T0186030

Modern Digital Radio Communication Signals and Systems

Sung-Moon Michael Yang

Modern Digital Radio Communication Signals and Systems

Second Edition, 2021

 Springer

Sung-Moon Michael Yang
BaycoreWireless.com
Irvine, CA, USA

ISBN 978-3-030-57708-7   ISBN 978-3-030-57706-3   (eBook)
https://doi.org/10.1007/978-3-030-57706-3

This Springer imprint is published by the registered company Springer Nature Switzerland AG
The registered company address is: Gewerbestrasse 11, 6330 Cham, Switzerland

Sung-Moon Michael Yang

Modern Digital Radio Communication Signals and Systems

Second Edition, 2021

 Springer

Sung-Moon Michael Yang
BaycoreWireless.com
Irvine, CA, USA

ISBN 978-3-030-57708-7   ISBN 978-3-030-57706-3 (eBook)
https://doi.org/10.1007/978-3-030-57706-3

This Springer imprint is published by the registered company Springer Nature Switzerland AG
The registered company address is: Gewerbestrasse 11, 6330 Cham, Switzerland

New Edition Preface

As said in the preface of the first edition, this book is initially written for practicing engineers and postgraduate students. However, most material, in particular, fundamentals in the book can be taught even at the undergraduate level. This was clear from the author's teaching experience. It is so as this field of digital communication systems has been matured. For this purpose, it is essential to have additional problems in each chapter and very useful as well. Thus the total of 298 problems are added in this new edition; most problems are basic to aid the understanding of the book material aiming at new-comer students but some are extending further the topics not covered in the main text and for useful practical applications.

In the first edition, there is an addendum of errata due to the mistranslations from the original manuscript format to a new one, i.e., to XMLT. In this edition it is eliminated. The structure of chapter organizations of the book is intact but minor technical errors are corrected as well as for clarity of expressions. Furthermore **the book support website: https://baycorewireless.com/ is organized and under development. It can be the arena of open communications for the users of this book.**

The book can be adapted as a textbook of three possible courses; (1) introductory communication systems course for junior and senior, (2) beginning graduate course of digital communication systems (3) wireless communication systems for graduate and practicing engineers. This author has experienced teaching all three courses, and selected appropriate chapters as below.

(1) Introductory communication course of one semester covers Chapters 9, 2, 3 intensively, and Chapters 8 and 1 are covered selectively and lightly. If there is time, Chapter 7 is covered lightly and selectively. Prerequisite is a course on signals and systems, in particular, Fourier transforms.

(2) Digital communication course for graduate students covers Chapters 9 and 1 quickly. And Chapters 2 and 3 are covered in depth, followed by selective coverage of Chapters 5, 6, and 7. For another course on error correction for digital communications, only Chapters 2 and 6 may be covered in depth.

(3) Wireless communication systems course covers Chapters 1, 2, 3, 4, 8, and
    selective topics from Chapters 5, 6, and 7, omitting 9.

Irvine, CA, USA   Sung-Moon Michael Yang

Preface of the First Edition

This book is written for practitioners of wireless digital communication systems - engineers, technical leaders, and managers for product and technology development - and for digital communication systems in general.

My goal is for this book to serve as an easily accessible reference, allowing the reader to learn and refresh a particular topic quickly. To this end, a number of figures, tables, examples, and exercises are included. A topic is described with a simple but non-trivial example, and then its variations and sophistications are presented. I believe it is the fastest and most effective way to learn and refresh. This does not mean only a collection of recipes. On the contrary, we emphasize that the fundamentals, when understood properly, are powerful in practice. For example, the performance of binary transmission system under additive Gaussian noise channel is fundamental, and also useful both in higher order modulations and in fading channels. A shaping pulse, matched and intersymbol inference free, is basic and applicable everywhere, e.g., it is insightful and practically useful, in orthogonal frequency division multiplex signal, to recognize its underlying pulse being rectangular. Thus this book emphasizes both practical problem solving and a thorough understanding of fundamentals and therefore should also be useful to newcomers to the area like graduate students, serious undergraduate students and others.

This book is the outgrowth of my involvement in telecommunication systems industry as a research and development engineer. The starting point of my career, in the early 80s, was also when the digitalization of the industry was at the starting point of its full swing. All of the topics in this book are, in one way another, related to my hands-on experience. Part of the book material was used in UCI and UCLA extension courses and at other universities.

I am indebted to the whole community of the technologies of communication systems, but in particular to the companies that gave me opportunities to work on the topics of this book. Technologies evolve continuously and incrementally through the work of many like biological evolution. And this author hopes this book to be a small thread in this continuing evolution, in addition to be useful to its intended audience, practitioners of digital communication systems. Thus it helps to push forward the field one small increment further.

Irvine, CA, USA   Sung-Moon Michael Yang

Contents

1   **Overview of Radio Communication Signals and Systems**   1
    1.1   Examples of Wireless Communication Systems   5
    1.2   Overview of Wireless Communication Systems   6
         1.2.1   Continuous Wave (CW) Signals   9
         1.2.2   Complex Envelope and Quadrature Modulation   10
         1.2.3   Digital Modulations   11
         1.2.4   Pulse Shaping Filter   11
         1.2.5   Channel Coding   13
         1.2.6   Demodulation and Receiver Signal Processing   14
         1.2.7   Synchronization and Channel Estimation   19
         1.2.8   More Modulation and Demodulation Processing   20
         1.2.9   Radio Propagation Channels   23
         1.2.10 Extension to Optical Fiber, and Other Systems   24
         1.2.11 Summary of the Overview   25
    1.3   The Layered Approach   26
    1.4   Historical Notes   27
    1.5   Organization of the Book   28
    1.6   Reference Example and its Sources   29
    1.7   Problems   30

2   **Digital Modulations**   33
    2.1   Constellation, Complex Envelope, and CW   37
         2.1.1   OOK, BPSK, and Orthogonal ($M = 2$)   38
    2.2   Power of Digitally Modulated Signals and SNR   42
         2.2.1   Discrete Symbol Average Power $\left(\sigma_s^2\right)$ Computation   43
         2.2.2   Power Spectral Density   43
         2.2.3   Signal Power and Noise Power Ratio (SNR)   44
    2.3   MAP and ML Detectors   45
         2.3.1   Symbol Error Rate of BPSK Under AWGN Channel . . .   47

	2.3.2	SER of OOK and Orthogonal signaling (M = 2) Under	
		AWGN Channel	49
2.4	PAM, QAM, and PSK		53
	2.4.1	M-PAM	53
	2.4.2	Square M- QAM	55
	2.4.3	PSK, APSK, and DPSK	62
2.5	BER and Different forms of SNR		66
	2.5.1	BER Requires a Specific Bit to Symbol Mapping	67
	2.5.2	A Quick Approximation of SER to BER conversion	67
	2.5.3	Numerical Simulations of BER with LLR	69
	2.5.4	SNR in Different Forms	70
2.6	Offset QAM (or Staggered QAM)		72
	2.6.1	SER vs. $\frac{E_s}{N_o}$ Performance of Staggered QAM	73
	2.6.2	CCDF of Staggered QAM vs. of 'Regular' QAM	74
	2.6.3	Other Issues	75
2.7	Digital Processing and Spectrum Shaping		75
	2.7.1	Scrambler	76
	2.7.2	Differential Coding or phase invariance coding	80
	2.7.3	Partial Response Signaling	84
2.8	Frequency Modulation – FSK, MSK, CPFSK		93
	2.8.1	Examples of FSK Signal Generation	94
	2.8.2	Non-coherent Demodulation of FSK	96
	2.8.3	FSK Signal Generations Using Quadrature	
		Modulator	97
	2.8.4	Binary CPFSK Example	98
	2.8.5	M-Level CPFSK	98
	2.8.6	MSK	101
	2.8.7	FSK with Gaussian Pulse	102
	2.8.8	Power Spectral Density of CPFSK, MSK,	
		and GMSK	103
	2.8.9	Partial Response CPFSK	105
	2.8.10	SER Performance Analysis of CPFSK	106
2.9	PSD of Digitally Modulated Signals		108
	2.9.1	Power Spectral Density of PAM Signal	109
	2.9.2	PSD of Quadrature Modulated Signals	111
	2.9.3	PSD of FDM and OFDM Signals	112
	2.9.4	PSD Numerical Computations and Measurements	112
	2.9.5	Numerical Computation of PSD Using FFT	114
	2.9.6	Example of PSD Computation by Using FFT	115
	2.9.7	PSD of Digital FM Signals –FSK, CPFSK	116
	2.9.8	PSD Computation Using Correlation	117
2.10	Chapter Summary and References		118
	2.10.1	Summary	118
	2.10.2	References	118
2.11	Problems		119

3 Matched Filter & Nyquist Pulse 131
   3.1 Matched Filters 134
       3.1.1 Matched Filter Defined and Justified 135
       3.1.2 Examples of Matched Filters 138
       3.1.3 Characteristic of Matched Filters 140
       3.1.4 SNR Loss Due to Pulse Mismatch 141
       3.1.5 A Mismatch Loss Due to Receive Noise Bandwidth 143
   3.2 Nyquist Criterion - ISI Free Pulse 145
       3.2.1 Nyquist Criterion - ISI Free End-To-End Pulse 145
       3.2.2 Frequency Domain Expression of Nyquist Criterion 147
       3.2.3 Band Edge Vestigial Symmetry and Excess
             Bandwidth 149
       3.2.4 Raised Cosine Filter 151
   3.3 Shaping Pulse (Filter) Design 153
       3.3.1 Practical Design Considerations 154
       3.3.2 A Practical Design Example of Analog Filter 154
       3.3.3 Design of Digital Matched Filters 156
   3.4 Performance Degradation Due to ISI 162
       3.4.1 A Design Case to Simplify the Shaping Filters 162
       3.4.2 A Quick Analysis of the Suggestion of Using a
             Rectangular Pulse 163
       3.4.3 A Discrete in Time Model for ISI Analysis
             in General 168
       3.4.4 A Discrete in Time Model for Simulation 170
       3.4.5 Summary for ISI Analysis Methods 171
   3.5 Extension to Linear Channel and Non-White Noise 172
       3.5.1 Linear Channels with White Noise 172
       3.5.2 Non- White Noise 174
   3.6 References 175
   3.7 Problems 175

4 Radio Propagation and RF Channels 183
   4.1 Path Loss of Radio Channels 186
       4.1.1 Free Space Loss 186
       4.1.2 Pathloss Exponent 187
   4.2 Antenna Basic and Antenna Gain 188
       4.2.1 Antenna Basics 188
       4.2.2 Antenna Pattern 189
       4.2.3 Directivity and Antenna Gain 189
       4.2.4 Aperture Concept and Antenna Beam Angle 190
   4.3 Path Loss Due to Reflection, Diffraction and Scattering 191
       4.3.1 Ground Reflection (2-Ray Model) 192
       4.3.2 Diffraction, Fresnel Zone and Line of Sight
             Clearance 194
       4.3.3 Examples of Empirical Path Loss Model 195
       4.3.4 Simplified Pathloss Model 195

	4.3.5	Shadow Fading: Variance of Path Loss	196
	4.3.6	Range Estimation and Net Link Budget	199
4.4	Multipath Fading and Statistical Models	202
	4.4.1	Intuitive Understanding of Multi-Path Fading	202
	4.4.2	Rayleigh Fading Channels	204
	4.4.3	Wideband Frequency Selective Fading	209
	4.4.4	Alternative Approach to Fading Channel Models	214
4.5	Channel Sounding and Measurements	220
	4.5.1	Direct RF Pulse	221
	4.5.2	Spread Spectrum Signal (Time Domain)	222
	4.5.3	Chirp Signal (Frequency Sweep Signal) for Channel Sounding	224
	4.5.4	Synchronization and Location	225
	4.5.5	Directionally Resolved Measurements (Angle Spread Measurements)	225
4.6	Channel Model Examples	226
	4.6.1	Empirical Path Loss Models	226
	4.6.2	M1225 of ITU-R Path Loss Models	227
	4.6.3	Multipath Fading Models in Cellular Standards	227
	4.6.4	Cellular Concept and Interference Limited Channels	228
	4.6.5	Channel Models of Low Earth Orbit (LEO) Satellite	236
4.7	Summary of Fading Countermeasures	240
4.8	References with Comments	242
4.9	Problems	243
5	**OFDM Signals and Systems**	247
5.1	DMT with CP – Block Transmission	252
	5.1.1	IDFT – DFT Pair as a Transmission System	253
	5.1.2	Cyclic Prefix Added to IDFT- DFT Pair	255
	5.1.3	Transmit Spectrum	256
	5.1.4	OFDM Symbol Boundary with CP	260
	5.1.5	Receiver Processing When the Channel Dispersion < CP	262
	5.1.6	SNR Penalty of the Use of CP	264
5.2	CP Generalized OFDM – Serial Transmission	265
	5.2.1	OFDM – Analog Representation	265
	5.2.2	Discrete Signal Generation of Analog OFDM	266
	5.2.3	Discrete Signal Reception	269
	5.2.4	Pulse Shape of DMT and its End to End Pulse	272
	5.2.5	Windowing	272
	5.2.6	Filter Method Compares with *DMT with CP*	273
5.3	Filtered OFDM	274
	5.3.1	Filtered OFDM Signal Generation	274
	5.3.2	Filtered OFDM Signal Reception	277
	5.3.3	Common Platform	279

	5.3.4	Impulse Response and Eye Pattern	280
	5.3.5	1-Tap Equalizer, Sample Timing and Carrier Phase Recovery, and Channel Estimation	282
	5.3.6	Regular FDM Processing with Filtered OFDM	283
5.4	OFDM with Staggered QAM		284
	5.4.1	Common Platform Structure for OFDM with Staggering	285
	5.4.2	Receiver Side of OFDM with Staggered QAM	289
	5.4.3	T/2 Base Implementation of Transmit Side	290
	5.4.4	Impulse Response, Eye Pattern, and Constellation of Staggered OFDM	293
5.5	Practical Issues		297
	5.5.1	Performance When a Channel Delay Spread > CP	297
	5.5.2	Digital Quadrature Modulation to IF and IF Sampling	299
	5.5.3	Modern FH Implementation with OFDM	301
	5.5.4	Naming of OFDM Signals	302
5.6	OFDM with Coding		304
	5.6.1	Coded Modulations for Static Frequency Selective Channels	304
	5.6.2	Coding for Doubly Selective Fading Channels	305
5.7	Chapter Summary		308
5.8	References and Appendix		309
5.9	Problems		313
6	**Channel Coding**		**321**
6.1	Code Examples and Introduction to Coding		323
	6.1.1	Code Examples – Repetition and Parity Bit	323
	6.1.2	Analytical WER Performance	326
	6.1.3	Section Summary	330
6.2	Linear Binary Block Codes		331
	6.2.1	Generator and Parity Check Matrices	331
	6.2.2	Hamming Codes and Reed-Muller Codes	334
	6.2.3	Code Performance Analysis of Linear Block Codes*	339
	6.2.4	Cyclic Codes and CRC	349
	6.2.5	BCH and RS Codes	361
	6.2.6	Algebraic Decoding of BCH	366
	6.2.7	Code Modifications – Shortening, Puncturing and Extending	369
6.3	Convolutional Codes		370
	6.3.1	Understanding Convolutional Code	370
	6.3.2	Viterbi Decoding of Convolutional Codes	385
	6.3.3	BCJR Decoding of Convolutional Codes	391
	6.3.4	Other Topics Related with Convolutional Codes	403

6.4	LDPC	407
	6.4.1 Introduction to LDPC Code	407
	6.4.2 LDPC Decoder	412
	6.4.3 Bit Node Updating Computation	412
	6.4.4 LDPC Encoder	420
	6.4.5 Useful Rules and Heuristics for LDPC Code Construction	425
	6.4.6 LDPC in Standards	433
6.5	Turbo Codes	437
	6.5.1 Turbo Encoding with G=15/13 RSC and Permutation	438
	6.5.2 G=15/13 Code Tables for BCJR Computation Organization	439
	6.5.3 The generation of 'extrinsic' information (E1, E2)	441
	6.5.4 Numerical Computations of Iterative Turbo Decoding	442
	6.5.5 Additional Practical Issues	447
6.6	Coding Applications	450
	6.6.1 Coded Modulations	451
	6.6.2 MLCM, TCM and BICM	461
	6.6.3 Channel Capacity of AWGN and of QAM Constellations	470
	6.6.4 PAPR Reduction with Coding	470
	6.6.5 Fading Channels	472
6.7	References with Comments and Appendix	479
	Appendix 6	480
	A.1 CM Decoding Example of Figure 6-36	480
	A.2 The Computation of p_0 (p_1), and LLR for BPSK	482
	A.3 Different Expressions of Check Node LLR of LDPC	483
	A.4 Computation of Channel Capacity	487
	A.5 SER Performance of Binary PSK, DPSK, FSK	489
6.8	Problems	490
7	**Synchronization of Frame, Symbol Timing and Carrier**	**505**
7.1	Packet Synchronization Examples	512
	7.1.1 PLCP Preamble Format of IEEE 802.11a	513
	7.1.2 RCV Processing of STS and LTS	516
	7.1.3 802.11_a Synchronization Performance	525
	7.1.4 DS Spread Spectrum Synchronization Example	529
7.2	Symbol Timing Synchronization	532
	7.2.1 Symbol Timing Error Detector for PAM/QAM	533
	7.2.2 Known Digital Timing Error Detectors	539
	7.2.3 Numerical Confirmation of S-Curve of Timing Error Detectors	545

	7.2.4	Timing Detectors with Differentiation or with Hilbert Transform	550
	7.2.5	Intuitive Understanding of Timing Detectors	550
	7.2.6	Carrier Frequency Offset Estimation	552
	7.2.7	Embedding Digital TED Into Timing Recovery Loop	554
	7.2.8	Resampling and Resampling Control	556
	7.2.9	Simulations of Doppler Clock Frequency Shift	561
7.3	Carrier Phase Synchronization		565
	7.3.1	Carrier Recovery Loop and Its Components	566
	7.3.2	Phase Locked Loop Review	566
	7.3.3	Understanding Costas Loop for QPSK	567
	7.3.4	Carrier Phase Detectors	569
	7.3.5	All Digital Implementations of Carrier Recovery Loop	571
7.4	Quadrature Phase Imbalance Correction		572
	7.4.1	IQ Imbalance Model	573
	7.4.2	$\hat{\theta}$, φ_i, and φ_d Measurements	576
	7.4.3	2-Step Approach for the Estimation of $\hat{\theta}$, φ_i, and φ_d	577
	7.4.4	Additional Practical Issues	578
	7.4.5	Summary of IQ Phase Imbalance Digital Correction	579
7.5	References with Comments		580
Appendix 7			581
	A.1	Raised Cosine Pulse and its Pre-filtered RC Pulse	581
	A.2	Poisson Sum Formula for a Correlated Signal	582
	A.3	Review of Phase Locked Loops	583
	A.4	FIR Interpolation Filter Design and Coefficient Computation	588
7.6	Problems		594
8	**Practical Implementation Issues**		**603**
8.1	Transceiver Architecture		605
	8.1.1	Direct Conversion Transceiver	606
	8.1.2	Heterodyne Conversion Transceiver	608
	8.1.3	Implementation Issues of Quadrature Up-Conversion	610
	8.1.4	Implementation Issues of Quadrature Down-Conversion	612
	8.1.5	SSB Signals and Image Cancellation Schemes	614
	8.1.6	Transceiver of Low Digital IF with Image-Cancelling	619
	8.1.7	Calibration of Quadrature Modulator I Demodulator	648
	8.1.8	Summary of Transceiver Architectures	650
8.2	Practical Issues of RF Transmit Signal Generation		650
	8.2.1	DAC	652
	8.2.2	Transmit Filters and Complex Baseband Equivalence	658
	8.2.3	TX Signal Level Distribution and TX Power Control	659
	8.2.4	PA and Non-linearity	662

	8.2.5	Generation of Symbol Clock and Carrier Frequency	670
	8.2.6	Summary of RF Transmit Signal Generation	672
8.3	Practical Issues of RF Receive Signal Processing	673
	8.3.1	ADC	673
	8.3.2	RX Filters and Complex Baseband Representation	677
	8.3.3	RCV Dynamic Range and AGC	678
	8.3.4	LNA, NF, and Receiver Sensitivity Threshold	682
	8.3.5	Re-generation of Symbol Clock and Carrier Frequency	684
	8.3.6	Summary of RF Receive Signal Processing	685
8.4	Chapter Summary and References with Comments	685
	8.4.1	Chapter Summary	685
	8.4.2	References with Comments	685
8.5	Problems	686

9   Review of Signals and Systems, and of Probability and Random Process . . . . | 697

9.1	Continuous-Time Signals and Systems	699
	9.1.1	Impulse Response and Convolution Integral – Time Domain	699
	9.1.2	Frequency Response and Fourier Transform	702
	9.1.3	Signal Power and Noise Power	705
9.2	Review of Discrete-Time Signals and Systems	713
	9.2.1	Discrete-Time Convolution Sum and Discrete-Time Unit Impulse	713
	9.2.2	Discrete Fourier Transform Properties and Pairs	715
9.3	Conversion Between Discrete-Time Signals and Continuous-Time Signals	717
	9.3.1	Discrete-Time Signal from Continuous-Time Signal by Sampling	717
	9.3.2	Continuous-Time Signals from Discrete-Time Signal by De-sampling (Interpolation)	718
9.4	Probability, Random Variable and Process	720
	9.4.1	Basics of Probability	721
	9.4.2	Conditional Probability	721
	9.4.3	Probability of Independent Events	723
	9.4.4	Random Variable and CDF and PDF	724
	9.4.5	Expected Value (Average)	725
	9.4.6	Some Useful Probability Distributions	726
	9.4.7	Q(x) and Related Functions and Different Representations	726
	9.4.8	Stochastic Process	730
	9.4.9	Stationary Process, Correlation and Power Density Spectrum	731
	9.4.10	Processes Through Linear Systems	732
	9.4.11	Periodically Stationary Process	732

9.5   Chapter Summary and References with Comments   733
        9.5.1   Chapter Summary   733
        9.5.2   References with Comments   733
9.6   Problems   734

Index   739

Author Bio

 **Sung-Moon Michael Yang** is a practicing communication systems engineer specializing in digital communication signals and systems, with an emphasis on wireless channels. In his career he worked in the communication industry in Silicon Valley for such companies as Hewlett-Packard Laboratory and Harris Microwave Communication Division. More recently he worked in Southern California for companies such as Boeing and SpaceX on the design and development of wireless system products. He has also taught digital signal processing (DSP), wireless communication systems, and analog circuits at various universities, e.g., 'DSP for communication systems' at UCI, DCE. He received a BS from Seoul National University, Seoul, Korea, and an MS and PhD from UCLA, all in Electrical Engineering.

Chapter 1
Overview of Radio Communication Signals and Systems

Contents

1.1	Examples of Wireless Communication Systems	5
1.2	Overview of Wireless Communication Systems	6
	1.2.1 Continuous Wave (CW) Signals	9
	1.2.2 Complex Envelope and Quadrature Modulation	10
	1.2.3 Digital Modulations	11
	1.2.4 Pulse Shaping Filter	11
	1.2.5 Channel Coding	13
	1.2.6 Demodulation and Receiver Signal Processing	14
	1.2.7 Synchronization and Channel Estimation	19
	1.2.8 More Modulation and Demodulation Processing	20
	1.2.9 Radio Propagation Channels	23
	1.2.10 Extension to Optical Fiber, and Other Systems	24
	1.2.11 Summary of the Overview	25
1.3	The Layered Approach	26
1.4	Historical Notes	27
1.5	Organization of the Book	28
1.6	Reference Example and its Sources	29
1.7	Problems	30

Abstract This chapter introduces the intent of the book briefly, and then overviews the content of it in a non-trivial manner thus a bit lengthy as an introduction. But, it is hoped that it stimulate a reader to the topics of the book. A new comer might find it a bit challenging for the first reading as an introduction, but one can move on since the topics will be elaborated later in detail.

General Terms AWGN · channel coding · channel estimation · complex envelope · constellations · CW signals · demodulation · digital modulations · DS spread spectrum · DSP · FH · layered approach · low noise amplifier · OFDM · path loss · PCM · power amplifier · pulse shaping filters · quadrature modulations · radio propagation channel · signals and systems · software defined radio · synchronization

© Springer Nature Switzerland AG 2020
S.-M. Yang, *Modern Digital Radio Communication Signals and Systems*,
https://doi.org/10.1007/978-3-030-57706-3_1

List of Abbreviations

1G, 2G, 3G	first generation, second generation, third generation
ADC	analog digital converter
AM	amplitude modulation
ASK	amplitude shift keying
AWGN	additive white Gaussian noise
BCH	Bose, Chaudhuri, Hocquenghem (names)
BCJR	Bahl, Cocke, Jelinek, Raviv (names)
BPSK	binary phase shift keying
BSC	binary symmetric channel
CDMA	code division multiple access
CM	correlation metric (decoding)
CP	cyclic prefix
CW	continuous wave
DAC	digital analog converter
DC	direct current
DFT	discrete Fourier transform
DMC	discrete memoryless channel
DMT	digital multi-tone
DS	direct sequence
DSB	ouble sideband
DSL	digital subscriber loop
DSP	digital signal processing
FDM	requency division multiplex
FEC	forward error correction
FH	frequency hopping
FM	frequency modulation
FSK	frequency shift keying
Gbps	Giga bit per second
GHz	Giga Hertz (10E9 Hz)
GPS	global positioning system
GSM	lobal system for mobile communications
HD	hard decision
IDFT	inverse discrete Fourier transform
IEEE	Institute of electrical electronics engineers
IoT	Internet of Things
LAN	local area network
LDPC	low density parity check code
LED	light emitting diode
LHS	left hand side
LNA	low noise amplifier
LOS	line of sight
LTE	long term evolution

MELP	mixed excitation linear predictive vocoder
MIMO	multiple input multiple output (antenna system)
MP3	MPEG-1 or 2 Audio Layer III
MPEG	motion picture expert group
MSK	minimum shift keying
OFDM	orthogonal frequency division multiplex
OOK	on-off keying
OSI	open system interconnect
PA	power amplifier
PAN	personal area network
PCB	printed circuit board
PCM	pulse coded modulation
QAM	quadrature amplitude modulation
QPSK	quad phase shift keying
RA	epeat accumulator
RF	radio frequency
RHS	right hand side
RS	Reed, Solomon
SD	soft decision
SDR	software defined radio
SERDES	serial, de-serial (parallel)
SNR	signal to noise ratio
SSB	single sideband
TDD	time division duplexing
VSB	vestigial sideband
WDM	wave division multiplex
Wi-Fi	Wireless Fidelity
WiMAX	wireless microwave access

The title of this book is *Modern Digital Radio Communication Signals and Systems*. We explain it first.

'*Modern*' is related with actual implementation of a system. Most systems take advantage of computational power being cheap to do ever more sophisticated discrete-time signal processing, often called digital signal processing (DSP), and thus achieve system performance close to theoretically possible optimum. Other than antennas, power amplifier (PA), low noise amplifier (LNA), mixers and oscillators, most digital radio systems are implemented by using DSP, i.e., numerical computations with digital hardware or with computer processors. The idealization of such implementations is sometimes called software defined radio (SDR). We do not follow its particular style in this book. If necessary, see Figure 1-2 wireless system block diagram.

'*Digital*' is related with the user messages. It means that a system will carry digital information. The simplest representation of it is to use binary number {0, 1}.

The user message is in the form of digital representation such as files, packet, or continuous stream of binary bits. Analog message signals such as picture, voice, and movie are converted to digital format, say using one of standard conversion formats like MPEG (motion picture expert group), MP3 (MPEG-1 or 2 Audio Layer III) and MELP (mixed excitation linear predictive) voice coder.

'*Radio*' is related with communication channels. It may be called wireless but 'radio' seems more specific than 'wireless'. Here we use 'wireless' exchangeable with 'radio'. One of the most ubiquitous wireless systems is cellular phone networks. In addition there are many wireless systems; Wi-Fi wireless LAN (local area network), WiMAX, satellite communication systems, deep space communication systems, and line of sight microwave radios, and short distance systems such Bluetooth, digital cordless phones. A long list is necessary to enumerate most of them. The communication channels have a large impact to the communication signal design, along with the user need and the performance requirement. For example, DSL (digital subscriber loop) and voice band modem, it is common practice to measure a channel response during the call setup, adjust the transmission rate accordingly. On the other hand, in radio systems, it is not possible since the channels are time varying due to the radio media (e.g., diffraction) or due to the mobility of the radios. Fading, fast and slow, is due to mobility, and is present for most of the time.

'*Communication signals and systems*' means here two way communications, unlike broadcasting which is one way. Radar is two way but it receives its own signal after delay and distortion. Because of digital signal processing implementations, a signal representation can be implemented numerically. For testing and simulation purpose, radio channels can be represented as a signal or system representation such as an impulse response or frequency response. In this situation, signals and systems are interchangeable. Historically some ingenious circuits were very useful at the time of invention for simplicity of implementations but are no longer the case since most of implementations are 'numerical'. They are not our focus here rather the ideas or their mathematical representations behind the circuits may be utilized for 'numerical' implementations. Thus the fundamentals represented mathematically are important and should be mastered. After that the circuit ingenuity may be appreciated as interesting examples.

'*Modern Digital Radio Communication Signals and Systems*' is written mainly for practicing communication system design engineers and managers. This book intends to be used as a reference book in practice as well as for quickly learning and refreshing. For this purpose we include examples, figures, and tables that are useful in practice so that they can be understood and remembered at a glance. Our approach is practical and pragmatic while we realize that the solid foundation in fundamentals is powerful in practice. This book helps to realize this complementary relationship between practice and theory. In this way, it is also useful to graduate students and even to senior undergraduate students in communication systems. In fact, part of this book material has been used for post graduate extension courses and for senior level undergraduate courses in communication systems. As said before, communication systems are heavily influenced by communication channels and user performance requirements. However, even though this book emphasizes radio channels, fundamentals of carrier modulation signals (CW signals) that are covered here will also be useful to wireline channels – DSL (twisted pair), SERDES (backplane trace), coax, power lines, and optical fibers.

1.1   Examples of Wireless Communication Systems

Cellular network systems are by far the largest digital radio communication system in use today and the 4th generation is here and next generation (5G) is brewing. Its user device was started as a mobile phone equipped for car in the trunk, and is now evolved into smart phones way beyond the initially envisioned phone service. They are not only phones but also perform multiple functions, especially any time Internet access without a call set up. Additional functions are camera, video, calendar and associated personal digital assistant functions. But the most useful functions come from the fact that a smart phone can access a larger number of applications on Internet with digital connectivity. This connectivity between Internet and a user device is accomplished by digital radio communication signals, through cellular phone networks. In order to support the user applications, the physical layer link utilizes radio propagation channels with appropriate communication signals and systems, which will be studied intensely and extensively in this book.

A major innovation for a high capacity cellular network (1G) is the spatial reuse of the same frequency over and over again. A base station is located in each cell of hexagonal geometric shape, and thus is called cellular networks. Next innovation (2G) is to use digital transmission using CDMA and GSM, which increase the network capacity by nearly a factor of 10. Then the 3G system is designed for Internet access in mind. The 4th generation is based on LTE (long term evolution) – OFDM (orthogonal frequency division multiplex) signaling format and on multiple antennas for even higher throughput. In Figure 1-1, examples of cellular base station antennas are shown. LHS is a close up view of antennas, and RHS is an example with camouflage as part of tree to be less visible to the environment.

The next most ubiquitous wireless system is Wi-Fi wireless LAN (local area network). One key innovation was to use unlicensed RF frequencies (2GHz, 5GHz). It is served as effective in-house wiring system connecting computers and network access for Internet. In the airport and at hotels, it is one prevalent way of providing Internet access, with least cost, without wiring up a whole building.

Related with wireless LAN, Bluetooth is a short wire replacement between earphones to smart phones, between earphones to mobile phone inside car. It is to replace a short wire within device. There are various similar devices called PAN (personal area network) such as ZigBee.

The cordless phones use the same unlicensed frequency as Wi-Fi. One can move around the house with a cordless phone. A telephone cord is replaced by a radio (mostly digitally these days) communication system.

LOS (line of sight) microwave systems were used extensively to carry long distance telephone traffics before the advent of optical fiber systems. It was designed

Figure 1-1: Examples of cellular base station antennas are shown. LHS is a close up view of antennas, and RHS is another example where antennas are blended as part of tree in an attempt to be less visible to the environment. Antennas are most visible part of the whole cellular infrastructure equipment

to be very reliable and highly bandwidth efficient systems (5–10 bps/Hz). This means that, as an example, a system with 10 bps/Hz bandwidth efficiency can transport 1Gbps with 100 MHz bandwidth. LOS microwave systems are still used extensively for backhaul connections of cellular base stations. Now it is declined substantially compared to its peak time, but they are useful where there is no optical fiber infrastructure.

Most of TV broadcasting uses digital communication signals. Satellite based system uses digital communication signals, which may be used for TV and for two-way communications. GPS (global positioning system) uses digital communication signals but it is one way from sky to the earth. Garage door opener uses a simple form of digital communication signals. Many sensor network devices use digital communication signals, where one important constraint may be the battery power. They are expected to proliferate with the spread of IoT perhaps with hundreds of thousands sensors which must be connected.

1.2   Overview of Wireless Communication Systems

Based on Figure 1-2 wireless communication system block diagram, we overview the whole system. To someone new to this wireless system area, the figure may look complicated, but it is a generic block diagram and contains only blocks of high level view. As we go through this book, the diagram will increasingly appear to be simple. The figure is drawn somewhat unconventionally yet intentionally. Typically it is drawn horizontally that the receiver is on the right-hand side with the transmitter on the left-hand side, and the channel in the middle. The intention of it will be clear as we go along; a brief justification follows.

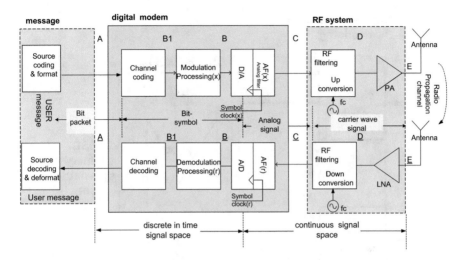

Figure 1-2: Digital radio communication system block diagram will be used throughout the book

In Figure 1-2, if one works on a pair of blocks, for example channel coding and decoding blocks, its right side can be considered as a channel (or resources to be utilized) and its left side is the user information to carry (or users to serve). On the channel side, from channel coding point of view, it can be abstracted from complicated transmission chain from B_1, B, C, D, E, antenna and radio channels back to $\underline{E}$, $\underline{D}$, $\underline{C}$, $\underline{B}$, and $\underline{B}_1$ (note underlines) as shown below;

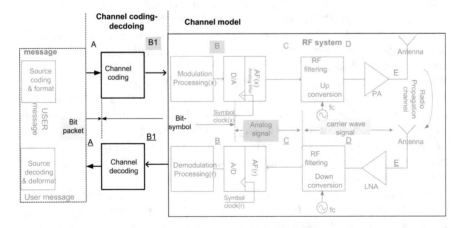

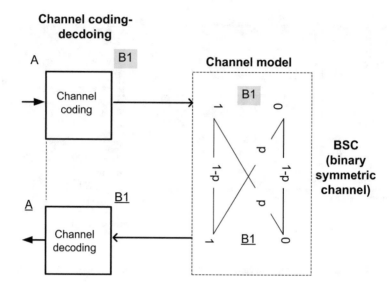

This abstraction, a digital channel, is often represented to have a probability of bit error (p). This abstraction as shown above is very useful and often used in practice and thus a specialist in coding can work without knowing many details on radio channel (coding is applicable to other channels like optical fibers and even for storage systems like magnetic tape and optical disks). The input A is a stream of bits and so is the output A. This channel model may be called binary symmetric channel (BSC). Different channel models (not shown here) are possible - discrete memoryless channel (DMC), erasure channel and AWGN channel.

Back to Figure 1-2, there are four large blocks in the figure; 1)user message, 2) digital modem, 3)RF sub-system, and 4) antennas and radio propagation channel (in the figure no block is drawn for radio channel). In this book our focus will be on the digital modem sub-system. However, it is critically influenced by other blocks, particularly by radio propagation channels. The right side of the digital modem side may be considered as a 'transmission channel' (C, D, E and radio channel, then back to E, D, C). Its left side is the user message, which can be a continuous stream of bits or frames / packets which consist of bits (A). At the receiver side (A), the recovered message should be delivered to the other user in the far end. It is not a loop which appears to be so in the figure.

Note also that signals are divided into two categories; discrete-time signals and continuous-time signals. Its boundary is between B-C at the transmitter side (DAC and transmit analog filter), and between C-B at the receiver side (receive analog filter and ADC). The function of a digital modem is to convert discrete-time signal to continuous-time signal and vice versa. At C and C, the signals are analog baseband and continuous-time (also called baseband complex envelope), and the signal at D/E and D/E is called CW (continuous wave) signals, i.e., carrier modulated sinusoidal waves. CW is real, not complex, in time domain. The bandwidth of analog baseband signals is limited, and typically much smaller than a carrier frequency, f_c. We explain it in more detail below starting from CW signals.

1.2.1   Continuous Wave (CW) Signals

A signal at D, after up-conversion with a carrier frequency, f_c, can be represented as a cosine wave being modulated by a baseband signal at C with its amplitude (A) and phase (θ). It is given by,

$$s_x(t) = A \cos\left(2\pi f_c t + \theta\right) \tag{1-1}$$

Note that the CW itself, without modulation, is given by $1.0 * \cos(2\pi f_c t)$, which is generated by a carrier frequency oscillator. A base band signal is represented by the amplitude ($A(t)$) and phase ($\theta(t)$), which is a function of time, carrying the information. The carrier itself does not carry the information. In order to increase the power of CW, additional amplifier such as PA (power amplifier) will be used but here we set aside the issue and the gain is set to be unity; $A(t)$ and $\theta(t)$ represent entirely baseband information carrying signals.

Historically CW modulation was a major innovation in communication signals in early 20th century. It was applied to AM broadcasting. $A(t) = 1.0 + k * a(t)$ where k is a constant called modulation index, and $a(t)$ is the modulating, analog baseband signal, i.e., voice. The phase $\theta(t)$ is not critical since it can be recovered non-coherently (i.e., without knowing the carrier phase of a signal). FM broadcasting followed, where the information is carried by frequency change, or $d\theta/dt$ while $A = $ constant. More sophisticated carrier modulations were developed in the context of TV and telephony with FDM; DSB, VSB and SSB with suppressed carrier. In particular, SSB is the most efficient system for analog voice telephony and was used until digital hierarchy systems, such as T1, were developed. If you are not familiar with these concepts, do not worry since all of these analog modulations will be explained briefly below, and later in detail. A key point is that CW modulation is still essential part of digital radio signals. And we emphasize the universality of CW represented by (1-1), and that it is real, not complex, in time. The bandwidth of analog baseband signals is limited, and typically much smaller than a carrier frequency, f_c. Thus from this point of view $A(t)$ and $\theta(t)$ may be slow compared to the carrier and thus sometimes treated them as nearly constant.

Abbreviations used in this section are DSB (double sideband), VSB (vestigial sideband) and SSB (single sideband), FDM (frequency division multiplex).

1.2.2   Complex Envelope and Quadrature Modulation

An equivalent, but different representation called complex envelope is useful, particularly in the context of digital radio communication signals. (1-1) above is represented as,

$$s_x(t) = Re\left\{\left[A_I + jA_Q\right]e^{j2\pi f_c t}\right\} \tag{1-2}$$

where $A_I = A \cos(\theta)$ and $A_Q = A \sin(\theta)$.

The equation (1-2) can be written as,

$$s_x(t) = A_I \cos(2\pi f_c t) - A_Q \sin(2\pi f_c t) \tag{1-3}$$

using the relationship of $e^{j2\pi f_c t} = \cos(2\pi f_c t) + j\sin(2\pi f_c t)$.

It is important to see that (1-1), (1-2), and (1-3) are different representations of the same CW signal.

The complex envelope, $C = A_I + jA_Q$, can be considered as a phasor representation as in Figure 1-3 (LHS). However, this phasor is not static, but dynamically changing, depending on the baseband signal. For example, in FM, the rotation of a phasor and its angular speed (frequency) represent the information. For DSB, VSB, and SSB, (A_I, A_Q) are related. In SSB, $A_Q =$ Hilbert transform (A_I). In DSB, $A_I=A$ and $A_Q=0$ (null). In VSB, (A_I, A_Q) are not independent but related even though it is not possible to state it simply. Here we emphasize again the universality of complex envelope and quadrature modulation represented equivalently by (1-1), (1-2) and (1-3); it can represent all different analog modulations. In particular, (1-2) or (1-3) are more often used in digital transmission. We cannot over-emphasize its universality for any kind of analog and digital modulations.

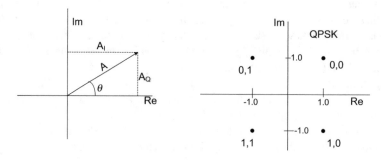

Figure 1-3: Phasor diagram (LHS) and QPSK constellation (RHS)

1.2.3 Digital Modulations

CW modulation (or quadrature modulation) with the complex envelope representation of a baseband signal in the form of (1-2) is general enough so that it is extended to digital transmission. For digital transmission it requires to generate a complex envelope $C = A_I + jA_Q$ (continuous signal) from discrete-time bit stream. There are many ways that can be done; QAM, QPSK, MSK, FSK, OOK, ASK and so on. This will be a major topic of this book. Abbreviations used here are QAM (quadrature amplitude modulation), QPSK (quad phase shift keying), MSK (minimum shift keying), FSK (frequency shift keying), OOK (on-off keying), ASK (amplitude shift keying); it is interesting to note that 'shift keying' is originated from early telegraph system, which is a form of baseband digital communication systems.

For actual signal generation, the equation (1-3) is often used, and it is called quadrature modulation, since two channels, in-phase (cosine) and quadrature phase (sine), are used and due to their 90° phase (quadrature) difference they are orthogonal. In digital transmission, rather than using SSB, two in-phase and quadrature phase channels are used to carry two independent data stream. This is equivalent to SSB in terms of bandwidth efficiency.

The symbols (discrete in time) to transmit can be represented by points on the complex plane. For example see Figure 1-3 right hand side for QPSK (RHS). A pair of bits are mapped into symbols as $(0, 0) \rightarrow \{1 + j\}$, $(0, 1) \rightarrow \{-1 + j\}$, $(1, 1) \rightarrow \{-1 - j\}$, and $(1, 0) \rightarrow \{1-j\}$. This bit assignment to symbol is called bit-to-symbol mapping. With Gray coding, adjacent symbol has one bit difference. In Figure 1-2, this bit-to-symbol mapping happens in modulation processing block (B1 –B in the figure). A sequence of digital modulation symbols is generated in this way.

Then from B – C digital modulation symbols are converted to continuous signal, $A_I + jA_Q$, called complex envelope, which is up-converted to CW using (1-2). This process of generating RF signals is summarized in Figure 1-4.

1.2.4 Pulse Shaping Filter

In order to generate A_I from the real part of, and A_Q from the imaginary part of, a complex symbol, digital to analog convertor (DACs) and analog filters (one set for the real part and the other set for imaginary part) will be used. Combined DAC and analog filter is a transmit pulse shaping filter. A simple pulse shape is a rectangular pulse. A pulse shape is an impulse response of a pulse shaping filter. In Figure 1-2, only one DAC is shown for simplicity, but for complex symbols, two sets of DAC are needed, and implicit in the figure.

Thus different baseband complex envelope signals are specified by the pulse shape and constellations on the complex plane. A constellation on the complex plane is a geometric representation of a signal, i.e., digital modulation of mapping bits into

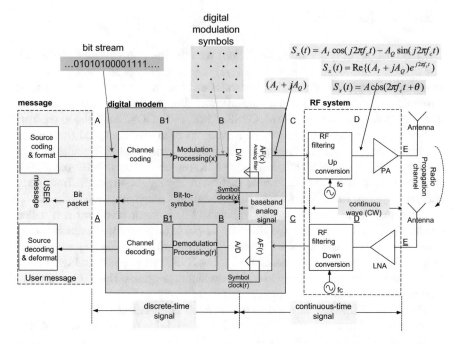

Figure 1-4: Summary of transmit signal generation from bit stream, modulation symbols, complex envelope and CW

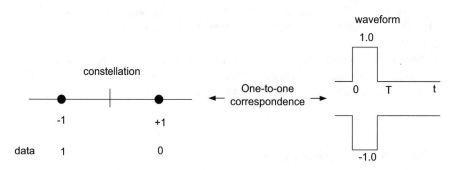

Figure 1-5: Constellation to pulse mapping; binary phase shift keying (BPSK) to rectangular pulse. Both pulse shape and constellation are necessary

discrete-time modulation symbols. The discrete-time modulation symbols are convolved in time with the impulse response of transmit filter (DAC and analog filter) to generate analog baseband signals. Another, perhaps more generic, view of this process of generating analog baseband signals is that a digital symbol is mapped to a pulse waveform with one to one correspondence. An example of BPSK is shown in Figure 1-5. In practice, a convolution is often used and in this book we use it exclusively otherwise stated.

1.2.5 Channel Coding

A channel coding is to add redundant bits to the information bits to be transmitted. There are infinitely many ways of how these redundant bits can be added and utilized in the receiver to correct or to detect errors. Thus the channel coding including FEC (forward error correcting codes) is a field by itself.

A repetition code is conceptually simple and so we use it as an example. A bit to be transmitted is simply repeated. For example, $\{1\} \rightarrow \{1,1,1\}$ and $\{0\} \rightarrow \{0,0,0\}$, repeated three times. And the code words are $\{1,1,1\}$ and $\{0,0,0\}$. All six other combinations are not code words; $\{0,0,1\}$, $\{0,1,0\}$, $\{1,0,0\}$, $\{0,1,1\}$, $\{1,1,0\}$, $\{1,0,1\}$. Thus at the receiver, when the received bit pattern is not from code words, it knows that errors occurred. This repetition code can correct a single bit error by using majority rule; two or more zeros decoded as $\{0,0,0\}$ / $\{0\}$ and two or more ones as $\{1,1,1\}$ / $\{1\}$. This decision rule is called hard decision decoding since received BPSK symbols are decided first (hard decision), and then the majority rule is applied for bit decoding.

The hard decision decoding can be improved by soft decision decoding, where the received sample of each bit is used for decision. For example, $\{0\} \rightarrow +1$, $\{1\} \rightarrow -1$ bit to symbol mapping, as in Figure 1-5, a received sample set is $\{+0.9, -0.2, -0.1\}$ after transmitting $\{0\}$. A hard decision will decode it incorrectly as $\{1\}$ since $\{0, 1, 1\}$ with each bit decision. With the soft decision rule, $0.9 - 0.2 - 0.1 = 0.6$, it correctly decodes it as $\{0\}$.

The soft decision rule can be obtained using a correlation metric decoding in general; the received sample $\{+0.9, -0.2, -0.1\}$ is correlated with all the code words (two in this example), $\{0,0,0\}$ i.e., $\{+1,+1,+1\}$ and $\{1,1,1\}$ i.e., $\{-1,-1,-1\}$ and then choose the maximum. The correlation metric with $\{-1,-1,-1\}$ will be -0.6 while with $\{+1,+1,+1\}$ it will be $+0.6$, only the sign difference. Obviously $+0.6 > -0.6$ thus it correctly decodes it as $\{0\}$.

Amazingly this correlation metric (CM) decoding works optimally for any linear binary block codes and for more. However, it may be quickly impractical as the size of code becomes even modest, say information bits of 20 or more. Thus it is directly usable only when the number of code words is small. But, it may still provide conceptual frame for understanding FEC and its performance characterization. It works for AWGN channels and fading channels as well as generating branch metrics of convolution code decoding.

Exercise 1-1: With the information bits of 20 in a code, how many codewords are there? Answer: $2^{20} = 1048576$ code words. In repetition code there is only one bit of information thus there is 2 codewords.

The hard decision decoding rule is obtained by applying the correlation metric decoding, left as an exercise. In this sense the difference between hard decision decoding and soft decision decoding can be understood clearly. A rule of thumb improvement of soft decision over hard decision is, in Gaussian noise channel, about 2 dB. In fading cases, it can be much larger than 2 dB.

There is no net coding gain with a repetition code. In order to see it our argument is as follows. In order to transmit three bits, after coding of one bit, the bandwidth required will be three times. Thus in the receiver, three time more noise must be allowed. In order to have a net coding gain, a code should be more sophisticated than the repetition. First of such code is Hamming code. For example, for 4 bits of information, 3 parity bits are added to have a single error correction.

Historically, algebraic codes such as Hamming, BCH and RS, were developed first and was important for simplicity of implementation and understanding. However, here we focus on the codes which can approach the channel capacity such as Turbo, LDPC and RA. These codes require iterative soft decoding. In wireless channels, the soft decision is important especially when FEC is used to cope with fading. Thus the channel coding is integral part of communication signal design. This will be explored in detail in Chapter 6 Channel Coding.

Abbreviations used in this section are LDPC (low density parity check code), RS (Reed, Solomon), BCH (Bose, Chaudhuri, Hocquenghem) and RA (repeat accumulator).

Exercise 1-2: A 3 bit repetition code may be modified as $\{1\} \rightarrow \{1,1,0\}$ and $\{0\} \rightarrow \{0,0,1\}$. Is it OK? Devise decoding schemes for HD and SD. Hint: It is OK. For HD choose $\{0\ 0\ 1\}$ if a received pattern after HD is one of $\{000\}$ $\{001\}$ $\{011\}$ $\{101\}$, i.e., data '0' is received. And $\{1\ 1\ 0\}$ if a received pattern after HD is one of $\{111\}$ $\{110\}$ $\{100\}$ $\{-010\}$, i.e., data '1' is received. For SD, use the correlation metric decoding.

1.2.6   Demodulation and Receiver Signal Processing

*This section might be skimmed through quickly without losing continuity if this type of material is new since it will be covered in later chapters.

Thus far we discussed a signal generation summarized in Figure 1-4, A-B1 (channel coding), B1 – B (digital modulation), B – C (baseband complex envelope with pulse shaping filter), and C – D (Complex envelope to CW signal with RF oscillator). In channel coding, we briefly discussed channel decoding as well.

Now we will discuss the receiver signal chain from $\underline{D} - \underline{C}$, $\underline{C} - \underline{B}$, and $\underline{B} - \underline{B}1$ in Figure 1-2. Essentially these are the inverse of transmission process; CW signal to complex envelope ($\underline{D} - \underline{C}$), analog complex envelope sampled ($\underline{C} - \underline{B}$) and then demodulated ($\underline{B} - \underline{B}1$) as shown in Figure 1-2.

1.2.6.1   CW Signal to Complex Envelope ($\underline{D} - \underline{C}$)

We explain two ways of recovering the complex envelope of A_I and A_Q from $s_x(t)$ represented by the equations of (1-1), (1-2) and (1-3).

First we use the equation (1-2). We need to find the corresponding imaginary part of (1-2), denoted as $\hat{s}_x(t)$.

$$s_x(t) + j\hat{s}_x(t) = \left[A_I + jA_Q\right]e^{j2\pi f_c t} \tag{1-4}$$

The imaginary part can be obtained from the real part, $s_x(t)$, by using Hilbert transform. It is a phase shift system by 90° (or multiplying j in the frequency domain; $-j$ for positive frequency and $+j$ for negative frequency).

$$\hat{s}_x(t) = Im\left\{\left[A_I + jA_Q\right]e^{j2\pi f_c t}\right\} \tag{1-5}$$

In order to remove CW modulation, (1-4) is multiplied by $e^{-j2\pi f_c t}$, i.e., demodulated. This is shown in Figure 1-6 (LHS). Note that the receiver should know the carrier frequency, f_c. In fact, the phase should be known as well, which may be absorbed into the complex envelope (A_I and A_Q). This carrier synchronization issue will be a major topic in Chapter 7.

Second we use the equation (1-3). We need $\cos(2\pi f_c t)$ and $\sin(2\pi f_c t)$, their phase is 90° apart. The equation (1-3) is multiplied by 2 $\cos(2\pi f_c t)$, and then remove the frequency component twice of the carrier by low pass filtering. The result is A_I. The equation (1-3) is multiplied by 2 $\sin(2\pi f_c t)$, and then remove the frequency component twice of the carrier by low pass filtering. The result is A_Q. In the figure the scale factor 2 is not shown for simplicity. In practice this is easy to accommodate by adjusting the gain of an oscillator. A single oscillator can generate cos () and sin () by 90° phase shift. This is shown in Figure 1-6 (RHS). Again note that the receiver should know the carrier frequency, f_c as well as carrier phase.

Both methods are used in practice. The second method was more common with analog implementations, but with the digital implementations, the first method is used since digital Hilbert transformer is not difficult to build numerically.

In practice, CW to complex envelope conversion (or quadrature demodulation) may happen at a convenient intermediate frequency (IF) after the down conversion (or frequency translation) from the carrier frequency.

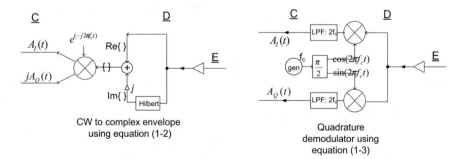

CW to complex envelope
using equation (1-2)

Quadrature
demodulator using
equation (1-3)

Figure 1-6: CW signal to complex envelope; two methods are displayed using equation (1-2) and (1-3) respectively

1.2.6.2 Analog Complex Envelope (A_I and A_Q) Is Sampled ($\underline{C}$ – B)

Here real and imaginary part of an analog complex envelope may be thought of as channels, commonly called in-phase channel and quadrature phase channels. Since the sampling process is identical, we consider only one channel (i-channel or q-channel). But it applies to both i- and q- channels.

In Figure 1-2, assuming that radio propagation channel does not distort the signal, and that RF system, both transmission direction (up conversion and PA) and receive direction (LNA and down conversion) are perfect, C → $\underline{C}$ is like a direct connection. Thus the signal at $\underline{C}$ is the same as one at C (or nearly so). To make it concrete, we use a rectangular pulse as shown in Figure 1-5. Then only thing that a receiver will do is to sample the signal at $\underline{C}$. The use of a rectangular pulse in this way happens in practice if C to $\underline{C}$ is connected by a short wire or by a PCB trace in a circuit board.

However, in radio channels, even if all things are perfect, one still need to consider thermal noise at the receiver front end (i.e., LNA). It is often (and accurately) modeled as noise with Gaussian distribution, called additive white Gaussian noise (AWGN). The white means that the spectrum of noise is flat in frequency domain. This channel model at C-$\underline{C}$ (toward right side in Figure 1-2) is shown in Figure 1-7, showing only i-channel, and the same figure is applicable to q-channel.

One needs to select a pulse shaping filter. If the transmit pulse shape is rectangular, the receive pulse shaping filter must be 'matched' to the transmit pulse shape. This is called a matched filter. This will maximize the signal to noise ratio at sampling instant. In general a matched filter pair is related as,

$$g_R(t) = g_X(-t) \tag{1-6}$$

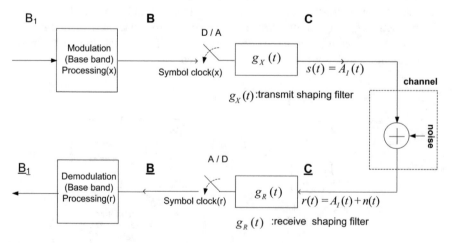

Figure 1-7: baseband AWGN channel model (i-channel only)

In case of a rectangular pulse shown in Figure 1-5, the receiver pulse is identical as transmit pulse. In order to make a shaping filter causal, or realizable, one needs to introduce a delay (D), which can be given by $g_R(t) = g_X(D - t.)$

There is one additional consideration for choosing a pair of pulse shaping filters. The criterion is called the Nyquist condition. A rectangular pulse of Figure 1-5 satisfies it. In fact any time limited pulse, less than or equal to symbol period T, will meet the condition. The end-to-end pulse is a convolution of transmit pulse with receive pulse, $g(t) = g_X(t) * g_R(t)$ ignoring the delay D for causality. Then the Nyquist condition says that a sampled end to end pulse is an impulse;

$$g(nT) = \delta(n) \tag{1-7}$$

where $n = .., -1, 0, 1, 2, ...$ a sampling index, and $\delta(n) = 1.0$ when $n = 0$, otherwise it is zero. The time origin of $g(t)$ is adjusted so that the peak of it happens at $t + D$ where a sampling occurs. With the rectangular pulse, $D = T$, where the end to end pulse is a triangle and its peak is at T.

With the Nyquist condition, there is no interference between symbols when they are synchronously transmitted symbol after symbol.

In passing we comment that Figure 1-7 may represent both i-channel and q-channel when modulations symbols are complex and noise is also represented by a complex random process (or 2-dimensional random process).

1.2.6.3 Digital Demodulation and Decision ($\underline{B} - \underline{B}1$)

We consider the case with Figure 1-7, where the received signal, $r(t)$, passes through the receiver filter, $g_R(t)$, and then sampled. Note that the noise, $n(t)$, also passes through the receiver filter and then sampled.

A sampled signal may not fall right on the transmitted constellation due to noise, but may be scattered around a constellation point as shown in Figure 1-8, where x denotes a received sample.

In Figure 1-8, a decision boundary for a sample is shaded by gray color. Any sample falls within the boundary will be decided 0 for BPSK (LHS of Figure 1-8) and {0,0} for QPSK (RHS of Figure 1-8) as an example. This decision boundary is intuitively satisfying and in fact it is optimal one if all the symbols are equally likely, which is typically the case. The decision boundary might be adjusted if symbols are not equally likely. For example, in BPSK, if the probability of {0} is more than ½, then the decision boundary should move to left, less than zero. The exact amount will depend on the statistics of noise.

The performance of error probability is often expressed by Q(x) function, which is a probability of a tail of Gaussian distribution with unity variance and zero mean.

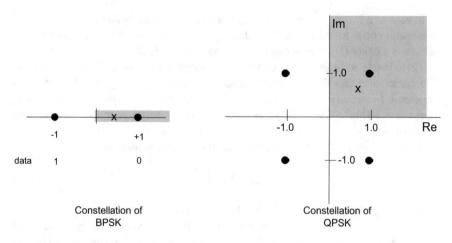

Figure 1-8:   Received sample (x) and decision boundary

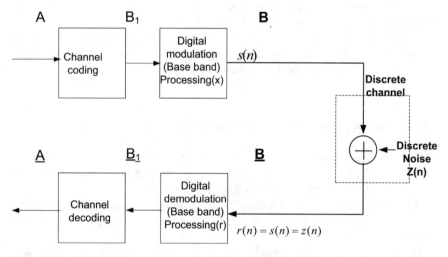

Figure 1-9: AWGN channel in discrete time; only constellations are necessary for performance computation with decision boundary (Figure 1-8)

$$Q(x) = \int_{x}^{\infty} \frac{1}{\sqrt{2\pi}} e^{-\frac{y^2}{2}} dy \qquad (1\text{-}8)$$

From digital modulation and demodulation point of view, RHS of (B – B) in Figure 1-2 can be abstracted as shown in Figure 1-9, where all the signals and noise are discrete in time, i.e., digital signals. DAC and filter, and ADC and filter are perfect and only thermal noise modeled by Gaussian probability distribution is present in the channel. The noise power is the variance of the Gaussian probability density function, σ^2. The signal power of BPSK is 1.0 in Figure 1-5 (LHS). It is convenient to use d rather than 1.0, i.e., $\{-d, d\}$ rather than $\{-1, +1\}$ and then the power is d^2. Its symbol error probability is given by $Q(\frac{d}{\sigma})$ assuming $\{0\}$ and $\{1\}$ are equally likely. This is a tail probability of Gaussian distribution with mean $= d$ and variance (σ^2). We will elaborate this derivation in a later chapter.

We relate digital signal to noise ratio $(\frac{d}{\sigma})$ with symbol energy (E_s) and two-sided noise density $(\frac{N_o}{2})$. We assume the transmit analog filter has a unity power gain, and the receive analog filter's noise bandwidth is optimal, i.e., the same as symbol rate $(1/T)$. And thus using $d^2 T = E_s$, i.e., the symbol energy (E_s) is the signal power (d^2) times symbol period, and noise power (σ^2) is noise density times noise bandwidth $(1/T)$, i.e., $\sigma^2 = \frac{N_o}{2} 1/T$. Thus $\frac{d}{\sigma} = \sqrt{2E_s/N_o}$.

$$Pe\{BPSK\} = Q\left(\sqrt{\frac{2E_s}{N_o}}\right) \tag{1-9}$$

In this case a symbol is a bit, thus symbol energy is bit energy.

With the same d for QPSK, the signal power is $2d^2$ and noise power is σ^2 in i-channel and in q-channel, total $2\sigma^2$. The symbol error rate for QPSK is given by $1 - (1 - Q(\frac{d}{\sigma}))^2$, is simplified to $2Q(\frac{d}{\sigma}) - Q(\frac{d}{\sigma})^2$. It is approximately $2Q(\frac{d}{\sigma})$. Here $2d^2 = E_s 1/T$. Thus $\frac{d}{\sigma} = \sqrt{\frac{E_s}{N_o}}$.

$$Pe\{QPSK \text{ symbol}\} \approx 2Q\left(\sqrt{\frac{E_s}{N_o}}\right) \tag{1-10}$$

This error performance is limited only by thermal noise. Additional imperfection due to filters and sampling errors must be taken into account separately.

1.2.7 Synchronization and Channel Estimation

So far we assumed that the receiver knows the carrier frequency and phase and sampling time. The receiver may know them roughly in advance, but the incoming carrier frequency and phase must be exactly synchronized with the locally generated frequency and phase for coherent demodulation. The symbol clock frequency and sampling phase must be known to the receiver precisely. Any small deviation will show as the degradation to the system, and worse it may not work at all. The receiver should recover the carrier frequency and phase, symbol clock frequency and phase from the receiving signal, and the process is called synchronization.

In addition, the receiver needs to know the gain and phase of a channel. When the channel is slowly time-varying, this may be done by decision directed way, i.e., by looking at errors after the symbol decision, one can estimate the gain and phase of a channel (actually including part of receiver circuits in addition to RF propagation).

When the channel is rapidly time-varying such as Rayleigh fading, a known data may be inserted, called pilot, to aid the channel estimation.

In addition to the carrier and symbol clock, a system may have a frame, a collection of symbols. In time division duplex (TDD) system, two channels of uplink and down link can be created by dividing time, i.e., a time for one directional transmission, and another time for other directional transmission. In this case there must be a TDD frame to distinguish the time boundary of transmission direction. Another example of a frame is to add preamble for pilots, and other control and housekeeping signals.

In direct sequence spread spectrum signals (or code division multiplex signals), the spreading code generated locally should line up with the incoming signal's code. This is similar to finding a frame boundary.

'Modern' implementations of a communication system uses digital signal processing as we discussed in the beginning of this chapter so that system performance is close to theoretically possible optimum. For the simplicity of description and ease of understanding, most of RF system and pulse shaping filters in the analog signal domain. However, in most of current implementations discrete signal to analog conversion (i.e., DAC and ADC and filtering) happens at IF frequency. Except for PA and LNA and antennas in Figure 1-2, all others can be implemented digitally. Thus the synchronization including carrier frequency and symbol clock recovery, and channel estimation may be done digitally.

1.2.8   More Modulation and Demodulation Processing

Spread spectrum signals are used for combating interference due to jamming or due to multi-users. Typically the bandwidth is much wider than information rate. Two forms are often used in practice; direct sequence (DS) spread spectrum and frequency hopping (FH). Spread spectrum signals have a rich history and were studied extensively in the past. Recently both forms of signals are used in cellular networks, the former as code division multiple access (CDMA) and the latter as orthogonal frequency division multiple access (OFDMA). Here we treat both of them as modulation processing and channel coding blocks, and corresponding receiver blocks, in Figure 1-2, i.e., all implemented and analyzed exclusively in digital domain or in discrete-in-time. This contrasts with traditional analog implementations.

1.2.8.1   DS Spread Spectrum System

DS expands k bits of information to $n = k L$ by the factor L, which is called spreading gain. It can be thought of (n, k) block code with block size n. A simple case is $k = 1$, $n = L$ with repetition code. Actual code words are changed for each bit with pseudo-noise (PN) sequence. Expanded transmission rate is often called chip rate. Assuming

BPSK, the rest of processing - DAC, filtering up conversion etc. in transmit side - follows by treating the chip rate as a symbol rate exactly same as 'normal' BPSK. The bandwidth expansion is just like one due to FEC. Many varieties are possible, and the performance analysis is similar to that of FEC. In the receiver side, PN sequence boundary must be recovered just as FEC block boundary should be recovered. The synchronization of PN sequence is one of major topics in direct sequence spread system

The spreading gain is 10 log (L) dB, and thus with spread spectrum system the receiver threshold can be improved (i.e., lowered) by the amount of the spreading gain. There is no net coding gain since the bandwidth is expanded by the same factor; essentially it may be a repetition code in this respect.

1.2.8.2   OFDM and FH Spread Spectrum

In a frequency hopping spread spectrum system, the available bandwidth is subdivided into a number of contiguous frequency slots. In any signaling interval, the transmitted signal occupies one or more of the available frequency slots. The selection of frequency slots in a signaling interval can be made pseudo random (PN). Typically this was implemented by frequency synthesizers switching the frequencies to right frequency slots. Due to this switching, FSK is often used since it is possible to demodulate it non-coherently, and furthermore, the signaling interval has a guard time for synthesizer switching, waste of precious bandwidth.

All these limitations can be eliminated by using OFDM which is efficiently implemented by DFT with cyclic prefix. Or DFT with commutating filters – this is one innovation that no textbook so far presented. (Figure 1-10 does not show CP or commutating filters for simplicity which will be amply elaborated in Chapter 5.) Any modulation scheme including coherent demodulation is usable and there is no synthesizer and thus there is no need of a switching guard time. Additional advantage of this approach is that all the sub-carrier channels (frequency slots) are demodulated simultaneously. This can be used for creating additional information channels at the expense of spreading gain, conversely increase the spreading gain by reducing information channels, seamlessly.

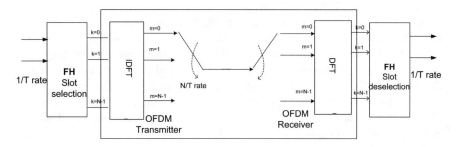

Figure 1-10:   FH implementation using OFDM

There are N sub carrier channels (frequency slots) and each slot carries 1/T rate symbols (complex for i-channel and q-channel). Depending on the FH pattern frequency slots are assigned to information channels (two channels are shown Figure 1-10). IDFT and DFT pairs are used to create frequency slots (k = 0, 1, ..., N–1).

Historically FDM was used long before OFDM. The major innovation came from the recognition that IDFT-DFT pair can be used a transmission system, and then CP is added as a gap between OFDM symbols in order to cope with channel impulse response dispersion. This signaling method is called DMT (digital multi tone) in DSL (digital subscriber line) area. This book, in Chapter 5, introduces DFT plus commutating filter structure to extend and generalize IDFT-DFT structure for any subcarrier shaping filter. DMT is limited for a rectangular pulse only.

1.2.8.3   Diversity Channels (Frequency, Time, Space)

Diversity is to send the same data over independent fading paths in space, or over different frequency bands, or over different time or over different codes, in general over independent communication channels.

In radio communication systems, polarization (horizontal and vertical) can be utilized as diversity as well. It may be considered as a special form of space diversity. The use of multiple antennas is specific to wireless communication channels whereas in wired communications such as optical fiber and DSL, more lines should be used. It is important to have uncorrelated channels to maximize the diversity effect. In spatial diversity, it means that the separation of antenna should be large enough. Similarly for frequency diversity and time diversity the separation should be sufficient to have uncorrelated channels. When one antenna is used in the transmit side and multiple antennas (e.g., 3) are used in the receive side, the entire signal is received through multiple paths. There will be diversity gain, which can be particularly large in radio fading situation such as Rayleigh fading. The essential receiver signal processing is to co-phase all the multiple signals and combine (e.g., add or select the best) them. This is essentially the same as repetition code. The performance improvement over fading can be analyzed the same way as repetition code; for example 3 repetitions with three receive antennas and the more repetitions, the more antennas. See Figure 1-11 LHS.

Spatial channels can be created by adding more transmit antennas at the same time taking advantage of space diversity. 3:3 MIMO (multiple input multiple output) is shown in Figure 1-11 RHS. When there is sufficient scattering in channels (9 paths in the figure), i.e., all the paths are uncorrelated, it can effectively create 3 spatial channels with the space diversity effect. The total transmit power is the same as a single antenna; the power of each antenna is 1/3 as shown in the figure. In practice, the required scattering (or no correlation) may not be good enough. Then the effective spatial channel may be less than 3.

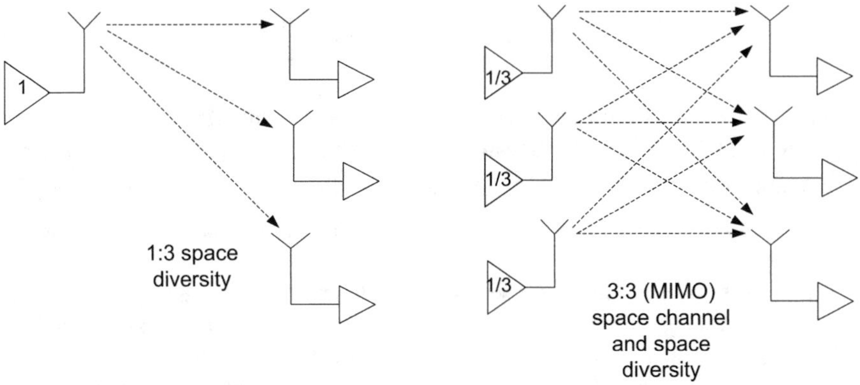

Figure 1-11: Space diversity and MIMO

Another way of using multiple antennas is beam forming, which effectively increase antenna gain or spatial filtering so that unwanted interference can be rejected.

1.2.9 Radio Propagation Channels

The variability of radio channels is tremendously large. For example, LOS (line of sight) microwave channel is clean and can carry high level modulation signals with the bandwidth efficiency of 5–10 bps/Hz where high gain antennas are in line of sight. On the other hand mobile wireless channels can carry less than 1.0 bps/Hz, plagued with interference and Rayleigh fading due to the mobility. In-door channels are dominated by multipath frequency selective fading.

The radio propagation phenomenon itself can be extremely complex. Yet radio propagation as communication channels are successfully modeled by using only several key characteristics. We consider two major characterizations of radio channels; large scale fading (simply path loss) and small scale fading (also called multipath fading).

The path loss is a complex function of carrier frequency, distance, and propagation profile. If there is no obstruction (i.e., clear LOS), then its signal power loss is the same as free space loss. Typically on the ground there are partial and full blockings of line of sight and then the path loss is much greater than that of the free space. The large scale fading is averaged locally and independent of signal structure and its bandwidth, only to the power of signal. There are a number of empirical models available in the literature. The path loss model can be simplified by three parameters; loss (A) at a reference distance (d_o), path loss exponent (γ), and shadowing (s). It is given by, with the distance d,

$$\text{Path loss (dB)} = A + 10\,\gamma\,\log\left(\frac{d}{d_o}\right) + s \qquad (1\text{-}11)$$

where the shadow fading s can be modeled by log normal distribution, i.e., Gaussian distribution in dB scale, of zero mean and s variance. Thus the path loss is log normal with the mean of $A+10\gamma\,\log\left(\frac{d}{d_o}\right)$. The path loss at a reference distance, A, may be modeled as a free space loss plus some additional corrections for a given carrier frequency.

The small scale fading is caused by rapid fluctuation of signal power due to movement which generate constructive and destructive wave front phasing, and by multiple signal paths with different delays and path attenuation (simply called delay spread) due to terrains. Doppler frequency spectrum should be specified, and its maximum frequency is related with the speed of movement. Its fading statistics are specified, Rayleigh or Rice. If there is no line of sight component it is Rayleigh and with LOS component, it is Rice.

Interference from other users or other networks may be treated, approximately, as thermal noise which comes from the receiver front end i.e., LNA. If there is knowledge on the structure of the interference, one may take advantage of it. For example a single tone type of interference may be reduced by using a narrow band notch filter (sometimes called excision).

1.2.10   Extension to Optical Fiber, and Other Systems

In this book we specifically focus wireless radio channels since the communication channel has, along with user need and performance requirements, a large impact to the signal and system design. However, most of CW signaling schemes are applicable to DSL and coax cable based system, where it is natural to take advantage of wire lines (twisted pair or coax cable); unlike mobile wireless channels, they are not rapidly changing.

Without CW modulation, a signaling scheme can be base band, and the frequency contents are near DC. This is the same as only i-channel (or q-channel) to make only one channel of complex envelope. Traditionally it is called PAM (pulse amplitude modulation) but many varieties are there, which may be used in short wire serial communications such as USB, Ethernet cable connections, and backplane connections between PCB boards. Even spectrum can be shaped. Bipolar pulse is used in PCM repeater to eliminate DC or a partial response pulse, called duo-binary, shapes the frequency response so that there is a notch at the half of the symbol rate. In this course we emphasize CW modulated signals since radio channels need them. However, we cover enough of base band signaling as well.

The evolution of optical fiber communication systems is spectacular in its transmission rate increase and the distance increase without repeaters. It was started as to transmit base band rectangular pulse (called NRZ); the presence of the pulse represents a bit one and the absence is a bit zero; very similar to OOK (on-off keying). Yet due to the large bandwidth of optical signals, it could be a high speed.

Wavelength division multiplexing (WDM) is a technology which multiplexes a number of signals onto a single optical fiber by using different wavelengths (i.e., colors) of laser light. This technique enables bidirectional communications over one strand of fiber by using two wavelengths. Furthermore it increases the capacity proportional to the number of colors of laser light. Note that the wavelength (λ) is closely related with the frequency ($\lambda = c/f$), where c is the speed of light, and it is essentially equivalent to CW signaling. However, the characteristic of optical fiber channel is different from the radio channel, and the generation of the optical signals and how to detect them are different as well. And fiber optical system is wired systems unlike radio. Thus even though the principles of communication signals and systems may apply to them, in practice they are treated separately. However, understanding the communication signals in radio channel will give some insights to the optical fiber communication systems as well.

Power line communication is systems for carrying data on a conductor also used for electric power transmission. It is a wired system with 'one' wire, and thus CW signaling with carrier frequency 100–200 kHz, is essential to utilize the capacity of the wire. Generally the power line channel is not clean, and thus all the principles of communication signals studied in this book will be useful. There are many different physical layer standards using pulse position modulation to OFDM. The applications include home networking, Internet access and automotive uses.

For SERDES using backplane trace connections, as its speed gets multiple Gbps, it may use multi-level, coded carrier modulation signals for ever higher transmission rate as in DSL and fiber optical cases.

Underwater acoustic communication is a technology to send and receive data under water. In underwater communication there are low data rates compared to terrestrial communication, since underwater communication uses acoustic waves rather than electromagnetic waves. Under water communication is difficult due to factors like multi-path propagation, time variations of the channel, small available bandwidth and strong signal attenuations. They use vector sensors, which is similar to multi-antenna systems. The underwater communication channels have their own peculiar features but the principles studied here should be useful to deal with these types of channels as well.

1.2.11 Summary of the Overview

This overview might be the most difficult part of the book to those who are new to digital radio communication signals and systems. However, it is intended as an insightful and structured summary of the topics of this book. Perhaps one may enjoy reading it after some progress into it or even after finishing it. In other words, we attempted this overview with key fundamentals and yet with significant key details, which make this overview somewhat difficult to newcomers. Another aim of the overview is to provide a frame for later development of detailed topics.

We introduce the complex envelope followed by quadrature modulation using (1-2) and (1-3) right from the beginning, rather than typical approach of first baseband and then passband, since it is essential for radio communication signals with a designated radio frequency to be used. Even for wired communication, the use of many frequencies in a single cable (i.e., FDM) is essential in order to increase the capacity. Then conventional analog modulations such as FM, AM and SSB are treated as how to generate the complex envelopes (A_I+j A_Q) and the relationship between them (A_I,A_Q).

Then the digital modem in Figure 1-2 consists of three main parts; channel coding, modulation processing and discrete to analog conversion in the transmitting side, and the corresponding inverse functions in the receiving direction.

The channel coding adds redundancy (parity bits) in information bits (or adds correlation among bits). There are many variety ways of adding the redundancy, and it is a field of its own (error correction codes).

Digital modulation processing includes bit-to-symbol mapping, and adding additional channels in frequency, time and space. This process is done in discrete in time i.e., digitally. It is also called geometric representations. Here there are many different ways possible to accommodate the different channel and user requirements.

Discrete modulation symbols must be converted to analog complex envelope (A_I, A_Q), which can be done by digital to analog converter followed by an analog filter. Here an end-to-end pulse should meet the Nyquist criterion of no inter-symbol interference and the receive filter pulse shape (in time domain) must match to that of the transmit filter pulse shape.

1.3   The Layered Approach

In passing we mention that providing a communication service to users is more than providing a connection by briefly introducing a layered model. Communication networks – Internet, telephone, and cellular networks are organized in hierarchy or in layer. The Open System Interconnection (OSI), standardized for computer communications, is defined with 7 layers. In terms of OSI model, the topics we cover in this book are related mostly with Physical layer and some with Link layer.

A detailed understanding of each layer is beyond the scope except for the physical layer which is a topic of this entire book. In order to give a rough idea of functions some layers are merged and indicted in Table 1-1 below. For example from a mobile user point of view, mobile apps installed in a smart phone with Internet service and thus see two layers.

The function of Physical layer and Link layer combined is to create a link (a visual analog may be a pair of connecting wires) and this link is utilized by Network layer. Higher layer utilize the low layer as resources. Each layer has its own protocol to communicate; e.g., Application in near side communication other side

Table 1-1: OSI model and its functions explained roughly

OSI 7-layer	Simplified functions	Mobile User perception
Application	User call / application	Mobile Apps
Presentation		Devices(e.g., smart phones with Internet service)
Session	Call set-up / find route	Infrastructure (Internet networks, cellular networks, telephone networks)
Transport		
Network	Switching / routing	
Link	Transmission / link	
Physical		

Application peer through application protocol. Physical layer this 'protocol' may be expressed packet format, transmission rate, modulation and coding, carrier frequency and etc.

A key point here is that making a reliable connection through physical layer is most complex task, and it requires substantial material resources, skilled human resources for the development of infrastructure and user gears. However, that is not the whole to provide a good service to users, more layers are necessary. The higher layers are typically implemented by software.

1.4 Historical Notes

Before the stable RF oscillators were developed, the electrical telegraphy, perfected and commercialized by Morse in the US in 1830s, used a single communication channel per one cable. It is essentially the base band transmission with four symbols of dot, dash, space between letters and space between words, the speed was about 50 words per minute. In order to increase the communication capacity a new cable should lay, which is very expensive.

In wireless telegraphy, spark gaps were used with crudely tuned circuits, one wire to the air and the other to the ground, and a center frequency was around 1 MHz. But due to the crude separation of dirty spark signals, it was difficult to have more channels. Marconi's wireless telegraphy demonstration in 1901 was using the spark gap signal.

The vacuum tube triode after the vacuum tube diode invention in 1904 by Fleming was the supreme device for the design of electronic amplifiers and oscillators. With the oscillators AM broadcasting was realized and later in 1936 FM radio was suggested by Armstrong.

The telephony, after its invention in 1875, started using frequency division multiplexing for efficient use of a cable, and by late 1930s combined with SSB, L-carrier could carry 600 voice channels and perfected further until digital T-carrier being deployed using PCM in 1970s.

It is interesting to note that in DSL, a twisted pair, initially designed for one telephone voice channel, now carries high speed data in the range up to a few 100 Mbps with carrier modulation and DMT (a form of FDM). A very similar pattern happened with optical fiber; initially baseband transmission was used with LED and multimode fiber, recently with WDM, the transmission capacity of one fiber is multiplied by a number of laser colors (wavelengths).

In 1928, Nyquist published a paper on the theory of signal transmission in telegraphy, and he obtained the Nyquist criteria to serially transmitting symbols without interfering (i.e., no inter-symbol interference condition) [1]. This is applicable to digital transmission. In 1943, D.O. North devised the matched filter for the optimum detection of a known signal in additive white noise in radar context [2]. The transmit pulse should be matched with the receive pulse shape. The minimum noise bandwidth of a receiver is the same as the transmission rate.

In 1948, Shannon developed a concept of channel capacity where with the transmission rate less than the capacity, it is possible to have an arbitrarily small error rate. He used a random coding, i.e., to choose codes randomly and thus without specifying any coding scheme, he developed the channel concept, which stimulated powerfully to search for practical coding schemes close to channel capacity. One such code is Turbo code discovered by C. Berrou, A. Glavieux and P. Thitimajshima in 1993. It turns out that the code is a clever combination of known schemes – convolutional codes, interleaver and BCJR iterative decoding. LDPC, invented in late 1960s, was rediscovered and its performance could be close to the capacity.

In the late 1990s, the access to Internet using telephone subscriber loop cable prompted to high speed connection. It is called DSL and its successful implementation was done by OFDM (called DMT). The heavy attenuation at high frequency in a twisted pair requires an equalizer, and by chopping a large bandwidth into smaller subcarriers the required equalizer in a subcarrier is simple; one tap equalizer with gain and phase of a sub-channel. In the beginning of an each connection, the cable loss is measured for each sub carrier, and thus SNR for each sub carrier is known at the transmit side. This can be utilized so that the higher SNR, the more bits are sent, called bit loading.

In wireless channel the bit loading is not typically practiced due to its variability of channels. OFDM is adapted in WiFi, WiMax and recently LTE in 4G cellular networks. For broadband systems, OFDM seems to be adapted everywhere. In conjunction with OFDM, multi-antenna technologies (MIMO) are used as well.

1.5 Organization of the Book

A list of chapters in this book is below.

1. Overview of radio communication signals and systems (this chapter)
2. Digital modulations
3. Matched filter and Nyquist pulse
4. Radio propagation and RF channels

5. OFDM signals and systems
6. Channel coding
7. Synchronization of frame, timing and carrier
8. Practical implementation issues
9. Review of Signals and Systems, and of probability and random process

   Chapters of 2, 3, 5, 6 and 7 are the core covering digital processing of signals. Chapter 1 and 9 cover overview, and basic review material. Chapter 4 is on propagation channel models, and Chapter 8 covers transceivers including PA, LNA and local oscillators.

1.6 Reference Example and its Sources

In this book there is no extensive list of reference papers and the reference is limited. The reason for this is that the availability of information on Internet. We still use the conventional reference for most relevant and specific cases, for example;

[1] Nyquist, H, "Certain Topics in Telegraphy Transmission Theory", AIEE Trans., vol.47 pp. 617–644, 1928
[2] North, D.O.,"An Analysis of the Factors which determine Signal / Noise Discrimination in Pulse-Carrier Systems", RCA Tech. Report No. 6 PTR-6C

Reference Sources
- For specific technical papers related with the topics of this book, we suggest to consult **IEEE.org | IEEE Xplore Digital Library**. For example, at Xplore search window, type 'marconi' and we find a paper titled "Marconi's experiments in Wireless Telegraphy", 1895.
- Another open source of information is **Wikipedia**; it is especially useful to grasp a rough, initial understanding of topics. For example;

  http://en.wkikpedia.org/wiki/morse_code   Morse code
  http://en.wkikpedia.org/wiki/wireless _telegraphy
  http://en.wkikpedia.org/wiki/AM_broadcasting   AM broadcasting
  http://en.wkikpedia.org/wiki/L-carrier   L carrier
  http://en.wkikpedia.org/wiki/T-carrier   T-carrier
  http://en.wkikpedia.org/wiki/OFDM   OFDM
  http://en.wkikpedia.org/wiki/Nyquist_ISI_criterion   Nyquist ISI free pulse
  http://en.wkikpedia.org/wiki/matched_filter   matched filter
  http://en.wkikpedia.org/wiki/shannon-hartley theorem   channel capacity
  http://en.wkikpedia.org/wiki/turbo_code   Turbo code
  http://en.wkikpedia.org/wiki/LDPC   LDPC

  These are addition reading materials for 1.4 Historical Notes above.
- Another information source is **Google scholar**; http://scholar.google.com/
  This generally contains a large data so one needs to get a hang of search skill.
- For books, **amazon.com /books** is a good place to begin to search.

1.7   Problems

P1-1: List the wireless devices that are used near you or by you; for example a smart phone (cell phone), Wi-Fi at home and in the airport, a garage door opener, GPS, car key, FM /AM radio, TV and so on. Identify them as many as possible.

P1-2: A channel model at B1 – $\underline{B1}$ in Figure 1-2 is modeled as BSC (binary symmetric channel). It is drawn horizontally (i.e., conventionally) as in Fig. P1-2 below. How many parameters are necessary to characterize BSC? And what is it?

P1-3: A block diagram at C – $\underline{C}$ in Figure 1-2 is drawn conventionally as in Fig. P1-3 below. Devise a simple test signal like sinusoid. If the channel is ideal except for thermal noise at receiver, sketch a simple test signal and its received signal.

P1-4: CW signal can be expressed three ways as below; express A and θ in terms of A_I, and A_Q.

$$s_x(t) = A\cos(2\pi f_c t + \theta),$$

$$s_x(t) = \mathrm{Re}\left\{\left[A_I + jA_Q\right]e^{j2\pi f_c t}\right\},$$

$$s_x(t) = A_I \cos\left(2\pi f_c t\right) - A_Q \sin\left(2\pi f_c t\right)$$

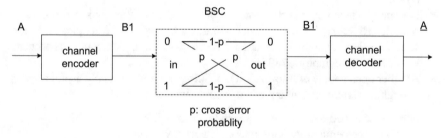

Fig. P1-2:   Channel coding end to end block diagram

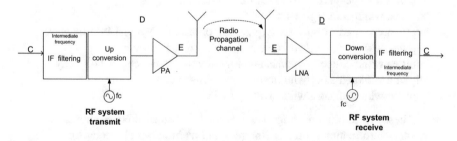

Fig. P1-3:   RF system + antenna +propagation channel end to end block diagram

Fig. P1-6: Orthogonal
signal set

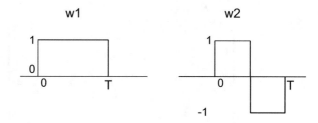

w1   w2

P1-5: The universality of CW signal generation is emphasized, i.e., all types of analog modulated signals can be done by using $s_x(t) = A_I \cos(2\pi f_c t) - A_Q \sin(2\pi f_c t)$. In particular for DSB, $A_I = A$, $A_Q = 0$. Draw a simplified block diagram for DSB generation.

P1-6: Rather than BPSK bit to symbol mapping, one may use a waveform as in Fig. P1-6, '0' $\rightarrow w_1$ and '1'$\rightarrow w_2$; it is called orthogonal signaling since $\int_0^T w_1(t)w_2(t)dt = 0$. By using the negation of each, i.e.,- w_1 and - w_2 along with w_1 and w_2, there are four waveforms. How many bits can be transmitted with this new signal set?

P1-7: A signal set of two, '0' $\rightarrow [+1,+1,+1]$ and '1'$\rightarrow [-1,-1,-1]$ is used in 3 bit repetition code. Received samples are $[+0.9, -0.2, -0.1]$. Show that CM can be computed in a matrix form and fill one blank.

$$\begin{bmatrix} +1 & +1 & +1 \\ -1 & -1 & -1 \end{bmatrix} \begin{bmatrix} +0.9 \\ -0.2 \\ -0.1 \end{bmatrix} = \begin{bmatrix} +0.6 \\ ? \end{bmatrix}$$

P1-8: Repeat the computation of CM with a quantized version of the received samples, i.e., $[+0.9, -0.2, -0.1] \rightarrow [+1, -1, -1]$ with a signal set of two, '0' $\rightarrow [+1,+1,+1]$ and '1'$\rightarrow [-1,-1,-1]$.

P1-9: 3-bit tuple of BPSK symbols can be detected by correlation metric (CM). It is an alternative to symbol by symbol detection and it requires more computation than the latter but nonetheless it is another way of detection.

3-bit tuple of all possible data and BPSK symbols in matrix form

$$X = \begin{bmatrix} 0 & 0 & 0 \\ 0 & 0 & 1 \\ 0 & 1 & 0 \\ 0 & 1 & 1 \\ 1 & 0 & 0 \\ 1 & 0 & 1 \\ 1 & 1 & 0 \\ 1 & 1 & 1 \end{bmatrix} \rightarrow \{`0` + 1, `1` - 1\} \rightarrow Y = \begin{bmatrix} +1 & +1 & +1 \\ +1 & +1 & -1 \\ +1 & -1 & +1 \\ +1 & -1 & -1 \\ -1 & +1 & +1 \\ -1 & +1 & -1 \\ -1 & -1 & +1 \\ -1 & -1 & -1 \end{bmatrix}$$

Received 3 BPSK symbols are $[-0.9 + 0.1 + 0.6]$. Compute CM for each 3-bit tuple; fill the blank and note that only 4 CM computation is good enough to find the largest.

$$\begin{bmatrix} +1 & +1 & +1 \\ +1 & +1 & -1 \\ +1 & -1 & +1 \\ +1 & -1 & -1 \\ -1 & +1 & +1 \\ -1 & +1 & -1 \\ -1 & -1 & +1 \\ -1 & -1 & -1 \end{bmatrix} \begin{bmatrix} -0.9 \\ +0.1 \\ +0.6 \end{bmatrix} = \begin{bmatrix} -0.2 \\ -1.4 \\ -0.4 \\ ? \\ +1.6 \\ +0.4 \\ +1.4 \\ +0.2 \end{bmatrix} \leftarrow \text{max}$$

Confirm the result of CM detection is the same as symbol by symbol detection.

P1-10: Draw the probability density function of Gaussian with mean $= 0$ and variance $= 1$. Then indicate $x = 1.0$, $Q(x)$ on the drawing.

Chapter 2
Digital Modulations

Contents

2.1	Constellation, Complex Envelope, and CW	37
	2.1.1 OOK, BPSK, and Orthogonal (M = 2)	38
2.2	Power of Digitally Modulated Signals and SNR	42
	2.2.1 Discrete Symbol Average Power $\left(\sigma_s^2\right)$ Computation	43
	2.2.2 Power Spectral Density	43
	2.2.3 Signal Power and Noise Power Ratio (SNR)	44
2.3	MAP and ML Detectors	45
	2.3.1 Symbol Error Rate of BPSK Under AWGN Channel	47
	2.3.2 SER of OOK and Orthogonal signaling (M = 2) Under AWGN Channel	49
2.4	PAM, QAM, and PSK	53
	2.4.1 M-PAM	53
	2.4.2 Square M- QAM	55
	2.4.3 PSK, APSK, and DPSK	62
2.5	BER and Different forms of SNR	66
	2.5.1 BER Requires a Specific Bit to Symbol Mapping	67
	2.5.2 A Quick Approximation of SER to BER conversion	67
	2.5.3 Numerical Simulations of BER with LLR	69
	2.5.4 SNR in Different Forms	70
2.6	Offset QAM (or Staggered QAM)	72
	2.6.1 SER vs. $\frac{E_s}{N_o}$ Performance of Staggered QAM	73
	2.6.2 CCDF of Staggered QAM vs. of 'Regular' QAM	74
	2.6.3 Other Issues	75
2.7	Digital Processing and Spectrum Shaping	75
	2.7.1 Scrambler	76
	2.7.2 Differential Coding or phase invariance coding	80
	2.7.3 Partial Response Signaling	84
2.8	Frequency Modulation – FSK, MSK, CPFSK	93
	2.8.1 Examples of FSK Signal Generation	94
	2.8.2 Non-coherent Demodulation of FSK	96
	2.8.3 FSK Signal Generations Using Quadrature Modulator	97
	2.8.4 Binary CPFSK Example	98
	2.8.5 M-Level CPFSK	98
	2.8.6 MSK	101
	2.8.7 FSK with Gaussian Pulse	102
	2.8.8 Power Spectral Density of CPFSK, MSK, and GMSK	103
	2.8.9 Partial Response CPFSK	105
	2.8.10 SER Performance Analysis of CPFSK	106

© Springer Nature Switzerland AG 2020

S.-M. Yang, *Modern Digital Radio Communication Signals and Systems*,

https://doi.org/10.1007/978-3-030-57706-3_2

2.9   PSD of Digitally Modulated Signals 108
        2.9.1 Power Spectral Density of PAM Signal 109
        2.9.2 PSD of Quadrature Modulated Signals 111
        2.9.3 PSD of FDM and OFDM Signals 112
        2.9.4 PSD Numerical Computations and Measurements 112
        2.9.5 Numerical Computation of PSD Using FFT 114
        2.9.6 Example of PSD Computation by Using FFT 115
        2.9.7 PSD of Digital FM Signals –FSK, CPFSK 116
        2.9.8 PSD Computation Using Correlation 117
2.10   Chapter Summary and References 118
        2.10.1 Summary 118
        2.10.2 References 118
2.11   Problems 119

Abstract We partition the communication signals into three – constellation, complex envelope (continuous I and Q baseband) and CW (real RF signal). Our focus in this chapter is on discrete-time representations but we have to relate them to complex envelope, and CW. This is done with binary cases – BPSK (2-PAM), OOK and orthogonal (a form of FSK) using a rectangular pulse for converting to complex envelope and quadrature modulator to CW. Then binary modulations are extended to multilevel ones – PAM, QAM, and PSK. Additional topics that arise due to practical reasons are covered; offset QAM, scrambler, 180° and 90° differential coding. PRS is an attempt to reduce the bandwidth. FSK is explored for non-coherent detection and its constant envelope property. The computation of PSD is explored analytically as well as numerically. Numerical simulation method is universally applicable even when no analytical solution is available.

Key Innovative Terms 1+D form of OOK

General Terms 1+D system · 8-QAM · 32-QAM · APSK · AWGN · BER · bit to symbol mapping · BPSK · complex baseband envelope · CPFSK · CW · digital constellation · DPSK · DSQ constellation · frequency modulation · FSK · Gaussian pulse · LLR · MAP · ML · M-PAM · M-QAM · OOK · orthogonal · phase invariance coding · power spectral density · PN · primitive polynomial · PRS · PSK · QAM · QPRS · QPSK · rotation invariance · scrambler · staggered QAM · SER · SNR

List of Abbreviations

A/D	analog digital converter
APSK	amplitude phase shift keying
AWGN	additive white Gaussian noise
BER	bit error rate
BPSK	binary phase shift keying
CCDF	complimentary cumulative density function
CM	correlation metric

CPFSK	continuous phase frequency shift keying
CW	continuous wave or carrier wave
D/A	digital analog converter
DFT	discrete Fourier transform
DPSK	differential phase shift keying
DSQ	double square (constellation)
FDM	frequency division multiplex
FFT	fast Fourier transform
FM	frequency modulation
FSK	frequency shift keying
IDFT	inverse discrete Fourier transform
I, Q	in phase, quadrature
LHS	left hand side
LNA	low noise amplifier
LPF	low pass filter
MAP	maximum a posterior probability
ML	maximum likelihood
MSK	minimum shift keying
OOK	n off keying
OFDM	orthogonal frequency division multiplex
PAM	pulse amplitude modulation
PA	power amplifier
PDA	personal digital assistant
PDF	probability density function
PN	pseudo noise
PRS	partial response system
PSD	ower spectral density
PSK	phase shift keying
QAM	quadrature amplitude modulation
QPRS	quadrature partial response system
QPSK	quadrature phase shift keying
RF	radio frequency
SER	symbol error rate
SNR	signal to noise ratio
TFM	tamed frequency modulation
USB	universal serial bus

Data stream in binary {1, 0} should be converted to modulation symbols, first discrete in time and then complex envelope (real and imaginary or also called I and Q) and then to continuous carrier wave (CW) by using a quadrature modulator for radio transmission. This process is summarized in Figure 2-1. The same figure in the overview chapter is repeated here in order to remind the reader that the signal to be transmitted in the air is a CW and of the partitioning of the overall signal formation process.

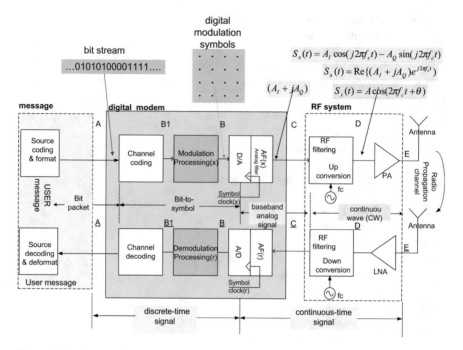

Figure 2-1: Digital radio communication system block diagram

In this chapter, our focus is on Modulation Processing (x and r) block – B_1 to B in Figure 2-1, where a stream of bits are converted to a stream of modulation symbols – called bit to symbol mapping – and then followed by complex envelope representations. Channel coding will not be considered in this chapter. It will be separately treated in Chapter 6. Note that it is essential in modern digital transmission. D/A and filtering (pulse shaping) will be treated as needed basis here using a simple pulse such as rectangular, and comprehensively described further in Chapter 3.

Note that the modulation symbols are complex numbers (real and imaginary part) in general. In modern wireless transmission we use a carrier wave practically in most situations. If not complex, we may consider it real, i.e., the imaginary is zero without loss of generality. In base band transmission (i.e., no carrier wave), it may be more convenient to consider the modulation symbols are real.

As we focus on the shaded blocks of Modulation Processing (both transmit and receive), the RHS from B-$\underline{B}$ may be considered as a channel. They can be represented in discrete time. Note that the channel may include the subsystems of PA and LNA as well as antennas and radio wave propagations. On the other hand LHS of $B_1 - \underline{B}_1$ may be abstracted to a stream (sequence) of random bits even though in practice it may have a complicated packet structure of many bits. We do not consider channel coding at this point. Later we develop the signaling schemes including coding combined with modulations. This is shown in Figure 2-2.

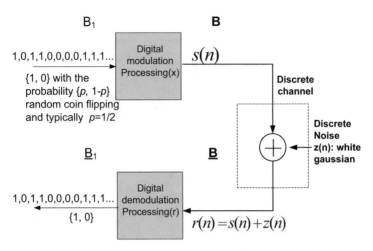

Figure 2-2: Digital modulations with discrete Gaussian noise channel

Abstracting Figure 2-1 into Figure 2-2, it is implicitly assumed that the implementations of the modulation and demodulation, D/A and sampling (A/D), and synchronization of carrier phase and symbol timing are all perfect. Only the thermal noise at the input of LNA is the channel impairment. It might not sound practical but it is important to understand the performance under additive white Gaussian noise (AWGN) and it provides practically important insight as well as reference point of a system error rate performance. The thermal noise is accurately represented by AWGN. Later we will include additional channel models such as fading (which may be represented by a time varying linear filter), and other circuit imperfections.

2.1   Constellation, Complex Envelope, and CW

In this section we consider digital modulation schemes that are represented on the complex plane (real and imaginary in mathematics) which is also called constellation plane or I-Q plane (in phase and quadrature phase) with radio communication terminology. Sometimes it is convenient to think I-Q plane to be two channels - I-channel and Q-channel. For example offset QAM two I-Q channels are in 90° apart in carrier phase but there is ½ symbol time difference in time. This case will be treated in later section. However, in most signaling schemes we consider here use a complex symbol. In other words, real and imaginary part is sent at the same time.

We give binary signal examples, OOK, BPSK, and Orthogonal (M = 2) in order to establish that RF signal can be decomposed into constellation, baseband complex envelope (I, Q) and CW. Then our focus will be on I-Q plane representation of constellations.

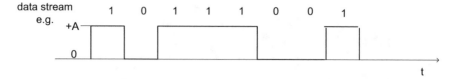

data stream
e.g.

Figure 2-3: A waveform example of OOK (without CW modulation)

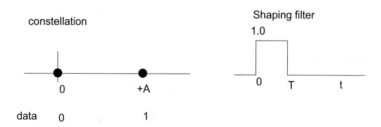

Figure 2-4: OOK and its constellation and a shaping filter

2.1.1 OOK, BPSK, and Orthogonal (M = 2)

In USB[1] or on the backplane of a computer it uses a simple signaling scheme since the cable is short and thus the channel is fairly clean. From the radio communications the same scheme sometimes is used, which is called **OOK** (on-off keying). An example waveform may look like Figure 2-3.

For our purpose it will be decomposed into two parts, and can be represented in Figure 2-4; the constellation and its shaping pulse being rectangular. Data 1 is represented by a voltage +A while data 0 by zero voltage. This is explained further below.

And the waveform in Figure 2-3 can be obtained by bit to symbol mapping as described in the constellation plane $\{0 \rightarrow 0$ and $1 \rightarrow +A\}$, and this impulse train of $\{0, +A\}$ pass through the shaping filter. The filtering can be described as a convolution; $\sum_{n=-\infty}^{n=+\infty} a(n)g(t - nT)$ where $g(t)$ is a rectangular pulse and $a(n)$ is modulation symbol and comes from $\{0, +A\}$. Since the duration of $g(t)$ is confined to the symbol period, the convolution is simple. It is a one to one mapping as shown in Figure 2-3; '0' corresponds to zero voltage and '1' corresponds to a rectangular pulse.

---

[1]Universal Serial Bus (USB) is an industry standard that defines the cables, connectors and communications protocols used in a bus for connection, communication and power supply between computers and electronic devices. USB was designed to standardize the connection of computer peripherals, such as keyboards, pointing devices, digital cameras, printers, portable media players, disk drives and network adapters to personal computers, both to communicate and to supply electric power. It has become commonplace on other devices, such as smart phones, PDAs and video game consoles.

We now change the OOK constellation to BPSK while keeping the same pulse shape.

BPSK (binary shift keying) is described by two parts; a constellation and a shaping filter as shown in Figure 2-5. The filtering can be described as a convolution; $\sum_{n=-\infty}^{n=+\infty} a(n)g(t-nT)$ where $g(t)$ is a rectangular pulse and $a(n)$ is modulation symbol and comes from $\{-1, +1\}$. Compared to an example of OOK in Figure 2-4, only the constellation is different while the pulse shape is the same.

An example data pattern (the same data as Figure 2-3) and its corresponding waveform is shown in Figure 2-6.

It is convenient to separate the constellation and the pulse shaping filter and furthermore one can describe many different possibilities of digital modulations. In practice a filter design can be done separately. In passing we mention that there are a variety of the pulse shape filters other than a rectangular pulse; e.g., an ideal low pass filter and a family of raised cosine filters. This will be covered in Chapter 3.

In both OOK and BPSK the complex envelope has only real part and the imaginary part is zero. It can be converted to a CW by a quadrature modulation where the input has only real part. In practice it is simpler than a full quadrature modulator. An example carrier wave for BPSK with the same data is shown in Figure 2-7. In Figure 2-7, the carrier frequency is about 1.41 times of symbol rate and there are 100 points per symbol period for the drawing purpose. During the data change there is a carrier phase discontinuity. The carrier phase carries the data information $\{0°, 180°\}$. For **coherent detection**, the carrier phase of the incoming signal must be synchronized with a local carrier phase. In passing we mention that the sensitivity of the carrier phase error is not high for OOK and BPSK. This issue will be explored later in great detail.

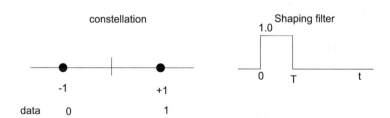

Figure 2-5: BPSK constellation and a shaping filter

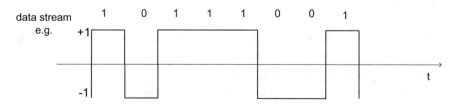

Figure 2-6: BPSK example with a rectangular pulse

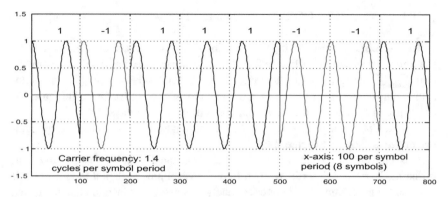

Figure 2-7:   Carrier wave of BPSK of Figure 2-6

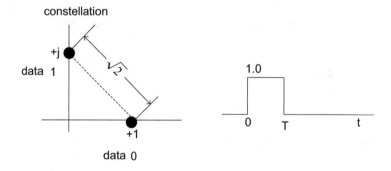

Figure 2-8:   RF waveform of OOK example in Figure 2-3

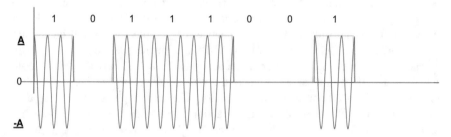

Figure 2-9:   Orthogonal signaling (M = 2) with shaping pulse

Exercise 2-1:   Plot CW of OOK in Figure 2-3 with carrier frequency is 3 times of symbol rate. Answer: see Figure 2-8.

Now we consider **orthogonal signaling** with M = 2 and its constellation diagram is shown in Figure 2-9. For data 1 its corresponding symbol is +j and for data 0, the symbol is +1. Here both I and Q channels are used but alternatively, i.e., when I-channel is in use for data 1, Q-channel is not used, and vice versa.

This orthogonal signaling might be considered as two OOK signaling combined; one in I-channel and other in Q-channel. Both I and Q channel may use a rectangular pulse filter (or other type of filter). See Figure 2-10 below.

In order to have a CW it should use a (full) quadrature modulator since its complex envelope contains both real part and imaginary parts; we need to compute

$$S_x(t) = A_I \cos (j2\pi f_c t) - A_Q \sin (j2\pi f_c t)$$

Exercise 2-2: Consider 90° rotated (counter-clockwise) BPSK constellation and draw it. Answer: Typically we consider bit symbol mapping in Figure 2-5 (i.e., constellation points on the real axis). What is the difference with the same shaping filter? CW signal is rotated by 90° and thus it is given by

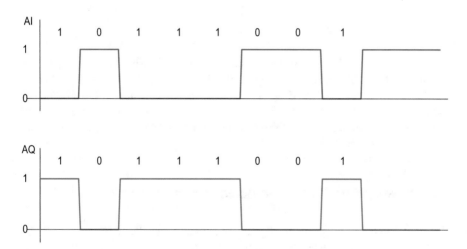

Figure 2-10: Complex envelope of M = 2 orthogonal signal; real (I) and imaginary (Q) part

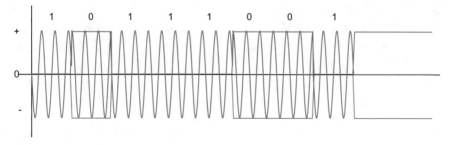

Figure 2-11: CW, output of quadrature modulator of Figure 2-10, $S_x(t)$

$$S_x(t) = -A_Q \sin (j2\pi f_c t)$$

A_Q is the same as in Figure 2-6.

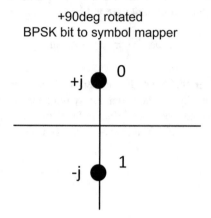

+90deg rotated
BPSK bit to symbol mapper

In this section, with binary signal examples, we established the decomposition into constellation and baseband complex envelope using a shaping pulse, CW by a quadrature modulation. This decomposition is very useful in many ways; implementation, performance analysis, understanding of a system, and even for inventing new ways. For example, for performance analysis we use constellations of a signal.

2.2   Power of Digitally Modulated Signals and SNR

We now compute the power of digitally modulated signals. It has two parts the power of symbol sequence and the power gain of a filter; the signal power (P) is shown in Figure 2-12, and the average power of a symbol sequence is denoted by σ_s^2. The filter power gain is given by $\frac{1}{T}\int_{-\infty}^{+\infty}|h(t)|^2 dt$ and it is 1.0 for the rectangular pulse. Figure 2-12 summarizes the process of digitally modulated complex envelope generation including a shaping filter.

Note that the signal power is determined by the discrete symbol average power σ_s^2 when the power gain of a shaping filter is unity. One can adjust the filter power gain to be unity. Thus we need to compute the discrete symbol power for different constellations.

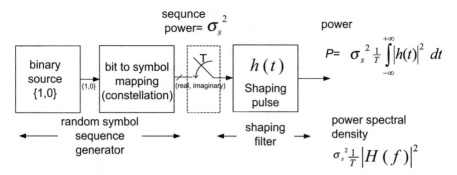

Figure 2-12: Power and power spectral density of digital modulated signals

2.2.1 Discrete Symbol Average Power $\left(\sigma_s^2\right)$ Computation

For BPSK constellation, $\sigma_s^2 = (+1)^2 p + (-1)^2 (1 - p) = 1$.

For OOK constellation, $\sigma_s^2 = (+A)^2 p + (0)^2 (1 - p) = A^2 p$. Assuming the probability of A, $p = 1/2$, what is A to make OOK power to be 1.0? The answer is $A = \sqrt{2}$.

For orthogonal signaling of M $= 2$, $\sigma_s^2 = (+1)^2 p + (+1)^2 (1 - p) = 1$. Note that the distance between the two constellation points is $\sqrt{2}$.

2.2.2 Power Spectral Density

The power spectral density is determined by the square of the frequency response of the shaping filter ($H(f)$) as shown in Figure 2-12, and is given by $\sigma_s^2 \frac{1}{T} |H(f)|^2$. The spectrum of these signals is essentially determined by the shaping filter. This is true if $\{1, 0\}$ is random and uncorrelated and there is no digital filter. In other words, the spectral density can be changed digitally or by adding correlation into random binary sequence. Here we consider the case depicted in Figure 2-12, and the modification of spectral density digitally will be treated separately.

Exercise 2-3: When the voltage level of d is assigned to each constellation points, the power is multiplied by d^2; $\sigma_s^2 = 1 \longrightarrow \sigma_s^2 = d^2$. Note that bit assignement can be opposite to that of Figure 2-5 as below; '0' $\rightarrow$ +d and '1' $\rightarrow$ -d.

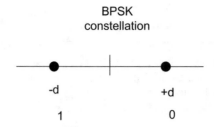

BPSK
constellation

2.2.3   Signal Power and Noise Power Ratio (SNR)

Consider BPSK first. In discrete model as in Figure 2-5 the digital signal power is denoted by $P_s = \sigma_s^2 = d^2$ for BPSK with d voltage. The noise power of AWGN is denoted by $P_n = \sigma^2$ per channel (I-channel only for BPSK). It is the same as its variance if its average (m) is zero. Thus for BPSK digital SNR is given by $\frac{P_s}{P_n} = \frac{d^2}{\sigma^2}$.

This digital SNR should be related with the thermal noise density (N_o) and the symbol energy (E_s), where it is related with signal power as $P_s = \frac{E_s}{T} = d^2$ and T is the symbol period.

$P_n = \sigma^2 = \frac{N_o}{2} B$; the noise power in I-channel only. When both I and Q-channels are used it will be twice.

$$\text{For BPSK,} \quad \frac{d}{\sigma} = \sqrt{\frac{\frac{E_s}{T}}{\frac{N_o}{2} B}} = \sqrt{2 \frac{E_s}{N_o}}.$$

The minimum noise bandwidth is the same as symbol rate, $B = 1/T$. This is justified in Chapter 3, and is explained in Chapter 9 (Section 9.1.3.4 noise power).

Note that in BPSK, $E_s = E_b$, i.e., bit energy is the same as symbol energy.

M-PAM is an extension of BPSK (or 2-PAM). M = 8 example.

8-PAM
constellation

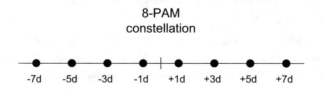

Exercise 2-4: Find the digital power of 8-PAM assuming symbols are equally likely ($p = 1/8$). Answer: $\{1^2 + 3^2 + 5^2 + 7^2 + (-1)^2 + (-3)^2 + (-5)^2 + (-7)^2\}$ $*1/8 = 168/8 = 21$, and thus $21d^2$.

Exercise 2-5: Derive the digital power of PAM, in general; $\frac{M^2-1}{3}d^2$.
Hint:$1^2 + 3^2 + 5^2 + \ldots + (M-1)^2 = \frac{M^2-1}{36}M$, using $1^2 + 2^2 + 3^2 + 4^2 + \ldots + k^2 = k(k+1)(2k+1)/6$.

The above argument of expressing $\frac{d}{\sigma}$ in terms of $\frac{E_s}{N_0}$ can be extended to M-PAM by noting that in M-PAM, $P_s = \frac{E_s}{T} = \frac{M^2-1}{3}d^2$, and $P_n = \sigma^2 = \frac{N_0}{2}B$.
In M-PAM,

$$\frac{d}{\sigma} = \sqrt{\frac{\frac{E_s}{T}}{\frac{N_0}{2}B}\frac{3}{M^2-1}} = \sqrt{\frac{3}{M^2-1}\left(2\frac{E_s}{N_0}\right)} \text{ if the noise bandwidth B} = 1/T.$$

Exercise 2-6: Express $\frac{P_s}{P_n}$ in terms of $\frac{E_s}{N_0}$.
Answer: $\frac{P_s}{P_n} = 2\frac{E_s}{N_0}\frac{1}{BT} = 2\frac{E_s}{N_0}$ if BT = 1, i.e. the noise bandwidth B = 1/T.
Note that in M-PAM, $E_b = \frac{E_s}{\log_2 M}$ since there are $\log_2 M$ bits per symbol.

2.3 MAP and ML Detectors

We consider AWGN channel in Figure 2-2. {+1, −1} symbols are transmitted sequentially one at a time and it is contaminated by thermal noise, modeled by additive white Gaussian, and thus the received sample is not exactly {+1, −1} but scattered around each symbol. Its distribution around a symbol, +1 or −1, is Gaussian with the mean +1 or −1. Thus intuitively, if a received sample is close to '+1' the decision is +1 is sent and if a received sample is close to '−1', the decision is −1 is sent. The decision boundary is zero; if a sample >0 then +1 sent, if a sample <0 then −1 sent. If it is exactly zero, coin flipping decision is probably fine. As it turns out that this intuition is exactly correct if +1 and −1 are equally likely, i.e., $p\{+1\} = p\{-1\} = 1/2$. If not equally likely then the decision boundary should be adjusted accordingly.

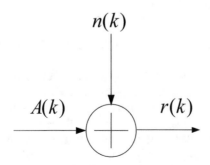

Discrete AWGN channel

We now need to be quantitative in order to compute symbol error rate probability. The detection strategy is the maximum conditional probability after observing received sample $r(k)$ where k is the sequence index. And $r(k) = A(k) + n(k)$ with $A(k) = \{+1, -1\}$ and $n(k)$ is a noise sample. This is shown in the figure (LHS).

Maximum a posterior probability (MAP) detection compares the conditional probabilities of

$$\Pr\{A(k) = +1|r(k)\} \geq \Pr\{A(k) = -1|r(k)\}; \text{then decision} : +1 \text{ sent}$$
$$\Pr\{A(k) = -1|r(k)\} > \Pr\{A(k) = +1|r(k)\}; \text{then decision} : -1 \text{ sent}$$

for each k.

The conditional probabilities are expressed as

$$\Pr\{A(k) = +1|r(k)\} = \frac{\Pr\{r(k)|A(k) = +1\}\,\Pr\{A(k) = +1\}}{\Pr\{r(k)\}}$$

$$\Pr\{A(k) = -1|\,r(k)\} = \frac{\Pr\{r(k)|A(k) = -1\}\,\Pr\{A(k) = -1\}}{\Pr\{r(k)\}}$$

It is important to see the meaning of these probability expressions. At first it might look complicated but once its 'physical' meaning is understood, it is straightforward. Note that for the decision process the actual computation of $\Pr\{r(k)\}$ is not necessary.

Exercise 2-7: Compute $\Pr\{r(k)\}$.

Answer: $\Pr\{r(k)\} = \Pr\{r(k)|A(k) = +1\}p + \Pr\{r(k)|A(k) = -1\}(1-p)$. It is a normalization factor so that the conditional probability is within 0 to 1.

$\Pr\{A(k) = +1\} = p$ and $\Pr\{A(k) = -1\} = 1 - p$ and the index k in probability expression may be not necessary since it does not depend on the index.

$\Pr\{r(k)|A(k) = +1d\}$ is a Gaussian with mean $= +1d$ and variance σ^2. Similarly $\Pr\{r(k)|A(k) = -1d\}$ is a Gaussian with mean $= -1d$ and variance σ^2. See Figure 2-13.

Maximum likelihood (ML) detection compares the conditional probability of

$$\Pr\{r(k)|A(k) = +1\} \geq \Pr\{r(k)|A(k) = -1\} \text{ decision} : +1 \text{ sent}$$
$$\Pr\{r(k)|A(k) = -1\} > \Pr\{r(k)|A(k) = +1\} \text{ decision} : -1 \text{ sent}$$

This ML is equivalent to MAP if $p = 1/2$, i.e., both symbols are equally likely. The decision boundary based on ML is shown in Figure 2-13.

Minimum distance detection compares the Euclidean distance from a received sample. The decision boundary is the same as ML detection.

Correlation metric (CM) receiver compares the correlation between a received sample to +1 and −1; choose the bigger from $\{r(k)(+1), r(k)(-1)\}$. This is the same as decide +1 if $r(k) > 0$, decide −1 otherwise. Thus the decision boundary is $r(k)$

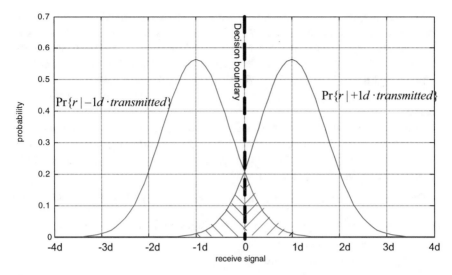

Figure 2-13: $\Pr\{\mathbf{r}(\mathbf{k})|A(k) = -1\mathbf{d}\}$ in red (LHS) and $\mathbf{Pr\{r(k)|}A(k) = +1\mathbf{d}\}$ in blue (RHS)

$(+1) = r(k)\ (-1)$ and thus $r(k) = 0$, which is the same as ML and minimum distance detector.

Example 2-1: $r(k) = 0.5$ find the Euclidean distance to $+1$ and -1 (with being $d = 1$).

   Answer: $(r-1)^2 = 0.25$ to $+1$, and $(r-(-1))^2 = 2.25$. Obviously it is closer to $+1$ than -1.

Example 2-2: $r(k) = 0.5$ find the correlation to $+1$ and -1.

   Answer: $0.5 * (+1) = 0.5$ with $+1$ and $0.5 * (-1) = -0.5$ with -1. It has a bigger correlation with $+1$ than -1. This CM receiver can be extended to a multi-bit symbol like M-PAM, which will be explored in channel coding context later.

   These different detections schemes will be explored in different situations later. But in BPSK they all point to the same as partitioning the decision boundary. When symbols $\{+1, -1\}$ are equally likely the decision boundary is zero. This is shown in Figure 2-13.

2.3.1   Symbol Error Rate of BPSK Under AWGN Channel

The **symbol error rate (SER)** performance of BPSK will be obtained from Figure 2-13 for a discrete channel representation of AWGN.

We mention why SER rather than bit error rate (BER). In BPSK case, a symbol is one bit thus SER is the same as BER. This is generally true with M = 2 for PAM, OOK and orthogonal signaling. For M-level modulation, e.g., M-PAM, SER is not the same as BER. In M-level signaling SER does not depend on the bit to symbol mapping and hence is straightforward. BER can be related to SER if bit to symbol relationship is known but, it may not be straightforward. Thus we consider SER to be generic.

SER is a function of signal power to noise power ratio (SNR). Intuitively it is clear that the higher the SNR the lower the SER. We first express it in terms of $\frac{d}{\sigma}$, which can be related to symbol energy (E_s) and noise power density (N_o) as shown in the previous section (Section 2.2.3).

Derivation of SER of BPSK $\Pr\{r(k)|A(k) = +1\text{d}\} = \frac{1}{\sigma\sqrt{2\pi}}e^{-\frac{(r-d)^2}{2\sigma^2}}$ and $\Pr\{r(k)|A(k) = -1\text{d}\} = \frac{1}{\sigma\sqrt{2\pi}}e^{-\frac{(r+d)^2}{2\sigma^2}}$ are displayed in Figure 2-13 with $\sigma^2 = 0.5$ (for concreteness) and d voltage.

Two possible errors; see Figure 2-13.

(1) SER happens when -1d sent but $r(k) > 0$ (red hatched area) with $1-p$ probability
(2) SER happens when $+1$d sent but $r(k) <= 0$ (blue hatched area) with p probability

In terms of equations, (1) and (2) are expressed by

$$\text{SER} = (1-p)\int_0^{+\infty} \frac{1}{\sigma\sqrt{2\pi}}e^{-\frac{(r+d)^2}{2\sigma^2}}dr + p\int_0^{-\infty} \frac{1}{\sigma\sqrt{2\pi}}e^{-\frac{(r-d)^2}{2\sigma^2}}dr$$

$$= (1-p)\int_{d/\sigma}^{+\infty} \frac{1}{\sqrt{2\pi}}e^{-\frac{y^2}{2}}dy + p\int_{d/\sigma}^{+\infty} \frac{1}{\sqrt{2\pi}}e^{-\frac{y^2}{2}}dy.$$

$$\text{SER of BPSK} = Q\left(\frac{d}{\sigma}\right) = Q\left(\sqrt{2\frac{E_s}{N_o}}\right) \text{ where } Q(x)$$

$$= \int_x^{+\infty} \frac{1}{\sqrt{2\pi}}e^{-\frac{y^2}{2}}dy \text{ is defined.}$$

For this SER, it does not depend on $p = \frac{1}{2}$. But the symbols have an equal energy, i.e., $\{+d, -d\}$ and the decision boundary is zero. ML, minimum distance, and correlation detection have the detection threshold to be zero for BPSK.

Exercise 2-8: Consider BPSK $\{+1, -1\}$ under AWGN. Find a decision boundary for MAP detection for a given p (probability of '0' data) and noise level (σ).

Answer: The decision boundary, for MAP, is given by $\Pr\{r(k)|A(k) = +1\}$ $p = \Pr\{r(k)|A(k) = -1\}(1-p)$. With AWGN, it is given by $e^{-\frac{(r-1)^2}{2\sigma^2}}p =$

$e^{-\frac{(r+1)^2}{2\sigma^2}}(1-p)$. It simplifies to be $e^{\frac{2r}{\sigma^2}} = \frac{1-p}{p}$ and thus $r|_{boundary} = \frac{1}{2}\sigma^2 \ln\left(\frac{1-p}{p}\right)$. Obviously $p = 0.5$ it is zero. When $p = 0.6$ then the decision boundary is $r = -0.2027\sigma^2$.

Exercise 2-9: LLR (log likelihood ratio) is defined as $LLR = \ln\left(\frac{\Pr\{r(k)|+1\}}{\Pr\{r(k)|-1\}}\right)$

assuming $p = \frac{1}{2}$. Find it for BPSK $\{+1, -1\}$. Answer: It is given by $LLR =$

$\ln \dfrac{\frac{1}{\sigma\sqrt{2\pi}}e^{-\frac{(r-1)^2}{2\sigma^2}}}{\frac{1}{\sigma\sqrt{2\pi}}e^{-\frac{(r+1)^2}{2\sigma^2}}} = \ln \dfrac{e^{-\frac{(r-1)^2}{2\sigma^2}}}{e^{-\frac{(r+1)^2}{2\sigma^2}}} = \frac{2r}{\sigma^2}$ .The decision boundary is $LLR = 0$; if $LLR > 0$, '+1' sent otherwise '−1' sent.

2.3.2 SER of OOK and Orthogonal signaling (M = 2) Under AWGN Channel

We consider equally likely symbols ($p = \frac{1}{2}$) and compare to BPSK. In summary OOK and orthogonal signaling require 3dB more SNR compared to BPSK, which will be explained in this section. SNR in dB, $\frac{E_s}{N_0}$ [dB], is defined as $\frac{E_s}{N_0}$ [dB] $\equiv \log_{10} \frac{E_s}{N_0}$. Note that we implicitly assume that OOK and orthogonal signaling use carrier coherent (i.e., the carrier phase of incoming signals and a locally generated carrier phase are synchronized) demodulation.

We mention that in practice both signaling schemes are amenable to non-coherent demodulation, which is simpler than the coherent demodulation. However, there is further SNR penalty (perhaps small) with non-coherent demodulation. On the other hand BPSK must be demodulated coherently. 3dB gain of BPSK is obtained with the coherent demodulation. In modern implementations this increase of complexity of coherent demodulation is readily handled.

Consider OOK (M = 2) and its SER performance.

We may go through a similar procedure for OOK as we did for BPSK in order to obtain SER of OOK. However, from Figure 2-14, SER of OOK can be obtained by comparing with Figure 2-13. Its distance to decision boundary is changed from d to $\frac{\sqrt{2}}{2}d$ i.e., the distance between symbols from $2d$ to $\sqrt{2}d$. Note that the signal power of OOK is the same as BPSK. We note that the translation of coordinates does not change the probability involved. This is also true for rotation assuming the independent Gaussian noise per channel (dimension).

Thus we conclude that

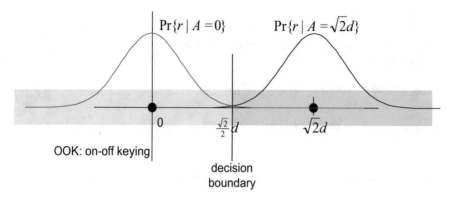

Figure 2-14: OOK decision boundary

$$\textbf{SER of OOK} = Q\left(\frac{\sqrt{2}}{2}d/\sigma\right) = Q\left(\sqrt{\frac{E_s}{N_o}}\right).$$

Consider orthogonal coherent signaling (M = 2). We argue and show next that

$$\textbf{SER of Orthogonal (M = 2)} = Q\left(\frac{\sqrt{2}}{2}d/\sigma\right) = Q\left(\sqrt{\frac{E_s}{N_o}}\right).$$

This is exactly the same as OOK.

We note that the translation or rotation of coordinates does not change the probability involved assuming that the Gaussian noise is independent in coordinates, i.e., I, Q channels. This is shown in Figure 2-15.

Exercise 2-10: Explain the SER of orthogonal signaling is equivalent to that of OOK; first find the decision boundary, and then obtain SER by computing the tail probability. Note that the probability density we need to use is 2 dimensional.

Answer: First the decision boundary, for MAP, is given by $\Pr\{r(k)|A(k) = +1\}$
$p = \Pr\{r(k)|A(k) = +j\}(1 - p)$. In AWGN case, it is given by
$\frac{1}{\sqrt{2\pi\sigma^2}}e^{-\frac{(x-1)^2}{2\sigma^2}}\frac{1}{\sqrt{2\pi\sigma^2}}e^{-\frac{(y)^2}{2\sigma^2}}p = \frac{1}{\sqrt{2\pi\sigma^2}}e^{-\frac{(y-1)^2}{2\sigma^2}}\frac{1}{\sqrt{2\pi\sigma^2}}e^{-\frac{(x)^2}{2\sigma^2}}(1 - p)$. After a bit of manipulation we obtain $y = x + \sigma^2\ln\left(\frac{p}{1-p}\right)$; it becomes $y = x$ if equally likely.

SER is related only with the computation of probability, and 2-dimensional independent Gaussian can be, due to circular symmetry, put into a single variable Gaussian for the probability computation. This is illustrated by the figure on LHS.

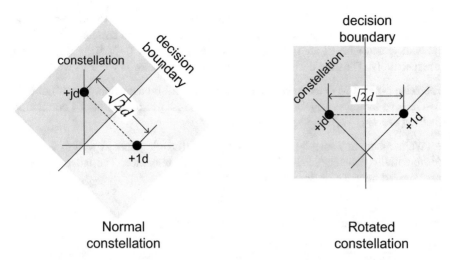

Figure 2-15: Orthogonal signaling (M = 2)

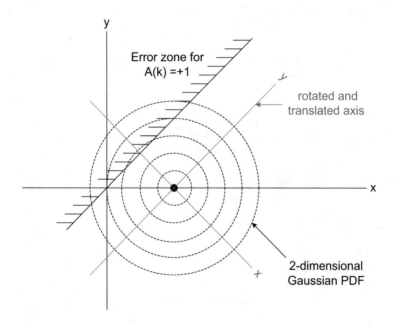

Exercise 2-11: Confirm the decision boundary in Figure 2-15 for orthogonal signaling with minimum distance detector. Assume d = 1 and equally likely. Answer: A received $r = (x, y)$ will have the decision boundary when the distance from +j to r and that from +1 to r is the same. $(y-1)^2 + x^2 = (x-1)^2 + y^2$. Solving it we obtain $y = x$.

Exercise 2-12: Confirm the decision boundary in Figure 2-15 for orthogonal signaling with correlation metric (CM) detector. Answer: The correlation $r^* = (x,-jy)$ with $+j$ and with $+1$ are $0^*x + (-jy)j = y$, and $(+1)^* x + (-jy)^*0 = x$, i.e., y and x respectively. Thus the decision boundary for CM detector is $y = x$ as well.

Exercise 2-13: CM detector can be extended to multi-bit detection. Works out with two receive samples r_1, r_2 for BPSK. Answer: Choose maximum out of 4 correlation metric $\{+r_1 +r_2, -r_1 -r_2, -r_1 +r_2, +r_1 -r_2\}$.

We tabulate, in Table 2-1, SER of BPSK, OOK, orthogonal as well as 2-DPSK (differentially coherent), 2-FSK (non-coherent) for reference. The latter two were not covered here. These are plotted in Figure 2-16.

Table 2-1: SER of BPSK, OOK, orthogonal, DPSK for $M = 2$

Mod	symbol error rate for $M = 2$				
	BPSK	**OOK**	**DPSK**	**FSK**	
Channel	Coherent	coherent	Differentially coherent	Orthogonal,	Non-coherent
AWGN	$Q\left(\sqrt{2\frac{E_s}{N_o}}\right)$	$Q\left(\sqrt{\frac{E_s}{N_o}}\right)$	$\frac{1}{2}\exp\left(-\frac{E_s}{N_o}\right)$	$Q\left(\sqrt{\frac{E_s}{N_o}}\right)$	$\frac{1}{2}\exp\left(-\frac{1}{2}\frac{E_s}{N_o}\right)$

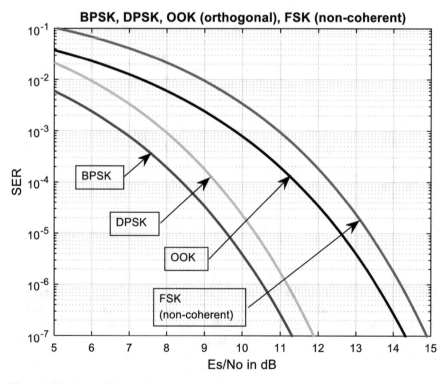

Figure 2-16: SER of BPSK, OOK, Orthogonal and DPSK, FSK (non-coherent)

2.4 PAM, QAM, and PSK

2.4.1 M-PAM

BPSK can be extended to multi-levels (M-level). Then the amplitude (and sign) carries the information, and BPSK may be considered as 2-**PAM (pulse amplitude modulation)**. 4-PAM constellation is shown in Figure 2-17, and it can be extended to M-level, where M = 2, 4, 8, 16, 32.. so on with M being the power of 2. M-level carries $\log_2 M$ bits of information, e.g., PAM with M = 32 carries 5 bits per symbol.

We explain more with M = 4 example in Figure 2-17.

We choose the constellation points $\{-3, -1, +1, +3\}$ so that the distance between the points are the same as BPSK, i.e., 2. For M-level it will be $\{-(M-1), -(M-3), \ldots, -3, -1, +1, +3, \ldots, +(M-3), +(M-1)\}$ and it ranges from $-(M-1)$ to $+(M-1)$. When a certain voltage level of d is assigned we may represent it as $\{-3d, -1d, +1d, +3d\}$ where d is the voltage level used.

For 4-PAM, assuming each symbol is equally likely (the probability is ¼ for each symbol), $\sigma_s^2 = 2\left[(+1)^2 + (+3)^2 + (-1)^2 + (-3)^2\right] * 1/4 = 5$.

For M-PAM, assuming each symbol is equally likely (the probability is 1/M for each symbol), $\sigma_s^2 = 2\left[(+1)^2 + (+3)^2 + .. + (M-1)^2\right] * 1/M = \frac{M^2-1}{3}$.

In this section SER of M-PAM using Figure 2-18 with M = 4 as an example. We assume that all symbols are equally likely $p = 1/M$.

For inner symbols $\{-1d, +1d\}$ it may make errors in both directions outside of the decision range. When $\{-1d\}$ symbol was sent, the received sample is the outside the ranges (hatched), symbol error will occur, and its probability is given by $2Q\left(\frac{d}{\sigma}\right)$. The same is true for $\{+1d\}$ and any inner symbols (in general M−2 inner symbols for M-PAM).

Figure 2-17: 4-PAM constellation

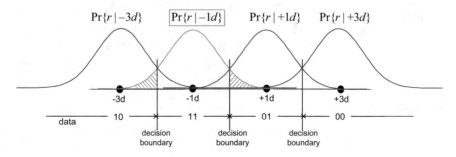

Figure 2-18: 4-PAM constellation and conditional PDF

2.4.1.1   SER of M-PAM Under AWGN Channel

For outer symbols $\{-3d, +3d\}$ it may make errors in only one side, and its probability is given by $Q(\frac{d}{\sigma})$. There are always two outer symbols for M-PAM.

Thus altogether SER is given by $2(\frac{M-1}{M})Q(\frac{d}{\sigma})$. Expressing $\frac{d}{\sigma} = \sqrt{\frac{6}{M^2-1}\frac{E_s}{N_o}}$ (see below)

$$\textbf{SER of M} - \textbf{PAM} = 2\Big(\frac{M-1}{M}\Big)Q\Big(\sqrt{\frac{6}{M^2-1}\frac{E_s}{N_o}}\Big).$$

Note that the above expression of SER, shown in Figure 2-19, is the same as BPSK with M = 2. In M-PAM, $P_s = \frac{E_s}{T} = \frac{M^2-1}{3}d^2$, and $P_n = \sigma^2 = \frac{N_o}{2}B$. Thus if the noise bandwidth B = 1/T.

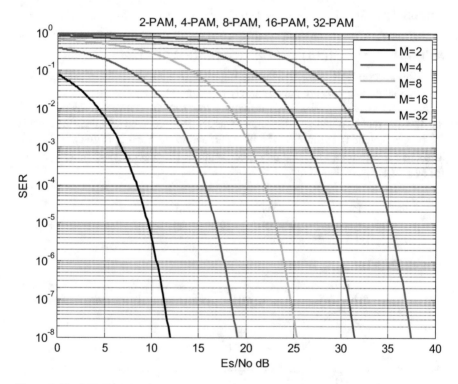

Figure 2-19:   M-PAM (M = 2, 4, 8, 16, 32) SER under AWGN

2.4.2   Square M- QAM

We now extend BPSK and M-PAM to QAM (quadrature amplitude modulation) with M constellation points on the complex plane. PAM has one dimension and QAM is of two dimensions. The naming of these different digital modulations comes from the history of their development and we continue to use them.

We consider QPSK and 16-QAM as in Figure 2-20. QPSK may be designated by 4-QAM equally well. The shade indicates the decision boundary. We consider only the equally likely symbols, i.e., each symbol probability the same as $p = \frac{1}{M}$.

The average power of QPSK is given by $P = 2d^2$. This is twice that of BPSK. It is shown below by following the definition of average for each symbol;

$$P = 2d^2 = \left[(1^2 + 1^2)\frac{1}{4} + \left((-1)^2 + 1^2 \right)\frac{1}{4} + \left((-1)^2 + (-1)^2 \right)\frac{1}{4} + \left(1^2 + (-1)^2 \right)\frac{1}{4} \right] d^2.$$

A similar process may be repeated for 16-QAM as above. However, it is not difficult to see the power of 16-QAM is the two times of 4-PAM $2\frac{m^2-1}{3}d^2 = 10d^2$ with m = 4, and thus the power of 16-QAM is $2\frac{M-1}{3}d^2 = 10d^2$ with M = 16. We use M or m consistently to be the number of constellation points.

This relationship is true for M-QAM with M being a number of constellation points; the power of rectangular M-QAM with M = 4, 16, 64, 256, ... = 2^{2k}(k = 1, 2, 3, ...) is $P = 2\frac{M-1}{3}d^2$. For example see Figure 2-21.

M-QAM with M = 8, 32, 128, 512, ... = 2^{2k+1}(k = 1, 2, 3, ...) are possible and used in practice. Two examples are shown in Figure 2-22.

8-QAM uses only black dots or includes 4 circles plus inner black dots and 12-QAM uses all points including 4 points of circles. Two consecutive 12-QAM

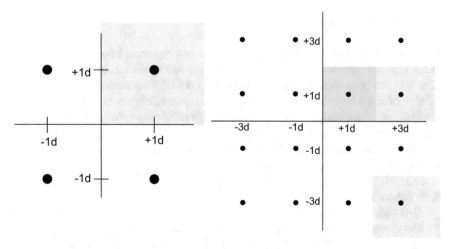

Figure 2-20: QPSK and 16-QAM constellations

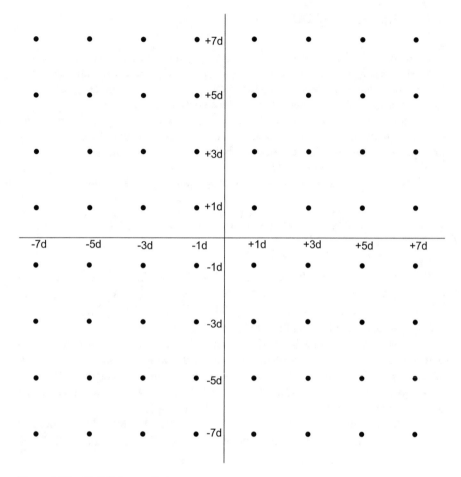

Figure 2-21: 64-QAM constellation

constellations may carry 5 bits since there are 144 equivalent constellation points (12 x 12). This is called 4 dimensional signaling and then each symbol carries 2.5 bits. This can be extended to three consecutive symbols (8-dimension). In this way a number of constellation points can be the integers other than power of 2. This type of constellations was used in very high bandwidth efficiency systems such as line of sight microwave radios as well as voice band modem.

The computation of power for these constellations is straightforward but some-what tedious for manual computations.

8-QAM: inner points power 2*4, and outer 10*4. Assuming equally likely, $p = 1/8$, its power is $(2*4 + 10*4)/8 = 6$.

32-QAM: inner 16 points; average 10 (from 16-QAM) *16 points, corner 4 sets of 4 points average 5 (from 4-PAM) *16, power due to shifting 5*5*16, and thus power = $(10*16 + 5*16 + 25*16)/32 = 20$. Note that d^2 part is dropped.

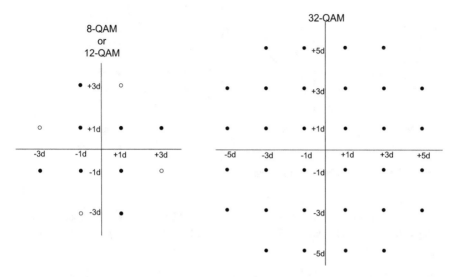

Figure 2-22: 8 /12 -QAM and 32-QAM

Exercise 2-14: Convince yourself this quick argument of 32-QAM power, or write a simple program to verify it.

Answer: List all the constellation points $C = \{+1+j1, +1+j3, 1+j5, \ldots \text{etc.}\}$. Then compute $P_{32QAM} = \frac{1}{32}\sum_{i=1}^{32}|c_i|^2$. This is generic, i.e., applicable to any constellation.

2.4.2.1 SER of Square QAM

We now consider the symbol error rate performance of M-QAM. We start with QPSK and 16-QAM in Figure 2-20. Decision boundaries are indicated in color shading; there are three kinds: inner points (4 sided), side (3 sided), and corner (2 sided). This is shown in Figure 2-23.

QPSK has only corner type constellations. Its symbol error probability is $1 - P_c$ and there are four of those and all symbols equally likely ($p = 1/4$). And $\frac{d}{\sigma} = \sqrt{\frac{E_s}{N_o}}$ since $P_s = \frac{E_s}{T} = 2d^2$ and $P_n = \sigma^2 = \frac{N_o}{2T}$. Thus

$$\textbf{SER of QPSK} = 2Q\left(\frac{d}{\sigma}\right) - Q^2\left(\frac{d}{\sigma}\right) = 2Q\left(\sqrt{\frac{E_s}{N_o}}\right) - Q^2\left(\sqrt{\frac{E_s}{N_o}}\right)$$

$$\approx 2Q\left(\sqrt{\frac{E_s}{N_o}}\right).$$

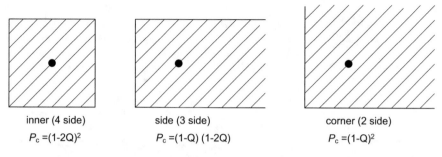

Figure 2-23: 3 kinds of constellation points and decision boundaries

Consider the SER of 16-QAM with Figure 2-20 (RHS). Q will be used for $Q\left(\frac{d}{\sigma}\right)$ and $\frac{d}{\sigma}$ will be expressed with $\frac{E_s}{N_o}$. Remember that all symbols are equally likely ($p = 1/16$).

inner points: 4: $1-(1-2Q)^2 = 4Q-4Q^2$
side points: 8: $1-(1-Q)(1-2Q) = 3Q-2Q^2$
corner points:4: $1-(1-Q)^2 = 2Q-Q^2$

$$\text{SER of 16-QAM} = \frac{1}{16}\left[(4Q-4Q^2)*4 + (3Q-2Q^2)*8 + (2Q-Q^2)*4\right]$$
$$= 3Q - \frac{9}{4}Q^2$$

From $P_s = \frac{E_s}{T} = 2\frac{M-1}{3}d^2$ and $P_n = \sigma^2 = \frac{N_o}{2T}$, we obtain $\frac{d}{\sigma} = \sqrt{\frac{E_s}{N_o}\frac{3}{M-1}}$ with $M = 16$.

$$\textbf{SER of 16-QAM} = 3Q\left(\sqrt{\frac{E_s}{N_o}\frac{3}{M-1}}\right) - \frac{9}{4}Q^2\left(\sqrt{\frac{E_s}{N_o}\frac{3}{M-1}}\right)$$
$$\approx 3Q\left(\sqrt{\frac{E_s}{N_o}\frac{3}{M-1}}\right).$$

For squareM-QAM, the above process can be used to obtain the SER.

1) inner points: $\left(\sqrt{M}-2\right)^2 * \left(4Q-4Q^2\right)$
2) side points: $\left(\sqrt{M}-2\right)*4*\left(3Q-2Q^2\right)$
3) corner points: $4*(2Q-Q^2)$

Add all three above and divide by M.

Exercise 2-15: Work out the computation above and simplify.

Answer: SER   of   M-QAM   (M = power of 2) = $\left(4 - \frac{4}{\sqrt{M}}\right)Q -$ $\left(4 - \frac{8}{\sqrt{M}} + \frac{4}{M}\right)Q^2$ after a tedious but straightforward simplification suggested above.

As M gets large, the inner points will dominate and, neglecting the squared terms, we obtain an approximate SER for M-QAM.

$$\textbf{SER of square M-QAM} \approx \left(4 - \frac{4}{\sqrt{M}}\right)Q\left(\sqrt{\frac{E_s}{N_o}\frac{3}{M-1}}\right).$$

For M = 4, 16, 64, 256, 1024, SER, using the above, is plotted for convenience in Figure 2-24.

A similar process of the above calculation can be applied to any constellation with rectangular decision boundaries.

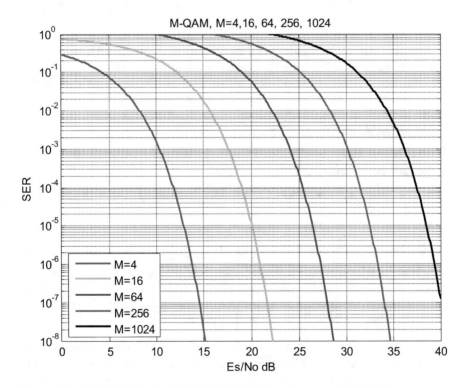

Figure 2-24: M-QAM (M = 4, 16, 64, 256, 1024) SER performance with AWGN

Exercise 2-16: Find SER of 32-QAM shown in Figure 2-22 assuming equally likely symbols.

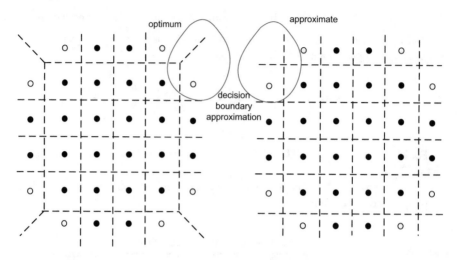

Answer: Decision boundaries are shown in the figure (LHS) with dotted lines. There are 16 points with 4 side (inner points), 8 points with 3-side (side points), and 8 points 'nearly' 3-side (4 circles). SER of 'nearly' 3-side points is a bit involved and so we approximate them 3-side. Then the approximate decision boundaries are shown below. Thus with the approximation, 16 points of 3-side.

For 32-QAM, we obtain the SER.

1) inner points: $16*(4Q - 4Q^2)$
2) side points: $16*(3Q - 2Q^2)$

by adding above and divide by 32. Thus SER is given by $[112 * Q - 96 * Q^2]/32 \approx 3.5 * Q$, dropping Q^2 term.

$P_s = \frac{E_s}{T} = 20d^2$ (32QAM digital power), and $P_n = \sigma^2 = \frac{N_0}{2T} * 2$ (noise with both I and Q). Thus $\frac{d}{\sigma} = \sqrt{\frac{E_s}{N_o} \frac{1}{20}}$ and SER $\approx 3.5 * Q\left(\sqrt{\frac{E_s}{N_o} \frac{1}{20}}\right)$.

We can form 'almost' square constellations of M = 128, 512, 2048 etc., knocking off corners; $144 (12^2) - 16 = 128$, $576 (24^2) - 64 = 512$, $2304 (48^2) - 256 = 2048$, and so on. In order to obtain SER we use an approximation such that all constellation points are either 4 sided or 3 sided. For example for almost square 128-QAM, there are 96 points of 4 sided and 32 points of 3 sided. Then $\frac{d}{\sigma} = \sqrt{\frac{E_s}{N_o} \frac{1}{digital\ power}}$ where '*digital power*' is an average digital power of a constellation with equal probability while the distance between points is 2.

Thus from the above, we conclude that SER of 'almost' square M-QAM $\approx \left(4 - \frac{4}{\sqrt{M}}\right)Q\left(\sqrt{\frac{E_s}{N_o} \frac{3}{M-1}}\right)$ is a useful approximation for even for 'almost' square

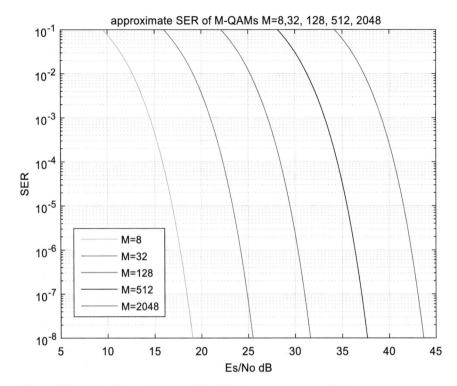

Figure 2-25: SER of M = 8, 32, 128, 512, 2048 almost square constellations

constellations. Note that we used an approximation of *digital power* $\approx \frac{M-1}{3}$. SER of M = 8, 32, 128, 512 and 2048 is shown in Figure 2-25.

2.4.2.2   Double Square (DSQ) constellations

Another form of constellation can be obtained from a square constellation. For example, from 16-QAM, we can obtain an 8-QAM as shown in Figure 2-26.

Exercise 2-17:   For DSQ 8-QAM in Figure 2-26, obtain SER in terms of E_s/N_o.

  Answer: SER $= \left[2 * \left(4Q\text{-}4Q^2\right) + 4 * \left(3Q\text{-}2Q^2\right) + 2 * Q\right]/8 = \frac{22}{8}Q\text{-}\frac{16}{8}Q^2$
where $Q = Q\left(\sqrt{\frac{E_s}{N_o}\frac{1}{5}}\right)$. Note that '5' is the digital power of DSQ 8QAM when the distance between constellation points is 2 (not $2\sqrt{2}$).

Exercise 2-18:   For DSQ 32-QAM in Figure 2-27, obtain SER in terms of E_s/N_o.

  Answer: SER $\approx \left[18 * \left(4Q\text{-}4Q^2\right) + 14 * \left(3Q\text{-}2Q^2\right)\right]/32 = \frac{114}{32}Q\text{-}\frac{100}{32}Q^2$ where
$Q = Q\left(\sqrt{\frac{E_s}{N_o}\frac{1}{21}}\right)$ and we approximated all outer constellation points as 3-side.

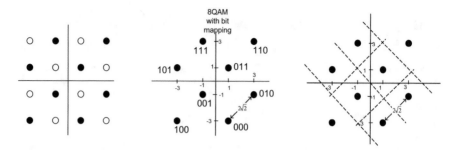

Figure 2-26: DSQ 8-QAM from square 16-QAM, and its decision boundary; compare this with one 8-QAM in Figure 2-22 (LHS). The digital power is 10, the same as that of square 16-QAM, i.e., doubled from 5. However, the distance between the constellation points is increased to $2\sqrt{2}$

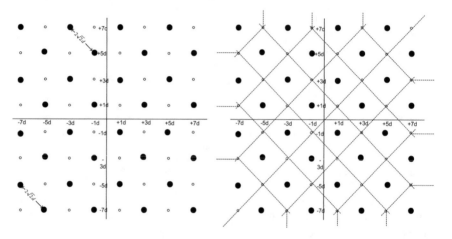

Figure 2-27: DSQ 32-QAM from 64-QAM and its decision boundary

Note that '21' is the digital power of DSQ 32-QAM when the distance between constellation points is 2 (not $2\sqrt{2}$), i.e., power is scaled by $\frac{1}{2}$.

From these two examples in the above, it is clear that DSQ constellation can be extended to the size of M = 128, 512, 2048 and so on.

We comment that DSQ constellation of 8-QAM is adapted by DOCSIS 3.1 and that of 128-QAM by 10 Gigabit Ethernet over Twisted-pair Copper.

2.4.3   PSK, APSK, and DPSK

We show two examples of PSK with M = 4 and 8 in Figure 2-28. Data bits are encoded into only the angle (phase), not into the amplitude. The radius of the circle is given by $1d$ and thus the power of M-PSK is $1d^2$ (with all symbols being equally likely or not). The shaded region is a decision boundary for a symbol, and similarly for all other symbols.

The SER performance of M-PSK will be discussed. M = 2 is BPSK whose SER is known from 2-PAM as well as BPSK before. M = 4 case is the same as QPSK except for 45° rotation, and thus the SER performance of 4-PSK is expected to be the same as QPSK (Note that we make a distinction between 4-PSK and QPSK).

$$\text{SER of 2-PSK} = \text{the same as BPSK} = Q\left(\sqrt{2\frac{E_s}{N_o}}\right)$$

$$\text{SER of 4-PSK} = \text{the same as QPSK} = 2Q\left(\sqrt{\frac{E_s}{N_o}}\right) - Q^2\left(\sqrt{\frac{E_s}{N_o}}\right)$$

For other M, the computation is slightly involved since we need to express the noise statistics in polar form. For M>4, it can be approximated by

$$\textbf{SER of M-PSK} \approx 2Q\left(\sqrt{2\frac{E_s}{N_o}\sin^2\left(\frac{\pi}{M}\right)}\right)$$

This approximation can be seen from Figure 2-28 (RHS); the symbol can be mistaken both sides of adjacent symbol if the noise moves a sample at least by $d *$ $\sin\left(\frac{\pi}{M}\right)$, i.e., BPSK double sided and decision boundary is $d * \sin\left(\frac{\pi}{M}\right)$. SER is plotted using this approximation in Figure 2-29.

Precise SER of M-PSK is expressed in an integral form as,

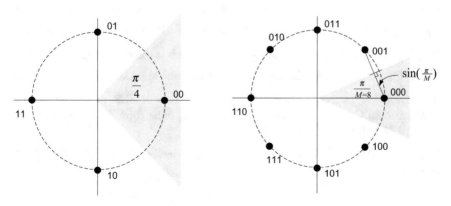

Figure 2-28: 4-PSK and 8-PSK with gray coding

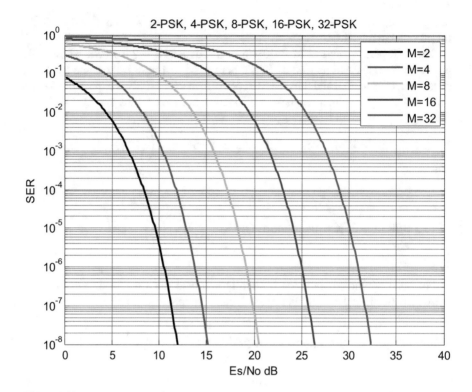

Figure 2-29: M-PSK SER performance with AWGN

$$\textbf{Precise SER of M-PSK } \frac{1}{\pi} \int\limits_{0}^{\pi-\frac{\pi}{M}} \exp\left(-\frac{\frac{E_s}{N_o} \sin^2\frac{\pi}{M}}{\sin^2\theta}\right) d\theta.$$

It is interesting to note that SER of 2-PSK in the above becomes $\frac{1}{\pi}\int_0^{\frac{\pi}{2}} \exp\left(-\frac{x^2}{2\,\sin^2\theta}\right)d\theta$ with $x = \sqrt{2\frac{E_s}{N_o}}$. This must be the same as SER of BPSK $= Q\left(\sqrt{2\frac{E_s}{N_o}}\right)$. Thus $Q(x) = \frac{1}{\pi}\int_0^{\frac{\pi}{2}} \exp\left(-\frac{x^2}{2\,\sin^2\theta}\right)d\theta$. This is the same as Craig's expression of $Q(x)$ function. For the derivation, it is outlined in pp.284 of [2]

Exercise 2-19: Derive the above integral expression of SER of M-PSK. (Hint: 2-dimensional Gaussian PDF may be expressed in polar form. Look for literature, e.g., pp284 of [2].)

Exercise 2-20*: Write a computer program to do the integration of SER M-PSK. Star (*) means a project.

2.4.3.1 APSK

Various variations of PSK, called APSK (amplitude phase shift keying), are used in DVB-S2 and –S2X (digital video broadcasting) standards for satellite communication systems. One example, 16-APSK, is shown in Figure 2-30.

A motivation of using this multi-ring constellation is to mitigate PA non-linearity, i.e., less backoff (more power) since the power at satellite is limited and precious. Even though SER of APSK requires more SNR than QAM without FEC, this can be recovered completely, or nearly completely, by a powerful error correction such as LDPC. This FEC is used in DVB standards.

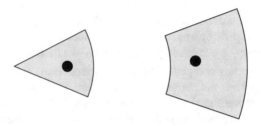

SER of APSK: SER in an integral form like PSK seems not available in the literature. The derivation of it requires the evaluation of Gaussian PDF (in 2-dimension) within the decision boundary shapes, two examples, as shown in the LHS figure. Even with an integral form of SER it requires numerical evaluations. In practice it can be done by numerical simulations.

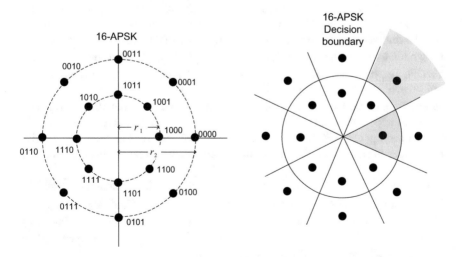

Figure 2-30: One form of 16APSK constellation example and its decision boundary for symbol detection. A specific bit to symbol mapping is shown as well (LHS). Note also the shapes of decision boundary in shade

In conjunction with FEC, code word error rate or bit error rate (BER) is most often computed by resorting to numerical simulations in practice. Note that the computation of SER requires only the decision boundary, but that of BER requires, in addition to the decision boundary, a bit to symbol mapping (an example is shown in Figure 2-30 at LHS), i.e., bits being assigned to a symbol. This will be elaborated in the next section.

2.4.3.2   Partially Coherent Detection (of DPSK) and Non-coherent Detection (of FSK)

Keep in mind that a channel we assume is AWGN, i.e., only thermal noise at the receiver front end is the impairment and all else - such as synchronization of carrier frequency and phase – is perfect. In particular, SER as a function of $\frac{E_s}{N_o}$ assumed that a perfect carrier frequency and phase is known (or recovered) at the receive side.

In order to minimize the complexity of a system, there was, especially in early days of communication signal and system development, an attempt was made to use a signal set so that it is possible to eliminate the need of carrier frequency and phase recovery at the receive side.

2.5   BER and Different forms of SNR

At this point, we summarize this chapter so far. As we open this chapter, a digital abstraction of Figure 2-2 from the block diagram of Figure 2-1, was presented immediately and followed by a variety of constellations, from 2 to many points on 1- and 2-dimension. The error performance of such signaling schemes was evaluated using SER as a function of SNR, in particular, $\frac{E_s}{N_o}$. Keep in mind that a channel is AWGN, i.e., only thermal noise at the receiver front end is the impairment and all else - such as filtering, synchronization of carrier frequency and phase, and of symbol timing frequency and carrier, and more – is perfect. Thus this performance is the best upper bound and with practical components it will be degraded except for channel coding, i.e., FEC may improve the performance. Thus in practice, we strive to minimize the degradation from this upper bound with actual implementations of a system.

We deliberately chose SER as a performance criterion in a function of $\frac{E_s}{N_o}$ since it depends only on the constellation and decision boundary. Traditionally, BER in terms of $\frac{E_b}{N_o}$ is often used in most textbooks. We will explore this topic and hopefully it will be clear why we stick with SER in terms of $\frac{E_s}{N_o}$ when we discuss constellations in this section. A short answer is that it is the simplest.

2.5.1   BER Requires a Specific Bit to Symbol Mapping

In BPSK, SER is the same as BER since one bit corresponds to one symbol. Similarly it is true with binary OOK and 2 level orthogonal signaling. However, in general BER is related with SER, not simple to find a precise relationship. First of all BER depends on a specific bit to symbol mapping detail. Additionally it is also a function of any additional processing like scrambling and differential coding, which will be covered in later section.

Example 2-3: bit to symbol mapping examples of binary case; BPSK {'0' → +1. '1' → −1}, OOK {'0' → +0, '1' → +$\sqrt{2}$}, orthogonal {'0'→ +1, '1'→ +j}.

Example 2-4: bit to symbol mapping of different constellations

constellation	4-PAM	8-QAM DOCSIS	4-PSK, 8-PSK	16 APSK
bit to symbol mapping	Figure 2-17	Figure 2-26	Figure 2-28	Figure 2-30

Note that we use Gray bit assignment, i.e., bit difference between adjacent symbols is only one. For example, for 4-PAM it is given by {'00' → +3, '01'→+1, '11'→−1, '10'→−3}.

Example 2-5: 16-QAM constellation bit to symbol mapping with Gray bit assignment

Exercise 2-21: A way of organizing bit to symbol mapping as a vector. Answer: A vector [+3, +1, −3, −1] may represent a bit to symbol mapping for 4-PAM while 2 bit data is implicit in the order of vector by ordering in this sequence{'0'→+3, '1'→+1, '2'→−3, '3'→−1}. See also RHS table of Figure 2-31 for 16-QAM.

Exercise 2-22: A digital power of a constellation can be computed from a bit to symbol mapping vector by squaring and summing the mapping vector divided by the number of constellation points assuming all points are equally likely.

2.5.2   A Quick Approximation of SER to BER conversion

In bit to symbol mapping Gray bit assignment is often used; adjacent symbols have only one bit difference. For example, in Figure 2-31, bit to symbol mapping of 16-QAM uses Gray coding. Most errors are made mistaken to be adjacent symbols. A symbol error tends to make one bit error but it is only approximately so most of time. Thus with Gray bit assignment, we may, somewhat crudely, approximate BER from SER as,

$$BER \approx \frac{SER}{\text{bit per symbol}}$$

16-QAM bit to symbol mapper b3b2b1b0

1000	1010	0010	0000
1001	1011	0011	0001
1101	1111	0111	0101
1100	1110	0110	0100

seq	binary	symbol
0	0 0 0 0	3+3i
1	0 0 0 1	3+1i
2	0 0 1 0	1+3i
3	0 0 1 1	1+1i
4	0 1 0 0	3-3i
5	0 1 0 1	3-1i
6	0 1 1 0	1-3i
7	0 1 1 1	1-1i
8	1 0 0 0	-3+3i
9	1 0 0 1	-3+1i
10	1 0 1 0	-1+3i
11	1 0 1 1	-1+1i
12	1 1 0 0	-3-3i
13	1 1 0 1	-3-1i
14	1 1 1 0	-1-3i
15	1 1 1 1	-1-1i

Figure 2-31: 16-QAM constellation with bit to symbol mapping and its vector representation

As noted before BER can get more complicated by additional processing such as differential coding and scrambler which makes one bit error tend to be multiplied. For example, differential coding makes one bit error into two.

Exercise 2-23: Obtain 64QAM bit to symbol mapping with Gray bit assignment; the first quadrant 16 points are bit assigned for LSB 4 bits ($b_3b_2b_1b_0$, note that this assignment is not the same as in Figure 2-31), and 2 MSB bits (b_5b_4) are '00' as shown below. Fill out the rest of quadrants. Hint: LSB bits ($b_3b_2b_1b_0$) between quadrant boundaries must be identical since 2 MSB bits (b_5b_4) are 1 bit different between quadrants.

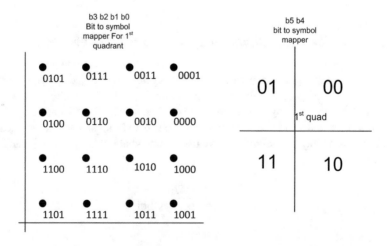

2.5.3 Numerical Simulations of BER with LLR

A quick estimation of BER from SER may not be good enough in some situations. There is a need of BER by simulations. This can be done in a straightforward manner from a simulation of SER by counting bit error after symbol detection and converting a symbol stream to a bit stream. A symbol decision is made from received symbols based on decision boundaries. This requires the least complex computation compared to LLR method we are about to describe. However, when a powerful FEC like LDPC or Turbo code is used, LLR computation of each bit, not each symbol, is necessary. We will describe how to compute LLR with a specific bit to symbol mapping constellation and then apply it to BER and SER simulations. Compared to symbol detection based on decision boundary, LLR computation is more computationally intensive. However, with the technology of these days it is well doable. Additional advantage is that there is no need of explicit decision boundary. It is implicit in LLR computation.

Example 2-6: LLR computation of 16-QAM whose bit to symbol mapping is the same as in Figure 2-31. LLR computation procedure is illustrated in Figure 2-32.

- $LLR(b_3|x) = \ln \dfrac{\sum\limits_{i \in S_0} pr(x|s_i : b_3=0)}{\sum\limits_{j \in S_1} pr(x|s_j : b_3=1)} = \ln \dfrac{\sum\limits_{i \in S_0} \exp\left(-|x-s_i|^2/2\sigma^2\right)}{\sum\limits_{j \in S_1} \exp\left(-|x-s_j|^2/2\sigma^2\right)}$

  - $S_0 = \{0, 1, 2, 3, 4, 5, 6, 7\}$ and $S_1 = \{8, 9, 10, 11, 12, 13, 14, 15\}$ for b_3. See Figure 2-31 table (RHS).

- $LLR(b_2|x) = \ln \dfrac{\sum\limits_{i \in S_0} pr(x|s_i : b_2=0)}{\sum\limits_{j \in S_1} pr(x|s_j : b_2=1)} = \ln \dfrac{\sum\limits_{i \in S_0} \exp\left(-|x-s_i|^2/2\sigma^2\right)}{\sum\limits_{j \in S_1} \exp\left(-|x-s_j|^2/2\sigma^2\right)}$

  - $S_0 = \{0, 1, 2, 3, 8, 9, 10, 11\}$ and $S_1 = \{4, 5, 6, 7, 12, 13, 14, 15\}$ for b_2. See Figure 2-31 table (RHS).

- And similarly for b_1 and b_0,

  - $S_0 = \{0, 1, 4, 5, 8, 9, 12, 13\}$ and $S_1 = \{2, 3, 6, 7, 10, 11, 14, 15\}$ for b_1.
  - $S_0 = \{0, 2, 4, 6, 8, 10, 12, 14\}$ and $S_1 = \{1, 3, 5, 7, 9, 11, 13, 15\}$ for b_0.

- There is a slight abuse of notation for S_0 and S_1 since for each bit this set is different.

Example 2-7: a numerical computation of SER and BER of 16-QAM

From Figure 2-33, we can observe following;

- SER is independent of the detail of bit to symbol mapping but depends on the decision boundaries.

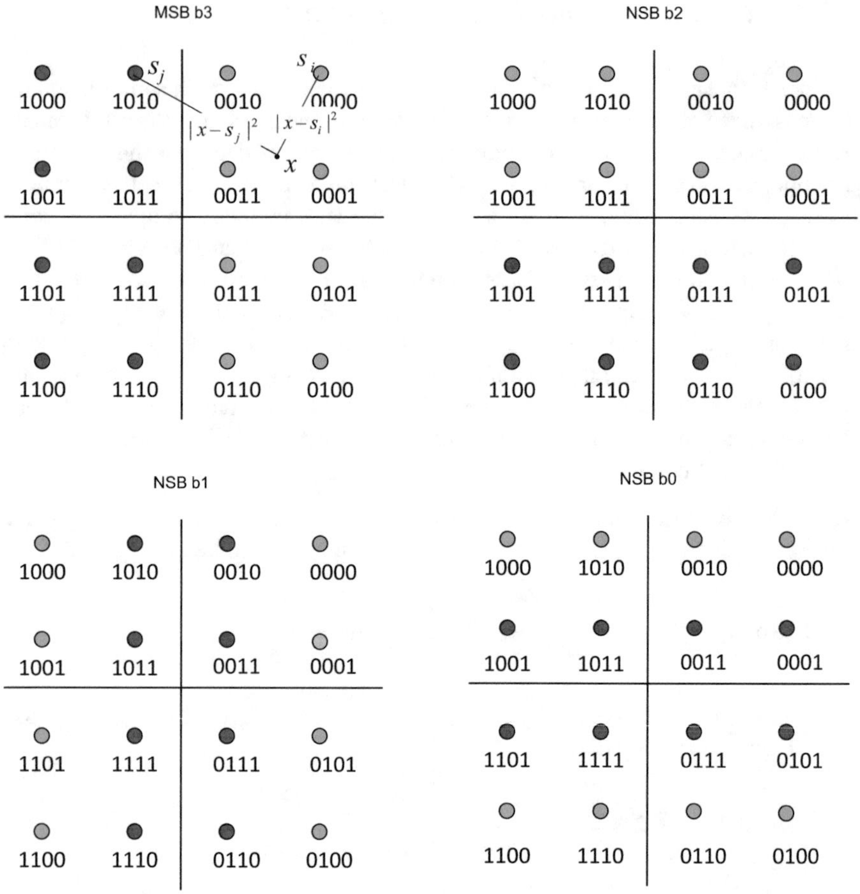

MSB b3

```
●       ● Sⱼ      ○        s ○
1000    1010     0010       0000
              |x-sⱼ|²   |x-sᵢ|²
●       ●        ○     ↘ x
1001    1011     0011    ○ 0001

●       ●        ○        ○
1101    1111     0111     0101

●       ●        ○        ○
1100    1110     0110     0100
```

NSB b2

```
○       ○    |   ○        ○
1000    1010 |   0010     0000

○       ○    |   ○        ○
1001    1011 |   0011     0001

●       ●    |   ●        ●
1101    1111 |   0111     0101

●       ●    |   ●        ●
1100    1110 |   0110     0100
```

NSB b1

```
○       ●    |   ●        ○
1000    1010 |   0010     0000

○       ●    |   ●        ○
1001    1011 |   0011     0001

○       ●    |   ●        ○
1101    1111 |   0111     0101

○       ●    |   ●        ○
1100    1110 |   0110     0100
```

NSB b0

```
○       ○    |   ○        ○
1000    1010 |   0010     0000

●       ●    |   ●        ●
1001    1011 |   0011     0001

●       ●    |   ●        ●
1101    1111 |   0111     0101

○       ○    |   ○        ○
1100    1110 |   0110     0100
```

Figure 2-32: LLR computation procedure of set partitioning

- BER depends on the detail of bit to symbol mapping, and Gray bit to symbol mapping is superior to random one, especially in the low SNR region. This has important consequence; it is common practice to use Gray bit assignment.
- SER is the upper bound for BER. And a quick approximation of BER, i.e., $BER \approx \frac{SER}{\text{bit per symbol}}$, works better with random bit to symbol mapping, and is poor in the low SNR region.

2.5.4   SNR in Different Forms

So far we used $\frac{E_s}{N_o}$ consistently and will continue so when SER is expressed. Here we comment on different types of SNR; $\frac{E_b}{N_o}$, $\frac{C}{N_o}$ and $\frac{C}{N}$. Our starting point is the ratio of

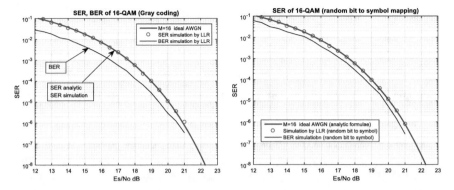

Figure 2-33: Gray bit assignment (LHS) to symbol and random bit to symbol mapping (RHS). Note that SER does not change with random bit to symbol mapping while BER changes dramatically, especially low SNR region. LLR simulations match with the analytical formula well

signal power to noise power measured at RF; $\frac{P}{N}$ where P is signal power and N is noise power. In terms of physical unit the power is given by [watt = joule /sec] and it is a flow of energy per unit time. For our purpose the unit time is symbol time or bit time. Thus

$$P = \frac{E_s}{T_s} = \frac{E_b}{T_b}$$

where E_s and E_b are symbol energy and bit energy, T_s and T_b are symbol period and bit period respectively. Symbol rate is $R_s = 1/T_s$ and bit rate $R_b = 1/T_b$.

Noise power is expressed in frequency domain. Its spectrum density is constant in frequency, compared to signal. Noise is mainly from thermal noise (also called kT noise) at the front end of a receiver (LNA: low noise amplifier) plus circuit noise expressed in terms of noise figure (NF). Its thermal spectrum density is denoted by $N_o = kT$ [watt /Hz] where k is Boltzmann constant and T is temperature in Kelvin. If the receiver noise bandwidth is BW then,

$$N = N_o(BW) = N_o R_s$$

where BW $= Rs$ which is optimum. If the noise band width (BW) of an actual system is bigger than the optimum Rs there will be the loss of SNR, i.e., more signal power is necessary to maintain the same performance.

$$\frac{P}{N} = \frac{E_s}{N_o} = \frac{E_b}{N_o} \frac{R_b}{R_s}$$

SNR is dimensionless and thus it can be expressed in dB, which is defined as $10 * \log_{10} \frac{P}{N}$.

Example 2-8: One symbol carries 4 bits (e.g., 16-QAM), and $10 \log_{10} \frac{E_s}{N_o} = 15[\text{dB}]$. What is $\frac{E_b}{N_o}$ in dB?

Answer: $10 \log 10 \frac{R_b}{R_s} = 10 \log_{10} \frac{4}{1} = 6$ dB. Thus $\frac{E_b}{N_o}$ in dB $= 15 - 6 = 9$ dB.

Example 2-9: A system has a symbol rate 1MHz. The receiver filter noise bandwidth is measured as 1.2 MHz. What is the SNR loss in dB?

Answer: $10 \log_{10} (1.2 / 1) = 0.8$ dB.

In some areas $\frac{C}{N_o}$ is used for convenience where C is a carrier power and N_o is noise power density per Hz, and thus it is not dimensionless but in [Hz] or dB-Hz. $N_o = kT$ it is -174 dBm at 27°C (300°K) and $k = 1.38 \times 10^{-23}$, where dBm means 0 dB with 1 mWatt. Thus it can be seen as carrier power compared to noise power density. This can be converted to $\frac{E_s}{N_o} : \frac{C}{N_o} = \frac{E_s}{N_o} R_s$. The ratio of carrier power to noise power, $\frac{C}{N}$, is the same as $\frac{P}{N}$.

Example 2-10: SNR $\frac{E_s}{N_o} = \frac{P}{N}$ is measured to be 10 dB, and symbol rate 1MHz. Find $\frac{C}{N_o}$ [dB-Hz]. Answer: $10 + 10 \log(10^6) = 70$ [dB-Hz].

2.6   Offset QAM (or Staggered QAM)

In the QAM we discussed so far we treat a symbol as a complex number on I-Q plane (complex plane) and its real part and imaginary part are transmitted at the same instant of a symbol period. However, it is possible that the real and imaginary parts are transmitted with a half symbol period apart. This signaling scheme is called offset or staggered QAM. At the time when this idea was invented, a motivation was to reduce the fluctuation of the signal envelope so that power amplifier (PA) can be used with better power efficiency, i.e., with less back-off. (If the signal has a large peak to average power ratio, PA must be used with less output power in order to be in a linear region. This is called power back-off. This advantage seems to be only in offset QPSK. With higher level modulations, it seems less so.)

We describe offset QAM signals. Since real part and imaginary part of a complex symbol are transmitted at different time by half a symbol time, the incoming bits stream is split into two sets, even and odd, and each rail is processed just like PAM. This is shown in Figure 2-34. For example, for offset QPSK, two bits of the incoming bits are split even bit and odd bit, and each rail goes through a bit to symbol mapping of 2 level PAM, followed by a shaping filter. The imaginary part or Q-channel is delayed by T/2.

This offset QAM can be extended to M^2-QAM. With more bits per even and odd partition, M-PAM per each, and combining both I and Q together this corresponds to offset M^2-QAM. Staggered 16-QAM example is shown in Figure 2-35.

In the receiver side, the inverse process will be used. Once the received carrier wave is converted to complex envelope ($A_I(t)$ and $A_Q(t)$), it should go through receive

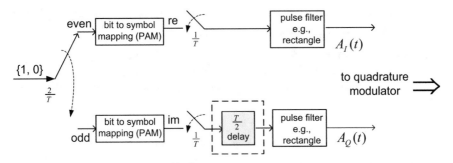

Figure 2-34: Offset QAM signal generation and $1/T$ is a symbol rate. $T/2$ delay is added by commutating between two (I, Q) channels or can be added as shown

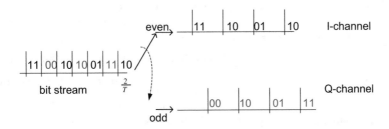

Figure 2-35: Staggered 16-QAM example and $1/T$ is a symbol rate

pulse filter (matched filter) then sampled. Each rail must be sampled at different phase of a symbol clock, half a symbol period, and processed like PAM. Then even and odd rails are combined into a single stream.

A staggered QAM may be considered to be two PAM signals combined by using I and Q channels separately, and this separation is due to transmission time staggering by a half symbol time. The idea of staggering transmission time will be visited again in conjunction with OFDM signaling as a way of making the system more bandwidth efficient. In practice, 'regular' QAM is more prevalently in use.

2.6.1   SER vs. $\frac{E_s}{N_o}$ Performance of Staggered QAM

In short, SER performance of staggered QAM is very similar to that of 'regular' QAM even though each rail (I, Q) is processed like PAM. The supporting argument for this is the same as QAM performance improvement over that of PAM. 16-QAM (4 bit per symbol) requires only 3dB more power over 4-PAM (2 bit per symbol) as can be seen in Figure 2-19 and in Figure 2-24; for example, 4-PAM requires about 17.5dB at SER 1E-6 and 16-QAM does about 20.5 dB at the same SER. Extending PAM to QAM, i.e., using only I-channel to using both I and Q requires 3dB more power at the same time bits per symbol double. Thus QAM improvement over PAM is more dramatic with more constellation points.

Exercise 2-24: Compare 8-PAM (3bits per symbol) and 64-QAM (6 bits per symbol). Here 3dB additional power 3 bit per symbol becomes 6 bits per symbol. Confirm this is the case from Figure 2-19 and Figure 2-24. Hint: check SNR at SER 1E-6.

2.6.2 CCDF of Staggered QAM vs. of 'Regular' QAM

A quick reminder of CCDF (complementary cumulative distribution function) is that CCDF = 1 – CDF. CDF is obtained by integration from PDF. A good example of PDF is a histogram. An example of Gaussian distribution is useful. PDF of a zero mean (m = 0), unity variance ($\sigma^2 = 1$) Gaussian, G(x), is given by $G(x) = \frac{1}{\sqrt{2\pi}}e^{-x^2/2}$. This is a well-known bell-shape curve centered on zero. Its integration is CDF as

$$CDF = \frac{1}{\sqrt{2\pi}}\int_{-\infty}^{x} e^{-y^2/2}dy = 1-\frac{1}{\sqrt{2\pi}}\int_{x}^{\infty} e^{-y^2/2}dy = 1\text{-}Q(x)$$ where Q(x) is defined as

shown above and it is a tail probability of Gaussian distribution. Thus CCDF of Gaussian (1-dimensional) is given by CCDF = Q(x).

In the context of communication systems of transmitting data, modulated RF signals are represented as $s_x(t) = Re\left\{ \left[A_I + jA_Q \right] e^{\, j2\pi f_c t} \right\}$

$$s_x(t) = A_I \cos\left(2\pi f_c t\right) - A_Q \sin\left(2\pi f_c t\right)$$
$$s_x(t) = A \cos\left(2\pi f_c t + \theta\right)$$

where $A = \sqrt{A_I^2 + A_Q^2}$ and $\theta = atan2(A_Q, A_I)$.

A_I, A_Q are formed by convolving discrete time modulated symbols with an interpolating filter. In order to calculate peak to average power ratio (PAPR), we need to know CCDF of $A_I + jA_Q$.

Example 2-11: Numerical computation of CCDF of 4-staggered QAM and that of 'regular' 4-QAM. This is plotted in Figure 2-36. The pulse shape we use is square root raised cosine with excess bandwidth parameter a = 0.2 to 0.9.

Example 2-12: Numerical computation of CCDF of 16- staggered QAM and that of 'regular' 16-QAM. This is shown in Figure 2-37.

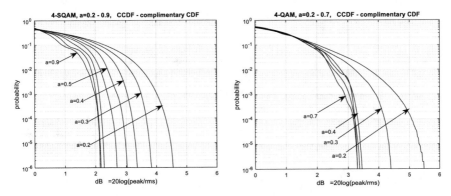

Figure 2-36: CCDF of staggered 4-QAM (LHS) and CCDF of regular 4-QAM (RHS). Compare a = 0.2 case 4.6dB at 1e-6 probability for staggered QAM vs. 5.5 dB for QAM

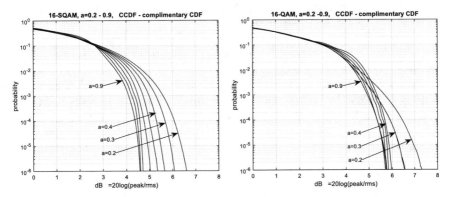

Figure 2-37: CCDF of staggered 16-QAM vs. of 'regular' 16-QAM. Clearly peak to average is less with staggering

2.6.3   Other Issues

There are additional practical issues like symbol timing recovery, carrier recovery, and differential coding carrier phase ambiguity and so on. These are different from those of regular QAM and should be addressed properly.

2.7   Digital Processing and Spectrum Shaping

Practical needs motivate people to invent. In this section we cover simple but effective practical solutions to the problems encountered in practice. Two additional blocks are often added as shown in Figure 2-38. Data bits to be transmitted are supposed to be random, but the random means occasionally a long stream of 0s or 1s

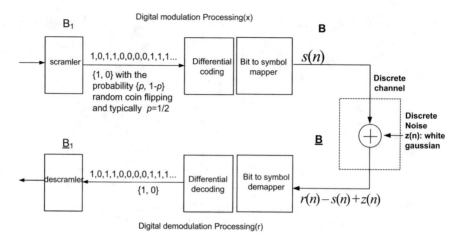

Figure 2-38: Scrambler and differential coding blocks are added to Figure 2-2, the additional blocks are part of digital modulation processing

are possible. To make sure randomness we use a scrambler, which is essentially a pseudo-random number generator.

Carrier wave may have phase ambiguity by $+90°$, $-90°$ or $180°$. In other words, rotating I-Q plane by $90°$ cannot be distinguished unless the receiver knows the absolute phase by some other means. A simple device called differential coding (or rotational invariance coding) will solve the problem or a reference known signal can be sent.

2.7.1   Scrambler

A scrambler is used to randomize the transmitted bits. Data bits to be transmitted are supposed to be random but it is necessary to make sure it is random or pseudo random at least. Pseudo random (PN) sequence generator is used for this purpose. As an example we use a 4 state pseudo random generator (without the input $d(n)$ in Figure 2-39) and is part of the scrambler. Its period is $2^4-1 = 15$ in this example, and PN with L state its period is 2^L-1; except for all zero, all other non-zero states occur.

A PN sequence generator is described by a primitive polynomial; $1 + x + x^4$ in this example. The input data sequence is recovered by an inverse process at the receiver side as shown in Figure 2-39.

One special input is a unit impulse $d(n) = 1$ when $n = 0$, $d(n) = 0$ otherwise, i.e., $\{1, 0, 0, \ldots\}$ sequence. And the impulse response will be a PN sequence. This will be repeated over and over again without any 1 input, i.e., the entire PN sequence is repeated. When there is another 1, a PN sequence in different time position (or phase) will be superimposed, i.e., added (modulo-2) to the previous impulse response.

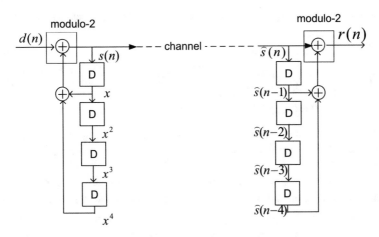

Figure 2-39: Scrambler and descrambler example with 4-state PN sequence

s(n) = 111101011001000 in one period in the example of Figure 2-39.

	1	2	3	4	5	6	7	8	9	10	11	12	13	14	15
s-1	1	1	1	1	0	1	0	1	1	0	0	1	0	0	0
s-2	0	1	1	1	1	0	1	0	1	1	0	0	1	0	0
s-3	0	0	1	1	1	1	0	1	0	1	1	0	0	1	0
s-4	0	0	0	1	1	1	1	0	1	0	1	1	0	0	1

The output example and the state of a PN sequence generator are in the above table. From this example we can observe its properties, which are summarized next.

2.7.1.1   Properties of PN Sequence

PN sequence has the following properties, which resemble to randomness of ideal coin-flipping experiment.

1. **The balanced property**: In a complete period 2^L-1 of a PN sequence, the number of 1's differs from the number of 0's by at most one. In the example 8 1's and 7 0's in a period. Relative frequency of 1's and 0's is about ½.
2. **The run lengths** of consecutive zeros and ones are similar to a coin flipping experiment; run length 1: 1/2, run length 2: ¼, and so on.
3. **The correlation property**: If a complete sequence is compared bit by bit with any shift of the sequence, the number of agreements minus the number of disagreements is always −1; that is, there is one more disagreement position than the agreement positions. This property is important in direct sequence (DS) spread spectrum.

Exercise 2-25: Write a computer program (e.g., Matlab script) to a unit impulse response a one bit delay, i.e., $d(n) = \{0, 1, 0, 0, 0, \ldots \ldots 0, 0\}$ $n = 0$ to 29 (i.e., two periods) sequence. And confirm this $d(n)$ is recovered (as $r(n)$ below) correctly after descrambling.

Answer:

n   0 1 2 3 4 5 6 .... 27 28 29
dn 0 1 0 0 0 0 0 0 0 0 0 0 0 0 0 0 0 0 0 0 0 0 0 0 0 0 0 0 0 0
sn 0 1 1 1 1 0 1 0 1 1 0 0 1 0 0 0 1 1 1 1 0 1 0 1 1 0 0 1 0 0
rn 0 1 0 0 0 0 0 0 0 0 0 0 0 0 0 0 0 0 0 0 0 0 0 0 0 0 0 0 0 0

Exercise 2-26: A type of scrambler in Figure 2-39 is self-synchronizing in the sense that the receiver descrambler does not need to know the state of transmit side scrambler.

Answer: after 4 (a number of states) receive samples without error would fill correct states since transmit symbols, s(n), are the states of scrambler.

2.7.1.2   Primitive Polynomials for PN Sequence

Additional primitive polynomials for PN sequence are listed in Table 2-2.

2.7.1.3   Another Type of Scrambler

PN sequence is used differently. Example 2-13 will be such an example.

Example 2-13: PN sequence may modify the data bits as in Figure 2-40. Discuss the difference with the scrambler in Figure 2-39. The same 4-statePN sequence generator is used.

Answer: Three different $d(n)$: all zeros, all ones and a single one followed by zeros. We set $s(4) = 1$. Then the transmit sequence, c(n) , is an impulse response with $d(n)$ all zeros and the inverse with $d(n)$ all ones. This is shown below.

dn 0 0 0 0 0 0 0 0 0 0 0 0 0 0 0 0
cn 1 1 1 1 0 1 0 1 1 0 0 1 0 0 0
rn 0 0 0 0 0 0 0 0 0 0 0 0 0 0 0
----
dn 1 1 1 1 1 1 1 1 1 1 1 1 1 1 1 1
cn 0 0 0 0 1 0 1 0 0 1 1 0 1 1 1
rn 1 1 1 1 1 1 1 1 1 1 1 1 1 1 1
----
dn 1 0 0 0 0 0 0 0 0 0 0 0 0 0 0 0
cn 0 1 1 1 0 1 0 1 1 0 0 1 0 0 0
rn 1 0 0 0 0 0 0 0 0 0 0 0 0 0 0

Table 2-2: Primitive polynomial list for PN sequence

degree	polynomial	degree	polynomial	degree	polynomial
2	$x^2 + x + 1$	12	$x^{12} + x^6 + x^4$ $+ x + 1$	22	$x^{22} + x + 1$
3	$x^3 + x + 1$	13	$x^{13} + x^4 + x^3$ $+ x + 1$	23	$x^{23} + x^5 + 1$
4	$x_4 + x + 1$	14	$x^{14} + x^{10} + x^6$ $+ x + 1$	24	$x^{24} + x^7 + x^2$ $+ x + 1$
5	$x^5 + x^2 + 1$	15	$x^{15} + x + 1$	25	$x^{25} + x^3 + 1$
6	$x^6 + x + 1$	16	$x^{16} + x^{12} + x^3$ $+ x + 1$	26	$x^{26} + x^6 + x^2$ $+ x + 1$
7	$x^7 + x^3 + 1$	17	$x^{17} + x^3 + 1$	27	$x^{27} + x^5 + x^2$ $+ x + 1$
8	$x^8 + x^4 + x^3$ $+ x^2 + 1$	18	$x^{18} + x^7 + 1$	28	$x^{28} + x^3 + 1$
9	$x^9 + x^4 + 1$	19	$x^{19} + x^5 + x^2$ $+ x + 1$	29	$x^{29} + x^2 + 1$
10	$x^{10} + x^3 + 1$	20	$x^{20} + x^3 + 1$	30	$x^{30} + x^{23} + x^2$ $+ x + 1$
11	$x^{11} + x^2 + 1$	21	$x^{21} + x^2 + 1$	31	$x^{31} + x^{28} + 1$

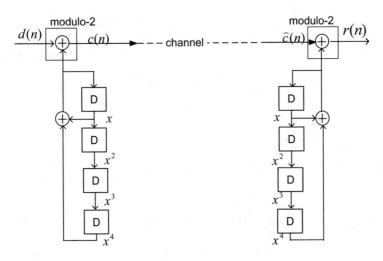

Figure 2-40: PN sequence used differently compared to Figure 2-39

Note that the scrambler in Figure 2-40 is not self-synchronizing. In order to descramble properly the descrambler needs to know the states of transmit scrambler.

This type of scrambler may be used for <u>direct sequence spread spectrum</u> <u>signal-</u><u>ing</u>; 15 zeros ($2^{\wedge\text{degree}}-1$) spread into a PN sequence of length 15, and 15 ones ($2^{\wedge\text{degree}}-1$) spread into its inverse. It is a repetition code of length 15 ($2^{\wedge\text{degree}}-1$); i.e., rather than sending 15 zeros or ones, a PN sequence of length 15 is sent.

For a long scrambling sequence, a higher degree PN polynomial in Table 2-2 may be used.

2.7.2   Differential Coding or phase invariance coding

During the signal transmission the polarity of a signal may change. In 2 dimensional signals, both in-phase (I) and quadrature phase (Q) may change its polarity, which is equivalent to 90° carrier phase rotation – 0°, 90°, 180° and 270°. In this subsection we explain how to handle this signal carrier phase change. For 180°, phase ambiguity 1 bit differential coding is necessary while 90° phase ambiguity 2 bit differential coding is necessary.

2.7.2.1   180° Phase Invariance Coding

In order to handle 180° phase ambiguity, a one bit differential coding as shown in Figure 2-41 may be used. In the encoding side, rather than sending data directly, it goes through a process of getting through the 'difference' of the data. It can be seen as a transfer function shown in Figure 2-41 (bottom). In the decoding side, the present and previous bits are summed. Its transfer function is shown in the figure as well. Modulo-2 arithmetic is defined as 0-0 = 0; 0-1 = 1; 1-1 = 0; 1-0 = 1 and it is an exclusive OR in logic circuit terminology. In this bit arithmetic + operation is the same as – operation.

Exercise 2-27: Compare modulo-2 operation and exclusive OR.

Example 2-14: 1-bit differential coding and decoding

d(n)	0	1	0	1	1	1	0	1	1
s(n)	0	1	1	0	1	0	0	1	0
s(n−1)	x = 0	0	1	1	0	1	0	0	1

At the receive side, the transmitted symbol may be inverted. Due to the differential coding the inversion does not matter in recovering the data as shown below.

s(n) normal	0	1	1	0	1	0	0	1	0
s(n)inverted	1	0	0	1	0	1	1	0	1
d(n)	x	1	0	1	1	1	0	1	1

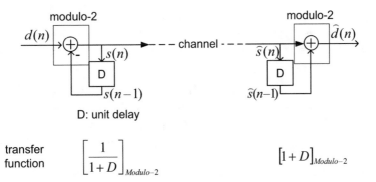

Figure 2-41: 180° phase invariance (differential) coding and decoding

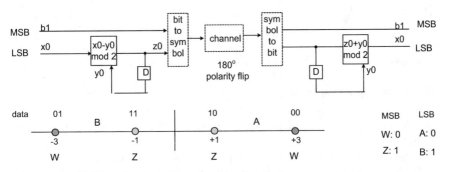

Figure 2-42: 2 bit example (4-PAM) of 180° phase invariance coding and decoding

2.7.2.2   Multi Bit Extension of 180° Phase Invariance Coding

In Figure 2-42, 4-PAM (2 bit) example of 180° phase invariance is shown, Gray bit assignment as well. After polarity inversion, MSB is the same while LSB is inverted.

Exercise 2-28: 8-PAM (3bit) example is shown below; bit assignment is partially done. Complete it. Is it unique?

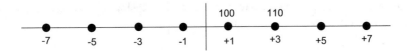

Answer: Yes it is unique as below.

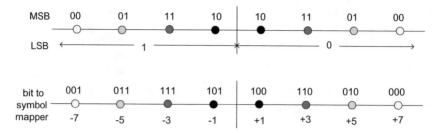

Table 2-3: Modulo 4 (x − y = z and x = y + z)

rotation	0°				+90°				180°				−90°			
x	0	1	2	3	0	1	2	3	0	1	2	3	0	1	2	3
y	0	0	0	0	1	1	1	1	2	2	2	2	3	3	3	3
z = x−y mod-4	0	1	2	3	3	0	1	2	2	3	0	1	1	2	3	0

2.7.2.3   90° Rotation Invariance

In order to resolve 90° phase ambiguity, a two-bit differential coding is necessary using the exactly the same structure as in Figure 2-41, the arithmetic is done in modulo-4, rather than modulo-2. We explain modulo-4 with table.

Modulo-4 operation example is shown in the above. [x−y] modulo-4 is done by normal subtraction, + or − 4 to it until the value is with [0−3]. For example 0 -1 = −1 + 4 = 3. Subtracting by −1 is the same as rotating +90° (counterclockwise rotation). From the above table x = [z + y]mod-4. This modulo-4 may be applied to 90° rotation invariance coding; let d(n) = x and s(n−1) = y , then s (n) = z = [x−y] mod-4.

Modulo arithmetic is not confined to integer numbers. It can be extended to real numbers. A familiar example is an angle on the plane. It is {0, .., 2π} [radian]. If an angle is more than2π, one can put it back to less than 2πby subtracting2π multiple times as needed. This can be seen easily an angle on a circle from zero. Thus the modulo arithmetic may be understood as a period extension of the range on the real number line which is shown in Figure 2-43.

2.7.2.4   Rotational Invariance of 16-QAM

We consider a bit to symbol mapping of 16-QAM to be rotationally invariant. With 90° rotation, the same color (W, X, Y, Z) position will coincide. Most significant bits of 2 are assigned as shown in Figure 2-44. This assignment is somewhat arbitrary but one can find an assignment such that bit error rate is the minimum when adjacent

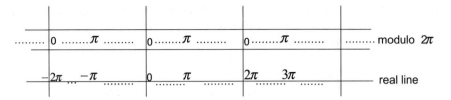

Figure 2-43: Example of modulo operation as periodic extension on real line

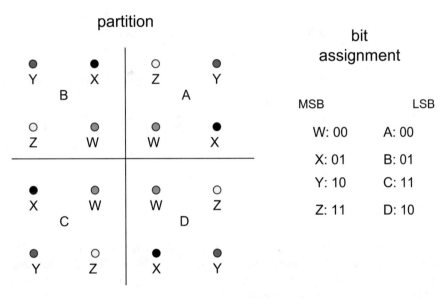

Figure 2-44: Rotation invariant bit to symbol mapping of 16-QAM

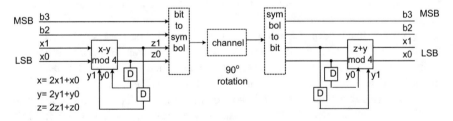

Figure 2-45: 16-QAM 90° rotation invariance coding and decoding

symbols are mixed up, like Gray assignment. Each quadrant bit (A, B, C, D) assignment uses Gray coding. Its block diagram implementation is shown in Figure 2-45.

Exercise 2-29: Complete a bit to symbol mapping of 16QAM, partially filled (LHS) as below as much with Gray coding, i.e., every adjacent symbol has one bit difference. Is it unique? Answer: RHS

partial bit assignment		bit to symbol assignment	

(LHS constellation)

●   ●   ○   ●
   0101

○   ●   ●   ●
     0000

●   ●   ●   ○

●   ○   ●   ●

(RHS constellation)

●   ●   ○   ●
1001   0101   1100   1000

○   ●   ●   ●
1101   0001   0000   0100

●   ●   ●   ○
0111   0011   0010   1110

●   ○   ●   ●
1011   1111   0110   1010

2.7.3   Partial Response Signaling

In certain situations DC component in frequency domain needs to be removed. This can be done digitally and used extensively in early PCM standard of digital telephony. We extend it to partial response signaling, generally shaping the spectrum of a signal digitally for practical reasons.

We discuss two important class of partial response signaling (PRS) systems. Let us call $1+D$ and $1-D^2$ system. The motivations of these schemes were to improve the bandwidth efficiency, i.e., more bits to be transmitted for a given bandwidth and to shape the frequency spectrum at DC.

To be more bandwidth efficient, one can create a null at half of a symbol rate thus the ideal low pass filtering is practically feasible. This can be accomplished by adding two consecutive symbols, $1+D$ filtering with D being a symbol time delay. Its frequency response is given by $1 + e^{-j2\pi fT} = 2 \cos (\pi fT)e^{-j\pi fT}$. Historically $1+D$ system is called duo-binary or correlative coding. This development was done 60's and 70's, used extensively. Practical situations sometimes necessitate removing DC components, which can be done by subtracting two consecutive symbols, $1-D$ filtering. This is combined with $1+D$ resulted in $1-D^2$ system, which has no DC and a null at the half of the symbol rate $\left(\frac{1}{2T}\right)$. Its frequency response is $1 - e^{-j2\pi f2T} = 2 \sin (2\pi fT)e^{-j\pi f2T}$.

Later it was generalized as partial response system with the filters being generalized. However, these two filters, $1+D$ and $1-D^2$, are the simplest and seem to be most useful. Furthermore a general filter seems to need to contain $1+D$ factor in order to be bandwidth efficient. For example, $(1+D)^2 = 1 + 2D + D^2$ and $1-D^2 = (1-D)(1+D)$.

2.7.3.1   1+D Duo-Binary digital filter

Most useful form of 1+D system for M-level (M = 2, 3, 4,..) is shown in Figure 2-46
. Before adding two consecutive symbols, a pre-coding in terms of transfer function
$\left[\frac{1}{1+D}\right]_{mod-M}$ is applied. And amazingly at the receiver mod-M operation recovers the
original symbol as shown in the figure. This precoding in the form of $\left[\frac{1}{G(D)}\right]_{mod-M}$
works with a shaping filter, $G(D)$, in general, and at the receiver modulo operation
will recover the original symbol.

1+D system often uses an ideal low pass filter as an interpolation, i.e., there is no
need of excess bandwidth as shown in Figure 2-47. Its impulse response is shown in
Figure 2-48 (RHS) along with the magnitude frequency response. When it is
sampled at −0.5T, +0.5T, its values are both 1.0 and zeroes at other T spaced points.

Referring to Figure 2-46, discrete time representation of 1+D, we give an example
below.

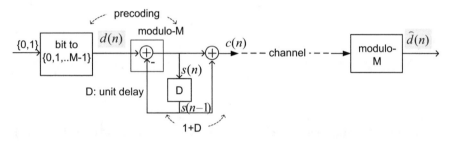

Figure 2-46: 1+D system with precoding

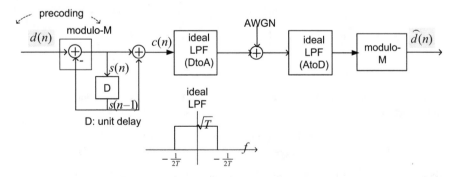

Figure 2-47: The implementation of digital spectrum shaping of 1+D with ideal low pass

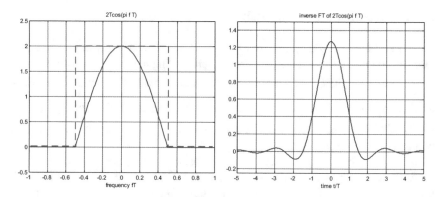

Figure 2-48: The magnitude of 1+D (LHS) and the impulse response with ideal LP (RHS); sampled at −0.5T, +0.5T its values are both 1.0 which indicate 1+D

Example 2-15: M = 4 with data $d(n) = [3\ 1\ 2\ 3\ 2\ 0\ 1\ 0\ 2\ 1\ 0\ 3\ 0\ 1]$

n	1	2	3	4	5	6	7	8	9	10	11	12	13	14
d(n)	3	1	2	3	2	0	1	0	2	1	0	3	0	1
s(n)	3	2	0	3	3	1	0	0	2	3	1	2	2	3
c(n)	3	5	2	3	6	4	1	0	2	5	4	3	4	5
d(n)	3	1	2	3	2	0	1	0	2	1	0	3	0	1

Note that at the receive side, mod(c(n), M) = d(n), i.e., mod operation on received sample recovers d(n) correctly. This is due to precoding. Note also that c(n) ranges from 0 to 6, 7 level from 4 level. In general M levels will be expanded to 2M-1 levels. For example, binary (M = 2) will expand to 3 levels. See an exercise below.

Exercise 2-30: M = 2 with data $d(n) = [0\ 1\ 0\ 1\ 1\ 0\ 1\ 0\ 1\ 1\ 0\ 1\ 0\ 1]$

n	1	2	3	4	5	6	7	8	9	10	11	12	13	14
d(n)	0	1	0	1	1	0	1	0	1	1	0	1	0	1
s(n)	0	1	1	0	1	1	0	0	1	0	0	1	1	0
c(n)	0	1	2	1	1	2	1	0	1	1	0	1	2	1
d(n)	0	1	0	1	1	0	1	0	1	1	0	1	0	1

Confirm that, mod M operation on c(n) recovers data d(n), and levels are increased to 3.

Exercise 2-31: As binary data is mapped into {0, 1} → {+1, −1}, 1+D expanded decimal {0, 1, 2}→ will be mapped {+2, 0, −2}. Hint: See Figure 2-49.

Exercise 2-32: With a mapping of 2 bit data {0, 1, 2, 3} → {+3, +1, −1, −3}, obtain 1+D expanded mapping. Hint: consider all combinations of two adjacent symbols. For M = 2 see Figure 2-49. Answer: {0, 1, 2, 3, 4, 5, 6} →{+6, +4, +2, 0, −2, −4, −6}.

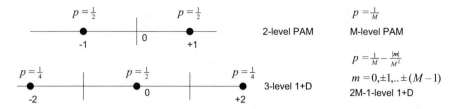

Figure 2-49: PAM and its 1+D system expansion

2.7.3.2 Performance of 1+D System with Symbol by Symbol Detection

In order to find the error rate performance we may compare the constellation of PAM and its expansion to 1+D system, which adds a present symbol to the previous to form 1+D symbol. We use $\{+1, -1\}$ for PAM rather than $\{0, 1\}$ in Figure 2-49. They are equivalent. The probability of 1+D system is a triangle distribution; $p = \frac{1}{M} - \frac{|m|}{M^2}$ with $m = 0, \pm 1, .. \pm (M - 1)$ while its underlying PAM symbols are equally likely. The number of symbols is enlarged to 2M-1 and their values $\{0, \pm 2, \pm 4, .. \pm 2(M - 1)\}$. M = 2 example is shown in Figure 2-49.

The power of 1+D system can be computed from its constellation in Figure 2-49 and is 2 ($=2^2/4*2 + 0/2$), which is twice, compared to 1 of 2-PAM. Note that the decision boundary distance from a symbol is the same, d, for both PAM and 1+D system. It can be shown that the power with M-level is also two times of PAM: $2\frac{M^2-1}{3}$.

Exercise 2-33: Work out the probability and power for M = 4, then for M in general.

Answer:

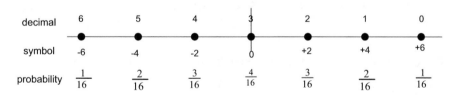

Power $= 2 * \frac{1}{16} \left[0^2 * 4 + 2^2 * 3 + 4^2 * 2 + 6^2 * 1\right] = 10$. The power of M = 4 PAM is 5.

The detection is done symbol by symbol like PAM. The SER of 1+D system can be represented by SER $= \frac{1}{2} 2Q\left(\frac{d}{\sigma}\right) + \frac{1}{4} 2Q\left(\frac{d}{\sigma}\right) = \frac{3}{2}Q\left(\frac{d}{\sigma}\right)$ for M = 2. (**Exercise 2-34:** Explain the process of obtaining the SER.)

In order to express SER in terms of signal to noise ratio of $\frac{Es}{No}$, we need to define the shaping filter to be an ideal low pass. And the receiver filter is matched to the transmit filter, i.e., the same ideal low pass filter. Thus the entire signal spectral shaping of 1+D is done by digitally adding two consecutive PAM symbols and its digital transmit power spectrum is in the shape of $|2T\cos(\pi fT)|^2$. In passing its power is obtained by integration; $P = d^2 \frac{1}{T} \int_{-\frac{1}{2T}}^{+\frac{1}{2T}} (2T\cos(\pi fT))^2 df = 2.0d^2$. Since $d = \sqrt{\frac{Es}{2T}}$ and $\sigma = \sqrt{\frac{No}{2T}}$, thus $\frac{d}{\sigma} = \sqrt{\frac{Es}{No}}$. Digital noise power is $\sigma^2 = \frac{No}{2T}$ since the noise bandwidth of ideal low pass filter is $1/T$. This situation is depicted in Figure 2-47.

SER of 1+D with M = 2 (symbol by symbol detection) is given by $\frac{3}{2}Q\left(\sqrt{\frac{Es}{No}}\right)$.

SER of 1 + D with M (symbol by symbol detection) $= 2\left(1 - \frac{1}{M^2}\right)$ $Q\left(\sqrt{\frac{3}{M^2-1}\frac{Es}{No}}\right)$

Exercise 2-35: Work out SER of 1+D with symbol by symbol detection for the case of M = 4. Then work out the general M case.

Answer: M = 4 case first

$$\frac{1}{16}Q + \frac{2}{16}2Q + \frac{3}{16}2Q + \frac{4}{16}2Q + \frac{3}{16}2Q + \frac{2}{16}2Q + \frac{1}{16}Q = \frac{30}{16}Q$$

For M in general, rearrange M = 4 case $\frac{4}{16}2Q + \frac{3}{16}2Q + \frac{2}{16}2Q + \frac{1}{16}Q + \frac{1}{16}Q + \frac{2}{16}2Q + \frac{3}{16}2Q$. For M in general, we see a pattern, $\frac{M+...+1}{16}2Q + \frac{(M-1)+...+1}{16}2Q - \frac{1}{16}2Q$. Thus it is expressed using $1 + 2 + ... + M = \frac{M(M+1)}{2}$ as $2\left(1 - \frac{1}{M^2}\right)Q$. Using the power of 1+D being : $2 * \frac{M^2-1}{3}$, SER $= 2\left(1 \frac{1}{M^2}\right)Q\left(\sqrt{\frac{3}{M^2-1}\frac{E_s}{N_o}}\right)$.

This 1+D system is 3dB worse than PAM but it is due to transmit filter and receive filter being not matched, and due to symbol by symbol detection. Since the transmitted symbol has a correlation, it can be exploited, and a detector can span a number of symbols, maximum likelihood sequence detection (MLSE) or simpler variation called AZD (ambiguity zone detection). We discuss the improvement with an analog matched filter with symbol by symbol next.

2.7.3.3   Performance of 1+D System with an Analog (1+D) Matched Filter

1+D spectrum shaping is done by two analog filters. Note that digital to analog conversion and low pass filter is part of transmit filter, and similarly at the receiver, an analog to digital conversion is followed by the analog filter. This is shown in Figure 2-50.

Rather than using the digital 1+D, 1+D part is combined with the ideal low pass filter. $s(n)$ and $s(n-1)$ in Figure 2-47 can be added by an analog filter. The pulse shape of the analog filter is, in frequency domain, $2T\cos(\pi fT)$ with $-\frac{1}{2T} \leq f \leq +\frac{1}{2T}$ and zero otherwise, and its pulse in time domain is given by $\frac{\sin(\pi t/T)}{(\pi t/T)} + \frac{\sin(\pi(t-T)/T)}{(\pi(t-T)/T)}$. Its peak value, at $t = T/2$, is $\frac{4}{\pi}$. This pulse, called an end-to-end pulse, is split into the

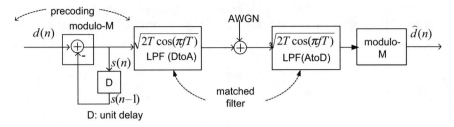

Figure 2-50: 1+D with an analog matched filter

transmit side and receive side to form a matched filter pair of $\sqrt{2T \cos(\pi f T)}$.

With this method, SNR penalty compared to PAM is $10\log_{10}\left(\left(\frac{\pi}{4}\right)^2\right) = 2.1$ dB rather than 3.0 dB (implemented as in Figure 2-47). In order to show it, $M = 2$ to be concrete, we compute the power at the output of the analog filter:

$$P = d^2 \frac{1}{T} \int_{-\infty}^{+\infty} |G_X(f)|^2 df = d^2 \frac{1}{T} \int_{-\frac{1}{2T}}^{+\frac{1}{2T}} 2T \cos(\pi f T) df = d^2 \frac{1}{T} \frac{4}{\pi}$$

And the noise power

$$\sigma^2 = \int_{-\infty}^{+\infty} S(f) |G_R(f)|^2 df = \frac{1}{2} N_o \int_{-\frac{1}{2T}}^{+\frac{1}{2T}} 2T \cos(\pi f T) df = \frac{1}{2} N_o \frac{4}{\pi}.$$

Thus $\frac{d}{\sigma} = \sqrt{2 \frac{Es}{No} \left(\frac{\pi}{4}\right)^2}$.

With M-level, it can be shown that $\frac{d}{\sigma} = \sqrt{\frac{6}{M^2-1} \frac{Es}{No} \left(\frac{\pi}{4}\right)^2}$.

Exercise 2-36: Find SER in terms of E_s/N_o for M-level 1+D system with matched filter.

1+D with matched filter SER $= 2\left(1 - \frac{1}{M^2}\right) Q\left(\sqrt{\frac{6}{M^2-1} \frac{E_s}{N_o} \left(\frac{\pi}{4}\right)^2}\right)$.

In practice this analog matched filter 1+D system (optimum symbol by symbol detection) is not often used. However, the implementation with 1+D digital shaping shown in Figure 2-47 (not matched case) is often implemented with a form of multi-symbol detection. SNR penalty can be compensated by multi-symbol detection and SNR recovery ranges from 1.4 dB (AZD) to 3.0 dB (ML). See Reference [3] pp. 511 - 515 for details.

Another interesting property of 1+D system is shown as a form of exercise below.

Exercise 2-37: Compare the constellation of binary 1+D with OOK. Remember that OOK can be demodulated non-coherently. The binary 1+D can be demodulated non-coherently.

Hint: Since {+2, −2} constellation points in Figure 2-49 carry the same data, it can be folded (rectified). This idea is discussed by Lender in Reference [4], pp.170.

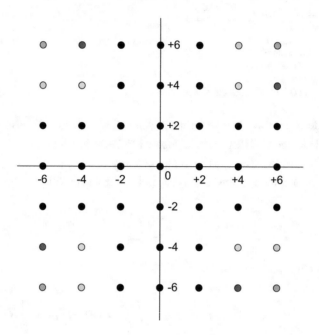

2.7.3.4   QPRS

Quadrature partial response systems (QPRS) can be formed by using PRS both in I and Q channels. An example constellation is on LHS. Its block diagram is shown in Figure 2-51. An example of the constellation, 49-QPRS, is shown on the left. Practical radio systems were built based on QPRS. See reference [5] for addition information where AZD is utilized for the improvement of receiver threshold, and for residual error control.

2.7.3.5 $(1-D^2)$ Modified Duo-Binary

Sometimes it is convenient to remove DC component of a signal. 1+D system can be extended to accomplish it, called $1-D^2$. The signal generation and reception can be essentially the same as 1+D system, other than the frequency shaping function is $2 \sin (2\pi fT)e^{-j\pi f2T}$. Thus Figure 2-46 and Figure 2-47 are applicable to this system except for replacing 1+D with $1-D^2$. Note that even the constellation is identical.

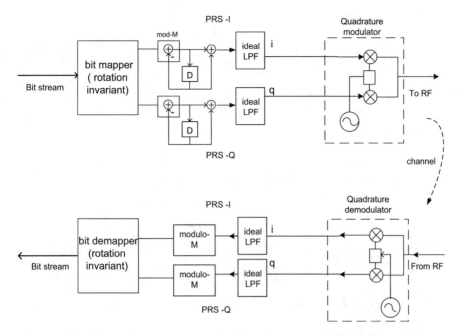

Figure 2-51: 1+D QPRS block diagram with 90° rotation invariant (optional) bit mapper and demapper. Care must be exercised for 90o rotation invariant mapper since the rotation of constellation occurs after PRS

The block diagram of this modified duo-binary is shown in Figure 2-52, and a numerical example (with the same data as $d(n) = [3\ 1\ 2\ 3\ 2\ 0\ 1\ 0\ 2\ 1\ 0\ 3\ 0\ 1]$) is shown in the table below. Notice that the symbol range is now $\{-3, -2, -1, 0, 1, 2, 3\}$.

n	1	2	3	4	5	6	7	8	9	10	11	12	13	14
d(n)	3	1	2	3	2	0	1	0	2	1	0	3	0	1
s(n)	3	1	1	0	3	0	0	0	2	1	2	0	2	1
c(n)	3	1	-2	-1	2	0	-3	0	2	1	0	-1	0	1
$\underline{d}$(n)	3	1	2	3	2	0	1	0	2	1	0	3	0	1

SER performance of the $1-D^2$ system is identical with 1+D

Exercise 2-38: Verify the above statement on SER performance of $1-D^2$. Use the integration of $\int_{-\frac{1}{2T}}^{+\frac{1}{2T}} 2T|\sin(2\pi fT)|df = \frac{4}{\pi}$.

$1-D^2$ system is very similar to 1+D system, including precoding and modulo M detection, except for the removal of DC. It is often used in magnetic recording context as partial response signaling, where a magnetic channel removes DC.

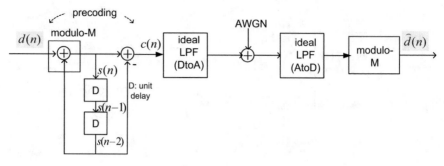

Figure 2-52: A block diagram of $1-D^2$ digital spectrum shaping system

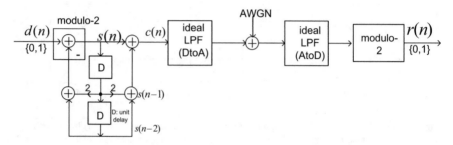

Figure 2-53: Exercise $1+2D+D^2$ system

2.7.3.6 Additional Example of PRS: $1+2D+D^2$

Exercise 2-39: $1+2D+D^2 = (1+D)^2$ system is depicted in Figure 2-53. Draw the system block diagram and find c(n) with $d(n) = [0\ 1\ 0\ 1\ 1\ 0\ 1\ 0\ 1\ 1\ 0\ 1\ 0\ 1]$. Show SER performance is given by SER $= \frac{7}{4}Q\left(\sqrt{\frac{1}{6}\frac{Es}{No}}\right)$.

n	1	2	3	4	5	6	7	8	9	10	11	12	13	14
d(n)	0	1	0	1	1	0	1	0	1	1	0	1	0	1
s(n)	0	1	0	0	1	0	0	0	1	1	1	0	1	0
c(n)	0	1	2	1	1	2	1	0	1	3	4	3	2	3
d(n)	0	1	0	1	1	0	1	0	1	1	0	1	0	1

n	1	2	3	4	5	6	7	8	9	10	11	12	13	14
d(n)	0	1	0	1	1	0	1	0	1	1	0	1	0	1
s(n)	0	1	0	0	1	0	0	0	1	1	1	0	1	0
c(n)	0	1	2	1	1	2	1	0	1	3	4	3	2	3
d̲(n)	0	1	0	1	1	0	1	0	1	1	0	1	0	1

decimal	4	3	2	1	0

power $=2*4^2/8 + 2*2^2*2/8 = 6$

symbol	-4	-2	0	+2	+4

probability	$\frac{1}{8}$	$\frac{2}{8}$	$\frac{2}{8}$	$\frac{2}{8}$	$\frac{1}{8}$

This system is used in 'tamed frequency modulation' (TFM) to control the spectrum of signal.

2.8 Frequency Modulation – FSK, MSK, CPFSK

FM uses the frequency (variation, change or difference) for carrying the data information. Its CW, in a general form, is represented by

$$s_x(t) = A \cos \left(2\pi \left[f_c t + f_d \int_0^t x_c(\tau) d\tau \right] + \theta_0 \right)$$

We explain the above equation; $x_c(t) = \sum_{n=-\infty}^{\infty} a_n g(t - nT)$ where a_n: data to be transmitted, $g(t)$: pulse shape. f_d is a parameter how much frequency changes to the waveform $(x_c(t))$ carrying data, called a peak frequency deviation, and θ_0 is an initial phase.

For example a binary frequency shift keying (FSK) can be generated by $a_n = \{1, -1\}$, and $g(t) =$ a rectangular pulse [0, T] in time and unity in amplitude. A numerical example is shown in Figure 2-54; binary FSK uses two tones (1200, 2400Hz) representing (0, 1) respectively with $f_c = 1800$ Hz and $f_d = 600$ Hz for symbol rate 1200 Hz. Thus one cycle or two cycles of carrier per symbol period. A rectangular pulse is implicitly used.

The characteristics of FSK are;

1. constant envelope (constant amplitude sinusoidal – low peak to average)

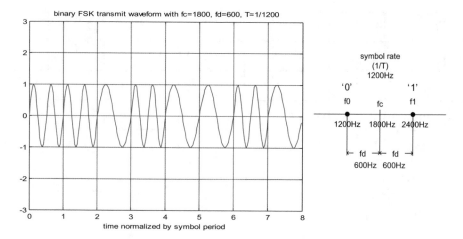

Figure 2-54: Binary FSK waveform example

2. non-coherent detection possible (see Figure 2-57)
3. frequency spectra of the signal is not the same as base band
4. the bandwidth tends to be larger than that of the base band as the peak frequency deviation gets larger (this is called Carlson's rule)

2.8.1   Examples of FSK Signal Generation

A simplified representation of a binary FSK example can be shown as in Figure 2-55 where two tones (f_0, f_1) generated by each oscillator are used representing {0,1}. This figure gives a simple visualization of binary FSK. Each oscillator phase may not be synchronized, and the switching between different frequencies create spectrum less clean, e.g., side lobe grows. This is not desirable. In modern digital hardware implementation we consider continuous phase implementations; the oscillators may be implemented by numerical oscillators (or equivalent like table of sine and cosine functions).

This binary FSK can be extended to multi-level FSK.

Another type of implementation is the use of VCO (voltage controlled oscillator). It is a common practice in analog radio implementation of high carrier frequency.

The instantaneous frequency of VCO changes proportional to the control voltage as shown in Figure 2-56.

$$f_i(t) = f_c + f_d x_c(t) = f_c + f_d \sum_{n=-\infty}^{\infty} a_n g(t - nT)$$

Figure 2-55: Binary FSK
generation using two
oscillators

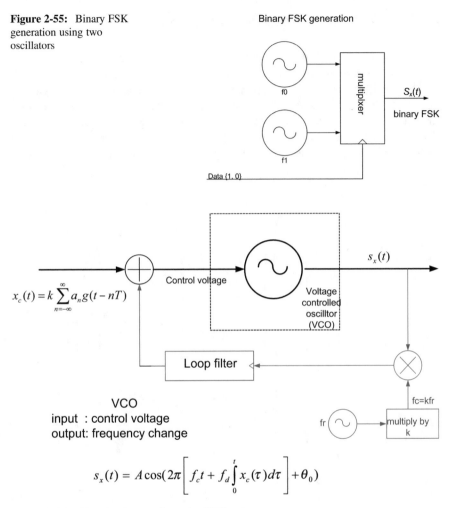

$$s_x(t) = A\cos(2\pi\left[f_c t + f_d \int_0^t x_c(\tau)d\tau \right] + \theta_0)$$

Figure 2-56: FSK signal generation using VCO

Note that the integration is built in part of VCO and that the peak frequency deviation is a gain parameter of the control voltage, part of VCO. The phase locked loop (PLL) is not essential in this picture, i.e., FM signal can be generated without it assuming VCO is centered on the desired carrier frequency. However, in practice, in order to stabilize VCO, PLL is almost always used. VCO is a device which generates a sinusoidal signal whose frequency is controlled by control voltage. PAM signal which carries the data, controls VCO directly.

Two examples of Figure 2-55 and Figure 2-56 are presented in analog form but it can be digital form using numerically controlled oscillator (NCO) in digital hardware. In this sense two examples may be useful in modern digital implementations.

2.8.2 *Non-coherent Demodulation of FSK*

One advantage of FSK is possible to demodulate it without carrier phase recovery at the receiver. We show this with an example shown in Figure 2-57.

The received waveform, perhaps contaminated by thermal noise but ignored, goes through a limiter, which is fine since the amplitude does not carry the information. Then measure the frequency variation by counting zero crossing – differentiation and rectifying – followed by a low pass filter. Note that it resembles to 2-level PAM signal. This process is summarized by a block diagram in Figure 2-58 as well.

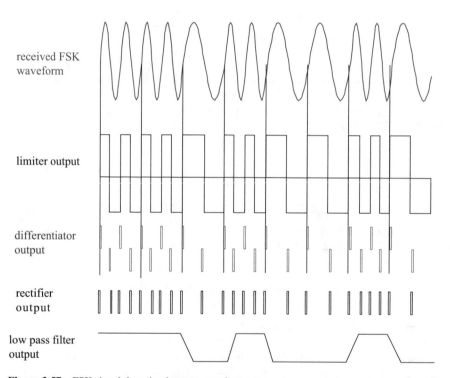

received FSK
waveform

limiter output

differentiator
output

rectifier
output

low pass filter
output

Figure 2-57: FSK signal detection by zero-crossing counts

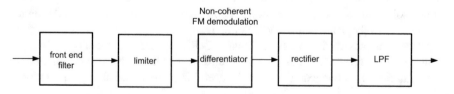

Non-coherent
FM demodulation

front end
filter → limiter → differentiator → rectifier → LPF

Figure 2-58: FSK signal detection with differentiator and rectifier

2.8.3 FSK Signal Generations Using Quadrature Modulator

Now back to the carrier wave, $s_x(t)$ can be put in exponential form as,

$$s_x(t) = A \operatorname{Re} \left\{ C(t) \cdot e^{j2\pi \cdot f_c \cdot t} \right\}$$

and the complex envelope $C(t)$ is given by

$$C(t) = \exp \left[j \left(2\pi \cdot f_d \int_0^t x_c(\tau) d\tau + \theta_0 \right) \right].$$

The complex envelope of FSK is complex, but real and imaginary parts are not independent while the real and imaginary parts of QAM complex envelope are independent and carrying independent data. Thus it may be considered to be one dimensional signal since $x_c(t)$ is PAM while QAM has two dimensions. Since FSK is of one dimension, its bit to symbol mapping can be done the same way as PAM.

A parameter, peak frequency deviation, may be expressed by $h = 2f_dT$, called modulation index, which is twice frequency deviation (in Hz) normalized to symbol rate. The name modulation index is somewhat a misnomer in the sense that it does not clearly indicate the degree of frequency deviation, and thus frequency deviation index might be more appropriate. However, we stick with the conventional name of modulation index.

We rewrite $C(t)$ with modulation index h, and inserting, $x_c(t) = \sum_{n=-\infty}^{\infty} a_n g(t - nT)$, we obtain $C(t) = \exp \left[j \left(2\pi \cdot \left[\frac{h}{2} \cdot \sum_{n=-\infty}^{\infty} a_n \int_{-\infty}^{t} \frac{1}{T} g(\tau - nT) d\tau \right] + \theta_0 \right) \right]$. Then define the integrated pulse shape as $q(t) = \int_{-\infty}^{t} \frac{1}{T} g(\tau) d\tau$. Now $C(t)$ is expressed using the modulation index as,

$$C(t) = \exp \left[j\theta(t) \right] = C_I(t) + jC_Q(t)$$

where $C_I(t) = \cos(\theta(t))$ and $C_Q(t) = \sin(\theta(t))$ and

$$\theta(t) = 2\pi \left[\frac{h}{2} \cdot \sum_{n=-\infty}^{\infty} a_n q(t - nT) \right] \cdot + \theta_0$$

With this representation the partitions into carrier wave, complex envelope, pulse filtering, and bit to symbol mapping (discrete time processing) are clearly identified. It is generic and thus different FSK signaling schemes can be studied within the

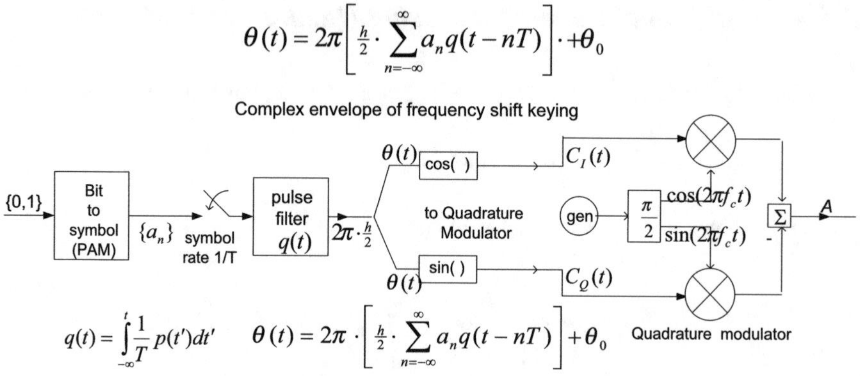

$$\theta(t) = 2\pi \left[\tfrac{h}{2} \cdot \sum_{n=-\infty}^{\infty} a_n q(t - nT) \right] \cdot + \theta_0$$

Complex envelope of frequency shift keying

$$q(t) = \int_{-\infty}^{t} \frac{1}{T} p(t') dt' \qquad \theta(t) = 2\pi \cdot \left[\tfrac{h}{2} \cdot \sum_{n=-\infty}^{\infty} a_n q(t - nT) \right] + \theta_0 \qquad \text{Quadrature modulator}$$

Figure 2-59: FSK signal generation using quadrature modulator

representation of complex envelope $C(t)$. Figure 2-59 is a literal implementation of FSK modulated carrier wave generation using complex envelope representation and followed by quadrature modulation.

2.8.4   Binary CPFSK Example

Consider the phase of $C(t)$, with the rectangular pulse shown in Figure 2-60, for $nT \le t \le (n + 1)T$,

$$\theta(t) = \pi h \left[\sum_{k=-\infty}^{n-1} a_k + a_n q(t - nT) \right] \mathrm{mod}(2\pi)$$

An example in Figure 2-54 is $h = 1 = 2f_d T$ example since $f_d = 600$ Hz for symbol rate 1200 Hz. Its phase change along with the waveform and frequency change is shown in Figure 2-61.

Phase trajectory moves linearly up or down during the symbol period T with $\{-1, +1\}\pi h$ for binary FSK. This is shown in Figure 2-61 (bottom), and for 4-level FSK, possible linear change during the symbol period are $\{-3, -1, +1, +3\}\ \pi h$.

2.8.5   M-Level CPFSK

M-level FSK can be represented similar to M-PAM but the distance between symbol constellation points is a frequency difference which can be given by modulation index $h = 2f_d T$. 4-FSK is shown in Figure 2-62. Keep in mind that $h = 2f_d T$ is a

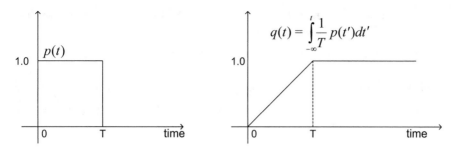

Figure 2-60:   Rectangular pulse and its integrated pulse

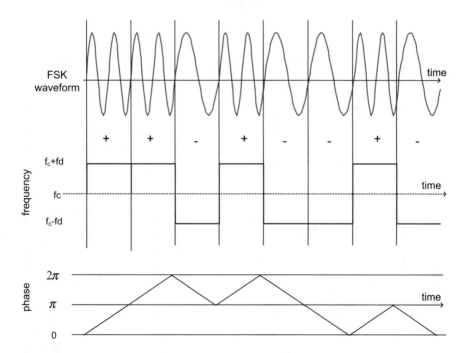

Figure 2-61:   CPFSK example with $h = 1$

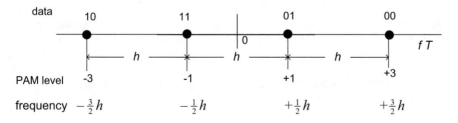

Figure 2-62:   4-level FSK discrete signal representation

frequency difference normalized by symbol rate. The frequency difference between symbols is given by h, modulation index.

Exercise 2-40: Draw a phase trajectory for 4-level FSK with the symbol pattern [...+1, −1, +3, −3, +3, −1, +1, −1, −1, +1, −3, +3..] with $h = 0.25$ and initial angle to be zero.

Figure 2-59 is a universal method of generating CPFSK of multi-level, different

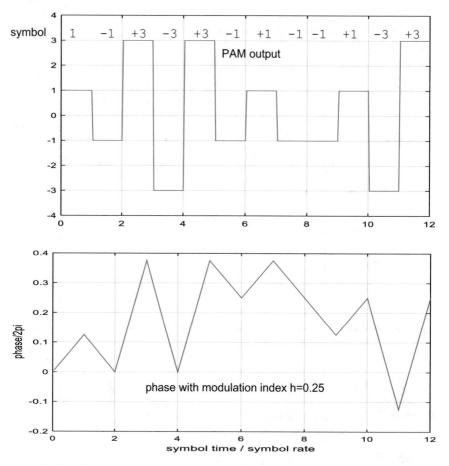

Figure 2-63: PAM output with rectangular pulse (above) is scaled by h and integrated to obtain phase, $\theta(t)$. The scaled phase, $\theta(t) / 2\pi$, is plotted and thus it may be reduced to modulo 1 or [−0.5, +0.5]

modulation index h and PAM pulse shape, with quadrature modulator.

2.8.6   MSK

The minimum h, still maintaining the orthogonality is $h = 1/2$. This case has a special name for it, minimum (frequency) shift keying (MSK) with 2-level (two tones). Since $h = 1/2$ means $f_d = \frac{1}{4T}$. The phase change during the symbol period is $\left\{+\frac{\pi}{2}\right\}$ or $\left\{-\frac{\pi}{2} = +\frac{3\pi}{2}\right\}$. Possible phase trajectory is shown in Figure 2-64.

A few observations on Figure 2-64 are made here. Positive phase change is in black color and negative change is in red color. The phase at even time (0, 2T, 4T, ..) is {0} or {π}, and the phase at odd time (T, 3T, ..) is $\left\{\frac{\pi}{2}\right\}$ or $\left\{-\frac{\pi}{2}\right\}$.

Using this even and odd separation and Figure 2-59 FSK generation, MSK signal can be generated as shown in Figure 2-65.

In-phase and quadrature phase has one bit delay (or half symbol clock) and thus one can recognize that this is offset QAM with the pulse shaping filter is given by

$$h(t) = \sin\left(\frac{\pi t}{2T}\right)$$
$$0 \le t \le 2T$$

This can be seen clearly by generating MSK signal with a computer program. An example is shown in Figure 2-66.

Exercise 2-41: Write a program routine (MATLAB) to generate MSK signal as Figure 2-65. Note that this task can be done by using Figure 2-59 for writing a program. Convince yourself that the pulse shaping filter is a half cycle of sine wave.

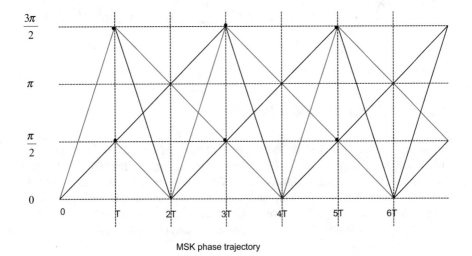

MSK phase trajectory

Figure 2-64: MSK phase trajectory

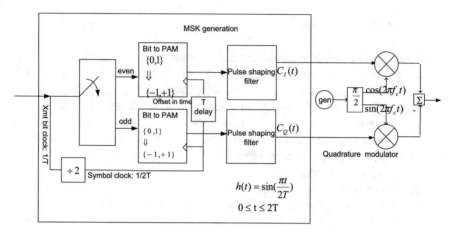

Figure 2-65: MSK generation using pulse shape filter. This is essentially the same as Figure 2-59 but is drawn with details of hardware implementation of clocking

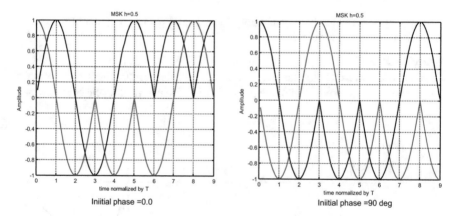

Figure 2-66: MSK waveform example of $C_I(t)$ and $C_Q(t)$

2.8.7   FSK with Gaussian Pulse

A special form of FSK, called GMSK, is used in GSM (cellular phone standard), DECT (digital European cordless telephone) and Bluetooth (IEEE801.15.1). The shaping pulse is not rectangular, but Gaussian. Thus the name is Gaussian MSK (GMSK) and MSK means modulation index $h = 1/2$. With the different pulse shape other than a rectangular, the spectrum of the signal is narrower than MSK and thus slightly more bandwidth efficient maintaining the advantages of FSK or MSK.

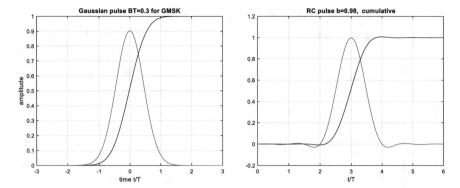

Figure 2-67: GMSK pulse example used in GSM cellular standard (LHS), and RC pulse is displayed (RHS)

$$h(t) = \frac{1}{\sqrt{2\pi\sigma^2}} \exp\left(-\frac{t^2}{2\sigma^2 T^2}\right)$$

$h(t)\frac{1}{T}$ is used for a proper scaling of q(t) $\rightarrow$ 1.0 as t $\rightarrow$ ∞; in other words, the scale for q(t) should be so that it becomes 1.0 as t become infinity. The pulse width is controlled by $\sigma = \frac{\sqrt{\ln 2}}{2\pi BT} \approx \frac{0.1325}{BT}$ where BT as a parameter BT = 0.3 is a typical standard. BT is in the range of [0, 1.0] ; 1.0 for narrow pulse, closer to 0 for wide pulse.

Example 2-16: For binary symbols $[-1 +1 -1 +1 -1 +1 +1 +1]$, PAM waveform, phase, and complex envelope of I and Q are shown in Figure 2-68.

2.8.8   Power Spectral Density of CPFSK, MSK, and GMSK

A different pulse filter may be used to contain the spectrum of FSK signal. Gaussian pulse used in GMSK is one such example. The spectrum of FSK signal is a function of peak frequency deviation f_d and the shaping pulse $h(t)$. However, it is not straightforward but involved and thus one in general resorts to numerical calculations in order to obtain the spectrum. When the pulse is rectangular with the pulse width being the same as symbol period as in Figure 2-60, Reference [1] (pp.203–207) computed the spectra as a function of level (M = 2, 4, 8) and h range of 0.125 to 1.5.

In Figure 2-69, we computed the spectrum of CPFSK signals - MSK, GMSK, and RC filtered with excess BW = 0.98. Numerical computation of the power spectral density (in short frequency spectrum) will be discussed in later subsection. It is conceptually straightforward and universal for any signal – modulation level, modulation index and pulse shape, and it is like using a spectrum analyzer instrument.

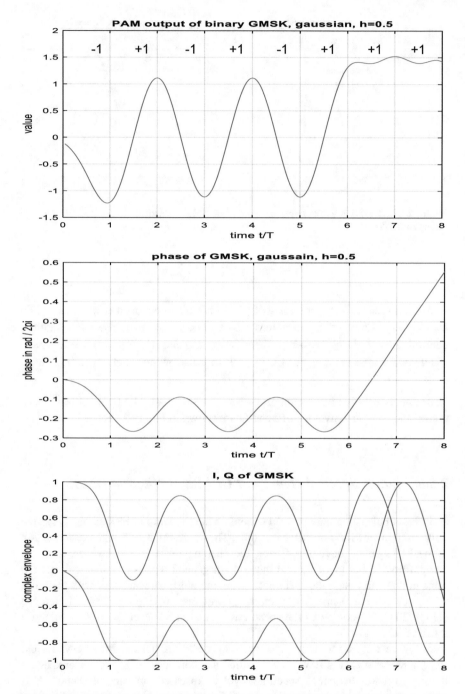

Figure 2-68: For binary symbol [−1 +1 −1 +1 −1 +1 +1 +1], PAM, phase and complex envelope are plotted

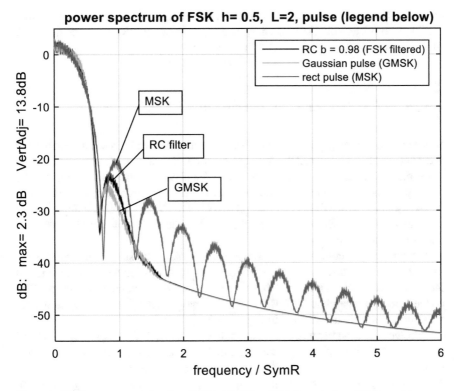

Figure 2-69: Power spectral density of MSK, GMSK and RC filtered where RC filter with excess BW = 0.98. Numerical simulations with random binary data are used

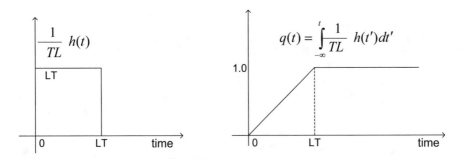

Figure 2-70: Extended pulse filter for FSK

2.8.9   Partial Response CPFSK

A base rectangular pulse width of T is extended to LT (L = 2, 3,..) as shown in Figure 2-70. This type of pulse is called partial response signaling, which can be done by discretely processing symbols $\{a_n\}$. The extended pulse in the above can be equivalently done by summing the past symbols of $\{a_n\}$ for example. In fact, entire partial response signaling systems may be applicable to CPFSK. For example, $1+2D +D^2$ in Figure 2-53 is applied and called tamed FM.

2.8.10   SER Performance Analysis of CPFSK

SER performance analysis of digital FM, in particular CPFSK, is complicated since it cannot be done with constellations but must be done at complex envelope level. Non-coherent case is even more complicated and will be treated separately. We resort to numerical simulations and our focus will be coherent demodulation, i.e., receiver knows carrier frequency and phase perfectly. This will give a upper bound performance of CPFSK since non-coherent demodulation will degrade, perhaps slightly, the performance.

We can draw a complex envelope simulation model as shown in Figure 2-71. Transmit side is a discrete version of Figure 2-59. A quadrature modulator is not included here since our interest is on the simulation of complex envelope. The conversion from baseband PAM to CPFSK may be considered as a numerical VCO in the model, as indicated in Figure 2-56.

Receive side is an inverse of transmit side with subtle difference. From baseband complex, $y(m\Delta t)$, its phase can be computed by $p(m\Delta t) = \frac{1}{2\pi} \tan^{-1}(im(y), re(y))$ with the range of $[-0.5 , +0.5]$. There are jumps (discontinuities), around $p (m\Delta t) = \pm 0.5$, due to the periodic nature of complex exponential (or trigonometric

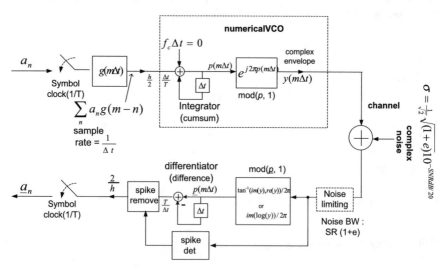

Figure 2-71: SER simulation model; transmit side is a discrete version of Figure 2-59 (without quadrature modulator) and receive side is a kind of inverse of transmit side, and noise limiting filter is at front end. This diagram implicitly assumes perfect carrier frequency and phase (coherent demodulation) since it is on complex envelope (baseband)

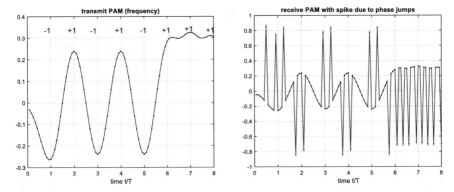

Figure 2-72: A numerical example of transmit side PAM (frequency) and receive PAM before removal of spikes, which must be removed before sampling and decision

function). These jumps become spikes after differentiation (approximated by differ-ence), which must be removed by spike detection / remove (shown in the figure) or by post detection filter (not shown in the figure). In addition to these systematic jumps, the noise may induce them randomly as well. An example is shown in Figure 2-72.

An idea of spike detection is simple and is to detect a large phase jump. In order to amplify the jump, additional interpolation may be used. As the interpolation is dense the jump is close to 1.0 of $p(m\Delta t)$. An idea of spike removal is to replace the peaking with an average of two adjacent (frequency) samples after differentiation. This spike detection and removal may be done by post detection filter to remove the spike of high frequency.

A noise limiting filter must be placed before receive processing as shown in Figure 2-71. In case of PAM (or PSK, QAM), the noise bandwidth of the filter is the same as symbol rate. However, in digital FM (CPFSK), it will be greater than the symbol rate due to the expansion of PSD (e.g., see Figure 2-69). In the model this can be done by adding more noise as indicated in the figure, $\sigma = \frac{1}{\sqrt{2}} \times \sqrt{(1+e)}10^{-SNRdB/20}$, $e = 0$ means that the noise bandwidth is the symbol rate. In our simulation we may use $e = 0$.

A simulation result is shown on the left. GMSK (h = 0.5, two level; binary, and Gaussian pulse) is simulated against BPSK under AWGN. Note that we assume that the noise limiting filter bandwidth is the same as symbol rate. GMSK is about 2dB worse than BPSK.

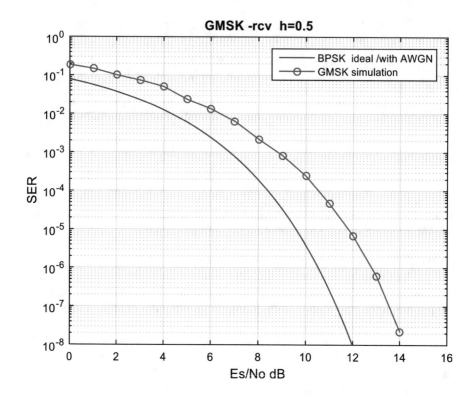

2.9   PSD of Digitally Modulated Signals

In this section we extend the power to its power spectral density; intuitively one can understand it as how the power of a random signal – digitally modulated signals are good examples of random signal - is distributed in frequency. This is reviewed in Chapter 9 but quickly repeated here again for convenience.

The power spectral density is Fourier transform of the correlation of a signal which is defined as $R(\tau) = E\{x(t)x(t + \tau)\}$ where $x(t)$ is a random signal (process); $S(f) = \int_{-\infty}^{+\infty} R(\tau)e^{-j2\pi f \tau}d\tau$. Its inverse is $R(\tau) = \int_{-\infty}^{+\infty} S(f)e^{+j2\pi f \tau}df$. This relationship is called Wiener-Khinchin theorem. We assume it is wide-sense stationary which means the statistics – average, variance, correlation - are not time dependent; not a function of t only of τ (time difference).

The power of a random process (signal) is $P_s = E\{|x(t)|^2\}$. From the definition of power spectral density, the power can be related to it as,

$$P_s = R(\tau)|_{\tau=0} = \int_{-\infty}^{+\infty} S(f)df.$$

In other words, the power is the sum (integral) of all power density in frequency. Intuitively it is satisfying.

2.9.1 Power Spectral Density of PAM Signal

With this definition of the power spectral density in the above, we would like to apply it to PAM signals and carry out the derivation. We use the work by Franks pp. 201–233 in this derivation (L.E. Franks," Signal Theory" Prentice-Hall, 1969). For more recent reference see [6] where a chapter is devoted to PSD.

Consider a random process defined by PAM signal, $x(t) = \sum_{n=-\infty}^{n=+\infty} a_n h(t - nT)$. We obtain its power spectral density. For discrete symbols of a_n, we assume they are stationary as, $E\{a_n\} = \mu$, $E\{a_n a_{n+m}\} = \alpha_m = \alpha_{-m}$, which do not depend on time index n. Then we compute the average and autocorrelation of $x(t)$ as,

$$E\{x(t)\} = \mu \sum_{n=-\infty}^{n=+\infty} h(t - nT)$$

$$E\{x(t)x(t + \tau)\} = \sum_{n=-\infty}^{n=+\infty} \sum_{k=-\infty}^{k=+\infty} E\{a_n a_k\} h(t - nT) h(t + \tau - kT).$$

It is convenient to change the index of the double summation $k = n + m$ as below

$$E\{x(t)x(t + \tau)\} = \sum_{m=-\infty}^{m=+\infty} \alpha_m \sum_{n=-\infty}^{n=+\infty} h(t - nT) h(t + \tau - nT - mT)$$

Clearly the random process $x(t)$ is periodically stationary (or cyclostationary) since $E\{x(t)\} = E\{x(t + T)\}$ and $E\{x(t + T)x(t + \tau + T)\} = E\{x(t)x(t + \tau)\}$. Thus the time dependence is due only to the periodic nature, i.e., in each period it is identical. This periodicity can be removed by averaging per symbol time with the uniform density.

$$x(t) = \sum_{n=-\infty}^{n=+\infty} a_n h(t - nT + \delta) \text{ with } 0 < \delta < T$$

$$E\{x(t)\} = \mu \frac{1}{T} \int_{-\infty}^{+\infty} h(t)dt$$

$$E\{x(t)x(t + \tau)\} = \frac{1}{T} \sum_{m=-\infty}^{m=+\infty} \alpha_m r(\tau - mT) \text{ with } r(\tau) = \int_{-\infty}^{+\infty} h(t)h(t + \tau)dt$$

Note that the Fourier transform of $r(\tau)$ is given by $R(f) = |H(f)|^2$.
Taking the Fourier transform of

$$S_{xx}(\tau) \equiv E\{x(t)x(t+\tau)\} = \frac{1}{T} \sum_{m=-\infty}^{m=+\infty} \alpha_m r(\tau - mT)$$

we obtain the power spectral density as,

$$S_{xx}(f) = \left[\sum_{m=-\infty}^{m=+\infty} \alpha_m e^{-j2\pi mfT} \right] \frac{1}{T} |H(f)|^2$$

This result is intuitively appealing. The first part in the parenthesis is the part from discrete time signal and the second part is the frequency response of a filter. The scaling by 1/T is due to the synchronous signaling with the symbol rate.

Example 2-17: $E\{a_n\} = 0$ and $E\{a_n a_{n+m}\} = d^2$ when $m = 0$, $E\{a_n a_{n+m}\} = 0$ when $m \neq 0$. BPSK is an example.

Answer: $S_{xx}(f) = d^2 \frac{1}{T} |H(f)|^2$. This result was already used in Figure 2-12 in order to relate the power of digitally modulated signals with analog filters.

Example 2-18: $E\{a_n\} = \mu$ and $E\{a_n a_{n+m}\} = \sigma^2 + \mu^2$ with $E\{(a_n - \mu)^2\} = \sigma^2$ variance when $m = 0$ and $E\{a_n a_{n+m}\} = 0$ when $m \neq 0$.

$$E\{x(t)x(t+\tau)\} = s_{xx}(\tau) = \frac{1}{T} \sigma^2 r(\tau) + \frac{u^2}{T} \sum_{m=-\infty}^{m=+\infty} r(\tau - mT)$$

We use a Fourier transform pair (the 22th of Table 9.1 in Chapter 9):

$$\sum_{m=-\infty}^{m=+\infty} r(t - mT) \Leftrightarrow \frac{1}{T} \sum_{k=-\infty}^{k=+\infty} R\left(f - \frac{k}{T} \right)$$

$$S_{xx}(f) = \sigma^2 \frac{1}{T} |H(f)|^2 + \frac{\mu^2}{T^2} \sum_{k=-\infty}^{k=+\infty} |H(f)|^2 \delta\left(f - \frac{k}{T} \right) \qquad \text{(Answer)}$$

There are power spectral lines at multiples of symbol rate if DC (average) of a signal is not zero.

Example 2-19: $1 - D^2$ system shown in Figure 2-52. Ignore the precoding part and focus on the formation of a new symbol by $b_n = a_n - a_{n-2}$. Notice that $E\{b_n\} = 0$ and $E\{b_n b_{n+m}\} = \beta_m = E\{(a_n - a_{n-2})(a_{n+m} - a_{n+m-2})\} = 2\alpha_m - \alpha_{m+2} - \alpha_{m-2}$

Note that $\alpha_0 = d^2$ and $\alpha_m = 0$ when $m \neq 0$.

Thus $\beta_0 = 2d^2$, $\beta_1 = \beta_{-1} = 0$, $\beta_2 = \beta_{-2} = -d^2$ and the power spectral density becomes $S_{xx}(f) = \left[\sum_{m=-\infty}^{m=+\infty} \beta_m e^{-j2\pi mfT} \right] \frac{1}{T} |H(f)|^2 = B(f) \frac{1}{T} |H(f)|^2$

where $B(f) = 4d^2 \sin^2(2\pi fT)$. We already obtained the same result by considering a transfer function of $|1 - e^{-j2\pi f 2T}|^2 = |2 \sin(2\pi fT)|^2$.

Example 2-20: Differential coding shown in Figure 2-41. It can be summarized as $b_n = (a_n - b_{n-1}) \bmod 2$ where $a_n = \{1, 0\}$. Intuitively it is clear that this differential coding will not change the frequency spectrum when the symbols are equally likely. What happens to the frequency spectrum when the probability of $a_n = 1$ is $p > 1/2$?

Answer: In order to evaluate the power spectral density for this differential coding signal, we need to know the mean and correlation of b_n. (1) mean: $\bar{b} = E(b_n) = \Pr(b_n = 1)1.0 + \Pr(b_n = 0)0.0 = \Pr(b_{n-1} = 0$ and $a_n = 1) + \Pr(b_{n-1} = 1$ and $a_n = 0) = (1 - \bar{b})p + \bar{b}(1 - p)$. Solving for $\bar{b} = E(b_n)$ it is ½. (2) correlation: $\beta_m = E\{b_n b_{n+m}\} = \Pr[b_n = 1$ and $b_{n+m} = 1] = \frac{1}{2}\Pr[b_{n+m} = 1 | b_n = 1]$ Now $2\,\beta_m = \Pr[b_{n+m} = 1 | b_n = 1]$ is expressed as the sum of two cases to make $b_{n+m} = 1$ due to differential coding; $2\beta_m = \Pr[b_{n+m-1} = 0$ and $a_{n+m} = 1 | b_n = 1] + \Pr[b_{n+m-1} = 1$ and $a_{n+m} = 0 | b_n = 1]$. Note that a_{n+m} and b_{n+m-1} are independent and that $\Pr[b_{n+m} = 0 | b_n = 1] = 1 - 2\beta_m$. We obtain a difference equation of $\beta_m - (1-2p)\,\beta_{m-1} = \frac{1}{2}p$. This has the solution of $\beta_m = \frac{1}{4}\left[(1 - 2p)^{|m|} + 1\right]$ since $\beta_m = \beta_{-m}$ with $\beta_0 = \frac{1}{2}$.

Thus spectral density is $S_{xx}(f) = \left[\sum_{m=-\infty}^{m=+\infty} \beta_m e^{-j2\pi mfT}\right] \frac{1}{T}|H(f)|^2$. We focus on the computation of $B(f) = \sum_{m=-\infty}^{m=+\infty} \beta_m e^{-j2\pi mfT}$ as below,

$$B(f) = \frac{1}{4}\sum_{m=-\infty}^{m=+\infty}(1 - 2p)^{|m|} e^{-j2\pi mfT} + \frac{1}{4}\sum_{m=-\infty}^{m=+\infty} e^{-j2\pi mfT}$$

$$= \frac{1}{4}M(f) + \frac{1}{4T}\sum_{k=-\infty}^{k=+\infty}\delta\left(f - \frac{k}{T}\right)$$

After some computation, using $\sum_{m=0}^{\infty}\gamma^m = \frac{1}{1-\gamma}$, we obtain $M(f) = \frac{p(1-p)}{p^2 + (1-2p)\sin^2 \pi Tf}$.

Finally, $S_{xx}(f) = \frac{1}{T}|H(f)|^2\left[\frac{1}{4}M(f) + \frac{1}{4T}\sum_{k=-\infty}^{k=+\infty}\delta\left(f - \frac{k}{T}\right)\right]$.

With this example it is shown that the probability has an impact to the power spectral density. However that the analog filters, typically low pass type, shape the overall band limiting effect and the digital shaping is periodic but limited within the bandwidth of a signal due to the low pass filtering. This example was taken from Franks, pp.216.

2.9.2 PSD of Quadrature Modulated Signals

Two baseband signals form a complex envelope of real and imaginary, and each gets into I-channel and Q-channel of a quadrature modulator in order to generate carrier wave for radio transmission. This process is practically an important part of implementations and corresponds to a frequency translation to a carrier frequency without altering signal spectral density. It provides a necessary amplification for feeding antenna. In practice the nonlinear effect of power amplification may cause inter-modulation distortion (IMD) which produces spectral growth. This will be treated as a separate topic. No analytical expression is available in this case, and thus we need to resort to numerical simulations or to measuring it by using an instrument called spectrum analyzer.

2.9.3 PSD of FDM and OFDM Signals

FDM signals are formed by adding different carrier (subcarrier) signals as used in AM, FM and TV broad casting. A special form of FDM is OFDM which will be covered intensively in Chapter 5. In OFDM subcarrier spacing is equally distributed and related with symbol rate. Schematically it is formed as in Figure 2-73.

In FDM and OFDM, PSD is a sum of PSD of each subcarrier, i.e., the squared magnitude of all subcarrier frequency response. We give an example of IEEE 802.11a/n shown in Figure 2-74. Here windowing is used to reduce the outband PSD. Windowing is to use a rectangular pulse smoothed by a window.

Note that practical OFDM implementation uses IDFT at transmit and DFT at receive side, rather than the schematic of Figure 2-73. Chapter 5 is devoted to OFDM issues.

2.9.4 PSD Numerical Computations and Measurements

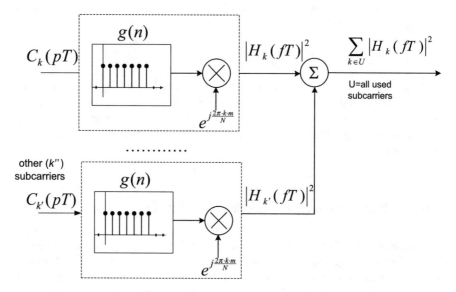

Figure 2-73: FDM signal formation by adding different carrier frequency signal

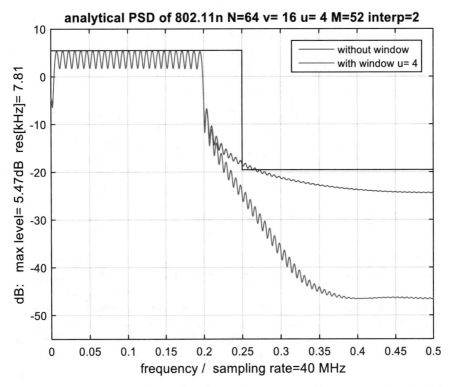

Figure 2-74: PSD of 802.11a/n signal with windowing (red) and without (blue); A mask of BW 20MHz is shown. Vertical scale is adjusted so that total power is $N = 64$

In practice it is important to know the power spectrum of a signal. In transmit side a signal should occupy a specified frequency band with a center frequency (carrier frequency) and bandwidth. It should occupy 'cleanly' without spilling over next adjacent channels. In receive side the signal is typically contaminated by noise and interference, which should be removed (limited) without distorting the signal itself. In this subsection we consider its practical aspects of measurements and computations.

A spectrum analyzer is a physical instrument for measuring power spectrum of arbitrary signals. This situation is quite different from finding a power spectral density for a given signal with all the signal structure we discussed so far. It is useful to understand the power spectral density intuitively since it provides an insight. Essentially a spectrum analyzer computes the following equation.

$$S_{xx}(f) = \lim_{C \to \infty} \left| \frac{1}{2C} \int_{-C}^{+C} x(t)e^{-j2\pi ft} dt \right|^2$$

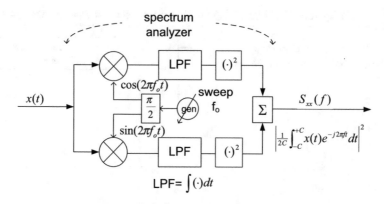

Figure 2-75: The principle of spectrum analyzer – sweep-tuned type - principle

The choice of the parameter C in the above equation is a frequency resolution. The integration can be implemented by a low pass filter with the appropriate cutoff frequency around 1/C. The 'carrier' frequency is variable over a frequency range of interest. It is called frequency sweeping range. Visualize one fixed frequency and it is the power at that frequency, and repeated to another frequency. Thus it is clear, with this formulation, that the power spectral density is the power distribution in frequency and by summing all it should be the (total) power.

Actual mechanization is similar to quadrature demodulator as shown in Figure 2-75.

This process can be done numerically; sample and store $x(t)$ and then compute Fourier transform (perhaps using fast Fourier transform: FFT) and squared magnitude. The computation of a signal spectrum has been a topic for a long time, which prompted to invent FFT. We cover this topic in detail next.

2.9.5 Numerical Computation of PSD Using FFT

We first summarize the idea of computing PSD using FFT,

1. Segment a long time signal $x(t)$ into a time segment of duration T, and compute Fourier transform of it and scaled by $\frac{1}{\sqrt{T}}$: $\widehat{X}_i(f) = \frac{1}{\sqrt{T}} \int_{iT}^{(i+1)T} x(t)e^{-j2\pi ft}dt$

2. Compute the squared magnitude of the Fourier transform for each segment, and then sum them and divide by the number of segments: $S_{xx}(f) \approx$

$\frac{1}{M}\sum_{i=0}^{M-1} \left|\widehat{X}_i(f)\right|^2$

3. As the number of segments becomes infinity, $S_{xx}(f) = E\left\{\left|\widehat{X}_i(f)\right|^2\right\}$.

The time length of a segment T is related to *frequency resolution*, i.e., frequency resolution $= \frac{1}{T}$. The number of segments, M, is related with the *smoothness* of PSD. It is called *video bandwidth* in spectrum analyzer jargon. In actual implementation, each segment may be generated with random data, rather than segmenting a long record of $x(t)$, if random signals can be generate readily. Note also that we use a rectangular window (or no windowing) for each segment. No windowing may be appropriate when each segment can be generated readily.

In order to utilize FFT we need to sample $x(t) \rightarrow x(n\Delta t)$, $T = N\Delta t$, i.e., there are N samples per segment $n = 0,1,..N-1$. We also need to limit frequency sample to N as well; $f \rightarrow k$ $\frac{1}{T}$ with $k = 0, 1, .. N-1$. Thus the frequency range is 0 to $\frac{1}{\Delta t} \left(= N\frac{1}{T}\right)$ or $-0.5\frac{1}{\Delta t}$ to $+0.5\frac{1}{\Delta t}$. The relationship of time sample and frequency range is fundamental to FFT computation.

Applying the above to $\widehat{X}_i(f) = \frac{1}{\sqrt{T}} \int_{iT}^{(i+1)T} x(t)e^{-j2\pi ft} dt$, we obtain $\widehat{X}_i\left(e^{j2\pi f}\right) = \frac{1}{\sqrt{N\Delta t}} \sum_{n=0}^{N-1} x(n\Delta t)e^{-j2\pi(f\Delta t)n}\Delta t$, i.e., the integration is replaced by summation. Using $f \rightarrow k\frac{1}{T} = k\frac{1}{N\Delta t}$, we obtain $\widehat{X}_i\left(e^{j2\pi k\frac{1}{T}}\right) = \frac{\sqrt{\Delta t}}{\sqrt{N}} \sum_{n=0}^{N-1} x(n\Delta t)e^{-j2\pi\left(\frac{k}{N}\right)n}$.

$S_{xx}\left(e^{j2\pi k\frac{1}{T}}\right) = E\left\{\left|\widehat{X}_i\left(e^{j2\pi k\frac{1}{T}}\right)\right|^2\right\}$ and inserting $\widehat{X}_i\left(e^{j2\pi k\frac{1}{T}}\right)$ we have

$S_{xx}\left(e^{j2\pi k}\right) = \Delta t\, E\left\{\frac{1}{N}\left|\sum_{n=0}^{N-1} x(n\Delta t)e^{-j2\pi\left(\frac{k}{N}\right)n}\right|^2\right\} = \frac{1}{N}E\left\{\frac{1}{N}\left|DFT\,(x(n))\right|^2\right\}$ note

that when $T = 1$, then $\Delta t = \frac{1}{N}$. The power of the segment in frequency should be the same as the power of time domain. For this the first scale factor $\frac{1}{N}$ is necessary. The next scale $\frac{1}{\sqrt{N}}$ is necessary to maintain the power level through DFT. It is useful to compute a PSD scaled by N as,

$$S_{xx}\left(e^{j2\pi k}\right)N = E\left\{\left|\frac{1}{\sqrt{N}}\,DFT\,(x(n))\right|^2\right\}$$

knowing that the total power is N times the power of signal, i.e., the computed PSD will be shifted vertically by factor of N. The averaging is done approximately by using.

$$S_{xx}\left(e^{j2\pi k}\right)N \approx \frac{1}{M}\sum_{i=0}^{M-1}\left|\frac{1}{\sqrt{N}}\,DFT\,\left(x(n); i^{th}\ \text{segment}\right)\right|^2$$

2.9.6 Example of PSD Computation by Using FFT

Using the method developed in the previous section, a computer program is written to compute PSD for different communication signals – QAM, OFDM, and FSK. We give an example of IEEE 802.11a/n and compare it with the analytical PSD shown in Section 2.9.3.

OFDM signal parameters are described in the IEEE standard in detail and we summarize them briefly here. IDFT size $N = 64$, CP $v = 16$, and thus 80 samples per OFDM symbol are transmitted serially with the sampling rate $(N + v)/T = 20$MHz. Out of 64 subcarriers M = 52 are used. OFDM symbol time $T = 4.0$ usec of which CP is 0.8 usec and data carrying portion is 3.2 usec. Sub-carrier frequency spacing is 312.5 kHz (= 1/3.2 usec).

The result of numerical calculation is shown in Figure 2-76. Frequency resolution 100Hz (2500 OFDM symbols in a segment) and averaging 30000 segments on LHS, and frequency resolution 7810Hz (32 OFDM symbols in a segment) and averaging 120 segments on RHS. Yet the difference appears minor – tail level −46dB vs. −41dB and peak level 5.55dB vs. 6.35dB. Both are overall resembling to the analytical one in Figure 2-74.

2.9.7   PSD of Digital FM Signals –FSK, CPFSK

The analytical PSD for CPFSK with rectangular pulse was done in [1] and further extended in [6]. In practical situations, however, numerical calculations are straightforward and can be accurate by increasing the length of each segment (reducing frequency resolution) and the number of segments for averaging (smoothness). The program written can compute all possible FSK signals but as a demonstration, we list several examples in Figure 2-77 – binary FSK, modulation index varying from $h = 0.5$, 0.7, 0.95, 1.0, 1.8, 2.0, 2.8 and 3.0. With h being integer or close to it, there is a tone of $\frac{h}{2}$ times symbol rate.

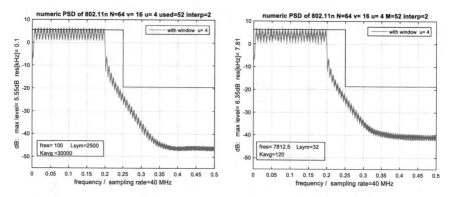

Figure 2-76:   Numerical PSD of 802.11a/n OFDM. Resolution 100Hz (2500 OFDM symbols in a segment) and averaging 30000 segments on LHS. Resolution 7810Hz (32 OFDM symbols in a segment) and averaging 120 segments on RHS. Yet the difference appears minor – tail level −46dB vs. −41dB and peak level 5.55dB vs. 6.35dB

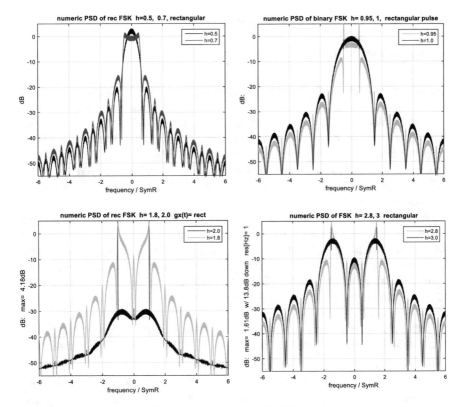

Figure 2-77: $h = 0.5, 0.7, 0.95, 1.0, 1.8, 2.0, 2.8$ and 3.0. With h being integer or close to it, there is a tone of $\frac{h}{2}$ times symbol rate

2.9.8 PSD Computation Using Correlation

Another way of measuring the power spectrum is to compute the correlation of a signal as defined as,

$$s_{xx}(\tau) = \lim_{C \to \infty} \frac{1}{2C} \int_{-C}^{+C} x(t)x(t + \tau)dt \dots$$

This might be considered as a generalization of the signal power since the power can be defined and measured by the equation $P_x = \lim_{C \to \infty} \frac{1}{2C} \int_{-C}^{+C} |x(t)|^2 dt = s_{xx}(0)$. Then the Fourier transform of $s_{xx}(\tau)$ will be the power spectrum using Wiener-Khinchin theorem. We do not pursue here.

2.10   Chapter Summary and References

2.10.1   Summary

We open this chapter with the overall block diagram of Figure 2-1 and immediately narrow down to discrete-time (digital) blocks – modulation and demodulation processing blocks – and abstract it rather quickly to Figure 2-2. The thermal noise at receiver front end, modeled accurately by AWGN, is the only channel impairment. In particular, we assume perfect carrier phase and frequency and symbol timing phase and frequency, and flat channel frequency response. Thus the performance of AWGN channel model is the best upper bound, which is a useful reference point in practical situations.

We partition the communication signals into three – constellation (digital repre-sentation), complex envelope (continuous I and Q baseband) and CW (real RF signals to transmit through antenna). Our focus in this chapter is on discrete-time representations but we have to relate them to complex envelope, and CW. This is done with binary cases – BPSK (2-PAM), OOK and orthogonal (a form of FSK) using a rectangular pulse for converting to complex envelope and quadrature modulator to CW.

The digital power of signal and noise is related with the corresponding analog one. For SNR we use $\frac{E_s}{N_o}$, which is the same as, $\frac{P}{N}$, measured signal power over noise power when the noise power measurement bandwidth is symbol rate. The error rate performance is measured by SER as a function of $\frac{E_s}{N_o}$. SER can be calculated from constellation and with decision boundary but BER needs a concrete bit to symbol mapping table and a function of addition digital processing – scrambler, rotational invariance coding.

Then binary modulations are extended to multilevel ones – PAM, QAM, and PSK. Additional topics that arise due to practical reasons are covered; offset QAM, scrambler, $180°$ and $90°$ differential coding. PRS is an attempt to reduce the bandwidth.

FSK is explored for non-coherent detection and its constant envelope property. The computation of PSD is explored both analytically and numerically. Numerical simulation method is universally applicable even when no analytical solution is available.

2.10.2   References

A vast amount of literature - textbooks, journal papers, conference papers, and technical presentations – is available on the topic of this chapter; digital modulations. Our reference is vanishingly small but it still can be a good starting point in search of additional references. For this purpose, we include seven text books [1] to [8] except for reference [5]. In particular [7] covers the topics of Section 2.1 to 2.5 but it is mathematically rigorous, opposite to our practical approach. Yet our practical approach is well supported mathematically as well according to [7]. A textbook of [8] is comparable to [2]. Reference [6] is an encyclopedic collection of modulations, a particular emphasis on CPFSKs.

[1]   Lucky, R.W., Salz, J., and Weldon, E.J., "Principles of Data communication", MacGraw-Hill, 1968, pp.203–207
[2]   Proakis, J. and Salehi,M., "Digital Communications" 5th ed, McGraw Hill 2008
[3]   Benedetto, S.,"Digital Transmission Theory" Prentice-Hall, 1987 pp. 511–515 on AZD and ML
[4]   Feher, Kamilo, "Digital Communications: Microwave Applications" Prentice – Hall, 1981. pp.170 on 1+D form of OOK by A. Lender
[5]   Yang, Sung-Moon,"A Digital Microwave Radio Based on 81-QPR Modulation" ICC 1988 Philadelphia, session 15.3.1-15.3.5, pp.478–482.
[6]   Xiong, Fuqin "Digital Modulation Technques", 2nd ed, Artech House, 2006, Ch.3, 5, 6 for FSK, MSK and CPM, Appendix A for PSD
[7]   Lapidoth, Amos "A Foundation in Digital Communication" Cambridge University Press, 2009
[8]   Barry, John, Lee, Edward, and Messerschmitt " Digital Communication", 3rd ed, Kulwer Academic Publishers, 2004

2.11   Problems

P2-1: A discrete-time transmission system is shown in Fig. P2-1. Binary data i $(n) \in \{0, 1\}$ is the input to the system and the channel is adding white Gaussian noise.

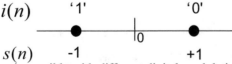

A variety of systems is possible with different digital modulation processing (x) at transmit side, and corresponding (digital) demodulation processing (r). Here we

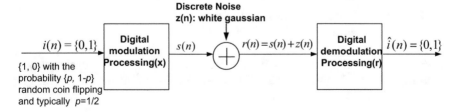

Fig. P2-1: Discrete-time transmission system with discrete AWGN

Table P2-1: BPSK example to fill blank

$i(n)$	1	1	0	1	0	0	0	1	1	1	0	0
$s(n)$	−1		+1									
$r(n)$	−0.49	−1.0	1.5	−1.1	0.91	2.05	1.99	0.03	−0.52	−1.85	1.51	2.1
$\widehat{i}(n)$												

Table P2-2: OOK example with A = 1.4

$i(n)$	1	1	0	1	0	0	0	1	1	1	0	0
$s(n)$	1.4		0									
$r(n)$	1.91	1.4	0.5	1.3	−0.09	1.05	0.99	2.43	1.88	0.55	0.51	1.1
$\widehat{i}(n)$												

specify a simple system of BPSK, whose bit to symbol mapping, i.e., digital modulation, is shown on the left; '0' → +1, and '1'→ −1. We consider what type of digital demodulation processing can be done to recover $i(n)$ at receive side from $r(n)$.

(1) Fill the blanks in Table P2-1 for transmit symbols $s(n)$ for the given $i(n)$.
(2) From received samples $r(n)$, obtain recovered data $\widehat{i}(n)$.
(3) Estimate noise samples of $z(n)$ using the recovered data $\widehat{i}(n)$. Keep in mind, at the receive side, only received samples are available.
(4) Estimate SNR in dB using the result of (3).

P2-2: We consider OOK constellation with data $i(n)$ in Table P2-2.

(1) Fill the blank of $s(n)$.
(2) Given $i(n)$ in Table P2-2, fill $\widehat{i}(n)$ with the threshold of 0.5A = 0.7.
(3) Estimate noise samples $z(n)$ when $\widehat{i}(n)$ is decided.
(4) Estimate the variance of $z(n)$ in (3) above.

P2-3: A pulse shape is trapezoidal as shown in Fig. P2-2.

(1) Find the transmit wave form of BPSK in P2-1.
(2) Find the transmit waveform of OOK in P2-2.

P2-4: A pulse shape is triangular with the two symbol span shown in Fig. P2-3.

(1) Find the transmit wave form of BPSK in P2-1.
(2) Find the transmit waveform of OOK in P2-2.

P2-5: The carrier frequency of CW is 3 times of a symbol rate; 3 cycles in one symbol period of a cosine wave as $S_x(t) = A_I \cos(j2\pi f_c t)$, and the shaping filter is shown Fig. P2-4;

(1) Sketch CW of BPSK in P2-1
(2) Sketch CW of OOK in P2-2.

Fig. P2-2: Trapezoidal
pulse

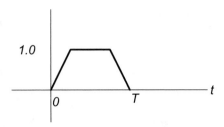

Fig. P2-3: Triangular pulse
with 2T span

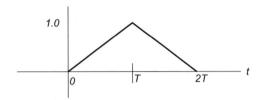

Fig. P2-4: Triangular pulse
with pulse span T

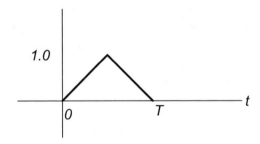

P2-6: Show the power of constellation of orthogonal signaling $(M = 2)$ is unity. See Figure 2-9.

P2-7: Show $1^2 + 3^2 + 5^2 + \ldots\ldots + (n\text{-}1)^2 = \frac{n(n+1)(n-1)}{6}$ when n is even. Hint: using $1^2 + 2^2 + 3^2 + \ldots\ldots + n^2 = \frac{n(n+1)(2n+1)}{6}$, show that $2^2 + 4^2 + 6^2 + \ldots\ldots + n^2 = \frac{n(n+1)(n+2)}{6}$ when n is even.

P2-8: Show $1^2 + 3^2 + 5^2 + \ldots\ldots + (n\text{-}1)^2 = \frac{n(n+1)(n-1)}{6}$ by mathematical induction, i.e., considering $1^2 + 3^2 + 5^2 + \ldots\ldots + (n\text{-}1)^2 + (n + 1)^2 = \frac{(n+2)(n+3)(n+1)}{6}$.

P2-9: Show that discrete symbol average power (variance) of OOK in Fig. P2-5, with '0' probability p, is given by $A^2 p$. When it is equally likely, we need to make it unity. What is A?

P2-10: A BPSK digital transmission system is shown in Fig. P2-6; the discrete power of BPSK output is d^2. And the rectangular pulse with the amplitude A is used. What is the output power? Hint: $P = d^2 \frac{1}{T} \int_{-\infty}^{+\infty} |h(t)|^2 dt$.

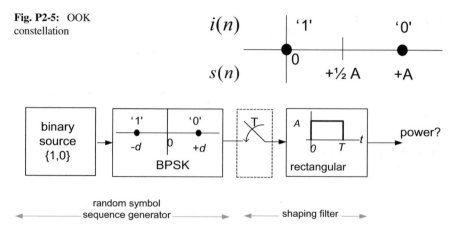

Fig. P2-5: OOK constellation

Fig. P2-6: BPSK with rectangular pulse with symbol period T

P2-11: In Fig. P2-6, BPSK is replaced by OOK ($p = 1/2$) with the discrete power is d^2. Show that the amplitude of non-zero symbol is A $= \sqrt{2}d$. What is the output power?

P2-12: Show that the power gain of a shaping filter, using Parseval's identity, is given by $P = \int_{-\infty}^{+\infty} |h(t)|^2 dt = \int_{-\infty}^{+\infty} |H(f)|^2 df$, which can be made unity by choosing a proper amplitude.

P2-13: SNR is defined as the ratio of signal power (P_s) over noise power (P_n), i.e., P_s/P_n. In Fig. P2-1, a BPSK discrete system, the variance of $s(n)$ is d^2, and the variance of AWGN, $z(n)$, is σ^2.

(1) Express SNR in terms of d^2 and σ^2.
(2) Signal symbol energy is given by $E_s = P_s T$ where T is a symbol period. Noise power at a receiver is expressed $P_n = \frac{N_0}{2} B$ where N_0 is thermal noise density and B is a noise bandwidth. Express SNR in terms of E_s, N_0, B, and T when $B T = 1.0$, which means the noise bandwidth is the same as the symbol rate ($1/T$).

P2-14: For 6-PAM constellation (Fig. P2-7), show that $\frac{d}{\sigma}$ is given by $\frac{d}{\sigma} = \sqrt{\frac{6}{35} \frac{E_s}{N_o}}$

when $B T = 1.0$.

P2-15: In Fig. P2-8, discrete BPSK with AWGN channel is shown along with conditional probabilities, e.g., $\Pr\{r(k)|A(k) = + d\}$, and so on.

(1) Assuming $\sigma^2 = 0.5$ and symbols are equally likely, roughly estimate from Fig. P2-8, $\Pr\{r(k)|A(k) = + d\}$ and $\Pr\{r(k)|A(k) = - d\}$ when $r(k) = + 0.6$.
(2) Assuming $\sigma^2 = 0.5$ and symbols are equally likely, compute $\Pr\{A(k) = + d|r(k)\}$ and $\Pr\{A(k) = - d|r(k)\}$ when $r(k) = + 0.6$.

Fig. P2-7: 6-QAM
constellation

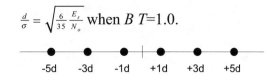

$$\frac{d}{\sigma} = \sqrt{\frac{6}{35}\frac{E_s}{N_o}} \text{ when } B\ T{=}1.0.$$

-5d   -3d   -1d   +1d   +3d   +5d

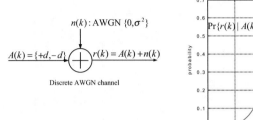

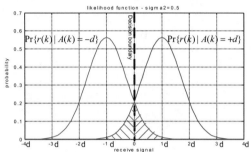

Fig. P2-8: Discrete BPSK with AWGN channel, and conditional probabilities

(3) For decision making of which symbol is transmitted, explain that there is no need of computing probabilities in this case but a decision boundary is enough.

(4) Compute the conditional probabilities of (1) and (2) using $Q(x)$ function.

P2-16: In Fig. P2-8, ML is a special case of MAP. What is the difference? When both are equivalent?

P2-17: In Fig. P2-8, minimum distance detection compares Euclidian distance from a received sample to constellation points, and choose a constellation point with the minimum distance to the received sample. Show the decision boundary is the same as ML.

P2-18: In Fig. P2-8, correlation metric (CM) decision the correlation between received sample to constellation points $\{+1, -1\}$, i.e., $r(k)(+1)$, $r(k)(-1)$. And thus decision boundary is $r(k) = 0$ which is the same as ML. For $r(k) = +0.6$, find the correlations.

P2-19: In BPSK of Fig. P2-8, three consecutive received samples are r_1, r_2, r_3. Show that these three bits can be detected simultaneously by CM, i.e., correlating with 8 possible three bit combinations. Choose the maximum out of $\{+r_1 + r_2 + r_3, +r_1 + r_2 - r_3, +r_1 - r_2 + r_3, +r_1 - r_2 - r_3, -r_1 + r_2 + r_3, -r_1 + r_2 - r_3, -r_1 - r_2 + r_3, -r_1 - r_2 - r_3\}$. What is the corresponding detected sample? And show that this is equivalent to sample by sample detection.

P2-20: In BPSK of Fig. P2-8, two consecutive received samples are r_1, r_2. Show that these two bits can be detected simultaneously by minimum distance decoding, i.e., by choosing the minimum out of $\{(r_1 + 1)^2, (r_2 + 1)^2, (r_1 + 1)^2 + (r_2 - 1)^2, (r_1 - 1)^2 + (r_2 + 1)^2, (r_1 - 1)^2 + (r_2 - 1)^2\}$, and that it is equivalent to sample by sample detection.

P2-21: 3-bit tuple of BPSK symbols can be detected by correlation metric (CM). It is an alternative to symbol by symbol detection and it requires more computation than the latter but nonetheless it is another way of detection.
3-bit tuple of all possible data and BPSK symbols in matrix form

$$
X = \begin{bmatrix} 0 & 0 & 0 \\ 0 & 0 & 1 \\ 0 & 1 & 0 \\ 0 & 1 & 1 \\ 1 & 0 & 0 \\ 1 & 0 & 1 \\ 1 & 1 & 0 \\ 1 & 1 & 1 \end{bmatrix} \rightarrow \{'0' \rightarrow +1, '1' \rightarrow -1\} \rightarrow Y = \begin{bmatrix} +1 & +1 & +1 \\ +1 & +1 & -1 \\ +1 & -1 & +1 \\ +1 & -1 & -1 \\ -1 & +1 & +1 \\ -1 & +1 & -1 \\ -1 & -1 & +1 \\ -1 & -1 & -1 \end{bmatrix}
$$

Received 3 BPSK symbols are $[-0.9 +0.1 +0.6]$. Compute CM for each 3-bit tuple; fill the blank and note that only 4 CM computation is good enough to find the largest.

$$
\begin{bmatrix} +1 & +1 & +1 \\ +1 & +1 & -1 \\ +1 & -1 & +1 \\ +1 & -1 & -1 \\ -1 & +1 & +1 \\ -1 & +1 & -1 \\ -1 & -1 & +1 \\ -1 & -1 & -1 \end{bmatrix} \begin{bmatrix} -0.9 \\ +0.1 \\ +0.6 \end{bmatrix} = \begin{bmatrix} -0.2 \\ -1.4 \\ -0.4 \\ ? \\ +1.6 \\ +0.4 \\ +1.4 \\ +0.2 \end{bmatrix} \leftarrow \text{max}
$$

Confirm the result of CM detection is the same as symbol by symbol detection.

P2-22: For two dimensional, independent, Gaussian distribution with the equal variance (σ^2) and the average m_x, m_y respectively, show that its probability density function is given by $f(x, y) = \frac{1}{2\pi\sigma^2} e^{-\frac{1}{2\sigma^2}\left[(x-m_x)^2 + (y-m_y)^2\right]}$, and it is circularly symmetric at $\{m_x, m_y\}$.

P2-23: Show that the rotation of a two-point constellation will have the SER performance expression $Q\left(\frac{d}{\sigma}\right)$ function. See some examples in Fig. P2-9. We assume that the noise of I and Q channel is independent Gaussian random variance with the equal variance.

(1) What about the translation? Can we say the same?
(2) What happens when SNR is expressed in terms of $\frac{E_s}{N_o}$? Explain.

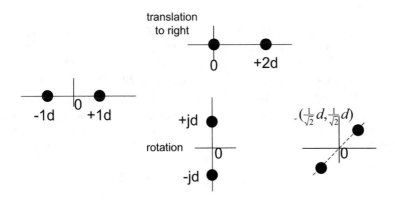

Fig. P2-9: The rotation and translation of two-point constellation

P2-24: In Fig. P2-7, 6-PAM, two consecutive symbols can be used to transmit 5 bits since $36 > 2^5$ and so 2.5 bits per symbol. Thus 4 out of 36 are not necessary. Which 4 combinations are to be eliminated in order to minimize average symbol energy, E_s? Show your choice and compute E_s with your choice.

P2-25: In Fig. P2-7, 6-PAM, two such symbols are transmitted by I-channel and Q-channel, at the same symbol period. This results in 32-QAM, which carries 5 bits per symbol. Out of 36 (6 x 6), four corner constellation points of {+5+j5, +5−j5, −5+j5, −5−j5} are not used. Compute the average power of 32-QAM constellation.

P2-26: Using Figure 2-19: SER curves of M-PAM estimate SNR in dB required to have 1E-3 and 1E-6 SER rate, and fill out the table below.

SER	2-PAM	4-PAM	8-PAM	16-PAM	32-PAM
1E-3	6.5 dB	14.0 dB			32.5 dB
1E-6	10.5 dB				

P2-27: Using Figure 2-24: SER curves of M-QAM (M = 4,16, 64, 256, 1024), estimate SNR in dB required to have 1E-3 and 1E-6 SER rate, and fill out the table below.

SER	4-QAM	16-QAM	64-QAM	256-QAM	1024-QAM
1E-3	10.2 dB	17.5 dB			36.2 dB
1E-6	13.5 dB				

P2-28: Using Figure 2-25: SER curves of M-QAM (M = 8, 32, 128, 512, 2048), estimate SNR in dB required to have 1E-3 and 1E-6 SER rate, and fill out the table below.

SER	8-QAM	32-QAM	128-QAM	512-QAM	2048-QAM
1E-3	14.5 dB	21.0 dB			39.1 dB
1E-6	17.5 dB				

P2-29: DSQ constellations (M-DSQ, $M = 8, 32, 128, 512, 2048$) are obtained from $2M$-QAM of rectangular QAM by selecting half of constellation points as shown in Figure 2-26. Show that the digital power is the half of $2M$-QAM $\{2M = 16, 64, 256, 1024, 4096\}$, i.e., $P = \frac{M-1}{3} d^2$ with $M = 8, 32, 128, 512, 2048$ where the distance between constellation point is $2d$, not $2\sqrt{2}d$.

P2-30: By partitioning inner points and 3-side points (approximately) as in Figure 2-27 (RHS), we can compute inner points and side points as shown below; fill out the blanks.

	32-DSQ	128-DSQ	512-DSQ	2048-DSQ
Inner points ($4Q-4Q^2$)	18	98	450	1922
Side points ($3Q-2Q^2$)	14 (4*2 + 3*2)	30(8*2 + 7*2)	62(16*2 + 15*2)	126(32*2 + 31*2)
SER	$\frac{114}{32} Q - \frac{100}{32} Q^2$			

P2-31: For a given system, the following statements on SER and BER are true or false. Hint: see Figure 2-33.

(1) BER can be approximated by SER / number bits in a symbol when a random bit to symbol mapping is used.
(2) For SER computation, a specific bit to symbol mapping is necessary.
(3) SER is in general an upper bound of BER.
(4) For BER computation, a specific bit to symbol mapping is necessary.

P2-32: For the following table, $\frac{Es}{No}$ and $\frac{Eb}{No}$ are converted for a given modulation; fill the blanks.

	8-QAM	32-QAM	128-QAM	512-QAM	2048-QAM
$\frac{Es}{No}$ symbol energy over noise density (dB)	17.5 dB	18.5dB	19.5dB		
$\frac{Eb}{No}$ bit energy over noise density (dB)	12.7 dB			20.5dB	20.5dB

P2-33: At the receiver front end, the power of signal is measured as 41dBm, noise figure is 5dB and the noise bandwidth is 50MHz. What is the thermal noise power at room temperature? What is SNR?

P2-34: In a staggered QAM system, Q-channel is delayed by half symbol period relative to I-channel. Discuss different ways of implementing this delay; one is to use a pulse shaping filter with the half symbol delay.

P2-35: It is argued that SER performance of staggered QAM is essentially comparable to that of 'regular' QAM. Show that this statement is true by comparing the required SNR of 8-PAM to that of 64-QAM at SER 1E-6.

P2-36: Examples of CCDF (complimentary cumulative distribution function) are shown in Figure 2-36. Explain the meaning of probability 1E-3 at 4dB.

P2-37: Draw a block diagram of a 3-state, self-synchronizing, scrambler and descrambler using the primitive polynomial of $x^3 + x + 1$ as shown in Table 2-2. An input to this scrambler is unit impulse, i.e., $\delta(n)$, obtain the output when $n = 0,1,2,\ldots.15$.

P2-38: In Fig. P2-10, 16-QAM constellation, partially filled, is shown

Complete it to make 90° rotation invariant and as close as Gray. Is it possible to be perfectly Gray? And generate a bit to symbol mapping table as shown below by filling the blanks;

0000	0001	0010	0011	0100	0101	0110	0111	1000	1001	1010	1011	1100	1101	1110	1111
3−j3					−3−j					1+j3					−1+j

P2-39: For 1+D system of M = 2 (see Figure 2-46), fill out the blanks below

n	1	2	3	4	5	6	7	8	9	10	11	12	13	14
d(n)	0	1	1	0	1	0	1	0	1	0	1	1	0	1
s(n)														
c(n)														
d(n)														

Fig. P2-10: Partially filled 16-QAM constellation

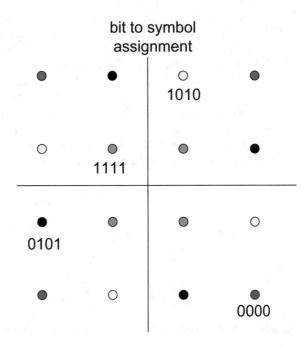

bit to symbol assignment

P2-40: The frequency response of $1+D$ system with an ideal lowpass is given by

$$Y(f) = 2T \cos(\pi fT) \quad |f| < \frac{1}{2T}$$

$$= 0 \qquad\qquad \text{otherwise}$$

which is shown in Figure 2-48 (LHS). Find its impulse response $y(t)$ by taking inverse Fourier transform.

P2-41: Show that M-level PAM becomes 2M-1 level $1+D$ system, e.g., M = 4 PAM becomes 7-level with $1+D$ system. Find the probability of each constellation point. (Hint: See Figure 2-49.)

P2-42: A modified $1-D^2$ is shown Fig. P2-11 (compare with Figure 2-52). Note that $1-D$ part may be considered as part of channel.

$d(n) = [3\ 1\ 2\ 3\ 2\ 0\ 1\ 0\ 2\ 1\ 0\ 3\ 0\ 1])$ is given as in the table below; find e(n), c(n) and $\underline{d}$ (n) fill the blanks.

n	1	2	3	4	5	6	7	8	9	10	11	12	13	14
d(n)	3	1	2	3	2	0	1	0	2	1	0	3	0	1
s(n)	3	1	1	0	3	0	0	0	2	1	2	0	2	1
e(n)	3													
c(n)	3	1												
$\underline{d}$(n)	3	1												

P2-43: 2-level FSK is shown in Fig. P2-12 (LHS); frequency deviation 600Hz, and symbol rate 1200Hz. Sketch CW waveform for the data [1 1 0 1 0 0 1 0]. Hint: consider how many cycles of carrier per symbol time.

P2-44: Consider 2-level FSK in Fig. P2-12 (RHS), and now carrier frequency is 7800Hz. How many cycles per symbol time for data bit '1' and '0'? What is the cycle difference per symbol time between data '1' and '0'?

P2-45: Non-coherent detection of FSK can be done by processing a received signal through the chain of limiter, differentiator, rectifier and LPF. A limiter makes a sine wave to rectangular; a differentiator differentiates a waveform; and a rectifier takes an absolute value. See Figure 2-57. Process the CW of P2-43 non-coherently by sketching the waveforms in each stage.

P2-46: FSK signals, specifically CW, can be generated by a VCO, which is shown in Fig. P2-13 along with its characteristic. The output of VCO is sinusoidal and is

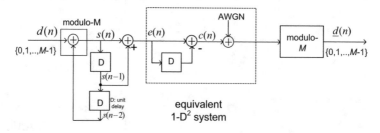

Fig. P2-11: A slight modification of $1-D^2$ system, which is sometimes called PRS 4 system

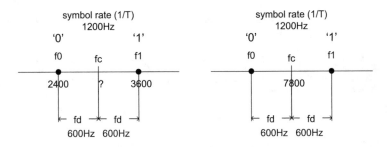

Fig. P2-12: 2-tone FSK

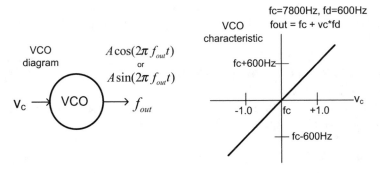

Fig. P2-13: A sinusoidal VCO and its characterization curve

given by $A\cos(2\pi f_{out}t)$. Obtain the output of VCO when the control voltage, V_c, is given by $V_c = \sin(2\pi 600t)$, i.e., 600Hz sine wave. Hint: write a Matlab script and plot.

P2-47: An FSK signal generation block diagram is captured in Fig. P2-14.

It consists of three major blocks; M-level PAM signal, FM complex envelope, and quadrature modulator. This is a generic FSK generation diagram. By using different M-level PAM, pulse filter and modulation index (shaded), many different CPFSK systems can be constructed.

(1) Express $x(t)$ in terms of $\{a_n\}$, T, and $p(t)$.
(2) Confirm that $\varphi(t)$ can be an angle in modulo 1 rather than in modulo 2π. Note that there is no need of taking modulo operation since the next complex exponentiation takes care of it.
(3) What is the impact to the signal if $x(t)$ is scaled by another factor, say 2.0, i.e., $x(t) \rightarrow 2.0x(t)$?
(4) Express $C(t)$, complex envelope, in terms of $x(t)$ and necessary parameters.
(5) Confirm $s_x(t)$ can be generated with this diagram.

P2-48: Referring to Fig. P2-14, obtain and sketch $x(t)$, and $\varphi(t)$ for the symbol pattern $\{\ldots -1, +1, -3, +3, -3, +1, -1, +1, +1, -1, +3, -3 \ldots\}$, i.e., $M = 4$. Assume that $p(t)$ is rectangular with pulse width being the same as symbol period, and amplitude being unity. h, modulation index, is 0.25.

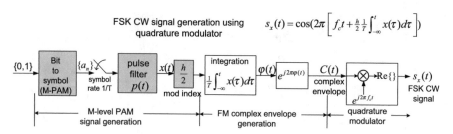

Fig. P2-14: FSK CW signal generation using quadrature modulator

P2-49: When $x(t)$, in Fig. P2-14, is continuous in time and given by$x(t) = \cos\left(2\pi \frac{0.25}{T} t\right)$, find the expressions of complex envelope,$C(t)$, and CW, $s_x(t)$.

P2-50: Show that $\mathrm{Re}\{C(t)\} = \cos(2\pi\varphi(t))$, and $\mathrm{Re}\{C(t)\} = \sin(2\pi\varphi(t))$ in Fig. P2-14. Given$C(t)$, $\varphi(t)$can be obtained; $\frac{1}{2\pi}\tan^{-1}(im\{C(t)\}, re\{C(t)\})$. Can you find another way?

P2-51: Given $\varphi(t)$in Fig. P2-14, is there an inverse to$x(t)$? Based on this problem and the previous (P2-49) draw a receiver block diagram. Hint: the inverse of integration is differentiation.

P2-52: Show that $E\left\{\sum_{n=-\infty}^{n=+\infty} a_n h(t - nT)\right\} = \mu \sum_{n=-\infty}^{n=+\infty} h(t - nT)$ is periodic with a period T. And show the identity of $\sum_{n=-\infty}^{n=+\infty} h(t - nT) = \sum_{k=-\infty}^{k=+\infty} \frac{1}{T} H\left(\frac{k}{T}\right) e^{j2\pi \frac{k}{T} t}$.

P2-53: $E\{x(t)\} = E\{a_n\} E\left\{\sum_{n=-\infty}^{n=+\infty} h(t - nT + \delta)\right\}$ assuming $a_n$and $\delta$are uncorrelated. With δbeing the uniform density, show that $E\left\{\sum_{n=-\infty}^{n=+\infty} h(t - nT + \delta)\right\} = \frac{1}{T}\int_0^T \left[\sum_{n=-\infty}^{n=+\infty} h(t - nT + \delta)\right] d\delta = \frac{1}{T}\int_{-\infty}^{+\infty} h(t) dt$.

P2-54: Consider $E\left\{\sum_{n=-\infty}^{n=+\infty} h(t - nT + \delta) h(t - nT + \delta + \tau - mT)\right\}$ with δbeing the uniform density, and show that it can be expressed as, using the result of P2-52, $E\left\{\sum_{n=-\infty}^{n=+\infty} h(t - nT + \delta) h(t - nT + \delta + \tau - mT)\right\} = \frac{1}{T}\int_{-\infty}^{+\infty} h(t) h(t + \tau - mT) dt$.

P2-55: A symbol time of OFDM is 4.0 μsec in 802.11n standard. In order to have a frequency resolution of 100 Hz, what is the time length of a segment in a number of OFDM symbols? Sampling rate being 20 MHz, what is the size of FFT to compute the spectrum?

Chapter 3
Matched Filter & Nyquist Pulse

Contents

3.1   Matched Filters 134
    3.1.1 Matched Filter Defined and Justified 135
    3.1.2 Examples of Matched Filters 138
    3.1.3 Characteristic of Matched Filters 140
    3.1.4 SNR Loss Due to Pulse Mismatch 141
    3.1.5 A Mismatch Loss Due to Receive Noise Bandwidth 143
3.2   Nyquist Criterion - ISI Free Pulse 145
    3.2.1 Nyquist Criterion - ISI Free End-To-End Pulse 145
    3.2.2 Frequency Domain Expression of Nyquist Criterion 147
    3.2.3 Band Edge Vestigial Symmetry and Excess Bandwidth 149
    3.2.4 Raised Cosine Filter 151
3.3   Shaping Pulse (Filter) Design 153
    3.3.1 Practical Design Considerations 154
    3.3.2 A Practical Design Example of Analog Filter 154
    3.3.3 Design of Digital Matched Filters 156
3.4   Performance Degradation Due to ISI 162
    3.4.1 A Design Case to Simplify the Shaping Filters 162
    3.4.2 A Quick Analysis of the Suggestion of Using a Rectangular Pulse 163
    3.4.3 A Discrete in Time Model for ISI Analysis in General 168
    3.4.4 A Discrete in Time Model for Simulation 170
    3.4.5 Summary for ISI Analysis Methods 171
3.5   Extension to Linear Channel and Non-White Noise 172
    3.5.1 Linear Channels with White Noise 172
    3.5.2 Non- White Noise 174
3.6   References 175
3.7   Problems 175

Abstract This chapter covers two major coupled topics; matched filter and Nyquist pulse. The former maximizes SNR at the sampling time, and the latter is intersymbol interference (ISI) free end to end pulse. Both considerations will guide how to design transmit and receive filters (shaping pulses). We cover two practical filter design cases; analog and digital hybrid, and purely digital ones.

Key Innovative Terms Analog and digital hybrid filter design · Purely digital filter design · Optimization of square root Nyquist filter · Window approach for digital filter design of square root Nyquist pulse · Frequency domain FIR filter

General Terms Matched filter · Nyquist pulse · Correlation by convolution · Root raised cosine pulse · Window · SNR loss due to pulse mismatch · Receiver noise BW · Nyquist criterion · ISI free · Band edge vestigial symmetry · Raised cosine, Kaiser window · Peak distortion · Eye pattern · Linear channel · Non-white Gaussian noise

List of Abbreviations

A/D	analog to digital conversion
AWGN	additive white Gaussian noise
BPSK	binary phase shift keying
CP	cyclic prefix
D/A	digital to analog conversion
EC	eye closure
FIR	finite impulse response
IEEE	Institute of electrical and electronics engineers
IF	intermediate frequency
ISI	intersymbol interference
LC	inductor and capacitor
LHS	left hand side
LAN	local area network
LNA	low noise amplifier
LPF	low pass filter
OFDM	orthogonal frequency division multiplex
PA	power amplifier
PD	peak distortion
PAM	pulse amplitude modulation
QAM	quadrature amplitude modulation
PD	peak distortion
PDF	probability density function
RC	raised cosine
RF	radio frequency
RHS	right hand side
SDR	software defined radio
SER	symbol error rate
SNR	signal power to noise power ratio
SQRT	square root

In this chapter we cover two major coupled topics; matched filter and Nyquist pulse. The matched filter maximizes SNR at the sampling time, and Nyquist criterion is intersymbol interference (ISI) free condition. Both considerations will guide how to design transmit and receive filters (shaping pulses). Discrete time to continuous signal conversion (D/A and analog filter) at the transmit side and the inverse process (analog filter and A/D) at the receiver are part of these filters. Using the system block diagram repeated here in Figure 3-1, the shaded blocks, (B–C) and (C–B), will be covered.

The block of Modulation processing (x) in Figure 3-1 generates modulation symbols, complex numbers in general, and then these discrete-time symbols are converted to analog complex envelope i.e., two continuous signals. These two continuous base band signals are the input to the quadrature modulator which up-converts them to RF frequency for radio transmission. In the receive direction essentially the inverse process happens. RF carrier wave is down converted to two continuous base band (complex envelope) signals, and then each base band is filtered and sampled to convert to discrete modulation symbols.

In this chapter, RHS of C–C in the figure is abstracted as two base band channels (I and Q) and each channel is perfect except for additive thermal noise modeled as AWGN. LHS of B–B in the figure is abstracted as a stream of complex modulation symbols (real and imaginary part). One channel, real or imaginary, can be represented as a baseband channel shown in Figure 3-2. Baseband processing (x) in Figure 3-2 generates a baseband modulation symbol stream. It is possible to interpret Figure 3-2 that $a(n)$ is complex, real and imaginary, symbols so that both channels are handled at the same time. However, we treat $a(n)$ to be real in Figure 3-2 unless otherwise stated.

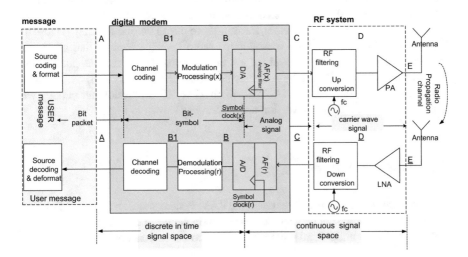

Figure 3-1: Digital radio communication system block diagram

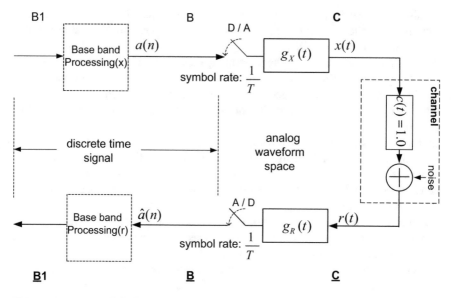

Figure 3-2: Baseband block diagram abstracted from Figure 3-1

In the figure the switches are used for representing D/A and A/D. The switch of D/A represents a filter with a unit impulse, i.e., that a discrete time symbol becomes an analog impulse with the corresponding amplitude. D/A is a part of shaping filter[1]. The switch of A/D is a sampling of analog signals to make it discrete-time signals.

Practical aspect of D/A and band limiting analog filter is very important, and is essential in any digital communication systems. In modern digital radio implementation, a quadrature modulator is often implemented digitally, and thus D/A and filtering is done at RF or IF frequency. In fact, the trend is to push D/A and filtering as close to PA and antenna. So called software defined radio (SDR)'s ideal is that all signal processing is done digitally except for PA and antenna. In the receive side a band limiting analog filter followed by A/D may be done at the baseband or IF/RF. SDR 's goal is to put A/D right after LNA. We will cover IF/RF conversion but initially we will focus on a base band conversion and filtering as shown in Figure 3-2.

3.1   Matched Filters

Now we confine our focus on the base band PAM system shown in Figure 3-2. It is a digital pulse transmission problem and is so fundamental to digital communications that it is critical to understand it thoroughly. We consider a channel with noise being

---

[1]In practice, D/A is not a unit impulse but a rectangular pulse. This can be compensated readily thus here we consider it a unit impulse.

the only impairment. It is the receiver front end noise. It is modeled well by a flat spectrum, i.e. white noise and Gaussian probability distribution, and thus AWGN. Actually the matched filter concept is applicable to non-white spectrum noise. It will be expanded in later section but here we first look at the simple case.

A question we try to answer is how to select the shaping filters ($g_X(t)$ and $g_R(t)$). There are two issues; (1) maximizing the signal to noise ratio (SNR) at the receiver sampling point, (2) no intersymbol interference (ISI). Both are important to symbol error rate (SER) performance. The second issue comes from the fact that data is transmitted consecutively in every symbol period (T), called synchronous transmission. A symbol in one symbol time should not interfere with other symbols. This ISI problem will go away if we suppose that there is a single pulse transmission.

A bit stream of $\{0, 1\}$ is mapped into a stream of symbols $a(n) \in \{+1, -1\}$ using a mapping $\{0 \rightarrow +1\}$ $\{1 \rightarrow -1\}$, where n is time index and integer. This is of course BPSK. With a proper bit to symbol mapping, $a(n)$ may be M-PAM symbols $\{-(M - 1), .. , -3, . -1, +1, +3, .. ,+(M - 1)\}$. Typically we may assume $a(n)$ to be M-level PAM symbols. In passing we mention that $a(n)$ may be generated from a Gaussian random number generator or a uniform random number generator as far as we discuss the problem of selecting the shaping filters. The point is that our discussion on the issue does not rely on the finite level of symbols.

A stream of symbols $a(n)$ is serially transmitted every T second of symbol time, and its inverse $\frac{1}{T}$ is called symbol rate. A symbol acts as an impulse to excite a filter represented by its impulse response $g_X(t)$ and its corresponding frequency response is denoted by $G_X(f)$. Baseband transmission means that the frequency response is centered on around the zero frequency; the filter is a low pass filter. A transmit signal at the filter output, $x(t)$ is expressed in terms of $a(n)$, $g_X(t)$, and T. It is a convolution of $a(n)$ with $g_X(t)$, and given by

$$x(t) = \sum_{n=-\infty}^{\infty} a(n) g_X(t - nT) \tag{3-1}$$

In the receive side $x(t)$ is contaminated by noise, and $x(t)$+ noise ($r(t)$ in Figure 3-2) is low pass filtered (shown as $g_R(t)$ in Figure 3-2) and sampled, and then a decision will be made which symbol is sent (not shown in Figure 3-2). This is the inverse of transmit side.

3.1.1   Matched Filter Defined and Justified

A question is how one selects transmit and receive filter pair. An optimum filter is a pair of matched filters in order to maximize SNR. In time domain the receive filter impulse response is a time-reversal of transmit filter impulse response, i.e. $g_R(t) = g_X(-t)$ [2]. They are exactly the same except for time reversed. Filtering operation in time domain is a convolution of incoming signal with the impulse response of a filter, and convolution is time reversed correlation, i.e., two signals are multiplied and summed (or integrated). Thus when two signals are matched in time there will be a peak. Thus a pair of matched filters maximizes SNR at the sampling point. Shortly we will show it mathematically.

Example 3-1: Correlation by a matched filter

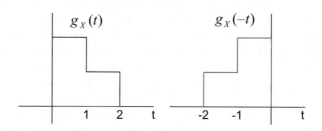

Find the convolution of $g_X(t) \otimes g_X(-t)$.

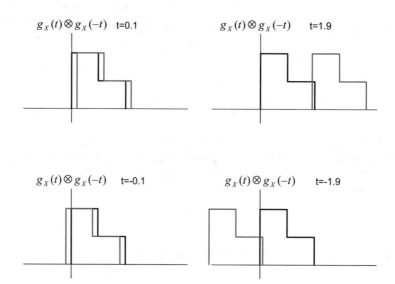

[2]We assume $g_X(t)$ to be real. When it is complex the matched filter is in general $g_X^*(-t)$. In other words, time reversal and conjugation of $g_X(t)$ will be a matched filter. Unless otherwise stated, we consider only a real impulse response. When $h(t)$ is real, $h(-t) = h^*(-t)$.

Note that the process of computing this convolution is the same as the process of finding auto-correlation of $g_X(t)$.

An impulse response has its corresponding frequency response and it is denoted as $g_X(t) \leftrightarrow G_X(f)$. And the time reversed conjugated impulse had a pair as $g_X^*(-t) \leftrightarrow G_X^*(f)$; its frequency response is the conjugation in frequency domain. This can be shown by evaluating Fourier transform of $g_X^*(-t)$, and since it is real, $g_X^*(-t) = g_X(-t)$. The frequency response of a matched filter is the complex conjugation[3] each other. Conjugation means that real part is the same and imaginary part is negated; it is the same as that the magnitude is identical and the phase is identical with the opposite sign.

Thus the time reversal of a real impulse response (no need of time conjugation) will give the conjugation in frequency response. Since $|G_X(f)| = |G_X^*(f)|$ the magnitude response of a pair of matched filters is identical for all the frequencies. The phase response is also identical for all the frequencies except for the sign being opposite.

Exercise 3-1: Discuss a frequency response of a pair of matched filters, $H(f)$, i.e., an end to end frequency response, $G_X(f)$ and $G_R(f) = G_X^*(f)$.

Answer: $H(f) = G_X(f)G_R(f) = G_X(f)G_X^*(f) = |G_X(f)|^2$. Note that the phase of the end to end frequency response, $H(f)$, is zero. This is non-causal. In order to make it causal $G_R(f) = G_X^*(f)e^{-j2\pi fD}$ where D is a delay in the time domain impulse response. When D is large enough, an impulse response can be made causal. The causality means an impulse response is zero if time is less than zero; $h(t) = 0, t < 0$.

In summary this receiver frequency response minimizes the noise power assuming the noise spectrum is flat since the magnitude of receive filter frequency is the same as that of the transmit one. Total noise power after receiver filter is the area of the squared magnitude receiver filter frequency response scaled by noise spectral density which is constant. A mathematical proof of this intuitive explanation is given below.

We will use the Schwarz inequality.

$$\left| \int f(x)g(x)dx \right|^2 \leq \int |f(x)|^2 dx \int |g(x)|^2 dx \qquad (3\text{-}2)$$

where the equality condition holds if and only if $f(x) = kg^*(x)$ with k=constant.

The noise power at the sampling point is given by $P_N = \frac{N_0}{2} \int_{-\infty}^{+\infty} |G_R(f)|^2 df$. The end to end transfer function is $G_X(f)G_R(f)$ and thus the received signal $r(t)$ without noise, at a sampling instant D, is expressed as $r(D) = \int_{-\infty}^{+\infty} G_X(f)G_R(f)e^{j2\pi fD} df$ so that it has a peak. Now we would like to maximize $\frac{r^2(D)}{P_N}$, a signal to noise ratio at the sampling point. In order to apply (3-2), let $f(x)$ and $g(x)$ be $G_R(f)$ and $G_X(f)e^{j2\pi fD}$

---

[3]This is generally true whether $h(t)$ is real or complex. Thus matched filter is also called conjugation filter.

respectively. Then applying (3-2) we get $\frac{r^2(D)}{P_N} \leq \frac{2}{No} \int_{-\infty}^{+\infty} |G_X(f)|^2 df$. The equality hold when $G_R(f) = G_X^*(f)e^{-j2\pi fD}$. In time domain it is expressed as $g_R(t) = g_X^*(D - t)$. We obtained the desired result.

Generally the time reversal makes a pulse non-causal but this can be made to be causal by introducing a sufficient delay. Thus sometimes the matched filter of $h(t)$ is denoted as $h(D - t)$ where D is a delay so that $h(D - t)$ to be causal; it is zero when $D - t < 0$. With the understanding that the causality can be solved by adding delay, we set $D = 0$ for simplicity.

With the matched filter pair, SNR is given by $\frac{r^2(D)}{P_N} = \frac{2}{No} \int_{-\infty}^{+\infty} |G_R(f)|^2 df$.

In summary a pair of matched filters maximizes SNR at the sampling point.

3.1.2   Examples of Matched Filters

We give a few examples of matched filters – a pair of filters which can be represented by pulse shape (impulse response of the filter) or by frequency response.

a)   Rectangular Pulse

This rectangular pulse filter is shown in Figure 3-3, and it is clearly a pair of matched filters, meeting requirement in time domain, reversal in time but requiring the delay of T. In fact it is often used in practice. A short distance interconnection uses it such as equipment backplane, Ethernet, serial port, circuit board and so on. Since the noise is not a problem in those cases the receiver is just sampling without receiver filter.

Overall end-to-end impulse response becomes a triangle pulse as shown in Figure 3-3.

Synthesizing a continuous filter (LC or any analog filter) with a rectangular impulse response is probably a non-trivial task, especially old days. In order to circumvent this problem of implementing rectangular pulse filter at the receiver, 'integrate and dump' method was used in the past. The incoming signal is integrated and dumped – emptying the integrator content (sampled). This process is repeated consecutively every T seconds. However, in modern times its implementation is simple once received signals are sampled, by using ADC, more frequently within a

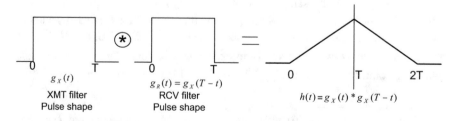

Figure 3-3:   A rectangular pulse as matched filters and its end-to-end pulse

symbol time, say 6 times per symbol period. Then this filter is simply adding all 6 samples. The more samples per symbol time the more accurately implementing the filter. In modern DSP implementation of communication systems it is simple to implement rectangular pulse receive filter in this way of adding samples.

More sophisticate use of this rectangular pulse filter is OFDM which is used in Wireless LAN IEEE 802.11a. All standards using OFDM with cyclic prefix uses rectangular pulses [6]. Cyclic prefix insertion is to elongate the rectangular pulse and elimination at the receiver requires shortening it by CP amount. This will be elaborated in Chapter 5.

b) **Ideal low pass filter**

The frequency response of an ideal low pass filter is rectangular in frequency which might be considered as dual to rectangular pulse in time domain as shown in Figure 3-3. Clearly a pair of ideal LPFs meets the matched filter requirement in frequency domain – conjugation filter in frequency, i.e., the magnitude response being the same.

A minimum bandwidth required for transmitting every T second is one half of symbol rate ($\frac{1}{2T}$) as indicated in Figure 3-4. Any ideal LPF with the bandwidth bigger than the minimum can form a matched filter. But it is not be used in practice since not only the waste of bandwidth but also due to ISI. This argument will be clarified later in this chapter after discussing Nyquist pulse.

We consider an ideal LPF with the bandwidth of one half of symbol rate. Its impulse response is shown in Figure 3-5 and its equation is given by

$$h(t) = \frac{\sin(z)}{z} \qquad \text{with } z = \frac{\pi \cdot t}{T} \qquad (3\text{-}3)$$

It is not causal therefore a delay is needed, and some approximations are necessary as well to make it implementable. This ideal LPF cannot be implemented exactly and is not used often in practice due to long 'ringing'. In practical situations a guard band is necessary for the practicality of filter design.

Figure 3-4: Frequency response of ideal low pass filterIdeal low pass filter (Brickwall filter)

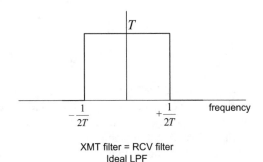

XMT filter = RCV filter
Ideal LPF

Figure 3-5: Impulse
response of ideal LPF

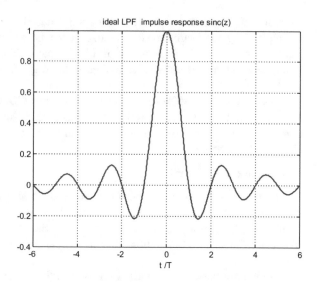

3.1.3   Characteristic of Matched Filters

A matched filter specifies a receive filter for a given transmit filter in order to maximize SNR at the sampling and vice versa. But it does not impose any more constraints. That means a lot of choices since for any given pulse $(g_X(t))$ one can construct its time revered version. A rectangular pulse is most commonly used in transmit side, and when the same pulse is used in the receive side, it is a matched filter.

A few comments are in order. The matched filters were discovered historically in radar context [3], and applied to communications. See [4] Turin (1960) for a good review.

A simple summary capture of matched filter condition: * means complex conjugation.

$$
\begin{array}{cc}
\text{Transmit} & \text{Receive} \\
g(t) & g^*(t) \\
\Updownarrow & \Updownarrow \\
G(f) & G^*(f)
\end{array}
$$

As mentioned briefly before, a matched filter is a filter method (convolution) of implementing correlation, i.e., multiplication of two signals and integration, in case of vectors multiply and sum, which is also called an inner product.

In modern digital implementations, it is often sampled much higher than symbol rate (e.g., 4 to 8 times) to make a receiver filter implemented digitally and an anti-aliasing analog filter can be implemented easily - a sharp cutoff analog filter is difficult. Once there is oversampling, a rectangular pulse filter can be implemented very easily by scaling and summing samples.

We now show that the end to-end pulse, $h(t) = g_X(t) \otimes g_X(-t)$ must be even symmetry, and its peak at $t = 0$.

Ignoring the delay D, the end to-end pulse, $h(t)$ can be expressed by

$$h(t) = \int_{-\infty}^{+\infty} G_X(f)G_R(f)e^{j2\pi ft}df = \int_{-\infty}^{+\infty} |G_X(f)|^2 e^{j2\pi ft}df.$$

It can also be expressed as,

$$h(t) = g_X(t) \otimes g_X(-t) = \int_{-\infty}^{+\infty} g_X(x)g_X(x+t)dx \qquad (3\text{-}4)$$

From (3-4) t replaced by $-t$, $h(t)$ does not change. Therefore it is even symmetry at zero. $h(t) \leq h(0)$ since $h(0) = \int_{-\infty}^{+\infty} |g_X(x)|^2 dx$. Note that $\int_{-\infty}^{+\infty} |g_X(x+A)|^2 dx$ has also the same value $h(0)$ for any constant A.

Exercise 3-2: With a delay, i.e., $g_R(t) = g_X(D - t)$, show that the peak of an end-to-end pulse, $h(t)$, is at $t = D$, and even symmetry around it.

Answer: Visualize the delay graphically, i.e., shifting it horizontally in time. Draw an arbitrary pulse. Find the matched pulse by time reversal and then delay. Convolve two.

Exercise 3-3: A chirp signal is defined as a complex envelope $e^{j2\pi kt^2}$, $0 \leq t \leq T$. It was studied extensively in radar context. Find the matched signal. Note this is a complex signal.

Answer: a matched filter is $e^{-j2\pi kt^2}$; time reversal and conjugation of $e^{j2\pi kt^2}$.

3.1.4 SNR Loss Due to Pulse Mismatch

The SNR does not depend on the details of the pulse shape as long as they are a pair of matched filters. However, there is SNR degradation if the match deviates.

We give examples of the mismatch loss computation which may give further insight into the nature of the matched filters. In order to calculate the mismatch loss we use Figure 3-6 which is a simplified form of Figure 3-2.

We need to fix the transmission power by inserting a gain block (a). Without loss of generality we can set $d^2 = 1.0$ and choose a so the transmission power to be unity; $Ps = 1.0$. And it is given by $Ps = \left[d^2 \frac{1}{T} \int |g_X(t)|^2 dt \right] a^2$.

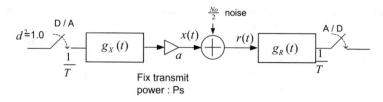

Figure 3-6: Pulse transmission system with AWGN for pulse mismatch loss calculation and digital power $d^2 = 1.0$ and a gain block (a) is used to make transmit power the same for different cases

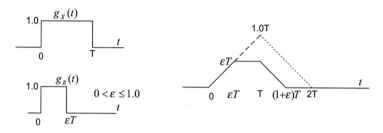

Figure 3-7:   Transmit pulse, receive pulse and the end-to-end pulse

The transmit pulse is a rectangular pulse with the width T but the receive pulse is a shortened rectangular with ε as a parameter. Its range is $0 < \varepsilon \leq 1.0$. This is shown in Figure 3-7. The overall (end to end) pulse is shown in the figure as well.

The signal at the receiver is given by $y(t) = dag_X(t) \otimes g_R(t)$, and sampled at the peak and its value is $y(T) = d\varepsilon Ta$. With $d = 1.0$ we choose $a = 1.0$ so that $Ps = 1.0$. Thus $y(T) = \varepsilon T$

The noise power after the receive filter is given by $P_N = \frac{No}{2} \int |g_R(t)|^2 dt = \frac{No}{2}\varepsilon T$.

SNR at the receiver $= \frac{y(T)^2}{P_N} = \frac{2}{No} T\varepsilon$. Compared with the matched case ($\varepsilon = 1.0$), the mismatch loss in dB is $10 \log \varepsilon$.

Another convenient normalization is to add a gain block (b) before the receive filter so that the peak is $1.0T$. This gain amplifies both the signal and noise so there is no effect on SNR but the peak is normalized. The end to end pulse is shown in Figure 3-8. Now compare only how much noise is entered in the system. In this example $b = 1/\varepsilon$. And $P_N = \left(\frac{1}{\varepsilon}\right)^2 \frac{No}{2}\varepsilon T$. Compare with the matched case, the noise power is $1/\varepsilon$. When $\varepsilon < 1.0$, more noise is getting into the system, losing SNR. The noise bandwidth concept is useful, which will be explained next section.

This example (Table 3-1) is not contrived as it may seem. The practical application will be in OFDM area. OFDM with cyclic prefix (CP) uses these rectangular pulses due to CP – lengthening at the transmit pulse and shortening at the receive one. This will cause the mismatch loss, which was pointed out in [7]. We will cover in more detail later in Chapter 5.

Figure 3-8: Receive filter gain adjustment so that the peak is 1.0T

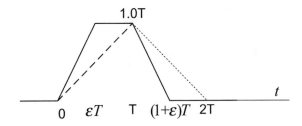

Table 3-1: Summary of mismatch example

Transmit pulse: $g_X(t)$	Receive pulse: $g_R(t)$	Mismatch loss
1.0 — 0 — T	1.0 — 0 — T	0 dB (matched)
1.0 — 0 — T	1.0 — 0 — εT, $0 < \varepsilon \leq 1.0$	$10 \log \varepsilon$ [dB]
1.0 — 0 — εT, $0 < \varepsilon \leq 1.0$	1.0 — 0 — T	$10 \log \varepsilon$ [dB]

Exercise 3-4: Exchange the pulses; the short pulse is used in the transmit side, and then show that the mismatch loss is the same $10 \log \varepsilon$ [dB]. Hint: consider a convolution of transmit pulse and receive pulse, i.e., end to end pulse, and the change its order does not change the end to end pulse.

3.1.5 A Mismatch Loss Due to Receive Noise Bandwidth

The source of AWGN is the receiver front end thermal noise. Its spectral density is flat and proportional to temperature in Kelvin (K). Its proportional constant is Boltzmann constant. Noise density is denoted as $N_o = kT$ [watt/Hz] where $k = 1.38\mathrm{E}^{-23}$, $T = 300$K (27 °C). Thermal noise power with bandwidth B [Hz] is given by kTB [watt][4] or $N_o B$. In the context of carrier wave signal (i.e., RF carrier), we use one sided bandwidth. In most of text books, the double sided noise power density, both positive and negative frequencies are considered, and thus to make the noise power equal we need to use $\frac{N_o}{2}$ with the twice of bandwidth.

A total noise power received by a receiver, after receiver filter described in frequency domain by $G_R(f)$, is given by

---

[4]T is used for temperature as well as for symbol time but no confusion with the context.

Figure 3-9: Noise
bandwidth concept:
arbitrary frequency shape
equivalent to rectangle
frequency shape (same area
under the curve)

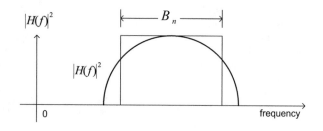

$$\frac{No}{2} \int_{-\infty}^{\infty} |G_R(f)|^2 df$$

Let $|G_R|^2_{max} = \max_{f} |G_R(f)|^2$. A noise bandwidth B_n is defined as

$$B_n = \frac{1}{|G_R(f)|^2_{max}} \int_{-\infty}^{\infty} |G_R(f)|^2 df$$

With B_n the total noise power is expressed by $\frac{N_o}{2} B_n |G_R(f)|^2_{max}$ and if I
$G_R|^2_{max} = 1.0$ then loosely used "bandwidth" is the same as noise bandwidth (B_n).
This noise bandwidth concept can be graphically illustrated in Figure 3-9.

A square root raise cosine filter with any excess bandwidth β has the noise
bandwidth of symbol rate ($\frac{1}{T}$). Any square root filter with vestigial symmetry will
have the noise bandwidth of $\frac{1}{T}$. This can be seen easily due to the vestigial symmetry
around half symbol rate; the area above $\frac{1}{T}$ is the same as the area below $\frac{1}{T}$ due to the
symmetry. RC pulse is listed in (3-10).

A rectangular pulse, with a pulse width T as a receiver filter, has the noise
bandwidth $\frac{1}{T}$. This can be seen by Parseval's theorem; a power measured in frequency
domain is the same as power in time.

$$\int_{-\infty}^{+\infty} |G_R(f)|^2 df = \int_{-\infty}^{+\infty} |g_R(t)|^2 dt = T$$

The second identity comes from the fact that the time domain integral of a
rectangular pulse with a pulse width T, which is the area under the curve is T. And
$|G_R|^2_{max} = T^2$ since $|G_R(f)| = T |\frac{\sin(\pi f T)}{(\pi f T)}|$ the noise bandwidth is the same as the
symbol rate ($\frac{1}{T}$). So called 'integrate and dump' is equivalent to rectangular pulse
matched filter, and thus the noise bandwidth of 'integrate and dump' is also $\frac{1}{T}$.

For a system with symbol rate ($\frac{1}{T}$), the minimum noise bandwidth is $\frac{1}{T}$. Note that
this is true that even with excess bandwidth as we discussed in the above. If it is
greater than $\frac{1}{T}$ then extra noise is added into the system and loss in SNR. On the other
hand if it is less than $\frac{1}{T}$ then ISI will be introduced, which will degrade SNR. ISI will
be discussed in Section 3.2.

Example 3-2: An ideal LPF is used in the transmit side, and a matched filter in the receive side is the ideal LPF. However, the bandwidth of receive filter is 10% wider than that of transmit side. What is SNR degradation due to this mismatch?

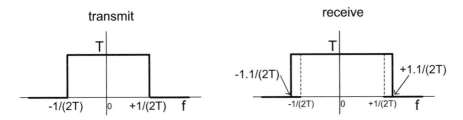

Answer: The noise bandwidth of the receive filter is increased by 10%, and thus more noise is admitted. SNR loss = 10*log(1.1) = 0.41 dB.

Exercise 3-5: What happens in Example 3-2 if one reduces the noise bandwidth less than the minimum (1/T)?

Answer: A short answer is to introduce a distortion to the signal, called ISI. A complete quantitative answer, hopefully, is clear at the end of this chapter.

3.2 Nyquist Criterion - ISI Free Pulse

In this section we consider another constraint – ISI free criterion. ISI happens due to the fact that a stream of symbols, $a(n)$, is transmitted sequentially, one symbol after another, in time. When only one symbol is transmitted in entire time, there is no ISI.

3.2.1 Nyquist Criterion - ISI Free End-To-End Pulse

Referring to Figure 3-2, a stream of symbols $a(n)$ is synchronously transmitted every T second. The signal at receiver sampling point is given by the convolution of transmit signal $x(t)$ with receiver filter $g_R(t)$ plus noise that is filtered by receiver filter, i.e., convolved with $g_R(t)$. $a(n)$ goes through a cascade of both transmit $g_X(t)$ and receive $g_R(t)$. A single equivalent filter is $h(t) = g_X(t) \otimes g_R(t)$ with $\otimes$ denoting convolution. This is called an end-to-end pulse. Thus this is equivalent to $a(n)$ being convolved with the end-to-end pulse, $h(t)$, for the signal part. And the noise goes through $g_R(t)$. Thus the received signal is expressed by

$$r(t) = \sum_{n=-\infty}^{\infty} a(n)h(t - nT) + n(t) \otimes g_R(t) \qquad (3-5)$$

Let us consider only the signal part at this point. Without noise we look for a condition such that a stream of $a(n)$ is recovered from $y(t) = \sum\limits_{n=-\infty}^{\infty} a(n)h(t - nT)$ with no ISI. This is called Nyquist criterion and $h(t)$ is called a Nyquist pulse if it meets the criterion.

We sample $y(t)$, every Tsecond, and $t = lT$ with $l = -\infty, .., 0, .., +\infty$.

$$y(lT) = \sum_{n=-\infty}^{\infty} a(n) h(lT - nT) \tag{3-6}$$

It can be written without T and is $y(l) = \sum\limits_{n=-\infty}^{\infty} a(n)h(l - n)$

Using the identity, $a(l) = \sum\limits_{n=-\infty}^{\infty} a(n) \delta(l - n)$, we can see that ISI free condition is

$$h(lT - nT) = \delta(l - n) \tag{3-7}$$

When (3-7) is met, $a(l)$ is recovered without any interference from other symbols, i.e., $a(n)$ where $n \neq l$. In other words, Nyquist criterion is that a sampled end-to-end pulse is a unit impulse (delta function); $\delta(l - n) = 1.0$ when $(l - n) = 0$, and $\delta(l - n) = 0 \, (l - n) \neq 0$.

This Nyquist criterion is intuitively satisfying and easily understood. Any arbitrary pulse in time with regular zero crossing at every symbol time will meet the requirement. For example, see a figure below; atypical one in order to emphasize the point.

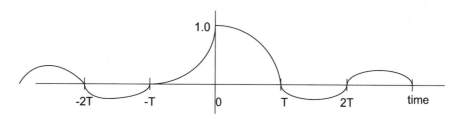

In practice things are not this chaotic, especially if it is combined with the matched filter requirement; a matched Nyquist pulse is of even symmetry and has a peak at zero (or at a delay in time to make it causal).

We mention some commonly used examples of ISI free, Nyquist pulse.

- Any time limited pulse less than two symbol times is a Nyquist pulse.
- A triangle pulse meets Nyquist criterion as well as an impulse response of an ideal LPF is qualified as a Nyquist pulse. In the previous section we showed that these pulses meet the matched filter requirement.
- Raised cosine (RC) pulses given by (3-9) and (3-10) are a set of Nyquist pulses with a parameter β called excess bandwidth.

3.2.2   *Frequency Domain Expression of Nyquist Criterion*

We now translate the time domain Nyquist criterion of (3-7) into frequency domain. For a given Fourier transform pair $h(t) \Leftrightarrow H(f)$, we express Nyquist criterion in frequency domain which is given by

$$\sum_{k=-\infty}^{+\infty} H\left(f - \frac{k}{T}\right) = T \qquad (3\text{-}8)$$

The equation (3-8) says that the frequency spectrum of a sampled $h(nT)$ must be flat.

Figure 3-10 shows an example of how (3-8) is computed. Remember that a sampled signal spectrum is periodic and thus specifying within one frequency period

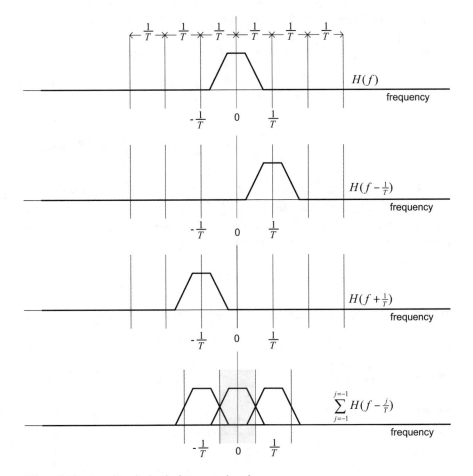

Figure 3-10: Nyquist criterion in frequency domain

$(\frac{1}{T})$ suffices. In Figure 3-10 one frequency period $(\frac{1}{T})$ around zero is shaded. This frequency region will be repeated. In the figure since $H(f)$ is band-limited less than twice of symbol rate (total positive and negative frequency) only three terms in the summation, i.e., $j = -1, 0, +1$ are necessary for specifying the frequency response of sampled end-to-end pulse, i.e., $h(nT)$.

Justification of Nyquist criterion in frequency domain in (3-8) follows. When continuous signal is sampled synchronously every T, its frequency spectrum will be expressed as shown below. This is one form of sampling theorem. A band-limited signal can be represented by its samples which can reconstruct the original continuous signal. Nyquist criterion may be interpreted as a re-statement of sampling theorem in digital pulse transmission point of view. It is suggested that for further understanding of sampling theorem consult textbooks on communication systems such as [1, 2]. The original paper [5] on Nyquist criterion was published in 1924 and was concerned with telegraph transmission.

The 2nd column of Table 3-2 above, a Fourier transform pair can be found in Chapter 9. See Table 9-1 Row 21 and the equation (9-21). It is also derived for convenience as follows:

$h(t) = \int_{-\infty}^{+\infty} H(f)e^{+j2\pi ft} df$ a Fourier representation of a signal.

$$h(nT) = \int_{-\infty}^{+\infty} H(f)e^{+j2\pi fnT} df = \sum_{m=-\infty}^{+\infty} \left[\int_{(2m-1)/2T}^{(2m+1)/2T} H(f)e^{+j2\pi fnT} df \right]$$

$$= \int_{-1/2T}^{+1/2T} \left[\frac{1}{T} \boxed{\sum_{m=-\infty}^{+\infty} H\left(f - \frac{m}{T}\right)} \right] e^{+j2\pi fnT} dfT$$

Thus we established a Fourier transform pair: $h(nT) \Leftrightarrow \frac{1}{T}\sum_{m=-\infty}^{+\infty} H\left(f - \frac{m}{T}\right)$.

Another identity is needed: $\delta(nT) = \int_{-1/2T}^{+1/2T}[1.0]e^{+j2\pi fnT} dfT$. This can be seen by directly evaluating the integration. If $n \neq 0$, the integration of sine and cosine in one period is zero, and if $n = 0$, the integration is 1.0. QED.

Table 3-2: Justification for Nyquist criteria using the Fourier transform of a sampled signal from continuous signal

	continuous	discrete	Nyquist criterion	
Time domain signal	$h(t)$	$h(nT)$	$h(nT) = \delta(nT)$	n:integer
	$\Updownarrow$	$\Updownarrow$	$\Updownarrow$	
Frequency domain	$H(f)$	$\frac{1}{T}\sum_{k=-\infty}^{+\infty} H\left(f - \frac{k}{T}\right)$	$\frac{1}{T}\sum_{k=-\infty}^{+\infty} H\left(f - \frac{k}{T}\right) = 1$	

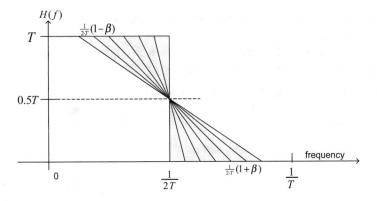

Figure 3-11: Vestigial symmetry at frequency band edge

3.2.3   Band Edge Vestigial Symmetry and Excess Bandwidth

The symmetry around the band edge is called vestigial symmetry. Remember the minimum bandwidth for $\frac{1}{T}$ rate signaling requires is $\frac{1}{T}$ (total both positive and negative). Thus the bandwidth needed more than the minimum is called excess bandwidth. The excess bandwidth parameter is denoted by β. Total bandwidth needed is given by $\frac{1}{T}(1 + \beta)$. For the frequency range of $\frac{1-\beta}{2T} < |f| \leq \frac{1+\beta}{2T}$ a Nyquist filter is vestigially symmetrical as shown in Figure 3-11; at the frequency of $0.5\frac{1}{T}$, its magnitude is half reduced from that of zero frequency.

In most of practical situations of digital transmission, bandwidth efficiency is important and thus we consider a case with the total bandwidth (positive and negative) of a signal is less than twice of symbol rate, a set of signals meeting the Nyquist criterion is shown in Figure 3-11. Trapezoidal shapes are shown in the figure, but any shape symmetrical around half of symbol rate will meet the Nyquist criterion including raised cosine (RC) shown in (3-9) and (3-10).

Example 3-3: Think of a different type vestigial symmetry. Draw it.

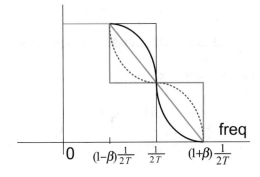

The figure above shows possible vestigial symmetries (4 cases are shown). The excess bandwidth is a parameter and the shape of the curves. There is a choice of a curve in the box from $(1-\beta)\frac{1}{2T}$ to $\frac{1}{2T}$ and then the other side $\frac{1}{2T}$ to $(1+\beta)\frac{1}{2T}$ be determined due to symmetry or vice versa.

Example 3-4: In Figure 3-11, a set of trapezoidal shapes, i.e., the transition is a straight line, is shown. Find closed form expressions for both time domain and frequency domain.

In frequency domain it is expressed as,

$$
H_{TZ}(f) = \begin{cases} T & \text{if } 0 \le |f| \le \dfrac{1-\beta}{2T} \\[2mm] T\dfrac{1}{\beta}\left(-|fT| + \dfrac{1+\beta}{2}\right)) & \text{if } \dfrac{1-\beta}{2T} < |f| \le \dfrac{1+\beta}{2T} \\[2mm] 0 & \text{if } |f| > \dfrac{1+\beta}{2T} \end{cases}
$$

In time domain it is expressed as,

$$
h_{TZ}(t) = \frac{\sin\left(\frac{\pi \cdot t}{T}\right)}{\frac{\pi \cdot t}{T}} \cdot \frac{\sin\left(\frac{\pi \cdot t \cdot \beta}{T}\right)}{\frac{\pi \cdot t \cdot \beta}{T}} .
$$

We explain the derivation of the above. A straight line in frequency domain has a slope of $-\frac{T}{\beta/T}$. This line is shifted horizontally such that zero crossing at $\frac{1+\beta}{2T}$.

For time domain impulse response, we note that a trapezoidal shape in frequency domain can be expressed as a convolution of two appropriate rectangular shapes.

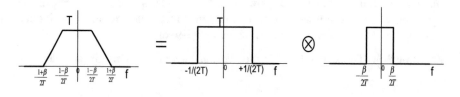

where $\otimes$ means convolution. Thus its impulse is a multiplication of each part.

This idea of decomposing into two parts of frequency response can be extended to other vestigial frequency shapes. For example,

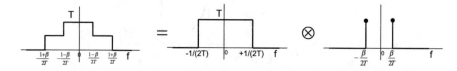

Thus its impulse response is given by,

$$h_{step}(t) = \frac{\sin\left(\frac{\pi \cdot t}{T}\right)}{\frac{\pi \cdot t}{T}} \cdot \cos\left(\frac{\pi \cdot t \cdot \beta}{T}\right).$$

In general, for impulse response, it is easy to notice that it is expressed in the form of

$$h_{vestigial}(t) = \frac{\sin\left(\frac{\pi \cdot t}{T}\right)}{\frac{\pi \cdot t}{T}} \cdot w\left(\frac{t}{T}\beta\right).$$

where $w\left(\frac{t}{T}\beta\right)$ comes due to vestigial part, and it is much 'wider' in time than $\frac{\sin\left(\frac{\pi \cdot t}{T}\right)}{\frac{\pi \cdot t}{T}}$ part. This form of multiplication of two impulse functions is called 'windowing', and used often in filter design. This windowing technique will be used in practical filter design situation and will be done later in this chapter.

With these examples, it is clear that there are many, infinitely many, possibilities of Nyquist pulse even within the vestigial symmetric case.

3.2.4   Raised Cosine Filter

A raised cosine filter in frequency domain is popular and used often and almost any textbook on communication systems. For reference see [1, 2].

It is defined in frequency domain (transmit and receive combined) as

$$RC(f) = \begin{cases} T & \text{if } 0 \le |f| \le \dfrac{1-\beta}{2T} \\[2mm] T\left(\dfrac{1}{2} + \dfrac{1}{2}\cos\left(\dfrac{\pi \cdot}{\beta}\left(|fT| - \dfrac{1-\beta}{2}\right)\right)\right) & \text{if } \dfrac{1-\beta}{2T} < |f| \le \dfrac{1+\beta}{2T} \\[2mm] 0 & \text{if } |f| > \dfrac{1+\beta}{2T} \end{cases} \tag{3-9}$$

And its corresponding impulse response is given by

$$rc(t) = \frac{\sin\left(\frac{\pi \cdot t}{T}\right)}{\frac{\pi \cdot t}{T}} \frac{\cos(\pi\beta t/T)}{1 - (2\beta t/T)^2} \tag{3-10}$$

Example 3-5: The plots of (3-9) and (3-10) are shown below; frequency response of (3-9) $\beta = 1.0, 0.5, 0.25$.

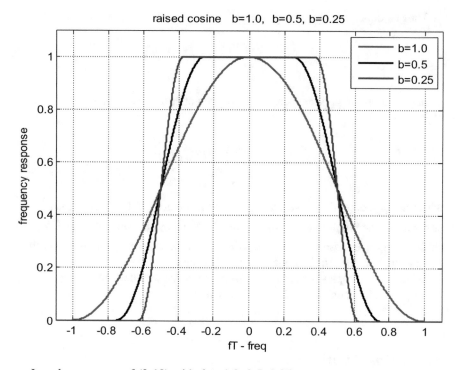

Impulse response of (3-10) with $\beta = 1.0, 0.5, 0.25$.

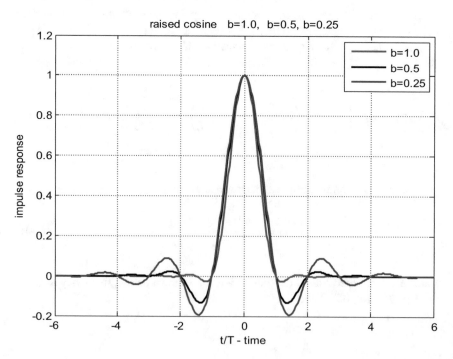

Now we need to find one side matched filter. We can meet its requirement in frequency domain by using square root of (3-9), i.e., $\sqrt{RC(f)}$. Its corresponding time domain pulse response is given by $rc_{sqrt}(t) \Leftrightarrow \sqrt{RC(f)}$ where $\Leftrightarrow$ means Fourier transform relationship. An example plot of square root RC filter is shown below.

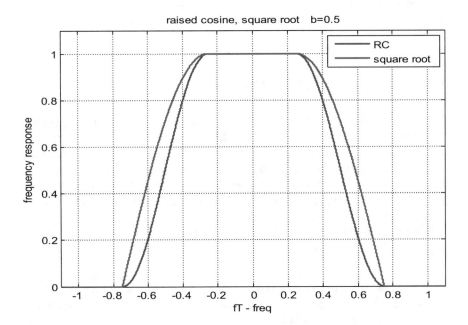

Numerical computation can be done; calculate $\sqrt{RC(f)}$ then find $rc_{sqrt}(t)$ using DFT. However, as mentioned in [8, 9], a closed form formula for $rc_{sqrt}(t)$ is available in the literature. It is also listed in Wikipedia under <Root raised cosine filter>. It is listed in (3-11) where we use it in designing digital matched filter.

3.3 Shaping Pulse (Filter) Design

So far the problem we posed in the beginning of this chapter is now solved; we need to select transmit and receive pulse (filter) to be both matched and ISI free. The ISI free condition specifies the end –to-end pulse through which a signal goes through. The pulse to meet the ISI free condition is called a Nyquist pulse. The matched filter specifies one side and the other side is determined since both of the impulse responses should be in mirror image (time reversal). In our digital communication, both conditions must be met; (1) $g_R(t) = g_X(-t)$ and (2) $h(t) = g_X(t) \otimes g_X(-t)$. We may call it a matched Nyquist pulse. Remember that the end to end pulse, $h(t)$, is even symmetry and the peak is around zero in time, and furthermore there are zero crossings uniformly with the symbol period.

As discussed briefly in RC filter context, any vestigial symmetry filter $H(f)$ can meet the ISI free criterion. In order to meet the matched filter condition one may choose the magnitude $|G_X(f)| = \sqrt{|H(f)|}$ and $|G_R(f)| = \sqrt{|H(f)|}$ and the angle so that $G_R(f) = G_X^*(f)e^{-j2\pi fD}$; $G_R(f)$ to be a complex conjugate of $G_X(f)$ and with a proper delay so that it is causal.

3.3.1 Practical Design Considerations

We now consider how to design pulse shaping filters in more detail. One first guiding principle is a matched filter to maximize SNR, which means that a pair of receiver and transmit filter must be 'matched'. Obviously SNR may be increased by increasing the transmit power so it is necessary to set the same transmit power to compare different situations. This was demonstrated when we calculated SNR loss due to the pulse mismatch in the previous section. The other second is the end to end pulse should be a Nyquist pulse, i.e., ISI free.

In wireless digital radio communication systems, we may state practical design considerations as follows

  i. End-to-end impulse response should be ISI free, i.e. meeting Nyquist criterion
 ii. A transmit filter must be matched to receive one in order to maximize SNR.
iii. At transmit side frequency spectrum mask must be met, i.e. bandwidth is limited. This is a frequency domain constraint.
 iv. A receiver should have minimum noise bandwidth and furthermore reject out of band interference, which is also a frequency domain constraint. It is particularly important to reject out of band interference as often in wireless systems band width is confined and adjacent channels are used by others.

Note that meeting four constraints above may be done easily in theory, e.g., ideal square root RC pulse may meet all four conditions. However, practical filter design - e.g., out of band rejection requirement- may entail the complexity of a filter to meet the conditions (iv above). We need practical tradeoff to reduce the complexity at the same time meeting the requirements.

3.3.2 A Practical Design Example of Analog Filter

In high speed digital radio systems, often the design of the pulse shaping filter is mixed with digital filters and analog filters, with multi-level modulations such as 64-QAM or even higher, the precision requirement of the shaping filters is

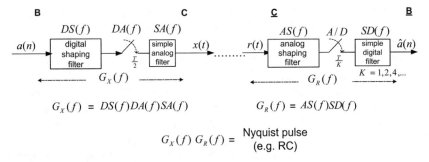

$$G_X(f) = DS(f)DA(f)SA(f) \qquad G_R(f) = AS(f)SD(f)$$

$$G_X(f) \, G_R(f) = \frac{\text{Nyquist pulse}}{\text{(e.g. RC)}}$$

Figure 3-12: Practical example of shaping filter design; a sqrt RC filter (or a sqrt Nyquist filter) is implemented by an analog filter at the receiver ($AS(f)$) approximating the magnitude, and delay distortion and the transmit side sqrt RC is done digitally at the transmit side ($DS(f)$). The digital filter also compensates D/A and a simple analog filter frequency response

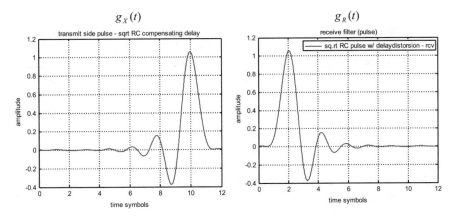

Figure 3-13: Receive filter is 40% sqrt RC with parabolic delay distortion; the sharper the receive filter (more out band rejection) the more delay distortion, and thus transmit digital filter gets more complicated. 40% means that the excess bandwidth $\beta = 0.4$ in equation (3-10)

demanding. In this situation, a following configuration was used successfully in actual design situations. It is shown in Figure 3-12.

A square root Nyquist filter is implemented by an analog filter at the receiver ($AS(f)$) approximating only the magnitude with the minimum phase (no delay equalizer), and the delay distortion and the transmit side sqrt RC are done digitally at the transmit side ($DS(f)$). The digital filter also compensates D/A and a simple analog filter frequency response. Thus the matched filter pulse shape is the inverse Fourier transform of the analog filter at the receiver ($AS(f)$), and the end to end pulse shape is a RC, or any Nyquist pulse.

A pulse example is shown in Figure 3-13, and the pulse shape of one side may be complicated as long as the end to end pulse is a Nyquist pulse. It appears to be a matched filter pair by cursory look. This is indeed a matched filter pair. A 'distorted' looking pulse shape is due to delay distortion of a sharp analog receiving filter. It is compensated by transmit side digital filter, and thus the delay of the end to end pulse is nearly flat.

This basic configuration in Figure 3-12 can be adjusted to a situation. If a higher sampling is possible, the analog filter can be simple and most of the shaping can be done by a digital filter at the receiver too. A sharp analog filter at the receiver saves a high speed sampling, symbol rate sampling or at most twice the symbol rate.

The configuration in Figure 3-12 can be made into a pair of standard components including digital filter D/A and analog filter in a package, and similarly another component is a combination of analog filter, A/D and digital filter. These may be implemented by a chip.

3.3.3   Design of Digital Matched Filters

In this section we discuss the design of digital matched filters. Due to high sampling we assume that the impact of analog filters and D/A is negligible.

As mentioned in the previous section, digital filters are used for both transmit and receive side when a high sampling is feasible. This is typically the case with the advent of very high speed D/A and A/D.

We consider two different methods - windowing in time domain, and optimization in frequency domain. In design process both may be used.

Peak distortion
The quality of end-to-end pulse is measured by peak distortion. We define it first.

First an end-to-end impulse response is sampled with symbol period. The result may be represented as

$$.., \left|h^0_{-3}\right|, \left|h^0_{-2}\right|, \left|h^0_{-1}\right|, \left|h^0_0\right|, \left|h^0_{+1}\right|, \left|h^0_{+2}\right|, \left|h^0_{+3}\right|, \ldots..$$

where h^0_{nT} is n-th sample. The value of n ranges over a filter length in symbol time. An example is shown in Figure 3-24.

A peak distortion (or 'eye closure' See the next subsection for eye pattern.) at k-th sampling offset, PD is defined as

$$PD = \sum_{n \neq 0} \left|h^k_n\right| / \left|h^k_0\right| \text{ with } k = 0 \text{ (i.e., no sampling offset case here)where } n \neq 0$$

means all of n except for zero, i.e. all ISI terms except for a main tap ($n = 0$). PD will be used in both methods. It is given as a design parameter.

3.3.3.1   Windowing Method

A simple method of obtaining a FIR form of sqrt RC digital filter is to compute samples of a continuous sqrt RC impulse and truncate them. An example of it is

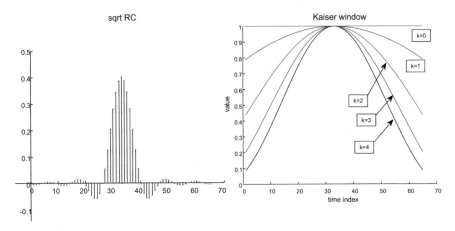

Figure 3-14: sqrt RC filter example and Kaiser Window

shown in Figure 3-14 (LHS); a number of taps N = 65, span of 8 symbols and 8 samples per symbol plus a center tap.

The truncation is implicitly a rectangular window. With a sufficient span of symbols, PD can be made small, say less than 0.01, but typically outband rejection may be not large enough with rectangular window. A windowing will improve the outband rejection at the expense of PD degradation. The Kaiser window is convenient for this purpose since it has a shape parameter, k, as shown in Figure 3-14 (RHS); the larger the k, the more outband rejection.

This windowing method has two control parameters: N_{span} (a number of symbol spans) and Kaiser window shape parameter k.

A number of sample per symbol ($N_{sampsym}$) determines the interpolation factor, and thus the higher the interpolation the more filter taps; Ntaps = $N_{span} * N_{sampsym} + 1$. In search of a properly windowed sqrt RC, $N_{sampsym}$ can be fixed (say 4 or higher) and not necessary to be varied.

A computer program (say a Matlab function) can be written as a function of excess bandwidth, β, N_{span}, and k.

For convenience we list a closed form formula of sqrt RC function, and Kaiser window function below.

The sqrt RC function is given by

$$g_{sqrtRC}(t) = \frac{1}{\sqrt{T_s}} \frac{\sin\left(\pi(1-\beta)\frac{t}{T_s}\right) + \frac{4\beta t}{T_s}\cos\left(\pi(1-\beta)\frac{t}{T_s}\right)}{\frac{\pi t}{T_s}\left(1 - \left(\frac{4\beta t}{T_s}\right)^2\right)} \tag{3-11}$$

with

$$g_{sqrtRC}(t = 0) = \frac{1}{\sqrt{T_s}}\left(1 - \beta + 4\frac{\beta}{\pi}\right)$$

$$g_{sqrtRC}\left(t = \pm\frac{T_s}{4\beta}\right) = \frac{1}{\sqrt{T_s}}\frac{\beta}{\sqrt{2}}\left(\left(1+\frac{2}{\pi}\right)\sin\left(\frac{\pi}{4\beta}\right) + \left(1-\frac{2}{\pi}\right)\cos\left(\frac{\pi}{4\beta}\right)\right)$$

Exercise 3-6: Find a sampled version of $g_{sqrtRC}(t)$.
Answer: A sampled version of it is given by

$$g_{sqrtRC}(n\Delta t) = \frac{\sin\left(\pi(1-\beta)\frac{n}{N}\right) + \frac{4\beta n}{N}\cos\left(\pi(1-\beta)\frac{n}{N}\right)}{\frac{\pi n}{N}\left(1 - \left(\frac{4\beta n}{N}\right)^2\right)}$$

$t = n\Delta t$ where $\Delta t = \frac{T_s}{N_{sampsym}}$ and $T_s = 1.0$, for easy looking $N_{sampsym} = N$
The power is $\sum_n \left(g_{sqrtRC}(n)\right)^2 = N_{sampsym} = N$. To make it unity power, a scale
factor $\frac{1}{\sqrt{N}}$ is necessary to add.

The Kaiser window, the length ($M+1$), is defined as

$$w(n) = \frac{I_0\left[k\sqrt{1 - \left[\frac{\left(n-\frac{M}{2}\right)}{\frac{M}{2}}\right]^2}\right]}{I_0(k)}, \quad 0 \le n \le M, \quad w(n) = 0, \quad \text{otherwise}$$

I_0 is the zeroth order modified Bessel function of the first kind. A shape parameter
is denoted as k.
A design example is shown in Figure 3-15; $\beta = 0.5$, $N_{span} = 8$, $N_{sampsym} = 8$,
PD = 0.0177 (1.77%), window shape parameter $k = 1.7$. In the figure stars (*) are
0.25, and 0.75 in frequency, and cross (+) is 3 dB at 0.5 in frequency.

3.3.3.2   Frequency Domain Optimization Method

This method utilizes a FIR digital filter design program – Parks-McClellan
algorithm.
    The filter design algorithm is optimal in the sense that it reduces its order,
equiripple in passband and stopband. One can specify passband ripple (δ_1), stopband
attenuation (δ_2), passband edge (f_p), stopband edge (f_s). The order of filter
(a number of taps) and transition from passband to stopband are not specified. The
order can be empirically estimated from given specifications. This LPF design
tolerance spec is shown in Figure 3-16 (LHS).

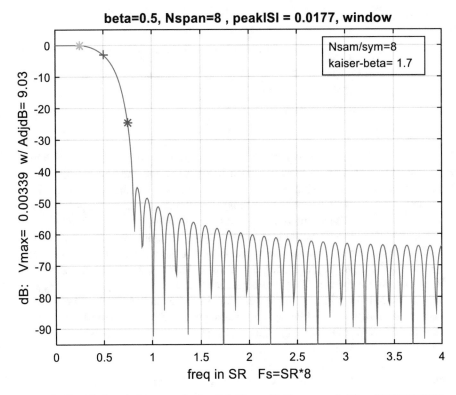

Figure 3-15: Window design example, $\beta = 0.5$, $N_{span} = 8$, $N_{sampsym} = 8$, PD $= 0.0177$ (1.77%), window shape parameter $k = 1.7$

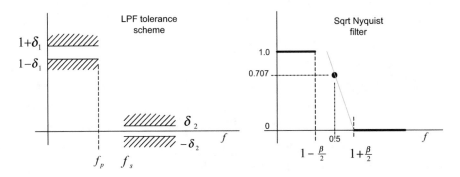

Figure 3-16: LPF design tolerance and square root Nyquist filter specification

For example, the program is available in Matlab, <*firpm.m*>;

```
rp = 0.05;rs=60;fp=0.364;  fs=0.75;FS=8;
f = [fp fs];                    % Cutoff frequencies
a = [1 0];                      % Desired amplitudes
dev = [(10^(rp/20)-1)/(10^(rp/20)+1)   10^(-rs/20)];
[n,fo,ao,w] = firpmord(f,a,dev,FS);
hcoeff = firpm(n,fo,ao,w);
```

For the design of a sqrt Nyquist filter it is important to control the transition from passband to stopband; the transition must be such that the end to end filter (equivalent to squaring of transition) must meet the vestigial symmetry. In particular it requires 3 dB point (0.707 in vertical axis) should occur at 0.5 in frequency in any transition of vestigial symmetry. Note that there are many possibilities of transition with vestigial symmetry but it must pass through 3 dB point at the frequency of 0.5. This observation may be exploited to use the FIR design program.

The FIR design program does not have a direct control over the transition. However, within the transition band, passband edge (f_p), stopband edge (f_s) may be adjusted to indirectly control the transition, in particular meeting 3 dB point at 0.5. In order to meet the given filter specifications, passband edge (f_p) must increase and stopband edge (f_s) must decrease. One possibility is to adjust only f_p while f_s is fixed. This idea of f_p adjustment is used in [9].

The f_p adjustment range should be limited since the sharper the transition the higher the order of filter. Thus the adjustment range is empirically given by

$$\left(0.5 - \frac{\beta}{2}\right) \le f_p \le \left(0.5 - \left(\sqrt{2} - 1\right)\frac{\beta}{2}\right)$$

This is shown in Figure 3-17.

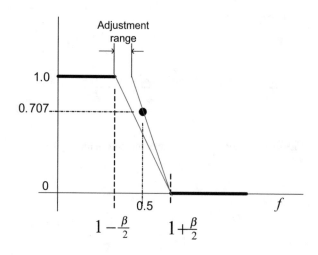

Figure 3-17: Square root Nyquist filter specification and adjustment range

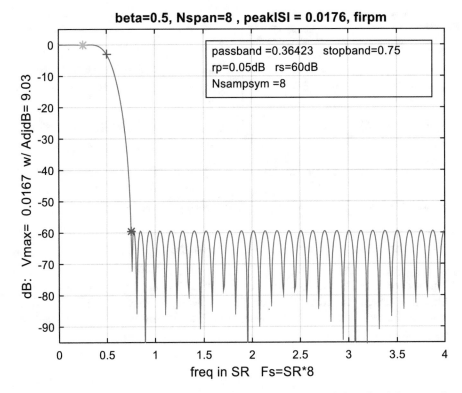

Figure 3-18: A design example of frequency domain method is shown below; $\beta = 0.5$, $N_{span} = 8$, $N_{sampsym} = 8$, PD = 0.0176 (1.76%), passband ripple = 0.05 dB

A design example of frequency domain method is shown in Figure 3-18; $\beta = 0.5$, $N_{span} = 8$, $N_{sampsym} = 8$, PD = 0.0176 (1.76%), passband ripple = 0.05 dB. In the figure stars (*) are 0.25, and 0.75 in frequency, and cross (+) is 3 dB at 0.5 in frequency

Exercise 3-7: Compare the two design examples in this subsection, and observe the difference.

Answer: Passband edge is different 0.25 vs. 0.364 for windowing and frequency domain method respectively. PD is nearly identical, 1.7%. Stopband loss at frequency 0.75 is 25 dB vs. 60 dB.

Example 3-6: Another example of design comparison, both the same parameters, is shown in Figure 3-19. $\beta = 0.2$, $N_{span} = 20$, $N_{sampsym} = 8$, PD = 0.0121 vs. 0.0245, loss at 0.6 = 26 dB vs. 60 dB

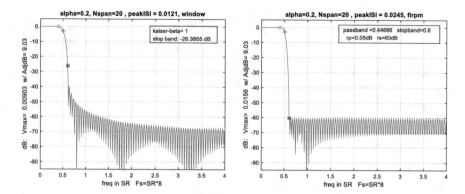

Figure 3-19:   Comparison of window method and frequency domain method

3.4   Performance Degradation Due to ISI

This section together will cover a case example with over simplifying the matched filter requirement in order to reduce the complexity. It may be skipped for first reading since the analysis is somewhat complicated. However, it is important in practice.

3.4.1   A Design Case to Simplify the Shaping Filters

There is a suggestion, in order to simplify the receiver filter implementation, rather than using a matched filter (square root RC), to use a rectangular pulse. In modern implementation it is essentially free to implement the rectangular pulse when there are multiple samples available in a symbol period; by simply adding those samples. In the transmit side there is no choice but given (square root RC) since it must meet the frequency spectrum mask; the spectrum of the signal must be within the specified limit.

A straightforward design would be to use square root RC pulse both at the transmit side and the receive side. To see the frequency response of the square root RC compared with RC it is shown in Figure 3-20.

The suggestion seems reasonable since the rectangular pulse matches with square root RC "reasonably well" as shown in Figure 3-21. Appropriately the suggestion is called 'no matched filter' since the complexity of the square root RC implementation is reduced to almost nothing. With the higher rate signaling, the complexity of the matched filter is higher.

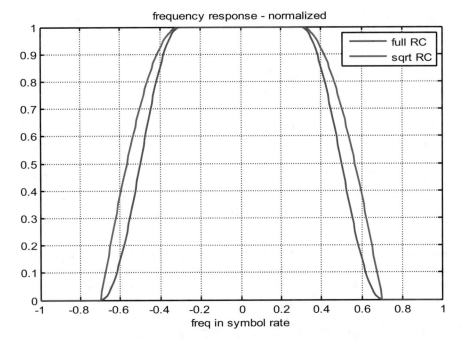

Figure 3-20: The frequency response of square root RC and full RC filter ($\beta = 0.4$)

3.4.2 A Quick Analysis of the Suggestion of Using a Rectangular Pulse

Now the task is to assess the situation quantitatively – analyzing the performance of the system with 'no matched filter'. We assume the rectangular pulse is perfectly done, i.e., sufficiently large number of samples in a symbol period. In practice 6 to 8 samples should be enough.

SNR degradation due to the pulse mismatch and due to ISI should be estimated. When there are many adjacent channels, the out of band interference must be estimated. This will be done when we consider OFDM. Here we focus on the mismatch loss and ISI. As for symbol rate ($\frac{1}{T}$), Figure 3-21 uses $\frac{1}{T} = 1.0$ scale or it may be considered to be normalized by the actual symbol rate in frequency and symbol time in time domain.

3.4.2.1 Mismatch Loss

Noise power after receive filter of a rectangular pulse can be seen easily by $P_N = \frac{No}{2} \int |g_R(t)|^2 dt = \frac{No}{2T}$ and for square root RC it can be done by $P_N = \frac{No}{2} \times \int |G_R(f)|^2 df = \frac{No}{2T}$. Both are the same. The peak amplitude of a signal is next; for

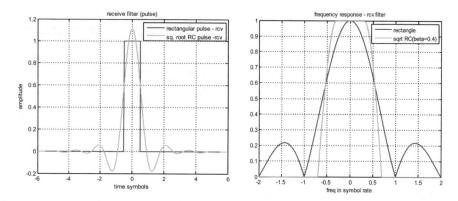

Figure 3-21: Rectangular pulse (blue) compares with $\beta = 0.4$ square root RC filter

the matched case it is 1.0 and for rectangular receive filter it can be computed by convolving a square root RC pulse with the rectangular pulse at time zero; in Figure 3-21 (LHS) it is the same as the (hatched) area under both the square root RC and the rectangular pulse. It is 0.9271. One can see the peak value from the end-to-end pulse. This is shown in Figure 3-22.

Thus the mismatch SNR loss is $10 \log_{10}(0.9271)^2 = -0.66$ dB.

3.4.2.2   SNR Loss Due to ISI:

ISI of rectangular pulse filter will be worse than square root RC filter since the end to end pulse does not meet Nyquist criterion while RC filter meet the criterion. ISI is a self-generated interference; other neighboring symbols interfere with the current symbol. This is due to the fact that the end-to-end pulse does not meet Nyquist criterion. The impact to the system performance is complicated and will be discussed next. A simplified expression of the degradation due to ISI may be expressed as SNR loss.

In order to estimate the impact of ISI we need an end-to-end impulse response, which is a convolution of transmit pulse with receive pulse as in Figure 3-22. In time domain Nyquist criterion is that a sampled end-to-end pulse is a delta function as discussed before. RC pulse has a completely regular zero crossings (green), thus there is no ISI. On the other hand, the end to end pulse with rectangular pulse receiver decays the zero crossings are not regular and thus there is ISI. This can be seen slightly easier when the peak values are normalized to be equal as shown in Figure 3-23.

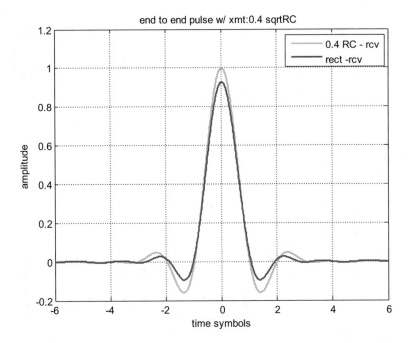

Figure 3-22:   End-to end pulses with two different receive filters; RC (green) and rect (blue)

3.4.2.3   Peak Distortion Estimation Method

We will explain how Figure 3-25 is obtained. First an end-to-end impulse response such as in Figure 3-23 is sampled with symbol period. The result may be represented as

$$.., \left|h^0_{-3}\right|, \left|h^0_{-2}\right|, \left|h^0_{-1}\right|, \left|h^0_0\right|, \left|h^0_{+1}\right|, \left|h^0_{+2}\right|, \left|h^0_{+3}\right|, ...$$

where h^{kdt}_{nT} is n-th sample with k-th time offset. The value of n ranges over a filter length in symbol time, and the range of k is the number of samples per symbol time period. An example is shown in Figure 3-24.

A peak distortion at k-th sampling offset, EC^k is defined as

$$EC^k = \sum_{n \neq 0} \left|h^k_n\right| / \left|h^k_0\right|$$

where $n \neq 0$ means all of n except for zero, i.e. all ISI terms except for a main tap ($n = 0$). Then the degradation in dB, C is given by

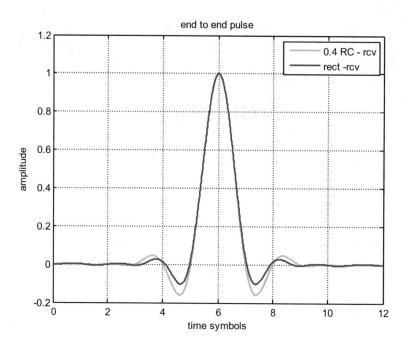

Figure 3-23: Two end to end pulses with different receive filters but normalized to have the same peak value

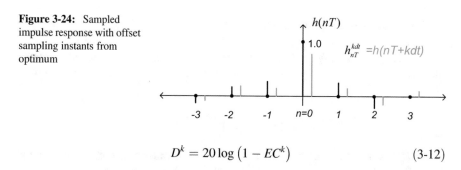

Figure 3-24: Sampled impulse response with offset sampling instants from optimum

$$D^k = 20 \log \left(1 - EC^k\right) \tag{3-12}$$

In Figure 3-25 D^k the ISI peak distortion in dB is displayed with k from zero to 0.2. This should be a good approximation, in particular, near optimum sampling point ($dt = 0$). In the figure mismatch loss is added to show the total SNR loss estimate.

Thus the total SNR loss because of the use of a rectangular pulse, rather than a matched filter, consists of the pulse mismatch loss (0.66 dB) and ISI loss (about 1 dB), total 1.66 dB around the optimum sampling point as shown in Figure 3-25. This approximate SNR loss in dB gives a quick intuitive feel with practically useful numbers.

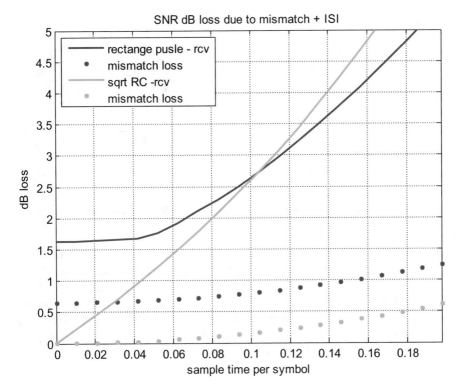

Figure 3-25: SNR dB loss due to mismatch and ISI

3.4.2.4 Eye Pattern

One way to see the impact of ISI is to use 'eye pattern', which is used often in conjunction with oscilloscope, particularly in the past. It is still useful when the signaling speed is very high like an optical transmission or high speed serial baseband connection. In oscilloscope display horizontal time base is synchronized with symbol rate then signal traces are traced repeatedly, and the result is overlapped pattern resembling 'eye shape', thus the name eye pattern. This can be done easily with a computer.

Two eye patterns are shown in Figure 3-26; one for RC filter (LHS) and the other is for rectangular pulse. The key point is at the sampling instant. At sampling point LHS (exact matched filter) of Figure 3-26 there is no degradation, i.e., no eye closure, and RHS (approximate matched filter) has amplitude degradation about 15%. This eye closure is due to ISI, i.e. adjacent symbols are affecting the current symbol. This reduces the immunity against noise.

The optimum sampling point is at the largest 'eye opening'. In Figure 3-26 time symbol '0', '1', '2' are such points. And the amplitude is to be +1 and −1. With the eye pattern one can see the quality of a system very quickly and thus useful. The 'eye

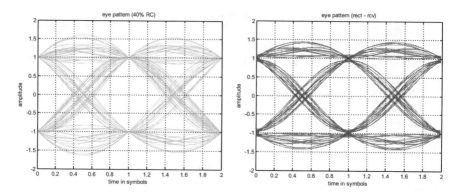

Figure 3-26: Eye patterns matched filter (LHS) and approximately matched filter by rectangular pulse (RHS)

closure' is how much amplitude reduction at the optimum sampling point. In Figure 3-26 (LHS), with the matched filter, no degradation is observable, a sharp point. On the other hand on RHS, one can easily see the degradation, about 15% or so. Note that the eye pattern does include both the reduction of the main tap of an end to end impulse response (mismatch loss) and uneven zero crossings (ISI loss) simultaneously. The eye pattern is still used extensively when a very high speed serial connection to estimate the quality of a system.

3.4.3 A Discrete in Time Model for ISI Analysis in General

The analysis done in the previous section, peak distortion model, may be refined further by finding a detailed probability distribution function of ISI. The peak distortion method may be viewed as a simplified version of ISI probability distribution. A sampled impulse response is first simplified to one ISI tap in addition to the main tap; a simplified impulse response has only two taps; $\left|h_0^0\right|$ and $\left|h_{+m}^0\right|$ where m may be arbitrary. A symbol at $\left|h_{+m}^0\right|$ may be one of symbols $\{+1, 1\}$ with the probability ½.

Before getting into further details, we develop a discrete model from Figure 3-2. RHS of B–$\underline{B}$ is a 'channel' for discrete symbols and LHS of B–$\underline{B}$ is a source of discrete symbols and a sink. Once the noise is added there is only essential discrete signal processing, i.e., to make a decision which symbols are transmitted. The amount of noise is limited by the receive filter ($g_R(t)$). Thus we can 'push' it toward the transmit side, and we obtain a discrete model as shown in Figure 3-27.

The noise samples, $w(n)$ are generated discretely but its variance, σ^2, should be related with SNR. For example M-PAM, $\frac{d}{\sigma} = \sqrt{\frac{6}{M^2-1}\frac{Es}{No}}$ and for a given $\frac{Es}{No}$, one set a discrete noise level. The noise variance is a function of the receive filter and it is given by $\sigma^2 = \frac{No}{2}\int|g_R(t)|^2 dt$.

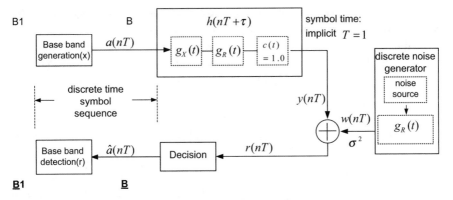

Figure 3-27: A discrete signal model of Figure 3-2. Once sampled the symbol time becomes implicit and thus one may set to be 1.0

Figure 3-28: A discrete time model of PAM with a linear filter

In discrete model only the end to end impulse response matters and it is sampled every symbol time as shown $h(n) = h(nT + \tau)$ where τ is a sampling phase within a symbol time. In fact a linear channel, represented by $c(t)$, can be incorporated as shown in the figure but at the moment we set it to be transparent. We need to compute the end to end impulse response as shown in Figure 3-27, and then it is sampled. Once the sampled impulse response is known, we may draw it simply as in Figure 3-28.

In order to compute SER for this system we need this discrete time model. In Chapter 2 we did the same except that there is no linear filter ($h(n)$). This linear filter may include the pulse mismatch and subsequent ISI, and sampling phase. Due to this filter things get much more complicated to analyze. This is the reason we tried to use a quick method.

Now back to the refinement of the peak distortion method. We drop the superscript for the value of $h(n)$ and use the notation $h(0) = h_0$, $y(n) = \sum_j a(j)h(n - j)$. The index for $h(n)$ is arranged so that the peak is at $h(0)$. Then the values are shown and its peak distortion approximation in the table below.

$h(n)$	$\ldots$	$\ldots$	h_{-3}	h_{-2}	h_{-1}	h_0	h_1	h_2	h_3	$\ldots$	$\ldots$
Approximation for peak distortion	0	0	0	0	0	h_0	$\sum_{k \neq 0} \lvert h_k \rvert$ $= \overline{h}_1$	0	0	0	0

With the approximation, we obtain $r(n) \cong a(n)h_0 + a(n - 1)h_1 + w(n)$ in order to estimate SER with the linear filter. The mismatch loss is the reduction of h_0 less than

1.0, which was 0.9271, which is equivalent to noise variance change (increase) to the same amount. And ISI term $\overline{h}_1/h_0 = e_1 =$ eye closure is not really immediate SNR reduction but the mean is no longer zero but add or subtract depending on the symbol at that moment; in the average half of time added and the other half time subtracted. With BPSK, SER is given by $Q\left(\frac{d}{\sigma}\right)$ and SER with the two tap approximation it will be given by

$$ \text{SER} = \frac{1}{2}Q\left(\frac{d-e_1}{\sigma}h_0\right) + \frac{1}{2}Q\left(\frac{d+e_1}{\sigma}h_0\right) \approx Q\left(\frac{d}{\sigma}\left(1-\frac{e_1}{d}\right)h_0\right) \qquad (3\text{-}13) $$

In the last equation, $\left(1-\frac{e_1}{d}\right)h_0$ is the SNR degradation due to both the mismatch and ISI. The quick method is thus somewhat justified and this process can be extended to make SER estimation further refined. Essentially more taps are considered with different possible past (future) symbols. If there are L taps in $h(n)$, there are $2^{(L-1)}$ combination of symbols possible. $L - 1$ terms of $h(n)$, except for h_0, are assigned to different combination of symbols, then multiplied and added (inner product). This is repeated all possible symbol combinations. These values are arranged to find a histogram. The frequency of each bin of the histogram is divided by $2^{(L-1)}$. This is a probability distribution of ISI.

$h(n)$	...	...	h_{-3}	h_{-2}	h_{-1}	h_0	h_1	h_2	h_3	...	...
Possible symbols	+-	+-	+-	+-	+-	h_0	+-	+-	+-	+-	+-

With BPSK, this method gives SER estimate as

$$ \text{SER} = \sum_i p_i Q\left(\frac{d-m_i}{\sigma}h_0\right) \qquad (3\text{-}14) $$

where p_i is the probability of ith bin with the value m_i. An example of probability density (PDF) of ISI is shown in Figure 3-29. A number of bins are16.

Even though conceptually simple, the computation of PDF of ISI is not trivial especially when there are many taps in a sampled impulse. This can be much worse with multi-level since a number of combination is $M^{(L-1)}$. The evaluation of (3-14) is another burden depending on the number of bins. However, with some experience it seems that the computation burden can be manageable.

3.4.4   A Discrete in Time Model for Simulation

Once a discrete system model is obtained as shown in Figure 3-27 or in Figure 3-28, a numerical simulation can be done to estimate SER. The computation burden for a

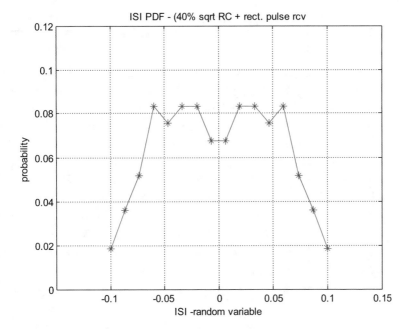

Figure 3-29:   PDF of ISI with rectangular pulse + 40% square root RC filter

brute-force simulation is manageable when SER required is not excessively low. However SER <1e-5 the computation becomes heavier since more symbols must be tried.

   Now compare all the methods we discussed so far. The result is summarized in Figure 3-30.

3.4.5   Summary for ISI Analysis Methods

A quick estimation of mismatch loss plus the peak distortion method is pessimistic but computationally the easiest. Thus in the early design stage this might be useful as an initial estimate. Computing PDF of ISI is computationally intense, but the result is accurate. It may be still less computation for low SER case than the straight simulation. A compromise might be to use 2-bin PDF (only peak distortion but both positive and negative) to compute SER using (3-13). As indicated earlier, the use of a rectangular pulse as a receive filter is very easy with modern implementation using digital hardware and DSP processors since it is just adding samples with the period of a symbol. However, the out band rejection may be an issue when there are many neighboring channels, which will be discussed when we discuss OFDM.

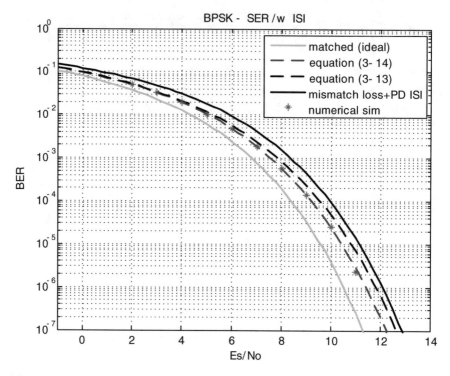

Figure 3-30: SER of BPSK with approximate matched filter. There are five cases: matched, PDF of ISI (3-14), 2-bin PDF of ISI (3-13), SNR dB loss method and numerical simulation. The results of PDF of ISI method line up well with the numerical simulation

3.5   Extension to Linear Channel and Non-White Noise

So far we considered a system where only AWGN is the impairment. We extend it to linear system channels and to noise without flat power spectrum density. In AWGN channel, an optimum solution can be described by a matched filter and ISI free Nyquist pulse where the noise passes through the noise limiting filter and the signal passes though both pairs of filters. This will maximize the SNR at the sampling time, which in turn minimizes SER. Once we open up linear channels and non-white, then things get much more complicated but the AWGN channel is a very useful reference point. In wireless channels, the linear channel is often randomly time-varying, which will be treated in Chapter 4 with ample details.

3.5.1   Linear Channels with White Noise

The system we consider is shown in Figure 3-31. The constraint for the optimization is that the linear channel is known both to the transmit side and to the receive side.

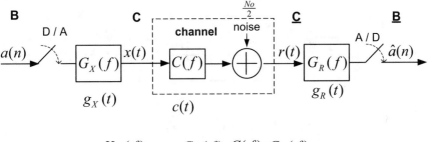

$$X_{rc}(f) \;\; = \;\; G_X(f)\, C(f)\, G_R(f)$$

Figure 3-31: PAM system with linear channel ($C(f)$)

One optimization constraint is to choose the end to end pulse to be ISI free – a Nyquist pulse. $X_{rc}(f)$ meets the Nyquist criterion; a sampled impulse response is a unit impulse $\delta(nT)$.

  The rest of optimization is how to split into a pair of matched filters. And then compute the noise power and the transmit power.

$P_x = \frac{d^2}{T} \int |G_X(f)|^2 df$ where d^2 is the power of discrete symbols of $a(n)$

$$P_N = \frac{No}{2} \int |G_R(f)|^2 df = \sigma^2$$

$$\frac{d^2}{\sigma^2} = \frac{2 P_x T}{No} \left[\int |G_X(f)|^2 df \int |G_R(f)|^2 df \right]^{-1}$$

When we choose $G_X(f) = \frac{\sqrt{X_{rc}(f)}}{C(f)}$ and $G_R(f) = \sqrt{X_{rc}(f)}$, and then $\frac{d^2}{\sigma^2} =$

$\frac{2 P_x T}{No} \left[\int \left| \frac{\sqrt{X_{rc}(f)}}{C(f)} \right|^2 df \right]^{-1}$. Thus compare to AWGN, there is SNR loss (or gain)

by $\int \left| \frac{\sqrt{X_{rc}(f)}}{C(f)} \right|^2 df$. We used that $\int |G_R(f)|^2 df = \int |X_{rc}(f)| df = 1.0$.

Exercise 3-8: If we choose $G_X(f) = \frac{\sqrt{X_{rc}(f)}}{\sqrt{C(f)}}$ and $G_R(f) = \frac{\sqrt{X_{rc}(f)}}{\sqrt{C(f)}}$, show that SNR

loss is given by $\left[\int |\frac{X_{rc}(f)}{C(f)}| df \right]^2$. And then this is smaller than or equal to

$\int \left| \frac{\sqrt{X_{rc}(f)}}{C(f)} \right|^2 df$ using Schwarz inequality.

Exercise 3-9: Another choice is $G_R(f) = \frac{\sqrt{X_{rc}(f)}}{C(f)}$ and $G_X(f) = \sqrt{X_{rc}(f)}$. Find

the SNR loss factor. (Hint: it is the same as $\int \left| \frac{\sqrt{X_{rc}(f)}}{C(f)} \right|^2 df$.)

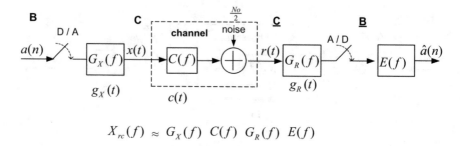

$$X_{rc}(f) \approx G_X(f) \; C(f) \; G_R(f) \; E(f)$$

Figure 3-32: More optimization with digital equalizer or MLSD at the receive side

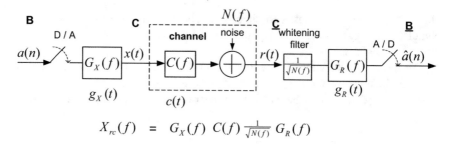

$$X_{rc}(f) = G_X(f) \; C(f) \; \frac{1}{\sqrt{N(f)}} \; G_R(f)$$

Figure 3-33: Non-white noise and the use of a whitening filter

By insisting no ISI, essentially the linear channel has to be inverted, which enhances noise. But one can relax the ISI requirement, allowing a smaller amount of ISI, and the noise may be minimized. The SNR loss due to ISI and noise may be simultaneously minimized, commonly by mean square error. An equalizer may be added at the receiver, and even more improvement is by adding maximum likelihood sequence detector (MLSD) or Viterbi detector (Figure 3-32). We do not get into any more here, but the linear channel as a fading in wireless channels will be treated later.

3.5.2   Non- White Noise

So far we consider white noise, whose power spectrum is flat in the frequency of interest. Is it possible to extend the matched filter and ISI free pulse idea to this situation? The answer is yes. We use the idea of whitening filter; essentially inversing the noise power spectrum, and then apply the matched filter and ISI free pulse idea (Figure 3-33).

Choose $G_X(f)C(f)\frac{1}{\sqrt{N(f)}} = \sqrt{X_{rc}(f)}$ and $G_R(f) = \sqrt{X_{rc}(f)}$. Thus this situation is equivalent to the channel linear filter becoming $C(f)\frac{1}{\sqrt{N(f)}}$. Then $G_X(f) = \frac{\sqrt{X_{rc}(f)}}{C(f)}\sqrt{N(f)}$ and the total receive side filtering is $\frac{\sqrt{X_{rc}(f)}}{\sqrt{N(f)}}$.

3.6 References

[1] Haykin, Simon ,"Communication Systems" 4th ed., John-Wiley & Sons, Inc., New York, 2001, Chapter 4. Baseband Pulse Transmission

[2] Carson, A. Bruce, Crilly, Paul B. and Rutledge, Janet C., "Communication Systems – An Introduction to Signals and Noise in Electrical Communication" 4th ed., McGraw Hill, Boston, 2002, Chapter 11. Baseband Digital Transmission

[3] North, D.O. , "An Analysis of the Factors Which Determine Signal/Noise Determination in Pulse-Carrier Systems," RCA Tech. Report No. 6 PTR-6C, 1943

[4] Turin, George, "An Introduction to Matched Filters", IRE Transaction on Information Theory, 1960, pp 311-329

[5] Nyquist, H., "Certain Factors Affecting Telegraph Speed," Bell Systems Technical Journal, vol.3, 1924, p.324

[6] IEEE Std 802.11™-2016 (Revision of IEEE Std 802.11-2012) Part 11: Wireless LAN Medium Access Control(MAC) and Physical Layer (PHY) Specifications

[7] Yang, Sung-Moon (Michael), "A New Filter Method of CP and Windowing in OFDM Signals" Military Communications Conference 2010 https://doi.org/10.1109/milcom.2010.5680210

[8] Lapidoth, Amos "A Foundation in Digital Communication" Cambridge University Press, 2009, pp.197 equation (11.31), which unfortunately has errors. See Wikipedia 'root raised cosine filter' for correct one.

[9] Harris, Frederic J. "Multirate Signal Processing for Communication Systems " Prentice Hall, 2004, pp.90 equation (4.18)

3.7 Problems

P3-1: In Fig. P3-1 a baseband digital transmission system under additive noise is captured. A shaded block is a model for the conversion from discrete-time to continuous signal at the transmit side. Another shaded block is a model for the conversion of continuous signal to discrete-time signal.

(1) When $a[n] = \delta[n]$, a single pulse of unit impulse, obtain $A_I(t)$.

(2) When $a[n] = \delta[n] + \delta[n - 3]$, find the simplest expression of $A_I(t)$.

(3) $A_I(t) = \sum_{n=-\infty}^{\infty} a[n]g_X(t - nT)$, i.e., continuous signal at the transmit side can be expressed as a convolution sum.

P3-2: Assuming that there is no noise, i.e., $n(t) = 0$, in Fig. P3-1,

(1) Obtain the output of the receive filter, $rr(t)$, when $a[n] = \delta[n]$.

(2) Find the simplest expression of $rr(t)$ when $a[n] = \delta[n] + \delta[n - 3]$.

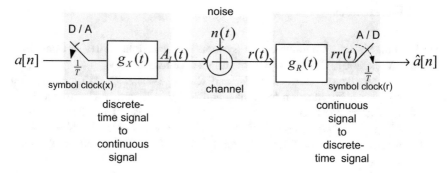

Fig. P3-1: Baseband digital transmission system under additive noise

Fig. P3-2: A transmit pulse

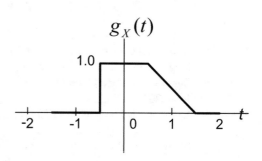

(3) What is $h(t)$ in the expression of $rr(t) = \sum\limits_{n=-\infty}^{+\infty} a[n]h(t - nT)$ in terms of $g_X(t)$ when a matched pair of filters is used?

P3-3: In Fig. P3-2 a transmit pulse is shown, which may be used in Fig. P3-1.

(1) Is this $g_X(t)$ causal? If not find a causal one and draw it.
(2) Find a matched filter for $g_X(t)$, and then obtain a causal one if not causal and sketch it.
(3) Find the end to end pulse for $g_X(t)$ and sketch it.

P3-4: A rectangular pulse is used in Fig. P3-1 as a transmit pulse. It can be expressed by unit step as $rec(t) = u(t) - u(t - T)$.

(1) Find a matched pulse for this rectangular pulse, $rec(-t)$, and sketch it. Is this matched filter causal? If not, find a causal one. Is it unique? If not, what other possibilities exist?
(2) Make $rec(-t)$ causal with the minimum delay.

Fig. P3-3: An example of mismatch loss and ISI

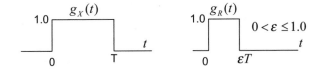

Fig. P3-4: DC zero pulse

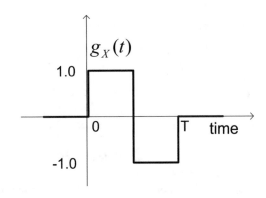

P3-5: In Fig. P3-3 a pair of pulses is shown.

(1) Find the mismatch loss in dB when $\varepsilon = 0.75$.
(2) A system has SNR = 45 dB. What is the SNR if the receive pulse width is 0.1T.
(3) Discuss what the problem is when the receive pulse is wider than T, say 10%?

P3-6: In Fig. P3-3, the shortened pulse ($g_R(t)$) is used at transmit side, and $g_X(t)$ at the receive side. Note that the transmit symbol rate is still 1/T.

(1) Find the mismatch loss in dB when $\varepsilon=0.75$.
(2) If $0 < \varepsilon \leq 1.0$, is there ISI ? Hint: consider the end to end pulse.

P3-7: An ideal LPF with the bandwidth 1/T is used at the transmit side, and an ideal LPF with the bandwidth 1.2 1/T (i.e., 20% wider) at the receive side.

(1) What is SNR loss compared to the matched case?
(2) Is there ISI in this case?

P3-8: Any time-limited pulse less than a symbol period and its time reversed version will form a matched filter pair. The convolution of this pair will be an end to pulse and will be even in time. Since it is time limited less than two symbol periods, it is a Nyquist pulse with even symmetry.

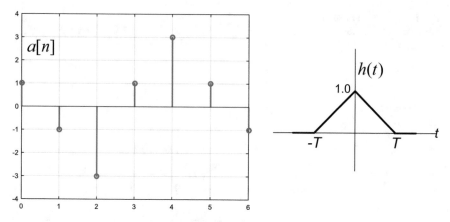

Fig. P3-5:   Input sequence and a triangular Nyquist pulse

(1) Obtain a causal matched filter with minimum delay corresponding to Fig. P3-4.

(2) Find the corresponding end to end pulse and sketch it.

P3-9: Any time-limited pulse less than two symbol periods is a Nyquist pulse. Is this assertion true? If yes, provide an example.

P3-10: An end to end impulse response, $h(t) = g_X(t) \otimes g_R(t)$, is a Nyquist pulse if its sampled version is $h(nT) = \delta[n]$, i.e., $h(nT) = 1.0$, $n = 0$ and $h(nT) = 0.0$ $n \neq 0$.

(1) This implies that $h(t)$ has periodic zeros every T (symbol period) except for $t = 0$. Construct examples of it.

(2) If $h(t) = h(-t)$, i.e., even, is imposed, construct examples.

P3-11: Show that a sampled impulse response can be expressed by $h(nT) = h(t) \sum\limits_{n=-\infty}^{+\infty} \delta(t - nT)$, and express its Fourier transform as a convolution of $H(f)$ and $FT\left\{ \sum\limits_{n=-\infty}^{+\infty} \delta(t - nT) \right\}$. Thus show that Fourier transform of a Nyquist pulse should meet the condition of $\sum\limits_{k=-\infty}^{+\infty} \frac{1}{T} H\left(f - \frac{k}{T} \right) = 1.0$.

P3-12: Show that $\sum\limits_{k=-\infty}^{+\infty} H\left(f - \frac{k}{T} \right)$ is periodic with $1/T$ where k is an integer.

P3-13: A triangular pulse, in Fig. P3-5 is an end to end Nyquist pulse.

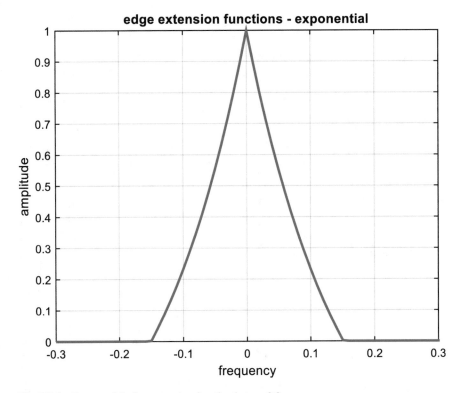

Fig. P3-6: Exponential edge extension function beta $= 0.3$

(1) Compute the output $x(t) = \sum_{n=-\infty}^{+\infty} a[n]h(t - nT)$

(2) Confirm that the output is a linear interpolation between points.

P3-14: If a Nyquist pulse is bandlimited by twice symbol rate, the frequency magnitude has a vestigial symmetry at the half of symbol rate, called Nyquist frequency. One example is a set of trapezoidal frequency response. Plot the frequency response with the excess bandwidth (β), $\beta = 0.2$, 0.4, 0.6, and 0.8, and corresponding impulse responses. Hint: see Fig. 3-11.

P3-15: Vestigially symmetrical Nyquist filters in frequency domain can be expressed by a convolution of the ideal LPF of bandwidth $1/T$, and an even symmetric function with bandwidth $\frac{\beta}{T}$, called edge extension function. An example is shown in Fig. P3-6, and it is expressed by an exponential function as,

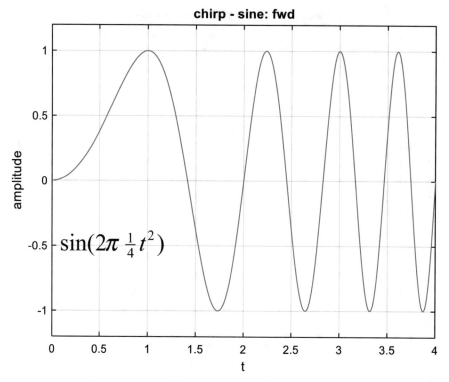

Fig. P3-7: Chirp signal

$$edge(f) = c_1 + c_2 \exp\left(-\frac{|fT|}{0.5\beta}\right)$$

where c_1 and c_2 are determined by $edge(|f| \geq 0.5\beta) = 0$, and $edge(0) = 1.0$.

(1) Find c_1 and c_2 so that to meet $edge(0) = 1.0$, and $edge(\pm 0.5\beta) = 0$.
(2) Plot (sketch) overall Nyquist pulses with $\beta = 0.2$, 0.4, and 0.6.

P3-16: PD (peak distortion) is defined as $\text{PD} = \sum_{n \neq 0} |h_n^k| / |h_0^k|$, sampled at kth sampling

offset phase. Find PD for 5% offset sampling of ideal LPF, $h(t) = \frac{\sin\left(\frac{\pi t}{T}\right)}{\frac{\pi t}{T}}$ with

$|t| \leq 3T$, i.e., 6 ISI terms.

P3-17: Repeat P3-16 with the raised cosine pulse $\beta = 0.5$ and compare with the

result of P3-16. Hint: $h(t) = \frac{\sin\left(\frac{\pi t}{T}\right)}{\frac{\pi t}{T}} \frac{\cos\left(\pi\beta t/T\right)}{1-(2\beta t/T)^2}$.

P3-18: A pulse at the transmit side is given by $g_X(t) = \sin\left(2\pi\frac{1}{4}t^2\right), 0 \leq t \leq 4$ which is shown in Fig. P3-7. Its symbol period is $T = 4$. This type of signal is called a chirp signal used in radar context. Find a matched pair for this signal and plot. Hint: use a computer program to plot $f(t) = \sin\left(2\pi\frac{1}{4}t^2\right)$ in the range of $-4 \leq t \leq 4$.

Chapter 4
Radio Propagation and RF Channels

Contents

4.1	Path Loss of Radio Channels	186
	4.1.1 Free Space Loss	186
	4.1.2 Pathloss Exponent	187
4.2	Antenna Basic and Antenna Gain	188
	4.2.1 Antenna Basics	188
	4.2.2 Antenna Pattern	189
	4.2.3 Directivity and Antenna Gain	189
	4.2.4 Aperture Concept and Antenna Beam Angle	190
4.3	Path Loss Due to Reflection, Diffraction and Scattering	191
	4.3.1 Ground Reflection (2-Ray Model)	192
	4.3.2 Diffraction, Fresnel Zone and Line of Sight Clearance	194
	4.3.3 Examples of Empirical Path Loss Model	195
	4.3.4 Simplified Pathloss Model	195
	4.3.5 Shadow Fading: Variance of Path Loss	196
	4.3.6 Range Estimation and Net Link Budget	199
4.4	Multipath Fading and Statistical Models	202
	4.4.1 Intuitive Understanding of Multi-Path Fading	202
	4.4.2 Rayleigh Fading Channels	204
	4.4.3 Wideband Frequency Selective Fading	209
	4.4.4 Alternative Approach to Fading Channel Models	214
4.5	Channel Sounding and Measurements	220
	4.5.1 Direct RF Pulse	221
	4.5.2 Spread Spectrum Signal (Time Domain)	222
	4.5.3 Chirp Signal (Frequency Sweep Signal) for Channel Sounding	224
	4.5.4 Synchronization and Location	225
	4.5.5 Directionally Resolved Measurements (Angle Spread Measurements)	225
4.6	Channel Model Examples	226
	4.6.1 Empirical Path Loss Models	226
	4.6.2 M1225 of ITU-R Path Loss Models	227
	4.6.3 Multipath Fading Models in Cellular Standards	227
	4.6.4 Cellular Concept and Interference Limited Channels	228
	4.6.5 Channel Models of Low Earth Orbit (LEO) Satellite	236
4.7	Summary of Fading Countermeasures	240
4.8	References with Comments	242
4.9	Problems	243

© Springer Nature Switzerland AG 2020

S.-M. Yang, *Modern Digital Radio Communication Signals and Systems*,

https://doi.org/10.1007/978-3-030-57706-3_4

Abstract This chapter covers RF channel models including propagation and antenna. From modem point of view all the components in RF, e.g., LO, LNA, and PA are part of channel. However, this chapter covers antenna, propagation media, frequency selective fading due to surrounding terrains and obstacles, and time selective fading due to the movement of transmitter and receiver. The measurement of RF channel is covered briefly as well. Channel models used in standards and LEO channel models are included as well.

Key Innovative Terms LEO fading channel model due to satellite motion · channel model with scattering description · use of repeaters and distributed antennas · summary of pathloss and fading channel counter measures

General Terms antenna gain · antenna pattern · aperture · beam angle · channel sounding · chirp signal · cellular frequency reuse · diffraction · directivity · Doppler spectrum · free space loss · fresnel zone · GSM channels · line of sight · link budget · pathloss exponent · Rayleigh fading · reflection · scattering · shadowing · solid angle · statistical models · thermal noise

Abbreviations

3G, 4G	3rd (4th) generation cellular system
AM	amplitude modulation
AWGN	additive white Gaussian noise
BICM	bit interleaved coded modulation
BW	bandwidth
FEC	forward error correcting codes
GSM	global system mobile
IEEE	institute of electrical and electronics engineers
LAN	local area network
LDPC	low density parity check code
LEO	low earth orbit
LNA	low noise amplifier
LOS	line of sight
LPF	low pass filter
MIMO	multi input multi output (antenna)
MLCM	multilevel coded modulation
NF	noise figure
OFDM	orthogonal frequency division multiplex
OOK	on off keying
PA	power amplifier
PDF	probability density function
PN	pseudo random noise
QAM	quadrature amplitude modulation
RCV	receive

RF   radio frequency
RHS   right hand side
RX   receive
TCM   trellis coded modulation
TX   transmit
VCO   voltage controlled oscillator
WiMAX   wireless microwave access

In this chapter we consider a major part of wireless systems, namely RF system and radio propagation channels and in particular, our focus is on radio propagation channel. However, from the digital modem point of view, RHS of (C –C) in Figure 4-1 may be treated as part of a channel. A real carrier wave signal, not complex, after power amplification, goes through an antenna, then passes through a radio channel and a receiving antenna captures it, amplified by LNA (low noise amplifier).

For example, a most basic channel model is additive Gaussian noise channel we often used in other parts of this book. This stems from the thermal noise due to random motion of electrons when a radio wave is converted to electric currents. In order to minimize additional thermal noise from the amplifying devices in a system, a signal should be amplified large enough so that the additional noise is relatively insignificant, i.e., LNA is necessary; inevitably LNA adds some noise of its own, which should be minimum possible. Other examples are PA's nonlinearity and the spectral purity of RF oscillators or phase noise of an oscillator. These issues will be treated in Chap. 8 and we focus on radio propagation channels including antennas.

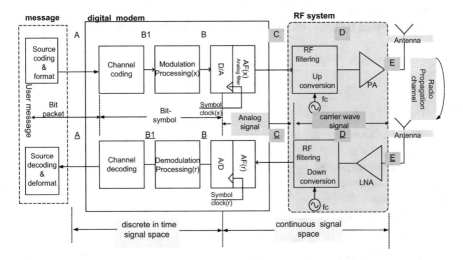

Figure 4-1: A wireless system block diagram emphasizing RF system and radio propagation channels with carrier frequency oscillator (f_c), antenna, PA and LNA

4.1 Path Loss of Radio Channels

Electromagnetic wave propagations and antenna is a wide, complex field of its own right, and there are many experts in the field. It can be very complex depending on the propagation environments. However, for our purpose, the mode of propagation is largely captured by reflection, diffraction and scattering. In wave propagations when there are the objects along the path of propagation, there it will interact with them. When the object is large compared to the wave length of a carrier, it will be reflected and when it is small, it will be scattered. If the object is partially blocking the path it will diffract. These will be discussed further in this chapter.

We need some models simple enough, yet capturing real situations accurately for our purpose of using radio channels for digital communications. Speaking of wireless channels, the underwater acoustic channels are also of no wire, but our focus will be wireless radio channels.

Even we focus on radio channels with RF frequency less than 100GHz (millimeter wave), there are tremendously diverse possibilities from high speed mobile (poor) to line of sight microwave (good), satellite, deep space communications and so on. Our major focus is on terrestrial mobile systems including cellular, 3G/4G, WiMAX, wireless LAN and LOS microwave but does not exclude other possibilities, e.g., LEO satellite channels.

4.1.1 Free Space Loss

We start with the simplest possible case, free space propagation. There is a transmit antenna and a receive antenna. It is shown in Figure 4-2.

It is described by Friis equation, which will be justified in the next section,

$$P_R = P_X\, G_X\, G_R\, \frac{\lambda^2}{(4\pi\, d)^2} \qquad (4\text{-}1)$$

where

d: distance from transmitter to receiver in meter [m]
P_X: transmit power [watts]
P_R: receive power [watts]
G_X: transmit antenna gain
G_R: receiver antenna gain
$\lambda = c\,\frac{1}{f}$ wave length of a carrier frequency
c: speed of light 3×10^8 m/s and f: carrier frequency in Hz.

The wave length is the distance that a wave propagates in one period, which is $T = \frac{1}{f}$. For example, $f = 1\,\text{GHz}(10^9\,\text{Hz}) \longleftrightarrow \lambda = 0.3\,\text{m}$ (30 cm).

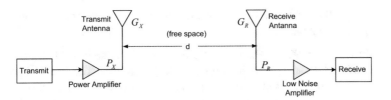

Figure 4-2: Free space propagation path loss with distance d, antenna gain G_X and G_R and with transmit power P_X and receive power P_R

The equation (4-1) captures the free space propagation in a simple but useful way by using the concept of antenna gain, which will be explained in the next section. The factor $\frac{\lambda^2}{(4\pi d)^2}$ in (4-1) is called 'path loss'. This loss is due to the fact that the transmit power is spread to the larger area with a longer distance. The area is the surface area of a sphere with the radius of d, i.e., $4\pi d^2$.

The dB of a ratio of power is defined as x [dB] $= 10\log_{10}$ (x). For example, x $= 10^6$, it is 60 dB. The use of dB is convenient when the range of x is very large. The dBm is often used as well, and its reference power is 1 mili-watt, i.e., 0 dBm $= 10^{-3}$ watts. 30 dBm is 1 watt. Antenna gain is also expressed in dB as well.

The path loss in dB is given by

$$\text{Pathloss [dB] in free space} = 20 \; \log \; (d) + 20 \; \log \; (4\pi/c) + 20 \; \log \; (f).$$

For example, 2 GHz carrier with the distance 1 km will have a path loss of 98.4 dB ($=60 - 147.6 + 186$).

Another example is a carrier 1 GHz with the distance 1 km will have a path loss of 92.4 dB ($=60 - 147.6 + 180$).

Exercise 4-1: find a path loss with the carrier frequency 1 GHz and distance 10 km. Answer: pathloss $= 92.4$ dB $+ 20 \log (10/1) = 102.4$ dB.

4.1.2   Pathloss Exponent

The path loss is expressed as

$$\text{Pathloss[dB]} = A + \gamma 10 \; \log \left(\frac{d}{d_o}\right) \tag{4-2}$$

where A $= 20 \log (d_o) + 20 \log (4\pi/c) + 20 \log (f)$, path loss at a reference distance (d_o) and for a given carrier frequency (f). In (4-2) γ is called a pathloss exponent. It is $\gamma = 2$ with free space. Other than free space the pathloss exponent is 2 or larger. This exponent concept will be used for propagation environments in general. For example, in cellular $\gamma = 3.8$ or 4 is typical and $\gamma = 5$ or so for indoor propagation. The bigger the exponent is the more loss with the same distance.

Example 4-1: $f = 900$ MHz, $d = 100$ m: omni-directional antenna at receiver ($G_R = 0$ dB) 18 dB antenna gain at transmit ($G_T = 18$ dB). The received power is $P_R = 1$ uW (-30 dBm) and then what is the transmit power P_T in dBm?

Answer: Express (4-1) in terms of dB.

$$P_R[\text{dBm}] = -30\text{dBm} = P_T[\text{dBm}] + 0 + 18 - \text{pathloss [dB]}$$

where pathloss [dB] $= 40 - 147.6 + 179 = 71.5$ dB. Thus $P_T = 23.5$ dBm.

4.2   Antenna Basic and Antenna Gain

The derivation of (4-1) requires the understanding of antenna gain.

4.2.1   Antenna Basics

Antenna is a transducer between a guided wave (cable, wire and wave guide) and free space (empty space) or vice versa. There are many, infinitely varied types are possible. For example, dipole, helical, micro strip (on printed circuit board), patch, horn, parabola and etc. In fact, an ear is an acoustic wave antenna, and it is more sensitive with a larger size. Of course, our focus is on electromagnetic propagation. A simple form of antenna is a dipole antenna shown in Figure 4-3.

In antenna description we often use a polar coordinate of (θ, φ, r). Its definition is shown in Figure 4-4.

Figure 4-3: A dipole antenna with a half wave length

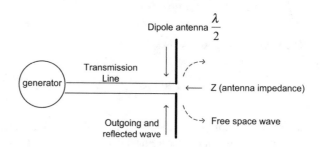

Figure 4-4: Polar
(spherical) coordinate
system

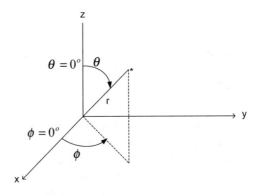

4.2.2 Antenna Pattern

Antenna patterns are described by field patterns, $E_\theta(\theta, \varphi)$, $E_\varphi(\theta, \varphi)$ or by power per unit area [watt/m^2], $S(\theta, \varphi)$. They are related as,

$$S(\theta, \varphi) = \left[E_\theta^2(\theta, \varphi) + E_\varphi^2(\theta, \varphi) \right] / Z_0 \qquad (4\text{-}3)$$

where Z_0 = free space impedance $377\Omega \left(= \sqrt{\frac{\mu_0}{\varepsilon_0}} \right)$.

It is sometimes convenient to use a normalized one as $P_n = S(\theta, \varphi)/S(\theta, \varphi)_{max}$.

4.2.3 Directivity and Antenna Gain

Although the radiation characteristics of an antenna involve 3-dimendsional patterns, many important radiation characteristics can be expressed in terms of single valued quantities such as beam width, main lobe area, directivity, antenna gain, aperture and so on. These will be explained as we go along.

Beam area in solid angle (Ω_A) is an extension of arc angle in a circle and summarized in a table below.

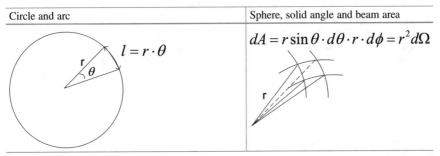

Circle and arc	Sphere, solid angle and beam area
$l = r \cdot \theta$	$dA = r \sin \theta \cdot d\theta \cdot r \cdot d\phi = r^2 d\Omega$

(continued)

Circle and arc	Sphere, solid angle and beam area
Circumference $= 2\pi \cdot r$	Area of sphere $= 4\pi \cdot r^2$
Angle of circle $= 2\pi$	Solid angle of sphere $= 4\pi$
$360°$ degree	$\left(\dfrac{180°}{\pi}\right)^2 \cdot 4\pi = 41253°$ Square degree

The directivity is defined and further antenna gain is defined as follows.

Directivity is defined as $D = 4\pi/\Omega_A$ where Ω_A is the total beam area in solid angle. Ω_A is further expressed as $\Omega_A = \Omega_M + \Omega_m$ with M means main lobe and m: minor lobe. Antenna efficiency factor is defined as $k = \Omega_M/\Omega_A$.

Antenna gain is defined as $G = k\,D$.

The smaller the beam area in solid angle (Ω_A) is the greater the directivity. When the power radiation is isotropic ($\Omega_A = 4\pi$), then $D = 1$ (0 dB). Note that Ω_A is not greater than 4π. Effective isotropic radiated power (EIRP) is given by $(EIRP) = P_X\,G_X$ with an antenna gain G_X and transmit power P_X.

Example 4-2: antenna beam angle of half power is given by $\theta_{HP}° = 3°$ and $\varphi_{HP}° = 120°$. What is the directivity?

Answer: $D = \dfrac{4\pi}{\Omega_A} \approx \dfrac{4\pi}{\theta_{HP}\cdot\varphi_{HP}} = \dfrac{41253}{\theta_{HP}°\cdot\varphi_{HP}°} = 114.6 (\text{or } 20.6 \text{dB}_i)$

Example 4-3: antenna beam angle of half power is given by $\theta_{HP}° = 3°$ and $\varphi_{HP}° = 120°$, the same as the previous example. Its antenna efficiency is $k = 0.6$ What is the antenna gain?

Answer: $G = k\,D = 0.6 * 114.6$ (18.3 dB). This is a typical 3-sector base station antenna gain.

4.2.4 Aperture Concept and Antenna Beam Angle

From Figure 4-5, the radiated power is Poynting vector [watt/m²], $\dfrac{|E_a|^2}{Z}$, times the area (A) and the same power is focused in the beam area ($r^2\Omega_A$) at the radial distance r. This is shown in (4-4) below.

Figure 4-5: Radiation from aperture area (A) with uniform field E_a through solid beam angle Ω_A

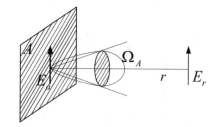

$$P = \frac{|E_a|^2}{Z} A = \frac{|E_r|^2}{Z} r^2 \Omega_A \tag{4-4}$$

It may be shown that $|E_r| = |E_a| \cdot \frac{A}{r\lambda}$ and it is not obvious but we omit a demonstration. It should be accepted as is.

Using this relationship we obtain the desired relationship. Thus the aperture area is related to the antenna pattern (solid angle):

$$\Omega_A = \frac{\lambda^2}{A} \tag{4-5}$$

Now the antenna gain is expressed as $G = kD = k4\pi/\Omega_A = kA\ 4\pi/\lambda^2$

$$G = A_e 4\pi / \lambda^2 \tag{4-6}$$

with $A_e = k\ A$.

Friis equation of (4-1) is obtained by using (4-6). The EIRP ($P_T\ G_X$) is reduced by $4\pi d^2$ with the distance d. This power density is picked up by a receiver antenna with effective aperture area of $A_e = G_R \lambda^2 / 4\pi$,

i.e., $P_R = P_T G_X / 4\ \pi d^2 A_e = P_T G_X / 4\ \pi d^2 G_R \lambda^2 / 4\pi = P_T G_X G_R \lambda^2 / (4\pi d)^2$.

Thus (4-1) is obtained.

Exercise 4-2: Show that the antenna size is proportional to the wave length or inversely proportional to the carrier frequency.

Answer: Use (4-6) or $A_e = G_R \lambda^2 / 4\ \pi \propto L^2$ where L is the size of an antenna, and $\lambda = \frac{c}{f}$.

4.3 Path Loss Due to Reflection, Diffraction and Scattering

Reflection occurs when a propagating electromagnetic wave impinges upon an object which has very large dimensions when compared to the wavelength of the propagating wave. Reflections occur from the surface of the earth and from buildings and walls; reflection from dielectrics and reflection from conductors:

Diffraction occurs when the path between the transmitter and receiver is obstructed by a surface that has sharp irregularities (edges). The secondary waves resulting from the obstructing surface are present throughout the space and even behind the obstacles, giving rise to a bending of wave around the obstacle, even when a line-of-sight path does not exist between transmitter and receiver. At high frequencies, diffraction like reflection depends on the geometry of the object, as well as the amplitude, phase, and polarization of the incident wave at the input of diffraction.

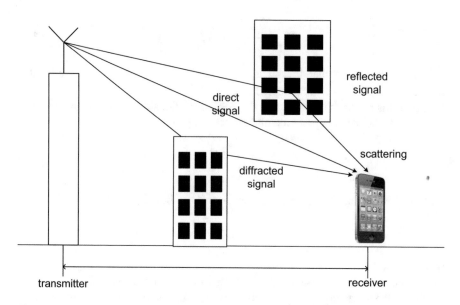

Figure 4-6: Direct signal, reflected signal, diffracted signal and scattering

Scattering occurs when the medium through which the wave travels consists of objects with dimensions that are small compared to the wavelength, and where the number of obstacles per unit volume is large. Scattered waves are produced by rough surfaces, small objects, or by other irregularities in the channel. In practice, foliage, street signs, and lamp posts induce scattering in a mobile communication system. These propagation mechanisms are schematically shown in Figure 4-6.

4.3.1   Ground Reflection (2-Ray Model)

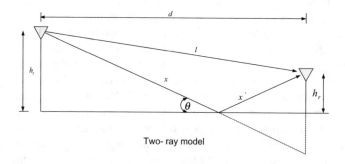

$$P_R = P_X G_X G_R \left[\frac{\lambda}{4\pi}\right]^2 \left|\frac{a_l}{l} + \frac{a_r e^{-j\Delta\varphi}}{x+x'}\right|^2 \tag{4-7}$$

$$\Delta\varphi = 2\pi(x + x' - l)/\lambda$$

$$x + x' - l = \sqrt{(h_t + h_r)^2 + d^2} - \sqrt{(h_t - h_r)^2 + d^2}$$

a_l and a_r are reflection coefficients with $a_l = 1$ and $a_r = -1$ for simplicity.

Exercise 4-3: Write a computer program (Matlab program and graph) for the computation of (4-7) with a graphical example of f = 900 MHz, $h_t = 50$ m, $h_r = 2$ m.

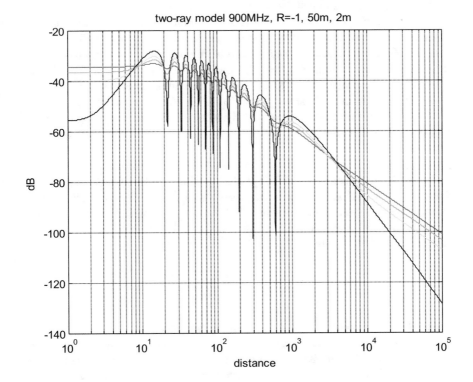

For large d, $x + x' \approx d$, $l \approx d$, $\theta \approx 0$, it can be shown that $P_R = P_X G_X G_R \left[\frac{\lambda}{4\pi \cdot d}\right]^2 \left[\frac{4\pi h_t h_r}{\lambda d}\right]^2$.

Exercise 4-4: work out the above in detail.
We obtain the desired result as,

$$P_R = P_X G_X G_R \left[\frac{h_t h_r}{d^2} \right]^2 \tag{4-8}$$

Compare (4-7) with (4-1). The path loss is independent of frequency and the path loss exponent is 4 rather than 2. In dB form, (4-8) can be put into $P_R\text{dBm} = P_X\text{dBm} + 10\log_{10}(G_X G_R) + 20\log_{10}(h_t h_r) - 40\log_{10}(d)$

4.3.2 Diffraction, Fresnel Zone and Line of Sight Clearance

Diffraction effects will be insignificant if obstructions are kept outside a volumes of revolution around a radio path know as a Fresnel zone.

Criteria for avoiding diffraction effects are normally based upon an exclusion volume in 3-dimensional space around the (normally line-of-sight) radio path of a fixed link. Such a volume is defined in terms of Fresnel zones. The n-th Fresnel is the locus of all points for which, if the radio signal traveled in a straight line from the transmitter to the point and then to the receiver, the additional path length compared to the straight transmitter-receiver path equals $n\lambda/2$, where λ = wavelength.

For large static obstructions, particularly terrain, a criterion requiring 0.6 of the first Fresnel zone radius to be unobstructed is commonly used. This should be calculated for the atmospheric refractivity gradient exceeded for perhaps 99% of an average year.

For the varying geometry it will be prudent to adopt a more conservative criterion than 0.6 of the 1st Fresnel Zone. It is suggested that to define a wind-turbine exclusion zone equal to the complete 2nd Fresnel zone would be realistic. The radius of this zone around the direct line-of-sight path of a radio link is given to an adequate approximation by:

$$R_{F2} = \sqrt{\frac{2\lambda d_1 d_2}{d_1 + d_2}} \tag{4-9}$$

where: d_1, d_2 = distances from each end of the radio path.

Figure 4-7 illustrates the general form of the zone produced by equation (4-9). The definition of Fresnel zone is based upon a fixed path difference between the direct and indirect paths between transmitter T and receiver R, which consists of an

Figure 4-7: Fresnel zone and line of sight clearance

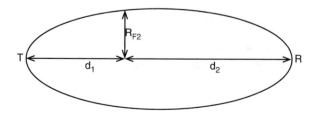

ellipse with T and R at the foci. As stated above, equation (4-9) is an approximation which clearly fails in the vicinity of the antennas. However this is not important since clearance from the antennas will be covered in any case by the other two criteria. Equation (4-9) is the normal method for computing Fresnel clearance around radio paths and is adequate for our present purposes.

Equation (4-9) thus provides a lateral clearance distance to be applied along a radio path. Although it should strictly be applied in 3-dimensional space, it will in most cases be adequate to apply it horizontally each side of the path of a fixed radio link.

It can be noted that the Fresnel clearance zone is a function of wavelength and path length only. It does not depend upon the antenna characteristics.

4.3.3 Examples of Empirical Path Loss Model

There are a number of empirical path loss models available in the literature. These are essentially obtained from field measurements followed by a modeling based on a theory and curve fitting.

Examples are (also repeated in Sect. 4.6):

- Hata: IEEE Trans. Veh. Tech. pp317 -325 August 1980
- Cost231: Euro –cost231 project 1991
- Erceg: IEEE J. Selected Areas of Communication pp. 1205-11 July 1999
- Walfish: IEEE Transaction Antenna and propagation pp. 1788-96, Oct 1988

4.3.4 Simplified Pathloss Model

$$\text{Pathloss} = A[dB] + 10\gamma \cdot \log_{10}\left(\frac{d}{d_o}\right) + s[dB] \qquad (4\text{-}10)$$

$A = -20\log_{10}\frac{\lambda}{4\pi\, d_o}$ free space loss in dB at distance d_o

d_o: reference distance, say 100 m or even smaller if necessary

γ: path loss exponent, derive from measurements by least mean square fit

s: shadow fading variance

The simplified model of pathloss in (4-10) is inspired by (4-2) form of free space loss. Additional term s, shadowing fading, is added to accommodate a random variation due to topographic blockings (say inside a building). The pathloss exponent, γ, is determined empirically, rather than 2 of free space.

Example 4-4: estimate a pathloss exponent γ from measured data

f = 900 MHz, d_o = 1 m, A = 20 log (3e8/900 e6)/4 π = − 31.54 dB

Data:

Distance	10 m	20 m	50 m	100 m	300 m
pathloss	70 dB	75 dB	90 dB	110 dB	125 dB

Solution: pathloss in $dB = A[dB] + 10\gamma \cdot \log_{10}\left(\frac{d}{d_o}\right)$

$$70 = 31.54 + \gamma * 10$$
$$75 = 31.54 + \gamma * 13$$
$$90 = 31.54 + \gamma * 16.9$$
$$110 = 31.54 + \gamma * 20$$
$$125 = 31.54 + \gamma * 24.7$$

Find γ for least square fit of the above data where ',' means transpose,

$$b = [38.46\ \ 43.46\ \ 58.46\ \ 78.46\ \ 93.46]'$$
$$a = [\,10.00\ \ \ 13.00\ \ \ 16.90\ \ \ 20.00\ \ \ 24.70\,]'$$
$$\gamma = x = a'b/a'a = 5815.2/1564.7 = 3.716 \ \text{(least square fit of γ)}$$

The error (variance) estimation since e = b − γ * a

$$e = [\,1.2950\ \ \ -4.8545\ \ \ -4.3488\ \ \ 4.1300\ \ \ 1.6625\,]'$$
$$e * e' = 63.97 \rightarrow \text{sqrt}(63.97/5) = 3.577\,\text{dB} \ \text{(variance)}$$

Least square fit in general

In general, A matrix [N, M] x vector [M] b vector [N] N > M

Ax = b → N > M means more equations than variable, called over-determined
Ax-b = error sum of (error)^2 is minimized
A'A x = A'b

solve for x in the above and it is a least square fit solution

4.3.5   Shadow Fading: Variance of Path Loss

PDF of shadow fading is represented by a log normal, which is Gaussian normal distribution but in dB (one dimensional)

A log-normal PDF is given by,

$$f_x = \frac{1}{\sqrt{2\pi\sigma^2}} e^{-\frac{(x-m)^2}{2\sigma^2}} \tag{4-11}$$

where x, m, σ are in dB and a random variable, mean, and variance respectively. In practice it is used often since it approximates real situations well for shadow fading.

As the name implies, this variance is largely due to shadowing because of some obstructions like buildings, or mountains. Inside a car or building will increase path loss, and additional margin for signal strength must be considered. This is called fade margin.

Review of Q(x)

$$Q(x) \equiv P_r(X \geq x) = \int_x^\infty \frac{1}{\sqrt{2\pi}} e^{-y^2/2} dy$$

$$Q(x) = \frac{1}{2} erfc\left(\frac{x}{\sqrt{2}}\right) = \frac{1}{2}\left[1 - erf\left(\frac{x}{\sqrt{2}}\right)\right]$$

$$Q(x) = 1 - Q(-x)$$

Example 4-5:

$$m = 5 \text{ dB} \quad s = 10 \text{ dB}$$

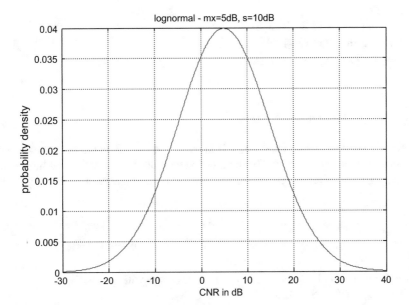

If $m = 5$ dB is used as SNR, then 50% of the time, SNR will be less than mean (5 dB)

In order to have a 90% of time what is the usable SNR dB?

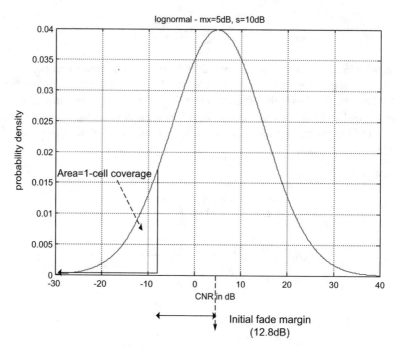

Solution:
fade margin = 12.8 dB (or 10* 1.28)
Solve x for $0.9 = Q\left(\frac{x}{\sigma}\right)$ with $\sigma = 10$ dB
$x = \sigma * Q^{-1}(0.9) = \sigma * 1.28$

4.3.5.1   Fade Margin and Reliability

When the shadow fading variance, s, is known, one can allow the fade margin with
the outage requirement as in Example 4-5. The table below will show how much
fade margin for different reliability. 90% reliability the fade margin needed is 1.28 s
[dB].

80%	84%	90%	99%	99.9%	99.99%	reliability
a = 0.845	1.0	1.282	2.33	3.08	3.72	a × s dB

Exercise 4-5: Find a fade margin for 99% coverage with s = 5 dB
     Answer: 3.08 × 5 = 15.4 dB

4.3.6   Range Estimation and Net Link Budget

A point to point link range will be estimated based on the pathloss model of (4-10) as expressed, $\text{pathloss} = A[dB] + 10\gamma \log_{10}\left(\frac{d}{d_o}\right) + s[dB]$. Specifically, we choose A [dB] to be free space loss at d_o and a model is adjusted to match observed data by using the pathloss exponent, γ, and the variance, s, for a given situation.

A system (end to end) should provide a gain to cope with the pathloss with adequate margin, fade margin, to meet reliability requirement (outage requirement).

Antenna, transmit and receive sides, can provide a gain. Antenna gain is essentially increased by narrowing the beam width, or increasing the antenna aperture area. It is given by $G = A_e 4\pi/\lambda^2$ and it is noted that the smaller the wavelength, i.e., the higher the carrier frequency is the greater the gain.

We introduce the concept of system gain as,

$$\text{system gain [dB]} = \text{transmit power[dBm]} - \text{receiver threshold [dBm]}.$$

It combines the transmit power and the receiver's sensitivity. For example, transmit power is 21 dBm, and the receiver can receive a signal with certain error probability as small as -59 dBm. Then the system gain is 80 dB. Transmit power is measured at the antenna port where it is connected, and it is a function of PA and other system considerations. The receiver threshold is a function of the receiver front end thermal noise and noise figure of LNA. Furthermore it is a function of modulation and coding as well as channel multipath fading. For our range estimation purpose we summarize it into a single number.

4.3.6.1   Thermal Noise

Thermal noise is present at the receiver front end once a signal is converted to electric current due to thermal random fluctuation of electrons. It may be characterized as a random process with the average being zero and its power spectrum is flat (white) with $S_n(f) = kT$ [watt/Hz] where k is Boltzmann constant 1.38×10^{-23} joule/degree [Kelvin] and T is temperature in $°K$ ($= °C + 273$). Thus the thermal noise per Hz at room temperature (20 $°C$), $293 \times 1.38 \, 10^{-23}$ watt/Hz, which is -174 dBm/Hz.

When the bandwidth of a system is 1 MHz, then the noise power is -114 dBm as used in the above example.

Exercise 4-6: Find out the definition of noise figure (NF). If an LNA has a NF of 6 dB, the thermal noise floor is increased by that amount, -174 dBm $+$ 6 dB per Hz.

Example 4-6: 100 Mbps data rate modem requires 20 dB SNR (Eb/No) at BER @1E-6 and the LNA noise figure is 5 dB. Thermal noise density is -114 dBm per MHz. What is the receiver threshold?
Answer: -114 dBm $+$ 20 dB (100 MHz) $+$ 20 dB (Eb/No) $+$ 5 (NF) $= -59$ dBm.

If there is a multipath fading in the channel due to delay spread and movement of devices, it may degrade the receiver threshold, i.e., higher SNR. The threshold degradation due to fading is captured as part of system gain degradation.

Net link budget = system gain (transmit power − receiver threshold) + antenna gain (transmit + receive) − fade margin (outage target) should balance out the pathloss.

Once net link budget is known from system gain, antenna gain and fade margin a link distance can be estimated.

4.3.6.2   Example of Range Estimation

Path loss model (ITU-R)

$$L = 40\left(1 - 4 \times 10^{-3}\Delta h_b\right) \log_{10}R - 18 \ \log_{10}\Delta h_b + 21 \ \log_{10}f + 80 \qquad \text{(a)}$$

Assuming the antenna height $\Delta h_b = 15$m becomes

$$L = 37.6 \ \log_{10}R + 58.83 + 21 \ \log_{10}f \qquad \text{(b)}$$

With f = 1850 (mobile test frequency 1850 MHz),

$$L = 37.6 \ \log_{10}R + 127.4 \qquad \text{(c)}$$

R in km.

There is another model by Erceg (IEEE JSAC vol. 17 No.1, July 1999 pp. 1205–1211)

$PL = 10\gamma \ \log_{10}\left(\frac{d}{d_o}\right) + A$ where d_o is reference distance so that.

$$A = 20\log_{10}\left(\frac{4\pi d_o}{\lambda}\right)$$

With 1850 MHz,

$$PL = 10\gamma \ \log_{10}\left(\frac{d}{d_o}\right) + 118 + s(= 10\text{dB})\text{with}\, d_o = 1 \text{ km}.$$

You can choose γ to approximate (c). *By choosing log-normal shadow fading variance 10 dB and $\gamma = 3.76$, the match is very good.*

We use (c) to estimate the distance.

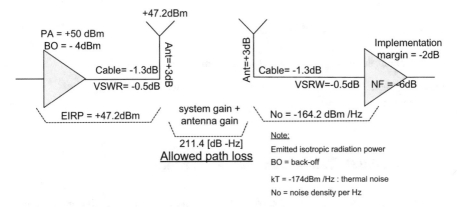

Figure 4-8: Net link budget including antenna gain, for a given RF blocks excluding the required SNR and data rate expressed C/No in [dB-Hz]

Net link budget

We assume the following parameters on transmit power and receiver noise figure and antenna gain as in Figure 4-8.

Transmit power + 47.2 dBm = 50 − 4 − 1.3 − 0.5 + 3

Receive threshold −164.2 dBm = −174 + 1.3 + 0.5 + 6 + 2

Total system gain (with antenna gain) 211.4 dB-Hz = 47.2 − (−164.2)

Estimated distance range

In order to estimate a distance we need to solve the following

$$L ==37.6 \log_{10}R + 127.4 = \text{total system gain} - C/No[dB\text{-}Hz] - \text{fade margin}$$
$$=211.4 - 80 - 0 \, (C/No = 80\,dB - Hz, 10MHz \text{ with } 10 \text{ dB SNR})$$

$$R = 1.27\,km.$$

Exercise 4-7: Estimate the distance with C/No =70 dB-Hz.

$$211.4 - 70 - 127.4 = 37.6 \log_{10}R, \text{ and } R = 2.35\,km.$$

Exercise 4-8: Estimate a distance with different C/No graphically.
   Hint: Draw L == 37.6 $\log_{10}$ R + 127.4 on semi-log paper.

4.3.6.3   Range Comparison with Different Pathloss Exponent

When the net link budget, i.e., system gain + antenna gain − fade margin, is the same for different pathloss exponent, then the range can be substantially different. A simple relationship can be found as below assuming the shadow fading is the same for both cases;

$$\left(\frac{d_2}{d_0}\right)^{\gamma_2} = \left(\frac{d_1}{d_0}\right)^{\gamma_1} \tag{4-12}$$

For example $\gamma_1 = 2$ to $\gamma_2 = 4$, $d_2 = 2.0$ km and $d_1 = 40$ km.

Exercise 4-9: $\gamma_1 = 2$ to $\gamma_2 = 4$, $d_2 = 1.2$ km and what is d_1?
   Answer: $d_1 = 15$ km

4.4 Multipath Fading and Statistical Models

In Sects. 4.2 and 4.3 we discussed pathloss and shadow fading, which may be called
a large scale fading since they are determined by large scale (compared to the
wavelength) features such as buildings and terrains along the propagation path. It
is a local average loss of different paths. In addition, there is a rapid local variation of
signal strength due to movement of transmitter, receiver and surroundings, and due
to many wave propagation paths, thus called multipath fading. This may be called
small scale fading since the signal strength fluctuate rapidly due to the constructive
and destructive combining of many waves of different phases. It is practically
convenient to use statistical models in order to characterize the multi-path fading
and to cope with it. The fading can be very destructive to the performance of a
system. Fortunately, there are various ways to handle the fading, and this topic is
beyond the scope of this chapter but summarized in Tables 4-5 and 4-6 at the end of
this chapter. Our focus here is to understand how the fading itself occurs and how to
model it. For further in-depth exploration see [2].

4.4.1 Intuitive Understanding of Multi-Path Fading

A plane wave in time and space (x and y-direction only), with a carrier frequency f_c
and a vehicle speed v_x and v_y can be represented by a complex exponential,

$$w(t) = \alpha \exp\left(j2\pi f_c t - j\frac{2\pi}{\lambda}(x + v_x t \cos\theta) + j\frac{2\pi}{\lambda}(y + v_y t \sin\theta)\right) \tag{4-13}$$

with $\lambda = \frac{c}{f_c}$ and c being the speed of light and α is an amplitude.
   When a vehicle speed is zero and ignoring y –direction, (4-13) becomes

$$w(t) = \alpha \exp\left(j2\pi f_c t - j\frac{2\pi}{\lambda}x\right) \tag{4-14}$$

From the above we can see the phase change due to time delay may be large when
$f_c \cdot t \gg 10$, and the phase change due to displacement (movement) may be large

when $x/\lambda \gg 1.0$. For example, the period of a carrier frequency 1 GHz is 1 nsec, and its wave length is $3 \times 10^8/1 \times 10^9 = 0.3$ m; at a given time, one wavelength displacement the phase will change from 0 to 360°.

A small scale fading is a local variation of signal due to many different multi-path components. In order to get intuitive feel, we will have a numerical example.

Example 4-7: A number of plane waves are combined with different phases and $\alpha_n \approx 1.0$, i.e., there is no dominant component (line of sight component). Incident angles θ are from all directions and time delay phases ϕ are uniformly distributed. Consider the range of about 5 times of wave length, i.e., $x, y \approx 5 \cdot \lambda$.

$$\mathrm{Re}\ (R) + j\ \mathrm{Im}(R) = \sum_n \alpha_n e^{\ j\phi_n} e^{-j2\pi\frac{x}{\lambda}\cos\theta_n - j2\pi\frac{y}{\lambda}\sin\theta_n}$$

α_n	θ_n	ϕ_n
1.0	17°	161°
0.7	126°	356°
1.1	343°	191°
1.3	297°	56°
0.9	169°	268°
0.8	213°	131°
0.7	87°	123
0.9	256°	22

1. compute PDF of Re (R): histogram
2. compute PDF of Im (R): histogram
3. compute PDF of magnitude (|R|): histogram
4. computer PDF of phase ($\angle$ R): histogram

   Write a program and display
   Answer:

1. approximately normal distribution
2. approximately normal distribution
3. approximately Rayleigh
4. approximately uniform

4.4.1.1   Doppler Frequency

(4-13) can be rearranged, ignoring y-direction for simplicity, and becomes

$$w(t) = \alpha \exp\left(j2\pi\left(f_c - \frac{v_x}{\lambda}\cos\theta\right)t - j\frac{2\pi}{\lambda}x\right) \qquad (4\text{-}15)$$

Thus the carrier frequency is changed by $f_D = \frac{v_x}{\lambda} \cos\theta$ with the maximum $\theta = 0$ and the minimum $\theta = \pi$. The apparent carrier frequency increases with the distance (between transmit and receive) decreasing movement while it decreases with the distance increasing movement.

Example 4-8: For 2 GHz carrier frequency with the speed of 60 km/hour, what is Doppler frequency? Assume the maximum (or minimum) Doppler.
    Answer: $60 \times 10^3/3600/3 \times 10^8 \ (2 \times 10^9) = 111.1$ Hz

4.4.2   Rayleigh Fading Channels

Many plane waves represented by (4-15) at a given location (x), vehicle speed (v_x) and carrier frequency and time are combined with different incident angles and antenna gain. We use a complex envelope representation and thus without loss of generality we can set $f_c = 0$. We focus on Doppler frequency with uniform scattering, i.e., the incident angle θ_n is distributed uniformly $(0, 2\pi)$ and (4-15) is represented as

$$C(t) = \lim_{N \to \infty} \sum_{n=1}^{N} a_n \exp\left(j\phi_n + j2\pi[\, f_{Dm} \cos\theta_n] \cdot t \right) \qquad (4\text{-}16)$$

with $\frac{v_x}{\lambda} = f_{Dm}$, $\bar{a}_n{}^2 = 1.0$ and one choice may be $a_n = \frac{1}{\sqrt{N}}$ for all n. In other words, there is no dominant component or it can be said there is no line of sight (LOS) component. We can show that real part of $C(t)$ and imaginary part of $C(t)$ are both normally distributed (Gaussian) by invoking the central limit theorem. This was demonstrated by a numerical example of Example 4-7. Due to statistical nature of θ_n (e.g., uniform distribution for Rayleigh fading) $C(t)$ is a random process, and we can compute its auto-correlation, $R(t_1, t_2) = E\{C(t_1)C(t_2)\}$ with the expectation respect to θ_n. With the simplifying assumption of no correlation between random variables, we obtain

$$R(t_1 - t_2) = \int_0^{2\pi} \exp\left(j2\pi(t_1 - t_2)f_{Dm} \cos\theta \right) p(\theta) d\theta \qquad (4\text{-}17)$$

with $p(\theta)$ being the distribution of incident angle (e.g., uniform). We dropped the index n and treated it as continuous random variable. With the uniform scattering, it is a Bessel function

$$R(\tau) = \int_0^{2\pi} \exp\left(j2\pi(\tau)f_{Dm} \cos\theta \right) \frac{1}{2\pi} d\theta = J_0(2\pi\tau \cdot f_{Dm}) \qquad (4\text{-}18)$$

By taking Fourier transform of (4-18), we can find Doppler power spectrum $S(f)$ as

$$R(\tau) \underset{FT}{\Leftrightarrow} S(f).$$

4.4.2.1 Doppler Spectrum with Uniform Scattering – Jakes Spectrum

Actual calculation of Doppler spectrum with uniform scattering, Fourier transform of (4-18), is given below. It is intuitively satisfying and easy mathematics is enough. It utilizes the physical meaning of power spectrum.

It is a PDF if the total power is normalized to be unity, and its variable is a frequency (f). The same power is distributed in incident angle (θ), and their relation is given by $f = f_{Dm} \cos \theta$.

$$P = \int S(f)df = \int G(\theta)p(\theta)d\theta$$

with $G(\theta)$ power distribution in angle (or antenna gain pattern) and $p(\theta)$ incident angle distribution. Thus this may be interpreted as a random variable change by a function $f(\theta) = f_{Dm} \cos \theta$. We can use <u>Fundamental Theorem of transformation of random variable</u>:

$$f_Y(y) = \frac{f_X(x_1)}{|g'(x_1)|} + \frac{f_X(x_2)}{|g'(x_2)|} + \ldots + f_X\left(\frac{x_{k)}}{|g'(xk)|}\right) \qquad (4\text{-}19)$$

if $y = g(x_1) = g(x_2) = \ldots = g(x_k)$ and $g'(x) = \frac{dg(x)}{dx}$.

To find a PDF in terms of $f(\theta)$, we need to solve $f(\theta) = f_{Dm} \cos \theta$. It is an even function and thus there are two solutions – positive and negative θ. Its derivative $df = f_{Dm}(-\sin\theta)d\theta = f_{Dm}\left(\sqrt{1 - \left(\frac{f}{f_{Dm}}\right)^2}\right)d\theta$. Thus

$$S(f) = \frac{G(\theta)p(\theta) + G(-\theta)p(-\theta)}{f_{Dm}\sqrt{1 - \left(\frac{f}{f_{Dm}}\right)^2}} \qquad (4\text{-}20)$$

With $G(\theta) = 1.0$ and $p(\theta) = \frac{1}{2\pi}$ (uniform scattering),

$$S(f) = \frac{1}{\pi \cdot f_{Dm}\sqrt{1 - \left(\frac{f}{f_{Dm}}\right)^2}} \qquad (4\text{-}21)$$

4.4.2.2   Rayleigh Fading Channel Implementation

The discussion so far can be used for implementing Rayleigh fading channel as in Figure 4-9. Two Gaussian noise generators and two Doppler filters are used.

In discrete time implementations of $C_I(t)$ and $C_Q(t)$, Doppler frequency is typically much smaller than the sampling rate of signal as shown in Figure 4-10. A possible efficient implementation may be that the sampling rate of Doppler is commensurate with Doppler frequency and then an interpolation will be used.

Another practical implementation of Figure 4-9 may be done as shown in Figure 4-11, where a frequency domain filtering, together with FFT –IFFT pair, is used and followed by interpolators.

Exercise 4-10: Jakes Doppler spectrum based on uniform scattering is called 'classic'. In real situations, this is replaced by a simple low pass type. Does this LPF spectrum will correspond to a uniform scattering too? Answer: No, a non-uniform scattering.

Figure 4-9: Complex envelope implementation of Rayleigh fading channel with two Gaussian noise generators and a filter obtained from Doppler spectrum $\sqrt{S(f)}$

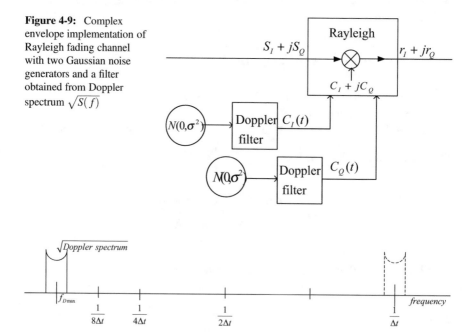

Figure 4-10: Doppler filter relative to the sampling rate

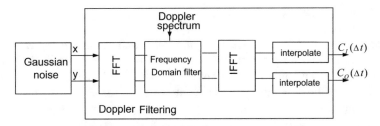

Figure 4-11: Rayleigh fading channel implementation

4.4.2.3 Rayleigh Fading of Power Distribution

The PDF of Rayleigh power, $Z^2 = X^2 + Y^2$, is simply exponential. x and y are independent Gaussian with zeros mean. It is given by

$$p_{Z^2}(x) = \frac{1}{P} e^{-x/\bar{P}} \tag{4-22}$$

It can be extended to higher order, so called Nakagami fading,

$$p_{Z^2}(x) = \left(\frac{m}{\bar{P}}\right)^m \frac{x^{m-1}}{\Gamma(m)} \exp\left[\frac{-mx}{\bar{P}}\right] \tag{4-23}$$

$m = 1$: Rayleigh same as (4-22)
$m = 2$: two Rayleigh added (equivalent to space diversity order $= 2$)
$\Gamma(x)$ is the gamma function, which is a generalization of factorial $\Gamma(x + 1) = x\Gamma(x)$
    and $\Gamma(1) = 1$. $\Gamma\left(\frac{1}{2}\right) = \sqrt{\pi}$. This, $p_{Z^2}(x)$ is the same as chi-square (χ^2) with even
    degree $m = 2n$.

4.4.2.4 Raleigh Envelope Distribution

Rayleigh envelope random variable is $Z = \sqrt{X^2 + Y^2}$ and its PDF is given by

$$p(x) = \frac{x}{\sigma^2} e^{-\frac{x^2}{2\sigma^2}}$$

This is not the same as Rayleigh power distribution. It can be generalized to $Z = \sqrt{\sum_{i=1}^{n} X_i^2}$.

4.4.2.5   Rice (LOS Component)

In many practical situations there is a dominant component in arriving waves, i.e.,
there is a direct, line of sight, path from transmitter to receiver. In Rayleigh fading all
incoming waves are uniformly distributed, i.e., scattered from 0 to 2π and thus we
need to modify it to accommodate LOS components. It can be done by making two
Gaussian noise generator non-zero mean, i.e., adding DC, which will be described
here as shown in Figure 4-12. This new case is called Rician fading.

The power of a LOS component is defined by K as shown in Figure 4-12. In the
figure Gaussian noise generators altogether have a unity variance. The power level is
controlled by σ and K. Doppler filter should have a unity power gain as well. In
practice K is often specified. For example $K = 0$ is equivalent to Rayleigh. $K = 1$
means that the power of LOS and non-LOS are equal.

$Z = X^2 + Y^2$ where X, and Y are Gaussian with variance σ^2 and mean m_i. Its PDF
is given by

$$p(x) = \frac{1}{2\sigma^2} e^{-\frac{s^2+x}{2\sigma^2}} I_0\left(\frac{s}{\sigma^2}\sqrt{x}\right) \tag{4-24}$$

with $s = \sqrt{m_x^2 + m_y^2}$

This can be extended with the degree of m, and is the same as the noncentral
chi-square (χ^2) random variable with the even degree ($m = n/2$).

$$p(x) = \frac{1}{2\sigma^2} \left(\frac{x}{s^2}\right)^{(m-1)/2} e^{-\frac{s^2+x}{2\sigma^2}} I_{m-1}\left(\frac{s}{\sigma^2}\sqrt{x}\right) \tag{4-25}$$

Rician envelope random variable is $Z = \sqrt{X^2 + Y^2}$ and its PDF is given by

$$p(x) = \frac{x}{\sigma^2} e^{-\frac{x^2+s^2}{2\sigma^2}} I_0\left(\frac{s \cdot x}{\sigma^2}\right)$$

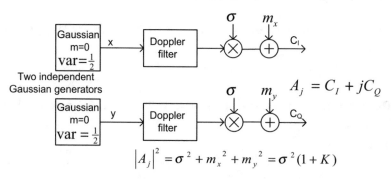

Rayleigh (K=0) and Rician statistics generator

Figure 4-12:   Rician fading channel implementation with the definition of K factor

This is not the same as Rician power distribution. It can be generalized to

$$Z = \sqrt{\sum_{i=1}^{n} X_i^2}.$$

4.4.2.6 Two Independent Zero Mean Gaussian Sample Generation

Two independent Gaussian generators may be implemented by using two uniform random number generators (one is shared in Figure 4-13) by taking advantage of Rayleigh as an intermediate step. In the figure R is Rayleigh envelope distribution (magnitude). This may be used for generating Gaussian random number generators with unity variance (x, and y together in the figure) and zero average.

Exercise 4-11: Modify Figure 4-13 so that two Gaussian generators with the variance σ^2 and mean m_i. Hint see Figure 4-12.

4.4.3 Wideband Frequency Selective Fading

An example of three tap delay line model of wideband frequency selective fading is given in Figure 4-14

Figure 4-13: Two independent Gaussian number generators with zero mean and unity variance (x, y separately)

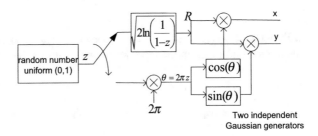

Figure 4-14: A tapped delay line wideband frequency selective fading channel model and the coefficients $A_j(t)$ are generated by power level and Doppler spectrum and fading statistics

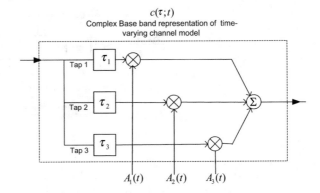

As we will show in the next section, most standards specify multipath fading model with a tapped delay line. Each tap is specified by relative power level, delay, Doppler spectrum, and statistics of signal envelope. Note that these are continuous signal models and can be implemented with RF circuitry, which can be used as an RF channel emulator in laboratory. These RF channel emulators are available commercially.

And its time-varying impulse response is represented by

$$c(\tau;t) = A_1(t)\delta(\tau - \tau_1) + A_2(t)\delta(\tau - \tau_2) + A_3(t)\delta(\tau - \tau_3).$$

where $A_m(t) = \alpha_m(g_{re}(t) + j \cdot g_{im}(t) + dc) \otimes Doppler(t)$.

We choose scales so that $\sum_m \alpha_m^2 = 1.0$ and the variance of a Gaussian noise generator plus dc, i.e., $(g_{re}(t) + i \cdot g_{im}(t) + dc)$, to be unity, and the power level of Doppler filter, $Doppler\ (t)$, is unity as well. In this way a whole multipath fading channel has a unity gain, and becomes a straight line connection when a channel is ideal. These time-varying coefficients are generated as shown in Figure 4-15.

Figure 4-14 can be implemented with RF circuits or in complex base band, then the resulting signal can be frequency translated to RF to do the same thing.

In general there are N taps time-varying impulse response of wideband frequency selective fading model is given by

$$c(\tau;t) = \sum_{i=1}^{N} A_i(t)\delta(\tau - \tau_i)$$

with $A_i(t) = \alpha_i(g_{re}(t) + j \cdot g_{im}(t) + dc) \otimes Doppler(t)$ and the same scaling as mentioned before. Each tap Doppler filter shape may be different.

Figure 4-15: Each tap coefficients are generated by narrow band fading channels as shown in Figure 4-12

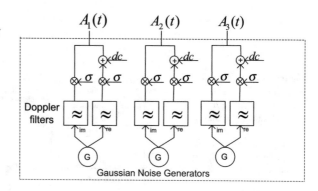

4.4.3.1 A Sampled (Discrete Time) Channel Model

We need a sampled channel model for simulations. In order to have a sampled version of this analog model as in Figure 4-14 there should be band limiting filters at both transmit side and receive side. This is depicted in Figure 4-16.

With the 3-tap example above an instantaneous channel end-to-end impulse response is given by

$$h(\tau; t) = A_1(t)g(\tau - \tau_1) + A_2(t)g(\tau - \tau_2) + A_3(t)g(\tau - \tau_3)$$

where we choose the gain of the combined filter, $g(\tau)$, is unity, and it may be a low pass filter with the cut-off frequency $0.5/\Delta t$ or a proper shaping filter such as raised cosine or a rectangular pulse in time. In this context Δt is a symbol period.

Note that in analog model the delay can be arbitrary and is of dimension in second but a sampled model uses a single sampling clock. Thus once a sampling period is known a discrete time signal can be considered as a sequence.

And another remark is that noise $n(t)$ is sometimes ignored when it is not immediately relevant. It is clear that the receive filter $g_R(\tau)$ should limit the noise while the signal passes it through without distortion. The minimum noise bandwidth for this optimum condition is the Nyquist rate, i.e., $0.5/\Delta t$.

4.4.3.2 Finding the Channel Coefficients of a Sampled Model

We need a clarification of two time notations (τ, t). If the channel is not time-varying, $h(\tau; t)$ does not depend on t and thus it can be represented by $h(\tau; 0) = h(\tau)$. For wireless channel it is typically time-varying though.

Now $h(\tau; t)$ is sampled with sampling parameters as $t = k\Delta t$ and $\tau = j\Delta t$ where $k = 0, 1, 2, \ldots \ldots \infty$ and $j = -\infty, \ldots -2, -1, 0, +1, +2, \ldots +\infty$ and its sampled version is shown below.

Notice that even though in analog model there is a finite number of taps (e.g., 3 taps), a sampled version may have an infinite number of taps and thus the index

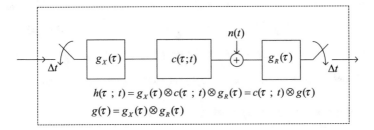

Figure 4-16: The end to end channel with band limiting filters

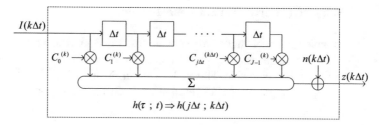

Figure 4-17: A sampled fading channel model

$j = -\infty, .. -2, -1, 0, +1, +2, .. +\infty$. For practical situations this should be a small number and it is related with the choice of the filter $g(\tau)$.

For simplicity of presentation we use the 3-tap example above to show how to obtain the tap coefficients of a sampled model. $h(\tau; t)$ is sampled with Δt and is denoted as $C_{j\Delta t}^{(k\Delta t)} = h(\, j\Delta t; k\Delta t) = h(\tau; t)$.

$$C_{j\Delta t}^{(k\Delta t)} = A_1(k\Delta t)g(\, j\Delta t - \tau_1) + A_2(k\Delta t)g(\, j\Delta t - \tau_2)$$
$$+ A_3(k\Delta t)g(\, j\Delta t - \tau_3) \tag{4-26}$$

As part of explanation of this equation (4-26), we try out an example $j = -1$, $0, 1, 2$ and $k = 0, 1$, total of 8 coefficients.

At $k = 0$, four coefficients are computed

$$C_{-1}^{(0)} = A_1(0)g(-\Delta t - \tau_1) + A_2(0)g(-\Delta t - \tau_2) + A_3(0)g(-\Delta t - \tau_3)$$
$$C_{.0}^{(0)} = A_1(0)g(\cdot 0 - \tau_1) + A_2(0)g(\cdot 0 - \tau_2) + A_3(0)g(\cdot 0 - \tau_3)$$
$$C_{+1}^{(0)} = A_1(0)g(+\Delta t - \tau_1) + A_2(0)g(+\Delta t - \tau_2) + A_3(0)g(+\Delta t - \tau_3)$$
$$C_{+2}^{(0)} = A_1(0)g(2\Delta t - \tau_1) + A_2(0)g(2\Delta t - \tau_2) + A_3(0)g(2\Delta t - \tau_3)$$

At $k = +1$, another four coefficients are computed

$$C_{-1}^{(1)} = A_1(1)g(-\Delta t - \tau_1) + A_2(1)g(-\Delta t - \tau_2) + A_3(1)g(-\Delta t - \tau_3)$$
$$C_{.0}^{(1)} = A_1(1)g(\cdot 0 - \tau_1) + A_2(1)g(\cdot 0 - \tau_2) + A_3(1)g(\cdot 0 - \tau_3)$$
$$C_{+1}^{(1)} = A_1(1)g(+\Delta t - \tau_1) + A_2(1)g(+\Delta t - \tau_2) + A_3(1)g(+\Delta t - \tau_3)$$
$$C_{+2}^{(1)} = A_1(1)g(2\Delta t - \tau_1) + A_2(1)g(2\Delta t - \tau_2) + A_3(1)g(2\Delta t - \tau_3)$$

For k in general it can be denoted in matrix form as,

$$
\begin{bmatrix} C_{-1}{}^{(k\Delta t)} \\ C_{\cdot 0}{}^{(k\Delta t)} \\ C_{+1}{}^{(k\Delta t)} \\ C_{+2}{}^{(k\Delta t)} \end{bmatrix} = \begin{bmatrix} g(-\Delta t - \tau_1) & g(-\Delta t - \tau_2) & g(-\Delta t - \tau_3) \\ g(0 - \tau_1) & g(0 - \tau_2) & g(0 - \tau_3) \\ g(+\Delta t - \tau_1) & g(+\Delta t - \tau_2) & g(+\Delta t - \tau_3) \\ g(+2\Delta t - \tau_1) & g(+2\Delta t - \tau_2) & g(+2\Delta t - \tau_3) \end{bmatrix} \begin{bmatrix} A_1(k\Delta t) \\ A_2(k\Delta t) \\ A_3(k\Delta t) \end{bmatrix}
$$

In frequency domain, this conversion from $A_i(k\Delta t)$ to $C_{j\Delta t}{}^{(k\Delta t)} = h(j\Delta t; k\Delta t)$ through $g(\tau)$ can be interpreted as picking up only frequency range of the filter $g(\tau)$ from much broader frequency response of an analog channel model.

4.4.3.3 Frequency Selective Channel Example

Example 4-9: $h(t) = A_0 + A_1\delta(t - D) + A_2\delta(t - 2D)$ and its frequency response is given by $H(f) = A_0 + A_1 e^{-j2\pi f \cdot D} + A_2 e^{-j2\pi f \cdot 2D}$ which is displayed below.

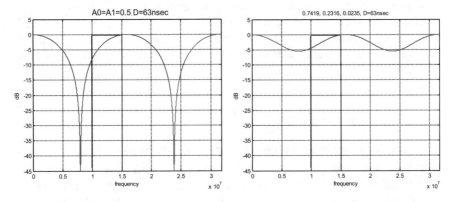

The parameters are $A_0 = A_1 = 0.5$, $A_2 = 0$, $D = 63$ nsec (LHS), and $A_0 = 0.7419$, $A_1 = 0.2316$, $A_2 = 0.0235$ (RHS). A_j may be random and time varying but choose them to be fixed at a given time.

4.4.3.4 On the Computation of $A_i(k\Delta t)$

In order to compute $A_i(k\Delta t)$ we need Doppler filters once two independent Gaussian noise generators available. Adding DC is necessary for Rician. A proper scaling is necessary; Doppler filter has a unity gain, and $\sigma^2(1 + K) = \alpha_m^2$ must be a corresponding to the gain of a tap in delay power profile. This is shown in Figure 4-12.

4.4.4   Alternative Approach to Fading Channel Models

As can be seen in the previous sections computational requirements for generating a sampled channel model is fairly heavy and thus some effort has been made to find an alternative to the approach discussed so far. Here we explore Monte Carlo method based on Reference [3].

4.4.4.1   A Channel Model with Scattering Description

We revisit the equation (4-16). With a slight different notation it is given by
$$C(t) = \lim_{N \to \infty} \sum_{n=1}^{N} \alpha_n \exp\left(j\theta_n + j2\pi \cdot f_{Dn} \cdot t \right) \text{ with } \alpha_n = \frac{1}{\sqrt{N}} \text{ for all } n.$$

Rather than implementing it with Doppler filters and Gaussian random number generators as shown in Figure 4-12 and in Figure 4-15, we implement the scattering description immediately as is. For example, rather than Doppler filtering, Doppler spectrum is implemented by random generation of $f_{Dn} = f_{D\max} \cos \theta_n$ and θ_n is uniformly distributed from 0 to 2π. With this scattering description the instantaneous channel impulse response, introducing delays for both wide band and narrow band, can be written as

$$c(\tau; t) = \lim_{N \to \infty} \frac{1}{\sqrt{N}} \sum_{n=1}^{N} \exp\left(j(\theta_n + 2\pi f_{Dn} t) \right) \delta(\tau - \tau_n) \qquad (4\text{-}27)$$

The factor $\frac{1}{\sqrt{N}}$ ensures convergence as $N \to \infty$, and its power level to be unity. Note that there is no amplitude modulation but only random phase modulations. Clearly real and imaginary part of $c(\tau; t)$ will be Gaussian with zero average, and the magnitude is Rayleigh distribution when θ_n is uniformly distributed $[0, 2\pi]$.

Note also that one needs to specify a scattering function -Doppler spectrum and delay power profile - to get a specific, single realization. Both Doppler spectrum and delay power profile are interpreted as and normalized to be a PDF.

For a single realization one needs N such samples for each random variable (θ_n, τ_n, and f_{D_n}) according to each probability distribution. With a portable uniformly distributed random number generator with $u_n = [0, 1)$ and to calculate v_n by a functional transformation

$$v_n = g_v(u_n) = P_v^{-1}(u_n) \qquad (4\text{-}28)$$

where v_n can be θ_n, τ_n, and f_{D_n} respectively and the memoryless non-linearity $g_v(u_n)$ is the inverse of the desired cumulative distribution function (CDF). In [4] this method is suggested.

A Jake spectrum discussed above, which models the Doppler spectrum with isotropic (uniform) scattering, is easy to implement with this method as $g_{f_D}(u_n) =$

$f_{D\max}\cos(2\pi \cdot u_n)$ where $f_{D\max} = \frac{v}{\lambda}$ the maximum Doppler frequency, which depends on the vehicular velocity v and the wavelength λ.

Then its Doppler spectrum becomes (4-21) which is $S(f) = \dfrac{1}{\pi \cdot f_{D\max}\sqrt{1 - \left(\frac{f}{f_{D\max}}\right)^2}}$.

Another example is given here to see that a portable uniform random number [0,1) generator can be used for generating any desired probability distribution by transformation. PDF and CDF of a Gaussian random variable with unity variance and zero mean is displayed along with the inverse of CDF which is the transformation. See Figure 4-18.

Once explicit transformation can be found by solving the inverse equation then this method is computationally efficient. Mathematically solving the inverse is symbolically given by

$$y = f(x) \quad \Rightarrow \quad x = f^{-1}(y).$$

An explicit expression may not exist. However, even if there is no explicit form, the inverse can be found numerically by interchanging $(x, y) \rightarrow (y, x)$ coordinates. In the case of the numerical inversion, a proper interpolation may be necessary.

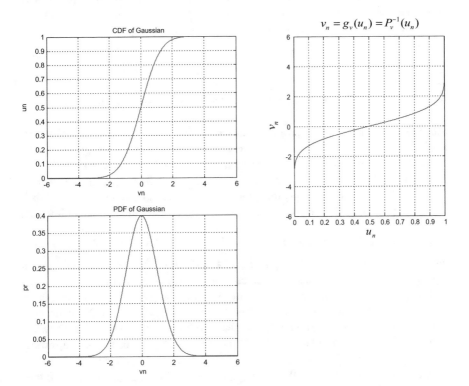

Figure 4-18: Realization of any random variable using CDF method from a portable uniform random number generator

4.4.4.2   A Channel Model with Transmit and Receive Filters

The channel model described by (4-27) is generic, and with the help of (4-28) it can be used for doubly selective fading channels – time selective and frequency selective. The same model may contain a band limiting filter. Combined impulse response of receive and transmit filters are denoted by $g(\tau)$. All representation here is based on a baseband complex envelope. Combining the representation of the channel with transmit and receive filters, the end-to-end channel impulse response is given by

$$h(\tau;t) = \lim_{N\to\infty} \frac{1}{\sqrt{N}} \sum_{n=1}^{N} \exp\left(j(\theta_n + 2\pi f_{Dn}t) \right) g(\tau - \tau_n) \qquad (4\text{-}29)$$

A single arbitrary realization of θ_n, τ_n, and f_{D_n} requires random number generators (three here) which meet respective PDF of each random variable. In particular θ_n is uniform, and τ_n, f_{D_n} are specified by a delay-Doppler scattering function. The scattering function is described by a Doppler spectrum such as (4-21) and a power delay model such as one sided exponential distribution as given by $p_\tau(\tau) = a \exp\left(-\frac{\tau}{b}\right)$ for $0 \le \tau \le c$ where a, b, c, are parameters and will be normalized to be a PDF.

Thus (4-27) or (4-29) with the help of (4-28) can generate any doubly selective channel model for a given statistics of θ_n, τ_n, and f_{D_n}. Before finding a discrete time model for (4-27) or for (4-29), we would like to introduce a simplification if a delay is discrete and fixed in a power delay model.

4.4.4.3   A Simplification when a Delay Profile Is Discrete in Time

Since typically delay power profile is discrete in time, i.e., echoes are impulses, we look at this case first. Finding the inverse of delay power profile given in time discrete form is not explicitly necessary and thus a simplification is possible.

In our applications a delay power profile is given in time-discrete as shown in Figure 4-19, and each tap is specified with power level (B_m) and delay (τ_m) and their underlying statistics (Rayleigh or Rician) and Doppler spectrum (classic or low pass type).

Consider the computation of $c(\tau;t) = \frac{1}{\sqrt{N}} \sum_{n=1}^{N} \exp\left(j(\theta_n + 2\pi f_{Dn}t) \right) \delta(\tau - \tau_n)$ under the condition that $\tau_n \in \{\tau_1, \tau_2, \tau_m, \tau_M\}$ is discrete. It means that only meaningful τ itself is discrete, i.e., the same as these discrete values of τ_n.

Using the example above we carry out to find the inverse of τ_n. First normalize relative power for each tap so that its sum becomes one, and thus PDF. In order to find the inverse we need to compute cumulative density function (CDF) as shown in Figure 4-19. Now we need to find the inverse of CDF. Numerically this is a sorting; when $u_n \le B_1$ then $\tau_n = \tau_1$, when $B_1 < u_n \le B_1 + B_2$ then $\tau_n = \tau_2$, when $B_1 + B_2 < u_n \le B_1 + B_2 + B_3$ then $\tau_n = \tau_3$, and when $B_1 + B_2 + B_3 < u_n \le 1.0$ then $\tau_n = \tau_4$.

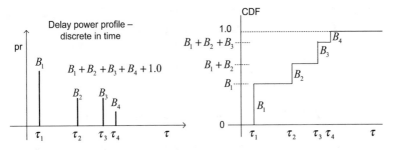

Figure 4-19: Discrete delay profile and its CDF

Now find how many samples belong to each and designate as N_1, N_2, N_3, and N_4, and these many samples are summed together for each τ_1, τ_2, τ_3, and τ_4. In other words, for given each τ_1, τ_2, τ_3, and τ_4 the exponential sum will be carried out, and it is denoted by,

$$c(\tau;t) = \sum_{m=1}^{M} \left[\frac{1}{\sqrt{N}} \sum_{n \in \{\tau_n = \tau_m\}} \exp\left(j(\theta_n + 2\pi f_{Dn}t) \right) \right] \delta(\tau - \tau_m)$$

Due to the nature of delta function the first summation (with m index) is not necessary, and $n \in \{\tau_n = \tau_m\}$ denotes it. This means that for mth tap, only θ_n and f_{D_n} random variables need a single realization of size N_m. Note that $N = \sum_{m=1}^{M} N_m$ where N_m is the number of samples in each delay tap. In other words, each tap has its own realization of θ_n and f_{D_n} with N_m rather than sorting θ_n and f_{D_n} following the index (n) since they are random variables.

Now it is clear that $\frac{N_m}{N} \approx B_m$ and in the limit of $N \to \infty$ the approximation becomes the equality. This suggests the use of B_m directly rather than following the process of the inverse of CDF. Choose $N_m = \frac{N}{M}$ and use $\sqrt{B_m}$ scaling factor for each tap. It is denoted as

$$c(\tau;t) = \sum_{m=1}^{M} \left[\frac{1}{\sqrt{N_m}} \sum_{n=1}^{N_m} \exp\left(j(\theta_n + 2\pi f_{Dn}t) \right) \right] \sqrt{B_m}\,\delta(\tau - \tau_m)$$

This is the simplification we sought, and can be extended to the case with the band limiting filters immediately as below. Notice here τ can be continuous rather discrete due to the filters, and thus the summation through m index is necessary.

$$h(\tau;t) = \sum_{m=1}^{M} \left[\frac{1}{\sqrt{N_m}} \sum_{n=1}^{N_m} \exp\left(j(\theta_n + 2\pi f_{Dn}t) \right) \right] \sqrt{B_m}\,g(\tau - \tau_m)$$

This simplification, i.e., only two random variables (θ_n and f_{D_n}, not τ_n) are needed, is useful when τ_n is specified discrete in time. However, it is also useful in numerical calculations even when τ_n is continuous since τ_n should be quantized as discrete for numerical calculations.

4.4.4.4 A Sampled Channel Model – Monte Carlo Method

We now describe how to convert an analog realization of channel impulse response into a sampled discrete time model, which is necessary for simulations.

For a sampling period Δt, each tap coefficient at time $t = k\Delta t$ is given by

$$C_{j\Delta t}^{(k\Delta t)} = h(\, j\Delta t; k\Delta t) = h(\tau; t)$$

$$C_{j\Delta t}^{(k\Delta t)} = \frac{1}{\sqrt{N}} \sum_{n=1}^{N} \exp\left(\, j(\theta_n + 2\pi f_{Dn} k\Delta t)\right) g(\, j\Delta t - \tau_n)$$

with $\tau = j\Delta t$ where $k = 0, 1, 2, \ldots \infty$ and $j = -\infty, \ldots, -2, -1, 0, .1, 2, \ldots + \infty$.

This tap coefficient computation can be done with the simplification discussed in the previous section.

$$h(\tau; t) = \sum_{m=1}^{M} \left[\frac{1}{\sqrt{N_m}} \sum_{n=1}^{N_m} \exp\left(i(\theta_n + 2\pi f_{Dn} t)\right) \right] \sqrt{B_m} g(\tau - \tau_m)$$

$$C_{j\Delta t}^{(k\Delta t)} = h(\, j\Delta t; k\Delta t) = h(\tau; t)$$

$$C_{j\Delta t}^{(k\Delta t)} = \sum_{m=1}^{M} \left[\frac{1}{\sqrt{N_m}} \sum_{n=1}^{N_m} \exp\left(i(\theta_n + 2\pi f_{Dn} k\Delta t)\right) \sqrt{B_m} \right] g(\, j\Delta t - \tau_m)$$

with $k = 0, 1, 2, \ldots \infty$ and $j = -\infty, \ldots, -2, -1, 0, 1, 2, \ldots + \infty$.

This can be put into the form as,

$$C_{j\Delta t}^{(k\Delta t)} = \sum_{m=1}^{M} [A_m(k\Delta t)] g(\, j\Delta t - \tau_m) \tag{4-30}$$

$$A_m(k\Delta t) = \sqrt{B_m} \frac{1}{\sqrt{N_m}} \sum_{n=1}^{N_m} \exp\left(i(\theta_n + 2\pi f_{Dn} k\Delta t)\right) \tag{4-31}$$

In a diagram a tap coefficient (4-31) can be shown in Figure 4-20.

This equation of (4-30) is the same as the one used in the filtering method of the previous section; see Figures 4-15 and 4-20. Only difference is how to compute $A_m(k\Delta t)$. The filtering method uses Gaussian noise which is filtered by Doppler filter to obtain $A_m(k\Delta t)$. Here with the Monte Carlo method there is no explicit Doppler filtering, rather imitating wave propagation directly. Both methods should be equivalent but a question is which is more efficient and convenient.

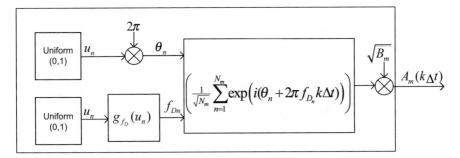

Figure 4-20: One tap implementation of fading model with alternative method

Accurately simulating Doppler spectrum is one key parameter. For the mobile with uniform scattering as we noted before it is given by $g_{f_D}(u_n) = f_{D\max} \cos(2\pi \cdot u_n)$ where $f_{D\max} = \frac{v}{\lambda}$.

For other shapes of Doppler spectrum one needs to find the explicit expression for scattering, i.e., the distribution of $g_{f_D}(u_n)$. However, it can be done off line before a simulation starts and so this should not be a burden during simulations.

Another question is how large N_m should be. An approximation to Gaussian does not require huge N_m, and it appear 10–20 might be good enough. However, in a long simulation a single arbitrary realization appears statistically not stable even though due to ergodicity it contains, on the average, the same statistics as any other realization of possible outcomes. In practice, for small N_m, it seems desirable to compute new random samples from time to time to improve the statistics. See the reference [3]. In a burst packet transmission case a natural boundary may be each packet; for a new packet a new realization is used to improve the statistics. For continuous transmission a periodic new realization will improve the statistics. This problem of choosing the size of N_m and how often a new realization will be computed needs to be investigated further in practice.

Note that this tap coefficient $C_{j\Delta t}^{(k\Delta t)}$ can be computed off-line and stored as a matrix of $J \times K$ since it is not involved with data, i.e., filter and statistics are known beforehand.

4.4.4.5 How to Incorporate Rician Fading Case

It is already indicated how to extend Rayleigh into Rician statistics in the Sect. 4.4.3.4 for the filtering method of computing $A_m(k\Delta t)$. By assuming the power level unity at the output of Doppler filter, the scaling factor of a tap should be $\sigma^2(1+K) = \alpha_m^2$. Thus when α_m and K is known, the scale factor σ and dc can be found. When K is zero (Rayleigh case), it reduces to $\alpha_m = \sigma$.

Now the same applies to the Monte Carlo method as well. $B_m = \sigma^2 + dc^2$. Its implementation is indicated in Figure 4-21.

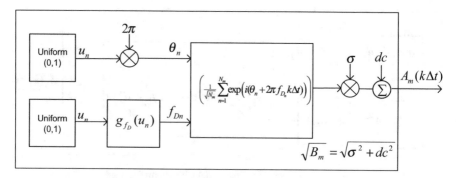

Figure 4-21: Rician fading implementation with alternative method

4.4.4.6   Further Development of Monte Carlo Method

By choosing proper statistical model of θ_n, τ_n, and f_{D_n} one can generate statistical channel samples for simulation of any doubly selective channel model. One immediate problem is to find a proper f_{D_n} so that it can meet any specified Doppler spectrum such a low pass type, rather than classical Jake of (4-21). A limited solution to this problem is hinted in (4-20) by applying non-uniform incident angle distribution. The relationship between different incident angle distribution and Doppler spectrum is needed to find. This may be a future research problem.

Exercise 4-12: Find an angle distribution for a given low pass type Doppler spectrum.
   Hint: Try a numerical example.

4.5   Channel Sounding and Measurements

Any channel model is based on measurement data. Measurement of wireless channel properties such as impulse responses is known as <u>channel sounding</u>. The word channel sounding gives a graphic description of such measurement process. A transmitter sends out a test signal that excites ("sounds") the wireless channel and the output of the channel is observed by a receiver, and stored. From the knowledge of the transmit and received signal, the impulse response of a channel is obtained.

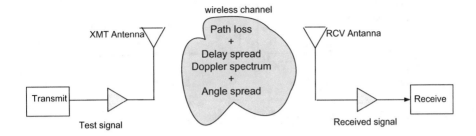

Test signals are:

1. direct RF pulse (time domain)
2. spread spectrum signal (time domain)
3. chirp signal (frequency sweep signal – frequency domain measurement)

The design of this test signal has similarity to digital data transmission problem, but not the same.

- Channel sounding: <u>known</u> = transmit signal + receive signal, <u>unknown</u> = channel impulse response
- Data transmission: <u>known</u> = receive signal (+channel, often estimated from received signals), <u>unknown</u> = transmit data

4.5.1   Direct RF Pulse

It can be interpreted as OOK (on off keying) signal or AM (amplitude modulation) signal.

A band-limited time-invariant channel can be identified by appropriate measurement methods without too much constraint to T_{rep}.

In a time-varying system, the repetition period T_{rep} should be smaller than the time over which the channel changes due to Doppler. This can be formalized by a sampling theorem – a minimum sampling rate of a band limited system is the twice the maximum frequency. There is a minimum temporal sampling rate to identify a time-varying process with a band limited Doppler spectrum.

$$\frac{1}{T_{rep}} = f_{rep} \geq 2f_{D\,\max} \qquad (4\text{-}32)$$

where $f_{D\,\max} = \frac{v}{\lambda} = $ maximum Doppler frequency

If a maximum delay spread for a channel is given as $\tau_{\max}$, then $T_{rep} \geq \tau_{\max}$ must hold, and combining with (4-32) (i.e., a sampling theorem applied), it is not hard to show, by multiplying the two, that the channels can be identified in an unambiguous way only if $2f_{D\max}\tau_{\max} \leq 1$.

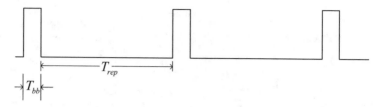

Figure 4-22: RF pulse as a test signal for channel sounding

Fortunately for almost all wireless channels this condition is met. In many cases $2f_{Dmax}\tau_{max} \ll 1$ is fulfilled. This means the channel is slowly time-varying.

Example 4-10: A channel sounder is in a car that moves along a street at 36 km/h. It measures the channel impulse response at a carrier frequency of 2 GHz. What is the repetition period? What is the maximum delay spread that can be measured?

Answer: v = 36 km/h = 10 m/s, wavelength = $3 \ 10^8/2 \ 10^9 = 0.15$ m

$T_{rep} = 0.15/10/2 = 7.5$ msec

Note that this corresponds to two samples per wavelength.

$\tau_{max} = \frac{1}{2f_{Dmax}} = T_{rep} = 7.5$ msec

Pulse width and time resolution

T_{bb}, pulse width of a test RF pulse, should be narrow, and that is the limit of time resolution of the measured impulse response. Ideally it should be an impulse, but a rectangular pulse, with carrier modulation, in practice should be used.

Measurement noise reduction

In actual measurements, measurement noise and interference from other should be taken into account. Thus the measurement should be repeated, and then average.

Example 4-11: the measurement noise is represented by Gaussian of zero mean and variance of σ^2, and with M sample measurements, SNR improves by a factor M.

$$y = \frac{1}{M} \sum_{i=1}^{M} x_i$$

$$m_y = \sum_{i=1}^{M} m_x = 0, \quad \text{and} \quad \sigma_y^2 = \frac{1}{M}\sigma^2$$

4.5.2   Spread Spectrum Signal (Time Domain)

4.5.2.1   Test Signal Design Problem

The characteristics of test signals in time domain and frequency domain will be described. Ideally the channel sounding signal should have the following properties.

i. Signal duration and repetition was discussed in the above and it is related with the fading rate (vehicle speed).

ii. A large bandwidth is necessary to have a proper delay resolution and this is the same as an impulse (small T_{bb}).

iii. The power spectrum of a test signal must be flat across the band and little outside the band. This may be accomplished by making autocorrelation close to an impulse.

iv. The peak to RMS ratio (also called crest factor) should be small so that a power amplifier is efficient.

Noise like signal meets the first three requirements but the crest factor is high. Closely related signal is a direct sequence spread spectrum signal with PN sequence. Schematically the test signal may be generated as in Figure 4-23.

PN (pseudo noise) sequence approximates a noise signal, but periodic while true noise is not periodic, thus pseudo noise.

A maximum length PN sequence (also called m-sequence) is often used in communication systems (scrambler, interleaver, CRC generation, spread-spectrum signal generation...). This was discussed in Chap. 2 Digital modulations in the context of scrambler. Here we give an example.

Example 4-12: n = 5 maximum length sequence
Period = $2^n - 1 = 2^5 - 1 = 31$ (except for 00000)
output = 0000101011101100011111001101001
Property:

1. balance property: 16 1s, 15 0s. In a complete period $P = 2n - 1$ of a PN sequence, the number of 1s differs from the number of 0s by at most 1.

2. run property: 16 runs of 1s and 0s, 8 runs of length 1, 4 runs of length 2, 2 runs of length 3, 1 run of length 4 (0000, 11111)

3. correlation property: If a complete sequence is compared bit by bit with any shift of the sequence, the number of agreements minus the number of disagreements is always -1; that is, there is one more disagreement position than the number of agreement positions.

Figure 4-23: PN sequence based test signal for channel sounding

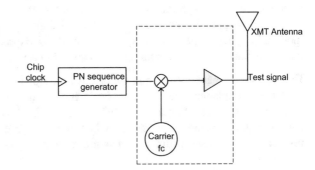

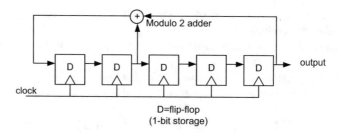

Figure 4-24: Maximum length sequence example with n = 5 and polynomial representation: $x^5 + x^2 + 1$

Exercise 4-13: Try out the property 3 with this example for yourself.

The period of a PN sequence corresponds to the pulse repetition (T_{rep}) and its pulse width i.e., signaling speed, is the same as T_{bb}. A RF carrier modulation is done on PN sequence. This is a BPSK. This is the same as direct RF pulse discussed in the previous section and is preferable in practice.

4.5.3   Chirp Signal (Frequency Sweep Signal) for Channel Sounding

Mathematically, a swept frequency oscillator will generate,

$$x(t) = A cos\left[2\pi\left(f_c + \frac{F_{range}}{T_{sweep}}t\right)t\right] \tag{4-33}$$

and an instantaneous frequency is $\left(f_c + \frac{F_{range}}{T_{sweep}}t\right)$ where f_c = carrier frequency, and F_{range} = sweep frequency range, T_{sweep} = sweep time. Its frequency is linear in the range, i.e., 'sweeping the freqeuncy range'.

A spectrum analyzer, vector signal analyzer and network analyzer are used in this sweep signal to do spectral analysis, frequency domain circuit analysis measurements. A complex envelope can be generated digitally followed by a quadrature modulator by RF carrier. Or it may be generated in RF directly using VCO. A set up using the chirp signal for sounding is shown in Figure 4-25.

Exercise 4-14: generate an example of this type signal of (4-33) using MATLAB.

Answer: implement the equation (4-33) directly in a program.

A time resolution of an impulse response is the inverse of F_{range}. The larger the sweep frequency is the larger the bandwidth of a test signal, which gives a finer time resolution. It is related with the pulse width of a test signal.

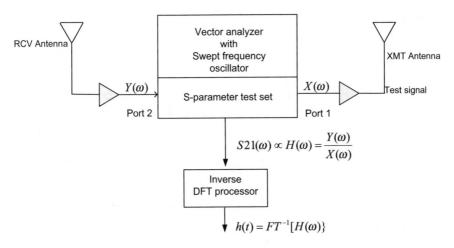

Figure 4-25: Chirp signal for sounding

T_{sweep} is equivalent to T_{rep}. It should be large enough to cover a delay spread but short enough to capture Doppler spectrum. A frequency resolution is inversely proportional to T_{sweep}.

4.5.4   Synchronization and Location

The synchronization of TX and RX is a key problem for wireless channel sounding. GPS (global positioning system) offers a way of establishing common time and frequency reference. Alternative is using rubidium clocks at the TX and RX. Its stability is 10^{-11}, and retain it for several hours

A location of TX and RX should be known accurately, and again GPS is very useful in this measurement.

4.5.5   Directionally Resolved Measurements (Angle Spread Measurements)

In order to measure incident angle spread, we need a number of antennas. And each can be highly directional (conceptually simple). Or antenna can be rotated mechanically for different angles. Schematically this is shown in Figure 4-26.

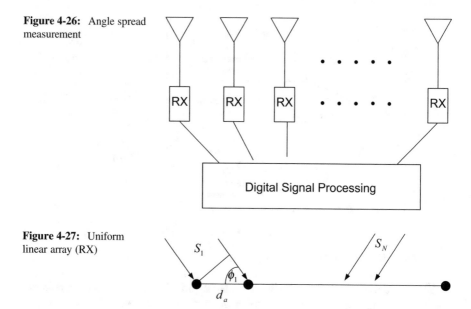

Figure 4-26: Angle spread measurement

Figure 4-27: Uniform linear array (RX)

4.5.5.1   Measurement with an Antenna Array

An array consists of a number of antenna elements, each of which is low directivity, which is spaced apart with a distance d_a that is on the order of wavelength.

4.5.5.2   More Topics with Multiple Antenna Channel Measurement <u>Not</u> Covered

- beam forming method (RX)
- high resolution algorithm (RX)
- extension to doubly directional (MIMO)

4.6   Channel Model Examples

4.6.1   Empirical Path Loss Models

Published examples are listed in 3.3, and repeated here

- Hata: IEEE Trans. Veh. Tech. pp317–325 August 1980
- Cost231: Euro –cost231 project 1991
- Erceg: IEEE J. Selected Areas of Communication pp. 1205–11 July 1999
- Walfish: IEEE Transaction Antenna and propagation pp. 1788–96, Oct 1988

4.6.2 M1225 of ITU-R Path Loss Models

For the terrestrial environments, the propagation effects are divided into three distinct types of model. These are mean path loss, slow variation about the mean due to shadowing and scattering, and the rapid variation in the signal due to multipath effects. Equations are given for mean path loss for each of the three terrestrial environments. The slow variation is considered to be log-normally distributed.

Equations are given for mean path loss as a function of distance for each of the terrestrial environments except the mixed-cell test environment. The slow variation is considered to be log-normally distributed. This is described by the standard deviation (dB) and the decorrelation length of this long-term fading for the vehicular test environment.

4.6.3 Multipath Fading Models in Cellular Standards

4.6.3.1 GSM Channels

Three tables for GSM multipath fading (doubly selective fading) channel models are listed below for HT 100, RA 250 and Urban 50. HT: hilly terrain 100 km/hour, RA: rural area 250 km/hour and urban with the vehicular speed of 50 km/hour. The carrier frequency is 1850 MHz.

Table 4-1: GSM HT 100

GSM Hilly Terrain, HT 100					
100 km/Hr, 1850 MHz mobile (Jakes spectrum)					
Tap	Power level relative (dB)	Delay (usec)	Doppler frequency (Hz)	Doppler spectrum	Statistics
1	−10.0	0.0	330	Jakes	Rayleigh
2	−8.0	0.1	330	Jakes	Rayleigh
3	−6.0	0.3	330	Jakes	Rayleigh
4	−4.0	0.5	330	Jakes	Rayleigh
5	0.0	0.7	330	Jakes	Rayleigh
6	0.0	1.0	330	Jakes	Rayleigh
7	−4.0	1.3	330	Jakes	Rayleigh
8	−8.0	15.0	330	Jakes	Rayleigh
9	−9.0	15.2	330	Jakes	Rayleigh
10	−10.0	15.7	330	Jakes	Rayleigh
11	−12.0	17.2	330	Jakes	Rayleigh
12	−14.0	20.0	330	Jakes	Rayleigh

Table 4-2: GSM RA 250

GSM Rural Area, RA 250

250 km/Hr, 1850 MHz mobile (Jakes spectrum)

Tap	Power level relative (dB)	Delay (usec)	Doppler frequency (Hz)	Doppler spectrum	Statistics
1	0	0.0	429	Jakes	Rician
2	−4.0	0.1	429	Jakes	Rayleigh
3	−8.0	0.2	429	Jakes	Rayleigh
4	−12.0	0.3	429	Jakes	Rayleigh
5	−16.0	0.4	429	Jakes	Rayleigh
6	−20.0	0.5	429	Jakes	Rayleigh

Table 4-3: GSM TU 50

GSM Typical Urban, TU50

50 km/Hr, 1850 MHz mobile (Jakes spectrum)

Tap	Power level relative (dB)	Delay (usec)	Doppler frequency (Hz)	Doppler spectrum	Statistics
1	−4.0	0.0	86	Jakes	Rayleigh
2	−3.0	0.1	86	Jakes	Rayleigh
3	0.0	0.3	86	Jakes	Rayleigh
4	−2.6	0.5	86	Jakes	Rayleigh
5	−3.0	0.8	86	Jakes	Rayleigh
6	−5.0	1.1	86	Jakes	Rayleigh
7	−7.0	1.3	86	Jakes	Rayleigh
8	−5.0	1.7	86	Jakes	Rayleigh
9	−6.5	2.3	86	Jakes	Rayleigh
10	−8.6	3.1	86	Jakes	Rayleigh
11	−11.0	3.2	86	Jakes	Rayleigh
12	−10.0	5.0	86	Jakes	Rayleigh

4.6.4   Cellular Concept and Interference Limited Channels

One of the major innovations in cellular mobile systems is the concept of frequency reuse; a set of carrier frequencies is used over and over again to cover the whole service area. In each cell a limited number of users are served and if it is moved to another cell, there should be handover. The technique of handover is now perfected but was one of hurdles to use the concept of frequency reuse.

Thus the channel model in this situation is effectively interference limited; other cell interference is a major source of impairment, typically not thermal noise. Fortunately the interference may be treated approximately as a noise and thus s system designed and optimized for noise should work well under the interference limited system. In fact, the cellular system capacity is sensitive to the required signal to interference plus noise. In this section we describe a cellular system concept and find the interference due to frequency reuse.

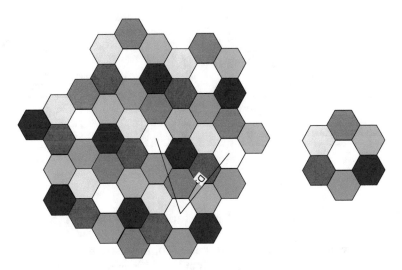

Figure 4-28: Frequency reuse with 7 frequencies (colors)

4.6.4.1 Cellular Frequency Reuse

A different color in Figure 4-28 means that a different frequency (or a set of different frequencies) is assigned. In this example there are 7 frequencies as shown on the right side of the figure. Each hexagonal is a cell to serve the users in it.

Focus on one color (say white). The same color cells are surrounded by 6 white color cells, and more white cells. The distance from a cell to the same color is the same. If there is infinite number of cells, there more cells of the same color. For a given cell, the first 6 cells of the same color (shown in the figure), the 12 cells of the same color (not shown in the figure) and so on.

Exercise 4-15: Show that a number of interfering (the same color) cells are 6, 12, 18, ... (n × 6 with n = 1, 2, 3, ...). Only 6 is shown in Figure 4-28.

Why hexagonal shape? A tessellation is possible and it is close to a circle. Tessellation is a tiling, i.e., filling up the plane without gap. Another such shape is rectangle and triangle. We focus on hexagonal shape.

Exercise 4-16: Can you think of any other shape that can be tessellated? Answer: No with a simple geometry of equal sides – regular polygon but infinite possibilities with complex geometrical shapes. For fun, google or Wikipedia 'tessellation' and then you will see.

We define radius of inner circle (R_i) and outer circle (R_o). We can see the hexagon is closer to circle.

shape	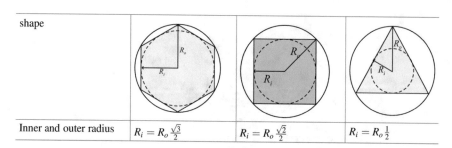		
Inner and outer radius	$R_i = R_o \frac{\sqrt{3}}{2}$	$R_i = R_o \frac{\sqrt{2}}{2}$	$R_i = R_o \frac{1}{2}$

4.6.4.2   Cluster Size (K) and Co-Channel Interference Distance (D)

We will show here the interference distance (D), the distance from the center to center with the same color (frequency), is expressed in terms of a cluster size, i.e., how many cells in a cluster ($K = 7$ with Figure 4-28). It is given by

$$\frac{D}{R_o} = \sqrt{3}\sqrt{K} \qquad (4\text{-}34)$$

We give two examples first. With $K = 3$, it is easy to see that $D = 3R_o$.

4.6.4.3   Cluster Size with Hexagons

Cluster size is possible only for certain integers as shown in Table 4-4. We show this shortly along with (i, j) coordinate.

Cluster size (K) possible with hexagons- 1, 3, 4, 7, 9, 12, 13, 19, ...

Exercise 4-17: Draw cell boundaries as Figure 4-28 for K = 4.

Exercise 4-18: Draw cell boundaries as Figure 4-28 for K = 19. Hint: From Table 4-4, $i = 2, j = 3$ ($i = 3, j = 2$) is required to be K = 19.

Figure. 4-29: $K = 3$
frequency reuse pattern and
$D = 3 R_o$

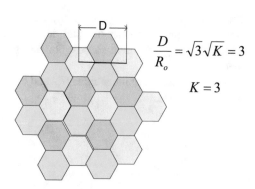

$$\frac{D}{R_o} = \sqrt{3}\sqrt{K} = 3$$

$$K = 3$$

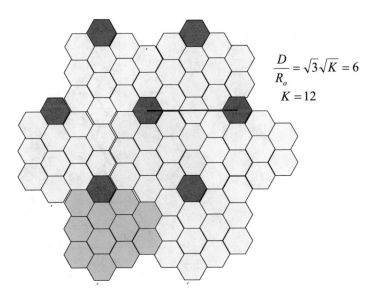

$$\frac{D}{R_o} = \sqrt{3}\sqrt{K} = 6$$

$$K = 12$$

Figure 4-30: $K = 12$ frequency reuse and $D = 6\,R_o$

Table 4-4: Cluster size in terms of i, j coordinates $K = i^2 + j^2 + ij$

j / i	0	1	2	3	4
0	0	1	4	9	16
1	1	3	7	13	21
2	4	7	12	19	28
3	9	13	19	27	37
4	16	21	28	37	48

From Figure 4-31, we can see that

$$d(i,j) = 2R_i\sqrt{i^2 + j^2 + ij} = \sqrt{3}R_o\sqrt{i^2 + j^2 + ij} \qquad (4\text{-}35)$$

We use one of trigonometric identity as below.

$$c^2 = a^2 + b^2 - 2ab\cos\theta$$

Exercise 4-19: See (4-35) is correct. One example is $i = 2$, $j = 2$ as shown in Figure 4-31

Another coordinate system (n, i) is convenient. This is shown in Figure 4-32. With Figure 4-32 coordinate system we can show that

Figure 4-31: Hexagonal (i, j) coordinate system and co-channel distance

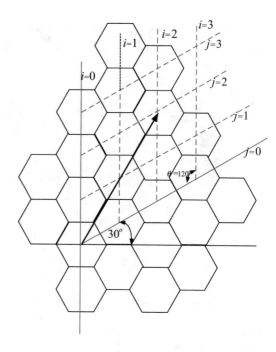

$$d(n, i) = 2R_i\sqrt{n^2 + i^2 - ni} = \sqrt{3}R_o\sqrt{n^2 + i^2 - ni} \qquad (4\text{-}36)$$

And allowed value of (n, i), for a center location of each cell,

$$n = 1, i = 1$$
$$n = 2, i = 1, 2$$
$$n = 3, i = 1, 2, 3$$
$$n = 4, i = 1, 2, 3, 4$$

and so on.

Exercise 4-20: Show for n = 1, 2, 3, 4, ... a number of cells is 6, 12, 18, 24, ...

Exercise 4-21: Work out the exact distance for 6, 12, 18, 24, ...

$$n = 1, i = 1$$
$$n = 2, i = 1, 2$$
$$n = 3, i = 1, 2, 3$$
$$n = 4, i = 1, 2, 3, 4$$

Answer: use (4-36).

Figure 4-32: (n, i)
coordinate for co-channel
interference

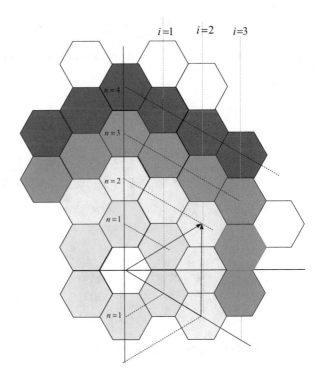

4.6.4.4   Co-channel Interference

Signal to noise plus co-channel interference can be expressed as

$$SINR = \frac{P_r}{kTB + P_I}$$

where P_r is the receiver power, and kTB thermal noise power, and P_I co-channel
interference. Typically the cellular system is designed to be interference limited
rather than noise limited, which means that co-channel interference power, is much
bigger than noise power. We ignore the thermal noise term.

So we consider, to a first order approximation, SIR as, with γ = pathloss exponent

$$\text{SIR} = \frac{P_r}{P_I} = \frac{P_t \cdot A_0 \cdot \left(\frac{1}{d_0}\right)^\gamma}{\sum\limits_{i=1}^{N_i} P_t \cdot A_0 \cdot \left(\frac{1}{d_i}\right)^\gamma} = \frac{\left(\frac{1}{d_0}\right)^\gamma}{\sum\limits_{i=1}^{N_i} \left(\frac{1}{d_i}\right)^\gamma}$$

Here we assume that down link so that transmit power of each cell base station is
equal, and the path loss power exponent is the same for all the cells. We are
interested in a first order estimation of SIR, relating with co-channel reuse distance.

Set the $d_0 = R_o$ the cell edge (the worst case), and $d_i = D$ for $n = 1$, and 2D for $n = 2$, and so on. This is again the worst case since for a given n, $i = 1$ case was considered here $i = 1$, $i = 2$ the distance is not exactly the same.

$$SIR = \frac{P_r}{P_I} = \frac{\left(\frac{1}{d_0}\right)^\gamma}{\sum_{i=1}^{N_i} \left(\frac{1}{d_i}\right)^\gamma} = \frac{\left(\frac{1}{R_o}\right)^\gamma}{\sum_{n=1}^{\infty} 6n\left(\frac{1}{nD}\right)^\gamma}$$

$$SIR = \left(\frac{D}{R_o}\right)^\gamma \frac{1}{6 \sum_{n=1}^{\infty} \left(\frac{1}{n}\right)^{\gamma-1}} = \left(\sqrt{3K}\right)^\gamma \frac{1}{6 \sum_{n=1}^{\infty} \left(\frac{1}{n}\right)^{\gamma-1}} \qquad (4\text{-}37)$$

Note that

$$\sum_{n=1}^{\infty} \left(\frac{1}{n}\right)^{\gamma-1} = \begin{cases} \infty & \gamma = 2 \\ 1.6449 & \gamma = 3 \\ 1.2021 & \gamma = 4 \\ 1.0823 & \gamma = 5 \end{cases}$$

This function is called Riemann zeta function.
For $\gamma = 4$ case,

$$SIR = \left(\frac{D}{R_o}\right)^\gamma \frac{1}{6 \sum_{n=1}^{\infty} \left(\frac{1}{n}\right)^{\gamma-1}} = \left(\sqrt{3K}\right)^\gamma \frac{1}{6 \sum_{n=1}^{\infty} \left(\frac{1}{n}\right)^{\gamma-1}} = 1.248K^2$$

$$= \begin{cases} 10.5\text{dB} & K = 3 \\ 13.0\text{dB} & K = 4 \\ 17.9\text{dB} & K = 7 \\ 20.0\text{dB} & K = 9 \\ 22.5\text{dB} & K = 12 \end{cases}$$

Sectoring:
Typically cellular systems use sectored antennas (3 sector is very common) rather than omni-antennas. Horizontal beam width 120°. Thus roughly it can expect to reduce the number of interferers by the same amount (1/3 in 3 sector case). For each n, interferers reduce to 2n rather than 6n. Thus the SIR improvement will be about 10 log (3.0) = 4.7 dB.

Exercise 4-22: Count the number of interferers for n = 1, K = 7. Answer = 6.

4.6.4.5   Use of Repeaters and Distributed Antennas:

In practice a cell shape is not hexagonal but irregular shape with different size. The hexagonal base cell may be used during the early planning stage. In practice there are many different techniques to improve the system. Repeaters and distributed antennas are often used in practice. Figure 4-33 is an example.

Exercise 4-23* (project): Discuss the advantage and cost of implementing Figure 4-33.

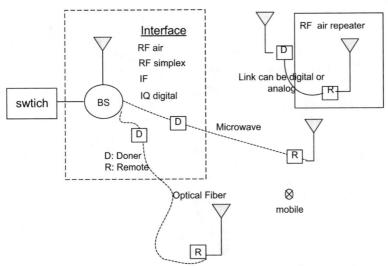

As a single cell, even with multiple repeaters or distributed antennas, it should behave just like a cell with a single antenna. When a mobile moves around in the cell, there is no need of hand-over.

Figure 4-33: Repeater and distributed antennas for cellular systems

4.6.5   Channel Models of Low Earth Orbit (LEO) Satellite

We use Figure 4-34 to derive a fading channel model of LEO - Doppler frequency and amplitude fading. The altitude of LEO satellite is 1000–2000 km, circling around the Earth every 3–4 hours.

4.6.5.1   Geometry of LEO

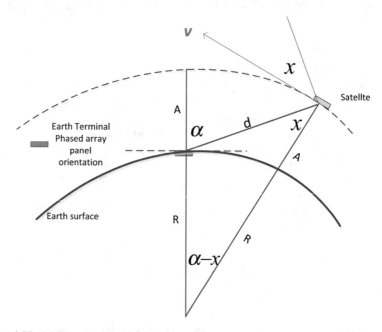

Figure 4-34:   LEO geometry used for deriving a channel model

4.6.5.2   Fading Channel Model Due to Satellite Orbiting

Complex baseband model of fading due to orbiting
For Figure 4-35 we summarize the fading model as below;

$$A_m = \frac{\sin \alpha}{\sin \left(\alpha - \sin^{-1} \left(\frac{R}{R+A} \sin \alpha \right) \right)} \frac{A}{R+A}$$

$$f_D = \frac{f_c}{c} \sqrt{\frac{GM}{R+A} \frac{R}{R+A}} \sin \alpha$$

Satellite revolution period $P = 2\pi \sqrt{\frac{R+A}{GM}} (R+A)$ [sec]

Satellite visible time: $t_v = P * (2 * 50 \text{ deg})/360$

Constants used in the model
Gravity constant: $G = 6.67384e\text{-}11$ [m^3/ (kg sec^2)]
Earth radius: $R = 6371\, e3$ [m]
Earth mass: $M = 5.9729\, e\, 24$ [kg]
α: Satellite elevation angle $[-50\ 50]$
c: speed of light 3e8m/sec
Altitude: A [m] (see the geometry of LEO)

Explanations
We use following trigonometric relation

$$\frac{\sin x}{R} = \frac{\sin \alpha}{R+A} = \frac{\sin (\alpha - x)}{d}$$

$$d = \frac{\sin (\alpha - x)}{\sin \alpha} (R+A) \text{ where } \sin (x) = \frac{R}{R+A} \sin (\alpha)$$

From the above we derive (4-38) below

Figure 4-35: LEO channel
model: Doppler by f_D and
A_m for amplitude fading

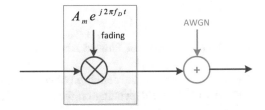

$$d = \frac{\sin\left(\alpha - \sin^{-1}\left(\frac{R}{R+A}\sin\alpha\right)\right)}{\sin\alpha}(R+A) \tag{4-38}$$

Exercise 4-24: The distance between satellite and ground terminal is expressed by (4-38). Can you see that? Hint: it can be easily derived from the trigonometric relationship of the law of sines.

From basic orbital mechanics compute v (speed of revolution), and find Doppler velocity (v_D) expressed in (4-39). From v, find the revolution period, (4-40), shown below.

$$v_D = v\frac{R}{R+A}\sin\alpha \text{ where } v = \sqrt{\frac{GM}{R+A}} \tag{4-39}$$

$$P = 2\pi\sqrt{\frac{R+A}{GM}}(R+A) \tag{4-40}$$

Exercise 4-25: From Figure 4-34, the velocity of satellite v has a component v_D which is directed toward the ground terminal. Obtain (4-39) from $v_D = v\cos(90° - x) = v\sin(x)$. Hint: v itself is not obtained from the figure but from other way.

Numerical examples and additional for derivation

Altitude of satellite: A = 1150 [km], Distance at 50 deg: d = 1625 [km] see the formula (1)

$$v_D = v\cos(90 - x) = v\sin(x) = v\frac{R}{R+A}\sin\alpha$$

$$= 7.284 * 6371/(6371 + 1150) * \sin(50\ deg * pi/180) = 4.7267\ [km/\ sec]$$

Period of revolution

Velocity: $v = \sqrt{\frac{GM}{R+A}}$ angular velocity $w = v/(R + A)$ [rad/ sec]

Period P $= 2\pi/w = 2\pi\sqrt{\frac{R+A}{GM}}(R+A)$

Satellite visible time: t_v = P* (2* 50 deg)/360

Also using Kepler's 3rd law: $\frac{P^2}{(a)^3} = \frac{4\pi^2}{G(M+m)}$ where $a = R + A$

Example plots with altitude A = 1150 km and max elevation angle=50° shown in Figure 4-36.

4.6.5.3   Cosine Loss of Phased Array Antenna

See LEO geometry for the definition of elevation angle of α. The gain of phased array antenna will change depending on the angle as follows;

Phased array ground terminal cosine amplitude = $\cos(\alpha)$
Phased array satellite cosine amplitude = $\cos(x)$ where $\sin(x) = \frac{R}{R+A}\sin(\alpha)$.

Figure 4-36: Numerical
example of Doppler
frequency, amplitude, and
elevation angle

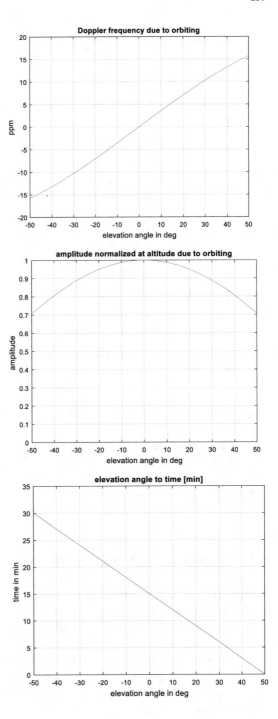

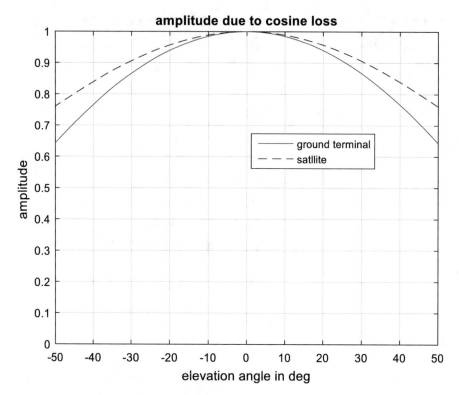

Figure 4-37: Numerical example of cosine loss of phased array antenna

This amplitude change is essentially the phased array antenna pattern change – looking each other with the elevation angle not being zero i.e., looking directly. This cosine loss is absent with mechanical tracking. A numerical example is shown in Figure 4-37.

4.7 Summary of Fading Countermeasures

In this section we summarize the discussion so far.

See Tables 4-5 and 4-6. Table 4-5 summarizes the fading – slow large scale fading (path loss) and fast small scale fading (Rayleigh) and their description along with typical examples.

Table 4-6 gives some specific instances and examples for large scale fading and small scale fading, and typical countermeasure techniques used in practice. The table is meant to give conceptual ideas, and their detailed explanations are found throughout this chapter.

Table 4-5: Pathloss and fading summary

		Outdoor mobile	In-door
Pathloss	Large scale path loss averaged locally independent of signal structure and bandwidth only to the power of signal	Pathloss (dB) = **A** + 10 **r** log (d/d$_o$) + **s** where **A** is the loss at a reference distance d$_o$ and **r** is called path loss exponent. **s** is a shadow fading standard deviation in dB and can be modeled by lognormal distribution. This model is generally accepted for different situations from fixed to mobile. Some additional correction factor can be added depending on terrain type and antenna height.	
		A = free space loss at d$_o$ (e.g., 100 m) **r** = 3 to 4 depending on terrain and surroundings **s** = 8 to 16 dB depending on terrain and building and car penetration distance.	**A** = free space loss at d$_o$ (e.g., 1 m) **r** = 3 to 5 depending on terrain and surroundings **s** = 10 to 14 dB depending on walls and structures
Delay spread Doppler spectrum Fade statistics	Small scale fading – rapid fluctuation of signal envelope due to movement and delay spread	Delay spread = typically tapped delay line model is used in standards; resulting frequency selective fading for wide band channel Doppler spectrum = spectrum shape of time varying channels; classical (jake), low pass type Fading statistics = probability distribution of signal power mainly due to arrival angle distribution and antenna gain; e.g., Rayleigh, Rice	
		Standard models available. Typically delay spread up to 20 usec, Doppler frequency 100 Hz with 2GHz 60 km per hour, jakes spectrum, Rayleigh or Rice fade statistics.	Some models available related with IEEE 802.11 and HIPERLAN-2. Delay spread = exponential decay 100–200 nsec Doppler spectrum up to 5 Hz with low pass filter type and Rayleigh statistics

Table 4-6: Countermeasure techniques for fading

		Available countermeasures
small scale 'fast' fading	time selective fading (e.g. Rayleigh)	Interleaving + FEC RCV space diversity + FEC Use of QAM, coherent demodulation with pilots large BW signal compared to fading rate
	frequency selective fading	Equalizers, rake receiver transmit side precoding for known channels OFDM or small BW sub-carrier signal
large scale 'slow' fading	AWGN	powerful FEC (iterative decoding –turbo, LDPC) for low SNR TCM, BICM, MLCM + Gaussian shaping for high SNR cases
	Shadow fading	Fade margin macro diversity (similar to soft HO) proposed by Jakes
	Multi-user interference	orthogonal signals multi-user detection (cancellation)
	Other-cell interference	avoidance by coordination, distance separation spatial filtering with multiple antennas (e.g. beam forming) spreading or averaging by using low bps/Hz signal (e.g. repetition, low rate FEC)

4.8 References with Comments

There are many textbooks available for antennas, and we selected classic one [1] which started being available from 50s. Our development of mobile fading channel was based on [3, 4] but vast amount of additional literature is available including [2]. For cellular and fixed channel model is covered in great detail through IEEE [5]. Our reference is vanishingly small compared to the open literature available, yet it may be a small opening to the additional reference as reader's needs arise.

Antenna areas
[1] Kraus, J. and Marhtfka, R. "Antennas" 4th ed, McGraw-Hill, 2001

Mobile fading channel model
[2] Parsons, J.D., "The Mobile Radio Propagation Channels", John Wiley & Sons, NY, 1992
[3] Hoeher, P., "A Statistical Discrete-time model for WSSUS multipath channel," IEEE Transaction on Vehicular Technology vol. 41 No.4 November 1992
[4] Coates, R.F.W., Jannacek, G.J., Lever, K.V., "Monte Carlo simulation and random number generation," IEEE J. Select. Areas Commun., vol. SAC-6, pp. 58066, Jan. 1988

More wireless channel model and references

[5] IEEE *J. Selected Areas Communication*," Special Issue on Channel and Propagation Modeling for Wireless System Design" April 2002, August 2002,
December 2002.

4.9   Problems

P4-1: What is the frequency of wavelength 1 mm assuming the speed of light
$c = 3 \times 10^8$ m/sec? How long does it take to move the wavelength?

P4-2: Express 20 dBm in terms of watts. Note that 0 dBm is defined as 1 miliwatt
in dB.

P4-3: A free space loss in ratio is expressed as $\lambda^2/(4\pi d)^2$ where λ wavelength [m] and
d distance [m]. Find the distance with 1 mm wave to have the loss of 60 dB, 90 dB
and 120 dB.

P4-4: Pathloss in dB model, inspired by free space loss, is given by

$$\text{Pathloss}\,[dB] = A\,[dB] + 10\gamma \cdot \log_{10}\left(\frac{d}{d_0}\right) + s[dB]$$

$A = -20\log_{10}\frac{\lambda}{4\pi\,d_0}$ free space loss in dB at distance d_0

d_0: reference distance, say 100 m or smaller if necessary

γ: pathloss exponent, ≥ 2, empirically determined from measurements

s: shadow fading variance with log-normal PDF, empirically determined from
measurements.

(1) Find $A = 20\log(d_0) + 20\log(4\pi f/c)$ when $d_0 = 1$ m, $f = 5$GHz.
(2) Find pathloss [dB] without shadow fading when d $= 100$ m with $\gamma = 2$ (free
space) and with $\gamma = 5$ (e.g., inside building)

P4-5: An approximate two ray model is described by (4-8). What is pathloss
exponent?

P4-6: Directivity of antenna is defined by the ratio of solid angle of sphere (4π) over
the total beam area in solid angle (Ω_A); $D = \frac{4\pi}{\Omega_A}$.

(1) What is the directivity of omnidirectional antenna?
(2) Antenna beam angle of half power is given by $\theta_{HP} = 1°$ and $\varphi_{HP} = 120°$.
What is the directivity of this antenna?

P4-7: The aperture area is related to the antenna pattern (solid angle) as $\Omega_A = \lambda^2/A$
(see the equation 4-5), and antenna gain (directivity*efficiency k) G $= A_e\,4\pi\,/\lambda^2$

where $A_e = kA$. Show that antenna size is proportional to wavelength, and compare the size difference with carrier frequency of 1 GHz and of 20 GHz.

P4-8: Fresnel zone (2nd) is shown in Figure 4-7, and its radius is given by (4-9). Assuming the distance between transmit and receive antenna is 1 km, find the radius R_{F2} when the carrier frequency is 1 GHz. Compare it with the case of 20 GHz.

P4-9: Measured path loss are given by a table below (as in Example 4-4). We try to use a model $PL = A + \gamma 10\log 10\, (d/d_o)$ where $d_o = 1.0$ m.

(1) Find A and γ using least square fit. Hint: $[A\ \gamma]' = (a'a)^{-1} (a'b)$.

Distance	10 m	20 m	50 m	100 m	300 m
pathloss	70 dB	75 dB	90 dB	110 dB	125 dB

(2) Find (square root) variance.
(3) Compare this result with Example 4-4.

P4-10: Consider the properties and computation of $Q(x)$.

(1) $Q(x)$ is expressed by $erfc()$ as $Q(x) = 0.5 erfc\left(x/\sqrt{2}\right)$ where $erfc$ $(x) = \frac{2}{\sqrt{\pi}} \int_x^\infty x^{-t^2} dt$. Use it computing $Q(x)$

(2) The inverse of $Q(x)$ is expressed by the inverse of $erfc()$ as $Qinv(x) = \sqrt{2} erfcinv(2x)$.
(3) Show that $Q(x) = 1 - Q(-x)$
(4) Fill the blanks

80%	84%	90%	99%	99.9%	99.99%	reliability
0.2	0.14	0.1	0.01	0.001	0.0001	outage
	1.0	1.282				Q^{-1}(outage)

P4-11: Define the terms. (1) System gain (2) Net link budget (3) Fade margin

P4-12: Using the equation (4-12) with path exponent $\gamma = 3.8$ the covered distance is 2.0 km. If it is applied to free space, what is the distance that can be covered?

P4-13: The PDF of Rayleigh fading in power is extended to (4-23) where $m = 1$ corresponds to Rayleigh fading. Plot $P_{Z^2}(x)$ when $m = 1$, 2, and 3 and compare them.

P4-14: Rician fading channel model is an extension of Rayleigh fading when there is LOS component. Its power distribution PDF is in (4-25). K factor is defined as ratio of the power of DC component over that of random component; $K = s^2/\sigma^2$. With $K = 0$ it becomes Rayleigh. Plot PDF (4-25) when $K = 5$ with $m = 1$, 2, and 3, and compare.

P4-15: In Figure 4-13, Gaussian random number generator, two (x, y) independent, is shown. Write a small program to confirm it, and compare with randn() of MATLAB.

P4-16: Look up a Fourier transform pair $J_0(t) \underset{FT}{\leftrightarrow} 2rect(\pi f)/\sqrt{1 - 4\pi^2 f^2}$ where $rect$ $(x) = 1.0$ when $-0.5 \leq x \leq 0.5$ and $rect\ (x) = 0.0$ otherwise. Then derive (4-21), Doppler spectrum with uniform scattering.

P4-17: A wideband frequency selective fading is often represented by discrete delay line model as, for example with two taps, where $A_1(t)$ and $A_2(t)$ from fading mechanism (e.g., Rayleigh) $c(\tau; t) = A_1(t)\delta(\tau - \tau_1) + A_2(t)\delta(\tau - \tau_2)$. When $A_1(t) = 1.0$ and $A_2(t) = -1.0$ for $0 \leq t < \infty$ and $\tau_1 = 0.0$ and $\tau_1 = 6.3nsec$. Find the magnitude of frequency response of $c(\tau)$.

P4-18: In direct RF pulse sounding system, a narrow pulse is transmitted repeatedly as shown in Figure 4-22. It is known that a channel is very slowly changing (practically time-invariant) and that its impulse response delay and spread is less than 10 nsec. In order to measure its frequency response what are the practically reasonable choice of a test pulse? Is it true that multiple measurements will improve measurement noise?

P4-19: Compare direct RF pulse method (Sect. 4.5.1) and spread spectrum method (Sect. 4.5.2), in particular, pulse width (T_{bb}) and chip clock period, pulse repetition (T_{rep}) and PN sequence length.

P4-20: A chirp signal, i.e., frequency sweep signal $x(t)$, can be used for channel sounding; $x(t) = A cos\left[2\pi\left(f_c + \frac{F_{range}}{T_{sweep}}t\right)t\right]$. Two parameters, F_{range} and T_{sweep}, are related with T_{bb} and T_{rep} of direct RF pulse method. Explain it.

P4-21: In Table 4-1 GSM HT 100 a multipath fading channel model is shown. What is the delay of a main tap? Sketch delay power spectrum.

P4-22: Cluster sizes (K) possible with hexagon are - 1, 3, 4, 7, 12, 13, etc. as in Table 4-4. Draw frequency reuse pattern with $K = 4$.

P4-23: In hexagonal pattern of frequency reuse co-channel interference distance (D) is related with cluster size (K); $D = R_o\sqrt{3K}$. Confirm it for $K = 7$ case using Figure 4-28.

P4-24: Show that when path loss exponent $\gamma = 4$, cluster size $K = 7$, SIR is about 18 dB. Find SIR when the cluster sized is reduced to $K = 3$. Use (4-37).

P4-25: Using (4-40) find LEO satellite revolution period when the altitude is 1150 km and 1150 km/2.

Chapter 5
OFDM Signals and Systems

Contents

5.1   DMT with CP – Block Transmission 252
     5.1.1 IDFT – DFT Pair as a Transmission System 253
     5.1.2 Cyclic Prefix Added to IDFT- DFT Pair 255
     5.1.3 Transmit Spectrum 256
     5.1.4 OFDM Symbol Boundary with CP 260
     5.1.5 Receiver Processing When the Channel Dispersion < CP 262
     5.1.6 SNR Penalty of the Use of CP 264
5.2   CP Generalized OFDM – Serial Transmission 265
     5.2.1 OFDM – Analog Representation 265
     5.2.2 Discrete Signal Generation of Analog OFDM 266
     5.2.3 Discrete Signal Reception 269
     5.2.4 Pulse Shape of DMT and its End to End Pulse 272
     5.2.5 Windowing 272
     5.2.6 Filter Method Compares with *DMT with CP* 273
5.3   Filtered OFDM 274
     5.3.1 Filtered OFDM Signal Generation 274
     5.3.2 Filtered OFDM Signal Reception 277
     5.3.3 Common Platform 279
     5.3.4 Impulse Response and Eye Pattern 280
     5.3.5 1-Tap Equalizer, Sample Timing and Carrier Phase Recovery, and Channel
         Estimation 282
     5.3.6 Regular FDM Processing with Filtered OFDM 283
5.4   OFDM with Staggered QAM 284
     5.4.1 Common Platform Structure for OFDM with Staggering 285
     5.4.2 Receiver Side of OFDM with Staggered QAM 289
     5.4.3 T/2 Base Implementation of Transmit Side 290
     5.4.4 Impulse Response, Eye Pattern, and Constellation of Staggered OFDM 293
5.5   Practical Issues 297
     5.5.1 Performance When a Channel Delay Spread > CP 297
     5.5.2 Digital Quadrature Modulation to IF and IF Sampling 299
     5.5.3 Modern FH Implementation with OFDM 301
     5.5.4 Naming of OFDM Signals 302
5.6   OFDM with Coding 304
     5.6.1 Coded Modulations for Static Frequency Selective Channels 304
     5.6.2 Coding for Doubly Selective Fading Channels 305

© Springer Nature Switzerland AG 2020

S.-M. Yang, *Modern Digital Radio Communication Signals and Systems*,

https://doi.org/10.1007/978-3-030-57706-3_5

5.7   Chapter Summary 308
5.8   References and Appendix 309
5.9   Problems 313

Abstract OFDM is a very special form of FDM signal applied to digital transmission by choosing the subcarrier spacing to be related with the subchannel symbol rate and the number of samples per symbol period. We need to choose the shaping filter of subcarrier to be the same. When it is a rectangular pulse it becomes implicit, and most often it results in DMT with CP. In this chapter we extend this basic concept to filtered OFDM, and staggered OFDM, and further show that all different OFDMs can be implemented by commutating filter (polyphase form) in conjunction with IDFT -DFT. The commutating filter is a special form time-varying FIR filter. We consider practical issues, applications of filtered OFDM, and applications to doubly selective fading channels.

Key Innovative Terms common platform with commutating filters · filter method of CP · filter method of windowing · serial transmission OFDM · naming of OFDM signals in use · IDFT – DFT pair as transmission system · SNR penalty of pulse mismatch with CP

General Terms 1-tap equalizer · bit loading · cyclic prefix · delay spread · DMT · eye pattern · FBMC · FMT · IF sampling · permutation · impulse response · iridium · Rayleigh fading · spectrum of OFDM · sub-channel spacing

List of Abbreviations

AM   amplitude modulation
AMPS   advanced mobile phone service
CATV   community antenna TV
CMT   cosine modulated multi-tone
CP   cyclic prefix
CW   continuous carrier wave
D/A   digital analog converter
DFT   discrete Fourier transforms
DMT   discrete multi-tone
DSL   digital subscriber line
DWMT   digital wavelet multi-tone
FBMC   filter bank multi-carrier
FDM   frequency division multiplex
FH   frequency hopping
FIR   finite impulse response
FFT   fast Fourier transforms
FM   frequency modulation
FMT   filtered multi-tone

IDFT	inverse discrete Fourier transform
IEEE	institute of electrical and electronics engineers
IF	intermediate frequency
LAN	local area network
LCD	least common divisor
LTE	long term evolution
OFDM	orthogonal frequency division multiplex
OQAM	offset QAM
PSD	power spectral density
QAM	quadrature amplitude modulation
QPSK	quadrature phase shift keying
RCV	receive
SMT	staggered multitone
SNR	signal to noise ratio
SQAM	staggered QAM
SSB	single side band
TV	television
VSB	vestigial side band
WiMAX	wireless microwave access
XMT	transmit

Let us first look at where OFDM signals belong in overall signal generation and reception. The OFDM signal processing belongs to the block modulation processing (x), B1 – B, and demodulation (r), <u>B1-B</u> in Figure 5-1 which is a repetition of the same block diagram. Once a discrete-time OFDM signal is generated and it will be

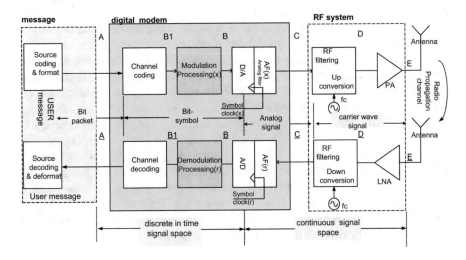

Figure 5-1: Digital radio communication system block diagram

made to complex envelope signals (I and Q), and then it will go through quadrature modulation to make it a CW, and this process is from B to C then D in Figure 5-1.

OFDM is a very special form of frequency division multiplex (FDM) signal applied to digital transmission. This will be explored, in great detail, in this chapter.

A conventional FDM signal is in fact everywhere and was started to be used as soon as CW signal i.e., $A(t) \cos (2\pi f_c t + \theta(t))$, was invented, by assigning different carrier frequencies (f_c) to create different channels. This invention, after inventing filters, amplifiers and oscillators, happened in the early 20th century. This was one of the most significant advancements in communication signals.

It was applied to first AM broadcasting, then FM broadcasting, and then TV broadcasting. Each broadcast station is assigned to a carrier frequency with a fixed bandwidth, 10 kHz, 200 kHz and 6 MHz respectively to AM, FM and TV. Another spectacularly successful use of FDM was in analog voice transmission with SSB modulation. A voice channel of 4 kHz is multiplexed into a group of 12 (48 kHz), then 5 groups (60 channels) are stacked in frequency – so called L-SSB carrier systems. SSB is as efficient as modern QAM where in-phase and quadrature phase are used in the same band while in SSB only one sideband (single sideband), upper or lower, is transmitted. See Appendix 5 A2.

Another example of FDM is the first generation high capacity cellular system, AMPS (advanced mobile phone service) which launched the service in early 80s. It uses 30 kHz channel for each user and its RF carrier frequency is in the range of 825 – 845 MHz or 870 – 890 MHz. With the frequency allocation, 666 voice channels are possible. The bandwidth 30 kHz is far wider than 4 kHz to accommodate FM modulation to deal with the interference, which is mostly due to spatial reuse of frequency. With 18 dB signal to interference ratio, it can meet the required (high) voice quality.

The orthogonality between the channels is attained by a sharp frequency domain filter. In other words, different channels occupy different carriers. The orthogonality means no mixing up between channels.

There were some attempts to use FDM in digital transmission in the late 50s to 70s. See [1], [2], [3], and [4]. I believe that a modern breakthrough came early 70s [1] and but was not widely used for nearly 20 years. The breakthrough is to choose sub-carrier frequency bandwidth (spacing), F, equal to the sub-channel symbol rate ($1/T$). By using a number of sub-carriers, N, the maximum transmission rate can be N/T. The sub carrier is numbered $k = 0, 1, 2, .., N - 1$, and thus each sub carrier frequency is k/T. At this point we mention that the word 'carrier' frequency is replaced by 'sub-carrier' since it is not RF carrier in order to distinguish the difference. The sub-carrier $k = 1$ has one cycle during the OFDM symbol period, and $k = 2$, has two cycles and so on. This seemingly simple idea is a breakthrough since the orthogonality does not rely on the frequency filtering but the orthogonality of $e^{j2\pi \frac{k}{T}t}$ with different $k = 0, 1...N - 1$. As we will see shortly below, its signal generation does not require separate oscillators but can be done numerically and efficiently by IDFT (inverse discrete Fourier transform).

Another choice of parameter is a number of samples per sub-channel symbol period (OFDM symbol time), or sampling rate $(1/\Delta t)$, $\Delta t = \frac{T}{N}$; there are N samples per period. This is the minimum sampling rate, and LN samples per period will do also, $L \geq 1$ as an interpolation factor. For our discussion we choose $L = 1$. With the choice of parameters above, the modulation (or multiplying by $e^{j2\pi\frac{kt}{T}}$) is done by $e^{j2\pi\frac{k}{N}n}$ in discrete time signal processing where n is time index and $-\infty < n < +\infty$.

With this particular choice of subcarrier spacing F and sampling rate Δt, it is possible to use inverse discrete Fourier transform (IDFT), and thus eliminating the need of oscillators (synthesizers or numerical oscillators), and possible to maintain the orthogonality without sharp frequency domain filters. This is the basis of OFDM. We will elaborate it in great detail.

Thus in OFDM sub-carrier spacing F and sample time Δt is related with the OFDM symbol rate as,

$$F = \frac{1}{T} \quad \text{and} \quad \Delta t = \frac{T}{N} \tag{5-1}$$

where N is the total number of sub-channels. By choosing N appropriately, say the power of 2, IDFT can be implemented by a fast Fourier transform (FFT).

As this idea of OFDM is applied to practical situations, soon it became apparent that the orthogonality is too fragile when the channel is dispersive (i.e., its impulse response is not a delta function), and thus sub-carrier spacing has to be increased and the number of samples per period to be increased as well, at the expense of transmission efficiency,

$$F = \frac{1}{T}\left(1 + \frac{v}{N}\right) \quad \text{and} \quad \Delta t = \frac{T}{N+v} \tag{5-2}$$

with $0 \leq v < N$ and in practice it is a small fraction of N. This is effectively creating a gap between OFDM symbols. This is called cyclic prefix (CP) and will be clear next section why it is named that way. When $v = 0$, (5-2) becomes (5-1)

We show the interdependency among sample rate, sub-channel spacing, and OFDM symbol rate as well as a number of sub-channels and cyclic prefix in the table below.

Sample rate, sub-channel spacing, and OFDM symbol rate interdependency

Sample rate $(\frac{1}{\Delta t})$	$\frac{1}{\Delta t}$	$\frac{1}{\Delta t} = FN$	$\frac{1}{\Delta t} = \frac{1}{T}(N+v)$
Sub-channel spacing (F)	$F = \frac{1}{\frac{\Delta t}{N}}$	F	$F = \frac{1}{T}\frac{(N+v)}{N}$
ODFM symbol rate $(\frac{1}{T})$	$\frac{1}{T} = \frac{\frac{1}{\Delta t}}{N+v}$	$\frac{1}{T} = F\frac{N}{(N+v)}$	$\frac{1}{T}$

As can be seen in the above table, sample rate, sub-channel spacing, and OFDM symbol rate are interdependent. For example, a sample rate is given along with N and v, sub-channel spacing, F, and OFDM symbol rate, $1/T$, are followed (column 1 in the table).

Example 5-1: $N = 64$ and $v = 16$ are given with OFDM symbol time $T = 4$ usec. What is sub-channel frequency spacing (F) and sample rate ($\frac{1}{\Delta t}$)?

Answer: The OFDM symbol rate $1/T = 250$ kHz, and $F = 250$ kHz $(80/64) = 312.5$ kHz, and sample rate $= 250$ kHz $(80) = 20$ MHz. These parameters are used in 802.11a shown in Table 5-1.

Example 5-2: $N = 2048$, $F = 15$ kHz, $T = 83.37$ usec are decided. Find out CP in samples and sample rate in MHz.

Answer: sample rate $= 2048 * 15$ kHz $= 30720$ kHz $= 30.72$ MHz, and $1/T = 1/$ $(83.37$ usec$) = 12$ kHz, and thus $N + v = 30720 / 12 = 2560$. $v = 512$. CP time $= 16.67$ usec. This set of parameters is used in LTE shown in Table 5-1.

We comment on the naming. Different names are used in different situations. Some examples are digital multi-tone (DMT) [5], multi-carrier [6], filtered multi-tone (FMT) [7], and filter bank multicarrier (FBMC). We explain different names as we go along. Following [8], we use OFDM generically to cover all the signals of FDM with the choice of sub-carrier spacing and sampling rate using (5-2). A summary of the naming is shown later in Table 5-5 which may be understood clearly after some development of this chapter.

In fact, we will show in this chapter that all the known OFDM signals can be represented by IDFT followed by a commutating filter in the transmit side, and at the receive side by DFT preceded by a matched commutating filter. The development of this 'universal' OFDM is based on Reference [14] and [15]. Furthermore this representation of IDFT – DFT with commutating filters has one-to-one correspondence to an analog system with the same OFDM parameters – symbol rate, number of sub-carriers, sub-carrier frequency spacing and pulse shape.

At the end of reading this chapter you may want to come back and to read this introduction again. You may enjoy it and feel insightful.

5.1 DMT with CP – Block Transmission

Most of OFDM signals used in practice today use CP. For example wireless LAN based on IEEE 802.11a standard, WiMAX based on IEEE 802.16e, 4G LTE, DSL, and digital broadcasting of video and audio. In order to understand the use of CP properly our departure point of exposition is IDFT – DFT pair as a transmission system, and then we explore why and how to add CP. We name this transmission scheme as 'DMT with CP' since it is widely used in practice.

5.1.1 IDFT – DFT Pair as a Transmission System

A pair of IDFT – DFT as a transmission system was explored by Weinstein and Ebert in 1971 [1] as shown in Figure 5-2. $C_k(pT)$ represents a complex modulation symbol such as QAM in the kth sub-channel and p is the sequence index of OFDM symbol. In every T, a set of N modulation symbols are transmitted. Thus the total transmission rate is $\frac{N}{T}$. After IDFT all the sub-channels are combined and its output is given by

$$D^{(m)}(pT) = \sum_{k=0}^{N-1} C_k(pT)\, e^{\,j\frac{2\pi \cdot k \cdot m}{N}} \tag{5-3}$$

Then this is sequentially transmitted from $m = 0, 1, 2, \ldots, N-1$. Of course in actual radio system this sequence should go through D/A, and analog filter followed by quadrature modulator. The real part of $D^{(m)}$ goes to I-channel, and the imaginary part goes to Q-channel of quadrature modulator as shown in Figure 5-3.

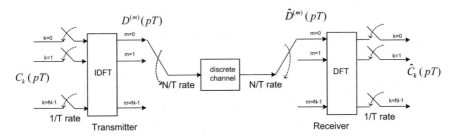

Figure 5-2: IDFT –DFT pair as a transmission system

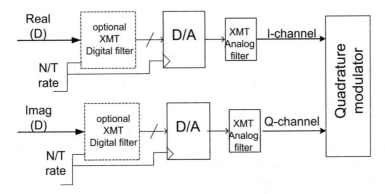

Figure 5-3: Further processing of $D^{(m)}$ to become complex envelope and input to a quadrature modulator

In the receiver essentially the inverse process must be done; quadrature demodulation, analog filtering and A/D to get the samples of $\widehat{D}^{(m)}(pT)$ in Figure 5-2.

Back to Figure 5-2, in order to recover the symbols sent, $\widehat{D}^{(m)}(pT)$ is processed by DFT, $\widehat{C}_k(pT) = \frac{1}{N} \sum_{m=0}^{N-1} \widehat{D}^{(m)} e^{-j\frac{2\pi \cdot k \cdot m}{N}}$,

Assuming that there is no noise, $\widehat{D}^{(m)}(pT) = D^{(m)}(pT)$, it is expressed as

$$\widehat{C}_k(pT) = \frac{1}{N} \sum_{m=0}^{N-1} \left[\sum_{l=0}^{N-1} C_l(pT) e^{j\frac{2\pi \cdot l \cdot m}{N}} \right] e^{-j\frac{2\pi \cdot k \cdot m}{N}} \tag{5-4}$$

After interchanging the summation, we get

$$\widehat{C}_k(pT) = \sum_{l=0}^{N-1} C_l(pT) \left\{ \frac{1}{N} \sum_{m=0}^{N-1} e^{j\frac{2\pi \cdot l \cdot m}{N}} e^{-j\frac{2\pi \cdot k \cdot m}{N}} \right\}$$

In DFT we know that $\frac{1}{N} \sum_{m=0}^{N-1} e^{j\frac{2\pi \cdot l \cdot m}{N}} e^{-j\frac{2\pi \cdot k \cdot m}{N}} = \frac{1}{N} \sum_{m=0}^{N-1} e^{j\frac{2\pi \cdot (l-k)m}{N}} = \delta(l-k)$. In fact this property is essential to define DFT. Thus each sub channel symbol is recovered perfectly, the condition of orthogonality. Keep in mind that we impose the condition $\widehat{D}^{(m)}(pT) = D^{(m)}(pT)$ for all $m = 0, 1, 2, \ldots, N-1$. It appears that the orthogonality is attained with no explicit filtering! In the subsequent sections we will show that there is an implicit filter. It is a rectangular pulse filter.

In practical applications of IDFT – DFT pair there are two immediate problems.

a) D/A and analog filtering need a guard band to make them practical
b) When a channel has a frequency response distortion, in particular delay dispersion, then the orthogonality condition cannot be met.

Fortunately, there is a solution to each problem.

For a), one cannot use all N sub-channels but some sub-channels near the band edge will not be used. Furthermore the DFT size N must be large. In 4G LTE, for example $N = 2048$, and used sub-channels are 1200. In WiMax, $N = 2048$ and used sub-channels are 1681. When N is not large enough, 'windowing' is used to make the transmit spectrum die out quickly at the edge of a channel. For example IEEE 802.11a, wireless LAN, uses windowing where $N = 64$.

For b) a solution is to introduce a gap in time between OFDM symbols; a number of samples per period are increased to $N + v$ from N. This is shown in Figure 5-4.

Additional v samples may be a gap. Then during the time there is no signal transmission, i.e., zero signals. Basically this will solve the problem of maintaining the orthogonality as long as the channel time dispersion (delay spread) is less than v, measured in samples. In practice, the gap is filled with a repetition of a part of the symbol. This can be seen if it is treated as a periodic signal. For a given set of $C_k(pT)$, the output of IDFT is periodic with N; as the index m increases to $N, N + 1, \ldots$ the output is the same as the value when $m = 0, 1, \ldots$ and so on.

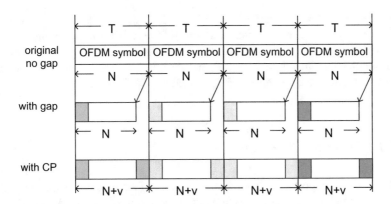

Figure 5-4: Time gap between symbols and addition of cyclic prefix

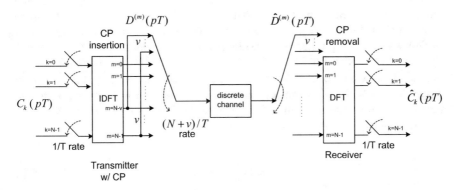

Figure 5-5: IDFT-DFT pair with cyclic prefix (CP)

This CP is useful in recovering the OFDM symbol boundary. However, there is a SNR penalty since it does not carry the information.

Exercise 5-1: Convince yourself that the problem b) can be solved by adding a gap or CP if the delay dispersion is less than CP. Hint: one can write a program, or use a mathematical argument based on equations, or use a graphical display, or a combination of all. This issue will be discussed again later section.

5.1.2   Cyclic Prefix Added to IDFT- DFT Pair

IDFT – DFT pair with CP is shown in Figure 5-5. Now the sample index ranges $m = 0, 1, 2, \ldots N - 1, N, \ldots N - 1 + v$.

OFDM with CP is a success story since virtually all standards using OFDM use this signaling format with different parameters. These are IEEE 802.11a, WiMAX [16], and 4G LTE in wireless area [17], and used in DSL [5] and TV and audio broadcasting. This is summarized in Table 5-1. In DSL area it is called DMT (digital multi-tone), where each sub-channel carries different data rate by changing coding rate, which is called bit loading. This is possible since its twisted pair cable channel does not change much during the call, and thus its frequency response is measured periodically, or in the beginning of a connection or off line measurement.

5.1.3   Transmit Spectrum

In practice it is important to confine the spectrum of a transmit signal to a specified frequency band. We would like to find out the spectrum of CP added OFDM signal. The spectrum must be measured at the antenna port where a transmit antenna is connected. For our purpose we can assume that the carrier frequency translation and amplification is ideal or nearly so. Then we consider the spectrum after D/A and analog filter and it will be good enough. We need to include the frequency response of D/A and an analog filter. We consider two identical sets for both I-channel and Q-channel as indicated in Figure 5-3 since $C_k(pT)$ is a complex. For simplicity of presentation and ease of visualization we consider only the real part of it or we treat $C_k(pT)$ to be real. We will use an example of $N = 5$ and $N + v = 7$, which is small but easy to be shown. The equation (5-3) is rewritten as, in order to see that the sampling time ($\Delta t = \frac{T}{N+v}$) and frequency spacing ($F = \frac{N+v}{N}T$),

$$D^{(m)}(pT) = \sum_{k=0}^{N-1} C_k(pT)\, e^{\, j2\pi \cdot kF \cdot m\Delta t} \tag{5-5}$$

For a given $C_k(pT)$, $m = 0, 1, \ldots, N + v - 1$, and $N + v$ samples will be generated. This is equivalent to the interpolating discrete time filter as shown below, which can be seen clearly for $k = 0$ sub-channel. In other words, the implicit pulse shape is rectangular.

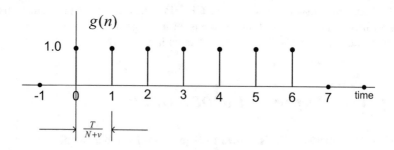

$g(n)$ of the figure above is a discrete version of a rectangular pulse – length $N + v$ samples. Its discrete Fourier transform is given by

Table 5-1: Examples of OFDM systems with CP

System	DFT size (N)	Used sub-channel	Sub carrier spacing kHz	Sample rate MHz	Bandwidth MHz	OFDM symbol time usec	Modulation coding and multi-antenna
IEEE 802.11a	64	52 (48 data + 4 pilot)	312.5	20	16.56	4.0 (3.2 + 0.8)	QPSK, 16QAM, 64QAM Convolution
WiMax	2048	1681	7.81	16	20	144 (128 + 16)	QPSK, 16QAM, 64QAM CTC
LTE	2048	1200	15	30.72	20	83.37 (66.67 + 16.7)	QPSK, 16QAM, 64QAM 2 × 2 MIMO
ADSL	256 (DL) 64(UL)	36 – 127 7-28	4.3125	1.104	1.104	231.9	0.64 – 8.192
DVB - T	2048	1712	4.464	9.174	7.643	224	0.68–14.92

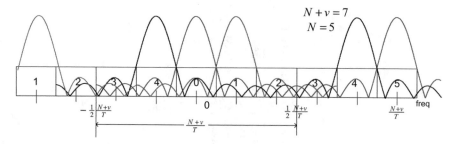

Figure 5-6: Sub-carrier digital spectrum with N = 5 and v = 2 example

$$\text{DFT of } g(n) \rightarrow G\left(e^{j2\pi f \frac{T}{N+v}}\right) = \frac{\sin\left(\pi fT\right)}{\sin\left(\pi fT(N+v)\right)} e^{-j(N+v-1)\pi f \frac{T}{(N+v)}}.$$

And kth sub-carrier translates the above discrete Fourier transform to the sub-carrier frequency $\frac{k \cdot (N+v)}{NT}$. Thus the magnitude of discrete Fourier transform of each sub-carrier is shown in Figure 5-6.

Exercise 5-2: In Figure 5-6, how many sub-carriers, out of N = 5, are used?

Answer: sub-channel number 2 and 3 (Figure 5-6) are <u>not</u> used so that filtering can be done.

Once the signal (5-5) is formed it will go through D/A and its impulse response is a rectangular with the pulse width $\Delta t = \frac{T}{(N+v)}$, the same as sample time. It will limit the spectrum further as shown Figure 5-7. Since the total sub-channel is only $N = 5$, thus D/A is not flat, which requires a compensation so that the frequency response is flat. There is an analog filter to remove high frequency side furthermore. An analog filter may be designed so that D/A and analog filter combined to be flat response LPF.

Another way is to make N large and enough number of subcarriers are not used so that the filtering is easy. As we remarked before, in practical systems, N is in the order of 1000 -2000. Not all sub-channels are usable. See Table 5-1.

So far the spectrum of DMT OFDM transmit signal is visually shown with a very simple example as in Figure 5-7, and we showed that each individual sub-channel spectrum is the same as that of a discrete rectangular pulse in time domain. Now we extend this idea of obtaining the actual transmit spectrum to practical cases. The idea is shown in Figure 5-8; a summary of the chain of signals for the spectrum computation is shown in the figure. The transmit spectrum before D/A is the squared sum of the spectrum of each sub-channel used in actual signal generation. This is indicated in the figure.

The measurement of the spectrum with a spectrum analyzer is equivalent to generating transmit signals, long enough to meet the spectrum resolution requirement, and then taking a discrete Fourier transform. To smooth out the computed spectrum, often multiple DTFs of different signals are averaged. It is equivalent to video filtering in spectrum analyzer measurement.

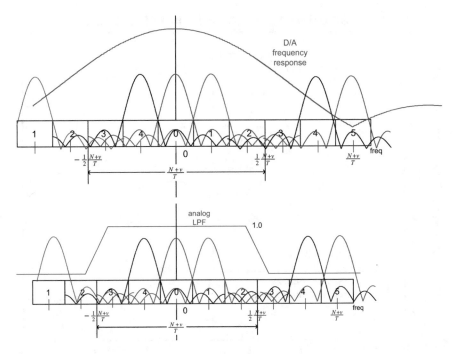

Figure 5-7: D/A frequency response and analog filter are imposed

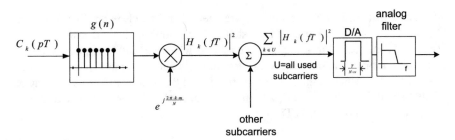

Figure 5-8: A signal chain in computing the signal spectrum and the spectrum before D/A is the squared magnitude sum of each sub-channel spectrum

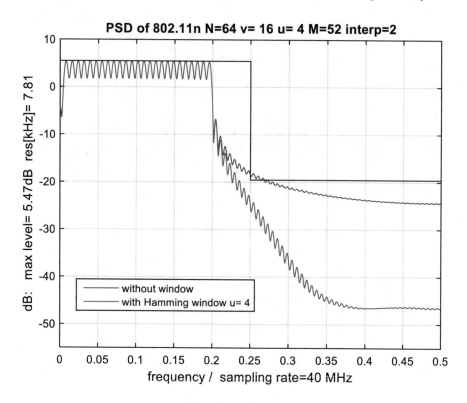

Example 5-3: IEEE 802.11a signal is used here. IDFT size $N = 64$, CP $v = 16$ and $M = 52$ sub-carriers are used. In the standard 'windowing' is used with $u = 4$ for spectrum confinement to a frequency band. This widowing will be explained later in this chapter. Note that the windowed spectrum dies out much more quickly and that DC channel is empty. A numerical confirmation of the method in Figure 5-8 may be done as well. See Chap. 2, Sect. 2.9, in particular Sect 2.9.3 for additional details of numerical computation of PSD.

5.1.4   OFDM Symbol Boundary with CP

As shown in Figure 5-5, the transmission system using IDFT-DFT pair with CP is a block transmission – one OFDM symbol as a block. But in the channel it is a serial transmission.one sample at a time. In the receiver the first process, after carrier demodulation and sampling, is to find the OFDM symbol boundary so that CP can be removed before getting into DFT. The details of this section may be skipped in the first reading without the loss of continuity.

As shown in Figure 5-4, CP is a repetition of part of an OFDM symbol. In order to find the symbol boundary it is utilized. In other words, it uses the fact that CP is repeated in an OFDM symbol, and separated by N samples. Our discussion is based [9], and see it for additional details. Another methods, not pursued here, are in [10], [11], and [12].

A correlation is computed in order to find OFDM symbol boundary. The received sample is denoted $r(n)$ where n is a sample time index. Note that the summation range is limited to the size of CP (v) and delay is N.

$$R(m) = \sum_{i=0}^{v-1} r(m+i)\, r^*(m+i+N) \tag{5-6}$$

We also need to compute the energy of the samples so that we have a discrimination function throughout the search span of samples.

$$\Phi(m) = \frac{1}{2} \sum_{i=0}^{v-1} r(m+i)r^*(m+i) + r(m+i+N)r^*(m+i+N) \tag{5-7}$$

One may use the discrimination function given by

$$D(m) = \frac{|R(m)|}{\Phi(m)} \tag{5-8}$$

A numerical example is shown in Figure 5-9

The correlation based on CP is also used to obtain the frequency difference, but here we do not concern with it.

In practice the computation of (5-6) and (5-7) can be done from stored samples and two blocks of samples are enough to find the peak.

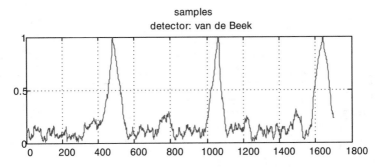

Figure 5-9: OFDM symbol boundary detection and its discrimination function ($D(m)$) with $N = 512$ and $v = 68$. The peak appears periodically per $N + v$ samples

5.1.5 Receiver Processing When the Channel Dispersion < CP

Once the OFDM symbol boundary is located, N samples, after discarding CP, get into DFT so that each sub-channel can be recovered. We show that if the channel impulse response delay spread is less than CP, one tap equalizer is enough for each sub-channel. One tap equalizer adjusts gain and phase.

First consider the following example.

Example 5-4: In Figure 5-10 it is a bit hard to show but within a symbol any consecutive N samples out of $N + v$, CP being removed, can get into DFT to recover the transmitted symbols properly. The phase of each sub-carrier must be appropriately adjusted to recover the transmitted symbols.

Answer: In order to recover a transmitted symbol we do DFT or demodulation. If there is an offset, ε, in sampling of N samples, then

$$\widehat{C}_k(pT) = \frac{1}{N}\sum_{m=0}^{N-1} \widehat{D}^{(m+\varepsilon)} e^{-j\frac{2\pi \cdot k \cdot m}{N}} \quad \text{where} \quad \widehat{D}^{(m+\varepsilon)}(pT) = \sum_{l=0}^{N-1} C_l(pT) e^{j\frac{2\pi \cdot l \cdot (m+\varepsilon)}{N}}.$$

Combining two together we obtain, after interchanging the summation sequence, using the identity $\frac{1}{N}\sum_{m=0}^{N-1} e^{j\frac{2\pi \cdot l \cdot m}{N}} e^{-j\frac{2\pi \cdot k \cdot m}{N}} = \delta(l-k)$,

$$\widehat{C}_k(pT) = \sum_{l=0}^{N-1} C_l(pT) e^{j\frac{2\pi \cdot l \cdot \varepsilon}{N}}\left\{\frac{1}{N}\sum_{m=0}^{N-1} e^{j\frac{2\pi \cdot l \cdot m}{N}} e^{-j\frac{2\pi \cdot k \cdot m}{N}}\right\} = \sum_{l=0}^{N-1} C_l(pT) e^{j\frac{2\pi \cdot l \cdot \varepsilon}{N}} \delta(l-k).$$

$$\widehat{C}_k(pT) = C_k(pT) e^{j\frac{2\pi \cdot k \cdot \varepsilon}{N}}$$

$$(5-9)$$

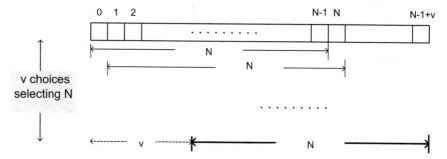

Figure 5-10: Different phases of N samples in a block of $N + v$ samples; the last one of v choices does not need subcarrier phase shift; in $\varepsilon = 0.0$ equation (5-9)

Remember that when CP, v, is added to $\{0, 1, 2, 3,\ldots .N - 1\}$index, $\epsilon=0$ in (5-9) corresponds to the vth delay in Figure 5-10 (last thick one) in order to eliminate subcarrier phase shift.

5.1.5.1   Gain and Phase Adjustment – "1-Tap Equalizer"

Now a channel impulse response is, as an example, represented by a main tap and another tap with a gain, g_1 and delay ε_1 samples, less than CP. From the above example, it can be shown that the symbol after DFT is given by

$$\widehat{C}_k(pT) = C_k(pT)\left[1.0 + g_1 e^{j\frac{2\pi \cdot k \cdot \varepsilon_1}{N}}\right] \qquad (5\text{-}10)$$

Thus by adjusting gain and phase out, i.e., multiplying $1/\left[1.0 + g_1 e^{j\frac{2\pi \cdot k \cdot \varepsilon_1}{N}}\right]$, the recovered symbol is the perfect replica of the transmitted one. This gain and phase adjustment is called 1-tap equalizer. See Figure 5-11 (top).

5.1.5.2   System Degradation if Delay Spread Is Bigger than CP

What happens if the impulse response of a channel has a delay spread beyond CP? A short and correct answer is that the orthogonality is destroyed. Next question is then how bad it is, or how much degradation.

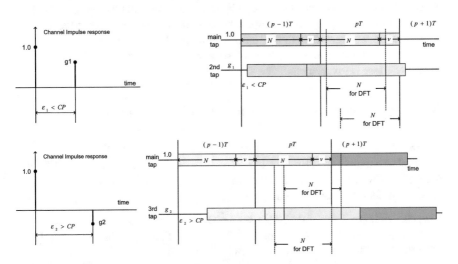

Figure 5-11: Illustrating the system when the delay spread is less than CP (top) and when it is greater than CP (bottom)

We answer the first question. Consider a case with the channel impulse response has three taps; the main tap, 2nd tap and 3rd tap. The delay of the 2nd tap is less than CP but the 3rd tap exceeds. Then there is interference due to 3rd tap. This is shown in Figure 5-11.

Note that N samples for DFT, out of $N + v$ samples per OFDM period, may have a range. But due to the 3rd tap the interference is unavoidable. Once DFT is done to N samples, it is not possible to get rid of the interference; no equalization is possible. This is called inter-channel interference (ICI). ICI can be equalized before DFT. An equalized impulse response may reduce or remove the ISI taps (in particular, 3rd taps). ISI of ODFM symbols become ICI in sub-channels.

For the 2nd question of how much degradation, the answer is involved. Essentially one must compute $N \times N$ impulse responses. We will cover this issue briefly later in Sect. 5.5. See Figure 5-38.

5.1.6   SNR Penalty of the Use of CP

In the literature it seems not explicitly mentioned but, the use of CP in OFDM has SNR penalty due to pulse mismatch in addition to the loss of bandwidth efficiency. This mismatch loss is covered in Chap. 3. In single carrier systems with an excess bandwidth there is no SNR loss due to the mismatch but, only the loss of bandwidth efficiency; it occupies more than the minimum bandwidth but there is neither increase in transmit power nor the noise bandwidth increase at the receiver.

CP is redundant and uses a part of transmit power, and thus there is a power loss. This transmit power loss can be avoided by sending zero signal instead of CP, which is not often done in practice. The noise bandwidth of a receiver is not the same as symbol rate but increased by the CP amount (mismatch loss). Let us denote DFT size be N which is the maximum number of subcarrier usable and v be CP in samples. Then SNR loss is given by

$$\text{SNR loss due to } CP = 10 \log_{10}\left(1 + \frac{v}{N}\right)^2 \text{ [dB]}$$

For example, with $N = 1024$ and $v = 136$, SNR loss = 1.1 dB.

Another example is IEEE 802.11a; $N = 64$, $v = 16$, and SNR loss = 1.94 dB. SNR loss is relative to without CP or to frequency filtered OFDM (which will be covered in Sect. 5.3) with the same parameters.

Exercise 5-3: In Table 5-1, LTE uses OFDM symbol time $T = 83.37$ usec and $CP = 16.67$ usec. See also Example 5-2. What is SNR loss due to CP in LTE? Answer: 1.94 dB = $20*\log_{10} (83.37 / (83.37-16.67))$.

One way to reduce this SNR loss is to use a large N and small CP. Some might argue that in a sense this SNR loss is well utilized in symbol synchronization. CP is indeed well utilized for recovery of OFDM symbol boundary discussed in 5.1.4 above.

5.2 CP Generalized OFDM – Serial Transmission

OFDM with CP we discussed so far is a successful signaling scheme adopted by various standards. Our starting point was IDFT – DFT pair and then inserted CP to maintain orthogonality under channel dispersion. When we computed the transmit spectrum, we notice that there is implicit pulse shaping, which is a discrete rectangular pulse. We called this type of OFDM, DMT with CP, here to distinguish it from the signaling scheme that will be developed here, similar but not exactly the same. Our scheme may be called CP generalized OFDM, or simply OFDM with CP. Our development is based on [15].

Our new starting point is a conventional FDM using filters and oscillators. However, we choose the subcarrier spacing (F) to be related with the subchannel symbol rate (OFDM symbol rate: $\frac{1}{T}$) and the number of samples per symbol period (T) is an integer ($N + v$). This is the same as (5-2), repeated here; $F = \frac{1}{T}\left(1 + \frac{v}{N}\right)$ and $\Delta t = \frac{T}{N+v}$. On top of it, we choose the shaping filter to be a rectangular pulse.

Jumping to the conclusion, with this approach, we arrive at a form of OFDM with CP, without explicitly inserting and removing it, but implicitly part of shaping filtering. An immediate application is to do 'windowing'. The windowing is used to smooth out the transmit spectrum, e.g., IEEE 802.11a, in an" awkward" way but both are equivalent.

This new method is called filter method since it is the result of interpreting the underlying pulse shaping filter is rectangular in time domain. However, in the filter method there is no explicit insertion and removal of CP, but implicitly part of filtering process. One immediate advantage of this filter method is that 'windowing' can be implemented as part of filtering as well. The windowing is necessary to meet transmit spectrum mask since the spectrum of a strictly rectangular pulse does not die out quickly. Thus the pulse edge must be smoothed out by windowing. With this new filtering approach the windowing is simply an extension of pulse to next OFDM symbol.

Furthermore this formulation will be extended to include a frequency domain filter, and thus provides a common platform for all known OFDM signals which can be implemented efficiently by using IDFT - DFT and commutating polyphase filters, which will be elaborated later in great detail.

5.2.1 OFDM – Analog Representation

An analog OFDM system, a pair of transmit and receive, is shown in Figure 5-12.

The transmit signal, $x(t)$, is given by

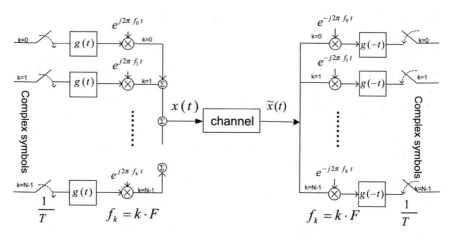

Figure 5-12: An analog OFDM system; filters are matched (time reversed impulse response) and oscillators are complex exponentials with subcarrier spacing $F = \frac{1}{T}\left(1 + \frac{v}{N}\right)$

$$x(t) = \sum_{k=0}^{N-1}\left[\sum_{l=-\infty}^{\infty} C_k(lT)g(t - lT)\right]e^{\,j2\pi \cdot k \cdot F \cdot t} \qquad (5\text{-}11)$$

This may be considered as a generic FDM, i.e., filtered base band signals inside the [ ] bracket are frequency modulated and summed. This generic FDM becomes OFDM with specific choice of subcarrier spacing as we discussed. We choose subcarrier spacing $F = \frac{N+v}{N \cdot T}$ with $N > v \geq 0$ and shaping pulse $g(t)$ to be 1.0 for $0 \leq t \leq T$ OFDM symbol time T. $C_k(lT)$ is a complex data symbol of kth sub-channel at lT OFDM symbol time.

5.2.2   Discrete Signal Generation of Analog OFDM

With the sampling time $\Delta t = \frac{T}{N+v}$, there are $N + v$ samples per OFDM symbol time T, and $t = n\Delta t = (p(N + v) + m)\Delta t$ and $m = 0, 1, .., N - 1, N, .., N + v - 1$. For a given set of N data C_k, $N + v$ samples are needed to represent the signal, and p tracks a number of OFDM symbols corresponding to a set of data C_k, and m is the index within one symbol time.

A sampled version of (5-11) becomes, after interchanging two summations,

$$x(n\Delta t) = \sum_{l=-\infty}^{\infty}\left[\sum_{k=0}^{N-1} C_k\left(\widehat{l}\right)e^{\,j\frac{2\pi \cdot k \cdot n}{N}}\right]g\left(\left(p - \widehat{l}\right)T + m\Delta t\right)$$

At this point it is slightly more convenient to change the summation index to $l = p - \widehat{l}$ and we obtain a sampled version

$$x(n\Delta t) = \sum_{l=-\infty}^{\infty} \left[\sum_{k=0}^{N-1} C_k(p-l)e^{j\frac{2\pi \cdot k \cdot n}{N}} \right] g(lT + m\Delta t) \qquad (5\text{-}12)$$

Since $g(t)$ is rectangular for the duration of $0 \le t \le T$, only $l = 0$ term in the summation remains. This single term for the shaping filter is called $L = 1$. When the filter has longer than one symbol period, L will be greater than 1. For windowing it is $L = 2$, which will be described later. Thus (5-12) becomes (5-13).

$$x(n\Delta t) = \left[\sum_{k=0}^{N-1} C_k(pT)e^{j\frac{2\pi \cdot k \cdot n}{N}} \right] g(m\Delta t) \qquad (5\text{-}13)$$

where $n = p(N + v) + m$, $m = 0, 1, \ldots, N + v - 1$, $p \in$ integer, and $g(m\Delta t) = 1.0$.

The equation (5-12) and (5-13) clearly indicates that IDFT can be used for generating transmit signal. Note that frequency index k in IDFT ranges 0 to $N - 1$. However, the time index n can be outside of the range. The result of IDFT for n bigger than $N - 1$ can be found by taking n mod N since IDFT is periodic with the period of N.

Thus the mth sample in a pth symbol of $x(n) = x(p(N + v) + m)$ can be computed from IDFT output of a pth symbol, index $jm = (p(N + v) + m)$ mod N. This is called address commutating or simply commutating. This is shown in Figure 5-13. While n can run indefinitely but in implementation only a finite number of OFDM symbols needs to be tracked. When $p = 0$, $jm = m$ and $p = N$, $jm = m$. In general when $p = lcm(N, N + v)/(N + v) = \widetilde{N}$, $jm = m$. Thus p ranges from 0 to $N - 1$ at most, i.e., $\widetilde{N} \le N$.

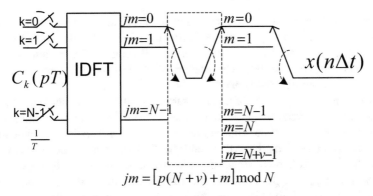

$$jm = [p(N+v)+m] \bmod N$$

Figure 5-13: Filter method of 'CP insertion'; no explicit insertion but N samples of IDFT output become $N + v$ samples by repeating a part of N samples every OFDM symbol time

Example 5-5: A numerical indexing example with $N = 9$, $v = 2$ is shown in a table below. Remember that $jm = (p(N + v) + m)$ mod$N$ and $n = p(N + v) + m$ where p OFDM symbol index and m sample index within OFDM symbol.

Address index example with $N = 9$, $v = 2$; p = OFDM symbol index

	p=0	p=1	2	3	4	5	6	7	8	p=9	10	11	
m=0	jm=0	2	4	6	8	1	3	5	7	jm=0	2	4	....
1	1	3	5	7	0	2	4	6	8	1	3	5	....
2	2	4	6	8	1	3	5	7	0	2	4	6	....
3	3	5	7	0	2	4	6	8	1	3	5	7	....
4	4	6	8	1	3	5	7	0	2	4	6	8	....
5	5	7	0	2	4	6	8	1	3	5	7	0	....
6	6	8	1	3	5	7	0	2	4	6	8	1	....
7	7	0	2	4	6	8	1	3	5	7	0	2	....
8	8	1	3	5	7	0	2	4	6	8	1	3	....
9	0	2	4	6	8	1	3	5	7	0	2	4	....
10	1	3	5	7	0	2	4	6	8	1	3	5	....

For every new $C_k(pT)$ or for a given symbol index, p, the index m runs from 0, 1, .. to $N + v - 1$ but, the starting point of jm is (pv) mod N and the ending point is $(pv + v - 1)$ mod$N$. Make sure this is the case in the table above.

On the other hand, the existing method of inserting CP in DMT is to force jm runs from 0, 1, .. to $N - 1$ for every new $C_k(pT)$. In fact, the starting point of jm can be fixed and it can be any from 0 to $N - 1$.

Exercise 5-4: Show that the last statement is true as an example $jm = 2, 3, \ldots$, $N - 1, 0, 1$ for a given $C_k(pT)$. Hint: Observe the table above when $p = 1$ case.

However, this fixed choice of jm index for all p makes the system as a block transmission. In other words, at the receiver, the OFDM symbol boundary must be known. There is no other choice.

With the new filter method of CP, the starting point of jm for each p changes and in this way the subcarrier phase is continuous. But the relative position of the repetition is the same in time. Thus at the receiver there is a choice. It can be treated as a block transmission if DFT is done once per OFDM symbol time i.e., the OFDM symbol boundary is known. However, if DFT is done more than once per OFDM symbol time, e.g., DFT per every sample, then there is no need of knowing OFDM symbol boundary before DFT. It can be recovered as a symbol timing recovery at each subchannel. This point will be clarified further after we discuss the receive side.

5.2.3 Discrete Signal Reception

Let us go back to Figure 5-12, analog representation of OFDM. In actual systems, there is a quadrature modulator to generate carrier wave for radio transmission and a corresponding quadrature demodulator at the receiver, which generates a complex envelope $\tilde{x}(t)$. We assume that the channel, including modulator and demodulator, is perfect and thus $\tilde{x}(t) = x(t)$. We also use the receive filter $h(t)$, which is not the matched filter shown in Figure 5-12 where the receive filter is a matched filter, $g(-t)$. In fact, for orthogonality, we do not use exactly a matched filter at the receiver. This will be clarified further soon.

An analog form of received signal without noise for the kth sub-channel is represented by

$$y_k(t) = \left[x(t)e^{-j2\pi\cdot k\cdot F\cdot t} \right] \otimes h(t) \tag{5-14}$$

where $\otimes$ denotes convolution by a receiver shaping filter, $h(t)$ rather than $g(-t)$ which is matched to transmit filter. This is shown in the receiver side of Figure 5-12.

A sampled version of (5-14), with $\Delta t = \frac{T}{N+\nu}$, is given by

$$y_k(n\Delta t) = \sum_{l=-\infty}^{\infty} x((n-l)\Delta t)\, e^{-j2\pi\cdot k\cdot F\cdot(n-l)\Delta t} h(l\Delta t)$$

where a convolution becomes a summation in discrete time domain, rather than an integral.

The summation can be decomposed into two, by setting $l = q \cdot N + i$ with $i = 0$, $1,...N - 1$, $q \in$ integer, and

$$u_i(n\Delta t) = \sum_{q=-\infty}^{\infty} x(n\Delta t - (qN + i)\Delta t)h((qN + i)\Delta t) \tag{5-15}$$

$$y_k(n\Delta t) = \sum_{i=0}^{N-1} u_i(n\Delta t)\, e^{-j\frac{2\pi\cdot k(n-i)}{N}} \tag{5-16}$$

These two (5-15) and (5-16) are key equations; (5-15) is a filtering and (5-16) is DFT. Clearly the receiver can be implemented pulse shaping filtering followed by DFT. Note that this derivation up to now does not depend on filter type. We need to clarify the index relationship.

(5-15) is a filtering of incoming signal with $x(n\Delta t)$, and that for one of its sample produces N outputs. These N outputs of filtering go through DFT using (5-16) to generate all sub-carrier channel signals. Before getting into DFT index must be properly aligned since for a given i of $u_i(n\Delta t)$, DFT time index is $im = (n - i) \bmod N$. This index alignment is called commutating.

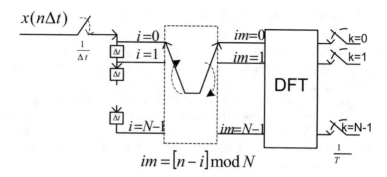

$$im = [n - i] \bmod N$$

Figure 5-14: Filter method of 'CP removal'; here for every sample of $x(n\Delta t)$, i.e., for every n, DFT is computed after 'address commutating' with $(n-i)$ mod N

We will explore in detail with a general receive filter, $h(t)$ later. Here we restrict it to be a rectangular pulse to match with the transmit filter. Since we need to drop the repetition part (or CP), $h(t)$ is rectangular pulse and its duration is shortened $\frac{N}{N+v}T$. This shortening of $h(t)$ is essential for orthogonality. Its sampled version $h((qN+i)\Delta t)$ is zero other than a single term with $q = 0$ in the summation for $u_i(n\Delta t)$.

(5-15) and (5-16) can be simplified and are given by,

$$u_i(n\Delta t) = x(n\Delta t - i\Delta t)h(i\Delta t) \tag{5-17}$$

$$y_k(n\Delta t) = \sum_{i=0}^{N-1} u_i(n\Delta t)\, e^{-j\frac{2\pi \cdot k(n-i)}{N}} \tag{5-18}$$

with $h(i\Delta t) = 1.0$ for $i = 0, 1,\dots N - 1$.

(5-17) and (5-18) can be represented (implemented) in Figure 5-14.

Let us look at (5-17) and (5-18) in conjunction with Figure 5-14. First address commutating. A case of $N = 6$, $v = 2$ is tabulated; for a given n and i, $(n - i)$ mod N is computed in the table below. As expected, $(n - i)$ is repeated per N. DFT is computed for every sample, i.e., $N + v$ DFT per symbol period.

Exercise 5-5: In Table 5-2, complete it, i.e., find im index when $n = 13, 14, 15$ cases.

Answer: $n = 13$; $im = 1, 0, 5, 4, 3, 2$ and $n = 14$; $im = 2, 1, 0, 5, 4, 3$ and $n = 15$; $im = 3, 2, 1, 0, 5, 4$.

There is no reason to compute DFT every sample. It can be done once per symbol period when OFDM symbol time boundary is known. Or it can be done twice per symbol time without knowing the boundary shown in Table 5-3; n jumps by 4.

Exercise 5-6: In Table 5-3, when $n = 1, 5, 9, 13, 17\dots$ cases find the corresponding im index. Think of other choices of n, and its meaning.

Answer: $im = [n - i] \bmod N$. There are half of $N + v$ choices ($N + v$ is even), and the signal delay (phase) will be different.

Table 5-2: Address commutating with N = 6, v = 2 where DFT is computed for every sample

$im = [n - i] \, mod \, N$	i					
n	0	1	2	3	4	5
0	0	5	4	3	2	1
1	1	0	5	4	3	2
2	2	1	0	5	4	3
3	3	2	1	0	5	4
4	4	3	2	1	0	5
5	5	4	3	2	1	0
6	0	5	4	3	2	1
7	1	0	5	4	3	2
8	2	1	0	5	4	3
9	3	2	1	0	5	4
10	4	3	2	1	0	5
11	5	4	3	2	1	0
12	0	5	4	3	2	1
13						
14						
15						

Table 5-3: N = 6, v = 2 address commutating where 2 DFT computations per OFDM symbol

$im = [n - i] \, mod \, N$	i					
n	0	1	2	3	4	5
0	0	5	4	3	2	1
4	4	3	2	1	0	5
8	2	1	0	5	4	3
12	0	5	4	3	2	1
16	4	3	2	1	0	5
20	2	1	0	5	4	3
24	0	5	4	3	2	1
28	4	3	2	1	0	5
32	2	1	0	5	4	3
36	0	5	4	3	2	1
40	4	3	2	1	0	5

Exercise 5-7: For Table 5-3 reduce it when 1 DFT computation per OFDM symbol. Hint: n jumps by 8.

This new filter method is a true digital implementation of analog representation in Figure 5-12. On the other hand CP method of DMT focuses on transfer in blocks of OFDM symbols $C_k(p)$. CP must be removed before DFT processing. This implies that OFDM symbol boundary should be known in order to remove CP. With the filter method there is a flexibility of sampling after DFT if there are more than one DFT per OFDM symbol.

5.2.4 Pulse Shape of DMT and its End to End Pulse

We obtained a pair of digital transmitter in Figure 5-13 and digital receiver Figure 5-14 with rectangular pulse filters, directly from Figure 5-12 which is an analog representation of OFDM. They are a literally true digital implementation of Figure 5-12. As it turns out that the result is a form of generalization of CP method. It is clear that DMT with CP uses rectangular pulses as shaping filters.

In Figure 5-15, transmit, receive, and end –to –end pulses are shown. The end -to-end pulse is observable from subchannel to subchannel. The pulse mismatch loss here is given by $10\log_{10}(1 + \frac{v}{N})$ [dB]. The same amount of transmission power loss since CP does not carry information. Thus the total SNR loss is twice the mismatch loss, i.e., $20\log_{10}(1 + \frac{v}{N})$ [dB]. This was discussed in Sect. 5.1.6.

5.2.5 Windowing

When there is a need to smooth out the transmit power spectrum, the 'windowing' is often applied. With the filter method it is natural to implement a 'windowing' filter. The pulse is extended to next symbol, say $L = 2$ case.

The windowing is a form of filter and the transmit side is implemented as shown in Figure 5-17.

The previous $((p - 1)T)$ output of IDFT is stored. Both current and stored DFT outputs are used for filtering. The address commutating is exactly the same as before.

Figure 5-15: Transmit pulse, receive pulse and end-to-end pulse

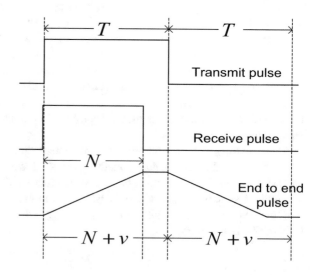

Figure 5-16: The edge of transmit pulse is smoothed out for windowing

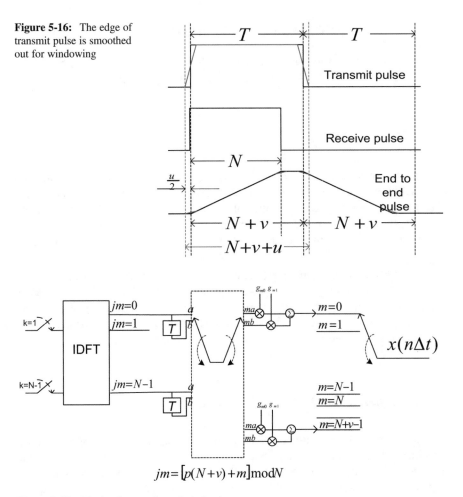

Figure 5-17: The implementation of windowing

5.2.6   *Filter Method Compares with* **DMT** *with* **CP**

The filter method of adding CP is a serial transmission while a typical OFDM with CP, *DMT with CP*, is a block transmission, which means OFDM symbol boundary must be recovered before DFT at receive side. On the other hand the filter method does not need to know OFDM symbol boundary before DFT but its sampling must be twice (2(N + v) sample per OFDM symbol) or higher. If it is sampled at N + v per OFDM symbol, OFDM symbol boundary must be known before DFT as well. The new filter method is more flexible at the receiver.

OFDM symbol boundary can be recovered by correlation discussed in 5.1.4. The exactly same method should be applicable to this new filter method.

We discussed SNR penalty in filter method is exactly the same as CP method.

The need of 1-tap equalizer discussed in 5.1.5.1 should be applicable to this filter method as well.

Both are essentially the same in many respects, which is not surprising since both are nearly the same except for the way CP is inserted and removed.

5.3   Filtered OFDM

So far we discussed the case where the shaping filter is rectangular in time domain, both DMT with CP and the filter method of CP. In this section we consider a frequency domain filter such as an ideal low pass or RC filters. The orthogonality depends on the frequency domain filtering similar to conventional FDM. We will show that by choosing the subcarrier spacing and the sample time for OFDM, $F = \frac{1}{T}\left(1 + \frac{v}{N}\right)$ and $\Delta t = \frac{T}{N+v}$ respectively, we obtain a common platform structure, i.e., IDFT followed by a commutating filter at the transmit side (Figure 5-18), and at the receiver a commutating filter followed by DFT (Figure 5-20). This is called a common platform since it can implement, with this structure, practically all OFDM systems of different pulse shaping filters including rectangular pulse. This is summarized in Figure 5-21. Our development is based on [14].

5.3.1   Filtered OFDM Signal Generation

When we derived the equation (5-12), Figure 5-12 is represented by discrete in time, and then we specialized it to a rectangular pulse (L = 1) to obtain (5-13). Another specific pulse was a windowed pulse (L = 2). For convenience, we repeat the equation (5-12) here, with a specialization of $g(lT + m\Delta t)$ to be a finite impulse response (FIR) of the length L, and $l = 0, 1, \ldots, L - 1$.

$$x(n\Delta t) = \sum_{l=0}^{L-1}\left[\sum_{k=0}^{N-1} C_k(p - l)e^{j\frac{2\pi \cdot k \cdot n}{N}}\right]g(lT + m\Delta t) \qquad (5-19)$$

where $n = p(N + v) + m$, $p \in$ integer, and $m = 0, 1, \ldots, N + v - 1$.

This equation (5-19) can be represented by IDFT followed by commutating filter structure as shown in Figure 5-18. In other words, the OFDM signal with a FIR filter can be obtained with this structure.

We explain Figure 5-18. The IDFT outputs at pT, are stored at L memory shift registers, and one set of these stored outputs indexed by jm is convolved with a set of the filter coefficients indexed by m. The matching between jm and m is given by $jm = (p(N + v) + m)$ mod$N$. For each set of IDFT, when $p = 0$ or a multiple of

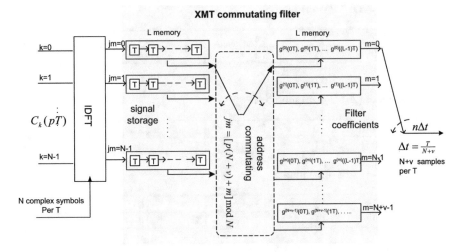

Figure 5-18: IDFT plus XMT commutating filter structure to implement (5-19)

N, $m = 0$ corresponds to $jm = 0$. Otherwise the position of jm is not zero but the subcarrier phase is continuous. This address commutating is exactly the same as Figure 5-13 or Figure 5-17.

Exercise 5-8: Show why for each set of IDFT, jm cannot start from 0 but the continuation from the previous set while m always starts from 0. Hint: v additional samples are not exactly the repetition but the interpolated to $N + v$ samples from N samples.

In Figure 5-18 the filter coefficients $g(lT + m\Delta t)$ is represented by $g^{(m)}(l)$ $m = [0, 1, .., N + v - 1]$ and $l = [0, 1, .., L - 1]$. It can be organized as an $(N + v)$ by L matrix as shown below.

$g^{(0)}(0T)$	$g^{(0)}(1T)$	....	$g^{(0)}((L-1)T)$
$g^{(1)}(0T)$	$g^{(1)}(1T)$	....	$g^{(1)}((L-1)T)$
....			
$g^{(N+v-1)}(0T)$	$g^{(N+v-1)}(1T)$	....	$g^{(N+v-1)}((L-1)T)$

Example 5-6: Show that Figure 5-18 is a true representation of (5-19) by working out the index relationship with a specific choice of N, v and L (say $N = 5$, $v = 2$, $L = 4$).

Answer:

$p = 0$: $jm = (p * 2 + m) \bmod 5$

jm	0	1	2	3	4	0	1
m	0	1	2	3	4	5	6

$p = 1: jm = (p * 2 + m) \bmod 5$

jm	2	3	4	0	1	2	3
m	0	1	2	3	4	5	6

$p = 2: jm = (p * 2 + m) \bmod 5$

jm	4	0	1	2	3	4	0
m	0	1	2	3	4	5	6

$p = 3: jm = (p*2 + m) \bmod 5$

jm	1	2	3	4	0	1	2
m	0	1	2	3	4	5	6

$p = 4: jm = (p*2 + m) \bmod 5$

jm	3	4	0	1	2	3	4
m	0	1	2	3	4	5	6

$p = 5: jm = (p*2 + m) \bmod 5, p = 5 \bmod 5 = 0$

jm	0	1	2	3	4	0	1
m	0	1	2	3	4	5	6

$p = 6: jm = (p*2 + m) \bmod 5, p = 6 \bmod 5 = 1$

jm	2	3	4	0	1	2	3
m	0	1	2	3	4	5	6

And so on. Thus it is clear that the range of p to track is 0 to $N - 1$ at most. Additional v samples per T, similar to cyclic prefix, are added as part of filtering.

5.3.1.1   Transmit Spectrum of Filtered OFDM

The transmit spectrum of filtered OFDM is shown in Figure 5-19 with $N = 5$ and $v = 2$. In filtered OFDM the orthogonality between subchannels is obtained by frequency filtering just like a conventional FDM. With the example of Figure 5-19, the excess bandwidth of an OFDM filter should be less than 0.4 $\left(= \frac{v}{N}\right)$. In the figure the subchannel filter is square root raised cosine shape with the excess bandwidth 0.4. Thus the subchannel bandwidth occupied by the signal is $F = \frac{1}{T}\left(1 + \frac{v}{N}\right)$, subchannel frequency spacing.

The length of FIR filter, $g\,()$, is $(N + v) * L = 7 * 4 = 28$ with this example. We may consider using a square root raised cosine filter for FIR filter, $g\,()$. However,

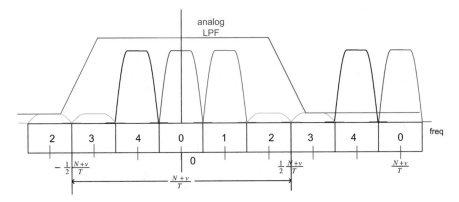

Figure 5-19: Filtered OFDM subcarrier digital spectrum with N = 5 and v = 2. OFDM filter has the excess bandwidth 0.4 (v/N). Compare this with Figure 5-6 of DMT with CP

optimizing it for filtered OFDM is very important in practice but beyond the scope here, and will be treated in Chap. 3.

Out of $N = 5$ subchannels, only 3 subchannels are used in Figure 5-19. And the same 3 subchannels are also used in Figure 5-6 for easy comparison. However, in filtered OFDM, since the subchannel filter is sharp at the channel edge frequency, it is considerably more flexible on the IDFT size of N. Additional subchannels are not needed to meet the frequency spectrum (band limiting) mask, but may be added for subsequent analog filtering to remove the high frequency of 'digital harmonics'.

Exercise 5-9: In Figure 5-19, the subchannel k = 2 is shown to be empty. If it is used what is the consequence? Hint: analog LPF has to be changed.

5.3.2 Filtered OFDM Signal Reception

For the receiver side, our starting point is the equations (5-15) and (5-16). Rather than specializing into (5-17) and (5-18) with a rectangular pulse ($L = 1$), we use a FIR filter with the length LN, which is not the same as the transmit filter length of $L (N + v)$. We overuse the same L to represent the filter length.

$$u_i(n\Delta t) = \sum_{q=0}^{L-1} x(n\Delta t - (qN + i)\Delta t)h((qN + i)\Delta t) \qquad (5\text{-}20)$$

$$y_k(n\Delta t) = \sum_{i=0}^{N-1} u_i(n\Delta t)e^{-j\frac{2\pi \cdot k(n-i)}{N}} \qquad (5\text{-}21)$$

with $i = 0, 1,..N - 1$, $q \in [0, 1,..,L - 1]$, and $n \in$ integer.

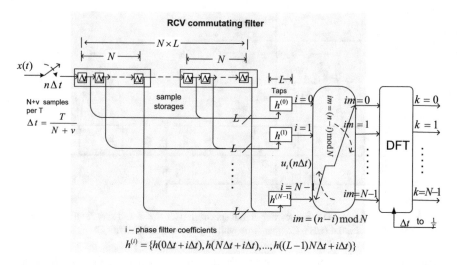

Figure 5-20: Receiver filter, address commutating and DFT

The FIR filter coefficients $h((qN + i)\Delta t)$ can be organized as an N by L matrix and are represented in Figure 5-20 as $h^{(i)}(q)$ as shown below, with $T' = N\Delta t < T$,

$h^{(0)}(0T')$	$h^{(0)}(1T')$	. . . .	$h^{(0)}((L-1)T')$
$h^{(1)}(0T')$	$h^{(1)}(1T')$	. . . .	$h^{(1)}((L-1)T')$
. . . .			
$h^{(N-1)}(0T')$	$h^{(N-1)}(1T')$	. . . .	$h^{(N-1)}((L-1)T')$

For a given sample $x(n\Delta t)$, ith phase of $u_i(n\Delta t)$ is computed using (5-20). This computation is organized, as in Figure 5-20, part of RCV commutating filter. For each increment of the index n, i.e., for a new sample, the output of the filter are N, indexed by i, which are matched to the input index of DFT, im. Their relationship is given by $im = (n - i) \bmod N$.

The frequency of DFT operation may range from one per OFDM symbol time (T) to every sample time (Δt), i.e., from one DFT computation per T to $N + v$ per T. In fact, not only DFT computation but the computation of RCV commutating filter may be done with the same frequency. Thus the whole receiver computation may range from one to $N + v$ per T. It is flexible. But in order to do one per T the OFDM symbol boundary is required before DFT. For filtered OFDM, 2 or 4 computations of DFT per T may be typical so that each subchannel may optimize for time sampling phase, gain and carrier phase. It is somewhat similar to the case that each channel is treated as a single carrier system.

Exercise 5-10* (project level): Figure 5-20 shows that the commutating filter part is done with every sample, i.e., every Δt while DFT can be done every sample ($N + v$

per *T*) or once per *T* or in between. This is programmed in Matlab in Appendix 5 A1. Confirm the Matlab program works and simplify, if possible, by doing the same processing rate for the commutating filter as DFT rate.

5.3.3 Common Platform

Figures 5-18 and 5-20 may be summarized, and shown in Figure 5-21.

The computational structure of Figure 5-21 is called a common platform of OFDM signals since it can implement all the known OFDM signals. This structure will also be used for staggered QAM based OFDM later. One condition is that the underlying shaping pulse is represented by a FIR filter. This is true any polyphase computational structure. One common shaping filter is used by all subchannels and with the common platform structure it is shared through all subchannels. A new part is a simple address commutating, which may be considered as a very special form of time-varying filter.

This common platform is a discrete domain computation of Figure 5-12. In fact a straightforward sampled version of Figure 5-12 may be used as a computational structure; this is shown in Figure 5-22. Thus Figure 5-21 may be considered as equivalent computational structure as far as the input and output are concerned; the complex symbols at the transmit side as the input, and the complex symbols at the receive side as the output. Compared to Figure 5-22, the common platform based on DFT is computationally far more efficient in general mainly due to the fast Fourier transform (FFT) and polyphase structure if all subchannels are computed simultaneously, which is the case with OFDM signal. This equivalence is useful conceptually. The straightforward structure is easy to visualize and to use in performance analysis due to channel fading.

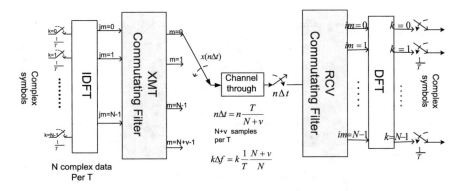

Figure 5-21: DFT based digital implementation of OFDM signals

5.3.4   Impulse Response and Eye Pattern

We look at the end-to-end impulse responses, eye patterns and constellations. Those are the ways to see the transmission quality, and very handy tools often used in practice to see the performance qualitatively even though the final performance measure is an error rate. Note that, with block transmission of DMT with CP, these tools do not apply. However, these tools give insights how a system behaves with the filter method, even if the underlying shaping pulse is the same rectangular pulse. Figure 5-21 of DFT with commutating filter, equivalently Figure 5-22, is a serial transmission system, and applicable to different shaping filters including a rectangular pulse.

An impulse response is obtained by sending an impulse $\delta(t)$ at a subchannel indexed by k and the impulse carries a complex symbol as $C_k(lT)\delta(t - lT)$ with $l = 0$. An example is shown in Figure 5-23. For the display purpose, $C_k(lT)= 1.0 + j\,0.5$ with subchannel $k = 2$. Real and imaginary impulse responses are displayed on the right hand side. The transmit pulse $N = 32$, $v = 8$ and $u = 8$; IDFT size, CP and window size in samples. There is a bit of roundness in the impulse response is due to windowing, and the trapezoidal shape is due to CP. The computation is done using the structure of Figure 5-21.

An adjacent subchannel $(k + 1)$ impulse response is displayed at the bottom of the left hand side. Notice that the sampling time it is zero.

Eye patterns are signal traces at the receiver before sampling synchronized by the symbol clock, i.e., folded and overlapped display in symbol clock. An oscilloscope can display eye patterns with horizontal trigger synchronized by the symbol clock.

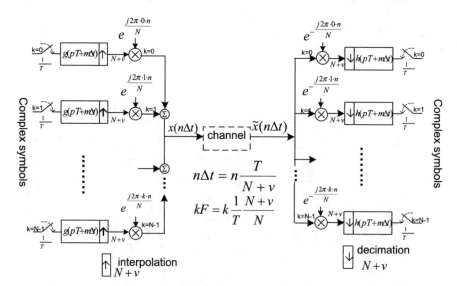

Figure 5-22:   A straightforward digital implementation of analog OFDM (Figure 5-12). In practice it may be used for regular FDM and efficient when one subchannel is of interest

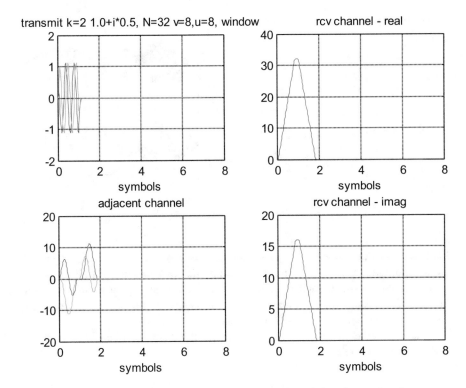

Figure 5-23: End to end impulse responses with a rectangular windowed transmit pulse

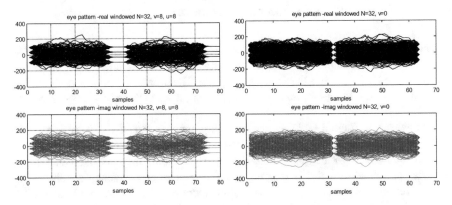

Figure 5-24: Eye patterns with 16 QAM, and the left hand side $N = 32$, $v == 8$ and $u == 8$; and the right hand side $N = 32$, $v = 0$ and $u = 0$

An example is shown in Figure 5-24. There are four levels at a sampling point (16 QAM). The left hand side of the figure shows the eye pattern with the pulse parameter as $N = 32$, $v = 8$ and $u = 8$. There is a margin in sampling (horizontal

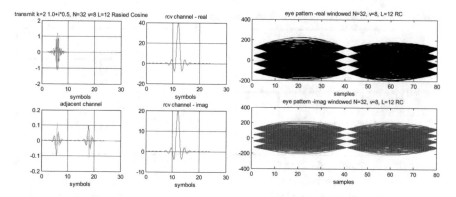

Figure 5-25: Impulse responses with frequency filter and eye patterns

line). The right hand side has the pulse parameter as $N = 32$, $v = 0$ and $u = 0$, i.e., no CP and no windowing. The optimum sampling time has no margin at all.

The impulse responses with a frequency domain filter and eye patterns are shown in Figure 5-25. Again the computations are done using the structure of Figure 5-21. It can also be done with Figure 5-22, but much more computations.

In order to compute impulse responses and eye patterns DFT must be computed every received sample.

5.3.5   1-Tap Equalizer, Sample Timing and Carrier Phase Recovery, and Channel Estimation

As discussed in Sect. 5.1.5, for DMT with CP, a 1-tap equalizer is enough if CP is greater than the channel delay spread and OFDM symbol boundary must be recovered before DFT, which is the same as sampling. The processing may be done for CP generalized OFDM (serial) transmission although there is flexibility to sample after DFT.

However, in filtered OFDMs, the sampling happens after DFT. A conceptually straightforward method is to treat each subchannel as a single carrier channel. A 1-tap equalizer might be good enough with proper sampling phase adjustment and carrier phase adjustment. It depends on the subcarrier frequency spacing. The equalizer adjustment and carrier phase adjustment may be done using known pilot symbols as well as decision directed. This issue will be handled in the future. There is not much literature available to take advantage of the fact that there are many subchannels, and thus it is one issue to do further research for filtered OFDM to be widely used.

5.3.6 Regular FDM Processing with Filtered OFDM

The regularity means that sub-carrier frequency spacing is uniform, and the symbol rate and the shaping filter are identical for each subchannel. We give a real life, but old, example. Iridium is a satellite based cell phone system to cover the globe. It was developed in early 1990s, and its signal format is FDM. In 10.5 MHz of bandwidth there are 252 sub-carrier channels. Each channel carries a single voice channel or data channel. Thus the allocated bandwidth of each sub-band is 41.67 kHz, and the symbol rate (with QPSK) in each sub-carrier is 25 kHz to make filtering easy. 40% RC signal occupies 35 kHz (=25 kHz (1 + 0.4)) bandwidth. Probably at that time a form of analog implementation might have been expected.

For a discrete time representation, a sampling rate for 10.5 MHz FDM signal may be 21 MHz (interpolation factor $L_i = 2$), i.e. $\Delta t = 1/21$ MHz. For higher sampling rate see Table 5-4, where it is listed up to $L_i = 8$. In practice for easy of RF / IF filtering it is not uncommon to use the sampling rate as high as 84 MHz ($L_i = 8$ case).

For the most right hand two columns of Table 5-4, we use the relationship $\Delta f = \frac{1}{\Delta t} \frac{1}{N}$ and $\frac{1}{T} = \frac{1}{\Delta t} \frac{1}{N+v}$.

In Table 5-4, we chose the sampling rate to match exactly the subchannel spacing and OFDM symbol rate. At the same time v can be chosen to be an even integer. However, N is not the power of 2. One can choose the sampling rate so that N is power of 2, and then OFDM symbol rate may be approximately 25 kHz.

Example 5-7: Choose N to be 256 ($=2^8$) and the subchannel spacing is exactly 41.67 kHz ($=10.5$ MHz /252). Find the sampling rate from $\Delta f = \frac{1}{\Delta t} \frac{1}{N}$. Then choose $\frac{1}{T} = \frac{1}{\Delta t} \frac{1}{N+v}$ close to 25 kHz and v to be a convenient integer. Choose the Interpolation factor $L_i \geq 2$.

Answer: $\frac{1}{\Delta t} = 10.5$ MHz $\quad x = \frac{256}{252} 10.667$ MHz . $256 + v = 426.667$ and $v = 170.667$. By choosing $v = 170$, $\frac{1}{T} = \frac{1}{\Delta t} \frac{1}{N+v} = 25.0391$ kHz.

Obviously the interpolation factor must be at least 2 or bigger to make the analog filtering feasible. For example, a sketch of transmit spectrum is shown in Figure 5-26. Out of 504 subcarriers, the number $k = 126$ to $k = 377$ are not used so that the subsequent filtering is easy or the filtering can be done digitally.

Table 5-4: Iridium example of sampling rate choice

Sampling rate (MHz) $\frac{1}{\Delta t} = \frac{N+v}{T}$	Interpolation factor (L_i)	FFT size (N)	Extra samples (v)	$N + v$	Sub-channel spacing (Δf)	OFDM symbol rate ($\frac{1}{T}$)
10.5	1	252	168	420	41.67 kHz = 10.5 MHz / 252	25 kHz = 10.5 MHz / 420
21	2	504	336	840		
42	4	1008	672	1680		
63	6	1512	1008	2520		
84	8	2016	1344	3360		

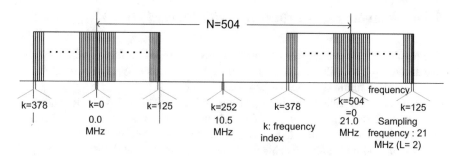

Figure 5-26: A sketch of transmit spectrum with the interpolation factor $L_i = 2$ and thus the subcarriers from 126 to 377 are empty (zero) to make the subsequent filtering easy

If the interpolation factor is 4 or bigger, then real IF samples, rather than complex baseband, can be generated. This is called IF sampling since the same can be done at the receiver, which will be treated later in this chapter.

5.4   OFDM with Staggered QAM

A staggered QAM was discussed previously in Chap. 2 (Sect. 2.6); we summarize it here briefly. Rather than a complex symbol from bit to symbol mapping, a bit stream is divided into even and odd bits (group of bits) as shown below, and thus real and imaginary parts are transmitted a half symbol time difference.

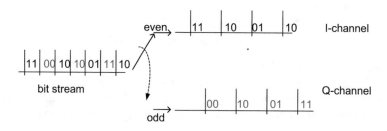

When the idea of staggering real and imaginary in time is applied to OFDM, then another form of OFDM is obtained and shown in Figure 5-27. It is represented in analog signal. Historically this form of OFDM was invented before the use of IDFT – DFT pair; see [2] and [3]. The result is that the filter can overlap with the neighboring subchannels, still maintain the orthogonality, and thus the most bandwidth efficient OFDM. There is no need of the gap between OFDM symbols, i.e., no CP, $v = 0$. The filter design is more flexible since it can overlap with the adjacent subchannels.

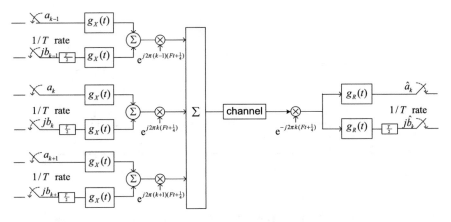

Figure 5-27: Analog representation of OFDM with staggered QAM

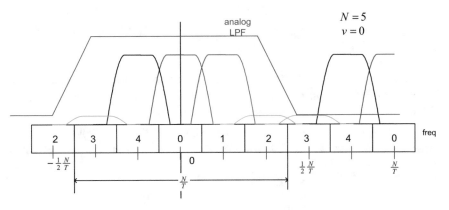

Figure 5-28: Staggered QAM based OFDM and frequency spectrum. The overlap with neighbor subchannels is allowed, maintaining the orthogonality

Its transmit spectrum example is shown in Figure 5-28. The excess bandwidth of the filter is 0.4 and it is square root raised cosine shape. The maximum excess bandwidth can be as large as 1.0, as shown in Figure 5-29

5.4.1   Common Platform Structure for OFDM with Staggering

In order to derive a common platform structure with IDFT-DFT our starting point is an analog representation of Figure 5-27. Our development is based on [14]. Alternative method, not pursued here, is in [13]

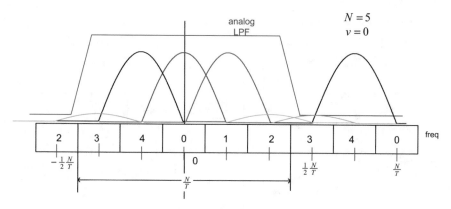

Figure 5-29: Staggered QAM based OFDM and frequency spectrum. The excess bandwidth is 1.0, the maximum possible while maintaining the orthogonality

$$x(t) = \sum_{k=0}^{N-1} \sum_{l=-\infty}^{\infty} \left[a_k(l) g_X(t - lT) + j b_k(l) g_X \left(t - lT \pm \frac{T}{2} \right) \right] j^k e^{j2\pi \cdot k \cdot Ft} \quad (5\text{-}22)$$

where $a_k(l)$, $b_k(l)$ are real, $F = \frac{1}{T}$ is a sub-channel frequency spacing, and $j^k = e^{j2\pi k \frac{1}{4}}$. $g_X(t)$ is a transmit shaping filter of OFDM which is band limited and its Fourier transform $G_X(f)$ is such that $G_X(f) = 0$ when $|f| \geq \frac{1}{T}$.

We obtain a sampled version of (5-22) by setting $t = n\Delta t = n\frac{T}{N}$ with $n = pN + m$ where n = sample index $\in$integer and p = OFDM symbol index $\in$integer, $m = [0, 1, 2, .. N - 1]$.

$e^{j2\pi \cdot k \cdot Ft} = e^{j2\pi \frac{k \cdot n}{N}}$ with $k = [0, 1, 2, .. N - 1]$.

$$x(n\Delta t) = \sum_{\widehat{l}=-\infty}^{\infty} \left\{ \sum_{k=0}^{N-1} a_k \left(\widehat{l} \right) e^{j2\pi \frac{k \cdot (n+N/4)}{N}} \right\} g_X \left(\left(p - \widehat{l} \right) T + m\Delta t \right)$$

$$+ \sum_{\widehat{l}=-\infty}^{\infty} \left\{ j \sum_{k=0}^{N-1} b_k \left(\widehat{l} \right) e^{j2\pi \frac{k \cdot (n+N/4)}{N}} \right\} g_X \left(\left(p - \widehat{l} \right) T + m\Delta t \pm \frac{T}{2} \right)$$

It is slightly more convenient to use the summation index $l = p - \widehat{l}$, and $g_X()$ is FIR with L taps.

$$x(n\Delta t) = \sum_{l=0}^{L-1} \left\{ \sum_{k=0}^{N-1} a_k(p - l) e^{j2\pi \frac{k \cdot (n+N/4)}{N}} \right\} g_X(lT + m\Delta t)$$

$$+ \sum_{l=0}^{L-1} \left\{ j \sum_{k=0}^{N-1} b_k(p - l) e^{j2\pi \frac{k \cdot (n+N/4)}{N}} \right\} g_X \left(lT + m\Delta t \pm \frac{T}{2} \right)$$

The inside of { } is represented as, with $m' = n + \frac{N}{4} = pN + m + \frac{N}{4}$,

$$A^{m'}(p-l) = \sum_{k=0}^{N-1} a_k(p-l)\, e^{\,j2\pi\frac{k \cdot m'}{N}} \tag{5-23}$$

$$B^{m'}(p-l) = \sum_{k=0}^{N-1} b_k(p-l)\, e^{\,j2\pi\frac{k \cdot m'}{N}} \tag{5-24}$$

We obtain a sampled transmit signal given by

$$x(n\Delta t) = \sum_{l=0}^{L-1} A^{m'}(p-l) g_X(lT + m\Delta t)$$

$$+ j \sum_{l=0}^{L-1} B^{m'}(p-l) g_X\left(lT + m\Delta t \pm \frac{T}{2}\right) \tag{5-25}$$

$$m' = \left(pN + m + \frac{N}{4}\right) \bmod N = \left(m + \frac{N}{4}\right) \bmod N \tag{5-26}$$

The equation (5-25), in conjunction with (5-23), (5-24) and (5-26), clearly shows that the computation of the transmit signal can be done by IDFT followed by a common transmit commutating filter.

The output phase (m') of IDFT must be circularly shifted to match the phase transmit time sample (m) as $m' = \left(pN + m + \frac{N}{4}\right) \bmod N = m\left(+\frac{N}{4}\right) \bmod N$ which is shown in (5-26). For filtering operation the output of IDFT is stored at m' position and those are used to compute m^{th} time sample.

The filter coefficients are organized as below, with the notation $g_X^{(m)}(lT) = g_X(lT + m\Delta t)$,

$g_X^{(0)}(0T)$	$g_X^{(0)}(1T)$	....	$g_X^{(0)}((L-1)T)$
$g_X^{(1)}(0T)$	$g_X^{(1)}(1T)$	....	$g_X^{(1)}((L-1)T)$
....			
$g_X^{(N-1)}(0T)$	$g_X^{(N-1)}(1T)$	....	$g_X^{(N-1)}((L-1)T)$

And for the half symbol delayed (or negative delayed) case, i.e., $\pm\frac{T}{2}$, the filter coefficients are organized, with the notation $g_X^{\left(m+\frac{N}{2}\right)}(lT) = g_X\left(lT + m\Delta t + \frac{N}{2}\Delta t\right)$, as below,

$g_X^{\left(\frac{N}{2}\right)}(0T)$	$g_X^{\left(\frac{N}{2}\right)}(1T)$	....	$g_X^{\left(\frac{N}{2}\right)}((L-1)T)$
$g_X^{\left(\frac{N}{2}+1\right)}(0T)$	$g_X^{\left(\frac{N}{2}+1\right)}(1T)$	....	$g_X^{\left(\frac{N}{2}+1\right)}((L-1)T)$
....			
$g_X^{\left(N+\frac{N}{2}-1\right)}(0T)$	$g_X^{\left(N+\frac{N}{2}-1\right)}(1T)$	....	$g_X^{\left(N+\frac{N}{2}-1\right)}((L-1)T)$

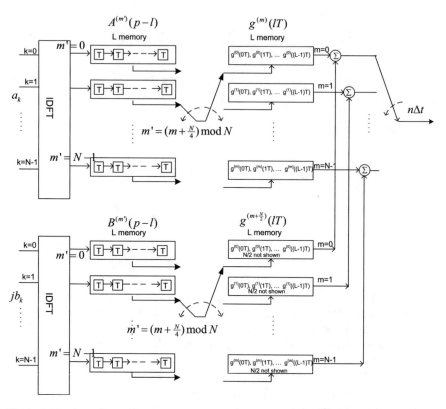

Figure 5-30: Staggered OFDM implementation with IDFT plus commutating filter

When the index $m + \frac{N}{2} \geq N$, then it is the same as the increment of l. This is shown as $g_X^{\left(m+\frac{N}{2}+N\right)}(lT) = g_X^{\left(m+\frac{N}{2}\right)}((l+1)T)$. For example, the last row of the coefficients can be written as,

$g_X^{\left(\frac{N}{2}-1\right)}(1T)$	$g_X^{\left(\frac{N}{2}-1\right)}(2T)$	$\cdots\cdots\cdots\cdots$	$g_X^{\left(\frac{N}{2}-1\right)}(LT)$

Thus the equation (5-25) can be 'implemented' as a block diagram shown in Figure 5-30.

Consider the implication of $m' = \left(m + \frac{N}{4}\right)$ mod N. Since we consider $v = 0$ case only for staggered OFDM, the range of m' is the same as that of m. Thus the commutating filter is not time-varying but the fixed phase difference only by $\frac{N}{4}$. Of course $v \neq 0$, it will be time-varying, i.e., commutating. There is no real motivation to use the case of $v \neq 0$.

5.4.2 Receiver Side of OFDM with Staggered QAM

Assuming that there is no distortion and noise in a channel, the received signal is the same as the transmit signal, the equation (5-22) written again for convenience,

$$x(t) = \sum_{k=0}^{N-1} \sum_{l=-\infty}^{\infty} \left[a_k(l)g_X(t-lT) + jb_k(l)g_X\left(t - lT \pm \frac{T}{2}\right) \right] e^{j2\pi \cdot k \cdot (Ft+0.25)} \quad (5\text{-}27)$$

As shown in Figure 5-27, $x(t)$ is processed first by demodulation, then filtering, finally sampling. In the derivation of a receiver structure we assume that $N + v$ samples per OFDM symbol period ($v \neq 0$), even though often $v = 0$ case in practice is used as we mentioned. The reason for this is that it turns out that nearly the same receiver front end structure (filtering and DFT) in Figure 5-20 may be used. Then we specialize the result with $v = 0$ for the possible simplification of this case.

A continuous form of the received signal without noise, is represented as for the kth channel,

$$y_k(t) = \left[x(t)e^{-j2\pi \cdot k \cdot (F \cdot t + 0.25)} \right] \otimes g_R(t) \quad (5\text{-}28)$$

where $\otimes$ denotes convolution.

A sampled version of (5-28), with $\Delta t = \frac{T}{N+v}$, is given by

$$y_k(n\Delta t) = \sum_{l=-\infty}^{\infty} x((n-l)\Delta t)\, e^{-j2\pi \cdot k \cdot (F \cdot (n-l)\Delta t + 0.25)} g_R(l\Delta t)$$

where a convolution becomes a summation in discrete time domain, rather than an integral. The summation can be decomposed into two parts, by setting $l = q \cdot N + i$ with $i = 0, 1, .. N - 1$, $q \in$ integer, and

$$u_i(n\Delta t) = \sum_{q=0}^{L-1} x(n\Delta t - (qN + i)\Delta t)g_R((qN + i)\Delta t) \quad (5\text{-}29)$$

$$y_k(n\Delta t) = \sum_{i=0}^{N-1} u_i(n\Delta t)\, e^{-j\frac{2\pi \cdot k \left(n + \frac{N}{4} - i\right)}{N}} \quad (5\text{-}30)$$

(5-29) is the same as (5-20) and (5-30) is the same as (5-21) except for the commutating index. We assumed that $g_R(l\Delta t)$ has L taps.

Clearly this shows that the receiver computation is decomposed into filtering followed by DFT. Again there is a need to match poly-phase of the filter represented as i and DFT input phase $\left(n + \frac{N}{4} - i\right)$. The ith polyphase of the filter output $u_i(n\Delta t)$ is computed and fed to DFT with $\left(n + \frac{N}{4} - i\right)$ modulo N address.

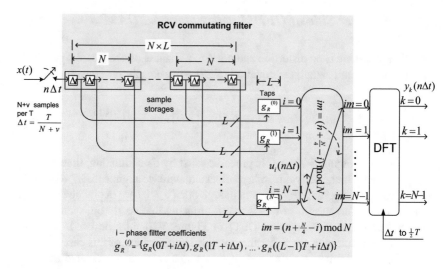

Figure 5-31:   Receiver structure of OFDM with staggered QAM

The implementation of (5-29) and (5-30) is essentially the same as (5-20) and (5-21) except for commutating. This is shown in Figure 5-31. In the figure for every sample of received signal, DFT is computed. Due to staggering the minimum number of DFT is 2 per period, rather than one.

After DFT, in practice, phase rotation, gain adjustment and sampling, equivalent to one –tap equalizer, must be performed in order to recover symbols correctly.

In deriving (5-29) and (5-30), there is no need of assuming $v = 0$. A simplification of Figure 5-31 with $v = 0$ may be considered in the future.

5.4.3   T/2 Base Implementation of Transmit Side

We organize the transmit side computation based on $\frac{T}{2}$. First, the filter coefficients are organized by $\frac{T}{2}$; two sets of filter coefficients in Figure 5-30 are interleaved as below

$g_X^{(0)}(0T)$	$g_X^{(0)}\left(\frac{1}{2}T\right)$	$g_X^{(0)}(1T)$	$g_X^{(0)}\left(\frac{3}{2}T\right)$	....	$g_X^{(0)}((L-1)T)$
$g_X^{(1)}(0T)$	$g_X^{(1)}\left(\frac{1}{2}T\right)$	$g_X^{(1)}(1T)$	$g_X^{(1)}\left(\frac{3}{2}T\right)$	......	$g_X^{(1)}((L-1)T)$
....					
$g_X^{(N-1)}(0T)$	$g_X^{(N-1)}\left(\frac{1}{2}T\right)$	$g_X^{(N-1)}(1T)$	$g_X^{(N-1)}\left(\frac{3}{2}T\right)$	......	$g_X^{(N-1)}((L-1)T)$

Note that the real part is transmitted on even $\frac{T}{2}$ and the imaginary part is transmitted on odd $\frac{T}{2}$. This can be displayed as in Figure 5-32.

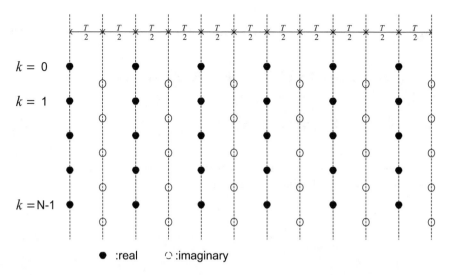

Figure 5-32: Symbol transmission arrangement; real at even of T/2,and imaginary at odd of T/2

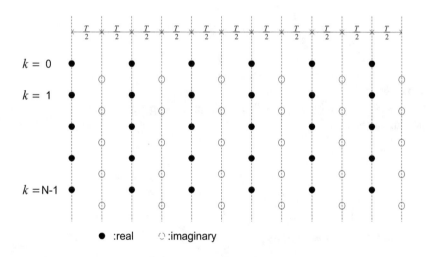

With the above observations in mind, our starting point is the equation (5-22) which is repeated as (5-27). Define a new symbol sequence $\widehat{a}_k$, $\widehat{b}_k$ from a_k, b_k as follows, with $l \in$ integer

$$\widehat{a}_k(2l) = a_k(l) \qquad \widehat{b}_k(2l+1) = b_k(l)$$
$$\widehat{a}_k(2l+1) = 0 \qquad \widehat{b}_k(2l) = 0$$

$$x(t) = \sum_{k=0}^{N-1} \sum_{l=-\infty}^{\infty} \left[\widehat{a}_k(2l)g_X\left(t - l\frac{T}{2}\right) + j\widehat{b}_k(2l+1)g_X\left(t - (2l+1)\frac{T}{2}\right)\right] e^{j2\pi \cdot k \cdot (Ft+0.25)}$$

Now $x(t)$ is expressed in $\frac{T}{2}$ base as below, with the index l for $\frac{T}{2}$ sequence,

$$x(t) = \sum_{k=0}^{N-1} \sum_{l=-\infty}^{\infty} \left[\widehat{a}_k\left(l\frac{T}{2}\right) + j\widehat{b}_k\left(l\frac{T}{2}\right)\right] g_X\left(t - l\frac{T}{2}\right) e^{j2\pi \cdot k \cdot (Ft+0.25)} \qquad (5\text{-}31)$$

The above (5-31) is converted to a sampled version following the same process, and we obtain, the length of the filter is $2L$ in $\frac{T}{2}$ and $n = pN + m$ with $m = [0, 1, \ldots, N-1]$,

$$x(n\Delta t) = \sum_{l=0}^{2L-1} \left\{C^{(m')}(p - l)\right\} g_X\left(l\frac{T}{2} + m\Delta t\right) \qquad (5\text{-}32)$$

where $m' = \left(m + \frac{N}{4}\right) \bmod N$, and

$$C^{(m')}(p - l) = \sum_{k=0}^{N-1} \left[\widehat{a}_k\left((p - l)\frac{T}{2}\right) + j\widehat{b}_k\left((p - l)\frac{T}{2}\right)\right] e^{j2\pi \frac{k \cdot (m+N/4)}{N}}.$$

(5-32) above can be implemented in a block diagram. It needs to computer every T, i.e., even $\frac{T}{2}$.

This is shown in Figure 5-33. The $\frac{T}{2}$ base processing requires the exactly the same amount as (5-25).

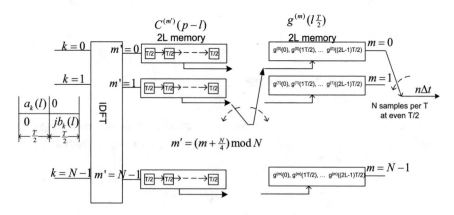

Figure 5-33:   T/2 based transmit organization of staggered OFDM

5.4.4 Impulse Response, Eye Pattern, and Constellation of Staggered OFDM

A major advantage of OFDM with staggered QAM is its spectral efficiency. It requires the minimum bandwidth yet the shaping filter can overlap with adjacent subchannels (both upper and lower sides). We need to understand how this is accomplished by examining the end to end impulse responses, eye patterns, and sampled constellations at the receiver.

As in Figure 5-27, we focus on only $k - 1, k, k + 1$ sub channels at the transmit side and k subchannel at the receive side. Or vice versa; k subchannel at the transmit side and $k - 1, k, k + 1$ sub channels at the receive side. The reason we can focus on three contiguous subchannels is that due to frequency filtering there is no inter-subcarrier interference (ICI). However, three adjacent channels are overlapped, yet maintaining the orthogonality.

We look at the end-to-end impulse responses, eye patterns and constellations. Those are the ways to see the transmission quality, and very handy tools often used in practice to see the performance qualitatively even though the final performance measure is an error rate.

Actual computation of an impulse response can be done by using Figure 5-30 (transmit side) and Figure 5-31 (receive side). We assume there is no noise and no channel distortion. An example is shown in Figure 5-34.

In the figure the impulse response of a main (k) subchannel is clearly a RC, and real and imaginary are separated by T/2 (right upper corner). The transmit impulse response sqrt RC modulated by a subcarrier ($k = 3$). The impulse responses of $k - 1$ and $k + 1$ subchannels are suppressed but not quite zero due to the overlap. But the orthogonality is obtained after sampling.

In terms of equations, the figures of Figure 5-34 may be shown as, when we consider received pulses ($r_k(t)$ and $r_{k \pm 1}(t)$) with $a_k(l) = \delta(l)$ and $b_k(l) = \delta(l)$;

$$r_k(t) = g_X(t) \otimes g_R(t) + j g_X\left(t \pm \frac{T}{2}\right) \otimes g_R(t)$$

$$r_{k\pm1}(t) = \left[g_X(t)e^{\pm j2\pi \cdot (Ft+0.25)}\right] \otimes g_R(t) + j\left[g_X\left(t \pm \frac{T}{2}\right)e^{\pm j2\pi \cdot (Ft+0.25)}\right] \otimes g_R(t)$$

In other words, in Figure 5-34, we displayed the impulse responses when the k subchannel transmits for clarity of display.

From the point of view of orthogonality between subcarriers, all three subchannels ($k - 1, k,$ and $k + 1$) transmits simultaneously as shown in Figure 5-27. Thus three receive impulse responses in Figure 5-34 are added together and are present in the k subchannel. From the figure the orthogonality (no interference between subchannels) comes in part from frequency domain filtering, in part from real and imaginary separation after sampling. In order to see it clearly, we may consider the case where all the three subchannels transmit an impulse, and group into

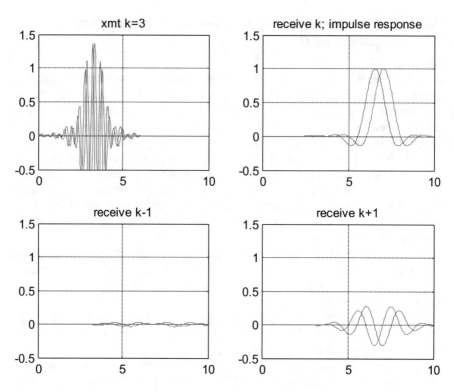

Figure 5-34: An impulse response of 3 adjacent subchannels, blue = i-channel and red = q-channel. The shaping filter is RC with excess bandwidth 0.5 and the filter length L = 6

two; one is real and sampled at even $\frac{T}{2}$ (a_T below) and the other is imaginary and sampled at odd $\frac{T}{2}$ ($b_{\frac{T}{2}}$ below);

$$a_T = g_X(t) \otimes g_R(t) + \left[g_X(t)e^{\pm j2\pi \cdot (Ft+0.25)} \right] \otimes g_R(t)$$

$$b_{\frac{T}{2}} = jg_X\left(t \pm \frac{T}{2}\right) \otimes g_R(t) + j\left[g_X\left(t \pm \frac{T}{2}\right)e^{\pm j2\pi \cdot (Ft+0.25)} \right] \otimes g_R(t)$$

The second terms are the interference from $k - 1$ and $k + 1$ subchannels while the first terms is the main carrying symbols. For the orthogonality, the second terms must be zero real and imaginary separation after sampling; for a_T taking real and for $b_{\frac{T}{2}}$ taking imaginary. Note that $\left[g_X(t)e^{\pm j2\pi \cdot (Ft+0.25)} \right] = g_X(t) \sin\left(2\pi \frac{t}{T}\right) \pm jg_X(t) \cos\left(2\pi \frac{t}{T}\right)$.

The eye pattern is another way, actually in practice used more often than the impulse response due to the fact that it can be observed with random symbol transmission. It is a synchronized display of signal traces with the symbol rate clock. This can be done easily with an oscilloscope; horizontal trigger comes from a symbol clock.

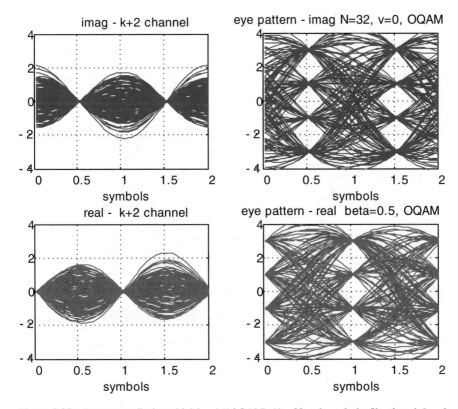

Figure 5-35:   Eye pattern display with M = 4 (16 QAM), $N = 32$ and $v = 0$, the filter length $L = 6$ (symbol time) and the excess bandwidth 0.5

An example of eye patterns is shown in Figure 5-35. In order to obtain the eye pattern in the figure, three contiguous subchannels are used in transmit side ($k - 1$, k, and $k + 1$) and the subchannel k is used for eye pattern at the receiver. For the interference we use $k + 2$. A clean eye pattern for real and imaginary separated by $\frac{T}{2}$. For the interference part, $g_X(t) \sin\left(2\pi \frac{t}{T}\right)$ pattern is clear; there is a zero crossing due to subcarrier modulation. Similarly for the imaginary part of the interference, $g_X(t) \cos\left(2\pi \frac{t}{T}\right)$ pattern is clear.

Compared with the case of no adjacent subchannel interference, i.e., a single carrier case, the eye pattern is narrower horizontally due to the presence of adjacent subchannels even though there is no interference at the sampling point.

The eye pattern can be obtained equivalently by transmitting one transmit subchannel and by adding all three subchannel eye patterns at the receiver. This will be elaborate further in later section.

For staggered QAM, strictly speaking, a constellation point is not a point since they are separated by the half symbol time. For the display purpose, one pair (real, imaginary) is designated as a point on a complex plane. This is shown in Figure 5-36.

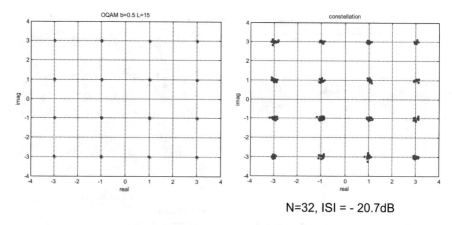

N=32, ISI = - 20.7dB

Figure 5-36: 16-QAM constellations, ideal and with ISI

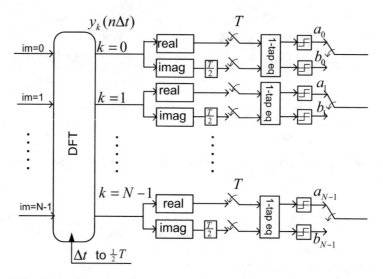

Figure 5-37: Further processing of the DFT outputs for recovering symbols

In order to obtain eye patterns and impulse responses, it is enough to compute $y_k(n\Delta t)$ in Figure 5-31 after DFT, and split into real and imaginary parts. If there is a distortion in channel – delay, and phase rotation, there may be an equalizer to see an eye pattern. For constellations, we need to sample the received signals. We show further signal processing after DFT in Figure 5-37.

The real and imaginary parts of $y_k(n\Delta t)$ are sampled even and odd time of $\frac{T}{2}$, every T and then followed by a one-tap equalizer (or more complex equalizer if necessary) and by decision. Two real and imaginary parts are interleaved to make them into one stream of symbols.

Exercise 5-11: Consider impulse responses in Figure 5-34 with three contiguous subchannels $k_T \in \{k-1, k, k+1\}$ and evaluate the impulse response of a receive sub channel $k_R = k_T$. The channel impulse response is denoted by $c(t)$, and consider four specific cases of it; 1) $c(t) = A\delta(t)$, 2) $c(t) = Ae^{i\theta}\delta(t)$, 3) $c(t) = A\delta(t-D)$ 4) $c(t) = Ae^{i\theta}\delta(t-D)$. Hint: Show that the impulse response can be represented by,

$$\sum_{k_T}\left[g_X(t)e^{j2\pi(k_R-k_T)\cdot(Ft+0.25)}\right] \otimes \left[c(t)e^{-j2\pi(k_R)\cdot(Ft+0.25)}\right] \otimes g_R(t)$$
$$+j\sum_{k_T}\left[g_X\left(t\pm\frac{T}{2}\right)e^{j2\pi(k_R-k_T)\cdot(Ft+0.25)}\right] \otimes \left[c(t)e^{-j2\pi(k_R)\cdot(Ft+0.25)}\right] \otimes g_R(t).$$

Evaluate four specific cases of $c(t)$ with a matched filter pair of $g_X(t)$ and $g_R(t)$.; say square root RC with $\beta = 0.5$.

5.5 Practical Issues

5.5.1 Performance When a Channel Delay Spread > CP

When a channel has a delay dispersion greater than CP, the performance analysis requires a channel impulse response, $c(n\Delta t; j\Delta t)$ to be taken into account. It is a time-varying impulse response $c(n\Delta t)$ at $j\Delta t$ where n and $j \in$ integer. A discrete time block diagram of an OFDM system is shown in Figure 5-38.

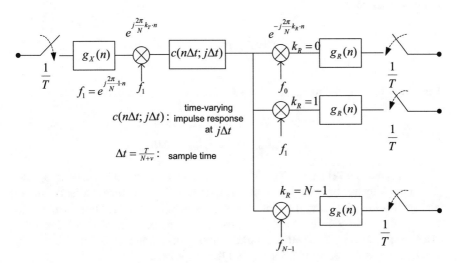

Figure 5-38: OFDM system with discrete time representation

5.5.1.1 Performance Analysis with a Linear Channel

Figure 5-38 is used to obtain $N \times N$ impulse responses for a given transmit subchannel k_T and receive subchannel k_R with $k_T = \{0, 1, .., N - 1\}$ and $k_R = \{0, 1, .., N - 1\}$. For this impulse response computation, it is convenient to use the following relationship.

Impulse response:

$$h(n) = \left[g_X(n) e^{\,j2\pi(k_R - k_T)\cdot\frac{n}{N}} \right] \otimes \left[c(n) e^{-j2\pi \cdot k_R \cdot \frac{n}{N}} \right] \otimes g_R(n) \qquad (5\text{-}33)$$

Frequency response:

$$H(f) = G_x(\, f + (k_T - k_R)\Delta f\,)C(\, f + k_R\Delta f\,) \cdot G_R(f) \qquad (5\text{-}34)$$

One can see that (5-33) is correct by first looking at the end to end frequency response (5-34). The end to end impulse response (5-33) can be modified slightly in order to accommodate staggered OFDM. See Exercise 5-11 (right after Figure 5-37).

In order to compute the performance degradation due to a linear channel (may be time-varying) the $N \times N$ impulse responses computed from (5-33) must be sampled in every T (or $N + v$ samples). These samples go through a 1-tap equalizer as follows; 1-tap equalizer: $\frac{1}{c(eq)} = h(n)$ sampled at every $N + v$ samples or

1-tap equalizer:

$$c(eq)_{re} + jc(eq)_{im} = \frac{1}{h_{re} + jh_{im}} = \frac{1}{h(n)} \qquad (5\text{-}35)$$

This is the same as the main tap to be 1.0 after the 1-tap equalizer;

$$c(eq)_{re}h_{re} - c(eq)_{im}h_{im} = 1.0$$
$$c(eq)_{re}h_{im} + c(eq)_{im}h_{re} = 0.0.$$

When there is no channel distortion or CP is greater than a channel delay spread (dispersion), then $N \times N$ sampled impulse responses to be unity matrix after the single tap equalization. However, if not the case there will ISI in diagonal terms and ICI in off-diagonal terms. For the ISI the analysis of performance degradation is the same as a single subcarrier. But with the multiple subcarriers, there are ICI terms. Further development is necessary but it is beyond the scope here.

5.5.2 Digital Quadrature Modulation to IF and IF Sampling

In Figure 5-13, at transmit side, a discrete complex envelope signal, $x(n\Delta t)$, may be considered as a complex baseband signal with a symbol (sample) rate $1/\Delta t$. It can be converted to a real carrier wave (analog), after D/A and analog filter for i-channel (real) and q-channel (imaginary), using an analog quadrature modulator, as shown in Figure 5-3. In order to make analog filtering easy, additional increase of sample rate, called interpolation, may be used. This is common practice since DSP is fast and cheap. In the receive direction, this corresponds to higher sampling rate, which again makes analog filtering easy, and subsequent processing reduces the sample rate, called decimation.

For example, the interpolation filter with a factor 4 can be implemented as shown in Figure 5-39, where $\{p_0, p_1, p_2, p_3, \ldots\}$ are the coefficients of a finite impulse response (FIR) interpolation filter. This is a polyphase implementation. Two sets of digital interpolators, D/As and LPFs, for real and imaginary, are necessary. The output will go to analog quadrature modulator for frequency translation to RF.

When IF frequency is chosen to be the sample rate $(1/\Delta t)$ and the interpolated sample rate is 4 times of the sample rate, then a specially simplified digital IF modulation and interpolation can be done at the same time.

$\cos(2\pi f_c t) = \cos\left(2\pi \frac{1}{\Delta t} \frac{\Delta t}{4} n\right) = \cos\left(\frac{\pi}{2} n\right)$ and $\sin(2\pi f_c t) = \sin\left(2\pi \frac{1}{\Delta t} \frac{\Delta t}{4} n\right) = \sin\left(\frac{\pi}{2} n\right)$, and thus the digital IF carrier has the value $1, 0, -1, 0$ or $0, 1, 0, -1$ for each channel (real and imaginary), only two samples are necessary per symbol, and

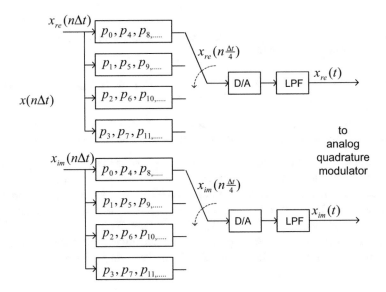

Figure 5-39: A factor of 4 interpolation and digital to analog converter and low pass filter and analog quadrature modulator for frequency translation

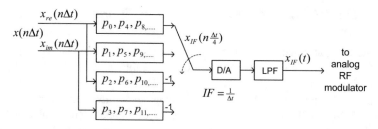

Figure 5-40: Digital IF generation with IF frequency = sample rate, and only one set of digital filter, D/A and LPF

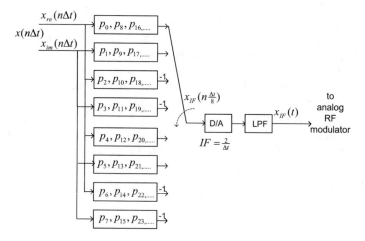

Figure 5-41: Exercise answer: digital IF generation with IF frequency = 2 sample rate, and only one set of digital filter, D/A and LPF

real and imaginary are alternating. With this special choice, digitally modulated IF signals are computed as shown in Figure 5-40.

Exercise 5-12: For a given IF = 2 sample rate and the interpolation factor 8, show the digital IF signal computation structure, shown in Figure 5-41.

The discussion so far is applicable to OFDM as well as to single carrier signals. It is a universal method of interpolations. However, in OFDM, there are other ways to generate digital IF signals. This was briefly discussed in Table 5-4 the interpolation factor is shown from 2 to 8; $L = 2$ is shown in Figure 5-26. When $L = 8$, DFT size (N) is increased by a factor 8. This is shown in Figure 5-42.

Digital IF signal may be generated directly by placing the subchannels differently with conjugate symmetry. This is shown, schematically in Figure 5-43.

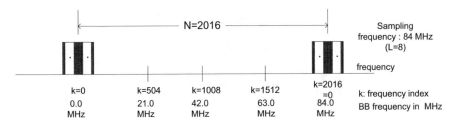

Figure 5-42: Use of IDFT size increase for interpolation $L = 8$ and many subcarriers are zero and subsequent analog filtering after D/A becomes easy

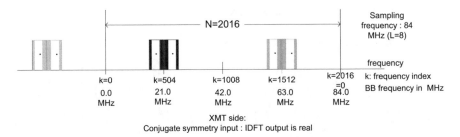

Figure 5-43: Digital IF generation using IDFT; $L = 8$ is shown. The IDFT output is real digital IF

5.5.2.1   IF Sampling at Receive Side

Typically at the receiver, RF carrier frequency is down converted to a convenient IF frequency, and then quadrature demodulated to baseband complex envelope and sampled, corresponding to $x(n\Delta t)$, with noise added and with channel distortion. In IF sampling, analog signal at IF frequency may be sampled and thus generated $x_{IF}\left(n\frac{\Delta t}{L}\right)$ with L oversampling factor. Remember that $x_{IF}\left(n\frac{\Delta t}{L}\right)$ is real. The subsequent processing can be done using the computational structure of Figure 5-20. DFT size is increased by the same factor L.

For example, IF $= 147$ MHz and sampling rate 84 MHz at IF the discrete signal is shown in Figure 5-44 and since IF is real the image in frequency is conjugate symmetric at 21 MHz and 63 MHz. These choices of IF is multiple and somewhat arbitrary. For example, IF $= 155$ MHz and sampling rate 88 MHz at IF yields frequency signal at 21 MHz and 65 MHz . After DFT one may use only one sideband.

5.5.3   *Modern FH Implementation with OFDM*

Frequency hopping is used in order to increase jamming (or interference) resistance. Out of many subchannels, only partial sets are used and its use pattern may change over time according to a random pattern. Before digital implementations, it requires synthesizers, filters and oscillators. In the receiver side the hopping pattern must be

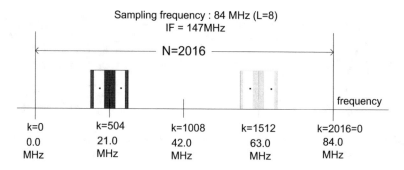

Figure 5-44: IF sampling example using Figure 5-43

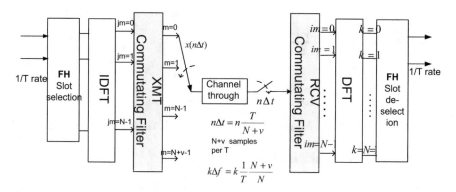

Figure 5-45: FH implementation with OFDM using DFT + commutating filter structure

synchronized by using a serial search. In digital implementations it requires numerical oscillators (NCO) and mixers (multipliers), and is computationally intensive. Only limited subchannels are available due to the heavy computational complexity.

With OFDM structure using DFT plus commutating filters, the implementation can be very efficient computationally and very fast hopping patterns are possible as shown in Figure 5-45. A multiple FHed subchannels are possible without any collision. Finding appropriate hopping pattern to implement FHed channels is a topic that can be handled separately. All the subchannels are available simultaneously. A subchannel may use any digital modulation, coherent or non-coherent. In the past, non-coherent modulations were used due to the switching of synthesizer for hopping.

5.5.4 Naming of OFDM Signals

OFDM signals discussed so far may be classified by a pulse shaping filter. In the literature there are different names used and here we summarize them. We use OFDM as a generic name for orthogonality condition in (5-2), i.e., subcarrier

Table 5-5: OFDM naming convention summarized

Name in DSL context		OFDM (generic name with the orthogonality condition)			
		DMT with CP	DMT with windowing	FMT	Staggered FMT
Pulse shaping filter	Transmit	Transmit pulse (T, $N+v$)	Transmit pulse (T, $N+v$)	(frequency filter)	(frequency filter)
	Receive	Receive pulse (T, N)	Receive pulse (T, N)	Matched frequency filter	Matched frequency filter
Other names used in signal processing context and comments		OFDM CP may be added as in Figure 5-5 for block transmission Or as in Figures 5-13 and 5-14 with commutating filter approach.	Filtered OFDM	FBMC Regular FDM	FBMC /OQAM DWMT SMT CMT

Note: FBMC: filter bank multicarrier, FMT: filtered multitone, DWMT: digital wavelet multitone, OQAM: offset QAM, SMT: staggered multitone, CMT: cosine modulated multitone. CMT is not the same as staggered FMT but it uses VSB rather than QAM, i.e., i-channel and q-channel are not independent but its efficiency is the same as staggered FMT frequency spacing and sampling time; $F = \frac{1}{T}\left(1 + \frac{v}{N}\right)$ and $\Delta t = \frac{T}{N+v}$. Any OFDM signal meets this choice of orthogonality with a common pulse shaping filter.

5.6   OFDM with Coding

The development of OFDM signaling was motivated by robust performance under frequency selective fading. For AWGN channels, OFDM signaling does not provide any advantage over a single carrier system. For frequency selective channels, i.e., dispersive impulse response, it simplifies the equalization since each subchannel is narrow and a single tap equalizer may be good enough. When the frequency selective channel does not change with time, i.e., not time selective, then each subchannel may be treated as AWGN channel with different SNR. The channel may be known a priori or measured. This static frequency selective channel is typical with cable system such as DSL or CATV. Each subchannel may have different modulation and coding classes, which may be implemented by different code rates with a common modulation depending on SNR of each subchannel. It is shown in Figure 5-46 as CM_i (coding and modulation).

5.6.1   Coded Modulations for Static Frequency Selective Channels

The real power of OFDM comes out with FEC. When the channel is known, by using Figure 5-46 with appropriate CMs, called *bit loading*, it can achieve close to the channel capacity. A single carrier system may use a pre-distortion of channel

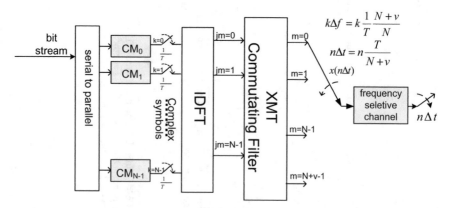

Figure 5-46: Bit stream is converted to a block of data and distributed to each sub channel followed by coding and modulations for static frequency channels such as DSL

frequency response but generally requires a sophisticated equalizer (overall more complex) at the receiver. The details of bit loading are <u>not</u> covered here.

5.6.2   Coding for Doubly Selective Fading Channels

Wireless channels are both time selective and frequency selective. Time selective fading is related with mobility and frequency selective is due to large scale objects along the propagation paths. In mobile channels, its fading rate is high and at the same time frequency selective. Thus the bit to symbol mapping must deal with this doubly selective fading situation. These fading channels are discussed in Chap. 4 in detail.

Due to the time varying nature of fading there is little chance to match coding and modulation to a subchannel. Instead we try to use coding to get the average channel throughout all subchannels, thus each subchannel is an average and for each subchannel the same coded modulation may be used; in Figure 5-46, $CM_0 = CM_1 = \ldots = CM_{N-1}$. This is much different from static frequency selective fading.

5.6.2.1   Subchannel Permutation (pi5)

OFDM system converts a wide band doubly selective channel into many narrow band subchannels with time selective fading. It can be shown that each subchannel is getting closer to Rayleigh even if the entire channel is Rician. Each subchannel is effectively a Rayleigh fading channel. To make sure each subchannel statistically identical or nearly so frequency subchannel permutation (π_5) is introduced as shown in Figure 5-47. One way of implementing this idea is to 'circularly shift' in each OFDM symbol time and thus a logical subchannel uses all physical subchannels. In order to have frequency diversity effect, a frequency separation between symbol times must be bigger than a correlation bandwidth, which is the inverse of delay spread. An example of π_5 is shown in Figure 5-48.

5.6.2.2   Coded Modulation (CM_R) for Rayleigh Fading Channels

Time selective fading like Rayleigh fading may be very damaging to the system error rate performance. A countermeasure is necessary and is in the form of diversity – frequency diversity, space diversity, time diversity and coding as a form of diversity. Diversity means that the same information is sent through different, independent channels. A typical solution, specifically for a single carrier system, uses a form of space diversity at the receiver (more antennas), and coding with interleaving.

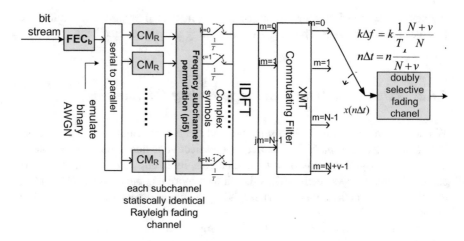

Figure 5-47: Coded bit to symbol mapping and modulation for doubly selective channels; each subchannel is a Rayleigh channel and coded modulation (CM$_R$) should handle it properly; binary FEC to handle AWGN while CM$_R$ converts Rayleigh channel to AWGN

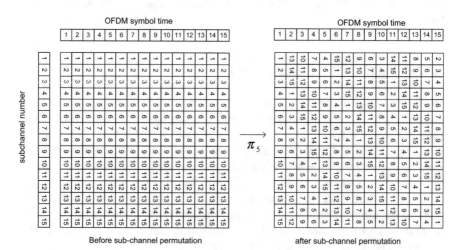

Figure 5-48: Subchannel permutation with 15 subchannels; permutation pattern, circular shift (column) each OFDM symbol time, is given by (0, 3, 6, 9, 12, 1, 4, 7, 10, 13, 2, 5, 8, 11, 14), and an ideal permutation should make a logical subchannel statistically identical

In OFDM system here we focus on the diversity based coding with interleaving. In Figure 5-47, CM$_R$ block is to handle Rayleigh fading. One key element is the interleaver. The use of it is to make a subchannel with Rayleigh fading to be like AWGN.

We partition the interleaver into two parts; one for time de-correlation π_3, and the other for modulation bit permutation π_1. Two together create binary AWGN like

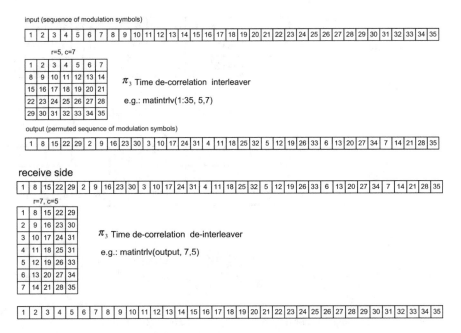

Figure 5-49: Symbol permutation for time de-correlation (π_3)

channel. With π_3, adjacent modulation symbols are transmitted with time separation so that they are de-correlated in time. The correlation time is the inverse of Doppler spectrum bandwidth. The larger the Doppler bandwidth, the faster the fading. For example, when the Doppler frequency spread is 2.0 Hz, the time separation must be 0.5 sec.

With π_1 each bit in a multi-bit modulation symbol is treated as a 'bit channel'. LSB, NSBs and MSB are used equally from a logical LSB,.., MSB bit channels. For example 45 bits are mapped into 15 symbols of 3 bit. The first row is MSB and the third row to be LSB. One symbol bit is transmitted in different time and different bit position in a symbol. This example is shown in Figure 5-50.

In practical implementations π_3 and π_1 may be implemented as a single block. Conceptually it is easy to see the permutation needs and thus the partition into two is correct and justified.

In Figure 5-51 the transmit side of CM_R is shown as well as the receive side. Each subchannel uses the same structure of the figure and same modulation and coding.

5.6.2.3   Binary FEC$_b$

In Figure 5-47 FEC$_b$ block is shown. This code should be good for handling AWGN channel and binary such as Turbo, LDPC and others. The details of such codes will be treated in Chap. 6. In Figure 5-51, as part of CM_R, there is another block for

Figure 5-50: An example of modulation bit permutation (π_1)

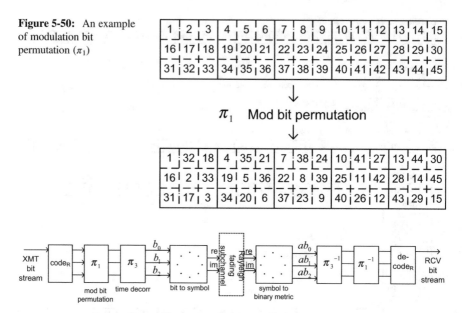

coding, denoted as code$_R$. This code should be efficient for converting Rayleigh (statistics) to AWGN, and can be a simple repetition code, or a convolution code. This two code scheme is called a two-step approach, and its aim it to achieve the channel capacity of a Rayleigh fading channel, ultimately a doubly selective channel.

In the literature it appears only one code scheme is described. This two-step approach should be experimentally verified; this is important in practice and will be treated in the future. Here we outlined a structure of overall OFDM system when it is combined with coding.

5.7 Chapter Summary

OFDM is a special form of frequency division multiplex (FDM) signal applied to digital transmission by choosing the subcarrier spacing (F) to be related with the subchannel symbol rate (OFDM symbol rate: $\frac{1}{T}$) and the number of samples per symbol period (T) is an integer ($N + v$) as $F = \frac{1}{T}\left(1 + \frac{v}{N}\right)$ and $\Delta t = \frac{T}{N+v}$. We need to choose the shaping filter to be the same for all subcarriers. In this chapter we apply to analog FDM and to sampled version of it. Thus we obtain different OFDM signaling formats depending on underlying pulse shape. One pulse shape is rectangular resulting in CP generalized OFDM and extend the pulse shape to frequency domain filters resulting in filtered OFDM, and staggered OFDM. In this context, a form of

OFDM used often currently, called DMT with CP, is a special case of CP generalized OFDM with rectangular pulse.

Furthermore we showed that all different OFDMs discussed in this chapter can be implemented by commutating filter (polyphase form) in conjunction with DFT (or IDFT). The commutating filter is a special form time-varying FIR filter. We emphasize that this commutating filter structure is derived for computational efficiency from analog / sampled prototype. And thus there is one to one correspondence between them, which may be utilized for analysis and other purposes.

We considered practical issues and applications of filtered OFDM. Additional practical issues are the design and optimization of matched filters – compact in time as well as in frequency even the excess bandwidth is very small, say less than 0.25 or so. A related one the design of window for 'windowing '. This is not covered here and will be treated in Chap. 3.

Practical system implementations require further detailed investigations. Additional future project and research problems related with them are suggested. For example, in Sect. 5.3.5, we need much more research is necessary for timing recovery, carrier recovery and channel estimation. We just briefly mentioned fast FH in 5.5.3, and need to compare other systems listed in 5.5.4, in particular Table 5-5.

We touched upon the issue of applying OFDM to frequency selective channel and to doubly selective channel, i.e., frequency selective and time selective (e.g., Rayleigh fading). Additional exploration, including two-step coding approach for doubly selective channel, will be very useful in practice.

5.8 References and Appendix

[1] S.B. Weinstein and P.M. Ebert, "Data Transmission by Frequency Division Multiplexing Using the Discrete Fourier Transform", IEEE Trans. Comm. Vol COM-19, No.5, October 1971.

[2] B. R. Saltzberg, "Performance of an Efficient Parallel Data Transmission System", IEEE Trans. Comm. Vol COM-15, No6, October 1967.

[3] R.W. Chang, "Synthesis of band-limited orthogonal signals for multichannel data," pp.1775–1797, BSTJ, Dec 1966

[4] B. Hirosaki, "An Orthothogonally Multiplexed QAM System Using the Discrete Fourier Transform", IEEE Trans. Comm. Vol COM-29, No.7, October 1981.

[5] J.S. Chow, J.C. Tu, and J.M. Cioffi, "A Discrete Multitone Transceiver Systems for HDSL Applications", IEEE J. of Sel. Area Commun., August 1991, pp.895–908

[6] J.A.C Bingham, "Multicarrier Modulation for Data transmission: an Idea whose time has come", IEEE Communications Magazine May 1990, No.5, pp.5–14

[7] G. Cherubini, E. Eleftheriou, S. Olcer, "Filtered Multitone Modulation for Ver High-Speed Digital Subscriber Lines" IEEE J. Sel. Areas of Communications, vol.20 No5., June2002

[8] L.J. Cimini,Jr., "Analysis and simulation of a Digital Mobile Channel Using Orthogonal Frequency Division Multiplxing", IEEE Trans. Comm. Vol COM-33, No.7, July 1985.

[9] J. van de Beek, M. Sandell, and P.O. Borjesson, "ML Estimation of Time and frequency Offset in OFDM Systems", IEEE Trans. Signal Processing Vol. 45, No.7, July 1995.

[10] J. Louveaux, L. Cuvelier, L. Vandendorpe, T. Pollet, "Baud Rate Timing Recovery Scheme for Filtered Bank-Based Multicarrier Transmission", IEEE Trans. Comm. Vol .51 NO.4, April 2003 pp.652–663

[11] K.E. Scott and E.B. Olasz, "Simultaneous clock phase and frequency offset estimation," IEEE Trans. Commun., vol.43 pp.2263–2270, July 1995

[12] V. Lottici, M. Luise, C. Saccomando, and F. Spalla, "Non-Data-Aided Timing Recovery for Filter-Bank Multicarrier Wireless Communications", IEEE Transaction on Signal Processing, Vol.54, No.11, November 2006, pp.4365–4375

[13] M.A. Aoude, R.A. Vallet, "Filtered OFDM/OQAM transmission system", IEEE 2003 Globecom pp. 1430–1432

[14] Sung-Moon Yang, "OFDM Signals Based on Shaping Filters" Military Communications Conference 2009,DOI: https://doi.org/10.1109/MILCOM.2009. 5380010

[15] Sung-Moon Yang, "A New Filter Method of CP and Windowing in OFDM Signals" Military Communications Conference 2010 DOI: https://doi.org/10. 1109/milcom.2010.5680210

[16] C. Eklund, R.B. Marks, K.L. Standwood, and S. Wang, "IEEE standard 802.16: A technical overview of the Wireless MAN 326 air interface for broadband wireless access," IEEE Comm. Mag., pp98–107, June 2002

[17] 3GPP specification series TS36.201:LTE physical layer; General description and TS 36.211 Physical channels and modulation 2007

Appendix 5
A.1 Matlab Program

```
% malab program - impulse response
N=64; v=16; u=4;   %DFT size, CP, window
L=2; L1=1;        %filter length xmt, rcv
%xmt filter pulse (RC window)
g=zeros(N+v, L);
a1 = -u/2/(N+v):1/(N+v): 0-1/(N+v);
g(1:u/2,1) = 1/2+1/2*cos(2*pi/((u+eps)/(N+v))*a1);
g(length(a1)+1:N+v,1)=1.0;
a2 = 0+1/(N+v):1/(N+v): +u/2/(N+v);
g(1:u/2,2) = 1/2+1/2*cos(2*pi/((u+eps)/(N+v))*a2);
g(length(a1)+1:N+v,2) = 0.0;
%rcv filter impulse
h = ones(N,L1);
```

```
%
MAX_iter=2; %no.of symbols to transmit
Dx=zeros(N, L); %IDFT output for filtering
xmt=zeros(N+v,MAX_iter);%time samples x_ser = zeros((N+v)*(MAX_iter
+1),1);%xmt
r_ser = zeros(1, N*L1); %receive shift register
Cxk = zeros(N,MAX_iter);%transmit symbols
Crk = zeros(N,length(x_ser)); %rcv symbols
% OFDM data symbol generator
loc = 2; %1--> k=0(DC) 2-> k=1
Cxk(loc,1) = 1 + 1i*0.5; %an impulse
%
iter = 1; p = 0;%iteration, symbols modulo N
while iter <= MAX_iter,
%-------IDFT----------
Dx_new = ifft(Cxk(:,iter),N)*N;
%shift and new IDFT output
for j = L:-1:2
  Dx(:,j) = Dx(:,j-1); %right shift
end
  Dx(:,1) = Dx_new;   %new stored
for m=1:N+v
  %commutatiing address: n mod N
  jm = mod((p*(N+v) + m-1), N)+1;
  %FIR filters (N+v)
  xmt(m,iter) = sum(Dx(jm,:).*g(m,:));
end
%next OFDM data symbol
iter=iter+1;
p = mod(p+1, N);
end   %while XMT
%channel: parallel to serial conversion
for j=1:MAX_iter
  x_ser((j-1)*(N+v)+1 :j*(N+v)) = xmt(:,j);
end
% Reciver processing
for n=1: length(x_ser)
  %shift for new input
  for k = N*L1 : -1: 2
    r_ser(k) = r_ser(k-1);
  end
  r_ser(1)=x_ser(n);
  %----filtering------
  i_N_set= zeros(1, L1);
  for ip=1:N
    i_N_set = ip:N:N*L1;
    ui(ip)= sum(r_ser(i_N_set).* h(ip,:));
  end
  %----commutating----
  n_i = mod((n-1 - (0:N-1)),N)+1;
  ui=ui(n_i);
  %-------DFT----------
  Crk(:,n)=fft(ui,N)/N;
```

```
%
end
%
%test display
plot((1:length(x_ser))/(N+v),real(Crk(loc,:)),'b',...
    (1:length(x_ser))/(N+v),imag(Crk(loc,:)),'r');
grid;axis([0 2 -1 1]);
title('rcv channel - real, imag')
xlabel('symbols')
```

A.2 FDM Signal Example (Figure 5-52)

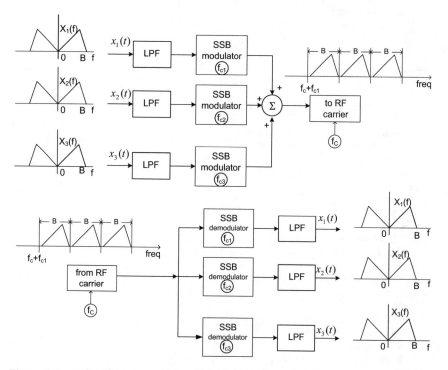

Figure 5-52: SSB FDM is most efficient in bandwidth efficiency, which was perfected in analog telephone systems. FDM in general used in many analog systems, e.g., 1st generation cellular phone – AMPS (advanced mobile phone system) uses DSB FDM

As can be seen in Figure 5-53, analog FDM system requires an oscillator and a filter to generate FDM signals and to recover them. Modern OFDM can be done with DFT. This is a major innovation since DFT can be implemented efficiently by FFT (fast Fourier transform). In order to do this the subcarrier frequency spacing (F) must be chosen such that it is related with symbol rate ($\frac{1}{T}$); $F = \frac{1}{T}\left(1 + \frac{v}{N}\right)$ and $\Delta t = \frac{T}{N+v}$ where $\frac{1}{\Delta t}$ is sampling rate and v is the size of cyclic prefix(CP).

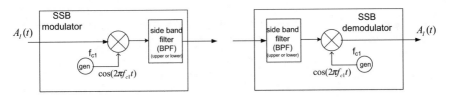

Figure 5-53: Single side band (SSB) modulator and demodulator where SSB filter is sharp to cut off one sideband (upper or lower)

5.9 Problems

P5-1: In FDM systems such as AM, FM and TV broadcastings, a block of frequency is divided with equal spacing and form a set of frequencies. Each frequency channel is used by one allowed station in a certain area. For example, in FM broadcasting in USA a frequency block from 87.8 MHz – 108 MHz is divided by 200 kHz for each channel. What is the maximum number of channels with this system? What are the carrier frequencies for each channel?

P5-2: Show that the condition of orthogonality, $\frac{1}{N}\sum_{n=0}^{N-1} e^{j\frac{2\pi\cdot(l-k)n}{N}} = \delta(l-k)$, where $l-k = 0, \pm N, \pm 2N, \ldots$ by taking DFT of $x[n] = e^{j\frac{2\pi\cdot ln}{N}}$ where $n = 0, 1, 2, \ldots N-1$, and $l = 0, 1, 2, \ldots N-1$.

P5-3: For a continuous periodic signal the Fourier series representation is given by $x(t) = \sum_{k=-\infty}^{+\infty} a_k e^{j\frac{2\pi kt}{T}}$. Express a_k in terms of $x(t)$, and show that $\frac{1}{T}\int_0^T e^{j\frac{2\pi\cdot(l-k)t}{T}}dt = \delta(l-k)$.

P5-4: OFDM symbol rate $(1/T)$ is 250 kHz and there are $N = 128$ sub-carriers. The size of CP in samples is 16.

(1) Obtain sub-carrier frequency (F) in kHz.

(2) Obtain the minimum sampling frequency $(1/\Delta t)$ in MHz.

(3) What is SNR loss due to CP in dB?

P5-5: A pair of IDFT – DFT can be used as a transmission system with the size of DFT being N. All sub-carrier channels are not practically usable. Explain why. Explain why CP is used in most practical systems.

P5-6: In Fig. P5-1, a block of N + v samples with a known OFDM symbol boundary is shown. There are v choices of selecting N samples. Is it OK to select 0 to $N-1$ and to recover transmit symbols correctly? What about selecting v to N − 1 + v?

P5-7: Show that in Fig. P5-1, all possible v choices of N sample before DFT can recover transmit data correctly. Explain how. Hint: see (5-9).

P5-8: When a channel has delay dispersion bigger than CP, there will be inter symbol interference as shown Fig. P5-2. Study the figure and explain why.

Fig. P5-1: A block of N + v samples with a OFDM symbol boundary before removing CP

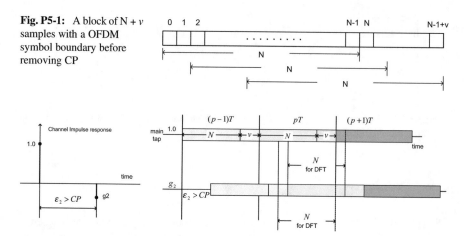

Fig. P5-2: When channel delay dispersion > CP, there is no N sample possible without ISI

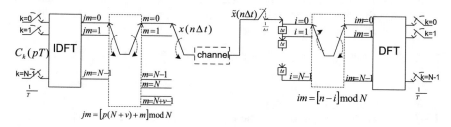

Fig. P5-3: IDFT – DFT commutating filter structure implementation when shaping pulses –g() and h()- are rectangular

P5-9: A block diagram of CP generalized OFDM is shown in Fig. P5-3. The underlying pulse shapes of this system are rectangular. A straightforward implementation of this system is shown in Fig. P5-4. There is one-to-one correspondence between the two.

(1) IDFT – DFT commutating filter structure of Fig. P5-3 is computationally more efficient compared to that straightforward structure of Fig. P5-4. Is it true if only one sub-carrier channel at the receive side is of interest?

(2) In IDFT – DFT commutating filter structure of Fig. P5-3, there is no explicit removal of 'CP', instead, the reduction of N + v samples to N samples happens. How often DFT at receive side happens per OFDM symbol period?

(3) When DFT at receive side happens once per OFDM symbol period, what is necessary condition? Compare this case with DMT with CP system.

(4) Is it possible to mix DMT with CP and CP generalized OFDM of Fig. P5-3? For example, at transmit side, IDFT with commutating filter structure is used and at receive side, (existing) DMT with CP system is used.

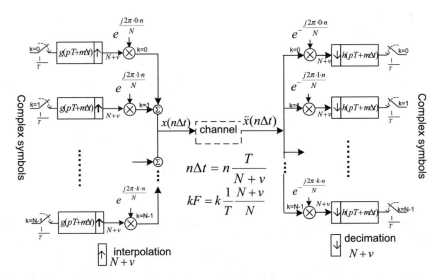

Fig. P5-4: Straightforward implementation of sampled OFDM

P5-10: In Fig. P5-3 commutating index is related as $jm = [p(N + v) + m]$ mod N where jm is the index of the output of IDFT and ranges from 0 to $N - 1$, and m ranges from 0 to $N - 1 + v$. What is the range of p to track by considering a concrete example of $N = 64$, $v = 16$?

P5-11: Consider a windowing at transmit side as shown in Figure 5-16 with $N = 128$, $v = 32$, $u = 8$.

(1) Sketch the transmit pulse assuming it is a trapezoidal pulse.
(2) How many samples in the main OFDM symbol period?
(3) How many samples of flat portion of the pulse?

P5-12: The input address (i) and the output address (im) of receive side commutating in Figure 5-14 (also Figure 5-20) are related as $im = [n - i]$ mod N.

(1) With $N = 6$, $v = 2$, generate the output address (im) when one DFT operation per OFDM symbol for 6 OFDM symbols starting $n = 0$.
(2) With $N = 8$, $v = 2$, generate the output address (im) when two DFT operations per OFDM symbol for 6 OFDM symbols starting $n = 0$.

P5-13: In Figure 5-18, the input address (jm) and output address (m) of transmit side commutating are related as $jm = (p(N + v) + m)$ mod$N$.

(1) Show that a simplification to $jm = (pv + m)$ mod N is OK.
(2) The maximum range that OFDM symbol index is limited by $\tilde{N} = lcm(N, N + v)/(N + v)$, i.e., $m = 0$ corresponds to $jm = 0$.
(3) Confirm (2) above numerically when $N = 8$, $v = 2$, i.e., $\tilde{N}=4$.
(4) Show that $\tilde{N} \leq N$ in general.

Fig. P5-5: Transmit filter
frequency response

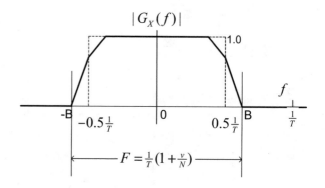

P5-14: A FIR filter is necessary in order to implement commutating filter in transmit
side (e.g., Figure 5-18). Its frequency response is sketched in Fig. P5-5 where
IDFT size = N, CP size = v and OFDM symbol rate = $1/T$.

(1) This filter design problem is the same as baseband filter design of Nyquist
pulse matched filter pair. What is the amplitude at $0.5/T$?
(2) Express an excess bandwidth β in terms of $N = 64$, and $v = 16$.
(3) Express B in Fig. P5-5 in terms of β, and of N, and v.

P5-15: A filter designed to approximate the frequency response shown in Fig. P5-5
has a symbol span of L OFDM symbols with zero phase, i.e., $g_x[n]$ is symmetrical
around $n = 0$. How many taps in the filter when $N = 64$, and $v = 16$? Sketch its
impulse response.

P5-16: A filter designed to approximate the frequency response shown in Fig. P5-5
has a symbol span of L OFDM symbols with being minimum phase and causal.
Explain the minimum phase. Express the receive side filter in terms of $g_x[n]$.

P5-17: When $g_x[n]$ is a rectangular pulse with its length being $N + v$;

$$g_X[n] = 1.0 \quad 0 \le n \le N+v-1$$

$$= 0 \quad \text{otherwise}$$

(1) Show that its frequency response is given by

$$G_X\left(e^{j2\pi f \frac{T}{N+v}}\right) = \frac{\sin(\pi f T)}{\sin\left(\pi f \frac{T}{N+v}\right)} e^{-j(N+v-1)\pi f \frac{T}{(N+v)}}$$

(2) Plot the frequency magnitude response when $N = 64$, and $v = 16$.

P5-18: When $g_R[n]$ as a matched filter to $g_X[n]$ in P5-17, is a rectangular pulse with its length being N;

$$g_R[n] = 1.0 \quad 0 \le n \le N-1$$
$$= 0 \quad \text{otherwise}$$

(1) Show that its frequency response is given by,

$$G_X\left(e^{j2\pi f \Delta t}\right) = \frac{\sin\left(\pi f N \Delta t\right)}{\sin\left(\pi f \Delta t\right)} e^{-j(N-1)\pi f \Delta t} \text{where } \Delta t = \frac{T}{N+v}$$

(2) Plot the frequency magnitude response when $N = 64$, and $v = 16$
(3) Compare the result of (2) with (2) of P5-17.

P5-19: In Figure 5-18 commutating index is related as $jm = [p(N + v) + m] \bmod N$ where jm is the index of the output of IDFT and ranges from 0 to $N - 1$, and m ranges from 0 to $N - 1 + v$.

(1) Show that, when $v = 0$, $jm = [p(N + v) + m] \bmod N = m$, i.e., there is no need of commutating since $jm = m$.
(2) Does a filter with ideal LPF meet the requirement such that $v = 0$?
(3) Consider 1+D system discussed in Chap. 2 (Sect. 2.7.3), in particular Figure 2-48 frequency response and impulse response. Does it meet the requirement of $v = 0$?

P5-20: Computationally efficient structure of Fig. P5-4 is shown in Fig. P5-6.

(1) Assume that the channel is ideal (no noise and through connection). Given $N = 32$, $v = 8$, and $u = 0$ (no windowing), at $k = 2$ sub-channel, an impulse $1.0 + j0.5$ at $n = 0$ is sent. Plot transmit side signal $x(n\Delta t)$. Hint: see Figure 5-23.
(2) Plot receive side signal at $k = 2$ sub-channel and $k = 3$ adjacent sub-channel. Hint: a computer program is necessary for the computation of (1) and (2).
(3) Use Fig. P5-4 structure for the computation of (1) and (2), and compare.
(4) In order to have an eye pattern send a random data of QPSK at $k = 2$ sub-channel. Plot eye pattern at $k = 2$ sub-channel and at $k = 3$ adjacent sub-channel.

P5-21: Analog representation of staggered OFDM system is expressed by (5-22) as,

$$x(t) = \sum_{k=0}^{N-1} \{A_k(t) + jB_k(t)\} j^k e^{j2\pi kFt}$$

where $\quad A_k(t) = \sum_{l=-\infty}^{+\infty} a_k(l)g_X(t - lT)$, $\quad$ and $\quad B_k(t) = \sum_{l=-\infty}^{+\infty} b_k(l)g_X\left(t - lT \pm \frac{T}{2}\right)$.

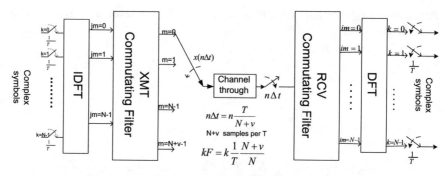

DFTbased digital implementation of OFDM signals

Figure P5-6: IDFT – DFT commutating filter structure is equivalent to Fig. 5-57 straightforward method of OFDM signal generation and reception

(1) Show that $j^k = e^{j2\pi k\frac{1}{4}}$ by evaluating $k = 0, 1, 2, 3$ and so on.

(2) What happens when j^k is removed? Hint: See Figures 5-28 and 5-29.

(3) One advantage of staggered OFDM is that there is no need of CP, i.e., $v = 0$.

(4) $g_X(t)$ is band limited and its Fourier transform $G_X(f)$ is such that $G_X(f) = 0$ when $|f| \geq \frac{1}{T}$. Thus there is overlap with adjacent sub-channels.

(5) Show that (3) and (4) are possible due the factor of j^k in addition to a half symbol period time staggering.

P5-22: Consider FH implementation with OFDM sketched in Figure 5-45. There are $N = 64$ sub-channels of which only 48 are used. If frequency diversity factor of 12 is necessary, how many FH channels are possible? When OFDM symbol rate $(1/T)$ is 1 MHz, what is the symbol rate of such FH channels?

P5-23: For frequency selective channel such as DSL, the concept of bit loading is used to achieve a channel capacity with OFDM systems. It means that modulation and coding class for each sub channel depends on SNR expected of each sub-channel, similar to 'water filling' idea. If SNR of one sub-channel is 3 dB more than that of other one, explain how many more bits per channel can be added.

P5-24: In doubly selective fading channels, the frequency response of a channel is time-varying, i.e., taps of tapped delay line model are changing in time. When OFDM systems are used in doubly selective fading channels, sub-channels are essentially Rayleigh fading. Argue to support this claim even under Rician fading channels.

P5-25: In OFDM systems with doubly selective fading channels, bit loading scheme is not feasible since there is little time to measure them. However, a statistical averaging can be used. One logical sub-channel uses all permutation of physical sub-channel using a permutation shown in Figure 5-48. When a random permutation of $N = 9$ is given by $\pi_5 = (0, 2, 7, 6, 3, 4, 8, 5, 1)$, obtain sub-channel matrix by down shifting a column vector of sub-channel number (1:9)' by a corresponding column in π_5.

P5-26: Create a 4 bit π_1 bit permutation similar to Figure 5-50 assuming 15 symbols of 4 bit, and thus 60 bits. Total symbol bits are organized as a matrix 4x 15, and then each bit in a symbol is down (circular) shifted by 0, 1, 2, 3.

P5-27: Symbol permutation for time de-correlation (π_3) is shown in Figure 5-49. For 60 symbols, obtain interleaved symbol sequence by r = 5, c = 12 matrix interleaver; write row by row and read column by column.

Chapter 6
Channel Coding

Contents

6.1   Code Examples and Introduction to Coding 324
     6.1.1 Code Examples – Repetition and Parity Bit 324
     6.1.2 Analytical WER Performance 328
     6.1.3 Section Summary 331
6.2   Linear Binary Block Codes 331
     6.2.1 Generator and Parity Check Matrices 332
     6.2.2 Hamming Codes and Reed-Muller Codes 334
     6.2.3 Code Performance Analysis of Linear Block Codes* 339
     6.2.4 Cyclic Codes and CRC 349
     6.2.5 BCH and RS Codes 361
     6.2.6 Algebraic Decoding of BCH 366
     6.2.7 Code Modifications – Shortening, Puncturing and Extending 369
6.3   Convolutional Codes 370
     6.3.1 Understanding Convolutional Code 370
     6.3.2 Viterbi Decoding of Convolutional Codes 385
     6.3.3 BCJR Decoding of Convolutional Codes 391
     6.3.4 Other Topics Related with Convolutional Codes 403
6.4   LDPC 407
     6.4.1 Introduction to LDPC Code 407
     6.4.2 LDPC Decoder 412
     6.4.3 Bit Node Updating Computation 412
     6.4.4 LDPC Encoder 420
     6.4.5 Useful Rules and Heuristics for LDPC Code Construction 425
     6.4.6 LDPC in Standards 433
6.5   Turbo Codes 437
     6.5.1 Turbo Encoding with G=15/13 RSC and Permutation 438
     6.5.2 G=15/13 Code Tables for BCJR Computation Organization 439
     6.5.3 The generation of 'extrinsic' information (E1, E2) 441
     6.5.4 Numerical Computations of Iterative Turbo Decoding 442
     6.5.5 Additional Practical Issues 447
6.6   Coding Applications 450
     6.6.1 Coded Modulations 451
     6.6.2 MLCM, TCM and BICM 461
     6.6.3 Channel Capacity of AWGN and of QAM Constellations 470
     6.6.4 PAPR Reduction with Coding 470
     6.6.5 Fading Channels 472
6.7   References with Comments and Appendix 479

© Springer Nature Switzerland AG 2020

S.-M. Yang, *Modern Digital Radio Communication Signals and Systems*,
https://doi.org/10.1007/978-3-030-57706-3_6

Appendix 6 480
    A.1   CM Decoding Example of Figure 6-36 480
    A.2   The Computation of p_0 (p_1), and LLR for BPSK 482
    A.3   Different Expressions of Check Node LLR of LDPC 483
    A.4   Computation of Channel Capacity 487
    A.5   SER Performance of Binary PSK, DPSK, FSK 489
6.8   Problems 490

Abstract This chapter covers the topics of channel coding applicable to digital communications – introduction with repetition codes and coding with parity, cyclic algebraic code of BCH and RS and more, convolutional codes with Viterbi decoding and with BCJR decoding, LDPC, turbocodes, applications of coding to higher order modulations as BICM and MLCM, to PAPR reduction, and to fading channels.

Key Innovative Terms correlation metric decoding · correlation metric used as branch metric · coding error performance with the consistent use of WER vs. Es/No · BICM and MLCM comparisons · PAPR reduction with coding · Fading channel counter-measure with coding

General Terms APP · bit node · BCJR · BCH · block code · linear code · convolution code · check node · coded modulation · CRC · cyclic code · algebraic decoding · girth · LDPC · message passing decoding · parity · parity check matrix · primitive polynomial · remainder · repetition · RSC · trellis · Turbo decoding

List of Abbreviations

APP	a posterior probability
AWGN	additive white Gaussian noise
BCH	Bose-Chaudhuri-Hocquenghem (names)
BCJR	Bahl-Cocke-Jelinek-Raviv (names)
BEC	binary erasure channel
BER	bit error rate
BICM	bit interleaved coded modulation
BLER	block error rate
BPSK	binary phase shift keying
BSC	binary symmetric channel
CCDF	complimentary cumulative density function
CM	correlation metric
CM	coded modulation
CRC	cyclic redundancy check
DPSK	differential phase shift keying
DSQ	double square quadrature (amplitude modulation)
DMC	discrete memoryless channel
DVB	digital video broadcasting

FEC	forward error correction
FSK	frequency shift keying
GF	Galois field
HD	hard decision
IEEE	institute of electrical and electronics engineers
LCM	least common multiple
LDPC	low density parity check
LLR	log likelihood ratio
LPF	low pass filter
MLCM	multi-level coded modulation
MLSD	maximum likelihood sequence detection
PA	power amplifier
PAPR	peak to average power ratio
PDF	probability density function
QAM	quadrature amplitude modulation
RA	repeat accumulate (code)
RS	reed solomon
RHS	right hand side
RSC	recursive systematic (convolution) code
SD	soft decision
SD	space diversity
SER	symbol error rate
SNR	signal (power) to noise (power) ratio
TCM	trellis coded modulation
WER	word error rate / block error rate

In modern implementations powerful error correction schemes such as LDPC, approaching channel capacity, may have profound practical impacts. In Figure 6-1 a wireless system block diagram is shown; from channel coding point of view, RHS of B1 – B1 can be considered as a 'channel'. Some useful such channels are BSC, DMC and AWGN, summarized in Table 6-15. We present FEC in practical point of view, e.g., consistently using Es/No in dB without including rate loss; thus a repetition code performance has 'coding gain' in Es/No, i.e., the more repetition the less Es/No required at receive side. We cover algebraic codes such as BCH but our emphasis is on modern coding such as LDPC and turbo form of convolutional codes. MLCM is covered in detail as a coding application to higher order modulations. The impact of coding to PAPR reduction and to fading channels is covered compactly. The literature on channel coding is vast. However, we believe that by studying this chapter thoroughly, especially examples, one may grasp the topic quickly and find useful in practice.

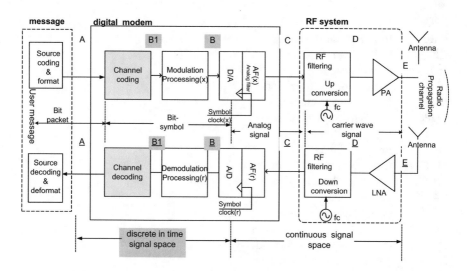

Figure 6-1: A wireless system block diagram; from channel coding point of view, RHS of B1-B1 can be considered as a 'channel'. In modern implementations, modulation processing can be intimately related with coding or the coding can be part of signal space including bit to symbol and vice versa

6.1   Code Examples and Introduction to Coding

We first introduce two basic examples of codes – repetition code and adding a parity bit. This is simple yet most fundamental in code construction.

6.1.1   Code Examples – Repetition and Parity Bit

Example 6-1: n = 3 repetition code with soft decision

Code word = {000, 111}, for {0, 1} respectably , which means three bits are transmitted for one bit of information, and will be transmitted by BPSK with $0 \rightarrow +1$ and $1 \rightarrow -1$ bit to symbol mapping.

Three received bits are contaminated by AWGN, and received signal y_i, $y_i = X + n_i$ where $i = 1, 2, 3$ and $X = \{+1, -1\}$, and n_i noise samples.

In order to recover the information bit, the three received bits are correlated to the code words $\{+1+1+1, -1-1-1\}$ in BPSK symbols, which correspond to {000, 111} in bits, and then compare and choose the maximum. Since there are only two code words, the bigger one will be chosen. This procedure is called correlation metric (CM) decoder with soft decision (SD). This is the same as adding y_i and taking the sign to make a decision, {*positive* $\rightarrow$ +1, *negative* $\rightarrow$ -1}, and thus {*positive* $\rightarrow$ 0, *negative* $\rightarrow$ 1}.

Note that SNR after addition, i.e., SNR of $\sum_i y_i = 3X + (n_1 + n_2 + n_3)$, is 3 times of $\frac{E_s}{N_o}$ i.e., $w_m \frac{E_s}{N_o}$, i.e., it is increased by code weight (w_m) . Thus the performance of the repetition code is obtained by effectively increasing SNR by the code weight. The code word error rate is given by $P_2(m) = Q\left(\sqrt{2\frac{Es}{No}w_m}\right)$ where the code weight w_m =3 and $\frac{E_s}{N_o}$ SNR measured in BPSK symbols. The *code rate*, R_c, is the ratio of the information bit over the bits in a codeword, $k/n = 1/3$.

There is only one non-zero code word, i.e., {111}, thus the average codeword error rate is trivially $P_w = P_2(m)$. This is peculiar to repetition codes but in general this is not the case. The computation of P_w for other than a repetition code is involved and complicated, which will be discussed in detail later. Furthermore since each code word represents one bit, the word error rate is also bit error rate as well.

There is no net <u>coding gain</u> since the transmit power is increased by a factor of 3, the inverse of code rate. Typically in a textbook, this factor is included for net coding gain. However, in our presentation, we do not explicitly include it in the error performance equation and plot. It is less confusing and from the receiver measurement point of view it is consistent but we keep in mind that the additional power is used for error correction. This can be easily adjusted by adding the code rate loss in dB, e.g., $10*\log(3) = 4.77$ dB.

Example 6-2: n = 3 repetition code with hard decision
The same repetition code of code word = {000, 111}, for {0, 1} is now decoded CM decoder with hard decision (HD). HD means that y_i is first quantized into $[y_i]$, i.e., $[y_i] = \{+1 \text{ or } -1\}$, then correlate with code words and then choose the maximum. Notice that the correlation computation can be done binary digits, rather than real number as in SD, and thus it is simpler than SD. This is the same as adding $[y_i]$, and then if $\sum_i[y_i] \geq 0$, +1 is decided otherwise -1 is decided.

The same decoding procedure can be more simplified if $[y_i] = \{0, 1\}$. Now $[y_1]$ $[y_2][y_3]$ is compared to {000, 111}. Then the distance, called Hamming distance, between code word and $[y_1][y_2][y_3]$. For example $[y_1][y_2][y_3] = 001$ then {0} is decided since Hamming distance is 1 to {000} and 2 to {111}. This is called minimum (Hamming) distance decoding. In this case it can be simplified into a majority decoding. For example, more {0}s in $[y_1][y_2][y_3]$ then it is decided into {0} and more {1}s in $[y_1][y_2][y_3]$ it is decoded into {1}.

In HD decoding of a repetition coding, CM decoding, minimum distance decoding and majority decoding are all equivalent.

Its performance can be analyzed. $p = Q\left(\sqrt{2\frac{Es}{No}}\right)$ is the error probability when y_i is decoded into $[y_i] = \{0, 1\}$. With the majority decoding, a single error will be corrected and double and triple errors are not corrected, and thus the word error probability is given by $P_w = \sum_{i=2}^{n}C(n,i)p^i(1-p)^{n-i} = 3p(1-p)^2 + p^3$ where $C(n, i)$ is a combination, $C(n,i) = \frac{n!}{(n-i)!i!}$.

Again this word error rate is the same as bit error rate.

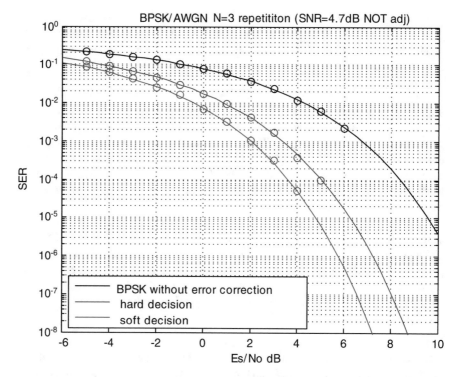

Figure 6-2: n=3 repetition code performance with SD and HD against without coding (black solid line). The circles are simulations and solid lines are analytical results discussed in Examples 6-1 and 6-2. Note that the rate loss (4.77dB, 1/3) is not included in this plot

The performance of n=3 repetition code with SD and HD against one without coding is shown in Figure 6-2. Note that the rate loss (4.77dB, 1/3) is not included in the figure. The circles on the figure are the result of simulations, and solid lines are analytical results discussed in Examples 6-1 and 6-2. The match between simulation and analytical analysis are excellent as expected since there is no approximation in the analytical results with these examples.

SD outperforms HD asymptotically about 1.6dB in this case.

Example 6-3: Hadamard H_4 (n=4) code (adding parity)

Info bits	00	10	01	11
Code word	0000 c_0	0101 c_1	0011 c_2	0110 c_3

H_{2m} can be constructed from H_m with the following recursion.

$$H_{2m} = \begin{bmatrix} H_m & H_m \\ H_m & \overline{H}_m \end{bmatrix} \text{ with } m = 2 \text{ as } H_2 = \begin{bmatrix} 0 & 0 \\ 0 & 1 \end{bmatrix} \text{ and } \overline{H}_2 = \begin{bmatrix} 1 & 1 \\ 1 & 0 \end{bmatrix}.$$

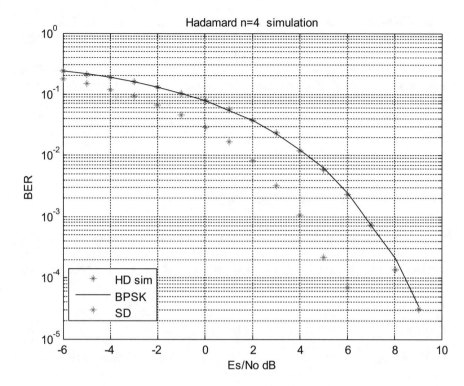

Figure 6-3: Hadamard n=4 code performance simulations with CM decoder. Blue star is with hard decision correlation metric (CM) decoder and red star with soft decision CM decoder and black solid line is BPSK symbol by symbol detector

The block size is $n = 2^k$ with k bit information and the minimum distance $d_{min} = n/2$. For example, with $n=4$, $d_{min} = 2$. The Hamming distance (different bits) between any two codewords is 2. This particular case of H_4 may be seen as adding one parity bit (to make even) into two information bits, and redundant {0} bit up front.

The precise analytical analysis of word error performance seems not straightforward for this code. However, it is simple to obtain a true performance by simulating with CM decoder since there are only 4 code words.

Figure 6-3 shows Hadamard n=4 code performance simulations with CM decoder. Blue star is with hard decision correlation metric (CM) decoder and red star with soft decision CM decoder and black solid line is BPSK symbol by symbol detector. Since the minimum distance of this code is $d=2$ there is no error correction and thus the performance of HD seems identical to that of BPSK. On the other hand SD has about 2dB improvement over HD.

In Figure 6-3, BER rather than word error rate (WER) is displayed. BER can be directly compared to the case without coding. BER rather than word error rate is displayed.

How WER can be converted to BER will be discussed briefly here. When there is k information bits in a code word ($k=2$ in Example 6-3), a corresponding BER p_b, assuming random bit errors, is related with WER P_w as $P_w = 1 - (1 - p_b)^k$. Its inverse relationship is given by $p_b = 1 - (1 - P_w)^{1/k}$. When P_w is very small, it may be approximated by $p_b \approx P_w/k$. This conversion may be used for a quick but approximate BER estimation. In general a precise BER is involved and requires simulations. In our presentation, we compute WER unless otherwise stated.

6.1.2 Analytical WER Performance

Example 6-3 can be slightly simplified as follows; for two bit information, one parity bit is added. There are 4 code words {000, 101, 011, 110}, which carry two information bits {00, 10, 01, 11} respectively. This is the simplest possible code with a parity bit and is summarized below.

Info bits	00	10	01	11
Code word	000 c_0	101 c_1	011 c_2	110 c_3

We did numerical simulations of Example 6-3, and displayed the result in Figure 6-3. Can we find analytical expressions for the simulation results? We focus on this simple code example, and later we may extend the insight obtained here to the cases of linear block codes.

We show that this code is linear, which means that zero, [000], is a code word and the sum of any code word is a code word. By adding [101], the code words {101, 000, 110, 011}. By adding [011], they become {011, 110, 000, 101}. By adding [110], they become {110, 011, 101, 000}. Thus the code is linear.

We assume that all code words are equally likely.

With the above two conditions, being a linear code and all code words occurring with the same probability ($1/2^k$), we can express WER P_w as follows.

$$P_w = \sum_{m=0}^{2^k-1} P_{e|m}p_m = \frac{1}{2^k} \sum_{m=0}^{2^k-1} P_{e|m} \quad \text{(equally likely)} \tag{6-1}$$

with p_m, the mth code word probability, $\frac{1}{2^k}$ when equally likely, and $P_{e\,|\,m}$ the error probability of the mth code word is transmitted and is in error.

With the condition of linearity, (6-1) simplifies to be

$$P_w = \sum_{m=0}^{2^k-1} P_{e|m}p_m = P_{e|m=0} \tag{6-2}$$

where $P_{e \mid m=0}$ = probability of $\{m \neq 0 \mid m = 0$ transmitted$\}$. In other words, WER can be computed by considering the code word zero being mistaken into other code words. In this case [000] is mistaken into other code words $\{101, 011, 110\}$. The event that the zero code word c_0 being mistaken into c_j codeword is denoted u_j with $j=1, 2, 3$.

$$P_w = P_{e \mid m=0} = p_r(u_1 + u_2 + u_3) = 1 - p_r(\overline{u_1 + u_2 + u_3}) = 1 - p_r(\overline{u_1} \cdot \overline{u_2} \cdot \overline{u_3}) \quad (6\text{-}3)$$

where u_j: $c_0 \rightarrow c_j$, the zero code word is mistaken into other code word and '+' means the union and '·' means the intersection. $p_r(u_1 + u_2 + u_3) = 1 - p_r(\overline{u_1 + u_2 + u_3})$ is based on the fact that $p_r(A) = 1 - p_r(\overline{A})$, and $1 - p_r(\overline{u_1 + u_2 + u_3}) = 1 - p_r(\overline{u_1} \cdot \overline{u_2} \cdot \overline{u_3})$ is based on De Morgan's rule of $\overline{A + B} = \overline{A} \cdot \overline{B}$

The equation (6-3) may be extended as a general expression of WER for linear block codes with equal code word probability with $M = 2^k$ code words as,

$$P_w = P_{e \mid m=0} = p_r(u_1 + \ldots \ldots + u_{M-1}) \quad (6\text{-}4)$$

In general, in order to evaluate (6-3) or (6-4), one needs a joint probability density function of $\{u_1, u_2, \ldots u_{M-1}\}$. For practical codes it is extremely complicated.

6.1.2.1   Independence and Mutually Exclusive Assumptions

Since the evaluation of (6-4) is complicated, a simplification is sought in terms of bound, which may provide insights and be useful in some situations where for example, SNR is high and only asymptotic behavior is of interest.

With the independence assumption, with no need of joint probability density, we can evaluate (6-3) using

$$P_w = 1 - p_r(\overline{u_1} \cdot \overline{u_2} \cdot \overline{u_3}) = 1 - (1 - p_r(u_1))\,(1 - p_r(u_2))\,(1 - p_r(u_3))$$

With the mutually exclusive assumption,
$P_w = p_r(u_1) + p_r(u_2) + p_r(u_3)$, which is often called a union bound. The term' union bound' is a misnomer since the precise codeword error rate is expressed as the probability of the union as in (6-4), but is stuck. We may use it as well.

These simplifying assumptions are not true for this simple code. Furthermore it is not generally true. However, it gives a very quick evaluation for a given code for high SNR case. With the independence assumption we compare the performance with simulated results (true performance) and it is shown in Figure 6-4. With BER better than 10^{-4}, the match is good. However, it is getting worse as SNR is smaller. With more powerful codes approaching the channel capacity, this discrepancy will progressively worsened such that the bound is not very useful in practice.

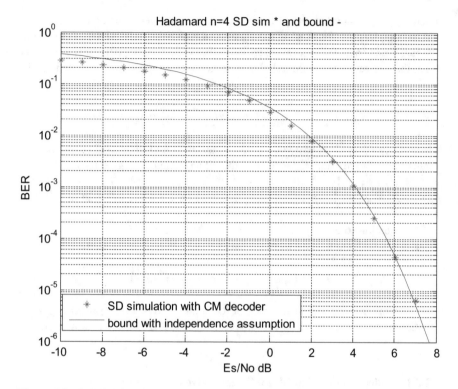

Figure 6-4: Soft decision with CM decoder (true, optimum performance) compared to a bound with assumption of independence (this is not a typical union bound in textbooks). Note that the bound matches very well with the true performance (simulation) when SNR is high but progressively worse when SNR is low. This discrepancy is due to the independence assumption. This 'small' discrepancy may be ignored for typical codes. However, with powerful codes whose performance approaching the capacity of a channel may be no longer ignorable. This issue will be addressed again in detail later in this chapter

For some practical situations, a union bound is not useful. Thus there is a need of more precise estimation of WER in practice. Typically one may resort to numerical simulations. In this chapter, we give an attempt to alleviate this problem and find analytical methods of predicting WER performance for a given code. It is a noble goal but very difficult. In fact, it is an open problem.

Exercise 6-1: Evaluate $p_r(u_1)$ and show that $p_r(u_1) = p_r(u_2) = p_r(u_3)$ with the code of Example 6-3 (Hadamard n=4).

Answer: $p_r(u_1) = Q\left(\sqrt{2\frac{Es}{No}w_m}\right)$ with $w_m = 2$. Note that u_1 is the event $\{000\}$ is mistaken as $\{101\}$. With the independence assumption, WER is given by $P_w \leq$

$1 - (1 - p_r(u_1))^3 = 1 - \left(1 - Q\left(\sqrt{2\frac{Es}{No}w_m}\right)\right)^3$. And BER is given by $p_b = 1 - (1 - P_w)^{1/2}$. This is plotted in Figure 6-4 (Blue solid).

6.1.2.2   Analytical Performance with HD

In Figure 6-3, the example code performance with HD is shown, with numerical simulations, to be the same as without adding parity bit. This makes sense since one parity bit there is no error correction possible; only the detection of one bit error. This can be extended to t error correcting codes.

$$P_w \le \sum_{i=t+1}^{n} \binom{n}{i} p^i (1-p)^{n-i} \tag{6-5}$$

In the example the minimum distance $d=2$, thus with $t=0$, the above (6-5) becomes $P_w = \sum_{i=1}^{n} C_i^n p^i (1-p)^{n-i} = 1 - (1-p)^n$ where $p = Q\left(\sqrt{2\frac{E_s}{N_o}}\right)$, and $C_i^n = \binom{n}{i}$ a combination. And BER is the same as the case without coding. For HD, (6-5) seems to be working reasonably well in general. A precise formulation requires the weight distribution of a code. This will be discussed later.

6.1.3   Section Summary

By some examples we introduced to channel coding in this section. Some concepts such as CM decoding, linearity and Hamming distance were used without formal definition s. This is partly intentional and will be further clarified in later sections. We mention that, for CM decoding, it is defined in Subsection 6.2.3 as well as in Chapter 2, and in Appendix 6 A.1 an example is given.

6.2   Linear Binary Block Codes

We introduced to coding by a few examples in the previous section. They are intuitive and thus easy to understand. Now we introduce the linear binary block codes. In fact all the examples – repetition and addition of parity are linear binary block codes. Our focus is on linear block codes with good reasons – in practice they are most often used and furthermore almost all codes including convolution codes and non-binary codes may be treated and converted to linear binary block codes.

A linear binary block code is usually put into (n, k), where k bit of information is expanded to n bit block by adding $n-k$ parity bits. There are 2^k code words. The linearity means the zero $(\mathbf{0})$ is a code word, and for any two codewords, $\mathbf{c}_1$ and $\mathbf{c}_2$, their modulo-2 addition, $\mathbf{c}_1 + \mathbf{c}_2$, is also a codeword.

It is referred to different names – group codes or generalized parity check codes. This includes polynomial generated codes such as BCH, RS, Reed-Muller codes, projective geometry codes, Euclidian geometry codes and quadratic residue codes, depending on how to construct the code and on how to decode efficiently. Almost all practical codes are linear. Interested readers for more details may consult a vast literature on coding.

Exercise 6-2: Show that the repetition code of n, $(n, 1)$, is a linear binary block code. Show also that a parity code $(n, n\text{-}1)$ is linear. Hint: apply the two conditions of linearity above.

Exercise 6-3: n-bit binary tuple may be considered as a code, (n, n). Show that it is linear.

Answer: As an example $n=2$ there are four {00, 01, 10, 11} codewords. It is easy to see $01+10 =11$, $10+11=01$ and so on. It is closed under modulo-2 addition, and thus linear.

6.2.1 Generator and Parity Check Matrices

A linear binary block code can be represented by a generator matrix, G $(k \times n)$. Codewords, c_m, can be generated from a binary vector length k denoting information bits, u_m;

$$c_m = u_m G \tag{6-6}$$

An example of $n=7$, $k=4$ is shown below.

$$\mathbf{G} = \begin{bmatrix} 1000111 \\ 0100011 \\ 0010101 \\ 0001110 \end{bmatrix} \quad \text{and} \quad \boldsymbol{u}_m = [0011], \quad \text{thus} \quad \boldsymbol{c}_m = [0011011].$$

Exercise 6-4: Find all, 16, codewords with above example.

Answer: all the code words can be found by linear combinations of the rows of G.
   Without loss of generality, a generator matrix can be put in a systematic form as

$$\mathbf{G} = [I_k | P] \tag{6-7}$$

Any non-systematic generator matrix can be converted to a systematic one by row combinations (a form of Gaussian elimination), and by possible column permutations (changing bit locations of a codeword).
   A linear block code can be described by a parity check matrix, H, $(n - k) \times n$, so that

$$c\,\mathbf{H}^{\mathrm{T}} = [\mathbf{0}] \tag{6-8}$$

In other words, a necessary and sufficient condition for c (dropping subscript m) to be a codeword is to meet the (6-8) conditions. $[\mathbf{0}]$ is a vector with $1 \times (n - k)$ dimension. Since the rows of G are codewords, we conclude that

$$\mathbf{G}\mathbf{H}^{\mathrm{T}} = [\mathbf{0}] \tag{6-9}$$

The row and column of (6-9) is given by $k \times (n - k)$ since there are k codewords in G.

A generator matrix example given above has a parity check matrix (3×7).

$$H = \begin{bmatrix} 1011100 \\ 1101010 \\ 1110001 \end{bmatrix}$$

If $u = (u_1, u_2, u_3, u_4)$ is an information sequence, the corresponding codeword $c = (c_1, c_2, c_3, c_4, c_5, c_6, c_7)$ is given by,

$$
\begin{aligned}
c1 &= u1, \\
c2 &= u2, \\
c3 &= u3, \\
c4 &= u4, \\
c5 &= u1 + u3 + u4, \\
c6 &= u1 + u2 + u4, \\
c7 &= u1 + u2 + u3
\end{aligned}
$$

From the above it is not hard to see that all the code words satisfy the equation (6-8). It is also noted that G in the form of (6-7) can be related with H as,

$$H = \left[P^{T} | I_{n-k} \right] \tag{6-10}$$

Exercise 6-5: We define a row vector, e.g., $[1\,0\,0\,0\,1]$ and its column vector $[1\,0\,0\,0\,1]^{\mathrm{T}}$. And c_m and u_m are row vectors. (6-6), (6-8) and (6-9) can be done with column vectors of $c_m{}^{\mathrm{T}}$ and $u_m{}^{\mathrm{T}}$.

$$c_m{}^{\mathrm{T}} = \mathbf{G}^{\mathrm{T}} u_m{}^{\mathrm{T}} \tag{6-6}$$

$$\mathbf{H} c^{\mathrm{T}} = [0]_{(n-k)\times 1} \tag{6-8}$$

$$\mathbf{H} \mathbf{G}^{\mathrm{T}} = [0]_{(n-k)\times k} \tag{6-9}$$

6.2.1.1   Dual Code

From (6-9) in the previous section, the role of generator matrix and parity check matrix is exchanged; consider a code H as a generator matrix and it will be $(n, n\text{-}k)$ code. This code is called dual of the original code.

6.2.1.2   Weight and Distance of Linear Block Codes

The weight of a codeword, c, is the number of nonzero components of it and denoted by $w(c)$. The Hamming distance between two codewords, c_1 and c_2, denoted by d (c_1, c_2) is the number of components at which c_1 and c_2 differ. Since the zero is a codeword of any linear block codes, it is clear that the weight of a codeword is its Hamming distance from 0 codeword.

The minimum distance of a code is the minimum of all possible distance between all distinct codewords,

$$d_{\min} = \min_{\substack{c1, c2 \in C \\ c1 \neq c2}} d(c_1, c_2) \tag{6-11}$$

The minimum weight of a code is the minimum weights of nonzero codewords.

$$w_{\min} = \min_{\substack{c \in C \\ c \neq 0}} w(c) \tag{6-12}$$

In linear block codes $d_{\min} = w_{\min}$.

6.2.2   Hamming Codes and Reed-Muller Codes

6.2.2.1   Hamming Codes

Hamming code is one of the first codes that sparked a search for better codes. It is simple but non-trivial single error correction code. When this code was invented and discovered, the soft decision computation was not feasible with the technologies at the time. As part of understanding the codes, we encourage a reader to re-discover it. How can you add parity bits so that a single error may be corrected?

A parity check matrix has rows and columns of $(n - k) \times n$. In order to identify a single error, the columns of H must be distinct and all possible combinations other than zero. This is possible when $n=7 =2^3\text{-}1$ and $n\text{-}k =3$. In general $n=2^m\text{-}1$ and $n\text{-}k=m$.

Since the column of H includes all nonzero sequence of length m, the sum of any two columns is another column, i.e. there always exist three columns that are linearly dependent. Thus for Hamming codes the minimum distance is 3 independent of m.

Once block size is identified and **H** matrix is known, the corresponding **G** matrix can be identified by using (6-10) and (6-7).

Example 6-4: Hamming code $n=7$ $k=4$, $d_{min}=3$
Hamming code is defined as $n=2^m-1$, $k=2^m-1-m$ for $m \geq 3$. And $(n, k) = (7, 4)$, $(15, 11)$, $(31, 26)$, $(63, 57)$, $(127, 120)$, $(255, 247)$, and so on. As the block size becomes larger, the code rate $(\frac{k}{n})$ gets closer to one.

We display $(7, 4)$ Hamming code in Table 6-1 in the standard array; $2^7=128$ bit patterns are organized so that syndrome based decoding process - hard decision decoding - can be seen. This will be explained below.

6.2.2.2 Hard Decision Decoding

Seven received bits in a block are contaminated by additive noise, and received signal y_i, $y_i = X + n_i$ where $i=1,2,\ldots,7$ and $X = \{+1, -1\}$, and n_i AWGN noise samples. For HD, y_i is quantized, $[y_i]$, to be $\{0\}$ if $y_i \geq 0$ and $[y_i] = \{1\}$ otherwise. In other words, a symbol by symbol (BPSK) decision is made.

$\mathbf{y} = [y_1][y_2][y_3][y_4][y_5][y_6][y_7]$ is correlated (compared) with all possible codewords, then choose a codeword with the minimum distance (maximum

Table 6-1: standard array for Hamming code $(7, 4)$ $d_{min}=3$

codeword	one bit error from a code word on the left – not codewords						
0000 000	1000 000	0100 000	0010 000	0001 000	0000 100	0000 010	0000 001
0001 110	1001 110	0101 110	0011 110	0000 110	0001 010	0001 100	0001 111
0010 101	1010 101	0110 101	0000 101	0011 101	0010 001	0010 111	0010 100
0011 011	1011 011	0111 011	0001 011	0010 011	0011 111	0011 001	0011 010
0100 011	1100 011	0000 011	0110 011	0101 011	0100 111	0100 001	0100 010
0101101	1101101	0001101	0111101	0100101	0101001	0101111	0101100
0110 110	1110 110	0010 110	0100 110	0111 110	0110 010	0110 100	0110 111
0111 000	1111 000	0011 000	0101 000	0110 000	0111 100	0111 010	0111 001
1000 111	0000 111	1100 111	1010 111	1001 111	1000 011	1000 101	1000 110
1001 001	0001 001	1101 001	1011 001	1000 001	1001 101	1001 011	1001 000
1010 010	0010 010	1110 010	1000 010	1011 010	1010 110	1010 000	1010 011
1011 100	0011 100	1111 100	1001 100	1010 100	1011 000	1011 110	1011 101
1100 100	0100 100	1000 100	1110 100	1101 100	1100 000	1100 110	1100 101
1101 010	0101 010	1001 010	1111 010	1100 010	1101 110	1101 000	1101 011
1110 001	0110 001	1010 001	1100 001	1111 001	1110 101	1110 011	1110 000
1111 111	0111 111	1011 111	1101 111	1110 111	1111 011	1111 101	1111 110

correlation becomes the minimum distance). This is called a minimum distance decoding. This CM based scheme works for any binary linear block codes in principle. However, it is severely limited when the block size is large (e.g., $k > 10$) due to computational burden.

Since a code is described by parity check matrix $\mathbf{H}$, it can be used to decode more efficiently. Suppose c_m is the transmitted codeword and $\mathbf{y}$, the received sequence after quantization, may be expressed as, e being an arbitrary error vector.

$$\mathbf{y} = c_m + e$$
$$s = \mathbf{y}\mathbf{H}^{\mathrm{T}} = c_m \mathbf{H}^{\mathrm{T}} + e\mathbf{H}^{\mathrm{T}} = e\mathbf{H}^{\mathrm{T}}$$

where (n-k) vector s is called syndrome of the error pattern. The vector s contains zero for all codewords and nonzero for error vector. For example, with error vector 1000000, the syndrome vector is $s=[111]^{\mathrm{T}}$ and with error vector 0100000, $s=[011]^{\mathrm{T}}$ and so on.

syndrome	Error pattern
111	1000 000
011	0100 000
101	0010 000
110	0001 000
100	0000 100
010	0000 010
001	0000 001

Exercise 6-6: syndrome decoding reduces correlation computations compared to CM method. Compare them.

Answer: syndrome computation, $s = \mathbf{y}\,\mathbf{H}^{\mathrm{T}}$, requires $n\text{-}k = m = 3$ CM computations while a number of codewords $2^k=16$ is required in general. With bigger block size of (31, 26) Hamming code, $n\text{-}k=m=5$ vs. $2^k=2^{26}$; the improvement is enormous.

6.2.2.3 HD Decoding Performance of (7, 4) Hamming

For t error correction code, WER can be bounded by $P_w \leq \sum\limits_{i=t+1}^{n} \binom{n}{i} p^i (1-p)^{n-i}$ where $p = Q\left(\sqrt{2\frac{Es}{No}}\right)$ with the assumption of BPSK under AWGN. This was shown in (6-5). It turns out that in this case it is not bound but exact equality since the code can correct exactly a single error pattern and no other pattern. In general, (6-5) is an upper bound since some other patterns (part of errors more than t) are corrected. Figure 6-5 shows the comparison of this WER against a simulation with syndrome decoding.

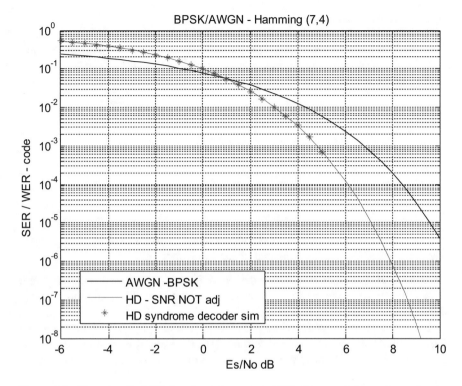

Figure 6-5: Hamming (7, 4) HD decoder performance in WER. Blue star is the syndrome decoder simulation and the solid blue line is based on WER of (6-5). The match is excellent as expected since the decoder corrects exactly a single bit error

6.2.2.4   Dual of Hamming Code is Maximal Length Code

A maximal length code is defined as $n=2^m-1$, $k=m$, and $n-k= 2^m-1-m$, for $m \geq 3$ and $(n, k) = (7, 3), (15, 4), (31, 5), (63, 6), (127, 7), (255, 8)$, and so on. It can be shown that $d_{\min} = 2^{m-1}$. Another property is that the all nonzero code word has the same code weight of 2^{m-1}. This property can be seen that the one's density is ½ of the length of a sequence. The maximal length sequence is used as a scrambler and as pseudo noise generators.

Exercise 6-7: For (7, 3) maximal length code, generator matrix and corresponding parity check matrix is shown below.

$$G = \begin{bmatrix} 1001011 \\ 0101101 \\ 0011110 \end{bmatrix} \qquad H = \begin{bmatrix} 1111000 \\ 0110100 \\ 1010010 \\ 1100001 \end{bmatrix}$$

6.2.2.5 Reed-Muller Code

The Reed-Muller codes are binary linear block codes that are equivalent to cyclic codes with an overall parity check added. For any value m the r-th order Reed-Muller code has parameters

$$n = 2^m$$

$$k = \sum_{i=0}^{r} \binom{m}{i}$$

$$d_{\min} = 2^{m-r}$$

Exercise 6-8: Consider $r=1$ RM codes. Show that $n=2^m$, $k=m+1$, $d_{\min} = 2^{m-1}$, e.g., $m=3$, $(n, k, d_{\min})$ =(8, 4, 4). Compare with a repetition code $(n, k, d_{\min})$ =(8, 4, 2). RM code has distance of 4 while that of repetition code is 2.

Exercise 6-9: Consider $r=1$ RM codes $m=4$; $(n, k, d_{\min})$ =(16, 5, 8). Compare it with 3 repetition code $n=15$, $k=5$, $d_{\min}$ =3. Hint: see the minimum distance difference.

Let v_0 vector whose components are all ones. Let $v_1, v_2, v_3, \ldots v_m$ be the rows of a matrix with all possible m-tuples as columns. The r-th order Reed-Muller code has as its basic vectors the vector $v_0, v_1, v_2, \ldots, v_m$ and all of their vector product r or fewer at the time (Table 6-2).

Example 6-5: The first order Reed-Muller code with block length 8 is an (8, 4) code with generator matrix

$$G = \begin{bmatrix} 1111 \cdot 1111 \\ 0000 \cdot 1111 \\ 0011 \cdot 0011 \\ 0101 \cdot 0101 \end{bmatrix} \tag{6-13}$$

The second-order Reed-Muller code with block length 8 has the generator matrix

Table 6-2: Basic vectors for the Reed-Muller codes of length-8

$v_0 = (1\ 1\ 1\ 1\ 1\ 1\ 1\ 1)$
$v_1 = (0\ 0\ 0\ 0\ 1\ 1\ 1\ 1)$
$v_2 = (0\ 0\ 1\ 1\ 0\ 0\ 1\ 1)$
$v_3 = (0\ 1\ 0\ 1\ 0\ 1\ 0\ 1)$
$v_1v_2 = (0\ 0\ 0\ 0\ 0\ 0\ 1\ 1)$
$v_1v_3 = (0\ 0\ 0\ 0\ 0\ 1\ 0\ 1)$
$v_2v_3 = (0\ 0\ 0\ 1\ 0\ 0\ 0\ 1)$
$v_1v_2v_3 = (0\ 0\ 0\ 0\ 0\ 0\ 0\ 1)$

$$G = \begin{bmatrix} 1111 \cdot 1111 \\ 0000 \cdot 1111 \\ 0011 \cdot 0011 \\ 0101 \cdot 0101 \\ 0000 \cdot 0011 \\ 0000 \cdot 0101 \\ 0001 \cdot 0001 \end{bmatrix}$$

and has the distance of 2.

Exercise 6-10: The first-order Reed-Muller can be obtained from a (7, 3) maximal length code in Exercise 6-9 by adding one more bit and adding overall parity check bit.

Answer:

$$G = \begin{bmatrix} 1 \cdot 1111111 \\ 0 \cdot 1001011 \\ 0 \cdot 0101101 \\ 0 \cdot 0011110 \end{bmatrix}$$

This G is not exactly the same as (6-13). But show that it is equivalent. Make it into a systematic form by row operations.

The Reed-Muller codes are of interest due to the flexibility and the existence of simple decoding algorithm in addition to be one of the earliest code, which is not covered here. If interested further, Wikipedia 'Reed-Muller code' or a textbook on classical coding may be consulted.

6.2.3 Code Performance Analysis of Linear Block Codes*

*This section is long and is an attempt to do analytical code performance analysis. We emphasize CM decoding. If you find it involved, you may skip or skim through quickly without loss of continuity.

We discuss the performance of linear binary block codes of (n, k) with the block length n and the k information bits. There are $M=2^k$ code words and designated by $m=0, 1, 2, \ldots, M-1$ and $m=0$ being zero (**0**) codeword. We assume all codewords are equally likely. Furthermore we assume BPSK i.e., each bit $\{0, 1\}$ is mapped into $\{+1, -1\}$ and AWGN channel here.

Our focus will be codeword error rate (WER), P_w, consistently since in coded systems a codeword is a symbol and thus the expression is simple in general, and furthermore, BER may be obtained, at least approximately, from WER with appropriate conversion.

$$P_w = \sum_{m=0}^{2^k-1} P_{e|m} \cdot P_m \qquad (6\text{-}14)$$

where p_m is the probability of m-th codeword and $p_{e\,|\,m}$ is the error probability when m-th code word is transmitted.

$$P_w = \frac{1}{2^k} \sum_{m=0}^{2^k-1} P_{e|m} \qquad (6\text{-}15)$$

when $p_m = \frac{1}{2^k}$ i.e., equally likely.

$$P_w = P_{e|m=0} \qquad (6\text{-}16)$$

with the code being linear. For the performance of linear codes it is enough to compute the error probability when the **0** codeword is transmitted.

In order to compute (6-16) we need to define the events u_j to be the (**0**) codeword, c_0, is mistaken into the j-th codeword, c_j; the event $u_j : c_0 \to c_j$ with $j=1, 2, \ldots, M\text{-}1$.

With this notation, (6-16) , the codeword probability, can be represented by the union of the event $u_j : c_0 \to c_j$ with $j=1, 2, \ldots, M\text{-}1$ and is given by

$$P_w = P_{e|m=0} = P_r(u_1 + u_2 + \cdots + u_{M-1}) \qquad (6\text{-}17)$$

Now with BPSK $P_r(u_j) = Q\left(\sqrt{\frac{2E_s}{N_o} w_j}\right)$ where w_j is the weight of j-codeword, a number of nonzero bits in the codeword. This situation is essentially the same as repetition code of {0 codeword, j-th codeword} except that the repetition is the same as the weight w_j. The knowledge of $P_r(u_j)$ is not enough to compute (6-17). It requires a joint probability density function of all the event $u_j : c_0 \to c_j$ with $j=1, 2, \ldots, M\text{-}1$, and this means all the codes must be enumerated and the decoding algorithm must be specified.

This is generally complicated especially when the block size is large. For HD case, the bound (6-5) seems to be good in general, and useful with the knowledge of minimum distance of a code. However, for SD case, usually asymptotic bound or union bound is often used. However, with powerful codes like LDPC and Turbo, it is not very useful. Thus numerical simulations seem to be only resort for these situations. Here we are interested in analytical methods of error performance analysis, accurate enough to be useful in practice.

6.2.3.1 Correlation Metric (CM) Decoding

With BPSK modulation, the received samples of a block, r_i with $i=1,2,\ldots, n$, are given by

$$r_i = X_i + n_i \qquad (6\text{-}18)$$

where $X_i = \{+1, -1\}$ BPSK transmitted symbols and n_i is AWGN samples after a receive filter. BPSK modulated (with $0 \rightarrow +1$, $1 \rightarrow -1$) version of all the codewords is denoted as $C_{mi} = 1 - 2c_{mi}$ with $m=0, 1, 2, \ldots, M\text{-}1$ and $i = 1,2,\ldots,n$.

CM decoding is defined by choosing a codeword, c_m, such that the correlation between a received vector with all the codewords in BPSK modulated form, is maximum

$$\max_m \sum_{i=1}^{n} C_{mi} \cdot r_i \qquad (6\text{-}19)$$

The correlation of (6-19) may be considered as a comparison of a received vector against all the codewords in BPSK symbols, just like measuring a distance is a comparison against a standard.

For HD, r_i is quantized first to $\{0\}$ if it is bigger than zero and $\{1\}$ otherwise and denoted as $[r_i]$. In this case of HD, (6-19) becomes a minimum distance decoding and given by,

$$\min_m \sum_{i=1}^{n} c_{mi} \cdot [r_i] \qquad (6\text{-}20)$$

We successfully used CM decoding for Examples 6-1, 6-2 and 6-3 in the previous section. Obviously when the block size is large, say $M > 2^{10}$, CM decoding is impractical. However it is conceptually simple and optimal in minimizing word error probability. For these purpose of demonstrating the optimality, we give examples of (n, n) codes, which is a binary tuple of length n without any parity bit added. Thus the code rate is unity and the minimum distance is 1.

Example 6-6: $(2, 2)$ code with codewords $\{00, 01, 10, 11\}$. We apply CM decoding to this code. We know that, bit by bit detection, its BER is given by $Q\left(\sqrt{\frac{2E_s}{N_o}}\right)$. Is there any advantage to use CM in this case, a code without any redundancy?

Clearly it is a linear code. Thus in order to find WER we can apply (6-17), and it is given by $P_w = P_r(u_1 + u_2 + u_3)$. This can be expanded as follows;

$$P_r(u_1 + u_2 + u_3) = P_r(u_1) + P_r(u_2) + P_r(u_3) - P_r(u_1u_2) - P_r(u_1u_3) - P_r(u_2u_3)$$
$$+ P_r(u_1u_2u_3)$$

The computation of each term is organized in a table below with $Q\left(\sqrt{\frac{2Es}{No}}\right) = Q(\sqrt{\gamma})$.

	$P_r(u_1)$	$P_r(u_2)$	$P_r(u_3)$	$P_r(u_1u_2)$	$P_r(u_1u_3)$	$P_r(u_2u_3)$	$P_r(u_1u_2u_3)$
Range of integration	1 $+n_2<0$	1 $+n_1<0$	2 $+n_1+n_2<0$	$1+n_2<0$ $1+n_1<0$	$1+n_2<0$ $2+n_1+n_2<0$	$1+n_1<0$ $2+n_1+n_2<0$	$1+n_2<0$ $1+n_1<0$ 2 $+n_1+n_2<0$
probability	$Q(\sqrt{\gamma})$	$Q(\sqrt{\gamma})$	$Q(\sqrt{\gamma2})$	$Q^2(\sqrt{\gamma})$	$\frac{1}{2}Q(\sqrt{\gamma2})$ $+$ $\frac{1}{2}Q^2(\sqrt{\gamma})$	$\frac{1}{2}Q(\sqrt{\gamma2})$ $+$ $\frac{1}{2}Q^2(\sqrt{\gamma})$	$Q^2(\sqrt{\gamma})$

Note that different symbols of signal to noise ratio $\gamma = \frac{1}{\sigma^2} = \frac{2Es}{No}$.

We explain the table above. $P_r(u_1)$ is the error probability of making mistake $00\rightarrow 01$, which happens when $1+n_2<0$. $P_r(u_1) = \int_{R\in 1+x<0} \frac{1}{\sqrt{2\pi\sigma^2}} \exp\left(-\frac{x^2}{2\sigma^2}\right)dx = Q(\frac{1}{\sigma})$. Similarly for $P_r(u_2)$. $P_r(u_3)$ is the error probability of making mistake $00\rightarrow 11$. This happens when $1+n1+1+n2<0$. $P_r(u_3) = \iint_{R\in 1+x+1+y<0} \frac{1}{\sqrt{2\pi\sigma^2}} \exp\left(-\frac{x^2}{2\sigma^2}\right)dx \frac{1}{\sqrt{2\pi\sigma^2}} \exp\left(-\frac{y^2}{2\sigma^2}\right)dy = Q\left(\frac{\sqrt{2}}{\sigma}\right)$.

$P_r(u_1u_2) = \iint_{\substack{1+x<0 \\ 1+y<0}} \frac{1}{\sqrt{2\pi\sigma^2}} \exp\left(-\frac{x^2}{2\sigma^2}\right)dx \frac{1}{\sqrt{2\pi\sigma^2}} \exp\left(-\frac{y^2}{2\sigma^2}\right)dy$ and this can be seen as $P_r(u_1u_2) = Q\left(\sqrt{\frac{2Es}{No}w_m}\right)$ with $w_m = 2$.

$$P_r(u_1u_3) = \iint_{\substack{1+x<0 \\ 1+x+1+y<0}} \frac{1}{\sqrt{2\pi\sigma^2}} \exp\left(-\frac{x^2}{2\sigma^2}\right)dx \frac{1}{\sqrt{2\pi\sigma^2}} \exp\left(-\frac{y^2}{2\sigma^2}\right)dy$$

The computation of $P_r(u_1u_3)$ is not obvious but can be seen geometrically as in Figure 6-6.

In Figure 6-6, the computation of $P_r(u_1u_3)$ is done geometrically and the hatched area is the integration and is given by $\frac{1}{2}Q\left(2\sqrt{2\frac{E_s}{N_o}}\right) + \frac{1}{2}Q^2\left(\sqrt{2\frac{E_s}{N_o}}\right)$.

$P_r(u_1u_2u_3)$ is the same as $P_r(u_1u_2)$ since $u_1u_2u_3$ is implied from u_1u_2.

$$P_w = P_r(u_1 + u_2 + u_3) = 2Q\left(\sqrt{\frac{2Es}{No}}\right) - Q^2\left(\sqrt{\frac{2Es}{No}}\right) = 1 - \left(1 - Q\left(\sqrt{\frac{2Es}{No}}\right)\right)^2$$

From P_w we can see BER is $p_b = Q\left(\sqrt{\frac{2Es}{No}}\right)$ which is identical to the case with bit by bit detection.

We now consider HD case. The computation process is summarized in the table below. Note that we use the same designation of probability for SD is used here even though the event u_i is not exactly the same; only SD becomes HD and thus there is no confusion.

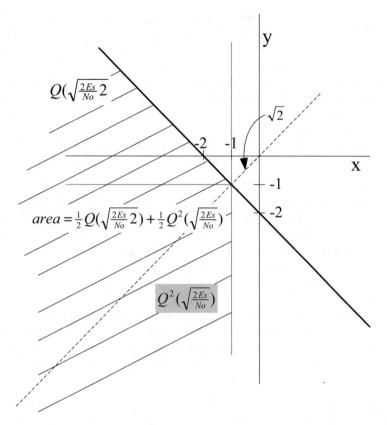

In the figure:

$$Q(\sqrt{\tfrac{2Es}{No}}\,2)$$

$$area = \tfrac{1}{2}Q(\sqrt{\tfrac{2Es}{No}}\,2) + \tfrac{1}{2}Q^2(\sqrt{\tfrac{2Es}{No}})$$

$$Q^2(\sqrt{\tfrac{2Es}{No}})$$

Figure 6-6: The computation of $P_r(u_1u_3)$

	$P_r(u_1)$	$P_r(u_2)$	$P_r(u_3)$	$P_r(u_1u_2)$	$P_r(u_1u_3)$	$P_r(u_2u_3)$	$P_r(u_1u_2u_3)$
Event with quantization $[1+n_i] = \{+1,-1\}$	$[1+n_2]$ <0	$[1+n_1]$ <0	$[1+n_1]+$ $[1+n_2]$ <0	$[1+n_2]$ <0 $[1+n_1]$ <0	$[1+n_2]<0$ $[1+n_1]+$ $[1+n_2]<0$	$[1+n_1]<0$ $[1+n_1]+$ $[1+n_2]<0$	$[1+n_2]<0$ $[1+n_1]<0$ $[1+n_1]+$ $[1+n_2]<0$
HD probability	$p =$ $Q(\sqrt{\gamma})$	p	p note1	p^2	$\tfrac{1}{2}p +$ $\tfrac{1}{2}p^2$ note2	$\tfrac{1}{2}p + \tfrac{1}{2}p^2$	p^2
Sum of HD	$P_w = 2p - p^2 = 1 - (1-p)^2$						
SD probability	$Q(\sqrt{\gamma})$	$Q(\sqrt{\gamma})$	$Q(\sqrt{\gamma 2})$	$Q^2(\sqrt{\gamma})$	$\tfrac{1}{2}Q(\sqrt{\gamma 2})$ $+$ $\tfrac{1}{2}Q^2(\sqrt{\gamma})$	$\tfrac{1}{2}Q(\sqrt{\gamma 2})$ $+$ $\tfrac{1}{2}Q^2(\sqrt{\gamma})$	$Q^2(\sqrt{\gamma})$

Each column of the above table is explained. $P_r(u_1)$ and $P_r(u_2)$ are identically given by $Q(\sqrt{\gamma})$, the same BER of BPSK. For $P_r(u_3)$, we need to consider 3 cases of $[1+n_1]\,[1+n_2] = \{ +-, -+, -- \}$. When the sum of $[1+n_1] + [1+n_2]$ is zero, it should be randomized, i.e. $\tfrac{1}{2}$ times. Thus the probability of $P_r(u_3)$ is $2\,p(1-p)\,1/2 + p^2 = p$ (note 1). $P_r(u_1u_2) = p^2$ since n_1 and n_2 is independent. $P_r(u_1u_3)$ and $P_r(u_2u_3)$ are the same.

We need to consider the cases of $[1+n_1] [1+n_2] = \{-+, --\}$ for $P_r(u_1u_3)$. Thus $P_r(u_1u_3) = p(1-p)1/2 + p^2$. For $P_r(u_1u_2u_3)$, it is the same as $[1+n_1] [1+n_2] = \{ -- \}$ and thus it is given by p^2 (note2). The sum of HD probability row in the above table, P_w, is given by $2p-p^2 = 1-(1-p)^2$. This is exactly the same as the case of SD. With (2,2) code, the bit by bit decision and CM decision with SD and HD are equivalent. Intuitively this is expected. However, formally going through the probability computation with CM is insightful. This shows that CM is optimum. With (2,2) code, $P_r(u_1 + u_2 + u_3) = 1 - P_r(\overline{u_1} \cdot \overline{u_2} \cdot \overline{u_3}) = 1 - P_r(\overline{u_1} \cdot \overline{u_2})$ and $P_r(\overline{u_1} \cdot \overline{u_2}) = P_r(\overline{u_1}) \cdot P_r(\overline{u_2}) = P_r^2(\overline{u_1})$ since u_1 and u_2 are independent. With this observation, we come to the conclusion that $P_w = 1 - (1 - P_r(u_1))^2$.

Example 6-7: (3, 3) code with codewords $\{000, 001, 010, 100, 011, 101, 110, 111\}$, which are denoted by $\{c_0, c_1, c_2, c_3, c_4, c_5, c_6, c_7\}$. Again the event u_j is defined as c_0 is mistaken as c_j; $u_j: c_0 \to c_j$ with $j \neq 0$. We assume CM decoding under AWGN channel. We need to compute $P_w = P_r (u_1+u_2+u_3+u_4+u_5+u_6+u_7)$. Following through the process of expansion requires many terms; $\binom{7}{1} = 7$ terms of $P_r(u_i)$, $\binom{7}{2}$ terms of $P_r(u_iu_j)$, $\binom{7}{3}$ terms of $P_r(u_iu_ju_k)$, $\binom{7}{4}$ terms of $P_r(u_iu_ju_ku_l)$, $\binom{7}{5}$ terms of $P_r(u_iu_ju_ku_lu_m)$, $\binom{7}{6}$ terms of $P_r(u_iu_ju_ku_lu_mu_n)$, $\binom{7}{7} = 1$ term of $P_r(u_1u_2u_3u_4u_5u_6u_7)$.

Even with this simple code it gets complicated for this formalized process. However, with this particular code, and generally for (n, n) code, it can be expressed much more simply. From the conclusion of Example 6-6, we make a critically important observation; the union of $u_1+u_2+u_3+u_4+u_5+u_6+u_7$ is equivalent to the union of $u_1+u_2+u_3$ and furthermore they are independent. By using the identity of $\overline{A+B} = \overline{A}\overline{B}$ repeatedly, we express it as $P_r(u_1 + u_2 + u_3) = 1 - P_r(\overline{u_1 + u_2 + u_3}) = 1 - P_r(\overline{u_1} \cdot \overline{u_2} \cdot \overline{u_3})$. With the independence $P_r(\overline{u_1} \cdot \overline{u_2} \cdot \overline{u_3}) = P_r(\overline{u_1})P_r(\overline{u_2})P_r(\overline{u_3})$. And $P_r(\overline{u_1}) = P_r(\overline{u_2}) = P_r(\overline{u_3})$.

Thus $P_w = 1 - (1 - P_r(u_1))^3$. This result applies to both HD and SD, and can be extended to (n, n) code.

6.2.3.2   Computation of WER of Linear Block Codes

We first consider a simple code with parity bit.

Example 6-8: We take Example 6-3 again. Consider a code; for two bit information $\{00, 10, 01, 11\}$, one parity bit is added, and thus there are 4 code words $\{000, 101, 011, 110\}$. This is the simplest possible code with parity. We use this example to consider the computation of the codeword error rate (WER).

Info bits	00	10	01	11
Code word	000 c_0	101 c_1	011 c_2	110 c_3

With $\{c_0, c_1, c_2, c_3\}$ we define the event of u_j as c_0 is mistaken as c_j, which is denoted by $u_j: c_0 \to c_j$ with $j = 1, 2, 3$.

Table 6-3: WER computation summary with HD

	$P_r(u_1)$	$P_r(u_2)$	$P_r(u_3)$	$P_r(u_1u_2)$	$P_r(u_1u_3)$	$P_r(u_2u_3)$	$P_r(u_1u_2u_3)$
Each [ ] has {+1, -1} after quantization	$[1+n_1]+[1$ $+n_3]<0$	$[1+n_1]$ $+[1$ $+n_2]$ <0	$[1+n_2]$ $+[1$ $+n_3]$ <0	$[1+n_1]$ $+[1+n_3]$ <0 $[1+n_1]$ $+[1+n_2]$ <0	$[1+n_1]$ $+[1+n_3]$ <0 $[1+n_2]$ $+[1+n_3]$ <0	$[1+n_1]$ $+[1+n_2]$ <0 $[1+n_2]$ $+[1+n_3]$ <0	$[1+n_1]+[1$ $+n_3]<0$ $[1+n_2]+[1$ $+n_3]<0$ $[1+n_2]+[1$ $+n_3]<0$
HD probability	$p = Q(\sqrt{\gamma})$	p	p	$\tfrac{1}{4}p + \tfrac{3}{4}$ p^2	$\tfrac{1}{4}p + \tfrac{3}{4}$ p^2	$\tfrac{1}{4}p + \tfrac{3}{4}$ p^2	$\tfrac{3}{4}p^2 + \tfrac{1}{4}$ p^3
Sum of HD	$P_w = 9/4p - 6/4\,p^2 + 1/4p^3$						

WER is given by $P_w = P_r(u_1 + u_2 + u_3) = 1 - P_r(\overline{u_1 + u_2 + u_3})$ as in (6-17). Note that the events of u_1, u_2, u_3 are not independent. Thus WER, the probability of the union of three events, is expressed in terms of intersections as follows; P_r $(u_1+u_2+u_3) = P_r(u_1) + P_r(u_2) + P_r(u_3) - P_r(u_1 u_2) - P_r(u_1 u_3) - P_r(u_2 u_3) + P_r$ $(u_1 u_2 u_3)$. And each term in the expression is computed and summarized in Tables 6-3 and 6-4 for HD and for SD respectively.

We now explain the computation of Table 6-3.

We need to go through a similar procedure of computation as in Example 6-6. For the probability of $P_r(u_1)$ in HD, there are four possible cases of $[1+n_1][1+n_3] = \{++, +-, -+, --\} = \{+2, 0, 0, -2\}$. Thus $P_r(u_1) = p(1-p)\,\tfrac{1}{2} + p(1-p)\,\tfrac{1}{2} + p^2 = p$. For the zero, we use the probability $\tfrac{1}{2}$ or randomizing. In practice this randomizing may happen by sending random bits of information or equally likely of all codewords. Exactly the same probability for $P_r(u_2)$ and $P_r(u_3)$ is obtained.

For $P_r(u_1 u_2)$, the computation of it is organized as the table below.

$[1+n1][1+n2][1+n3]$	$[1+n_1]+[1+n_3]$, $[1+n_1]+[1+n_2]$	probability	sum
+ + +	+, +	0	
+ + -	0, +	0	
+ - +	+, 0	0	
+ - -	0, 0	$p^2(1-p)\,\tfrac{1}{4}$	$\tfrac{1}{4}p + \tfrac{3}{4}\,p^2$
-+ +	0, 0	$p\,(1-p)^2\,\tfrac{1}{4}$	
-+-	-, 0	$p^2(1-p)\,\tfrac{1}{2}$	
--+	0, -	$p^2(1-p)\,\tfrac{1}{2}$	
---	-, -	p^3	

Exactly the same probability for $P_r(u_1 u_3)$ and $P_r(u_2 u_3)$ is obtained.
For $P_r(u_1 u_2 u_3)$ computation it is shown below.

$[1+n1][1+n2][1+n3]$	$[1+n_1]+[1+n_3]$, $[1+n_1]+[1+n_2]$ $[1+n_2]+[1+n_3]$	probability	sum
+ + +	+, +, +	0	
+ + -	0, +, 0	0	

(continued)

-+ +	0, 0, +	0	
+ - +	+, 0, 0	0	$\frac{3}{4}p^2 + \frac{1}{4}p^3$
-- +	0, -, 0	$p^2(1-p)^2 \frac{1}{4}$	
-+-	-,0, 0	$p^2(1-p) \frac{1}{4}$	
+--	0,0, -	$p^2(1-p) \frac{1}{4}$	
---	-, -, -	p^3	

Exercise 6-11: Consider a code of adding parity bit, $(n, n-1)$ and thus there are 2^{n-1} code words. For example, with $n=4$, there are 8 codewords. In order to compute P_r $(u_1+u_2+\ldots+u_7)$, show that a number of terms is $C_1^7+C_2^7+C_3^7+C_4^7+C_5^7+C_6^7+C_7^7$ where $C_1^7=7$, $C_2^7=21$, etc., i.e. combination. Show that the independence bound of WER is given by $P_w= 1- (1-p)^n$.

Now we explain the computation WER with SD summarized in Table 6-4.

With the integration region, R: $1+n_1+1+n_3<0$,

$$P_r(u_1) = \iint_{R \in 1+x+1+y<0} \frac{1}{\sqrt{2\pi\sigma^2}} \exp\left(-\frac{x^2}{2\sigma^2}\right) dx \frac{1}{\sqrt{2\pi\sigma^2}} \exp\left(-\frac{y^2}{2\sigma^2}\right) dy = Q\left(\frac{\sqrt{2}}{\sigma}\right).$$

With the integration region; R: $1+n_1+1+n_3<0$ and $1+n_1+1+n_2<0$, $P_r(u_1u_2)$ is given by,

$$\iiint_{\substack{1+x+1+z<0 \\ 1+x+1+y<0}} \frac{1}{\sqrt{2\pi\sigma^2}} \exp\left(-\frac{x^2}{2\sigma^2}\right) dx \frac{1}{\sqrt{2\pi\sigma^2}} \exp\left(-\frac{y^2}{2\sigma^2}\right) dy \frac{1}{\sqrt{2\pi\sigma^2}} \exp\left(-\frac{z^2}{2\sigma^2}\right) dz$$

$$(6\text{-}21)$$

We could not reduce this integration in terms of $Q(\cdot)$ and conjectured it as shown in the table. But the numerical computation shows it is not quite correct but is close.

With the integration region; R: $1+n_1+1+n_3<0$ and $1+n_1+1+n_2<0$ and $1+n_2+1$ $+n_3<0$, $P_r(u_1u_2u_3)$ is given by,

$$\iiint_{\substack{1+x+1+z<0 \\ 1+x+1+y<0 \\ 1+z+1+y<0}} \frac{1}{\sqrt{2\pi\sigma^2}} \exp\left(-\frac{x^2}{2\sigma^2}\right) dx \frac{1}{\sqrt{2\pi\sigma^2}} \exp\left(-\frac{y^2}{2\sigma^2}\right) dy \frac{1}{\sqrt{2\pi\sigma^2}} \exp\left(-\frac{z^2}{2\sigma^2}\right) dz$$

$$(6\text{-}22)$$

Again we could not reduce this integration in terms of $Q(\cdot)$ and conjectured it as shown in the table. But the numerical computation shows it is not quite correct but is close.

The type of integration shown in (6-21) and (6-22) should be reduced to a form of $Q(\cdot)$ in order to evaluate WER analytically. This problem is not solved at this point and any insight of how to do it will be very useful.

The numerical computation of Table 6-3 is shown in Figure 6-7.

Table 6-4: WER computation with SD

	$P_r(u_1)$	$P_r(u_2)$	$P_r(u_3)$	$P_r(u_1u_2)$	$P_r(u_1u_3)$	$P_r(u_2u_3)$	$P_r(u_1u_2u_3)$
Range of integration with CM decoding	$1+n_1+1$ $+n_3<0$	$1+n_1+1$ $+n_2<0$	$1+n_2+1$ $+n_3<0$	$1+n_1+1+n_3<0$ $1+n_1+1+n_2<0$	$1+n_1+1+n_3<0$ $1+n_2+1+n_3<0$	$1+n_1+1+n_2<0$ $1+n_2+1+n_3<0$	$1+n_1+1+n_3<0$ $1+n_1+1+n_2<0$ $1+n_2+1+n_3<0$
SD probability	$Q(\sqrt{2\gamma})$	$Q(\sqrt{2\gamma})$	$Q(\sqrt{2\gamma})$	$\tfrac{1}{4}Q(\sqrt{2\gamma})+\tfrac{3}{4}Q^2(\sqrt{2\gamma})$	$\tfrac{1}{4}Q(\sqrt{2\gamma})+\tfrac{3}{4}Q^2(\sqrt{2\gamma})$	$\tfrac{1}{4}Q(\sqrt{2\gamma})+\tfrac{3}{4}Q^2(\sqrt{2\gamma})$	$\tfrac{3}{4}Q^2(\sqrt{2\gamma})+\tfrac{1}{4}Q^3(\sqrt{2\gamma})$
Sum of SD	$P_w = 9/4\,Q(\sqrt{2\gamma}) - 6/4\,Q^2(\sqrt{2\gamma}) + 1/4\,Q^3(\sqrt{2\gamma})$						

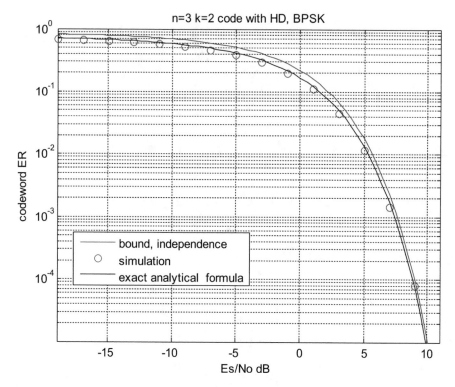

Figure 6-7: The code of Example 6-8 using Table 6-3 is analyzed and its numerical values are computed. The black solid is the formula in Table 6-3 and the blue solid is the bound with the assumption of independence and red circle is numerical simulations with CM decoding (HD). The bound is tight and useful. However, we strive to come to the exact analytical expression and succeeded. We hope that the same method is applicable to more complex codes

The numerical computation of Table 6-4 (SD) is shown in Figure 6-8. The conjectured formulas are exactly correct in this case. However, overall the match is better than the bound with independent assumption.

6.2.3.3   Additional Comments on the WER Computation for Linear Block Codes

As shown in Example 6-8 the code is the simplest with parity bit. Yet the exact computation of the codeword error rate (WER) is involved, and thus in the literature often the union bound is a resort. However, there is a practical need to know how to compute WER without resorting to a union bound. The method described in Example 6-8 is general enough and additional work is necessary, especially the integration of (6-21) type. This will be sought as a research project. MathCAD might a tool for this problem. Furthermore the method will be extended to fading channels.

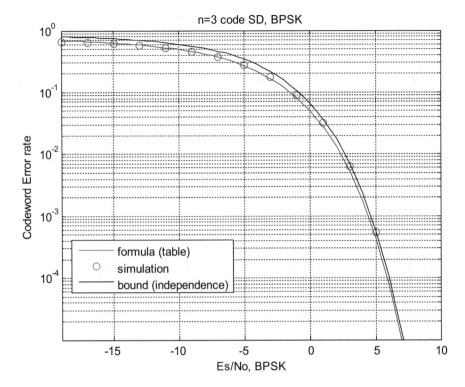

Figure 6-8: The code of Example 6-8 using Table 6-4 is analyzed and its numerical values are computed. The black solid is the formula in Table 6-4 and the blue solid is the bound with the assumption of independence and red circle is numerical simulations with CM decoding (SD). The bound is tight and useful. However, we strive to come to the exact analytical expression and not succeeded with SD case. We need to solve the integration problem

6.2.4 Cyclic Codes and CRC

Cyclic codes are linear binary block codes with an important additional property; a cyclic shift of any codeword is also a codeword. They can be extended to non-binary codes as important as RS (Reed Solomon) codes. Other than RS codes, however, our focus is on binary codes as before. This cyclic property can be exploited in the simplification of encoding and decoding, and thus large block size codes are implemented in practice. These cyclic codes are designed with, and amenable to, HD decoding in mind.

6.2.4.1 Cyclic Codes

If a codeword is represented by a n-tuple, $\mathbf{c} = (c_0, c_1, c_2, \ldots, c_{n-1})$, and its components are cyclically shifted one place to the right, we obtain another n-tuple, $\mathbf{c}^{(1)} = (c_{n-1}, c_0, c_1, c_2, \ldots c_{n-2})$, which is called a cyclic shift of $\mathbf{c}$. When they are cyclically shifted two places to the right, the resulting n-tuple would be $\mathbf{c}^{(2)} = (c_{n-2}, c_{n-1}, c_0, c_1, c_2, \ldots c_{n-3})$. Cyclic shift to the left is possible. For example, cyclically shifting 3 places to the right is equivalent to cyclically n-3 places shifting to the left. However, we use the right shift for consistency.

A formal definition of a cyclic code is that (n, k) linear code C is called cyclic if every cyclic shift of a codeword in C is also a codeword in C. Note that it is every possible cyclic shift. A simple parity bit code {000, 101, 011, 110} is cyclic, and a repetition code {000, 111} is cyclic too. (n, n) code, without redundancy, is also cyclic.

Exercise 6-12: Show that (4, 4) code is a cyclic code. Hint: enumerate all possible 4-tuples; 16 of them.

A codeword may be represented by a polynomial as follows;

$$\mathbf{c}(X) = c_0 + c_1 X + c_2 X^2 + \ldots + c_{n-1} X^{n-1}$$

and thus each code vector corresponds to a polynomial of degree n-1. '+' means addition in modulo 2; e.g., 1+1=0. Then the shifted codewords are represented as,

$$\mathbf{c}^{(1)}(X) = c_{n-1} + c_0 X + c_1 X^2 + c_2 X^3 + \ldots c_{n-2} X^{n-1}$$

$$\mathbf{c}^{(2)}(X) = c_{n-2} + c_{n-1} X + c_0 X^2 + c_1 X^3 + \ldots c_{n-3} X^{n-1}, \text{and so on.}$$

Now we relate the cyclic shifted codewords above with $X \mathbf{c}(X)$ and $X^2 \mathbf{c}(X)$ as follows;

$$X\mathbf{c}(X) = c_{n-1} (X^n + 1) + \mathbf{c}^{(1)}(X)$$
$$X^2\mathbf{c}(X) = (c_{n-1} X + c_{n-2}) (X^n + 1) + \mathbf{c}^{(2)}(X) \text{ and so on.}$$

The above can be seen easily by a long division of $X \mathbf{c}(X)$ and $X^2 \mathbf{c}(X)$ by (X^n+1).

Exercise 6-13: compute $X^2 \mathbf{c}(X)/ (X^n+1)$ and the quotient is $(c_{n-1} X + c_{n-2})$ and the remainder is $\mathbf{c}^{(2)}(X)$.

Exercise 6-14: compute $X^i \mathbf{c}(X)/ (X^n+1)$ and show the remainder is $\mathbf{c}^{(i)}(X)$, i.e.,

$$X^i \mathbf{c}(X) = q(X) (X^n + 1) + \mathbf{c}^{(i)}(X) \tag{6-23}$$

6.2.4.2   Generator Polynomial

A cyclic code is uniquely defined by a generator polynomial. We summarize it as follows. In an (n, k) cyclic code, there exists one and only one code polynomial (i.e., codeword) of degree n-k, called generator polynomial,

$$g(X) = 1 + g_1 X + g_2 X^2 + \ldots + g_{n-k} X^{n-k} \tag{6-24}$$

Every code polynomial is a multiple of $g(X)$ and every binary polynomial of degree n-1 or less that is a multiple of $g(X)$ is a code polynomial (i.e., codeword).

From the above definition of (6-24), every code polynomial $c(X)$ in an (n, k) cyclic code can be expressed in the following form;

$$c(X) = u(X)\, g(X) = \left(u_0 + u_1 X + u_2 X^2 + \ldots + u_{k-1} X^{k-1} \right) g(X) \tag{6-25}$$

If $u_0, u_1, u_2, \ldots u_{k-1}$ are the k information bits to be encoded, $c(X)$ is the corresponding codeword. Hence the encoding can be done by multiplying the message $u(X)$ by $g(X)$

Another important property of a cyclic code is that the generator polynomial g (X) of (n, k) cyclic code is a factor of (X^n+1).

This can be seen by using (6-23); $i = k$ and $c^{(i)}(X) = g^{(k)}(X)$. Since $X^k g(X)$ is of degree n we obtain, $X^k g(X) = (X^n+1) + g^{(k)}(X)$ where $g^{(k)}(X)$ is the remainder, and $g^{(k)}(X)$ is a code polynomial obtained by shifting $g(X)$ to the right cyclically k times. Thus $g(X)$ is a factor of X^n+1.

Another related property is that if $g(X)$ is a polynomial degree n-k and is a factor of X^n+1, then $g(X)$ generates a cyclic code of (n, k).

Example 6-9: The polynomial X^7+1 is factored as

$$X^7 + 1 = (X + 1)\left(X^3 + X + 1\right)\left(X^3 + X^2 + 1\right)$$

Four examples of cyclic code of length 7 are, depending on the choice of $g(X)$, given by,

$g(X) = (X^3+X+1);$ $(n, k) = (7, 4)$ Hamming code
$g(X) = (X+1);$ $(n, k) = (7, 6)$ parity check code
$g(X) = (X+1)\,(X^3+X+1)$ $(n, k) = (7, 3)$ maximal length code
$g(X) = (X^3+X+1)\,(X^3+X^2+1)$ $(n, k) = (7, 1)$ repetition code

A $(7, 4)$ cyclic code generated by $g(X) = (X^3+X+1)$, using (6-25).

Note that $u_0 u_1 u_2 u_3 = (0001)$ generates a codeword (0001101), and in polynomial notation it is $X^3 g(X)$, three position cyclic shift to the right. $u_0 u_1 u_2 u_3 = (1000)$ generates a codeword (1101000), which is $X^0 g(X)$ with no shift to the right. Similarly $u_0 u_1 u_2 u_3 = (0010)$ generates a codeword (0011010) and $u_0 u_1 u_2 u_3 = (0100)$ generates a codeword (0110100).

A codeword can be obtained by the operation of (6-25), multiplying the message polynomial and $g(X)$. It can be obtained binary addition of 4 highlighted (yellow) codewords in Table 6-5.

Notice that this code is not exactly the same as one listed in Table 6-1 even though both $(7, 4)$ code. A straightforward multiplication using (6-25) generates a non-systematic code.

Table 6-5: (7, 4) cyclic code with $g(X) = (X^3+X+1)$

$u_0u_1u_2u_3$	$c_0c_1c_2c_3c_4c_5c_6$	$u_0u_1u_2u_3$	$c_0c_1c_2c_3c_4c_5c_6$
0000	000 0000	1000	1101 000
0001	000 1101	1001	1100 101
0010	001 1010	1010	1110 010
0011	001 0111	1011	0010 111
0100	011 0100	1100	1011 100
0101	011 1001	1101	1010 001
0110	010 1110	1110	1000 110
0111	010 0011	1111	1001 011

Exercise 6-15: Show that a generator matrix for the code in Table 6-5.

Answer:

$$G = \begin{bmatrix} g_0 & g_1 & g_2 & g_3 & 0 & 0 & 0 \\ 0 & g_0 & g_1 & g_2 & g_3 & 0 & 0 \\ 0 & 0 & g_0 & g_1 & g_2 & g_3 & 0 \\ 0 & 0 & 0 & g_0 & g_1 & g_2 & g_3 \end{bmatrix} = \begin{bmatrix} 1 & 1 & 0 & 1 & 0 & 0 & 0 \\ 0 & 1 & 1 & 0 & 1 & 0 & 0 \\ 0 & 0 & 1 & 1 & 0 & 1 & 0 \\ 0 & 0 & 0 & 1 & 1 & 0 & 1 \end{bmatrix}$$

Clearly this G is not systematic.

Exercise 6-16: Make the G above systematic. Hint: use a different combination of rows, which may be considered as Gaussian elimination of a set of linear equations. The answer will be shown in problem P6-13.

6.2.4.3   Transmission Sequence of a Codeword and its Code Polynomial

A (n, k) code generator matrix of the form $\mathbf{G} = [I_k | P]$ as shown in (6-7) presupposes a block of information and parity bits is organized as, k information bits followed by $n-k$ parity bits as shown in Figure 6-9;

We need to find a systematic cyclic code polynomial, consistent with this transmission sequence: $(c_0, c_1, c_2, \ldots, c_{k-1}, c_k, c_{k+1}, \ldots, c_{n-1})$ where $(c_0, c_1, c_2, \ldots, c_{k-1}) = (u_0, u_1, u_2, \ldots, u_{k-1})$ information bits and $(c_k, c_{k+1}, \ldots, c_{n-1})$ are parity bits. In other words, we transmit $u_0, u_1, u_2, \ldots, u_{k-1}$ followed by $c_k, c_{k+1}, \ldots, c_{n-1}$. For this purpose we define

$$v(X) = u_0X^{k-1} + u_1X^{k-2} + \quad +u_mX^{k-1-m} + \quad +u_{k-1}X^0 \qquad (6\text{-}26)$$

$$X^{n-k}v(X) = u_0X^{n-1} + u_1X^{n-2} + \quad +u_mX^{n-1-m} + \quad +u_{k-1}X^{n-1-(k-1)} \qquad (6\text{-}27)$$

Using this definition, a codeword will be transmitted, with the higher polynomial exponent first and its polynomial is expressed by

Figure 6-9: A codeword sequence designation of a linear systematic block code

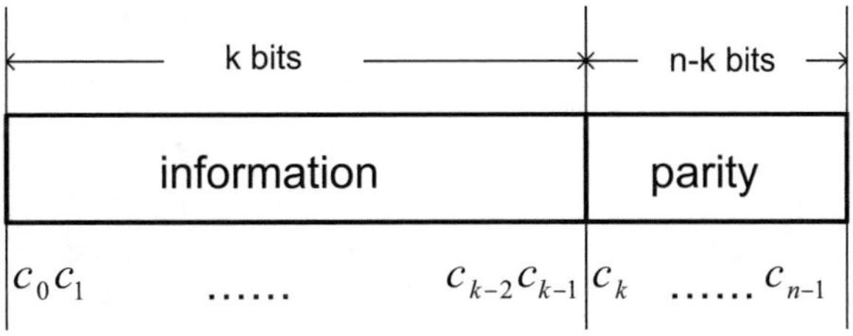

$$c(X) = X^{n-k}v(X) + c_k X^{n-1-k} + c_{k+1}X^{n-1-(k+1)} + \quad +c_{n-1}X^0 \qquad (6\text{-}28)$$

Now we need to compute n-k parity bits of $(c_k, c_{k+1}, \ldots, c_{n-1})$ which are the remainder of the division $X^{n-k}v(X)/g(X)$ since

$$X^{n-k}v(X) + \text{remainder} = (\text{quotient})\, g(X) \qquad (6\text{-}29)$$

Example 6-10: Consider (7, 4) code with $g(X) = (X^3+X+1)$. Let $(u_0, u_1, u_2, u_3) = (1\ 0\ 0\ 0)$. $v(X) = X^3$ and $X^{n-k}v(X) = X^6$ and thus $X^6/(X^3+X+1) = X^3+X+1$ and remainder $= X^2+1$. See the long division below

$$
\begin{array}{r}
X^3 + X + 1 \\
\hline
X^3 + X + 1 \,\big)\, X^6 \\
X^6 + X^4 + X^3 \\
\cdots\cdots X^4 + \cdots\cdots X^2 + X \\
\cdots\cdots\cdots\cdots X^3 + \cdots\cdots X + 1 \\
\hline
\text{remainder}\cdots\cdots\cdots\cdots X^2 \cdots\cdots + 1
\end{array}
$$

$c(X) = X^6 + X^2+1$ or codeword is (1000101).

Exercise 6-17: Consider (7, 4) code with $g(X) = (X^3+X+1)$. Let $(u_0, u_1, u_2, u_3) = (0\ 0\ 0\ 1)$. Show that the remainder is $X+1$. Hint: follow the rule of long division.

Exercise 6-18: Consider (7, 4) code with $g(X) = (X^3+X+1)$, and the transmission sequence is $u_0, u_1, u_2, u_3, c_4, c_5, c_7$ and find the corresponding generator matrix, G. Hint: By a long division we obtain; $X^6=(X^3+X+1)\, g(X)+X^2+1$, $X^5=(X^2+1)g(X) +X^2+1$, $X4=X\, g(X)+X^2+X+1$, and $X3=1\, g(X) +X+1$. Thus we obtain codeword polynomials $X^6+X^2+1=(X^3+X+1)\, g(X)$, $X^5+X^2+1=(X^2+1)\, g(X)$, $X4 +X^2+X+1=X\, g(X)$, and $X3+X+1=1\, g(X)$. Note that codeword polynomials are divisible by $g(X)$. Putting these 4 codeword polynomials in matrix form, a generator matrix G is obtained.

$$G = \begin{bmatrix} 1 & 0 & 0 & 0 & 1 & 0 & 1 \\ 0 & 1 & 0 & 0 & 1 & 1 & 1 \\ 0 & 0 & 1 & 0 & 1 & 1 & 0 \\ 0 & 0 & 0 & 1 & 0 & 1 & 1 \end{bmatrix}.$$

In the literature and textbooks, the designation of code bit and transmission sequence is not consistent, and there is a lot of confusion when the material is exposed for the first time. Here we try to be consistent; the lower the polynomial degree it is transmitted first. The definition of $c(X)$ in (6-28) is an efficient scheme of getting $c(X)$ by computing remainders. This polynomial convention is consistent with long division and the remainder computation. However, it is opposite to the sequence index convention, i.e., the bigger exponent coefficients for later transmission.

$c(z^{-1}) = c_0 + c_1 z^{-1} + c_2 z^{-2} + \ldots + c_{n-1} z^{-(n-1)}$. If we interpret $z^{-1} = X$ as a one sample delay, c_0 is transmitted first; the index designation of $c(z^{-1})$ is the transmission sequence, lower first and higher later. There is a way to do maintain this convention in code polynomials. This will be explored briefly. We define codeword polynomial as $c(X) = (u_0 + u_1 X + u_2 X^2 \ldots, u_{k-1} X^{k-1}) X^{n-k} g(X^{-1})$, a multiplication of information polynomial by another generator polynomial $X^{n-k} g(X^{-1})$. When it is applied to (7, 4) code with $g(X) = (X^3 + X + 1)$, $X^{n-k} g(X^{-1}) = g_3 + g_2 X + g_1 X^2 + g_0 X^3$.

A non-systematic $\underline{G}$ is shown below.

$$\underline{G} = \begin{bmatrix} g_3 & g_2 & g_1 & g_0 & 0 & 0 & 0 \\ 0 & g_3 & g_2 & g_1 & g_0 & 0 & 0 \\ 0 & 0 & g_3 & g_2 & g_1 & g_0 & 0 \\ 0 & 0 & 0 & g_3 & g_2 & g_1 & g_0 \end{bmatrix} = \begin{bmatrix} 1 & 0 & 1 & 1 & 0 & 0 & 0 \\ 0 & 1 & 0 & 1 & 1 & 0 & 0 \\ 0 & 0 & 1 & 0 & 1 & 1 & 0 \\ 0 & 0 & 0 & 1 & 0 & 1 & 1 \end{bmatrix}.$$

After Gaussian elimination, by only row operations, we can find a systematic G. It turns out the systematic G is the same as the one obtained by remainder computation. In this way the transmission sequence is the lower polynomial coefficient first then higher. The definition of $c(X)$ in (6-28) may be interpreted as a scheme for computing remainders. The transmission sequence is to transmit c_0 first, then c_1, c_2, $\ldots$, and c_{n-1} the last. This designation is in line with a typical sequence designation in digital signal processing.

6.2.4.4 Remainder Computation with Shift Register Circuits

By now it is clear that the encoding of a cyclic binary code can be accomplished by appending a remainder of $X^{n-k} v(X) = u_0 X^{n-1} + u_1 X^{n-2} + u_m X^{n-1-m} + u_{k-1} X^{n-1-(k-1)}$ as shown in (6-27) for a given information sequence of bits $(u_0, u_1, u_2, \ldots,$

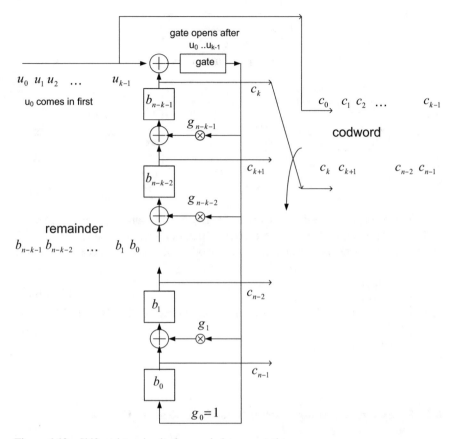

Figure 6-10: Shift register circuits for remainder computation

u_{k-1}). We discussed a long division by a generator polynomial $g(X)$, and constructing a generator matrix. There is another way to accomplish it by using shift register circuits. It is essentially the same as the long division. The circuit is shown in Figure 6-10. The remainder is computed by passing through all k information bits starting from u_0. Then the contents of the shift registers are the remainder. This can be shifted out one at a time with gate open, or unload the shift register contents and appended to the information bits at the tail.

Example 6-11: Consider $(7, 4)$ code with $g(X) = (X^3+X+1)$, and the transmission sequence is u_0, u_1, u_2, u_3, c_4, c_5, c_6. In the previous subsection we found the corresponding generator matrix G. Now generate codewords with shift register circuits in Figure 6-10 and redrawn for this case below.

In the shift register circuits the initial contents are set to be zero, and with each input the content is shifted. A shift register is one bit memory clocked sequentially with each input.

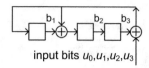

input bits u_0, u_1, u_2, u_3

input	$b_1b_2b_3$	input	$b_1b_2b_3$	input	$b_1b_2b_3$	input	$b_1b_2b_3$
	000		000		000		000
1	110	0	000	0	000	0	000
0	011	1	110	0	000	0	000
0	111	0	011	1	110	0	000
0	101	0	111	0	011	1	110

Note that the transmission sequence of parity bits (remainder) is $b_3b_2b_1$, i.e., b_3 first. In other words codewords are 1000101, 0100111, 0010110, and 0001011 with the above table. This is the same as G, i.e., another way to find a generator matrix,

$$G = \begin{bmatrix} 1000101 \\ 0100111 \\ 0010110 \\ 0001011 \end{bmatrix}. \text{ And the corresponding parity check matrix } H = \begin{bmatrix} 1110100 \\ 0111010 \\ 1101001 \end{bmatrix}.$$

6.2.4.5   Syndrome Computation with the Shift Register Circuits

A syndrome can be in general computed by using a parity check matrix H using the relationship $s = H\,e^T$ where e is a row vector of length n, and s is a column vector of length $(n-k)$.

In cyclic codes, however, the syndrome can be computed with the essentially same shift register circuits as in Figure 6-10. The table below shows the computation.

input	b_1b_2 b_3	input	b_1b_2 b_3	input	b_1b_2 b_3	input	b_1b_2 b_3	input	b_1b_2 b_3	input	b_1b_2 b_3	input	b_1b_2 b_3
	000		000		000		000		000		000		000
1	110	0	000	0	000	0	000	0	000	0	000	0	000
0	011	1	110	0	000	0	000	0	000	0	000	0	000
0	111	0	011	1	110	0	000	0	000	0	000	0	000
0	101	0	111	0	011	1	110	0	000	0	000	0	000
0	100	0	101	0	111	0	011	1	110	0	000	0	000
0	010	0	100	0	101	0	111	0	011	1	110	0	000
0	001	0	010	0	100	0	101	0	111	0	011	1	110

$$s = \begin{bmatrix} 1001110 \\ 0100111 \\ 0011101 \end{bmatrix}$$ computed using Figure 6-10 does not correspond to the parity

check matrix obtained from generator matrix G, i.e., s is cyclically shifted version of the parity matrix H, 3 positions to the right.

In order to obtain the same H, the input position has to be cyclically shifted 3 positions to the left, which can be done as below.

Exercise 6-19: Write a computer program to generate H matrix from Figure 6-11.

A MATLAB program:

```
% syndrome computation --> H matrix
n=7; k=4; nmk = n - k;
g = [1 1 0 1];      %g0, g1, ..., gn-k; generator polynomial
list =[1 0 0 0 0 0 0; 0 1 0 0 0 0 0; 0 0 1 0 0 0 0;...
       0 0 0 1 0 0 0; 0 0 0 0 1 0 0; 0 0 0 0 0 1 0; 0 0 0 0 0 0 1];
hmat = zeros(nmk,n);
for k =1: length(list(:,1))
  u = list(k,:);
  b = zeros(1, nmk);
  for i=1 : length(u)
    temp = b(nmk);                 %syndrome
    for j=1:nmk-1
      b(nmk+1-j) = g(nmk+1-j)*temp + b(nmk-j);
      b(nmk+1-j) = mod(b(nmk+1-j), 2);
    end
    b(1) = mod(temp + u(i),2);     %syndrome
  end
  hmat(:,k) = b(nmk:-1:1)';
end
disp(hmat)   % parity check matrix
```

6.2.4.6   CRC

A very common use of cyclic codes is called CRC for error detection, rather than error correction. It is a code of $(n, k) = (k + r, k)$; r bits of redundancy are added at the end of k bits of information, and r bits of remainder are computed using Figure 6-10 for a given generator polynomial (or CRC polynomial) of degree r. In the receiving side, k bits of information are divided by the same generator polynomial for error detection, and add the result with the received CRC bits (this

Figure 6-11: For (7, 4) cyclic code, receiver syndrome computation

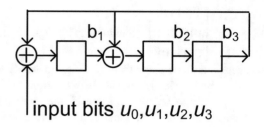

input bits u_0, u_1, u_2, u_3

is the same as comparing the two) and zero if there is no error and non-zero if there is any error in $k + r$ bits including CRC of r bits.

A list of the most popular CRC polynomials is given below:

CRC name	$n=2^m$	Polynomial g(X)
CRC-12	2^{12}	$X^{12} + X^{11} + X^3 + X^2 + X + 1$
CRC-16	2^{16}	$X^{16} + X^{15} + X^2 + 1$
CRC-CCITT	2^{16}	$X^{16} + X^{12} + X^5 + 1$
CRC-32	2^{32}	$X^{32} + X^{26} + X^{23} + X^{22} + X^{16} + X^{12} + X^{11} + X^{10} + X^8 + X^7 + X^5 + X^4 + X^2 + X + 1$

- See Wikipedia 'cyclic redundancy check' for more CRCs.

The choice of the generator polynomial is dictated by the undetected error probability, which depends on the minimum distance (weight) of the code. When there are A_i codewords with i weight in the code, then the undetected probability is given by

$$P_u = \sum_{i=d\,\min}^{n} A_i \, p^i (1-p)^{n-i} \leq \sum_{i=d\,\min}^{n} C_i^n \, p^i (1-p)^{n-i} \qquad (6\text{-}30)$$

where p is the bit error probability and C_i^n is a combination i from n.

Typically g(X) is of the form g(X) = $(1+X)$ $g_h(X)$ where $g_h(X)$ is the generator polynomial of a cyclic Hamming (BCH) code $n=2^m-1$. This is the same as adding an additional parity bit.

Exercise 6-20: Draw the shift register circuits for CRC-CCITT.
Answer: similar to Figure 6-12 with the tap position will be different.

Exercise 6-21: Write a computer program to compute CRC for a given file length less than 2^m. To be specific, work out the CRC-16 case. Hint: study the Matlab program of Exercise 6-20 related in the previous subsection.

Note that the when the length of the file including CRC is less than 2^m then the code is not cyclic but it does not matter for error detection. To make it cyclic, think of it as known 0 bits are not transmitted, and adding those 0 bits may be added to make it cyclic. This is called shortening of a code.

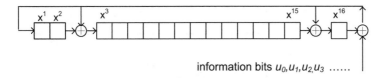

information bits u_0, u_1, u_2, u_3 ......

Figure 6-12:   CRC-16 implementation of CRC bits computation, which will be at the shift registers after k bits of information entered. The same structure can be used in the receiver

6.2.4.7   Decoding Cyclic Codes

We show the decoding of cyclic codes in general. Decoding of cyclic codes consists of syndrome computation, association of the syndrome to an error pattern and error correction. The syndrome computation of cyclic codes can be accomplished by dividing received n bits, or its polynomial r(X) by a generator polynomial g(X). n-k bit syndrome pattern is generated from the computation. The n-k syndrome bits are associated with the error pattern, which can be implemented by logic circuits or lookup table of 2^{n-k} inputs and n output bits of error pattern. This error pattern can be added (modulo 2) to the received bits to correct errors. This is summarized in Figure 6-13, and Figure 6-14 shows a specific example.

As the block size gets larger, in particular n-k, the complexity increases exponentially. For example, BCH code $n=255$, $k=231$ can correct t=3 errors, and n-$k=24$. A number of cases are $2^{24} \approx 16.77$ e^6. This is called complete decoding, which means correct all correctable errors including the weight higher than 3 errors.

One of reduction of the complexity is to store only t=1, 2 and 3 correctable error patterns and the required memory size $= 255 + 255 * 254/2 + 255 * 254 * 253/ 6 \approx 2.764$ e^6. This reduced complexity decoding can correct any error pattern up to t=3 and is called bounded distance decoding.

Exercise 6-22: BCH code $n=255$, $k=223$ can correct t=4 errors. Find the complexity (memory size) of complete decoding and of bounded distance decoding.

Answer: 2^{32} for complete decoding, and 255+ 255*254/2 + 255*254*253/6 + 255*254*253*252/24 for bounded distance decoding in order to enumerate t=1, t=2, t=3 and t=4 error patterns.

Additional reduction of the complexity is possible by using the cyclic property. We explain Figure 6-13 and the decoding operation is described step by step:

Step1. The syndrome is formed by shifting the entire received n bit block in to the syndrome computation circuit, and at the same time the received block is stored into the buffer register. In Figure 6-13, gate 1 is closed and gate 2 is open.

Step2. The syndrome, n-k bits, is tested for the corresponding error pattern. The error pattern detector output is '1' if and only if the syndrome in the syndrome register corresponds to a correctable error pattern with an error at the highest-order position. That is, if a '1' appears at the output of the detector, the received symbol in the rightmost stage of the buffer register is assumed to be erroneous and must be corrected, and if a '0' appears the received symbol at the rightmost stage of the

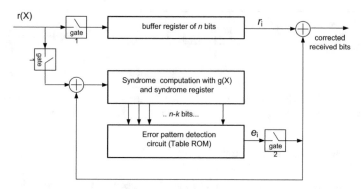

Figure 6-13: Decoding of cyclic codes. When the block size is large, the complexity is on error pattern detection circuit, which may be implemented by a look up table memory. Syndrome can be computed by division circuit with *n-k* registers

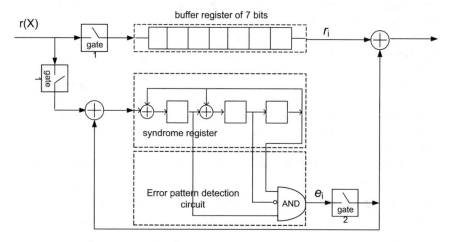

Figure 6-14: Decoding of (7,4) Hamming code using Figure 6-13

buffer register is assume d to be correct and no correction is necessary. Thus, the output of the detector is the estimated error value for the symbol to come out of the buffer.

Step3. The first received symbol is read out of the buffer. If the received symbol is detected to be erroneous symbol, it is then corrected by the output of the detector. The output of the detector is also fed back to the syndrome register to modify the syndrome, i.e., to remove the error effect from the syndrome. These results in a new syndrome, which corresponds to the altered received vector, shifted one place to the right.

Step4. The new syndrome formed in step3 is used to detect whether or not the second received symbol (now at the rightmost stage of the buffer register) is an erroneous symbol. Step 2 and 3 are repeated. The second received symbol is corrected in exactly the same manner as the first received symbol was corrected. This step is repeated until the entire received n bits are read out of the buffer.

The complexity of error pattern detection is now reduced since the rightmost bit position is in error. For example, BCH code $n=255$, $k=231$ can correct $t=3$ errors needs the size of memory: $1+ 254+254*253/2 =32,386$. This is much smaller than $255+ 255*254/2 + 255*254*253/6 = 2,763,775$.

Exercise 6-23: BCH code $n=255$, $k=223$ can correct $t=4$ errors. Find the memory size for serial decoding (using the above step1 to step4).

Answer: $1+254+ 254*253/2 + 254*253*252/6$. Compare with $255+ 255*254/2 + 255*254*253/6 + 255*254*253*252/24$.

6.2.5   BCH and RS Codes

6.2.5.1   BCH Codes

A binary, t-error correcting, BCH code exists with the following parameters for $m \geq 3$:

Block length n: $n = 2^m-1$
Number of parity check bits: $n - k \leq mt$
Minimum distance: $d_{min} \geq 2t + 1$

For example, $m = 4$, we list possible BCH codes in a table

Block length n	Info bits k	Parity bits n-k	Errors correctable t	Generator polynomial $g(X)$	binary representation of $g(X)$	Octal $g(X)$
15	11	4	1	X^4+X+1	10011	23
	7	8	2	$X^8+X^7+X^6+X^4+1$	111010001	721
	5	10	3	$X^{10}+X^8+X^5+X^4+X^2+X$ $+1$	10100110111	2467

In order to generate a codeword, i.e., to compute parity bits, we use a generator polynomial $g(X)$, which will be shown shortly. And we treat a codeword as a polynomial; $c(X) = c_0X^{n-1} + c_1X^{n-2} + \ldots + c_{n-2}X + c_{n-1}$.

The $c(X)$ above is a polynomial with GF (2) coefficients, simply a polynomial with GF (2). GF (2) contains 2 elements $\{0,1\}$. And thus the coefficients are from $\{0, 1\}$. In order to find zeros of $c(X)$, i.e., $c(a) = 0$, we need GF (2^m), an extension of GF (2). GF (2^m) contains 2^m elements of m-tuple (i.e., m-bits). A question is how to find GF (2^m).

First we give an example of GF (2^m) with $m=4$;

Power representation	Polynomial representation	4-tuple representation
0	0	[0000]

(continued)

1	1	[1000]
a^1	a	[0100]
a^2	a^2	[0010]
a^3	a^3	[0001]
a^4	$1+a$	[1100]
a^5	$a+a^2$	[0110]
a^6	a^2+a^3	[0011]
a^7	$1+a+a^3$	[1101]
a^8	$1+a^2$	[1010]
a^9	$a+a^3$	[0101]
a^{10}	$1+a+a^2$	[1110]
a^{11}	$a+a^2+a^3$	[0111]
a^{12}	$1+a+a^2+a^3$	[1111]
a^{13}	$1+a^2+a^3$	[1011]
a^{14}	$1+a^3$	[1001]
$1=a^{15}$	1	[1000]

For the above representation, we need $1+a+a^4=0$, and $a^{15}+1=0$.

In polynomial forms $X^4+X+1 = p(X)$, and $X^n+1 = X^{15}+1$, and thus a is a zero of these polynomials. Thus we found an extension to GF (2^m) in this case with p(X). Is this possible with any other m as well? It turns out the answer is yes. We need a polynomial – irreducible one. Any lower order polynomial, degree less than m, cannot divide this irreducible polynomial. It sounds very similar to a prime number.

In order to have GF (2^m), we need an irreducible polynomial degree m, called primitive polynomial. Then if $p(X) = 0$ with $X = a$, this element, a can generate all non-zero elements in GF (2^m). For a given m, there is a list of irreducible polynomials tabulated.

In Matlab form, a table is given by

```
function p = primpol(m)
% list of primitive polynomials
% use: p = primpol(m) where m=2,.., m=32
% [p0 p1 p2 p3......pm] with pi ={0,1}
% ref: pp.42 Shu Lin 2ed,  S. Wicker pp.460

if m == 2,  p = [1 1 1];              end      %m=2
if m == 3,  p = [1 1 0 1];           end      %m=3
if m == 4,  p = [1 1 0 0 1];         end      %m=4
if m == 5,  p = [1 0 1 0 0 1];       end      %m=5
if m == 6,  p = [1 1 0 0 0 0 1];              end      %m=6
if m == 7,  p = [1 0 0 1 0 0 0 1];            end      %m=7
if m == 8,  p = [1 0 1 1 1 0 0 0 1];          end      %m=8
if m == 9,  p = [1 0 0 0 1 0 0 0 0 1];        end      %m=9
if m == 10, p = [1 0 0 1 0 0 0 0 0 0 1]; end      %m=10
if m == 11, p = [1 0 1 0 0 0 0 0 0 0 0 1];              end      %m=11
if m == 12, p = [1 1 0 0 1 0 1 0 0 0 0 0 1];            end      %m=12
if m == 13, p = [1 1 0 1 1 0 0 0 0 0 0 0 0 1];          end      %m=13
if m == 14, p = [1 1 0 1 0 1 0 0 0 0 0 0 0 0 1];        end      %m=14 DVB-S2
if m == 15, p = [1 0 1 1 0 1 0 0 0 0 0 0 0 0 0 1];      end      %m=15 DVB-S2
if m == 16, p = [1 0 1 1 0 1 0 0 0 0 0 0 0 0 0 0 1];          end      %m=16 DVB-S2
if m == 17, p = [1 0 0 1 0 0 0 0 0 0 0 0 0 0 0 0 0 1];        end      %m=17
if m == 18, p = [1 0 0 0 0 0 0 1 0 0 0 0 0 0 0 0 0 0 1];      end      %m=18
end
```

With $m=4$ above, note that it is listed as $[1\ 1\ 0\ 0\ 1]$ which represents for $1+X+X^4$.
Another question is how to find generator polynomials.

Before answering the question we need to establish one important property related with GF (2^m). In the above example of m=4 case we use $1+a+a^4=0$. Now try to find

$$p(a) = 1 + a + a^4$$
$$p(a^2) = 1 + a^2 + a^8 = 1 + a^2 + 1 + a^2 = 0,$$
$$p(a^4) = 1 + a^4 + a^{16} = 1 + 1 + a + a^{15}a = 0,$$
$$p(a^8) = 1 + a^8 + a^{32} = 1 + 1 + a^2 + a^{30}a^2 = 0$$
$$p(a^{16}) = p(a)$$

In other words, $p(a)=0$ implies that $p(a^2)=0$, $p(a^4)=0$, and $p(a^8)=0$ with $m=4$.

Consider $g_1(X) = (X + a)(X + a^2)(X + a^4)(X + a^8)$. From the above we claim $g_1(X) = p(X)$ since a 4th degree polynomial has four zeros, and both polynomials have exactly the same zeros.

Now we are back to a question how to find generator polynomials. If a code polynomial, $c(X)$, contains a factor $g_1(X)$, it must be divisible by $g_1(X)$, i.e., $c(X) = q(X) g_1(X)$. We may choose $q(X)$ be a polynomial for information bits. The degree of $c(X)$ is $n-1$ $(=14)$, and the degree of $q(X)$ is $k-1$ (10) since the degree of $g_1(X)$ is m $(=4) = n-k$. This $g_1(X)$ is a generator polynomial for $t=1$ case.

Now consider another polynomial

$$g_2(X) = (X + a^3)(X + a^6)(X + a^{12})(X + a^{24})$$
$$= (X + a^3)(X + a^6)(X + a^{12})(X + a^9).$$

With GF (2^m) arithmetic, it can be shown that $g_2(X) = X^4 + X^3 + X^2 + X^1 + 1$.

For $t=2$, a generator polynomial is $g(X) = g_1(X) g_2(X)$. Again using GF (2^m) arithmetic, $g(X) = X^8 + X^7 + X^6 + X^4 + 1$.

Now consider another polynomial,

$$(X + a^5)(X + a^{10})(X + a^{20})(X + a^{40}) = (X + a^5)(X + a^{10})(X + a^5)(X + a^{10})$$

Thus we choose $g_3(X) = (X + a^5)(X + a^{10})$.

With GF (2^m) arithmetic, it can be shown that $g_3(X) = X^2 + X^1 + 1$.

For $t=3$, a generator polynomial is $g(X) = g_1(X) g_2(X) g_3(X)$. Again using GF (2^m) arithmetic, $g(X) = X^{10} + X^8 + X^5 + X^4 + X^2 + X + 1$.

We completed BCH code table with $m=4$, $t=1$, 2, and 3. To generate a codeword, k information bits, in polynomial form with $k-1$ degree, $u(X)$, may be multiplied by a generator polynomial with degree $n-k$. However, this cannot maintain a codeword format of k information bits followed by n-k parity bits, i.e., not systematic.

In order to keep a codeword systematic, i.e., k information bits appended by n-k parity bits, we need to X^{n-k} u(X), i.e., the degree of u(X) information polynomial is increased to n-1 and followed by n-k zeros. Now X^{n-k} u(X) is divided by g(X) and the remainder (n-k bits) is appended (replaced zeros of tail end). The GF(2) polynomial division implementation is shown below.

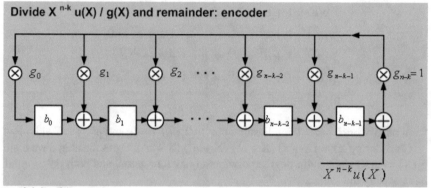

Matlab implementation

```
%u = [1 0 0 0]; g =[1 1 0 1]; k = length(u); nmk = length(g);n = n+nmk;
for i=1 : length(u)
    temp = mod(b(nmk) + u(i),2);
    for j=1:nmk-1
        b(nmk+1-j) = g(nmk+1-j)*temp + b(nmk-j);
        b(nmk+1-j) = mod(b(nmk+1-j), 2);
    end
    b(1) = temp;
end
codeword =[u b(nmk:-1:1)])   %correct transmission seq
```

Exercise 6-24: To design a double error correcting BCH code with block length n=15 (m=4), we need the minimal polynomial of α and α^3. The minimal polynomial of α is $\phi_\alpha(X) = X^4+X+1$, $\phi_{\gamma=\alpha^3}(X) = \prod_{i=0}^{3}\left(X + \gamma^{2^i}\right) = (X + \alpha^3)(X + \alpha^6)(X + \alpha^9)$ $(X + \alpha^{12}) = X^4+X^3+X^2+X+1$ and thus $g(X) = \phi_\alpha(X)\phi_{\gamma=\alpha^3}(X) = X^8+X^7+X^6+X^4+1$. In octal representation it is given by 111 010 001 = 7 2 1. In Table 6-6 g(X) is expressed by octal notation.

6.2.5.2   RS Codes

Reed-Solomon (RS) codes are probably one of the most widely used codes in practice. These codes are used in communication systems as well as storage systems such as CD.

Table 6-6: BCH code n=7 to n=127 with generator polynomial

n	k	t	g(X): octal representation	n	k	t	g(X): octal representation
7	4	1	13	63	18	10	1363026512351725
15	11	1	23		16	11	6331141367235453
	7	2	721		10	13	472622305527250155
	5	3	2467		7	15	5231045543503271737
31	26	1	45	127	120	1	211
	21	2	3551		113	2	41567
	16	3	107657		106	3	11554743
	11	5	5423325		99	4	3447023271
	6	7	313365047		92	5	624730022327
63	57	1	103		85	6	130704476322273
	51	2	12471		78	7	26230002166130115
	45	3	1701317		71	9	6255010713253127753
	39	4	166623567		64	10	120653402557077310045
	36	5	1033500423		57	11	33526525205705053517721
	30	6	157464165547		50	13	54446512523314012421501421
	24	7	17323260404441		43	14	17721772213651227521220574343

One way of looking at RS codes is to interpret as t error correcting 2^m-ary (non-binary) BCH codes with block length $N=2^m-1$ symbols, i.e., $m*N$ bits, and $N-K=2t$ where t symbols of m bits can be corrected. RS is called MDS (minimum distance separable), i.e., $D_{min} = N-K +1 = 2t+1$.

To specify a RS code, we need a generator polynomial g(X) of the form, where $\alpha \in GF(2^m)$ is a primitive element,

$$g(X) = (X + \alpha)(X + \alpha^2)(X + \alpha^3) \ldots (X + \alpha^{2t})$$
$$= X^{2t} + g_{2t-1}X^{2t-1} + \ldots + g_1 X + g_0 \tag{6-31}$$

where $g_i \in GF(2^m)$ for $0 \leq i \leq 2t - 1$ and g(X) is a divisor of $X^{2m-1} +1$.

Exercise 6-25: A triple error RS code (thus $t=3$) with $m=4$ has parameters of $N=15 = 2^4-1$, $N-K = 2t+1=7$, and $K=8$ (symbols of 4 bits). $\alpha \in GF(2^4)$ to be a primitive element. $g(X) = (X + \alpha)(X + \alpha^2)(X + \alpha^3)(X + \alpha^4)(X + \alpha^5)(X + \alpha^6) = X^6 + \alpha^{10}X^5 + \alpha^{14}X^4 + \alpha^4 X^3 + \alpha^6 X^2 + \alpha^9 X + \alpha^6$

This is a (15, 8) triple error correcting RS code over $GF(2^4)$. A block length of this code is 15 symbols of 4-bit, thus 60 bits.

A popular RS is (255, 223) over $GF(2^8)$ and $D_{min} = N-K +1 = 2t+1 = 33$ and thus it is capable of correcting 16 symbol errors. In bits the block length is $255*8 = 2040$ bits.

The performance of the HD decoder for RS code may be characterized by (6-5), interpreting p in (6-5) to be symbol error rate (p_S) rather than bit error rate (p), and its codeword error probability is bounded by,

$$P_w \leq \sum_{i=t+1}^{N} \binom{N}{i} p_s^{\,i} (1 - p_s)^{N-i} \qquad (6\text{-}32)$$

where p_S is RS code symbol (m bits) error rate. If BPSK is used, p_S may be obtained from bit error rate p, assuming random errors, $p_S = 1 - (1 - p)^m$. When other modulations are used, its expression may be found accordingly.

6.2.6   Algebraic Decoding of BCH

In Figure 6-13, we showed the decoding of any cyclic codes. The complexity of decoding is reduced substantially, as shown in the exercises. However, by utilizing the cyclic nature of code, algebraic decoding is possible, and thus it extends the usefulness of such codes. The algebraic decoding of BCH and RS was intensely studied during 60s to 70s. Even though there are a variety of techniques available, here we outline the decoding method by showing a double error ($t = 2$) correcting BCH as an example. The same method is applicable to RS codes. Extending a single error correcting code to multiple error correcting codes is a non-trivial jump, considerable sophistication of code construction using finite field arithmetic, as shown briefly in the previous sections, and furthermore in decoding as well. However, this example contains all the necessary steps for algebraic decoding.

Example 6-12: We use (15, 7) double error correction BCH as an example to describe the algebraic decoding process. We need to go back to the construction of generator polynomial, g(X). The elements in GF (2^4) with p(X) = $1+X+X^4$, and its corresponding minimal polynomial is listed in the table below.

Conjugate roots	Minimal polynomial $\phi_i(X)$	i
0	X	
1	X+1	
$\alpha, \alpha^2, \alpha^4, \alpha^8$	X^4+X+1	1
$\alpha^3, \alpha^6, \alpha^{12}, \alpha^{24} = \alpha^9$	$X^4+X^3+X^2+X+1$	3
α^5, α^{10}	X^2+X+1	5
$\alpha^7, \alpha^{14}, \alpha^{28} = \alpha^{13}, \alpha^{56} = \alpha^{11}$	X^4+X^3+1	7

Then g(X) = LCM{$\phi_1(X), \phi_2(X)$, , $\phi_{2t}(X)$}. It can be reduced to g(X) = LCM {$\phi_1(X), \phi_3(X)$, , $\phi_{2t-1}(X)$} since we need to take only least common multiple. For t=2, g(X) = $\phi_1(X)\phi_3(X)$ and $n-k = 8$, and for t=3, g(X) = $\phi_1(X)\phi_3(X)\phi_5(X)$. The parity check matrix, H, of (15, 7) code is given by,

$$H = \begin{bmatrix} 1 & \alpha & \alpha^2 & \alpha^3 & \alpha^4 & \alpha^5 & \alpha^6 & \alpha^7 & \alpha^8 & \alpha^9 & \alpha^{10} & \alpha^{11} & \alpha^{12} & \alpha^{13} & \alpha^{14} \\ 1 & \alpha^3 & \alpha^6 & \alpha^9 & \alpha^{12} & \alpha^{15} & \alpha^{18} & \alpha^{21} & \alpha^{24} & \alpha^{27} & \alpha^{30} & \alpha^{33} & \alpha^{36} & \alpha^{39} & \alpha^{42} \end{bmatrix}$$

Any codeword polynomial v(X) should meet v H^T=0. For the received codeword polynomial r(X) = v(X) + e(X) and thus $\mathbf{r}\,\mathbf{H}^T = \mathbf{S} = \mathbf{e}\,\mathbf{H}^T = (S_1, S_3, \ldots S_{2t-1})$.

And the syndrome can be evaluated from received polynomial as $S_i = r(\alpha^i)$. With t=2, there are S_1, S_3.

Say there are two errors with location j_1 and j_2, and then

$$S_1 = \alpha^{j1} + \alpha^{j2}$$
$$S_3 = \left(\alpha^{j1}\right)^3 + \left(\alpha^{j2}\right)^3$$

There are two unknowns of j_1 and j_2 and two equations. We need to solve these equations. The algebraic decoding is essentially to reduce the problem to the equation and solving it.

6.2.6.1   A Block Diagram of Algebraic Decoding Process

For a large block size code, the above formulation gives an overview but is not practical enough. Now we outline a general algebraic decoding of BCH codes applicable to a large block size, e.g., m=16, n = 65535.

Decoding process in summary, as shown in Figure 6-15:

1) Syndrome computation from r(X)
2) Find error location polynomial degree t or less $\sigma(X)$
3) Find zeros of $\sigma(X)$
4) Correct errors and check of a codeword

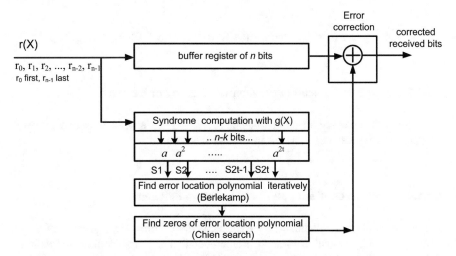

Figure 6-15: Algebraic decoding of BCH codes. Compare this figure with Figure 6-13, and its overall structure is similar. However, here its key component is to find an error location polynomial iteratively (Berlekamp), and then solve it

6.2.6.2   Syndrome Computation From r(X) – Method A

A polynomial evaluation can be used syndrome computation since $S_i = r(\alpha^i)$ with $i=1, 2,\ldots,2t$ where $r(X)$ is code packet in polynomial form; higher degree comes first.

Given r(X),

$$r(X) = r_0 X^{n-1} + r_1 X^{n-2} + \ldots\ldots + r_{n-3} X^2 + r_{n-2} X + r_{n-1}$$

compute u(α), s $= 0$ initialize to be zero,

$$r(\alpha) = r_0 \alpha^{n-1} + r_1 \alpha^{n-2} + \ldots\ldots + r_{n-3} \alpha^2 + r_{n-2} \alpha + r_{n-1}$$

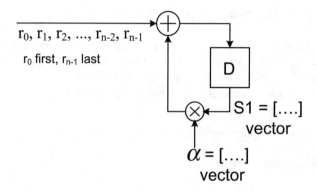

The above process is repeated for $r(\alpha^2)$, $r(\alpha^3)$,$\ldots r(\alpha^{2t})$ for S_2, S_3, $\ldots S_{2t}$.

6.2.6.3   Find Error Location Polynomial Degree t or Less $\sigma(X)$

Since it is a bit involved it is not explained here; see [11]. A Matlab routine has been written but not presented here but in another medium.

6.2.6.4   Find Zeros of $\sigma(X)$ an Error Location Polynomial

Using the polynomial evaluator, try each element at a time to see if it is zero, i.e., s=0 when each evaluation is done. Remember that an error location polynomial degree is limited to t or less, and trial locations may be limit to after the number of shortening bits.

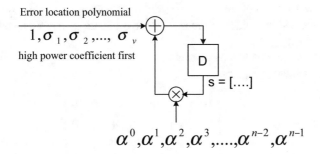

6.2.7   Code Modifications – Shortening, Puncturing and Extending

The block length of linear block code appears to be inflexible. For example, in BCH and Hamming codes the block length is 2^m-1 for m is an integer. With m=4, a (15, 11) code exists. However, there are ways to modify a code. In this section we consider shortening and lengthening, puncturing and extending of block codes.

Shortening: An (n, k) linear block code may be shortened to $(n - s, k - s)$ by s information bits. This may be considered as the leading information bits of s are known to be '0', thus not transmitted. The shortened code generator matrix, G_s, is of $(k - s, n - s)$ rows and columns, is obtained from the original code generator matrix, G $= [I_k \mid P]$, by removing s columns of I_k and the s rows that correspond to where the selected columns are nonzero.

For example, (7, 4) Hamming code is shortened to be (5, 2) with s=2.

$$G = \begin{pmatrix} 1 & 0 & 0 & 0 & 1 & 0 & 1 \\ 0 & 1 & 0 & 0 & 1 & 1 & 1 \\ 0 & 0 & 1 & 0 & 1 & 1 & 0 \\ 0 & 0 & 0 & 1 & 0 & 1 & 1 \end{pmatrix} \rightarrow \text{shortening} \rightarrow G_s = \begin{pmatrix} 1 & 0 & 1 & 1 & 0 \\ 0 & 1 & 0 & 1 & 1 \end{pmatrix}$$

Exercise 6-26: Find the shortened parity check matrix from the original parity check matrix.

$$H = \begin{pmatrix} 1 & 1 & 1 & 0 & 1 & 0 & 0 \\ 0 & 1 & 1 & 1 & 0 & 1 & 0 \\ 1 & 1 & 0 & 1 & 0 & 0 & 1 \end{pmatrix} \quad \text{and remove the leading two } (s = 2) \text{ columns.}$$

Extending: An (n, k) block code may be extended by adding more parity bits to $(n + e, k)$. The extended code has minimum distance is greater than or equal to that of the original code, $d_{ext} \geq d_{min}$.

A $(7, 4)$ Hamming code may be extended to $(8, 4)$. The parity check matrix, H, of $(7, 4)$ can be extended H_{ext} as,

$$H_{ext} = \begin{bmatrix} 1 & 1 & 1 & 0 & 1 & 0 & 0 & 0 \\ 0 & 1 & 1 & 1 & 0 & 1 & 0 & 0 \\ 1 & 1 & 0 & 1 & 0 & 0 & 1 & 0 \\ 1 & 1 & 1 & 1 & 1 & 1 & 1 & 1 \end{bmatrix}$$

Puncturing: It is the opposite of extending a code. A block code (n, k) is punctured to $(n - p, k)$ by removing p parity bits.

A puncturing technique is often used in convolution code to adjust the code rate, simple yet very useful technique. This will be discussed again in convolution codes.

We summarize this section in a table.

Technique	Action	Code parameters
Shortening	Removing information symbols	$(n - s, k - s)\, d_s \geq d$
Extending	Adding parity check symbols	$(n + e, k)\, d_{ext} \geq d$
Puncturing	Removing parity check symbols	$(n - e, k)\, d_{pun} \leq d$

6.3   Convolutional Codes

Linear block codes are discussed up to now. There is another class of linear codes that use a form of digital filter in order to generate parity bits. Convolutional codes are widely used in communication systems and its decoding may be done by hard decision or by soft decision where block codes are designed often for HD decoding. When the memory of the filter is small, Viterbi decoding, maximum likely sequence detector (MLSD) is almost universally used and efficient in implementations. We also discuss BCJR decoding as well which maximizes a posteriori probability.

6.3.1   Understanding Convolutional Code

6.3.1.1   Example of G=15/13 RSC

Parity bits (A1) of a convolution code are generated by a filter as shown in Figure 6-16. The operation of a linear filter is a convolution of input bit with an impulse response of it. Thus the name is convolution code. An input to the filter is binary

Figure 6-16: Systematic convolutional code G=15/ 13 used in 3GPP as a constituent code. It is a RSC

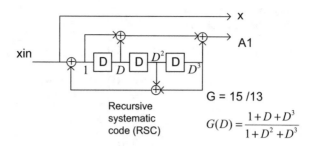

Recursive systematic code (RSC)

$$G = 15/13$$

$$G(D) = \frac{1+D+D^3}{1+D^2+D^3}$$

(0, 1) and output (x and A1) also binary. D represents one bit delay and addition is modulo 2 (or exclusive or). The output (x, A1) is two bits for one input bit; this is the rate ½ code. A short hand notation using octal form is common and in this case it is shown as G= 15/13 = 001 101 / 001 011, which shows which taps are connected in a filter. This also can be represented as a transfer function $G(D) = \frac{1+D+D^3}{1+D^2+D^3}$ with D is one bit delay. This particular convolution code is used by 3GPP cellular phone industry as a constituent code of a turbo code.

From the filter description of a code, other descriptions are derived and they can be useful. A state transition table can be generated. A state is defined by memories of a filter; $D^1D^2D^3$, and there are 8 possible states in this case since there are three positions and each one bit. When a new input bit arrives, it generates the output (x, A1) and moves to next state from current state. For example from 000 (0) state to 100 (1) when xin = 1 while from 000 (0) to 000 (0) when xin= 0. Try out all possible cases; 8 states with xin =0 or 1 thus 16 possibilities. This is shown in Table 6-7 below.

Table 6-7 can be represented as a state diagram as in Figure 6-17. The circled numbers are states, and a / bc means input bits / output bits; 1/10 means input bit =1 and output bits are x=1, A1= 0. State transitions are represented by arrowed lines. It can begin in any state and end in any state; there is no termination.

For a linear code, all zero code must be a code word. The code can be represented by a trellis starting from zero state and terminating at zero state. Figure 6-18 shows with 9 consecutive input bits starting from zero state and terminated at zero state by adding terminating bits (not information bits). There are 512 possible paths in this trellis diagram. When all zero code word is transmitted, any other code word - any one of all other paths (there are 511 of such) - is selected at receiver it is in error (resulting in code word error). In a trellis, MLSD - Viterbi algorithm - can be visualized and implemented. This will be amply explored later in this chapter. Dotted line means xin =1 while solid line means xin=0 in the figure.

A convolution code can be used without termination. A continuous stream of input bits are coming and the output bits are generated. At the receiver there is no pre-determined termination, but when a most likely path (series of states) is emerged, then a path is selected, or when multiple paths are competing simply a path with the least metric (i.e., most likely at the moment) may be selected. In a turbo code case a termination is clearly specified thus there is no ambiguity when the decision will be made.

Table 6-7: State transition table of 3GPP convolution code G=15/13

current state $D^1D^2D^3$		xin	A1	next state $D^1D^2D^3$	
0 0 0	(0)	0	0	0 0 0	(0)
0 0 0	(0)	1	1	1 0 0	(1)
1 0 0	(1)	0	1	0 1 0	(2)
1 0 0	(1)	1	0	1 1 0	(3)
0 1 0	(2)	0	1	1 0 1	(5)
0 1 0	(2)	1	0	0 0 1	(4)
1 1 0	(3)	0	0	1 1 1	(7)
1 1 0	(3)	1	1	0 1 1	(6)
0 0 1	(4)	0	0	1 0 0	(1)
0 0 1	(4)	1	1	0 0 0	(0)
1 0 1	(5)	0	1	1 1 0	(3)
1 0 1	(5)	1	0	0 1 0	(2)
0 1 1	(6)	0	1	0 0 1	(4)
0 1 1	(6)	1	0	1 0 1	(5)
1 1 1	(7)	0	0	0 1 1	(6)
1 1 1	(7)	1	1	1 1 1	(7)

Figure 6-17: State diagram for G=15/13 convolutional code with input bit / output bits

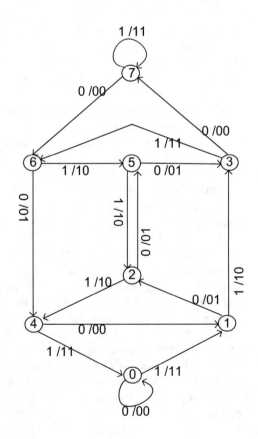

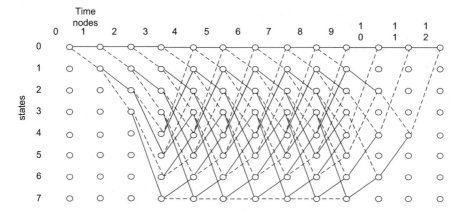

Figure 6-18: Trellis representation of G=15/13 code (terminated). The termination requires different patterns of input bits (000, 011, 110, 101, 100, 111, 010, 001) depending on the states. This is due to feedback and different from the non-recursive case where all zeroes are necessary

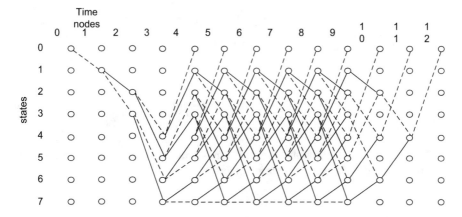

Figure 6-19: Transfer function (open 'zero' state) trellis of G=15/13 code

When there is no determined termination we consider the trellis diagram as in Figure 6-19, i.e., zero state is not extended. This reduces a number of possible paths considerably. Any one of these non-zero paths will cause an error assuming a zero code word is transmitted. This consideration of zero code word is general enough for performance analysis of linear codes since a zero is a code word by the definition of linear codes. Any path diverged from the zero path is called a single error event. The state diagram corresponding to this trellis (Figure 6-19) is the same as the zero state is 'opened' as in Figure 6-20.

Figure 6-20: State diagram for transfer function of G=15/13 convolutional code, where zero state is open

$$1/11 \to w/z^2$$
$$0/00 \to 1/1$$
$$1/10 \to w/z$$
$$0/01 \to 1/z$$

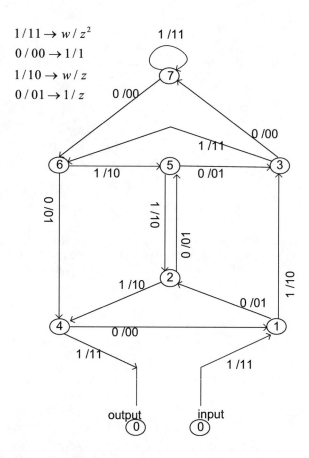

In order to find a symbolic form of transfer function, each branch on a state diagram will be replaced by dummy variables;

$$1/11 \to w \cdot z^2$$
$$0/00 \to 1 \cdot 1$$
$$1/10 \to w \cdot z$$
$$0/01 \to 1 \cdot z$$

When the number of state is small, a transfer function may be found manually. However, when it is large, a symbolic analysis of a state diagram (sometimes it is called flow diagram in the literature) is needed. It is like circuit analysis, solving linear simultaneous equations, not numerically but symbolically. And the resulting transfer function will be a function of (w, z). Even for Figure 6-20, it is laborious for manual calculation of transfer function. Later we will give a simpler example.

6.3.1.2 G =7 I5 Nonsystematic Convolutional code Example

Applying the basic transfer functions shown in Figure 6-23, a transfer function of this state diagram on Figure 6-22 is given by

$$T(w, z) = \frac{w \cdot z^5}{1 - 2w \cdot z} = w \cdot z^5 + 2w^2 \cdot z^6 + 4w^3 \cdot z^7 + 8w^4 \cdot z^8 + \dots\dots\dots$$

This transfer function says that there is one path of output distance 5 with input weight 1, two paths of output distance 6 with input weight 2, four paths of output distance 7 with input weight 3, and so on. One can identify these paths by looking at the state diagram (zero state open); state transitions from 0 to 1, 2 back to 0 again. This may be represented by a trellis shown in Figure 6-24.

0 - 1 - 2 - 0   (distance =5, input weight =1)
0 - 1 - 2 - 1 - 2 - 0 (distance =6, input weight =2)
0 - 1 - 3 - 2 - 0   (distance =6, input weight =2), and so on.

6.3.1.3   Recursive form of G=7/5 convolutional Code

A non-systematic code can be transformed to be a recursive systematic code (RCS) as shown in Figure 6-25.

And its state transition table is shown below, and is not the same as Tables 6-8 and 6-9.

This systematic recursive form is important for turbo codes as a constituent convolution code of it. On the other hand when a convolution code is used alone it is often the case that non-systematic non-recursive form is used. Given a one form it can be changed to other form.

6.3.1.4   Optimum Convolutional Code Tables

In Table 6-10, an example of $m=2$ is shown in Figure 6-21 G=7I5 code. All filter coefficients in the table are represented by octal G=7I5 $= 111$ I 101.

Figure 6-21: Non-systematic non-recursive convolutional code G=7I5

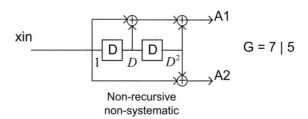

Non-recursive non-systematic

Table 6-8: G=7|5 non-systematic recursive code state transition table

current state	xin /A1A2	next state
0 0 (0)	0 /0 0	0 0 (0)
0 0 (0)	1 / 11	1 0 (1)
1 0 (1)	0 /1 0	0 1 (2)
1 0 (1)	1 /0 1	1 1 (3)
0 1 (2)	0 /1 1	0 0 (0)
0 1 (2)	1 /0 0	1 0 (1)
1 1 (3)	0 /0 1	0 1 (2)
1 1 (3)	1 /1 0	1 1 (3)

0 1 (2)   1 / 0 0   1 0 (1)

1 1 (3)   0 / 0 1   0 1 (2)

1 1 (3)   1 /1 0   1 1 (3)

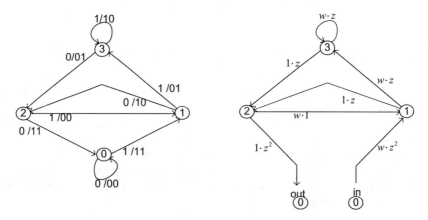

Figure 6-22: State diagram (LHS) and transfer function (RHS) for G=7|5 code

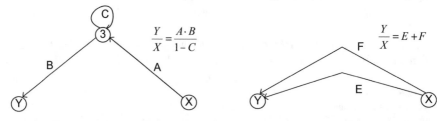

Figure 6-23: Basic transfer function of flow diagrams

Note that these are non-systematic, and converting to systematic can be done as shown in Figure 6-25.

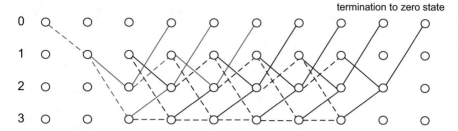

Figure 6-24: Trellis representation of (open zero state) diagram

Figure 6-25: A recursive form of Figure 6-21 code

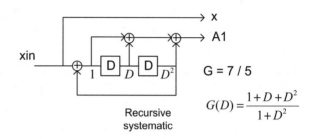

$G = 7 / 5$

$$G(D) = \frac{1+D+D^2}{1+D^2}$$

Recursive systematic

Table 6-9: Recursive G=7/5 code state transition table

current state x A1		next state
0 0 (0)	0 0	0 0 (0)
0 0 (0)	1 1	1 0 (1)
1 0 (1)	0 1	0 1 (2)
1 0 (1)	1 0	1 1 (3)
0 1 (2)	0 0	1 0 (1)
0 1 (2)	1 1	0 0 (0)
1 1 (3)	0 1	1 1 (3)
1 1 (3)	1 0	0 1 (2)

Finding these codes in Tables 6-10 and 6-11 can be done by trial and error using a computer, and see [10]. More codes are available in the literature.

Minimum free distance (d_{free}): One important parameter of a convolutional code is free distance. It is briefly shown in Figure 6-24 or can be obtained by transfer function method. But it is limited with a small number of states. One can write a computer program to compute free distance of a convolutional code. It is, essentially, systematically enumerating codewords with trellis. d_{free} is defined as $d_{free} \equiv$ min {d (v, v'): u $\neq$ u'} where v, v' are the codewords corresponding to the information sequence u, and u', respectively. Since it is a linear code, d_{free} is the minimum weight code word of any length produced by a nonzero information sequence.

Table 6-10: Rate ½ maximum free distance codes

Memory : m	Generator s in octal form		d_{free}
2	5	7	5
3	15	17	6
4	23	35	7
5	53	75	8
6	133	171	10
7	247	371	10
8	561	753	12
9	1 167	1 545	12
10	2 335	3 661	14
11	4 335	5 723	15
12	10 533	17 661	16
13	21 675	27 123	16

Table 6-11: Rate 1/3 maximum free distance codes

Memory : m	Generator s in octal form			d_{free}
2	5	7	7	8
3	15	15	17	10
4	25	33	37	12
5	47	53	75	13
6	133	145	175	15
7	225	331	367	16
8	557	663	711	18
9	1 117	1 365	1 633	20
10	2 353	2 671	3 175	22
11	4 767	5 723	6 265	24
12	10 533	10 675	17 661	24
13	21 645	35 661	37 133	26

Exercise 6-27: Show that the minimum free distance of $G=7|5$ code is 5, from Figure 6-24. Hint: input bits of $\{1, 0, 0\}$ produce the output $\{11, 10, 11\}$ with the state transition $\{1, 2, 0\}$.

6.3.1.5 Block Codes from Convolutional Codes

Since in practice a convolutional code is used to transmit a block of information bits, its trellis must be terminated. Using $G=7|5$ of non-recursive code in Figure 6-21 its terminated trellis is shown in Figure 6-26.

The encoder state initially is assumed to be zero and then the final state to be zero by adding appropriate bits (zeroes in the case of non-recursive example here). Note that the block length of $(N + m)/R$ bits is flexible with different choice of N.

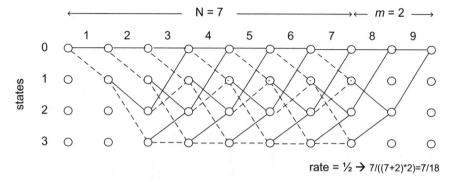

Figure 6-26: Terminated trellis of G=7|5 non-recursive code in Figure 6-21 where N=7 data bits and m=2. And the code rate is changed to 7/ (9*2) = 7/18 < 1/2. This may be considered as a block code of $(n, k) = (9*2, 7) = (18, 7)$. All possible codewords (or paths in trellis), including the tail of 2 zeroes in the example, are represented by the trellis. The information bits are $\{u_1, u_2, u_3, u_4, u_5, u_7, 0, 0\}$ in this example; i.e., two known bits $\{0, 0\}$ are appended

In passing we remind that in Figure 6-18, a terminated trellis of RSC G=15/13 has different patterns of input bits (called termination input bits) necessary to drive the final state to be zero due to feedback.

Exercise 6-28: Termination of RSC to zero state requires termination input bits that are not zero in general. Convince that this is true by studying Figure 6-18 and apply to Figure 6-25. Hint: to drive states to zero, xin + 'feedback bit' in the figure, not just xin being zero, must be zero.

When the data block size, N, is much larger than the memory (m) of the code, the rate reduction is small since it is given by $\frac{N}{N+m}$, and thus overall code rate is the original code rate (R=1/2 in this example) times the reduction, i.e., $\frac{R \cdot N}{N+m}$. In practice, this termination scheme is used often. However, there are ways to avoid the rate reduction. One simple way is not to terminate at all, no termination bits are added. But in practice, the Viterbi decoder needs to be terminated once in a while even if the transmission is continuous, i.e., semi-infinite.

Tail biting termination: The idea is to maintain the initial state and final state the same without adding termination bits. The final state depends on the last m input bits out of N bits. The block length does not increase with tail-biting termination and thus it is used when the data block size is small at the expense of decoder complexity. The complexity increases since the decoder does not know the initial and final state, which depend on the last m input data bits $\{u_{N-m+1}, \ldots, u_{N-1}, u_N\}$ out of $\{u_1, u_2, \ldots, u_{N-m}, u_{N-m+1}, \ldots u_{N-1}, u_N\}$.

We explain the tail-biting termination by an example shown in Figure 6-28. We use the same G=7|5 code in Figure 6-21, and the same input bits of Figure 6-27 with zero termination are used for comparison $\{1\ 0\ 0\ 0\ 1\ 1\ 0\}$. The corresponding tail-biting output bits are $\{11\ 00\ 00\ 00\ 11\ 01\ 01\}$. The figure explains how to obtain them.

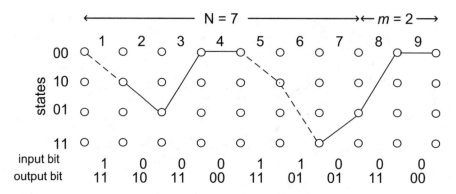

Figure 6-27: One particular codeword (path in trellis) with the input bits (1000110 +00) and its output bits are shown. The initial state and final state are both zero state

Exercise 6-29: For the same code of Figure 6-28, compute the tail-biting output bits for given input bits of {1 1 1 1 0 0 1}
Answer: {01 01 01 10 01 11 11} For the details of the solution see Figure 6-29.

6.3.1.6 Input Weight and Output Distance of Convolutional Codes

We discussed the performance of linear block codes of (n, k) and came to a conclusion that the codeword error probability can be expressed by (6-17), as repeated here for convenience, $P_w = p_{e \mid m = 0} = P_r(u_1 + u_2 + \cdots + u_{M-1})$, where $p_{e \mid m = 0}$ means that the zero codeword is transmitted and decoded into non-zero codeword, and since there are $M = 2^k$ codewords including the zero, it may be mistaken into any one of M-1 codewords. And u_j is defined as the event $C_0 \rightarrow C_j$ with $j = 1, 2, \ldots, M$-1. In BPSK transmission, $P_r(u_j) = Q\left(\sqrt{\frac{2E_s}{N_0} w_j}\right)$ where w_j is the weight of the codeword C_j.

Another useful variation of the expression of P_w is given by

$$P_w = 1 - P_r(\overline{u_1 + u_2 + \cdots + u_{M-1}}) = 1 - P_r(\overline{u_1} \cdot \overline{u_2} \cdots \cdots \overline{u_{M-1}}) \qquad (6\text{-}33)$$

The events of u_j are typically not independent, but when the independence is assumed, then (6-33) may be considerably simplified, and P_w is upper-bounded, called independence bound, by the simplified expression of

$$P_w = 1 - P_r(\overline{u_1} \cdot \overline{u_2} \cdots \cdots \overline{u_{M-1}}) \leq 1 - \prod_{j=1}^{M} \left[1 - P_r(u_j)\right] \qquad (6\text{-}34)$$

The above (6-34) can be further simplified to $P_w \leq \sum_{j=1}^{M} P_r(u_j)$ when the events of u_j are assumed to be mutually exclusive. This expression, called union bound, is

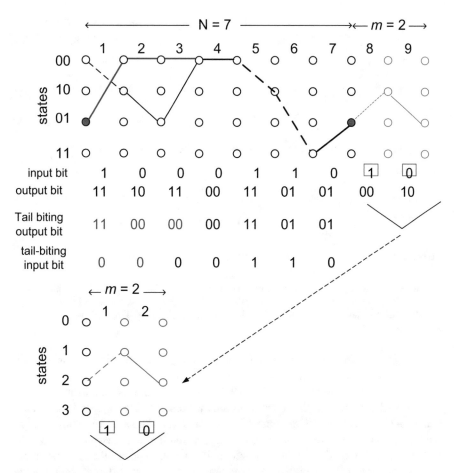

Figure 6-28: Tail-biting termination example using code G=7l5 of Figure 6-21. With the initial state of zero and the input bits {1000110 + 10} path are computed as shown and the final state is 01. The m (2 in this example) last input bits are repeated and the output bits are {00, 10}, and these are added to the first m (2 in this example) bits and resulted {11, 00}. Now the initial state is 01 to produce {11, 00}. Note that the third output bits are changed to 00 with the input bit the same 0. At the decoder the path (thick line) is recovered from the output bits , i.e., tail-biting bits, of { 11 00 00 00 11 01 01}. Then the tail-biting information bits are recovered as {0 0 0 0 1 1 0}. And the original information bits are {1 0 0 0 1 1 0}

often used in textbooks and in the literature, but it is a misnomer. It gives some indication of the code performance, but is not very useful for powerful codes like Turbo codes and LDPC in practical situations.

Convolutional codes are linear codes and may be converted to block codes by termination. In the previous section was discussed three ways of termination – tailbiting, the final state to be zero, and direct termination without controlling the final state. The convolutional codes may be used without termination, where the block length may be considered to be the infinite. Note that one important parameter of

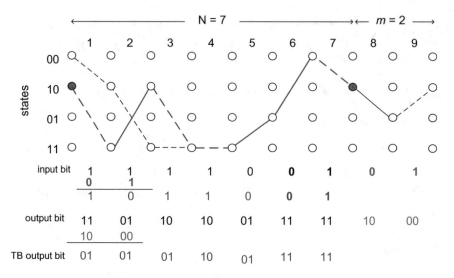

Figure 6-29: The solution to the above exercise; find the tail-biting output bits for the input bit of {1 1 1 1 0 0 1} and the answer is {01 01 01 10 01 11 11}. The process is the same as Figure 6-28; the repeated tail bits are added to the initial bits of the block, modifying the path (codeword) accordingly

convolutional codes is the minimum free distance (d_{free}) and that it is obtained when the block length is arbitrarily large, i.e., no termination.

As we briefly discussed in Section 6.2.3, the precise computation using (6-33) is complicated even for simple codes. And thus we resort to the independence bound using (6-34) where we group some of codewords into the same Hamming weight (called distance), distance d. We also classify, within the same distance codewords, further depending on the input (information bits) weight. The input of zero produces the zero codeword. The weight one of the input may produce different weight at the output (codeword), and the weight two of the input may produce different weight at the output, and so on. This was discussed in Section 6.3.1.2 using a transfer function. As we mentioned earlier that the transfer function method is limited to simple codes, especially by manual calculations, we present it with a table. This distance distribution with the input weight is called *input weight and output distance*, A (w, d), and code distance distribution (spectrum) is given by $B(d) = \sum_w A(w, d)$. Then (6-34) can be expressed as,

$$P_w \leq 1 - \prod_{j=1}^{M} \left[1 - P_r(u_j)\right] = 1 - \prod_{d}[1 - \Pr(u_d)]^{B(d)} \qquad (6\text{-}35)$$

In order to find the independence upper-bound, we need the code distance spectrum, $B(d)$, which can be obtained from *input weight and output distance* table, A (w, d) using $B(d) = \sum_w A(w, d)$.

Input weight and output distribution examples:
We give two examples of A(w, d), input weight and output distribution. No entry means zero.

In Table 6-12, A(w, d) of G=7|5 non-recursive, non-systematic code in Figure 6-21 is displayed. Similarly in Table 6-13, A(w, d) of RSC G=7/5 in Figure 6-25 is shown. Both codes have the same B(d) even though A(w, d) is not exactly the same. And thus the upper-bound of the codeword error rate, P_w, is the same, which is not surprising since both are the 'same' code in this regard.

A(w, d) contains more information than that of B(d) since, for a given distance d, the input weight distribution is shown in detail. For a limited input weight and output distance enumeration, we need to consider only small input weight codewords as shown in the above two tables. This trend is true for convolutional codes with more states.

Codeword error rate independence bound:
Using (6-35) and with B(d) in Table 6-12, we computed WER for G=7 | 5 code, and is shown in Figure 6-30.

Table 6-12:   A(w, d) in table form for G=7|5 code in Figure 6-21

		Input weight w									B(d)
		w=1	2	3	4	5	6	7	8	9	
Output distance d	d=5	1									1
	6		2								2
	7			4							4
	8				8						8
	9					16					16
	10						32				32
	11							64			64
	12								128		128
	13									256	256

Table 6-13:   A(w, d) in table form for RSC G=7/5 code in Figure 6-25

		Input weight w										B(d)
		w=1	2	3	4	5	6	7	8	9	10	
Output distance d	d=5		1		0		0					1
	6		1		1		0					2
	7		1		3		0					4
	8		1		6		1					8
	9		1		10		5					16
	10		1		15		15		1			32
	11		1		21		35		7			64
	12		1		28		70		28		1	128
	13		1		36		126		84		9	256

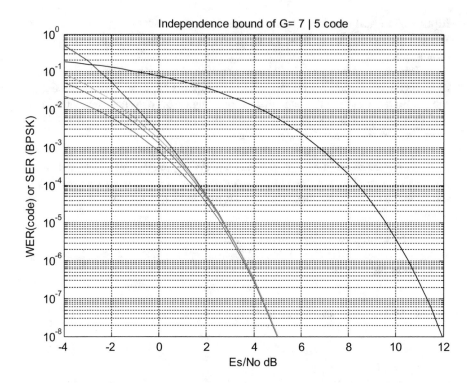

Figure 6-30: WER performance of $G = 7 \mid 5$ code with (6-35) and B(d) of Table 6-12

We explain Figure 6-30. There is no dB reduction due to code rate (1/2, 3dB) in the curves. The red curve shows WER with only one term of d=5 and B(d) =1, and the blue curve takes all the distance from d=5 to 13 into account. The magenta and green curve are for 2 and 3 terms respectively. Note that all WER curves are merge as SNR increases, i.e., d=5 indicates the asymptotic performance well, and for low SNR it needs more terms. SER of BPSK is shown too as a reference.

Exercise 6-30: Write a computer program to generate the curves of Figure 6-30 using the upper bound of (6-35).

Exercise 6-31: Write a simulation program to check the tightness of the bound in the example in Figure 6-30. Choose N=12 with zero state termination to make it a block code, and then use CM decoding discussed in Section 6.2.3.1.

6.3.1.7   Optimized Computation of A (w, d)

A limited code distance (input weight, output distance) distribution, A (w, d), can be generated by a computer program in table form as shown in Tables 6-12 and 6-13. An optimized enumeration method may be developed. It is optimized in the sense that we do not explore more than the number of codewords within the bound of $d \leq$ distance target, $w \leq$ input weight target. Target means the maximum output distance and input weight to consider computing. We do not pursue any further here.

6.3.2 Viterbi Decoding of Convolutional Codes

Viterbi decoding is maximum likelihood sequence detection (MLSD); a sequence in trellis representation of convolutional codes is a path. It is an optimal path finding algorithm. Viterbi decoding was very successful in coding applications.

6.3.2.1 Review of CM and its Use as Branch Metric

We covered CM decoding in subsection 6.2.3.1 but review it here again here for convenience. In Viterbi decoding CM is applied to branch metric computation.

With BPSK modulation, the received samples of a block, r_i with $i=1,2,\ldots,n$ are given by $r_i = X_i + n_i$ where $X_i = \{+1, -1\}$ BPSK transmitted symbols and n_i is AWGN samples after a receive filter. BPSK modulated (with $0\rightarrow+1$, $1\rightarrow-1$) version of all the codewords is denoted as $C_{mi} = 1 - 2c_{mi}$ with $m=0, 1, 2, \ldots, M\text{-}1$ and $i =1,2,\ldots,n$ where M is the total number of codewords.

$$\max_m \sum_{i=1}^{n} C_{mi} \cdot r_i \qquad (6\text{-}36)$$

CM decoding is defined by choosing a codeword, c_m, such that the correlation between a received vector, r_i with $i=1,2,\ldots,n$, with all the codewords in BPSK modulated form, is maximum, which is given by (6-36).

The correlation in (6-36) may be considered as a 'similarity' comparison of a received vector against all the codewords in BPSK symbols, just like the measurement of a distance is a comparison against a standard stick; the more 'similar' the bigger the correlation.

For HD, r_i is quantized first to be $\{0\}$ if it is bigger than zero, $\{1\}$ otherwise, and denoted by $[r_i]$. In this case of HD, (6-36) becomes a minimum distance decoding, say using Hamming distance, and given by,

$$\min_m \sum_{i=1}^{n} c_{mi} \cdot [r_i] \qquad (6\text{-}37)$$

In Viterbi decoding, CM is not applied to the entire received vector but to each branch between states in trellis representation of a code. If it is applied to the entire received vector, the number of codewords becomes exponentially large, 2^k with k being information bits, and thus computationally impractical. The saving grace of

Viterbi decoding is the reduction of computation to $2^{(s+1)} k$ where s is the number of states in the code (why not 2^s?). This is possible because the code is represented by a trellis. In (6-36) and (6-37), the summation goes up to the number of bits in each branch; $n =$ the number of bits per branch; $n=2$ for ½ rate code, 3 for 1/3 codes etc.

Exercise 6-32: When r_i is quantized to be $[r_i] = \{+1, -1\}$ rather than $\{0, 1\}$, then the equation (6-37) may be represented by $max_m \sum_{i=1}^{n} C_{mi}[r_i]$. Show that this is equivalent to (6-37). Thus (6-36) and (6-37) are qualitatively the same except for the quantization level difference. In other words, (6-36) and (6-37) are the same form of computation with different quantization level.

Answer: The computation of $max_m \sum_{i=1}^{n} C_{mi}[r_i]$ is the same as (6-36) with $[r_i]$ being quantized. When it is replaced C_{mi} with c_{mi}, and $\{+1, -1\}$ with $\{0, 1\}$ in the equation, it becomes (6-37).
   Here we use (6-36) as a branch metric for SD, and (6-37) for HD. There is no need of max or min when it is used as a branch metric since m is known for each branch.

6.3.2.2   Viterbi Decoding Explained With an Example of HD

We now describe the algorithm of Viterbi decoding. We explain Viterbi decoding with an example of G = 7|5 convolutional code in Figure 6-31. Note that there are 2^N codewords (paths), i.e., N=7, information bits, and represented by a trellis of states and branches. We consider HD, i.e., received bits are quantized to $\{0\}$ or $\{1\}$.

Viterbi decoding Algorithm: The computation proceeds from the time index 1, 2, ..., to N+s, N=7 s=2, In each time index, each state has a state branch metric

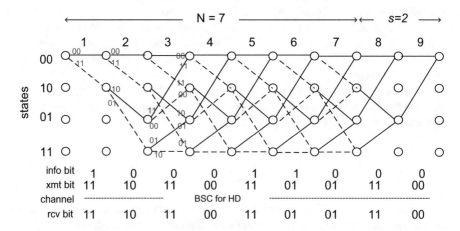

Figure 6-31:   Trellis diagram of G = 7|5 Non-systematic code with N=7 and zero state termination

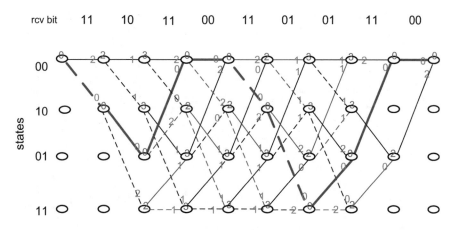

Figure 6-32: Example of Viterbi decoding algorithm with G=7⎸5 code

accumulated up to the time index, and $2^{(s+1)}$ branch metrics need to be computed, which are added to the corresponding state branch metric in order to obtain new state branch metrics. In each state two branch metrics are compared and the minimum is selected and assigned to a new state branch metric for the next time index. This is repeated for $N+s$ received symbols. Note that for the initial two (s in general) and the last two symbols, less than $2^{(s+1)}$ branch metrics need to be computed and there is no need of comparison and selection.

We give a numerical example in Figure 6-32. Consider a case without error in the received bits. Certainly this should work!

There is no error in received bits. Red number above the state circle is the state metric, and the green number is the branch metric (Hamming weight). Selected branch is in blue color. Branches of traced back states are in thick blue.

Exercise 6-33: Confirm all the branch metrics and state metrics in Figure 6-32. Hint: the receiver knows that the beginning state is zero and so is the end state. Patiently try out all branch metrics and state metrics for forward direction, perhaps manually, in conjunction with Figure 6-31.

Then for reverse direction, from zero state at the end of trellis back to beginning, choose only one survived path, with branches of the minimum branch metric at each survived state. This is called *traceback*.

Exercise 6-34*: Write a computer program to carry out the algorithm described in Figure 6-32, and make sure to include trace back and to find the information bits. The star (*) means project exercise.

6.3.2.3   Viterbi Decoding Explained With an Example of SD

We give another example with the organization of computing in Table 6-14. This time with SD. This means that received symbols are NOT quantized to be binary,

Table 6-14: Viterbi decoding algorithms with the same transmit data of Figure 6-32

time	1			2			3			4		
XMT	-1,-1			-1,+1			-1,-1			+1,+1		
RCV	-1	0.1		-1	1		-1	0.1		-0.2	1	
state	sm	bm		sm	bm		sm	bm		sm	bm	
00	0	-0.9	-0.9	-0.9	0	-0.9	-0.9	-0.9	-1.8	3.8	0.8	4.6
		0.9	0.9		0	-0.9		0.9	0		-0.8	3
10				0.9	2	2.9	-0.9	1.1	0.2	2	1.2	3.2
					-2	-0.9		-1.1	-2		-1.2	0.8
01							2.9	0.9	3.8	0.2	-0.8	-0.6
								-0.9	2		0.8	1
11							-0.9	-1.1	-2.2	0	-1.2	-1.2
								1.1	0		1.2	1.2

time	5			6			7			8		
XMT	-1,-1			+1,1			+1,-1			-1,-1		
RCV	-1	-1		1	-1		1	-1		-1	-1	
	sm	bm		sm	bm		sm	bm		sm	bm	
00	4.6	-2	2.6	5.2	0	5.2	5.2	0	5.2	5.2	-2	3.2
		2	6.6		0	5.2		0	5.2			
10	3	0	3	6.6	-2	4.6	5.2	-2	3.2	5.2	0	5.2
		0	3		2	8.6		2	7.2			
01	3.2	2	5.2	3	0	3	5	0	5	10.6	2	12.6
		-2	1.2		0	3		0	5			
11	1.2	0	1.2	3	2	5	8.6	2	10.6	7.2	0	7.2
		0	1.2		-2	1		-2	6.6			

time	9					
XMT	+1,+1					
RCV	1	1				
	sm	bm				
00	12.6	2	14.6		14.6	
10	0					
01	7.2	-2	5.2			
11	0					

thus SD decoding and errors (violet color in Table 6-14) and will be corrected through decoding process. Yellow is the state metric (maximum at the time index) and the corresponding branch is the state transition these states with the maximum state metric. The metric is CM defined in (6-36). The initial state is the zero state and the final state is zero too. sm= state metric, and bm= branch metric. Additional explanation follows after Table 6-14.

We explain the computation of Table 6-14. Take 'time index 3' case as an example. Each state has a state metric (sm) from the previous time index 2. Branch metrics (bm), total 8, are computed and the result is (-0.9, 0.9, 1.1, -1.1, 0.9, -0.9, -1.1, 1.1) and these are added to the current sm and result (-1.8, 0, 0.2, -2, 3.8, 2, -2.2,

0). Then pair them (-1.8, 3.8), (0,2), (0.2,-2.2) (-2,0) and choose the larger one to update the state metric. This process is repeated. The corresponding branch is a state transition from (01) to (00) and its value is (11).

time			3			4
XMT			11			
RCV		-1		0.1		
state	sm		bm			sm
00	-0.9	+1,+1	-0.9	-1.8		Max(-1.8, 3.8)
		-1,-1	0.9	0		
10	-0.9	-1,+1	1.1	0.2		Max(0, 2)
		+1,-1	-1.1	-2		
01	2.9	-1,-1	0.9	3.8		Max(0.2, -2.2)
		+1,+1	-0.9	2		
11	-0.9	+1,-1	-1.1	-2.2		Max(-2, 0)
		-1,+1	1.1	0		

Exercise 6-35: Study Table 6-14 in detail with the help of 'time' $=3$ case in the above, and convince yourself that the table is correct. Hint: this requires intense attention and patience.

6.3.2.4 Viterbi Decoding with Quantized SD

Received samples are quantized to be $\{+1\}$ or $\{-1\}$ so for branch metric we use (6-36), the same as SD case, and thus the calculation of branch metric is simpler and thus less complex. It will be exactly the same organization as Table 6-14 with different branch metrics. We put this case as an exercise and the computation is shown on a trellis shown in Figure 6-33.

Exercise 6-36: Repeat the same computation of Table 6-14 with quantized to +1 when received value is positive and -1 otherwise. Note that the performance is worse than SD of Table 6-14.

Answer: Using Figure 6-33 we explain this case. Unlike SD case, due to errors (red mark on received BPSK symbol), there are ties in two branch comparison (state nodes gray filled). In trace back of path (thick blue line), there are two possible paths due to tie at '00' state. Note that one path is correct and the other path is incorrect, which results in two bit errors. Because of these ties it is clear that the performance of quantized SD is worse than SD. Other ties are not important.

6.3.2.5   Why Viterbi Decoding is Effective?

The number of branch metric calculations in Viterbi decoding for a given convolutional code represented by a trellis is $2 \times 2^s \times k$, where 2^s is a number of states, k is a number of information bits in a block (e.g., the same as N in Figure 6-31). This is possible because the code is represented by a trellis.

Suppose that a code has no structure, i.e., no trellis representation, and then an optimal CM decoding is applied for entire block. The number of CM calculations is 2^k in this case.

For Figure 6-31 example, $2 \times 2^s \times k = 56$, $2^k = 128$. Now increase $k = 20$ (small block size), $2 \times 2^s \times k = 160$, $2^k = 1048576$! CM decoding is optimal and general but completely impractical.

6.3.2.6   Summary of Viterbi Decoding Algorithm

For a given received symbol sequence, it is an algorithm to find an optimum (the maximum CM or the minimum Hamming distance) path with the trellis representation of a code.

1. Given state metrics from the previous time (or initialize)
2. For a received symbol (BPSK symbols or binary digits), compute branch metric using CM or hamming distance
3. Add the branch metric to state metric
4. Compare and select the maximum (CM) or minimum (Hamming distance) branch and update each state metric
5. Go back to 2 until all received symbols are done
6. Trace back a survived path with termination
7. Convert the decided path to a bit sequence.

In step 4 a tie break method is necessary if there is no distinct maximum (or minimum) branch to choose; random selection may be one. In Figure 6-33, one can choose a branch originated from the greater state metric in the previous time. Is this better in general? A minor improvement may be possible but it needs to be confirmed.

Exercise 6-37*: Write a computer program of Viterbi decoder with CM branch metric for a given convolutional code. The star * means the exercise may be a project.

6.3.2.7   Branch Metric Revisited

<u>Other branch metric</u> may be Euclidean distance, $\min_{m} \sum_{i=1}^{n} (C_{mi} - r)^2$ for AWGN channel. No min operation since m is known for each branch. The Euclidean distance is a poor metric for fading channels.

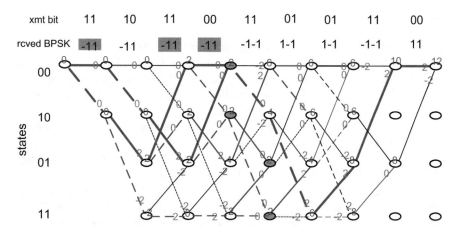

Figure 6-33: Trellis representation of quantized SD, i.e., quantized to +1 when received value is positive and -1 otherwise

Exercise 6-38: In the computation of Table 6-14 replace the branch metric of CM with that of Euclidean distance, $\min_m \sum_{i=1}^{n} (C_{mi} - r)^2$. No min operation when $m=1$. Hint: the same computational organization as Table 6-14 but branch metric is different.

In information theoretic general description of MLSD, log of probability is often suggested as a branch metric, b_m as, $b_m = \log \prod_{i=1}^{n} p_r(r_i|c_{mi}) = \sum_{i=1}^{n} \log (p_r(r_i|c_{mi}))$. Note that max operation is dropped here. This metric is generic enough to be used for AWGN and fading channels.

Here we use only (6-36) for SD and (6-37) for HD.

6.3.3 BCJR Decoding of Convolutional Codes

BCJR is the abbreviation of four names of Bahl-Cocke-Jelinek-Raviv, who invented the decoding scheme in early 70s' [1] . It is an optimal, in the sense of minimizing symbol error rate rather than codeword error rate, decoding of linear codes that are described by a trellis. When it is applied to a convolutional code, computationally it is much more complex than Viterbi decoding which optimizes for codeword error rate or detecting a codeword with the maximum CM. Thus the method was not often used for convolutional codes, but it became a major, important, decoding block in Turbo codes, and in general for iterative decoding.

By examining the received sequence of an entire block, BCJR decoding computes a posteriori probability (APP) of states and branches of a trellis which represent the code, and from them, APP of symbols and of transmitted digits is computed. Thus the decoding output is the probability or soft-information, which can be used to make a final decision or which can be used for the next iteration in iterative decoding.

6.3.3.1   Notations

Here we describe BCJR with the same example of G= 7 | 5 convolution code used in Figures 6-31, 6-32, and 6-33. We draw it again with additional notations in Figure 6-34. There are M distinct states which are indexed by the integer m, $m=0$, 1, 2, ..., M-1. In Figure 6-34 bit representation is in parenthesis, e.g., 0 (00), 1 (10) etc. The state at time t is denoted by S_t, and a state sequence from 1, 2, ...t is denoted by $S_1^t = S_1 S_2 ... S_t$. Similarly their associated transmit digit at time t is denoted by X_t, and a transmit digit sequence from 1,2, ... t is denoted by $X_1^t = X_1 X_2$ ... X_t, and the corresponding information is denoted by i_t. For the received digit similarly $Y_1^t = Y_1 Y_2 ... Y_t$. Thus the entire received sequence is denoted by Y_1^τ.

With Figure 6-34, time index $t=1, 2,..., \tau$, where $\tau=9$, and a state at time t is designated as S_t. Thus $S_0(00)$ is the initial state and $S_9(00)$ is the ending state. States are designated by 0(00), 1(10), 2(01), 3(11). Transmit digit at time t is X_t and its corresponding information bit is i_t and received digit is Y_t, which may be a set of real numbers (BPSK symbols + AWGN). The branch is designated as $\boxed{1}, \boxed{2}, ..., \boxed{8}$ with square and its values (branch bit) are {00, 11, 10, 01, 11, 00, 01, 10}.

First the decoder is to compute the APP of the states and of branches, i.e., the conditional probabilities, conditioned on Y_1^τ, an entire received block (with transition probabilities for BSC).

$$\Pr\left\{S_t = m | Y_1^\tau\right\} \tag{6-38}$$

$$\Pr\left\{S_{t-1} = m', S_t = m | Y_1^\tau\right\} \tag{6-39}$$

where $m=\{0, 1, 2, ...M\text{-}1\}$ of states and, $t=\{1, 2, 3, ... \tau\}$ of time index.

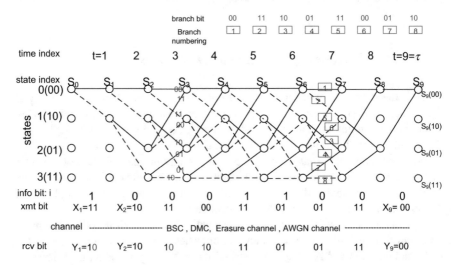

Figure 6-34: Trellis of G=7|5 convolutional code with time index, state index, states, information bit, transmit digits, receive digits and branch designation

Then using (6-38) or (6-39), the APP of information bit, i_t, and transmission digit X_t can be computed. From the APP a decision on information bit and transmission digit can be made.

Conditional probability: A few identities on the conditional probability are used extensively in deriving the algorithm. We list them here as a reminder.

$$\Pr{(AB)} = \Pr(A|B)\,\Pr(B) = \Pr{(B|A)}\,\Pr(A) \tag{6-40}$$

$$\Pr(AB|C) = \Pr(A|BC)\,\Pr(B|C) = \Pr{(B|AC)}\,\Pr(A|C) \tag{6-41}$$

$$\Pr(A|B) = \Pr(AB)/\Pr(B) = \Pr{(B|A)}\,P(A)/\sum_j \Pr(B|A_j)\Pr(A_j) \tag{6-42}$$

(6-40) and (6-41) show that joint probabilities are expanded by conditional probabilities. (6-42) is called *Bayes Theorem* and useful in computing conditional probabilities. Additional details are reviewed in Chapter 9 of this book.

Exercise 6-39: Try to express (6-40) with a diagram (Venn diagram). Hint: see Chapter 9.

6.3.3.2 BCJR Algorithm (Forward – Backward Algorithm)

It is simpler to use the joint probabilities of states and branches at time t, instead of (6-38) and (6-39),

$$\lambda_t(m) = \Pr\left\{S_t = m, Y_1^\tau\right\}$$

$$\sigma_t(m', m) = \Pr\left\{S_{t-1} = m', S_t = m, Y_1^\tau\right\}$$

In order to obtain the conditional probabilities of (6-38) and (6-39), conditioned on Y_1^τ, we can divide $\lambda_t(m)$ and $\sigma_t(m', m)$ by $\Pr\{Y_1^\tau\}$. Alternatively, we can normalize $\lambda_t(m)$ and $\sigma_t(m', m)$ to add up to 1.0 for all states and for all branches at time t respectively. The normalization is often used in practice rather than explicitly computing $\Pr\{Y_1^\tau\}$.

We now derive a computational algorithm for obtaining the probabilities $\lambda_t(m)$ and $\sigma_t(m', m)$, which are conditioned on entire received bits Y_1^τ. We need additional intermediate probabilities in order to compute $\lambda_t(m)$ and $\sigma_t(m', m)$.

Define the intermediate probabilities for forward and backward iterations as,

$$\alpha_t(m) = \Pr\left\{S_t = m, Y_1^t\right\} \qquad \text{\% forward iteration}$$
$$\beta_t(m) = \Pr\left\{Y_{t+1}^\tau \,|\, S_t = m\right\} \qquad \text{\% backward iteration}$$
$$\gamma_t(m', m) = \Pr\left\{S_t = m, Y_t \,|\, S_{t-1} = m'\right\} \quad \text{\% branch probabilities at } t$$

We now express $\lambda_t(m)$ and $\sigma_t(m', m)$ in terms of the above probabilities.

$$\lambda_t(m) = \alpha_t(m)\,\beta_t(m) \tag{6-43}$$

$$\sigma_t(m', m) = \alpha_{t-1}(m')\,\gamma_t(m', m)\,\beta_t(m) \tag{6-44}$$

with $t = 1, 2, \ldots \tau$ (time index) and $m = 0, 1, 2, \ldots, M\text{-}1$ (states).

The derivation of (6-43) and (6-44) is shown below.

$$
\begin{aligned}
\lambda_t(m) &= \Pr\left\{S_t = m, Y_1^\tau\right\} && \text{\% definition} \\
&= \Pr\left\{S_t = m, Y_1^t, Y_{t+1}^\tau\right\} && \text{\% } 1,2,\ldots t, \text{ and } t+1,t+2,\ldots \tau \\
&= \Pr\left\{S_t = m, Y_1^t\right\}\Pr\left\{Y_{t+1}^\tau | S_t = m, Y_1^t\right\} && \text{\%P(AB)=P(A)P(B|A)} \\
&= \alpha_t(m)\Pr\left\{Y_{t+1}^\tau | S_t = m\right\} && \text{\%Once } S_t = m \text{ is known, } Y_1^t \text{ is not} \\
&&& \text{relevant} \\
&= \alpha_t(m)\,\beta_t(m) && \text{\% } \beta_t(m) \text{ definition}
\end{aligned}
$$

QED.

$$
\begin{aligned}
\sigma_t(m',m) &= \Pr\left\{S_{t-1} = m', S_t = m, Y_1^\tau\right\} && \text{\% definition} \\
&= \Pr\left\{S_{t-1} = m', S_t = m, Y_1^{t-1}, Y_t, Y_{t+1}^\tau\right\} && \text{\% P(AB)=P(A)P(B|A) A =} \\
&&& S_{t\text{-}1} = m', \, Y_1^{t-1} \\
&= \Pr\left\{S_{t-1} = m', Y_1^{t-1}\right\} && \text{\% Pr(AB|C)=Pr(A|BC) Pr} \\
&\quad \Pr\left\{S_t = m, Y_t, Y_{t+1}^\tau | S_{t-1} = m', Y_1^{t-1}\right\} && \text{(B|C)} \\
&= \alpha_{t-1}(m')\Pr\left\{S_t = m, Y_t | S_{t-1} = m', Y_1^{t-1}\right\} && \text{\% Once } S_{t\text{-}1} = m \text{ is known,} \\
&\quad \Pr\left\{Y_{t+1}^\tau | S_{t-1} = m', Y_1^{t-1}, S_t = m, Y_t\right\} && Y_1^{t-1} \text{ is not relevant} \\
&&& \text{\% Once } S_t = m \text{ is known,} \\
&&& S_{t\text{-}1} = m', Y_t, Y_1^{t-1} \text{ are not} \\
&&& \text{relevant} \\
&= \alpha_{t-1}(m')\Pr\left\{S_t = m, Y_t | S_{t-1} = m'\right\}\Pr\left\{Y_{t+1}^\tau | S_t = m\right\} \\
&= \alpha_{t-1}(m')\gamma_t(m',m)\beta_t(m)
\end{aligned}
$$

QED

<u>Iterative computation</u> of $\alpha_t(m')$, $\beta_t(m)$, $\gamma_t(m',m)$

They are given by

$$
\alpha_t(m) = \sum_{m'}\alpha_{t-1}(m')\gamma_t(m',m) \tag{6-45}
$$

$$
\beta_t(m) = \sum_{m'}\beta_{t+1}(m')\gamma_{t+1}(m',m) \tag{6-46}
$$

$$
\begin{aligned}
\gamma_t(m',m) &= \Pr\{S_t = m, \ Y_t | S_{t-1} = m'\} && \text{\%definition, } \Pr(AB|C) \\
&= \Pr\{S_t = m | S_{t-1} = m'\}\Pr\{Y_t | S_t = m, S_{t-1} = m'\} \\
&&& \text{\%Pr(AB|C) = Pr(B|C)Pr(A|BC)} \\
&= p_t(m|m')\ \Pr\{Y_t | S_t = m, S_{t-1} = m'\} \\
&= p_t(m|m')\sum_X \Pr\{Y_t | X, S_t = m, S_{t-1} = m'\}\Pr\{X | S_t = m, S_{t-1} = m'\} \\
&&& \text{\%apply the identity } \Pr(B|C) = \sum_j \Pr(B|A_jC)\Pr(A_j|C)
\end{aligned}
\tag{6-47}
$$

We explain (6-47), first, term by term and then the derivation of (6-45) and (6-46) follows.

$p_t(m|m') = \frac{1}{2}$ in Figure 6-34 example with equally likely information bit.

$\Pr\{Y_t | X, S_t = m, S_{t-1} = m'\}$ is a branch metric of a received digit Y_t at time t and is expressed in terms of probability, rather than CM or Hamming distance. For BSC (binary symmetric channel) it is represented by a cross over probability, p. For branch $\boxed{1}$, m'=0 $\rightarrow$ m=0, and Y_t=00, Pr $\{$ Y_t $|X$=00, S_t=0, S_{t-1}=0$\}$=$(1\text{-}p)(1\text{-}p)$, Pr $\{$ Y_t $|X$=01, S_t=0, S_{t-1}=0$\}$=$(1\text{-}p)p$, Pr $\{$ Y_t $|X$=10,S_t=0, S_{t-1}=0$\}$=$p(1\text{-}p)$, and Pr $\{$ Y_t $|$ X=11,S_t=0, S_{t-1}=0$\}$=$p\,p$. DMC and other channel models will be explained further in the next section.

$\Pr\{X | S_t = m, S_{t-1} = m'\}$ represents the branch connection of $X = \{00, 01, 10, 11\}$. For branch $\boxed{1}$, m'=0 $\rightarrow$ m=0, Pr $\{X$=00$|$ S_t=0, S_{t-1}=0$\}$=1, Pr $\{X$=01$|$ S_t=0, S_{t-1}=0$\}$=0, Pr $\{X$=10$|$ S_t=0, S_{t-1}=0$\}$=0, and Pr $\{X$=11$|$ S_t=0, S_{t-1}=0$\}$=0.

Exercise 6-40: Find $p_t(m|m')$ and Pr $\{X | S_t = m, S_{t-1} = m'\}$ for branch $\boxed{2}$.
Answer: $p_t(m|m') = 1/2$, Pr $\{X$=00$|$ S_t=1, S_{t-1}=0$\}$=0, Pr $\{X$=01$|$ S_t=1, S_{t-1}=0$\}$=0, Pr $\{X$=10$|$ S_t=1, S_{t-1}=0$\}$=0, and Pr $\{X$=11$|$ S_t=1, S_{t-1}=0$\}$=1.

We now show the validity of (6-45) and (6-46).

Derivation of (6-45) for $\alpha_t(m)$, and (6-46) for $\beta_t(m)$
We now derive (6-45) and (6-46) of iterative computation of forward state probability $\alpha_t(m)$ and backward state probability $\beta_t(m)$.

$$
\begin{aligned}
\alpha_t(m) &= \Pr\{S_t = m, Y_1^t\} & &\%\text{definition}\\
&= \Pr\{S_t = m, Y_1^{t-1}, Y_t\} & &\%\text{divide into two parts}\\
&= \sum_{m'}\Pr\{S_{t-1} = m', Y_1^{t-1}, S_t = m, Y_t\} & &\%P(AB) = P(A)P(B|A)\\
&= \sum_{m'}\Pr\{S_{t-1} = m', Y_1^{t-1}\}\Pr\{S_t = m, Y_t|S_{t-1} = m', Y_1^{t-1}\}
\end{aligned}
$$

$\%$Once $S_{t-1} = m'$ is known, is not relevant Y_1^{t-1}

$$
\begin{aligned}
&= \sum_{m'}\Pr\{S_{t-1} = m', Y_1^{t-1}\}\Pr\{S_t = m, Y_t|S_{t-1} = m'\} & &\%\text{definitions}\\
&= \sum_{m'}\alpha_{t-1}(m')\gamma_t(m', m) & &\text{QED}
\end{aligned}
$$

$$
\begin{aligned}
\beta_t(m) &= \Pr\{Y_{t+1}^{\tau}|S_t = m\} & &\%\text{definition}\\
&= \Pr\{Y_{t+2}^{\tau}, Y_{t+1}|S_t = m\} & &\%\text{divide into two parts}\\
&= \sum_{m'}\Pr\{Y_{t+2}^{\tau}, Y_{t+1}, S_{t+1} = m'|S_t = m\}\\
&= \sum_{m'}\Pr\{Y_{t+2}^{\tau}|\,Y_{t+1}, S_{t+1} = m', S_t = m\}\Pr\{Y_{t+1}, S_{t+1} = m'|S_t = m\}\\
&= \sum_{m'}\Pr\{Y_{t+2}^{\tau}|S_{t+1} = m'\}\gamma_{t+1}(m, m')\\
&= \sum_{m'}\beta_{t+1}(m')\gamma_{t+1}(m, m') & &\text{QED}
\end{aligned}
$$

(6-45), (6-46), and (6-47) are computed for all states of $m = \{0, 1, 2, \ldots M\text{-}1\}$ at each time index, and for entire time index $t=\{1, 2, 3, \ldots \tau\}$. A major portion of decoding computation is on this forward – backward iterative computation. Thus it is critical to understand this iterative computation thoroughly.

6.3.3.3   Computation of APP of Information and Transmit Digits

The APP of the information bit, i_t, and transmission digit, X_t, can be computed from the APP of states $\lambda_t(m)$ and that of branches $\sigma_t(m', m)$.

We need to define a set of states and branches. A_t is a set of states such that $i_t=0$; $A_t=\{S_t=00, S_t=01\}$ in Figure 6-34. B_t is a set of branches such that the first digit of X, $X_t^{(1)}=0$; $B_t=\{\boxed{1}, \boxed{4}, \boxed{6}, \boxed{7}\}$ in Figure 6-34.

Exercise 6-41: For Figure 6-34 find a set of states such that $i_t=1$. Answer $=$ $\{S_t=10, S_t=11\}$. Find a set of branches such that the second digit of X, $X_t^{(2)}=0$. Answer $= \{\boxed{1}, \boxed{3}, \boxed{6}, \boxed{8}\}$. See the summary below.

Set of states A_t such that $i_t=0$	$S_{i0} = \{00,01\}$
Set of states A_t such that $i_t=1$	$S_{i1} = \{10,11\}$
Set of branches B_t such that $X^{(1)}=0$	$\boxed{1}$ $\boxed{4}$ $\boxed{6}$ $\boxed{7}$
Set of branches B_t such that $X^{(2)}=0$	$\boxed{1}$ $\boxed{3}$ $\boxed{6}$ $\boxed{8}$
Set of branches C_t such that $X=00$	$\boxed{1}$ $\boxed{6}$

The joint probability of the information, $i_t=0$ for a received digit Y_1^τ, is given by

$$\Pr\{ i_t = 0, Y_1^\tau \} = \sum_{m \in S_{i0}} \lambda_t(m)$$

And the conditional probability is obtained by normalizing by $\Pr\{Y_1^\tau\}$ as

$$\Pr\{ i_t = 0|Y_1^\tau \} = \frac{1}{\Pr\{Y_1^\tau\}} \sum_{m \in S_{i0}} \lambda_t(m) \qquad (6\text{-}48)$$

We decode $i_t=0$ when $\Pr\{ i_t = 0|Y_1^\tau \} \geq 0.5$ otherwise $i_t=1$.
Note that $\Pr\{ i_t = 1|Y_1^\tau \} = 1 - \Pr\{ i_t = 0|Y_1^\tau \}$.
The joint probability of the transmit digit, first digit at t, is similarly given by

$$\Pr\left\{ X_t^{(1)} = 0, Y_1^\tau \right\} = \sum_{(m', m') \in B_t} \sigma_t(m', m)$$

which can be normalized to give the conditional probability $\Pr\{X_t^{(1)} = 0|Y_1^\tau\}$.

We can obtain the probability of any event that is a function of the states by summing the appropriate $\lambda_t(m)$; likewise, the $\sigma_t(m', m)$ can be used to obtain the probability of any event which is a function of branches.

6.3.3.4 Channel Model and Branch Metric

As shown in (6-47) the branch metric, $\gamma_t(m', m)$, is expressed in 3 factors

$$\gamma_t(m', m) = \Pr\{S_t = m| S_{t-1} = m'\} \sum_X \Pr\{Y_t|X, S_t = m, S_{t-1} = m'\} \Pr\{X| S_t = m,$$

$S_{t-1} = m'\}$; the first term, $\Pr\{S_t = m| S_{t-1} = m'\}$, is a probability of state transition, typically ½ with equally likely in binary input and the third term, $\Pr\{X| S_t = m, S_{t-1} = m'\}$, is 0 or 1 probability of the alphabet X. If an alphabet happens in the branch then it is 1.0, if not 0. In other words we consider the only alphabet in the branch, typically only one unless there are parallel transitions. With no parallel transition and a single random binary input the branch metric is simplified as

$$\gamma_t(m', m) = \frac{1}{2} \sum_{X \in (m,m')} \Pr(Y_t |X, S_t = m, S_{t-1} = m')$$
$$= \frac{1}{2} \Pr(Y_t |X = X_t, S_t = m, S_{t-1} = m')$$

(6-49)

For example (0,0) branch $\boxed{1}$, X=00, and (0,1) branch $\boxed{2}$, X=11.

We now consider $\Pr(Y_t |X = X_t, S_t = m, S_{t-1} = m')$. It means that, when X_t is transmitted, the probability of $Y_t = r$, received value which is not explicitly shown in the expression. For BSC, $\Pr(Y_t^{(1)}=0 | X_t^{(1)}=0) = 1-p$ and $\Pr(Y_t^{(1)}=1 | X_t^{(1)}=0) = p$.

Exercise 6-42: Find the probability of $\Pr(Y_t^{(1)} Y_t^{(2)}=00 | X_t^{(1)} X_t^{(2)}=00)$. It is $(1-p)$ ^2. Find the probability of $\Pr(Y_t^{(1)} Y_t^{(2)}=01 | X_t^{(1)} X_t^{(2)}=00)$. Answer is $(1-p)p$.

We now consider more channel models and specifically relate them with AWGN channel. In particular we look at Figure 6-1 and the RHS of B1-$\overline{B1}$, from coding and decoding sub-blocks, can be considered as a 'channel', say $\overline{AWGN}$ channel. In coding context BSC and DMC (discrete memoryless channel) are often used. Here we also include one more channel model - AWGN channel specified by conditional probabilities. All three models are considered as a translation of the received value r, to a conditional probability.

BSC may be obtained from BPSK with AWGN channel and its transition probability, p, is computed as shown in Table 6-15. Note that p is a function of SNR.

DMC with 4 –level quantization is shown in the table and the transition probabilities may be obtained as shown in Table 6-15.

The received signal r of AWGN channel is converted to the conditional probability p_r, which in turn is obtained as shown in the figure of Table 6-15, from PDF of AWGN for a transmit symbol, +1, $p_r(r|+1)=p_0$ for data '0', and similarly of a symbol,-1, $p_r(r|-1) =p_1$ for data '1'.

Table 6-15: Channel model of BSC, DMC and AWGN in coding context

	Graph representation		Computation of the probability p

BSC

Graph representation:

in: 0 —1-p— 0 out, 1 —1-p— 1, with cross paths p, p

$\Pr\{r \mid -1\}$ $\Pr\{r \mid +1\}$

−1, +1, 0 ; 1−p, p: cross over probability, 1−p

Computation of the probability p:

$$p = Q\left(\tfrac{1}{\sigma}\right)$$
$$\frac{1}{\sigma} = \sqrt{\frac{2Es}{No}}$$

Note that the cross-over probability depends on SNR.

DMC

Graph representation:

in: 0, 1 → out: A, B, C, D with transitions p_A, p_B, p_C, p_D

$\Pr\{r \mid -1\}$ $\Pr\{r \mid +1\}$

−1, +1, 0

$p_D \rightarrow p_C \; p_B \; p_A$

Computation of the probability p:

$$p_A = 1 - Q\left(\frac{1/2}{\sigma}\right)$$
$$p_B = Q\left(\frac{1/2}{\sigma}\right) - Q\left(\frac{3/2}{\sigma}\right)$$
$$p_C = Q\left(\frac{1}{\sigma}\right) - Q\left(\frac{3/2}{\sigma}\right)$$
$$p_D = Q\left(\frac{3/2}{\sigma}\right)$$
$$\frac{1}{\sigma} = \sqrt{\frac{2Es}{No}}$$

AWGN

p_0(black), p_1(blue) graphs

conditional probability distribution of AWGN ; probability ; received r

$\Pr\{r \mid -1\}$ $\Pr\{r \mid +1\}$

−1, +1, 0 ; 0.8, 0.2, 0.8, 0.5, 0.2

Computation of the probability p_0:

$$p_0 = 1/\left(1 + e^{-2r/\sigma^2}\right)$$
$$p_1 = 1/\left(1 + e^{+2r/\sigma^2}\right)$$
$$\frac{1}{\sigma} = \sqrt{\frac{2Es}{No}}$$

Note that the probability depends on SNR.

$$p_0 = \frac{\Pr(r|+1)}{\Pr(r|+1) + \Pr(r|-1)} \text{ and } p_1 = \frac{\Pr(r|-1)}{\Pr(r|+1) + \Pr(r|-1)} \tag{6-50}$$

Exercise 6-43: Use (6-50) and show $p_0 = 1/\left(1 + e^{-2r/\sigma^2}\right)$ with AWGN, and find the expression for $1 - p_0 = p_1 = 1/\left(1 + e^{+2r/\sigma^2}\right)$.

Answer: remember that $\Pr(r|+1) = \frac{1}{\sqrt{2\pi\sigma^2}} \exp\left(-\frac{(r-1)^2}{2\sigma^2}\right)$ and $\Pr(r|-1) = \frac{1}{\sqrt{2\pi\sigma^2}} \exp\left(-\frac{(r+1)^2}{2\sigma^2}\right)$. These are plugged into (6-50) to obtain the expression of p_r.

See additional elaboration in Appendix A.2 if necessary.

6.3.3.5 Numerical Example of BCJR Decoding of Figure 6-34

We show the iterative computation of (6-45), (6-46) and (6-47) with numerical example using Figure 6-34, which is the same as Figure 6-31 G=7|5 convolutional code. We assume BSC channel with the transition probability p=0.2. And the initial conditions are $\alpha_0(0) = 1.0$, and zero for all other states with t=1,…,9 and τ=9. Similarly for $\beta_9(0) = 1.0$, and zero for all other states. There are 4 states and 8 branches. Received bits are {10, 10, 10, 10, 11, 01, 01, 11, 00} and red bits in error. See Tables 6-16, 6-17, and 6-18.

Now using (6-43) we obtain $\lambda_t(m)$, not scaled APP of states, and using (6-44) we obtain $\sigma_t(m', m)$, not scaled APP of branches. APP with proper scale can be obtained by normalizing with the sum (0.00079) to make a total APP to be unity.

Exercise 6-44: There is a practical issue of representing $\alpha_t(m)$ and $\beta_t(m)$ with finite precision. With more iteration (t increases for $\alpha_t(m)$ while t decreases for $\beta_t(m)$), the number representing them becomes smaller. One solution is to use a constant scaling at each time of iteration. The scaling may be a power of 2, which can be implemented by shifting positions. Alternatively each time index t, $\alpha_t(m)$ and $\beta_t(m)$ are normalized through all states.

Note that the normalizing factor $\Pr\{Y_1^\tau\}$ in (6-48) is 0.00079 as shown in Table 6-19. The conditional probability of (6-48) is obtained by normalizing, i.e., the sum of $\lambda_t(m')$ for all m' must be unity.

Using $\lambda_t(m)$ and (6-48), the state probability, the first digit of state, i.e., the information bit, may be obtained. It is shown in Table 6-20.

By summing appropriate $\sigma_t(m', m)$, relevant to a branch (i.e., a set B_t in Section 6.3.3.3), APP of transmit symbols, X_t, may be obtained, and is shown in Table 6-21. The probability is normalized, and a decision may be made by choosing the largest.

Table 6-16: Forward iteration $\alpha_t(m)$ with $t=1, \ldots, 9$ and $\alpha_0(0) = 1.0$ BSC **channel** $p=0.2$

	t= 1			t= 2			t= 3	
$\alpha_0(m')$	$\gamma_1(m',m)$	prod	$\alpha_1(m')$	$\gamma_2(m',m)$	prod	$\alpha_2(m')$	$\gamma_3(m',m)$	prod
1.0	0.16	0.16	0.16	0.16	0.03	0.0256	0.16	0
	0.16	0.16		0.16	0.03		0.16	0
			0.16	0.64	0.1	0.0256	0.64	0.02
				0.04	0.01		0.04	0
						0.1024	0.16	0.02
							0.16	0.02
						0.0064	0.04	0
							0.64	0

	t= 4			t= 5			t= 6	
$\alpha_3(m')$	$\gamma_4(m',m)$	prod	$\alpha_4(m')$	$\gamma_5(m',m)$	prod	$\alpha_5(m')$	$\gamma_6(m',m)$	prod
0.0205	0.16	0	0.0059	0.04	0	0.00876	0.16	0
	0.16	0		0.64	0		0.16	0
0.0205	0.64	0.01	0.0059	0.16	0	0.00433	0.04	0
	0.04	0		0.16	0		0.64	0
0.0166	0.16	0	0.0133	0.64	0.01	0.00161	0.16	0
	0.16	0		0.04	0		0.16	0
0.0051	0.04	0	0.0041	0.16	0	0.00161	0.64	0
	0.64	0		0.16	0		0.04	0

	t= 7			t= 8			t= 9		
$\alpha_6(m')$	$\gamma_7(m',m)$	prod	$\alpha_7(m')$	$\gamma_8(m',m)$	prod	$\alpha_8(m')$	$\gamma_t(m',m)$	prod	$\alpha_9()$
0.0017	0.16	0	0.00046	0.04	0	0.0012	0.64	0.00078	0.00079
	0.16	0							
0.0017	0.04	0	0.00046	0.16	0	0			
	0.64	0							
0.0012	0.16	0	0.00188	0.64	0	0.0003	0.04	1E-05	
	0.16	0							
0.0028	0.64	0	0.00117	0.16	0	0			
	0.04	0							

6.3.3.6   Summary of BCJR Algorithm

A summary of BCJR algorithm of computing APP of info bits and of transmit bits is summarized in Table 6-22.

Once $\alpha_t(m)$ and $\beta_t(m)$ for all time $t=1,\ldots,\tau$, and for all states m $=0, 1, \ldots$ M-1 are available $\sigma_t(m',m) = \alpha_{t-1}(m')\, \gamma_t(m',m)\, \beta_t(m)$ can be computed for all trellis branch.

Table 6-17: Backward iteration $\beta_t(m)$ with $t=0, 1, \ldots, 8$ and $\beta_9(0)$

0	t= 1			t= 2			t= 3	
$\beta_0(0)$	prod	$\gamma_1()$	$\beta_1(m)$	prod	$\gamma_2()$	$\beta_2(m)$	prod	$\gamma_3()$
0.00079	0.0004	0.16	0.00244	0.0006	0.16	0.0038	0.0028	0.16
	0.0004	0.16		0.0018	0.16		0.001	0.16
			0.00252	0.0024	0.64	0.0114	0.0113	0.64
				8E-05	0.04		9E-05	0.04
						0.0038	0.0028	0.16
							0.001	0.16
						0.0021	0.0007	0.04
							0.0014	0.64

t= 4			t= 5			t= 6	
$\beta_3(m)$	prod	$\gamma_4()$	$\beta_4(m)$	prod	$\gamma_5()$	$\beta_5(m)$	prod	$\gamma_6()$
0.0177	0.0172	0.16	0.1078	0.0002	0.04	0.0041	0.0008	0.16
	0.0005	0.16		0.1076	0.64		0.0033	0.16
0.0061	0.006	0.64	0.0029	0.0007	0.16	0.16814	0.0002	0.04
	0.0001	0.04		0.0022	0.16		0.1679	0.64
0.0177	0.0172	0.16	0.0093	0.0026	0.64	0.0041	0.0008	0.16
	0.0005	0.16		0.0067	0.04		0.0033	0.16
0.0022	0.0004	0.04	0.0029	0.0007	0.16	0.01377	0.0033	0.64
	0.0018	0.64		0.0022	0.16		0.0105	0.04

t= 7			t= 8			t= 9		
$\beta_6(m)$	prod	$\gamma_7()$	$\beta_7(m)$	prod	$\gamma_8()$	$\beta_8(m)$	prod	$\gamma_9()$	$\beta_9(0)$
0.0051	0.0041	0.16	0.0256	0.0256	0.04	0.64	0.64	0.64	1
	0.001	0.16		0	0.64				
0.0205	0.0164	0.04	0.0064	0.0064	0.16	0			
	0.0041	0.64		0	0.16				
0.0051	0.0041	0.16	0.4096	0.4096	0.64	0.04	0.04	0.04	
	0.001	0.16		0	0.04				
0.2624	0.2621	0.64	0.0064	0.0064	0.16	0			
	0.0003	0.04		0	0.16				

Table 6-18: State probability $\lambda_t(m)$ state

	$\lambda_0()$	$\lambda_1()$	$\lambda_2()$	$\lambda_3()$	$\lambda_4()$
0	0.00079	0.00039	9.7E-05	0.00036	0.00064
1		0.0004	0.00029	0.00012	1.7E-05
2			0.00039	0.00029	0.00012
3			1.4E-05	1.1E-05	1.2E-05
sum	0.00079	0.00079	0.00079	0.00079	0.00079
	$\lambda_5()$	$\lambda_6()$	$\lambda_7()$	$\lambda_8()$	$\lambda_9()$
0	3.6E-05	8.5E-06	1.2E-05	0.00078	0.00079
1	0.00073	3.4E-05	2.9E-06	0	
2	6.6E-06	6.1E-06	0.00077	1E-05	
3	2.2E-05	0.00074	7.5E-06	0	
sum	0.00079	0.00079	0.00079	0.00079	0.00079

Table 6-19: Branch probability $\gamma_k(m', m)$

branch	$\sigma_1()$	$\sigma_2()$	$\sigma_3()$	$\sigma_4()$	$\sigma_5()$	$\sigma_6()$	$\sigma_7()$	$\sigma_8()$	$\sigma_9()$
1	0.00039	9.7E-05	7.3E-05	0.00035	9.7E-07	7.2E-06	6.8E-06	1.2E-05	0.00078
2	0.0004	0.00029	2.5E-05	9.4E-06	0.00064	2.9E-05	1.7E-06	0	
3		0.00039	0.00029	0.00012	3.9E-06	8.9E-07	2.7E-05	2.9E-06	
4		1.4E-05	2.3E-06	2.3E-06	1.3E-05	6.8E-06	6.8E-06	0	
5			0.00029	0.00029	3.5E-05	1.3E-06	4.9E-06	0.00077	1E-05
6			1E-04	7.6E-06	9E-05	5.3E-06	1.2E-06	0	
7			4.5E-06	1.9E-06	2.7E-06	5.3E-06	0.00074	7.5E-06	
8			9E-06	9.4E-06	9E-06	1.7E-05	7.3E-07	0	
sum	0.00079	0.00079	0.00079	0.00079	0.00079	0.00079	0.00079	0.00079	0.00079

6.3.4 Other Topics Related with Convolutional Codes

We showed so far that once a code is represented by a trellis, both Viterbi and BCJR decoding algorithms can be applied. For convolution codes, it is natural to represent them by a trellis. Here we demonstrate that block codes may be represented by a trellis even though its trellis structure is not regular unlike convolution codes. In this way Viterbi and BCJR decoding can be applied to block codes at least in principle. This idea of trellis representation of block codes was part of BCJR's original paper, and this trellis representation is another connection between block codes and convolutional codes.

6.3.4.1 Trellis Representation of Block Codes

Any linear block codes may be represented by a generator matrix, G, or a parity check matrix, H. Here we use parity check matrix to demonstrate a trellis representation of a block code.

Example 6-13: Hamming (7, 4) code with generator matrix G and corresponding parity check matrix H. $\mathbf{G} = \begin{bmatrix} 1000111 \\ 0100011 \\ 0010101 \\ 0001110 \end{bmatrix}$ and they are related as $\mathbf{G}\,\mathbf{H}^{\mathrm{T}} = 0$

Here we are using H matrix for constructing a trellis. Since there are $n\text{-}k$ parity check bits, we need $2^{(n-k)}$ states including the '000' state.

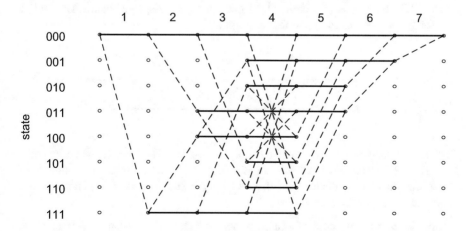

Table 6-20: The probability of state, first digit and shading the larger

	t=1	2	3	4	5	6	7	8	9
s(1)=i='0'	0.49136	0.61444	0.82837	0.96383	0.05351	0.01845	0.98683	1	1
s(1)=i='1'	0.50864	0.38556	0.17163	0.03617	0.94649	0.98155	0.01317	0	0

The figure above, a trellis diagram of Hamming (7, 4) using H matrix specified in the example, is explained. For t=1, 2, 3, 4 the information bits are 1 or 0 but t=5, 6, 7 parity bits are determined by the code. Thus there are 16 paths (2^k). Dotted line is for the output bit being 1 and solid line for being 0.

$$\mathbf{H} = \begin{bmatrix} 1011100 \\ 1101010 \\ 1110001 \end{bmatrix} = [\mathbf{h}_1\mathbf{h}_2 \ldots \mathbf{h}_7] \text{ and } \mathbf{h}_1 \text{ is the first column vector } \begin{bmatrix} 1 \\ 1 \\ 1 \end{bmatrix}.$$

With t=1, 2,...,n (n=7) and S_0=000, c_t being a bit at t, a trellis is constructed by a state transition described as in (6-51) below

$$S_t = S_{t-1} + c_t\mathbf{h}_t \tag{6-51}$$

For t=1, 2, 3, 4 (in general up to k information bits), the information bits are 1 or 0 but for t=5, 6, 7, parity bits (in general n-k bits) are determined by the code. Thus there are 16 paths (in general 2^k).

Exercise 6-45: For (n, n-1) code, one bit parity being added, construct a trellis diagram for it.

Answer: since H =[1 1 1 ...1], there are two states including 0. We designate them 0 and 1.To be concrete the parity check code of n=7 is displayed below. Dotted line is for '1' output and solid line for '0' output.

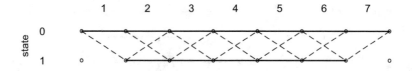

The above diagram is obtained by applying (6-51) to H of a single bit parity code.

The trellis construction based in **H** needs $2^{(n-k)}$ states. When the number of information bits is smaller than that of parity check bits, **G**, a generator matrix instead of **H**, may be used.

Exercise 6-46: Dual code of a single parity bit code is a repetition code. In this case we may use **G** for trellis construction since G = [1 1 ...1].

Table 6-21: The probability of transmit symbols and shading the largest

	t=1	2	3	4	5	6	7	8	9
X=00	0.49136	0.12289	0.21733	0.45482	0.1141	0.01568	0.01011	0.01476	0.98683
X=11	0.50864	0.36847	0.39711	0.37355	0.84973	0.03783	0.00834	0.97207	0.01317
X=10	0	0.49155	0.377	0.16626	0.01629	0.02237	0.03516	0.00369	0
X=01	0	0.01709	0.00856	0.00537	0.01988	0.92412	0.94638	0.00948	0

Table 6-22: Summary of BCJR algorithm

	Computation summary	remarks
APP of info bits	$\Pr\{ i_t=0 \mid Y_1^\tau \} = \frac{1}{\Pr\{Y_1^\tau\}} \sum_{m \, \in \, s_{i0}} \lambda_t(m)$ $\Pr\{ i_t=1 \mid Y_1^\tau \} = \frac{1}{\Pr\{Y_1^\tau\}} \sum_{m \, \in \, s_{i1}} \lambda_t(m)$	Y_1^τ: entire block of received values S_{i0}: a set of states such that i_t $='0'$ S_{i1}: a set of states such that i_t $='1'$
State probability given Y_1^τ received	$\lambda_t(m) = \Pr\{S_t = m, \; Y_1^\tau\}$ $= \alpha_t(m) \; \beta_t(m)$	State probability after forward and backward iteration done
APP of transmit bits	$\Pr\{X_t^{(1)}=0, \; Y_1^\tau\} = \sum_{(m,m') \in B0t} \sigma_t(m', m)$ $\Pr\{X_t^{(1)}=1, \; Y_1^\tau\} = \sum_{(m,m') \in B1t} \sigma_t(m', m)$ $\Pr\{X_t^{(1)}=0 \mid Y_1^\tau\}$, $\Pr\{X_t^{(1)}=1 \mid Y_1^\tau\}$ can be obtained by scaling of $\frac{1}{\Pr\{Y_1^\tau\}}$	B_{0t}: a set of branches such that $X_t^{(1)} = '0'$ B_{1t}: a set of branches such that $X_t^{(1)} = '1'$
Branch probability given Y_1^τ	$\sigma_t(m', m) = \alpha_{t-1}(m') \gamma_t(m', m) \; \beta_t(m)$	$\sigma_t(m', m) = \Pr\{S_{t-1}=m', S_t=m, Y_1^\tau\}$: definition
Forward and backward state probability at t	$\alpha_t(m) = \sum_{m'} \alpha_{t-1}(m') \; \gamma_t(m', m)$ $\beta_t(m) = \sum_{m'} \beta_{t+1}(m') \; \gamma_{t+1}(m', m)$	$\alpha_t(m) = \Pr\{S_t = m, Y_1^t\}$ $\beta_t(m) = \Pr\{Y_{t+1}^\tau \mid S_t = m\}$
	$\alpha_0(m) = \beta_\tau(m) = 1.0$ if $m = 0$ $\alpha_0(m) = \beta_\tau(m) = 0.0$ if $m \neq 0$	Initialize: start from zero state and end at zero state
Trellis branch metric probability with Y_t	$\gamma_t(m', m) = \Pr\{ Y_t \mid S_t=m, S_{t-1}=m'\} \Pr \{S_t=m \mid S_{t-1}=m'\}$	$\gamma_t(m', m) = \Pr\{S_t = m, Y_t \mid S_{t-1} = m'\}$: definition
	$\gamma_t(m', m) = \sum_X \Pr\{Y_t \mid X, S_t = m, S_{t-1} = m'\}$ $\Pr\{X \mid S_t = m, S_{t-1} = m'\} \Pr\{ St = m \mid St\text{-}1 = m'\}$	$\Pr\{ S_t=m \mid S_{t-1}=m'\} = 1/2$ $\Pr\{X \mid S_t = m, S_{t-1} = m'\}$ $= 0$ or 1: connected or not $\Pr\{Y_t \mid X, S_t = m, S_{t-1} = m'\}$: connected branch probability (metric)

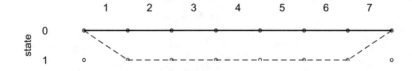

Thus the representation of block codes by a trellis requires more states and the resulting trellis is not regular. Thus, in practice, it is not often used. Instead a convolutional code may be used.

6.3.4.2   Other Decoding Techniques of Convolutional Codes

Historically before the advent of Viterbi decoding and BCJR, other decoding techniques were invented and explored, in late 50s to through 60s; they are sequential decoding, majority logic decoding, and threshold decoding. They are useful in certain situations, e.g., sequential decoding may be used when the memory of a code, m, is large, say $m > 10$. $m+1$ is called constraint length of a code.

A major motivation of developing such decoding methods is to reduce the complexity of implementation; the decoding techniques should be implementable with available technologies at the time of decoding technique development. For example, in [6], a special form of convolutional code, called canonical self orthogonalizing convolutional code which is decodable by majority logic, was constructed for this. It was used in order to meet the signaling speed requirement of the radio. This is much simpler than MLSD at the expense of small performance loss, and thus suboptimum. For additional details of these suboptimal decoding techniques, see the reference [11].

6.4 LDPC

LDPC stands for low density parity check code, and a forward error correcting code (FEC) can be specified by a parity check matrix (H). LDPC is defined by H but to be LDPC it must be low density, a small number of ones. After a gentle introduction to LDPC with an example, we cover its decoding by message passing algorithm, and encoding. Then we introduce some rules and heuristics for LDPC code construction followed by surveying some practical LDPC codes used in industry along with its error rate performance.

6.4.1 Introduction to LDPC Code

A parity check matrix is defined as a null space for codewords, i.e.,

$$Hc^{T} = [0]$$

with every codeword, e.g., $c = [c_1 \, c_2 \, c_3 \, c_4 \, c_5 \, c_6]$ and ^{T} means transpose.

For example, $c = [1 \, 1 \, 1 \, 1 \, 0 \, 0]$ and with H below,

$$H = \begin{bmatrix} 1 & 0 & 0 & 1 & 1 & 0 \\ 1 & 1 & 0 & 0 & 0 & 1 \\ 0 & 1 & 1 & 0 & 1 & 0 \\ 0 & 0 & 1 & 1 & 0 & 1 \end{bmatrix} \begin{bmatrix} 1 \\ 1 \\ 1 \\ 1 \\ 0 \\ 0 \end{bmatrix} = \begin{bmatrix} 0 \\ 0 \\ 0 \\ 0 \end{bmatrix}.$$

Multiplication and additions are done in GF (2); modulo 2 additions $1+1=0$, $1+0=1$, etc. For example, for the first row, $[1\ 0\ 0\ 1\ 1\ 0] \times [1\ 1\ 1\ 1\ 0\ 0]^T = 1 + 0 + 0 + 1 + 0 + 0 = 0$.

We emphasize the importance of low density, i.e., small number of ones in H compared to total number of entries ((no of rows) × (no. of columns) of H) and thus the name. The complexity of decoder is proportional to the density and the code performance depends on it as well. With this simple example it is not really 'low' density but as the block size becomes large the density can be made very low (say much less than 1%).

With the above parity check matrix, we have 4 parity check equations;

$$c_1 + c_4 + c_5 = 0$$
$$c_1 + c_2 + c_6 = 0$$
$$c_2 + c_3 + c_5 = 0$$
$$c_3 + c_4 + c_6 = 0$$

where + means modulo 2 addition i.e., in GF(2).

Exercise 6-47: Solve $c_4\ c_5\ c_6$ in terms of $c_1\ c_2\ c_3$. This can be used for encoding when c_1, c_2, c_3 are information bits. Answer: $c_4 = c_1 + c_2 + c_3$, $c_5 = c_2 + c_3$, $c_6 = c_1 + c_2$.

Note that the block size of this code is $n=6$, and information bits are $k=3$, and thus (n, k) = (6, 3) code. Parity bits are (n – k) =3 even though there are 4 parity check equations. In LDPC, parity check matrix may have more than (n-k) equations.

These, derived and equivalent, parity check equations, $c4 = c_1 + c_2 + c_3$, $c_5 = c_2 + c_3$, $c_6 = c_1 + c_2$, can be put into matrix form as

$$H_r = \begin{bmatrix} 1 & 1 & 1 & 1 & 0 & 0 \\ 0 & 1 & 1 & 0 & 1 & 0 \\ 1 & 1 & 0 & 0 & 0 & 1 \end{bmatrix}$$

With this form of parity check matrix, encoding is straightforward; given information bits one can obtain parity bits to find a code word. When a parity check matrix is not in this form one may convert to this form by using row operations – Gaussian elimination. This equivalent parity check matrix, H_r, is useful for encoding but may no longer 'low density' in general. This example is so small and in fact a number of ones are reduced. For decoding, keeping low density form of parity check matrix is essential. In the design of LDPC codes, one tries to construct a low density H matrix.

6.4.1.1 Graphical Representation of LDPC Code

$$H = \begin{bmatrix} 1 & 0 & 0 & 1 & 1 & 0 \\ 1 & 1 & 0 & 0 & 0 & 1 \\ 0 & 1 & 1 & 0 & 1 & 0 \\ 0 & 0 & 1 & 1 & 0 & 1 \end{bmatrix}$$

$$\begin{array}{c} \cdot i \cdot = 1 \quad 2 \quad 3 \quad 4 \quad 5 \quad 6 \\ H = \begin{bmatrix} 1 & 0 & 0 & 1 & 1 & 0 \\ 1 & 1 & 0 & 0 & 0 & 1 \\ 0 & 1 & 1 & 0 & 1 & 0 \\ 0 & 0 & 1 & 1 & 0 & 1 \end{bmatrix} \begin{array}{l} j = 1 \\ 2 \\ 3 \\ 4 \end{array} \end{array}$$

We designate column numbers and row numbers as shown on the right side of the above; the column numbers correspond to codeword bits and the row number is parity check equation number.

This example parity check matrix, H, may be represented by a graph as in Figure 6-35, called Tanner graph, honoring the inventor of such graphs to represent LDPC codes or code in general.

In this H matrix there are 12 ones ('1'), and each column has two entries, which is called column weight $w_c=2$, and each row has three ones, which is called row weight $w_r= 3$. These ones in H correspond to connections in the graph. When these column and row weights are uniform for each column and for each row, then it is called *regular* LDPC. The weights can be *irregular*, i.e., each row and column may have different weights. A number of columns, $n=6$, are a block size of a code, and a number of rows, $m=4$, are the number of parity check equations.

In a regular LDPC, the number of connections is $w_c n = w_r m$. When $m=n - k$, a number of parity bits are equal to the number of parity check equations, $w_c n = w_r(n - k)$.

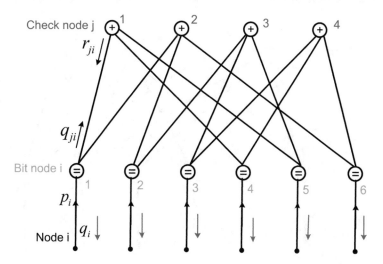

Figure 6-35: Example LDPC code in Tanner graph with check node and bit node numbers designated

Thus $k/n = 1 - w_c/w_r$, which is a code rate. This relationship does not hold with the above example since $m \neq n - k$, i.e., $m=4$, $n - k =3$.

Exercise 6-48*: For a given H with m row and n columns $m \geq n-k$, find its code rate k/n. Hint: we need to find a number of information bits k. This requires finding a reduced and equivalent H_r as we did with the example. The star * means a project exercise.

6.4.1.2   A numerical Example of Tanner Graph as a Decoder

We demonstrate the use of the graphic representation as a decoder. For example, received samples are given by

$$\mathbf{r} = [1.3802 \quad 2.2968 \quad -0.5972 \quad 1.6096 \quad 1.2254 \quad 0.0753]$$

and decoded as **rbit** $=[0\ 0\ 1\ 0\ 0\ 0]$ by using if $r>0$ rbit $=0$, if $r<= 0$, rbit$=1$. This is called hard decision (HD). But we know that it is not a codeword. Why? Check $H\, \mathbf{rbit}^T = [0\ 0\ 1\ 1]^T \neq [\mathbf{0}]$. See Appendix A.1 where we can do optimum decoding by using correlation metric (CM) decoding for this small code.

Here we pursue the idea that by using this graphic representation of a code, i.e. parity check matrix H, as in Figure 6-36, a decoding algorithm may be developed.

The check node 3 and 4 does not satisfy parity check. We notice that it is due to node 3 bit being '1', and by changing to '0' all check nodes are satisfied. Decoding is done with a single error being corrected. For simple case above, this heuristic approach is doable. However, as the block size (a number of bit nodes) gets large,

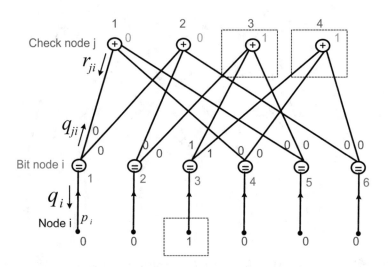

Figure 6-36: Example LDPC code as in Figure 6-35 with binary input

say 100, one needs to be systematic. At a check node, parity equation must be met, and at a bit node, all the branch bits must be equal; these are called constraints.

(1) Can we use $\mathbf{r} = [1.3802\ 2.2968\ -0.5972\ 1.6096\ 1.2254\ 0.0753]$ without HD to bit? The answer is yes. We will elaborate it below. Iterative message-passing (MP) algorithm can be applied using $\mathbf{r}$.

(2) As it turns out that we need to formulate the problem by assigning probability to input bit nodes; given $\mathbf{r}$ we assign probabilities of '1', p_1 and of '0', p_0 with $p_1 + p_0 = 1.0$.

$$p_1 = [0.0040\ \ 0.0001\ \ 0.9160\ \ 0.0016\ \ 0.0074\ \ 0.4253]$$
$$p_0 = [0.9960\ \ 0.9999\ \ 0.0840\ \ 0.9984\ \ 0.9926\ \ 0.5747]$$

We use Table 6-15 and the result of an exercise followed; a probability may be assigned from the received sample value, r, as $p_0 = \frac{1}{1+\exp(-2r/\sigma^2)}$ and $p_1 = 1 - p_0 = \frac{1}{1+\exp(+2r/\sigma^2)}$ assuming AWGN (additive white Gaussian noise). See Appendix A.2 *The computation of p_0 and LLR for BPSK*, for additional details.

Graphically it is shown below. Note that it is a function of noise. We assumed $\sigma = 0.5$ and $\sigma = 1.0$ respectively.

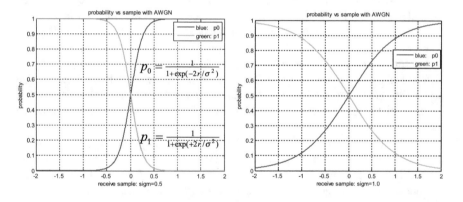

(3) LLR (log likelihood ratio), defined as $\log \frac{p_0}{p_1} = +\Lambda$, is often used in practice. It is related with received samples as $\frac{2r}{\sigma^2} = \Lambda$, i.e., a scaled version of received samples. By using LLR as given here, it is equivalent to (2) where a probability is assigned to each bit.

(4) Yet there is another way assigning a probability to input bit nodes. After hard decision to bit, a probability to be correct is assigned as,

$$\text{rbit} = [0\ 0\ 1\ 0\ 0\ 0]$$
$$\mathbf{p_{1hd}} = [0.079\ \ 0.079\ \ 0.921\ \ 0.079\ \ 0.079\ \ 0.079]$$
$$\mathbf{p_{0hd}} = [0.921\ \ 0.921\ \ 0.079\ \ 0.921\ \ 0.921\ \ 0.921]$$

This is the case that the channel model is binary symmetric channel (BSC) with cross over probability p where this example $p=0.079$. See Table 6-15.

(5) Given **rbit** $=[\,0\ 0\ 1\ 0\ 0\ 0\,]$, at a check node, parity equation must be met, and all the branch bits, at a bit node, must be equal; these are called constraints. This can be done, iteratively, by so called *bit-flipping algorithm*. It is a form of *message-passing (MP) algorithm*.

We will focus on (2) and (3). The computational algorithms, iterative computations of APP under the constraints of check nodes and of bit nodes, will be developed using (2), and then the results will be converted to LLR forms of (3).

Additional comments: all the five different cases above are related. (1) is related with (2); (1) is an approximation of (2) without noise scaling. In (4) AWGN channel of (2) is converted to BSC. (5) is a HD version of (3).

6.4.2   LDPC Decoder

It is typical to discuss encoder first, and then decoder. Here we cover decoding first since LDPC code is described by parity check matrix H, a generic description of code. The decoding scheme should work for any H even though for practical reasons of implementation and performance, additional constraints are imposed on H.

The decoding algorithm can be formulated as the computation of APP given received samples ($\mathbf{r}$). APP is a conditional probability of transmit data conditioned on (or after observing) the received samples. The computation is iterative, and with different names, *message-passing* (MP), *belief propagation* (BP) in general. The probability based formulation is called *sum-product algorithm* (SPA). This is soft decision (SD) decoding and our focus will be SD decoding for better performance than HD.

We develop bit node updating (computing q_{ji} from r_{ji}), check node updating (computing r_{ji} from q_{ji}), and updating (q_i, APP version of input p_i). See Figure 6-35 for naming of branches and nodes. In practice, LLR will be used often for the computational advantage. Thus the useful results will be expressed in LLR.

6.4.3   Bit Node Updating Computation

In order to explain decoding we use the example of Figure 6-35 in the previous section. And then a pattern of generalization is provided.

Bit node updating rule is simply the summation of LLRs of only neighbors; for example, LHS of Figure 6-37, LLR (q_{11}) = LLR (r_{21}) + LLR (p_1), and RHS of it LLR (q_{21}) = LLR (r_{11}) + LLR (p_1). Below we derive this conclusion using the concept of conditional probability.

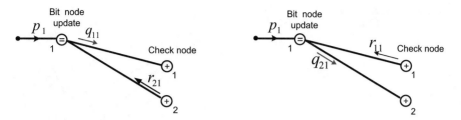

Figure 6-37: Bit node updating with probability (or LLR); q_{11} updating from p_1 and r_{21} (LHS) and q_{21} updating from p_1 and r_{11} (RHS)

We need to compute the probability of q_{11} being '1', $pr(q_{11} = \,'1' \mid p_1 = \,'1'$, $r_{21} = \,'1')$ with the condition of knowing $p_1 = \,'1'$ (the probability of receive sample p_1 being '1') and the probability of r_{21} being '1', $r_{21} = \,'1'$, which is denoted as $\mathrm{p}^1 (r_{21})$ below. Note that the probability of its own edge, r_{11} is NOT included. It is formulated as a conditional probability. An abbreviated notation of pr $(q_{11} = $ '1'$\mid p_1, r_{21}) = pr(q_{11} = $ '1'$\mid p_1 = $ '1', $r_{21} = $ '1') is used without confusion. At a bit node all branch must be the same '1' or the same '0'; it is bit node constraint. Using the properties of conditional probability, it is expanded as,

$$pr(q_{11} = \,'1'|p_1, r_{21}) = \frac{pr(q_{11} = \,'1', p_1, r_{21})}{pr(p_1, r_{21})} = \frac{pr(p_1, r_{21}|q_{11} = \,'1')pr(q_{11} = \,'1')}{pr(p_1, r_{21})}$$

$$pr(q_{11} = \,'0'|p_1, r_{21}) = \frac{pr(q_{11} = \,'0', p_1, r_{21})}{pr(p_1, r_{21})} = \frac{pr(p_1, r_{21}|q_{11} = \,'0')pr(q_{11} = \,'0')}{pr(p_1, r_{21})}$$

where the scaling factor $pr(p_1, r_{21})$ is obtained by

$$pr(p_1, r_{21}) = pr(p_1, r_{21}|q_{11} = \,'0')pr(q_{11} = \,'0')$$
$$+ pr(p_1, r_{21}|q_{11} = \,'1')pr(q_{11} = \,'1').$$

And a priori probability is equally likely, i.e., $pr(q_{11} = $ '1'$)=pr(q_{11} = $ '0'$)= \frac{1}{2}$, and assuming p_1 and r_{21} are independent (we assume all the edges are independent),

$$pr(p_1, r_{21}|q_{11} = \,'1') = \mathrm{p}^1(r_{21})\, \mathrm{p}_1('1')$$

$$pr(p_1, r_{21}|q_{11} = \,'0') = (1 - \mathrm{p}^1(r_{21}))\,(1 - \mathrm{p}_1('1')) = \mathrm{p}^0(r_{21})\, \mathrm{p}_1('0')$$

Thus, updating q_{11}, in terms of probability, is given by

$$\mathrm{p}^1(q_{11}) = pr(q_{11} = \,'1'|p_1, r_{21}) = \mathrm{p}^1(r_{21})\mathrm{p}_1('1')/(\mathrm{p}^1(r_{21})\mathrm{p}_1('1') + \mathrm{p}^0(r_{21})\mathrm{p}_1('0'))$$

$$\mathrm{p}^0(q_{11}) = pr(q_{11} = \,'0'|p_1, r_{21}) = \mathrm{p}^0(r_{21})\mathrm{p}_1('0')/(\mathrm{p}^1(r_{21})\mathrm{p}_1('1') + \mathrm{p}^0(r_{21})\mathrm{p}_1('0'))$$

$(\mathrm{p}^1(r_{21})\, \mathrm{p}_1('1') + \mathrm{p}^0(r_{21})\, \mathrm{p}_1('0'))$ is a scaling factor to make a conditional probability whose value is within 0 to 1.

Figure 6-38: Bit node updating with K check node connections

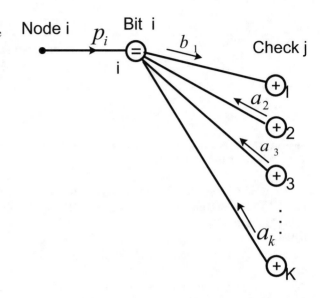

Updating q_{11}, in terms of LLR, is the sum of LLRs of r_{21} and p_1.

$$\text{LLR } (q_{11}) = \log \left[p^0(q_{11})/p^1(q_{11}) \right] = \log \left[p^0(r_{21}) \, p_1('0')/p^1(r_{21}) \, p_1('1') \right]$$
$$\text{LLR } (q_{11}) = \text{LLR } (r_{21}) + \text{LLR}(p_1).$$

Similarly we can obtain the update of q_{21}, in probability and in LLR,

$$p^1(q_{21}) = pr(q_{21} = '1'|p_1, r_{11}) = p^1(r_{11}) \, p_1('1')/(p^1(r_{11}) \, p_1('1') + p^0(r_{11}) \, p_1('0'))$$
$$p^0(q_{21}) = pr(q_{21} = '0'|p_1, r_{11}) = p^0(r_{11}) \, p_1('0')/(p^1(r_{11}) \, p_1('1') + p^0(r_{11}) \, p_1('0'))$$
$$\text{LLR}(q_{21}) = \text{LLR}(r_{11}) + \text{LLR } (p_1).$$

The similar update computation will be repeated for all bit nodes 2, 3, 4, 5, 6.

This bit node update can be generalized into a bit node with K edges plus 1 input node as shown in Figure 6-38, assuming the independence of edges and a prior probability being equal,

$$pr(b_1 = '1'|p_1, a_2, a_3, \ldots, a_K) = p_i('1')pr(a_2 = '1')pr(a_3 = '1'), \ldots, pr(a_K = '1')/\text{scale}$$
$$pr(b_1 = '0'|p_1, a_2, a_3, \ldots, a_K) = p_i('0')pr(a_2 = '0')pr(a_3 = '0'), \ldots, pr(a_K = '0')/\text{scale}$$

with scale $= p_i('1')pr(a_2 = '1')pr(a_3 = '1'), \ldots, pr(a_K = '1') + p_i('0')pr(a_2 = '0')pr(a_3 = '0'), \ldots, pr(a_K = '0')$.

$$\text{LLR } (b_1) = \text{LLR}(p_i) + \sum_{k=2, k\neq 1}^{K} LLR(a_k) \qquad (6\text{-}52)$$

This is repeated for all other edges 2, 3, K-1, K.

Similarly all the bit nodes, i=1, 2, 3,....n will be updated.

Note that LLR requires addition and probability requires multiplication and scaling, and thus LLR is used in practice. However, the concept behind can be understood with that of conditional probability, and LLR is equivalent to probability based computation, e.g., convergence behavior.

6.4.3.1 Check Node Updating Computation

We need to compute the probability of r_{11} being '0' with the condition of knowing the probability of q_{14} and q_{15} and of meeting a parity check.

$$r_{11} = \,'0' : \text{if } \{q_{14} = \,'0' \text{ and } q_{15} = \,'0'\} \text{ or } \{ q_{14} = \,'1' \text{ and } q_{15} = \,'1'\}$$
$$r_{11} = \,'1' : \text{if } \{q_{14} = \,'0' \text{ and } q_{15} = \,'1'\} \text{ or } \{ q_{14} = \,'1' \text{ and } q_{15} = \,'0'\}$$

With the notation of, noting that p_i is temporary notation <u>not</u> input node probability i=1, 2,

$$p(q_{14} = \,'0') = 1 - p_1, p(q_{14} = \,'1') = p_1, \text{and}$$
$$p(q_{15} = \,'0') = 1 - p_2, p(q_{15} = \,'1') = p_2$$

the probabilities of r_{11}='0' and r_{11}='1' are given by

$$\text{pr} \left(r_{11} = \,'0' \mid q_{14}, q_{15} \right) = p_1 p_2 + (1 - p_1)(1 - p_2) = \tfrac{1}{2} [1 + (1 - 2 p_1)(1 - 2 p_2)]$$
$$\text{pr} \left(r_{11} = \,'1' \mid q_{14}, q_{15} \right) = (1 - p_1) p_2 + p_1 (1 - p_2) = \tfrac{1}{2} [1 - (1 - 2 p_1)(1 - 2 p_2)]$$

The second equality can be checked out by expanding both sides.

Similarly pr $(r_{14}$='0'| $q_{11}, q_{15})$ and pr $(r_{14}$='1'| $q_{11}, q_{15})$ can be computed with the same form of the expressions.

Similarly again pr $(r_{15}$='0'| $q_{11}, q_{14})$ and pr $(r_{15}$='1'| $q_{11}, q_{14})$ can be computed with the same form of the expressions.

This computation can be repeated to all other check nodes 2, 3, 4.

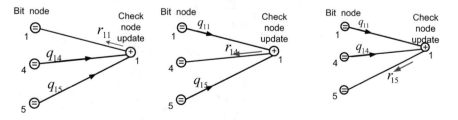

Figure 6-39: Check node updating; r_{11}, r_{14} r_{15} from left to right

In terms of LLR,

$$\Lambda(r_{11}) = \log \frac{1 + (1 - 2p_1)(1 - 2p_2)}{1 - (1 - 2p_1)(1 - 2p_2)} \quad \text{where } p(q_{14} = {}'1') = p_1 \text{ and } p(q_{15} = {}'1') = p_2.$$

Since $(1 - 2p) = \tanh\left(\frac{1}{2}\Lambda\right)$, with the definition of $\log \frac{p_0}{p_1} = +\Lambda$, (See Appendix A.2.)

$$\Lambda(r_{11}) = \log \frac{1 + \tanh\left(\frac{1}{2}\Lambda(q_{14})\right) \tanh\left(\frac{1}{2}\Lambda(q_{14})\right)}{1 - \tanh\left(\frac{1}{2}\Lambda(q_{14})\right) \tanh\left(\frac{1}{2}\Lambda(q_{14})\right)},$$

$$\Lambda(r_{14}) = \log \frac{1 + \tanh\left(\frac{1}{2}\Lambda(q_{11})\right) \tanh\left(\frac{1}{2}\Lambda(q_{15})\right)}{1 - \tanh\left(\frac{1}{2}\Lambda(q_{11})\right) \tanh\left(\frac{1}{2}\Lambda(q_{15})\right)}$$

$$\Lambda(r_{15}) = \log \frac{1 + \tanh\left(\frac{1}{2}\Lambda(q_{11})\right) \tanh\left(\frac{1}{2}\Lambda(q_{14})\right)}{1 - \tanh\left(\frac{1}{2}\Lambda(q_{11})\right) \tanh\left(\frac{1}{2}\Lambda(q_{14})\right)}$$

We generalize the case of 3 edges to K edges. We omit the detailed derivation and a general expression for all other edges j = 2, 3, ... K as shown in Figure 6-40.

$$\Pr(a_1 = {}'0') = \frac{1}{2}\left[1 + \prod_{k=2, k\neq 1}^{K} (1 - 2(\Pr(b_k = {}'1')))\right]$$

$$= \frac{1}{2}\left[1 + \prod_{k=2, k\neq 1}^{K} (\Pr(b_k = {}'0')) - (\Pr(b_k = {}'1'))\right]$$

$$\Pr(a_1 = {}'1') = \frac{1}{2}\left[1 - \prod_{k=2, k\neq 1}^{K} (1 - 2(\Pr(b_k = {}'1')))\right]$$

$$= \frac{1}{2}\left[1 - \prod_{k=2, k\neq 1}^{K} (\Pr(b_k = {}'0')) - (\Pr(b_k = {}'1'))\right]$$

$$\Lambda(a_1) = \log \frac{1 + \prod_{k=2, k\neq 1}^{K} \tanh\left(\frac{1}{2}\Lambda(b_k)\right)}{1 - \prod_{k=2, k\neq 1}^{K} \tanh\left(\frac{1}{2}\Lambda(b_k)\right)}$$

$$= 2\tanh^{-1}\left[\prod_{k=2, k\neq 1}^{K} \tanh\left(\frac{1}{2}\Lambda(b_k)\right)\right] \tag{6-53}$$

Figure 6-40: Check node updating with K edges

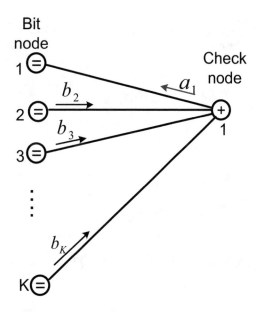

We used the identity of $2\tanh^{-1}[y] = \log\frac{1+y}{1-y}$, and $\tanh\left(\frac{x}{2}\right) = \frac{e^x-1}{e^x+1}$.

Check node update in (6-53) may have different expressions, which are explored in Appendix A.3. In particular, note that

$$\Lambda(a_1) = \prod_{k=2,k\neq1}^{K} sign(\Lambda(b_k))\log\tanh 2\left[\sum_{k=2,k\neq1}^{K}\log\tanh 2(\Lambda(b_k))\right] \quad (6\text{-}54)$$

$$\Lambda(a_1) \approx \prod_{k=2,k\neq1}^{K} sign(\Lambda(b_k))\min_{k=2,k\neq1}^{K}[\Lambda(b_k)] \quad (6\text{-}55)$$

For simplicity, (6-55) may be used in practice at the expense of 'small' performance degradation. (6-54) may be used with a look up table.

6.4.3.2 Updating q_i; Computation of APP of p_i

We compute a posteriori probability (APP) of p_i as shown in Figure 6-41. This is essentially the same as bit node updating except that all the edges from check nodes are used.

$$pr(q_i = '1'|r_{1i}, r_{2i}, r_{3i}, \ldots, r_{Ki}) = \prod_{k=1}^{K} pr(r_{ki} = 1)/scale$$

$$pr(q_i = '0'|r_{1i}, r_{2i}, r_{3i}, \ldots, r_{Ki}) = \prod_{k=1}^{K} pr(r_{ki} = 0)/scale$$

Figure 6-41: Bit node APP
computation with K edges
with check nodes

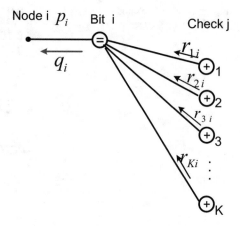

with $scale = \prod_{k=1}^{K} pr(r_{ki} = 1) + \prod_{k=1}^{K} pr(r_{ki} = 0).$

$$\text{LLR}(q_i) = \sum_{k=1}^{K} LLR(r_{ki}) \qquad (6\text{-}56)$$

6.4.3.3   The Computational Organization with H ➜ edge tables

Given receive samples (**r**) or LLRs (or probability) from demapper of BPSK and
even high order modulations, as input to decoder, it needs to update all the edge
values of LLR (or probability) until stopping condition (e.g., H $\mathbf{q}^T = \mathbf{0}^T$) is met.

For iterative computation of q_{ji}, r_{ji}, and q_i, there are steps of *initialization*,
updating, checking for *stopping* criterion. After some iterations, when stopping
criterion is met, such as syndrome H $\mathbf{q}^T = \mathbf{0}^T$, then a codeword is reached and
iteration stops, where $\mathbf{q} = [q_i]$ at the iteration time (omitted the time index for
simplicity). This stopping criterion (H $\mathbf{q}^T = \mathbf{0}^T$) is universally applicable to all
decoding schemes.

(1) *Initialization* and set Max iterations
(2) *Updating* q_{ji} from the previous r_{ji}
(3) *Updating* r_{ji} from the previous q_{ji}
(4) *Updating* q_i from the current r_{ji}
(5) *Computing* stopping criterion such as syndrome (H $\mathbf{q}^T$)
(6) Terminate iteration if stopping criterion met or if Max iterations are reached, if
not going back to 2).

In the previous three sub-sections (6.4.2.1, 6.4.2.2, 6.4.2.3), we obtained updating
rules for bit nodes, check nodes and a posteriori probability (LLR) for input bits. In
other words, the items of (2), (3), and (4) of the above were explained in detail.

H → edge tables (q, r)

Now we need to organize decoding computation for a given H matrix. There are many possible organizations but we illustrate one way of converting H to edge tables of q-table, r-table. We use the table names as q and r; nothing to do with APP and receive samples but to do with the branch names of q_{ji} and r_{ji}.

In order to organize the decoding computation, we create a list of connecting edges in two forms of tables, called the table of q-table and r-table.

Example 6-14: The tables, q and r, are shown below when H matrix is given by Figure 6-35.

q table

jr	ic	r indx	br v
1	1	1	0
2	1	4	0
2	2	5	0
3	2	7	0
3	3	8	0
4	3	10	0
1	4	2	0
4	4	11	0
1	5	3	0
3	5	9	0
2	6	6	0
4	6	12	0

r table

jr	ic	qindx	br v
1	1	1	0
1	4	7	0
1	5	9	0
2	1	2	0
2	2	3	0
2	6	11	0
3	2	4	0
3	3	5	0
3	5	10	0
4	3	6	0
4	4	8	0
4	6	12	0

The length of tables (row count) is the number of edges, the first column is row number (parity check node number) of H, the second column is column number (bit node number) of H, the third column cross-reference address for easy lookup, and the fourth column is the values (LLR, probability, or '1'/'0' bit) of edge. In case of probability, is added one more column for the probability of '0', p_0, which can be obtained by $p_0 = 1$-p_1 for convenience and clarity of script.

(jr, ic) pair in the table is a location of (row, column) in H where there is '1', i.e., connection / branch.

q-table is used for bit node updating, and cross-reference address is where r values are located. Updated values are stored at column 4 (br v)

r-table is used for check node updating, and cross-reference address is where q values are located. Updated values are stored at column 4 (br v)

Matlab m-script for creating the tables from H is given below.

```
% --------------H --> edge table (q, r)--------
% h : parity check matrix H
row_cnt = length(h(:,1));    %row dimension
col_cnt = length(h(1,:));    %column dimension - block size
con_cnt = length(find(h));   %connections. edges
% row and column location of parity check matrix h
jrow = mod(find(h), row_cnt); jrow(jrow==0) = row_cnt;
icol = ceil(find(h)/row_cnt);
%
seq1 = [jrow icol];
[v, iii] = sort(jrow);              %iii = q_adress
seq2 = seq1(iii,:);
[vv, ii] = sort(seq2(:,2));         %ii = r_address
q = [seq1 ii];
r = [seq2 iii];
% ----- initilialze ---------------
q(:,4) = 0;      % q(:,5) = 0;
r(:,4) = 0;      % r(:,5) = 0.5;
% [1   2      3            4      ] column number
% [j   i      x_address    LLR    ]
%   j; check node number
%   i: bit node number
```

The above m-script is universal in the sense that it can be used for any type of LDPC H matrix. In fact a simplified version of q will be used for encoding as well as will be seen later.

Exercise 6-49*: A universal message passing decoder may be organized by a program for a given H with programmable parameters such as Max iterations, stopping criterion and output format. Write a program for AWGN channel, and generate curves of error performance with BPSK. The star * means a project. Hint: this is an advanced project exercise.

6.4.4 LDPC Encoder

A transposed version of (6-9), $\mathbf{HG}^T = [\mathbf{0}]_{(n-k)\times k}$, is slightly more convenient, and will be used to describe encoding process.

If a parity check matrix H can be reduced to $H_r = [H_0 | I_{n-k}]$ as in (6-10), G^T can be expressed as $G^T = [I_k; H_0]_{(n \times k)}$.

Thus parity bits are generated by $[\mathbf{p}]_{(n-k) \times 1} = [H_0]_{(n-k) \times k} [u^T]_{k \times 1}$, which will be appended after u^T information bits.

Example 6-15: We use the same H used in Figure 6-35.

$$H = \begin{bmatrix} 1 & 0 & 0 & 1 & 1 & 0 \\ 1 & 1 & 0 & 0 & 0 & 1 \\ 0 & 1 & 1 & 0 & 1 & 0 \\ 0 & 0 & 1 & 1 & 0 & 1 \end{bmatrix} \rightarrow H_r = \begin{bmatrix} 1 & 1 & 1 & 1 & 0 & 0 \\ 0 & 1 & 1 & 0 & 1 & 0 \\ 1 & 1 & 0 & 0 & 0 & 1 \end{bmatrix} = [H_0 | I_3]$$

$$G^T = \begin{bmatrix} 1 & 0 & 0 \\ 0 & 1 & 0 \\ 0 & 0 & 1 \\ 1 & 1 & 1 \\ 0 & 1 & 1 \\ 1 & 1 & 0 \end{bmatrix} \quad G^T u^T = c^T, \text{ and e.g.,} \quad \begin{bmatrix} 1 & 0 & 0 \\ 0 & 1 & 0 \\ 0 & 0 & 1 \\ 1 & 1 & 1 \\ 0 & 1 & 1 \\ 1 & 1 & 0 \end{bmatrix} \begin{bmatrix} 1 \\ 1 \\ 1 \end{bmatrix} = \begin{bmatrix} 1 \\ 1 \\ 1 \\ 1 \\ 0 \\ 0 \end{bmatrix}$$

Now we discuss how to obtain H_r from H. It is a form of Gaussian elimination method with mod 2 row operations, which is used for solving linear simultaneous equations. A row is added to another and replaced; which will be repeated until a desired form is reached. Similar to Gaussian elimination for linear equations the reduction process needs to go to a lower triangular form first and then identity matrix of (n-k) dimension. As in the above example, a number of rows might be reduced if the rank of H is less than the number of rows. This form of H, H_r, is only for encoding, but we must use H for decoding since H_r is not in general low density.

Exercise 6-50*: Write a program to convert H to H_r. Then generate G^T. (*) means a project exercise. Hint: use row swapping only. If column swapping is used, it will be a different code due to the fact that bit locations are swapped.

6.4.4.1 Encoder with a Special Form of H – Band Diagonal

Now we consider a special form of H, as shown in Figure 6-42; parity bit part of H is band diagonal. This form of H is called repeat-accumulator (RA) code; its naming will be clear later this subsection. It is often used in practical standards since its encoding is possible without converting to a diagonal form. This will be discussed below. Its encoder implementation can be done as in Figure 6-43, which is a 'most' serial implementation, i.e., 'least' hardware.

As for Figure 6-43, encoder for systematic RA codes is shown. H matrix column vector for connect (inside dotted line) is time varying connections to a set of 1-tap integrators (pre-coder). After k shifts of info bits, the temporary stored parity bits are accumulated again to generate final parity bits. This is least 'parallel', i.e., most 'serial', implementation.

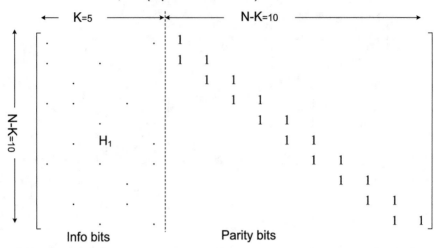

Figure 6-42: RA form of H,and definition of H_1 in H of LDPC; information part of H and its dimension is (N-K, K) = (10, 5) in this example

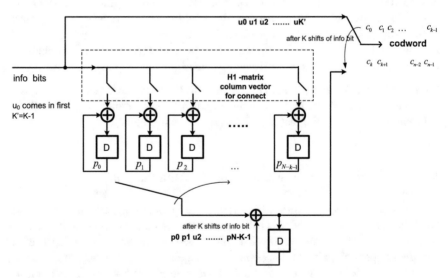

Figure 6-43: Encoder for systematic RA codes; structural 'similarity' to BCH encoder with generator polynomial

6.4.4.2   Different Encoder Implementation of RA Form

Another encoder implementation is shown in Figure 6-44, with an example *H* matrix below.

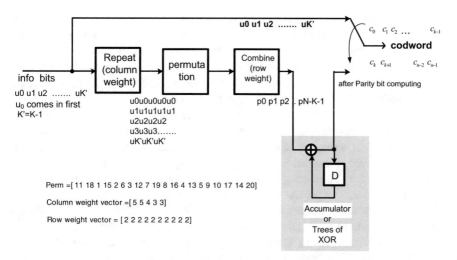

Figure 6-44: Permutation based encoder implementation of LDPC in RA form. For easy understanding we give a concrete H. We need permutation vector, column weight vector and row weight vector to represent H matrix. Parity side is done by accumulator. The output of this implementation is the same as Figure 6-43

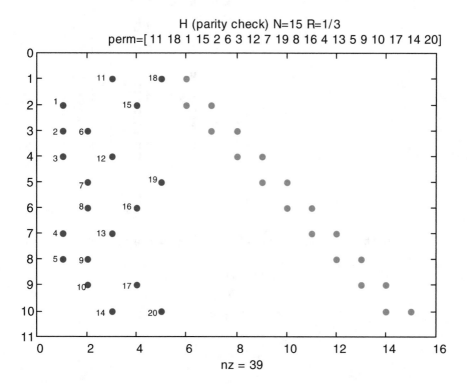

In Figure 6-44, we need a permutation vector, a column weight vector and a row weight vector to represent H matrix. Parity side of H is done by accumulator (trees of XOR). The output of this implementation is the same as Figure 6-43. The original RA code was invented as a simple turbo code and decoded by turbo decoding. This encoding scheme was used. Later it is interpreted as a LDPC code and developed further with additional refinement, e.g., AR4JA codes.

We explain, with an example in Figure 6-44, how parity bits are computed. The input info bits are expanded by repeat block according to column weight, i.e., K (5 in this example) bits are expanded to the number of ones in H_1 (info part of H), nz (20 in this example). This expanded nz bits are permuted by a permutation sequence, which can be obtained by numbering ones in H_1 column wise as shown in the figure and reading row wise. The permuted sequence, nz in length, is 'combined' (i.e., exclusive ORed) when the bits are in the same row. After combing the number of bits becomes the same as the number of rows in H_1. Note that this is the same as $H_1 * U^T$ $= P^T$ where $U=[u_0 \ u_1 \ \ldots \ u_{K-1}]$ and $P=[p_0 \ p_1 \ldots p_{N-K-1}]$ with the addition is the same as XOR, i.e., GF(2) addition. With this example it is given by

$$
H_1 * U^T = P^T \quad \leftarrow \rightarrow \quad
\begin{bmatrix}
0 & 0 & 1 & 0 & 1 \\
1 & 0 & 0 & 1 & 0 \\
1 & 1 & 0 & 0 & 0 \\
1 & 0 & 1 & 0 & 0 \\
0 & 1 & 0 & 0 & 1 \\
0 & 1 & 0 & 1 & 0 \\
1 & 0 & 1 & 0 & 0 \\
1 & 1 & 0 & 0 & 0 \\
0 & 1 & 0 & 1 & 0 \\
0 & 0 & 1 & 0 & 1
\end{bmatrix}
\begin{bmatrix}
u_0 \\ u_1 \\ u_2 \\ u_3 \\ u_4
\end{bmatrix}
=
\begin{bmatrix}
p_0 \\ p_1 \\ p_2 \\ p_3 \\ p_4 \\ p_5 \\ p_6 \\ p_7 \\ p_8 \\ p_9
\end{bmatrix}
$$

Final parity bits, represented by $Q=[q_0 \ q_1 \ldots q_9]$, are computed as

$$q_0 = p_0$$
$$q_1 = q_{0+}p_1$$
$$q_2 = q_{1+}p_2$$
$$\ldots\ldots\ldots$$
$$q_9 = q_{8+}p_9$$

which is called 'accumulator' and + is XOR. We note that this computation may be seen as back substitution using parity side of H (a band matrix) in solving a linear simultaneous equation.

6.4.5 Useful Rules and Heuristics for LDPC Code Construction

In this subsection useful definition, theorems, and corollaries for code construction and for code performance analysis are presented with some examples.

6.4.5.1 Definition of Cycle, Trapping (stopping) Set

Definition of cycle: A cycle is a path from a variable node, v_i, back to itself, if any edge in the path is used only once. The length of the cycle is the number of edges contained in this cycle. The cycle involving two variable nodes and two check nodes is the smallest and its length is 4. The cycle involving three variable nodes and check nodes is 6. This can be generalized to 4, 5, 6, variable nodes and check nodes. Girth is defined as a minimum cycle for a given H.

Definition (1) of stopping set: A stopping set, S, is a subset of v, the set of variable (bit) nodes, such that all neighbors of S, i.e., all check nodes which are connected to S, are connected to S at least twice. The support of any codeword is a stopping set. But it is not limited to codewords, i.e., the cardinality of stopping set is larger than that of codewords. The entire set of nodes is a stopping set. It should contain cycles. It is also called trapping set.

Definition (2) of stopping set: A stopping set in a Tanner graph is a set of message (bit) nodes such that the graph induced by these message nodes has the property that no check node (connected to the set) has degree one.

We elaborate these terms with examples.

It is useful to represent cycles and stopping sets, even codewords by parity check matrix and its Tanner graph. A few parity check matrix examples are given below;

$$H_4 = \begin{bmatrix} 1 & 1 & 0 & 0 & 1 & 0 \\ 1 & 0 & 1 & 1 & 0 & 0 \\ 0 & 1 & 0 & 0 & 1 & 1 \\ 0 & 0 & 1 & 1 & 0 & 1 \end{bmatrix} \quad H_6 = \begin{bmatrix} 1 & 0 & 1 & 1 & 0 & 0 \\ 1 & 1 & 0 & 0 & 0 & 1 \\ 0 & 1 & 1 & 0 & 1 & 0 \\ 0 & 0 & 0 & 1 & 1 & 1 \end{bmatrix} \quad H_8 = \begin{bmatrix} 1 & 0 & 0 & 1 & 1 & 0 \\ 1 & 1 & 0 & 0 & 0 & 1 \\ 0 & 1 & 1 & 0 & 1 & 0 \\ 0 & 0 & 1 & 1 & 0 & 1 \end{bmatrix}$$

Their graphical representations, Tanner graph, are shown below.

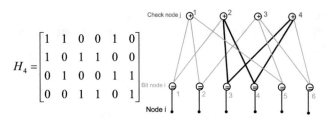

$$H_4 = \begin{bmatrix} 1 & 1 & 0 & 0 & 1 & 0 \\ 1 & 0 & 1 & 1 & 0 & 0 \\ 0 & 1 & 0 & 0 & 1 & 1 \\ 0 & 0 & 1 & 1 & 0 & 1 \end{bmatrix}$$

One with black line is (2,4, 3,4). Can you see more patterns of 4-cycle? Answer (1,3, 2,5), and that is all. Girth of H_4 is 4.

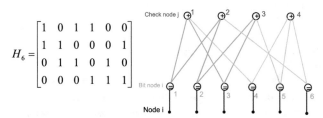

$$H_6 = \begin{bmatrix} 1 & 0 & 1 & 1 & 0 & 0 \\ 1 & 1 & 0 & 0 & 0 & 1 \\ 0 & 1 & 1 & 0 & 1 & 0 \\ 0 & 0 & 0 & 1 & 1 & 1 \end{bmatrix}$$

One with green is (1,2,3, 1,2,3). Can you see more? Answer (1,3,4, 3,5,4), (2,3,4 ,2,5,6), (1,2,4, 1,4,6). Girth of H_6 is 6.

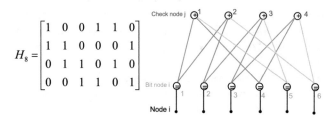

$$H_8 = \begin{bmatrix} 1 & 0 & 0 & 1 & 1 & 0 \\ 1 & 1 & 0 & 0 & 0 & 1 \\ 0 & 1 & 1 & 0 & 1 & 0 \\ 0 & 0 & 1 & 1 & 0 & 1 \end{bmatrix}$$

One with red is (1,2,3,4, 1,2,3,4). Can you see more? Answer: (1,2,4,3, 2 6 4 5), (1,3,4,2, 1,5, 3,6)

H_4 has 4-cycles. Look for this pattern, in H_4, $\begin{bmatrix} . & . & . & . & . \\ . & 1 & . & 1 & . \\ . & . & . & . & . \\ . & . & . & . & . \\ . & 1 & . & 1 & . \end{bmatrix}$ or any column

permuted patterns or row permuted patterns.

H_6 has 6-cycles. Look for the pattern
$$\begin{bmatrix} . & 1 & . & 1. & . \\ . & 1 & 1 & . & . \\ . & . & 1 & 1 & . \\ . & . & . & . & . \\ . & . & . & . & . \end{bmatrix}$$
or any column per-

muted patterns or row permuted patterns.

H_8 has 8-cycles. Look for the pattern
$$\begin{bmatrix} . & 1 & . & . & 1 \\ . & 1 & 1 & . & . \\ . & . & 1 & 1 & . \\ . & . & . & 1 & 1. \\ . & . & . & . & . \end{bmatrix}$$
or column permuted

patterns or row permuted patterns.

Now in order to check all 4-cycles in a given H, one needs to look all two column combinations, i.e., $\binom{N}{2} = N(N-1)/2$, and then by adding two columns (not in GF (2) but in real), to see if at least two rows have the value equal to 2. If there are 2 of such rows, then one 4-cycle exists. If there are 3 of such rows, then three 4-cycles exist. This pattern continues, denoting the number of such rows (i.e., with two column addition row by row the value equals 2) is M_R then a number of 4-cycles is $\binom{M_R}{2} = M_R(M_R - 1)/2$ with the two nodes (i.e. columns).

To check all 6-cycles, one needs to look all three column (node) combinations, i.e., $\binom{N}{3}$, by adding the three columns (not in GF(2) but in real), to see if, at least three rows have the value equal to 2 or 3. If there are 3 such rows, there is one 6-cycle. If there are more than 3 such rows, say, $M_R > 3$, then choose sets out of $\binom{M_R}{3}$ combinations such that the sum of selected 3 rows (i.e., row addition column by column) should be equal to 2 or 3.

Exercise 6-51: For a given H below, how many 6-cycles are there?

$$H = \begin{bmatrix} . & 1 & . & 1 & . \\ . & 1 & 1 & . & . \\ . & . & 1 & 1 & . \\ . & . & 1 & 1 & . \\ . & 1 & 1 & . & . \end{bmatrix} \begin{matrix} j=1 \\ 2 \\ 3 \\ 4 \\ 5 \end{matrix}$$

Answer: There are 4 of 6-cycles in H. We explain how to obtain it below.

There are $\binom{M_R}{3} = 10$ combinations of 3 rows since $M_R = 5$. Not all 10 form 6-cycles but only 4 of 3 row combinations form 6-cycle as shown below.

	j rows			3 row sum			6-cycle?
1	1	2	3	2	2	2	yes
2	1	2	4	2	2	2	yes
3	1	2	5	3	2	1	No
4	1	3	4	1	2	3	No
5	1	3	5	2	2	2	yes
6	1	4	5	2	2	2	yes
7	2	3	4	1	3	2	No
8	2	3	5	2	3	1	No
9	2	4	5	2	3	1	No
10	3	4	5	1	3	2	No

This pattern will continue for 8-cycle, i.e., to check all 8-cycles, one needs to look all four column (node) combinations, i.e., $\binom{N}{4}$, by adding the four columns (not in GF(2) but in real), to see if, at least four rows have the value equal to 2 or 3 or 4. . If there are 4 such rows, there is one 8-cycle. If there are more than 4 such rows, say, $M_R > 4$, then choose sets out of $\binom{M_R}{4}$ combinations such that the sum of selected 4 rows (i.e., row addition column by column) should be equal to 2 or 3 or 4.

The enumeration of 4-cycles (i.e., actually programming) is relatively straight-forward for a given H. However, for 6-cycles and beyond, it gets more involved computationally. In any case, it is computationally intense as the column size (i.e., block size) of H gets large.

Stopping set examples are shown for a parity check matrix H below.

$$
\begin{array}{c}
\cdot i\cdot = 1\ \ 2\ \ 3\ \ 4\ \ 5\ \ 6 \\
H = \begin{bmatrix} 1 & 0 & 0 & 1 & 1 & 0 \\ 1 & 1 & 0 & 0 & 0 & 1 \\ 0 & 1 & 1 & 0 & 1 & 0 \\ 0 & 0 & 1 & 1 & 0 & 1 \end{bmatrix} \begin{array}{l} j = 1 \\ 2 \\ 3 \\ 4 \end{array}
\end{array}
$$

Column number set of $(1, 2, 5)$ is a stopping set and so is column set of $(1, 2, 3, 4)$. Their column sums are $[2, 2, 2, 0]^T$ and $[2, 2, 2, 2]^T$ respectively. Note that these are also the bit node set of 6-cycle and 8-cycle; these stopping sets contain cycles. Note that these two sets are codewords as well (Can you see it? It will be explained as Theorem 6.1 in sub-section 6.4.5.2 later).

Column number set, entire column, $(1, 2, 3, 4, 5, 6)$ is a stopping set since their column sum, row by row, is $[3, 3, 3, 3]^T$, all bigger than one. On the other hand the column set of (1, 2, 3) is not a stopping set since its column sum $[1, 2, 2, 1]^T$. Note that the value less than 2 (i.e., one) in the sum of column vectors, it is not a stopping set.

Now in order to check stopping sets in an H, one needs to look all ρ column combinations, i.e., $\binom{N}{\rho}$ to see if by adding ρ columns (in real) and all rows (to be precise all non-zero rows) have the value greater than equal to 2, i.e. no 1s. These combinations will $\binom{N}{\rho}$ grow quickly when N is large similarly to finding cycles. Here note that ρ is not known in advance.

Additional comments on bit node degree: To have a cycle through a bit (variable) node it should have at least two connections to the node to leave and to return. That is, the bit node degree is assumed to be at least 2. This should apply to stopping set definition as well. This assumption of bit node degree greater than 1 is practically necessary.

Some useful heuristics on column weight distribution and row distance
Definition of row distance

If, in a bit node of info bit, two adjacent check nodes are connected to the bit node (i.e., two consecutive rows in H are assigned as '1'), a 4-cyle is created with one variable node in parity bit of RA). See Figure 6-45 with red dots. Two consecutive rows have the <u>row distance</u> 1.

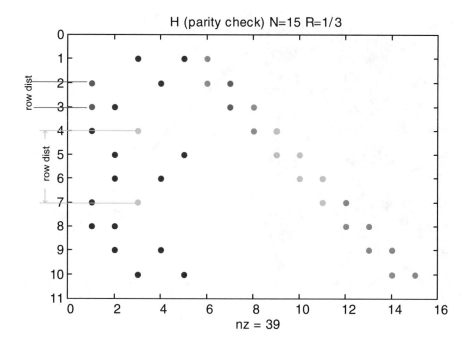

Figure 6-45: Examples of row distance

If the row distance in a variable node is 2, a 6-cycle is created with two variable nodes in parity bit of RA. This pattern can be generalized to the row distance 3, 4, with the cycle 8 (= (3+1)*2), 10 (= (4+1)*2), and so on.

Thus the small cycles with the bit nodes in parity bits with one bit node in info bit can be prevented by increasing the row distance in info bit variable node.

And the low weight variable nodes should be assigned, in edge assignment, such that no small cycles should be avoided. We elaborate this idea below.

It seems desirable that the cycles should go through bit nodes with high degree of connectivity, i.e., large column weight. The measure of this is referred to as <u>ACE</u> <u>(approximate cycle extrinsic message node degree)</u>. This ACE can be increased by sorting the variable nodes, low to high column weight, and in the beginning of edge assignment (e.g., PEG: progressive edge assignment algorithm – good for small block (100s) size but for large block size (>1k) computationally too intense), the edges of low weight node are assigned first in order to prevent having small cycles between variable (bit) nodes with low degree.

For example (DVB-S2 standard listed in Table 6-23), variable nodes in parity check part of RA code have degree 2. But all the info bits degree (column weight) is 3 or more, and some bit node have much higher degree. For example N=16200, R=2/5 case, 6(12) and 12(3) are shown in the DVB-S2 standard. This means that K1 =6 blocks have column weight 12, and K2=12 blocks have column weight 3 where a total info blocks are 18 (i.e. 6480 bits = 18*360) shown below.

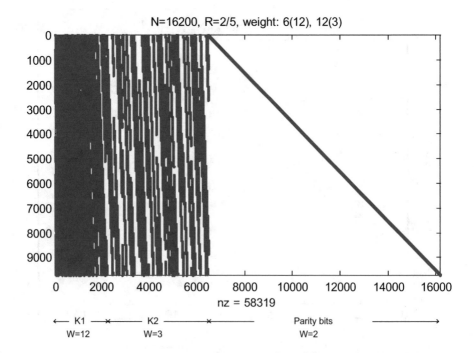

N=16200, R=2/5, weight: 6(12), 12(3)

6.4.5.2 Properties of H Related with and Useful to Code Construction

A useful theorem on the column sums of H matrix is given below.

Theorem 1: If there are l columns such that the vector sum of these l columns is zero in GF(2) i.e., all even integer, then the codeword has distance (from zero codeword) l, and the inverse of this is also true, i.e., when a codeword has a weight l, then there exist l columns whose vector sum is zero in GF(2).

One simple example of this theorem is that when two columns of a parity check matrix H are identical, then the distance of the code will be two since adding the two columns in GF(2) will be zero.

Corollary 1: If no d-1 or fewer columns of H add to zero in GF(2), the code has minimum distance at least d.

Corollary 2: The minimum distance of a linear code is equal to the smallest number of columns of H that sum to zero in GF(2).

Corollary 3: If d columns of H of a linear code sum to zero in GF(2), the minimum distance of the code is upper bounded by d, i.e., not greater than d.

Corollary 4: Consider a special form of parity check matrix, i.e., H with the parity side being diagonal. When $H = [P^T \ I_{n-k}]$, i.e., the parity part is diagonal (identity matrix), each column of P^T corresponds to parity bits when the info bit of the column is one.

For example, when info bits u = [1 0 0 ... 0] then the parity bits are the first column of P^T, and u = [0 1 0 ... 0] the corresponding parity bits are the second column of P^T, and so on. When u = [1 1 0 ... 0] the corresponding parity bits are the sum of the first column and the second column in GF(2).

It is easy to see the argument with a concrete example.

$$H_r = \begin{bmatrix} 1 & 1 & 1 & 1 & 0 & 0 \\ 0 & 1 & 1 & 0 & 1 & 0 \\ 1 & 1 & 0 & 0 & 0 & 1 \end{bmatrix}.$$

With this H, a code word, the first info bit is 1, is given by [1 0 0 1 0 1]. Note that the weight of this codeword is 3. And its row sum for information part is given by [2 3 2]. Thus the minimum distance of this code cannot be greater than 2+1=3.

Notice that three columns (1, 2, 5) are added to be zero in GF (2). Then the minimum distance of the code is upper-bounded by the smallest sum of each column +1 (in integer) since it is the weight of codewords when one info bit location is '1' and the rest is '0'.

This corollary may be used to generate the codewords when the info bit has one of '1' in all possible locations (K locations). It can generate the codewords when the info bit has 2 of '1's in all possible locations, i.e., (K, 2). It can be extended to the case of 3 bits and 4 bits. This may be useful to find the upper bound of minimum distance and enumeration of codewords with the low weight codewords when the info bits contain a small number of '1's. Note that this is only the upper bound of the minimum distance and a true minimum distance requires much larger enumeration of codewords, which requires a literally huge computation.

If a code of (n, k) is described by a check matrix H, and furthermore its minimum distance d_{min} is known, can we relate it with the girth (i.e., minimum cycle) of H? Corollary 5 partially answers to this question.

Corollary 5: A codeword weight d leads to a cycle of 2 d or less.

Eliminating small cycles is similar to eliminating codewords of small weight in the code. This relationship between the weight of codewords and cycle is very tight but only indicative. Nevertheless removing small cycles, especially 4-cycle, improves the code.

In order to see the corollary is true, we take $d = 5$ example.

$$
H = \begin{bmatrix}
. & . & . & . & . & . & . \\
. & 1 & . & . & . & 1 & . \\
. & 1 & 1 & . & . & . & . \\
. & . & 1 & 1 & . & . & . \\
. & . & . & 1 & 1 & . & . \\
. & . & . & . & 1 & 1 & . \\
. & . & . & . & . & . & .
\end{bmatrix}
\quad \text{adding more 1's} \rightarrow \quad
H' = \begin{bmatrix}
. & . & . & . & . & . & . \\
. & 1 & . & \bar{1} & \bar{1}. & 1 & . \\
. & 1 & 1 & . & . & . & . \\
. & . & 1 & 1 & . & . & . \\
. & . & . & 1 & 1 & . & . \\
. & . & . & . & 1 & 1 & .
\end{bmatrix}
$$

If all the dots are zero in H (left hand side), this pattern generates 2 $d = 10$ cycles.

From Corollary 3, all the row weights in a collection of nodes (d) must be even, i.e., zero in GF (2), to be a codeword. Additional connection i.e., '1's can be added in a row by even number within d nodes (here $d=5$). As an example, see H' on the right hand side where 2 of '1's are added. This clearly generates 4-cycles. Thus Corollary 5 is true in general.

Exercise 6-52*: Based on Theorm1 and its corollaries, write a program to enumerate 4-cycles (and 6-cycles) for a given H. The star * means a project exercise.

Exercise 6-53*: Write a program to estimate minimum distance of a code. Hint: first find H from a code description, and then apply the corollaries.

Another useful theorem related with error correction and erasure capabilities of a code and with its minimum distance is summarized and given below.

Theorem 2: A code with minimum distance d_{min} is capable of correcting and any pattern of v errors and e erasures with the following condition is satisfied;

$$d_{min} \geq 2v + e + 1$$

A receiver may be designed to decide one, zero with hard decision or to declare a symbol to be erased when it is received unreliably. In this case, the received sequence consists of ones, zeros and erasures. A code with minimum distance is capable of handling errors and erasures at the same time with the above condition satisfied.

In order to see the above theorem, delete all the codewords with the e bit positions which the receiver has declared erasures. This deletion results in a shortened code of block length $n - e$. The minimum distance of this shortened code is at least $d_{min} - e$ which is, i.e., $d_{min} - e \geq 2v + 1$. Hence the v errors can be corrected in unerased positions. Still $d_{min} \geq e + 1$, there is one and only one original codeword that agrees with the unerased components. So the entire codeword can be recovered. When the erasures are reliable (i.e., $v=0$), a code with d_{min} can correct up to $d_{min} - 1$ erasures.

ML decoder vs. MP decoder

ML (maximum likelihood decoder): errors due to decoding to wrong codewords

MP (message passing decoder): error due to decoding to wrong codewords plus pseudo codewords (near to codewords) as well. However, when the cycle is large, it will be rare for the latter (decoding to pseudo codewords) and converge to ML performance.

6.4.6 LDPC in Standards

This is a survey of most used and known LDPC codes in industry. In cellular industry Turbo code is adapted early on so it is not included. This motivated the rediscovery of LDPC codes with essentially similar performance with Turbo codes approaching Shannon channel capacity. Other than cellular industry, LDPC is widely used now.

Table 6-23 below is the list of LDPC codes that I encountered so far, probably most often used in industry at this point of time. My starting point was DVB-S2, -S2X where LDPC was used early on (chips began to be available in 2006).

The most salient feature through the list is quasi-cyclic in structure of H (parity check matrix). This means that if any code word is shifted by the number of quasi-cyclic block size in bits then it becomes another codeword. It is called quasi-cyclic since it is cyclic in block but not in bit. A cyclic code like BCH and RS is cyclic in bit.

This quasi-cyclic structure is very useful for easy implementation of encoder as well as decoder. It is also useful in code description as well as code construction. Practically it can approach Shannon capacity as the block size becomes large. No random code like McKay code seems to become a standard even though it is often used for performance comparison.

The block size ranges from N=64K to less than 1K. The larger the block it gets closer to the channel capacity at the expense of implementation complexity.

The code rate ranges from ¼ to 9/10.

The code description varies depending on the standards. We need to convert it to parity check matrix H. For this conversion, the m-scripts may be written.

The description of codes in each standard is not exactly same but it can be converted to H parity check matrix. One common feature is the use of quasi-cyclic blocks and each block is a circulant (except for 10GBase –T, each block is not circulant but permutation matrix). And the parity part is band diagonal for ease of encoding. (DVB-S2 or like is slightly deviated but can be made band diagonal with row and column permutations.)

<u>Circulant defined</u>
An identity matrix may be circularly shifted to generate a circulant matrix. It is easy
to show with an example; an identity matrix (5, 5) to circulant

$$
\begin{bmatrix}
1 & 0 & 0 & 0 & 0 \\
0 & 1 & 0 & 0 & 0 \\
0 & 0 & 1 & 0 & 0 \\
0 & 0 & 0 & 1 & 0 \\
0 & 0 & 0 & 0 & 1
\end{bmatrix}
\rightarrow
\begin{bmatrix}
0 & 0 & 0 & 0 & 1 \\
1 & 0 & 0 & 0 & 0 \\
0 & 1 & 0 & 0 & 0 \\
0 & 0 & 1 & 0 & 0 \\
0 & 0 & 0 & 1 & 0
\end{bmatrix}
$$

identity matrix   column circular permutation:
                                     2 3 4 5 1

eye(5)   circshift(eye(5), [0, -1])
                                 circshift(eye(5), [0, +4])

In this example columns are circularly shifted to left by 1; this can be obtained by
permutation of columns with a permutation vector of [2 3 4 5 1]. Note that this can be
done circular shift right by 4 as well.

A convention of left shift is used in standards. Equivalently a row circular shift is
equivalently possible but we stick with column left shift to avoid confusion.

A circulant matrix is a special case of permutation matrix; random permutation of
row or column applied to an identity matrix is called permutation matrix. We show
examples;

$$
\begin{bmatrix}
1 & 0 & 0 & 0 & 0 \\
0 & 1 & 0 & 0 & 0 \\
0 & 0 & 1 & 0 & 0 \\
0 & 0 & 0 & 1 & 0 \\
0 & 0 & 0 & 0 & 1
\end{bmatrix}
\quad
\begin{bmatrix}
0 & 0 & 1 & 0 & 0 \\
0 & 0 & 0 & 0 & 1 \\
1 & 0 & 0 & 0 & 0 \\
0 & 1 & 0 & 0 & 0 \\
0 & 0 & 0 & 1 & 0
\end{bmatrix}
\quad
\begin{bmatrix}
1 & 0 & 0 & 0 & 0 \\
0 & 0 & 0 & 0 & 1 \\
0 & 0 & 0 & 1 & 0 \\
0 & 1 & 0 & 0 & 0 \\
0 & 0 & 1 & 0 & 0
\end{bmatrix}
$$

    identity matrix   row random permutation: column random permutation:
                                3   5   1   2   4   1   4   5   3   2

Exercise 6-54:   Check the above permutation matrix examples.

We give an example; IEEE 802.11n LDPC codes are specified by *block circulant*
as shown in Figure 6-46 and in Figure 6-47.

The notation of circulant block, e.g., '57', in Matlab language, circshift (eye (81),
[0, -57]). '0' means eye (81), identity matrix of size 81, in Figure 6-46. '-1' means no
connection. Code size = (n, k)=(1944, 972) =(24*81, 12*81).

Table 6-23: LDPC codes used in industry

system	Block length n	Code Rate	Parity check matrix construction	Remarks **Matlab script to obtain H**
DVB-S2, -S2X	N= 64800 N= 32400 N=16200	Many rates staring from ¼ to 9/10	Irregular repeat – accumulate (RA), Quasi-cyclic size =360	Optimized for N= 64800 with APSK constellations
GMR	N=8K, 4K, 2K, 1K	½, 2/3, ¾, 4/5, 9/10	Similar to DVB-S2 but 3 column weight sets	Rate is related with constellations similar to DVB-S2
DOCSIS 3.1 uplink	N=16200, 5940, 1120	8/9. 0.845, 0.75	Quasi-cyclic size: 360, 180, 56 similar to DVB-S2	N=480, 160 for ranging
Wi-Fi (IEEE 802.11n)	N=1944, 1296, 648	½, 2/3, ¾, 5/6	Quasi-Cyclic with circulant blocks Circulant block parity check – band or low triangle	Wireless LAN, IEEE 802.11n, 11ac
Wi-Fi (IEEE 802.11ad)	N=336, 252, 118, 126	½, 5/8, ¾, 13/16		
WiMAX (IEEE 16e)	N=2304, 1920, 1248,to 672	½, 2/3, ¾, 5/6		Wirless MAN, IEEE 802.16e, m
NASA CCSDS	(8160, 7136) LDPC + RS (255, 223, 33)	0.7648 (7136 / 8160 * 223 / 255)	QC with two circulant blocks per one location of proto H	Called C2, and concatenation with RS!
	K=16384, 4096, 1024	½, 2/3, 4/5	Essentially irregular RA codes	AR4JA codes
10GBase-T	(2048, 1723)	0.841	RS-based LDPC, (6,32) regular LDPC lifted by 64 x 64 permutation matrix (not circulant)	Not quasi-cyclic but close to random.

```
57  -1  -1  -1  50  -1  11  -1  50  -1  79  -1   1   0  -1  -1  -1  -1  -1  -1  -1  -1  -1  -1
 3  -1  28  -1   0  -1  -1  -1  55   7  -1  -1  -1   0   0  -1  -1  -1  -1  -1  -1  -1  -1  -1
30  -1  -1  -1  24  37  -1  -1  56  14  -1  -1  -1  -1   0   0  -1  -1  -1  -1  -1  -1  -1  -1
62  53  -1  -1  53  -1  -1   3  35  -1  -1  -1  -1  -1  -1   0   0  -1  -1  -1  -1  -1  -1  -1
40  -1  -1  20  66  -1  -1  22  28  -1  -1  -1  -1  -1  -1  -1   0   0  -1  -1  -1  -1  -1  -1
 0  -1  -1  -1   8  -1  42  -1  50  -1  -1   8  -1  -1  -1  -1  -1   0   0  -1  -1  -1  -1  -1
69  79  79  -1  -1  -1  56  -1  52  -1  -1  -1   0  -1  -1  -1  -1  -1   0   0  -1  -1  -1  -1
65  -1  -1  -1  38  57  -1  -1  72  -1  27  -1  -1  -1  -1  -1  -1  -1  -1   0   0  -1  -1  -1
64  -1  -1  -1  14  52  -1  -1  30  -1  -1  32  -1  -1  -1  -1  -1  -1  -1  -1   0   0  -1  -1
-1  45  -1  70   0  -1  -1  -1  77   9  -1  -1  -1  -1  -1  -1  -1  -1  -1  -1  -1   0   0  -1
 2  56  -1  57  35  -1  -1  -1  -1  -1  12  -1  -1  -1  -1  -1  -1  -1  -1  -1  -1  -1   0   0
24  -1  61  -1  60  -1  -1  27  51  -1  -1  16   1  -1  -1  -1  -1  -1  -1  -1  -1  -1  -1   0
```

Figure 6-46: IEEE 802.11n specifies LDPC (N=1944, R=1/2, circulant block size=81) where '-1' means no connection, and the numbered blocks are circulant. '0' means the identity matrix of size 81. Note that the parity part is band diagonal

Figure 6-47: IEEE
802.11n specifies LDPC
(N=1944, R=1/2, circulant
block size =81) where dot
means '1'and no dot means
no connection. Thus note
that the column weight is
1 and row weight is 1 in
each circulant block

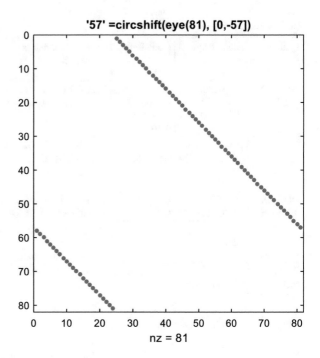

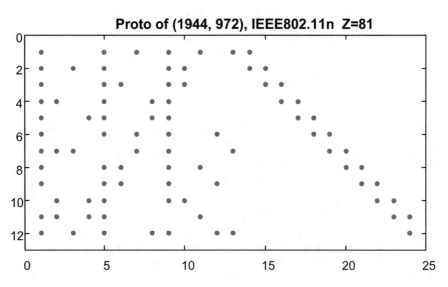

Figure 6-48: IEEE 802.11n specifies LDPC (N=1944, R=1/2, Z=81) by *proto* before lifting

The code described in Figure 6-46, (n, k) = (1944, 972), can start from a *proto* (24, 12) and then it is _lifted_ by circulant blocks of size Z=81. The *proto* is shown in Figure 6-48 where there is dot means a circulant will be inserted, and no dot means no connection, i.e., zero matrix of size Z will be inserted. This process is called *lift*; a small size proto becomes a larger, by factor of Z, parity check matrix.

6.5   Turbo Codes

A paper on parallel concatenated convolutional coding scheme was published in 1993 [2] by C. Berrou, A. Glavieurx and P. Thitimajshima at a conference of IEEE demonstrating that the possibility of a coding scheme approaching channel capacity is well within the technology at the time. It was one of seminal, land marking events in digital communication systems technology development. A rush of new efforts to find other possibilities was ensued, for example, resulting in the rediscovery of LDPC [3].

It was an ingenious combination of all known components – BCJR decoding of convutional codes, permutation (or interleaver), concatenation, and feedback – with added new piece of ideas – 'turbo decoding' called by the authors; two decoders are coupled through extrinsic information (E1, E2 in Figure 6-49), and thus an iterative decoding is possible.

We use Figure 6-49 throughout this section to explain Turbo codes in detail. It uses G=15/13 RSC which is covered before, and BCJR decoder is shown in log domain, i.e., log of probability and of its ratio (LLR), rather than probability directly. A block $\boxed{\pi}$ is permutation and $\boxed{\pi^{-1}}$ is its inverse (or depermutation).

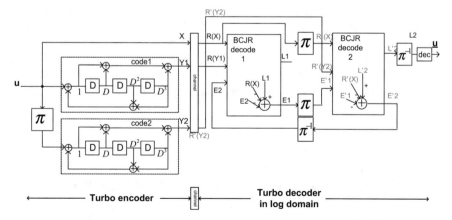

Figure 6-49:   Turbo encoder and decoder using G=15/13 RSC constituent codes, and decoding side is shown in log domain, rather than probability directly, and E1 and E2 coupling (or feedback) is called 'turbo'

Table 6-24: Summary of BCJR decoding in probability and in log probability

	Probability domain decoder computation	log domain decoder computation
APP of transmit bits	$\Pr\{X_t^{(1)}{=}0, Y_1^\tau\} =$ $\sum_{(m,m')\in B0t}\sigma_t(m',m)$ $\Pr\{X_t^{(1)}{=}1, Y_1^\tau\} =$ $\sum_{(m,m')\in B1t}\sigma_t(m',m)$ $\Pr\{X_t^{(1)}{=}0|Y_1^\tau\}, \Pr\{X_t^{(1)}{=}1|Y_1^\tau\}$ can be obtained by scaling of $\frac{1}{\Pr\{Y_1^\tau\}}$	$LLR\big(Xt^{(1)}\big|Y_1^\tau\big) = \log \frac{\Pr\{Xt^{(1)}=0\mid Y_1^\tau\}}{\Pr\{Xt^{(1)}=1\mid Y_1^\tau\}}$ $= \log \frac{\sum_{(m,m')\in B0t}e^{D_t(m',m)}}{\sum_{(m,m')\in B1t}e^{D_t(m',m)}}$ $= \log \sum_{(m,m')\in B0t}e^{D_t(m',m)}$ $\quad - \log \sum_{(m,m')\in B1t}e^{D_t(m',m)}$
Branch probability given Y_1^τ	$\sigma_t(m',m)= \alpha_{t-1}(m')\gamma_t(m',m)\,\beta_t(m)$	$D_t(m',m) = A_{t-1}(m') + \Gamma_t(m', m) + B_t(m)$
Forward and backward state probability at t	$\alpha_t(m) = \sum_{m'}\alpha_{t-1}(m')\,\gamma_t(m',m)$ $\beta_t(m) = \sum_{m'}\beta_{t+1}(m')\,\gamma_{t+1}(m',m)$ $\alpha_0(m) = \beta_\tau(m) = 1.0$ if $m = 0$ $\alpha_0(m) = \beta_\tau(m) = 0.0$ if $m \neq 0$	$A_t(m) = \log\sum_{m'}e^{A_{t-1}(m')+\Gamma_t(m',m)}$ $B_t(m) = \log\sum_{m'}e^{B_{t+1}(m')+\Gamma_{t+1}(m',m)}$ $A_0(m) = B_\tau(m) = 0.0$ if $m = 0$ $A_0(m) = B_\tau(m) = -\infty$ if $m \neq 0$
Trellis branch metric probability with Y_t	Single branch of X $\gamma_t(m',m)= \Pr\{ Y_t|X, S_t{=}m, S_{t-1}{=}m'\}$ $\Pr\{S_t{=}m \mid S_{t-1}{=}m'\}$	$\Gamma_t(m', m) = \log \Pr\{ Y_t\mid X, S_t = m, S_{t-1} = m'\} + \log \Pr\{S_t{=}m \mid S_{t-1}{=}m'\}$ when a branch is single (i.e., not of parallel connection) and X is branch transmit bits. There are 2 times total states of $m' \to m$ transitions, from t-1 to t.
	Parallel branches of X $\gamma_t(m',m) = \sum_X \Pr\{Y_t|X, S_t = m,$ $S_{t-1} = m'\}\ \Pr\{X|S_t = m,$ $S_{t-1} = m'\}\ \Pr\{S_t{=}m \mid S_{t-1}{=}m'\}$	When $Y_t = \{Y_t^{(1)}, Y_t^{(2)}\}$ at receive and corresponding $X = \{X_t^{(1)}, X_t^{(2)}\}$ at transmit, then $\Gamma_t(m', m) = \log \Pr\{ Y_t^{(1)} \mid X_t^{(1)}, S_t{=}m, S_{t-1}{=}m'\} + \log \Pr\{Y_t^{(2)}\mid X_t^{(2)}, S_t = m, S_{t-1} = m'\} + \log \Pr\{S_t{=}m \mid S_{t-1}{=}m'\}$

BCJR decoding is described in detail as a decoding of convolutional code. A summary of BCJR algorithm of computing APP is shown in Table 6-24 for Turbo code applications where the computation may be done in log domain for implementation practicality.

It is essential to familiarize oneself with RSC type convolutional codes and BCJR decoding in order to understand Turbo codes. It is recommended to review the above table thoroughly or better to review entire BCJR decoding of Section 6.3.3 if necessary.

6.5.1   Turbo Encoding with G=15/13 RSC and Permutation

An input to encoder is u=[1 1 0 1 0 1 0 1 1] and permutation π =[2 8 1 5 9 4 7 6 3]. The 1st encoder (Figure 6-49) will generate X, Y1 as shown in Figure 6-50. It adds 3 more bits [1 1 1] for termination to zero state for the 1st encoder.

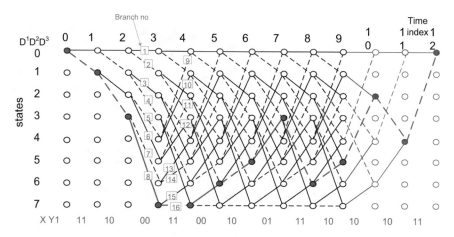

Figure 6-50: X Y1 generation from encoder 1 on trellis diagram; solid line means '0' input and dotted line means '1' input. Termination bits [1 1 1] are necessary to terminate to zero state

Exercise 6-55: $u=[1\ 1\ 0\ 1\ 0\ 1\ 0\ 1\ 1]$ permuted by $\pi =[2\ 8\ 1\ 5\ 9\ 4\ 7\ 6\ 3]$ becomes u $(\pi)=[1\ 1\ 1\ 0\ 1\ 1\ 0\ 1\ 0]$, i.e., order of transmission is altered by π .What is the inverse (or depermutation) π^{-1} ?

Answer: $\pi^{-1}= [3\ 1\ 9\ 6\ 4\ 8\ 7\ 2\ 5]$. Confirm that $u(\pi)$ returns to u after applying π^{-1}
    Y2 is generated from permuted u, shown in Figure 6-51, and thus state transition is entirely different from encoder 1 and thus termination bits, [0 1 1], are different as well. These termination bits may or may not be transmitted; in our numerical example later, no transmission. In practice, some of parity bits of Y1 and Y2 may not be transmitted. This is called puncturing, which is not shown in Figures 6-50 and 6-51, for the time being we may assume there is no puncturing. Puncturing is often used in order to change code rate.

6.5.2   G=15/13 Code Tables for BCJR Computation Organization

A visual representation of a convolution code can be done as a trellis, e.g., Figures 6-50 and 6-51. Another representation is by a code table as shown in Figure 6-52 where <table_bwd> is obtained by listing all the branches in a trellis with all the branch number sequence, from 1 to 16. Then <table_fwd> is obtained by row sorting of <s(t)> of <table_bwd>. Thus forward iterative computation, $\alpha_t(m)$, of BCJR decoding can be organized conveniently. <table_bwd> is organized for backward iterative computation of $\beta_t(m)$. However, it turns out that in a Matlab program only <table_bwd> is used for BCJR decoding organization.
    The first column of <table_bwd> is branch numbers, <br_no>; it is implicit in the table, not stored as part of it. <br_no> is designated in Figure 6-50 as $\boxed{1}, \boxed{2}, \ldots, \boxed{16}$.

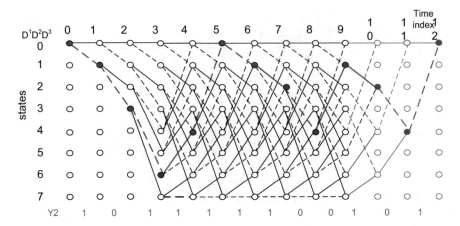

Figure 6-51: Y2 generation from encoder 2 with permuted input of u. Termination bits are [0 1 1] which may not be transmitted. The permuted u (or X) is not transmitted

	table_bwd						table_fwd			
br no.	st(t-1)	xin	X	Y	st(t)		st(t-1)	xin	X	Y	st(t)
1	0	0	0	0	0		0	0	0	0	0
2	0	1	1	1	1		4	1	1	1	0
3	1	0	0	1	2		0	1	1	1	1
4	1	1	1	0	3		4	0	0	0	1
5	2	1	1	0	4		1	0	0	1	2
6	2	0	0	1	5		5	1	1	0	2
7	3	1	1	1	6		1	1	1	0	3
8	3	0	0	0	7		5	0	0	1	3
9	4	1	1	1	0		2	1	1	0	4
10	4	0	0	0	1		6	0	0	1	4
11	5	1	1	0	2		2	0	0	1	5
12	5	0	0	1	3		6	1	1	0	5
13	6	0	0	1	4		3	1	1	1	6
14	6	1	1	0	5		7	0	0	0	6
15	7	0	0	0	6		3	0	0	0	7
16	7	1	1	1	7		7	1	1	1	7

Figure 6-52: Two code tables of G=15/13 code ; branch numbers are designated in <table_bwd> and shown in Figure 6-50. <table_fwd> is obtained by row permutation of <table_bwd>. In a Matlab program only <table_bwd> is used for BCJR decoding

In the table the actual first column is <st(t-1)> which is the beginning state, and the 5th column <st(t)> will be the next state. An encoder will generate <X> and <Y> depending on state transition. In the table <xin> which is the same as <X> and thus redundant; however, it is deliberately done so in order to emphasize the 3 input decoder as shown in Figure 6-49. The other decoder will give an equivalent <xin> as 'extrinsic' information E_2 and E_1 as shown in Figure 6-49.

6.5.3   The generation of 'extrinsic' information (E1, E2)

In Figure 6-49 BCJR decoder 1 has three inputs, $R(X)$, $R(Y1)$ and E2, and then generates APP of u, information bits, designated as L_1. In log domain it is a log of APP. In addition to L_1, an 'extrinsic' information E_1 is generated and it is symbolically $E_1 = L_1 - R(X) - E_2$ in the figure and literally correct if each term is LLR. Thus E_1 is APP of u only using $R(Y1)$ while L_1 is APP of u using all three inputs, $R(X)$, $R(Y1)$ and E2. Note that u and <xin> are the same.

   In Figure 6-49 BCJR decoder 2 has three inputs, $R'(X)$, $R'(Y2)$ and E_1', and then generates APP of *permuted* information bits, designated as L_2'. $R'(X)$ means permuted $R(X)$, and similarly for $R'(Y2)$ and E_1'. In addition to L_2', an 'extrinsic' information E_2' is generated and it is symbolically $E_2' = L_2' - R'(X) - E_1'$ in the figure. Again E_2' is APP of permuted u only using $R'(Y2)$ while L_2' is APP of permuted u using all three inputs, $R'(X)$, $R'(Y2)$ and E_1'. When E_2' is used by 1st decoder it must be depermuted to E_2 by using $\boxed{\pi^{-1}}$.

Exercise 6-56: When E_2' is depermuted to E_2 by using $\boxed{\pi^{-1}}$, note that termination part (3 bits in our example) cannot be used. Can you see why?

Answer: The state transitions of 1st encoder and 2nd encoder are different and thus termination bits are not the same. This is shown in Figure 6-50 and 6-51. The termination bits can be treated as erased, i.e., probability $=0.5$ or LLR $=0.0$. Similarly it is necessary to use erased termination bits in the case of permutation from $R(X)$ to $R'(X)$ for the same reason.

   The understanding of how to generate E_1 and E_2 is important to the coupling of two decoders, i.e., turbo decoding; a key innovation. We explain the computation of 'extrinsic' information in probability domain, and that of log domain will follow in a straightforward manner. We need to revisit trellis branch probability defined in (6-47) and its simplified version is listed in Table 6-24 as $\gamma_t(m',m) = \Pr\{\ Y_t \mid X,\ S_t=m,\ S_{t-1}=m'\}\ \Pr\{S_t=m \mid S_{t-1}=m'\}$. It can be, assuming that equally likely information bits, simplified to, where Y_t is received and X_t is transmitted,

$$\gamma_t(m',m) = \Pr\{Y_t \mid X_t, S_t = m, S_{t-1} = m'\}$$

where $Y_t = \{R(X), R(Y1), E_2\}$ at time t for decoder 1, and $X_t = \{X, Y1, xin\}$ at time t. Similarly for decoder 2, Y_t and X_t are understood. This is a bit of notational abuse but justified with no confusion. Thus for decoder 1, $\gamma_t(m',m) = \Pr\{R(X) \mid X, S_t = m, S_{t-1} = m'\} \Pr\{R(Y1) \mid Y1, S_t = m, S_{t-1} = m'\} \Pr\{E2 \mid xin, S_t = m, S_{t-1} = m'\}$ at t, assuming independence, i.e., no correlation. Similarly for decoder 2.

   This trellis branch metric in probability will be used for computing $\alpha_t(m)$, $\beta_t(m)$ and will be used for computing APP of branches, $\sigma_t(m',m) = \alpha_{t-1}(m')\gamma_t(m',m)\ \beta_t(m)$, and for computing APP of X, designated as $\Pr\{X_t^{(1)}=0 \mid Y_1^{\tau}\}$, $\Pr\{X_t^{(1)}=1 \mid Y_1^{\tau}\}$ in Table 6-24. It will be L_1 for decoder 1, and L_2 for decoder 2 in Figure 6-49 after taking log.

Now for E_1 computation at decoder 1, branch metric probability will be

$$\gamma_t(m',m)|_{Y1} \Pr\{R(Y1)|Y1, S_t = m, S_{t-1} = m'\}$$

And its corresponding APP of branches, in order to compute APP of X, i.e., E_1

$$\sigma_t(m',m)|_{Y1} = \alpha_{t-1}(m')\, \gamma_t(m',m)|_{Y1}\quad \beta_t(m)$$

$$\Pr\{E1 = 0 \text{ at } t\} = \frac{1}{scale} \sum_{(m,m')\in C0} \sigma_t(m',m)\Bigg|_{Y1}$$

$$\Pr\{E1 = 1 \text{ at } t\} = \frac{1}{scale} \sum_{(m,m')\in C1} \sigma_t(m',m)\Bigg|_{Y1}$$

where C0 is a set of branches so that transmit data being '0' and C1 a set of branches so that transmit data being '1'. From <table_bwd>, reading column 2, C0= {1,3,5,7,9,11,13,15} and C1={2,4,6,8,10,12,14,16}. *scale* is for the sum to be unity. Similarly E2 for decoder 2 can be computed.

6.5.4   *Numerical Computations of Iterative Turbo Decoding*

We continue on with the same example and hope turbo coding to be understood quickly. The information bit (infoK =9) u=[1 1 0 1 0 1 0 1 1] using permutation π =[2 8 1 5 9 4 7 6 3] will generate X, Y1 and Y2 as shown in Figures 6-50 and 6-51 with termination bits. No puncturing, thus 36 bits are transmitted. After BPSK modulation and added noise (Es/No= 2dB, i. e. , $\sigma^2 = 0.3155$), i.e., AWGN channel, we obtain 36 received samples as shown in Figure 6-53.

The goal of turbo decoder is to process received samples and compute APP of u, or of <xin> or <X> since we use RSC. A Matlab script is written for it. In hardware implementation log domain processing might be used more often. However, here we use probability domain computation with the hope of understanding decoding structure quickly and in computer simulation multiplications are fast.

We use (6-50), the result of sub-section 6.3.3.4, in order to convert received samples to probability (Figure 6-54).

$$p_0 = 1/\left(1 + e^{-2r/\sigma^2}\right), \quad p_1 = 1/\left(1 + e^{+2r/\sigma^2}\right), \quad \frac{1}{\sigma} = \sqrt{\frac{2Es}{No}}$$

Pr (r I x=0) = 1/(1+exp(-2r / sgm^2))
Pr (r I x=1) = 1/(1+exp(+2r / sgm^2)) where sgm^2 = noise variance =0.3155.

transmit

receive

1-2X	1-2Y1	1-2Y2
-1	-1	-1
-1	1	1
1	1	-1
-1	-1	-1
1	1	-1
-1	1	-1
1	-1	-1
-1	-1	1
-1	1	1
-1	1	-1
-1	1	1
-1	-1	-1

channel
$\longrightarrow$

Xr	Y1r	Y2r
-0.6980	-0.5926	-0.7254
0.0300	0.9646	1.5812
-0.2687	1.4015	-0.5917
-0.5157	-1.1151	-1.1704
1.1790	0.9303	-0.8349
-1.7345	1.8367	-1.4422
0.7565	-0.2086	-0.5010
-0.8076	-0.2040	0.3557
1.0099	1.3772	0.3996
0.5555	0.3218	-1.4547
-1.7582	1.4029	-0.6537
0.7046	-0.0843	-0.1921

Figure 6-53: Example of transmit BPSK symbols and received samples after AWGN channel. No puncturing here, i.e., all parity bits are transmitted. If punctured, punctured samples are inserted with the value being zero at the receiver

$Pr(r|x=0) = 1/(1+exp(-2r/sgm^2))$
$Pr(r|x=1) = 1/(1+exp(+2r/sgm^2))$

where sgm^2 = noise variance = 0.3155

Xr	Pr(Xr|x=0)	Pr(Xr|x=1)	Y1r	Pr(Y1r|Y1='0')	Pr(Y1r|Y1='1')	Y2r	Pr(Y2r|Y2='0')	Pr(Y2r|Y2='1')
-0.6980	0.0118	0.9882	-0.5926	0.0228	0.9772	-0.7254	0.0100	0.9900
0.0300	0.5475	0.4525	0.9646	0.9978	0.0022	1.5812	1.0000	0.0000
-0.2687	0.1540	0.8460	1.4015	0.9999	0.0001	-0.5917	0.0229	0.9771
-0.5157	0.0366	0.9634	-1.1151	0.0009	0.9991	-1.1704	0.0006	0.9994
1.1790	0.9994	0.0006	0.9303	0.9973	0.0027	-0.8349	0.0050	0.9950
-1.7345	0.0000	1.0000	1.8367	1.0000	0.0000	-1.4422	0.0001	0.9999
0.7565	0.9918	0.0082	-0.2086	0.2104	0.7896	-0.5010	0.0401	0.9599
-0.8076	0.0059	0.9941	-0.2040	0.2153	0.7847	0.3557	0.9051	0.0949
1.0099	0.9983	0.0017	1.3772	0.9998	0.0002	0.3996	0.9265	0.0735
0.5555	0.9713	0.0287	0.3218	0.8849	0.1151	-1.4547	0.0001	0.9999
-1.7582	0.0000	1.0000	1.4029	0.9999	0.0001	-0.6537	0.0156	0.9844
0.7046	0.9886	0.0114	-0.0843	0.3694	0.6306	-0.1921	0.2283	0.7717

Figure 6-54: Received samples are converted to conditional probabilities to compute branch metric probabilities

For example branch metric probability of <X>, the above is arranged in a matrix form.

Similarly for Y1, and Y2, $\gamma_t(m', m)$ can be arranged as in Figure 6-55. This arrangement is useful in Matlab programming taking advantage of speedy matrix operations.

For E2 input for 1st decoder, it will be initialized to be 1/2 with the same format as Figure 6-55. Thus $\gamma_t(m', m)$ for all three inputs Xr, Y1r and E_2 can be computed by a matrix (16 x 12) multiplication; in Matlab code it is $prx.*pry1.*prE_2$.

Using the recursion for $\alpha_t(m)$ and $\beta_t(m)$ (see Table 6-24), we obtain, for 1st decoder, after iteration 1 as in Figure 6-56.

$$\gamma_t(m',m) \quad \text{for X}$$

X	t=1	t=2	t=3	t=12
0	0.0118	0.5475	0.1540	0.9886
1.0000	0.9882	0.4525	0.8460	0.0114
0	0.0118	0.5475	0.1540	0.9886
1.0000	0.9882	0.4525	0.8460	0.0114
1.0000	0.9882	0.4525	0.8460	0.0114
0	0.0118	0.5475	0.1540	0.9886
1.0000	0.9882	0.4525	0.8460	0.0114
0	0.0118	0.5475	0.1540	0.9886
1.0000	0.9882	0.4525	0.8460	0.0114
0	0.0118	0.5475	0.1540	0.9886
1.0000	0.9882	0.4525	0.8460	0.0114
0	0.0118	0.5475	0.1540	0.9886
0	0.0118	0.5475	0.1540	0.9886
1.0000	0.9882	0.4525	0.8460	0.0114
0	0.0118	0.5475	0.1540	0.9886
1.0000	0.9882	0.4525	0.8460	0.0114

Figure 6-55: Branch metric probabilities for X, similar branch probabilities for Y1, and Y2

$$\alpha_t(m) \quad \text{for 1}^{\text{st}} \text{ decoder, iteration 1}$$

	t=1	t=2	t=3	t=4	t=5	t=6	t=7	t=8	t=9	t=10	t=11	t=12
S=0	1.0000	0.0003	0.0003	0.0003	0.0145	0.0145	0.0000	0.0000	0.0000	0.0000	0.0018	0.0002	1.0000
S=1	0	0.9997	0.0000	0.0000	0.0003	0.0000	0.0000	0.0000	0.0000	0.1484	0	0	0
S=2	0	0	0.0027	0.0000	0.0000	0.0000	0.0000	0.0022	0.0000	0.0000	0.3724	0	0
S=3	0	0	0.9970	0.0000	0.0000	0.0000	0.0000	0.9978	0.0000	0.0000	0	0	0
S=4	0	0	0	0.0145	0.0000	0.0000	0.0000	0.0000	0.0006	0.0397	0.6256	0.9998	0
S=5	0	0	0	0.0000	0.0000	0.0000	1.0000	0.0000	0.0000	0.4071	0	0	0
S=6	0	0	0	0.0007	0.0000	0.9855	0.0000	0.0000	0.9977	0.4047	0.0002	0	0
S=7	0	0	0	0.9845	0.9851	0.0000	0.0000	0.0000	0.0016	0.0000	0	0	0

$$\beta_t(m) \quad \text{for 1}^{\text{st}} \text{ decoder, iteration 1}$$

	t=1	t=2	t=3	t=4	t=5	t=6	t=7	t=8	t=9	t=10	t=11	t=12
S=0	1.0000	0.0000	0.0000	0.0000	0.0008	0.0008	0.0000	0.0000	0.0056	0.0046	0.0007	0.9808	1.0000
S=1	0	1.0000	0.0003	0.0000	0.0000	0.6521	0.9640	0.0000	0.0000	0.8055	0	0	0
S=2	0	0	0.0132	0.0015	0.0000	0.0000	0.0001	0.9351	0.0000	0.0013	0.9923	0	0
S=3	0	0	0.9865	0.0001	0.0000	0.0000	0.0067	0.0013	0.0000	0.0000	0	0	0
S=4	0	0	0	0.0023	0.6582	0.0016	0.0000	0.0193	0.9871	0.0000	0.0070	0.0192	0
S=5	0	0	0	0.0000	0.0000	0.0077	0.0035	0.0000	0.0000	0.1830	0	0	0
S=6	0	0	0	0.0681	0.0000	0.3377	0.0199	0.0204	0.0004	0.0056	0.0000	0	0
S=7	0	0	0	0.9280	0.3409	0.0000	0.0058	0.0239	0.0069	0.0000	0	0	0

Figure 6-56: State probabilities of 1st decoder after iteration 1. Starting from zero state to end state. Note that there are errors compared to Figure 6-50 (encoder)

Exercise 6-57: APP of states can be obtained using $\lambda_t(m) = \Pr\{S_t = m, \ Y_1^\tau\} = \alpha_t(m)\,\beta_t(m)$ and dividing $\lambda_t(m)$ by $\Pr\{Y_1^\tau\}$. It can be done by $\alpha_t(m)\,\beta_t(m)$ and proper scaling. How can one do the scaling?

Answer: Scale so that after the multiplication the sum of each column should be unity. In our example we do not need $\lambda_t(m)$ but $\alpha_t(m)$ and $\beta_t(m)$ separately in order to compute outputs of decoder (L1, E1). See Figure 6-57.

outputs of 1st decoder, iteration 1

		t=1	t=2	t=3	t=4	t=5	t=6	t=7	t=8	t=9	t=10	t=11	t=12
L1	Pr(X=0|rcvd)	0.0000	0.0000	0.9999	0.0001	1.0000	0.0000	0.3911	0.0116	0.6205	0.6205	0.0000	0.0116
	Pr(X=1|rcvd)	1.0000	1.0000	0.0001	0.9999	0.0000	1.0000	0.6089	0.9884	0.3795	0.3795	1.0000	0.9884
E1	Pr(X=0|Y1r)	0.0000	0.0000	1.0000	0.0015	0.9999	0.0001	0.0053	0.6621	0.0027	0.0461	0.1971	0.0001
	Pr(X=1|Y1r)	1.0000	1.0000	0.0000	0.9985	0.0001	0.9999	0.9947	0.3379	0.9973	0.9539	0.8029	0.9999

termination to zero state

Figure 6-57: Outputs of 1st decoder after iteration 1 in probabilities

Permutation: [2   8   1   5   9   4   7   6   3]

For feedback, permuted E1 of 1st decoder, iteration 1

π (E1)	Pr(X=0|Y1r)	0.0000	0.6621	0.0000	0.9999	0.0027	0.0015	0.0053	0.0001	1.0000	0.5000	0.5000	0.5000
	Pr(X=1|Y1r)	1.0000	0.3379	1.0000	0.0001	0.9973	0.9985	0.9947	0.9999	0.0000	0.5000	0.5000	0.5000

Permuted termination is
not the same

$\alpha_t(m)$ for 2nd decoder, iteration 1

	t=1	t=2	t=3	t=4	t=5	t=6	t=7	t=8	t=9	t=10	t=11	t=12
S=0	1.0000	0.0000	0.0000	0.0000	0.0000	0.9918	0.0000	0.0000	0.0000	0.0000	0.0005	0.0005	1.0000
S=1	0	1.0000	0.0000	0.0000	0.0000	0.0082	1.0000	0.0000	0.0000	0.9995	0	0	0
S=2	0	0	0.0000	0.0000	0.0000	0.0000	0.0000	0.9390	0.0000	0.0000	0.9995	0	0
S=3	0	0	1.0000	0.0000	0.0000	0.0000	0.0000	0.0610	0.0000	0.0000	0	0	0
S=4	0	0	0	0.0000	1.0000	0.0000	0.0000	0.0000	0.9932	0.0005	0.0000	0.9995	0
S=5	0	0	0	0.0000	0.0000	0.0000	0.0000	0.0000	0.0000	0.0000	0	0	0
S=6	0	0	0	1.0000	0.0000	0.0000	0.0000	0.0000	0.0068	0.0000	0.0000	0	0
S=7	0	0	0	0.0000	0.0000	0.0000	0.0000	0.0000	0.0000	0.0000	0	0	0

$\beta_t(m)$ for 2nd decoder, iteration 1

	t=1	t=2	t=3	t=4	t=5	t=6	t=7	t=8	t=9	t=10	t=11	t=12
S=0	1.0000	0.0000	0.0008	0.0000	0.0021	0.2344	0.0008	0.0001	0.0000	0.0000	0.0036	0.2283	1.0000
S=1	0	1.0000	0.0000	0.0007	0.0000	0.0002	0.0819	0.0008	0.0002	0.0120	0	0	0
S=2	0	0	0.0000	0.0000	0.0002	0.0000	0.0000	0.1254	0.0000	0.0000	0.0120	0	0
S=3	0	0	0.9992	0.0019	0.0062	0.0000	0.6450	0.0003	0.0003	0.7596	0	0	0
S=4	0	0	0	0.0000	0.2321	0.0023	0.0001	0.0000	0.0405	0.0036	0.2248	0.7717	0
S=5	0	0	0	0.0260	0.0000	0.0001	0.0055	0.0000	0.2027	0.0000	0	0	0
S=6	0	0	0	0.9713	0.0038	0.0000	0.0000	0.6279	0.0010	0.2247	0.7596	0	0
S=7	0	0	0	0.0000	0.7556	0.7630	0.2666	0.2454	0.7554	0.0001	0	0	0

Figure 6-58: State probabilities of 2nd decoder after iteration 1. Starting from zero state to end state. Note that there are errors compared to Figure 6-51 (encoder)

E_1 should be permuted and add erased termination bits in order to be an input to 2nd decoder. This is shown below.

For 2nd decoder three inputs, R'(X), R'(Y2) and E_1', all permuted, will be used. Y2 permuted at transmit side, and thus proper termination bits transmitted. For permutation of X, we need to use erased (i.e., unknown) termination bits similar to E1. Again using the recursion for $\alpha_t(m)$ and $\beta_t(m)$, we obtain, for 2nd decoder, after iteration 1 as in Figures 6-58 and 6-59.

Now for the 2nd iteration 1st decoder has three inputs R(X), R(Y1) and E2 (depermuted), and then generates outputs L_1 and E_1. Then updated, permuted version of E_1 along with , R'(X), R'(Y2) will be the inputs to 2nd decoder. And this iteration may continue until converged or stop after fixed number of iterations (Figures 6-59 and 6-60).

In Figure 6-49, depermuted (π^{-1}) L2 is shown as the output of Turbo decoder. From this example it is clear that L1 can be the output as well.

outputs of 2nd decoder, iteration 1

		t=1	t=2	t=3	t=4	t=5	t=6	t=7	t=8	t=9	t=10	t=11	t=12
L2	Pr(X=0|rcvd)	0.0000	0.0000	0.0000	1.0000	0.0000	0.0000	0.9998	0.0000	1.0000	0.9998	0.0002	0.0002
	Pr(X=1|rcvd)	1.0000	1.0000	1.0000	0.0000	1.0000	1.0000	0.0002	1.0000	0.0000	0.0002	0.9998	0.9998
E2	Pr(X=0|Y2r)	0.0000	0.0000	0.0000	1.0000	0.0000	0.0000	0.9999	0.6347	1.0000	0.9998	0.0002	0.0002
	Pr(X=1|Y2r)	1.0000	1.0000	1.0000	0.0000	1.0000	1.0000	0.0001	0.3653	0.0000	0.0002	0.9998	0.9998

termination to zero state

Inverse permutation [3 1 9 6 4 8 7 2 5]

For feedback, inverse permuted E2 of 2nd decoder, iteration 1

		t=1	t=2	t=3	t=4	t=5	t=6	t=7	t=8	t=9	t=10	t=11	t=12
$\pi^{-1}(E2)$	Pr(X=0|Y2r)	0.0000	0.0000	1.0000	0.0000	1.0000	0.6347	0.9999	0.0000	0.0000	0.5000	0.5000	0.5000
	Pr(X=1|Y2r)	1.0000	1.0000	0.0000	1.0000	0.0000	0.3653	0.0001	1.0000	1.0000	0.5000	0.5000	0.5000

Inverse permuted termination is not the same

Figure 6-59: Output of 2nd decoder L2 and E2 in probabilities and depermuted E2

outputs of 1st decoder, iteration 2

		t=1	t=2	t=3	t=4	t=5	t=6	t=7	t=8	t=9	t=10	t=11	t=12
L1	Pr(X=0|rcvd)	0.0000	0.0000	1.0000	0.0000	1.0000	0.0000	1.0000	0.0000	0.0000	0.0000	0.0000	0.0000
	Pr(X=1|rcvd)	1.0000	1.0000	0.0000	1.0000	0.0000	1.0000	0.0000	1.0000	1.0000	1.0000	1.0000	1.0000
E1	Pr(X=0|Y1r)	0.0000	0.0000	1.0000	0.0000	1.0000	0.0000	0.9992	0.0000	0.0000	0.0000	0.0000	0.0000
	Pr(X=1|Y1r)	1.0000	1.0000	0.0000	1.0000	0.0000	1.0000	0.0008	1.0000	1.0000	1.0000	1.0000	1.0000

outputs of 1st decoder, iteration 3

		t=1	t=2	t=3	t=4	t=5	t=6	t=7	t=8	t=9	t=10	t=11	t=12
L1	Pr(X=0|rcvd)	0.0000	0.0000	1.0000	0.0000	1.0000	0.0000	1.0000	0.0000	0.0000	0.0000	0.0000	0.0000
	Pr(X=1|rcvd)	1.0000	1.0000	0.0000	1.0000	0.0000	1.0000	0.0000	1.0000	1.0000	1.0000	1.0000	1.0000
E1	Pr(X=0|Y1r)	0.0000	0.0000	1.0000	0.0000	1.0000	0.0000	1.0000	0.0000	0.0000	0.0000	0.0000	0.0000
	Pr(X=1|Y1r)	1.0000	1.0000	0.0000	1.0000	0.0000	1.0000	0.0000	1.0000	1.0000	1.0000	1.0000	1.0000

outputs of 2nd decoder, iteration 2

		t=1	t=2	t=3	t=4	t=5	t=6	t=7	t=8	t=9	t=10	t=11	t=12
L2	Pr(X=0|rcvd)	0.0000	0.0000	0.0000	1.0000	0.0000	0.0000	1.0000	0.0000	1.0000	1.0000	0.0000	0.0000
	Pr(X=1|rcvd)	1.0000	1.0000	1.0000	0.0000	1.0000	1.0000	0.0000	1.0000	0.0000	0.0000	1.0000	1.0000
E2	Pr(X=0|Y2r)	0.0000	0.0000	0.0000	1.0000	0.0000	0.0000	0.9999	0.3443	1.0000	1.0000	0.0000	0.0000
	Pr(X=1|Y2r)	1.0000	1.0000	1.0000	0.0000	1.0000	1.0000	0.0001	0.6557	0.0000	0.0000	1.0000	1.0000

outputs of 2nd decoder, iteration 3

		t=1	t=2	t=3	t=4	t=5	t=6	t=7	t=8	t=9	t=10	t=11	t=12
L2	Pr(X=0|rcvd)	0.0000	0.0000	0.0000	1.0000	0.0000	0.0000	1.0000	0.0000	1.0000	1.0000	0.0000	0.0000
	Pr(X=1|rcvd)	1.0000	1.0000	1.0000	0.0000	1.0000	1.0000	0.0000	1.0000	0	0.0000	1.0000	1.0000
E2	Pr(X=0|Y2r)	0.0000	0.0000	0.0000	1.0000	0.0000	0.0000	0.9999	0.3443	1.0000	1.0000	0.0000	0.0000
	Pr(X=1|Y2r)	1.0000	1.0000	1.0000	0.0000	1.0000	1.0000	0.0001	0.6557	0.0000	0.0000	1.0000	1.0000

Figure 6-60: 2nd and 3rd iteration results. Note that L1 recovers correctly transmit bits at the 2nd iteration, and that E2 at $t=8$ does not change after 2nd iteration

Exercise 6-58*: Write a program for decoding example in Section 6.5.4 in log / LLR domain. * means a project exercise.

Exercise 6-59: In order to appreciate the inventiveness of turbo part (feedback and iteration), we consider a concatenated decoder as shown in Figure 6-61. Hint: there is no feedback thus no iteration possible.

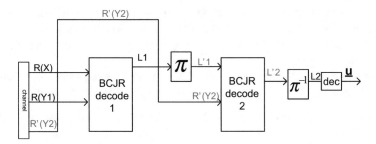

Figure 6-61: A concatenated decoder to use all the information of three inputs for discussion exercise

6.5.5   Additional Practical Issues

So far we use one short block (N=12, R ≈ 1/3, G=15/13) Turbo code to understand how it works. It is a small code but good enough to learn the workings of Turbo code in general. In this section we discuss practical issues of Turbo codes; log domain decoding, permutation (interleaver) and more examples of constituent convolution codes.

6.5.5.1   Decoding Computation in Log Domain

As shown in Table 6-24, log domain computation is in practice often used to use addition operation instead of multiplication operations. However, we need to go back to probability in case of e.g., $A_t(m) = \log \sum_{m'} e^{A_{t-1}(m') + \Gamma_t(m',m)}$.

This is called Jacobi logarithm as $\max^*(x,y) = \log (e^x + e^y) = \max (x,y) + \log (1 + e^{-|x-y|}) = \max (x,y) + f_c (|x-y|)$ where function f_c can be approximated.

One such approximation is given by

$\max^*(x,y) \approx \max (x,y) + 0$ (if $|x-y| > 1.5$) and
$\max^*(x,y) \approx \max (x,y) + 0.5$ (if $|x-y| \leq 1.5$).

The same organization of program - e.g., using <table_bwd> can be done as in our example in Section 6.5.4.

6.5.5.2   Permutations (Interleaver)

The permutation is used to indicate the order of transmission, i.e., which bit is transmitted in which time. If a block of information data is numbered from 1, 2, ..., K, a trivial example is a permutation of [K, K-1, ..., 2, 1], which means transmitting in completely reverse order. Another example is called a uniform interleaving, e.g., for K=12, [1, 5, 9, 2, 6, 10, 3, 7, 11, 4, 8, 12]. This is equivalent write in row and read in column with 3 rows and 4 columns. A uniform interleaving is often used to combat fading in time, in order to spread adjacent symbols far apart so that after fading erased (or bad) symbols appear random. Spreading is related with the duration of fading.

Sometimes a structured way of obtaining permutation sequence, with desirable randomness and separation, may be done by a permutation polynomial.

Example 6-16: quadratic permutation polynomial $y = 6x^2 + x \ mod \ 12$

0 1 2 3 4 5 6 7 8 9 10 11

$\pi \downarrow$

0 7 2 9 4 11 6 1 8 3 10 5

The permutation inside Turbo code, also called interleaving by tradition, does not have the same need of adjacent symbol separation, but, rather it should be 'random' so that extrinsic information is uncorrelated with (or independent of) other signal, say X itself, and thus the branch metric probabilities of three inputs (X, Y, E) can be multiplied with the assumption of independence, or at least very little correlation. Thus the choice has a big influence to the error performance of Turbo codes. Another one is with the choice of a constituent convolution code.

With a block size K, there are K factorial (K!) permutations possible, a huge number with any practical block size. Unfortunately there is no known way of selecting an optimum permutation sequence. It is done empirically (i.e., trial and error) in practice.

3GPP interleaver with G=15/13 constituent code is given as a standard; Multiplexing and channel coding (FDD) T_1TRQ3GPP 25.212-330 ver. 330, pp. 16 -20.

The range of K (a number of information bits): K=40 – 5114. A Matlab script is written based on the standard.

Example 6-17: K=41. In this example 41st bit is transmitted first and then 31st so on.

infoK=41 bit sequence   Altered transmission order

1	2	3	4	5	6	7
8	9	10	11	12	13	14
15	16	17	18	19	20	21
22	23	24	25	26	27	28
29	30	31	32	33	34	35
36	37	38	39	40	41	

Permutation (interleaver)

π

41	31	21	11	1	37	22
18	7	35	24	19	5	32
28	16	2	33	25	14	3
40	30	20	10	34	29	13
4	36	27	12	6	39	23
15	9	38	26	17	8	

In the receiver permuted sequence is restored by depermutation.

1	2	3	4	5	6	7
8	9	10	11	12	13	14
15	16	17	18	19	20	21
22	23	24	25	26	27	28
29	30	31	32	33	34	35
36	37	38	39	40	41	

dePermutation
(deinterleaver)
←
π^{-1}

Restore original order

5	17	21	29	13	33	9
41	37	25	4	32	28	20
36	16	40	8	12	24	3
7	35	11	19	39	31	15
27	23	2	14	18	26	10
30	6	38	34	22	1	

Exercise 6-60*: Write a program for Example 6-17 for all possible K. * means a project exercise.

6.5.5.3 Constituent Convolution Codes

In IEEE 802.16 standard two bit input (duo-binary) convolution code is used. We do not list here; see Figure 6-20 in Std 802.16.1-2012. A motivation for duo-binary convolution code is to improve error performance, in particular in the region of error floor.

Exercise 6-61: Typically in convolution code alone, a non-systematic code is often used in practice. However, in Turbo code situation, systematic convolution codes are used. Why?

Answer: In Turbo decoding two BCJR decoders need three inputs (X, Y, E). For the 2nd decoder X can be obtained by permutation without transmitting permuted X except for termination bits.

A number of good convolution codes are listed in Tables 6-10 and 6-11, typically non-systematic form, which can be converted to RSC form as shown in Figures 6-21 and 6-25. In general in this way one can start with a good constituent code.

Another example constituent convolution code is shown in Figure 6-62.

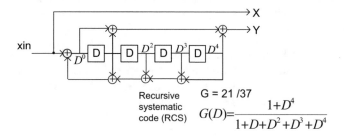

Recursive
systematic
code (RCS)

G = 21 /37

$$G(D) = \frac{1+D^4}{1+D+D^2+D^3+D^4}$$

Figure 6-62: Convolution code of G=21/37 in Berrous's original paper [2]

Exercise 6-62: Obtain a code table, similar to Figure 6-52 (<table_bwd>), for G=21/37 RSC. Answer: See below. There are 16 states and 32 branches.

Code table for G=21/37 code

br no.	st(t-1)	xin	X	Y	st(t)	br no.	st(t-1)	xin	X	Y	st(t)
1	0	0	0	0	0	17	8	1	1	0	0
2	0	1	1	1	1	18	8	0	0	1	1
3	1	1	1	0	2	19	9	0	0	0	2
4	1	0	0	1	3	20	9	1	1	1	3
5	2	1	1	1	4	21	10	0	0	1	4
6	2	0	0	0	5	22	10	1	1	0	5
7	3	0	0	1	6	23	11	1	1	1	6
8	3	1	1	0	7	24	11	0	0	0	7
9	4	1	1	1	8	25	12	0	0	1	8
10	4	0	0	0	9	26	12	1	1	0	9
11	5	0	0	1	10	27	13	1	1	1	10
12	5	1	1	0	11	28	13	0	0	0	11
13	6	0	0	0	12	29	14	1	1	0	12
14	6	1	1	1	13	30	14	0	0	1	13
15	7	1	1	0	14	31	15	0	0	0	14
16	7	0	0	1	15	32	15	1	1	1	15

6.6   Coding Applications

So far we covered binary linear error correcting codes. These binary codes are most often used. In decoding of such code, the received samples are assumed to use binary modulations, often BPSK. Further conversion to their probabilities and log likelihood ratios are based on binary modulations. In this section we cover coding applications to higher order modulations. In order to make our discussion concrete and simple we use specific LDPC codes – IEEE 802.11n LDPC codes. See [7] IEEE Std 802.11n -2009 Annex R.

During code development it is a common practice to use BPSK with AWGN channel. In some situations other channel models like BSC, DMC in addition to AWGN may be used. Other models are related with AWGN, and may be considered as additional simplification for binary code development. Our focus will be on AWGN channel. As a reminder on channel model we provide Exercise 6-63 below.

Exercise 6-63: In Table 6-15, BSC, DMC and AWGN channel models are listed. A three level quantization of DMC is called an erasure channel. Show this is the case and find an erasure probability. Answer: See below. Note that P_B can change depending on the threshold (e.g., ½) and SNR (σ).

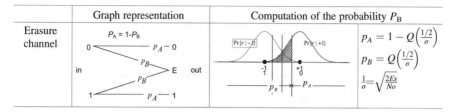

	Graph representation	Computation of the probability P_B
Erasure channel	$P_A = 1 - P_B$ in $0 \rightarrow P_A \rightarrow 0$ $P_B \rightarrow E$ out P_B $1 \rightarrow P_A \rightarrow 1$	$P_A = 1 - Q\left(\frac{1/2}{\sigma}\right)$ $P_B = Q\left(\frac{1/2}{\sigma}\right)$ $\frac{1}{\sigma} = \sqrt{\frac{2E_s}{N_o}}$

Back to Figure 6-63, a code word (or block) of N bits is generated by encoder and mapped into BPSK symbols {+1, -1}. A channel is AWGN adding noise. And received samples {r} are demapped into LLR, which is particularly simple in BPSK, as shown in the figure, LLR=$2r/\sigma^2$. The decoder needs to process LLR to generate a posterior LLR (or APP) and then decision to get K information bits. See A.2 for additional detail of LLR.

We rely on numerical simulations of Figure 6-63; sending enough number of codewords and counting errored blocks i.e., block error rate (or word error rate, WER). The result is shown in Figure 6-64. We use LDPC N=1944 R=2/3 of IEEE 802.11n.

It is a dramatic improvement; by reading the graph, 5% SER of BPSK is reduced to 1E-6 BLER after LDPC decoding. For horizontal scale of SNR is Es/No in dB. Note that we use Es/No measurable at receive, i.e., we did not discount transmit power increase of 1.76dB (or rate loss).

6.6.1 Coded Modulations

In applying binary codes to higher order modulations (multi bits), we update Figure 6-63 slightly to Figure 6-65; mapper and demapper need to be specified, and modulation symbols are complex (numbers) rather than real.

We now apply the same LDPC code to 8 level (3bit) constellations; specifically 8-PSk and 8-DSQ (double squared QAM) shown in Figures 6-66 and 6-67 respectively.

A mapper table is shown (far LHS); constellation number, 3 bits, and outputs (complex number). This is a specific assignment of an input bit pattern into a constellation point. S_0 and S_1 are a set of constellation points such that S_0 set carries '0' of a specific bit and S_1 does '1'; e.g., for b_2, S_0 ={0, 1, 2, 3} carries '0' and S_1={4, 5, 6, 7} does '1'.

Exercise 6-64: For b_1 find a set of constellation points which carry '0'. Answer: S_0 = {0, 1, 4, 5}. For b_0, find a set of constellation points which carry '0'. S_0 = {0, 2, 4, 6}. Note that there is slight abuse of notation by using S_0 for both.

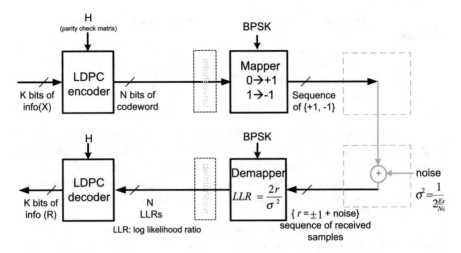

Figure 6-63: LDPC code performance simulation with BPSK

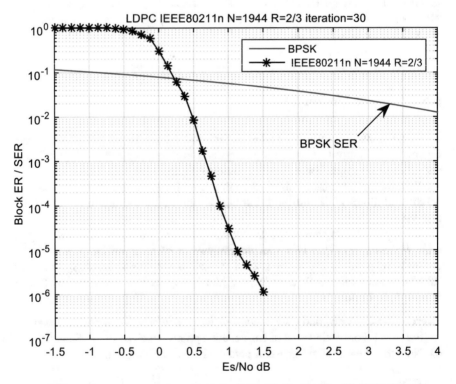

Figure 6-64: Performance simulation of LDPC with BPSK; N=1944 R=2/3 (IEEE 802.11n)

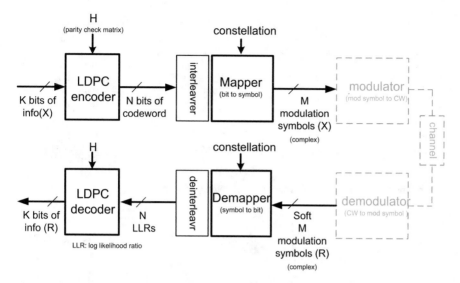

Figure 6-65: Binary LDPC codes (specified by parity check matrix H) are applied to higher order modulations specified by constellation. Interleaver (deinterleaver) is explicitly shown

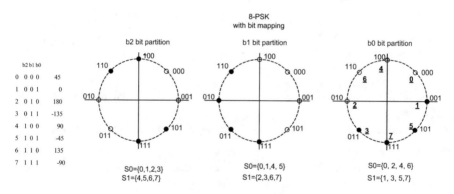

Figure 6-66: 8-PSK constellation and mapper, demapper

6.6.1.1   LLR Computations of Multi-Level Constellations

From LDPC decoder point of view, it needs LLR of each bit. In BPSK it is simply a scaled version of received sample; $LLR = 2r / \sigma^2$. We need to compute LLR of each bit from samples for high order constellations. We use the same constellation of Figures 6-66 and 6-67 for concreteness. We present the result first and then followed by additional explanations.

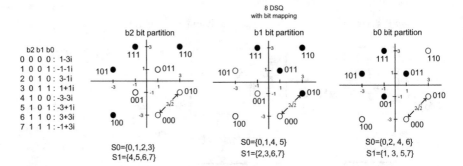

Figure 6-67: 8-DSQ constellation and mapper, demapper

$$LLR(b_2|r) = \ln \frac{\sum_{p_i \in S_0} \exp\left(-\frac{|r-p_i|^2}{2\sigma^2}\right)}{\sum_{p_j \in S_1} \exp\left(-\frac{|r-p_j|^2}{2\sigma^2}\right)} \text{ where } S_0 = \{0,1,2,3\} \text{ and } S_1 = \{4,5,6,7\}$$

$$LLR(b_1|r) = \ln \frac{\sum_{p_i \in S_0} \exp\left(-\frac{|r-p_i|^2}{2\sigma^2}\right)}{\sum_{p_j \in S_1} \exp\left(-\frac{|r-p_j|^2}{2\sigma^2}\right)} \text{ where } S_0 = \{0,1,4,5\} \text{ and } S_1 = \{2,3,6,7\}$$

$$LLR(b_0|r) = \ln \frac{\sum_{p_i \in S_0} \exp\left(-\frac{|r-p_i|^2}{2\sigma^2}\right)}{\sum_{p_j \in S_1} \exp\left(-\frac{|r-p_j|^2}{2\sigma^2}\right)} \text{ where } S_0 = \{0,2,4,6\} \text{ and } S_1 = \{1,3,5,7\}$$

Note that this result has been shown in Section 2.5.3 (Chapter 2) for 16QAM example.

Additional explanation on how to obtain the above result is given below;

$$p(b_2 = '0'|r) = \frac{p(r|p_0)p(p_0) + p(r|p_1)p(p_1) + p(r|p_2)p(p_2) + p(r|p_3)p(p_3)}{P(r)}$$

$$p(b_2 = '1'|r) = \frac{p(r|p_4)p(p_4) + p(r|p_5)p(p_5) + p(r|p_6)p(p_6) + p(r|p_7)p(p_7)}{P(r)}$$

Taking the ratio, we have, assuming all constellation points are equally likely,

$$\frac{p(b_2 = '0'|r)}{p(b_2 = '1'|r)} = \frac{p(r|p_0)p(p_0) + p(r|p_1)p(p_1) + p(r|p_2)p(p_2) + p(r|p_3)p(p_3)}{p(r|p_4)p(p_4) + p(r|p_5)p(p_5) + p(r|p_6)p(p_6) + p(r|p_7)p(p_7)}$$

$$\times \frac{p(r|p_0) + p(r|p_1) + p(r|p_2) + p(r|p_3)}{p(r|p_4) + p(r|p_5) + p(r|p_6) + p(r|p_7)}$$

$$LLR(b_2|r) = \ln \frac{p(b_2 = '0'|r)}{p(b_2 = '1'|r)} = \ln \frac{p(r|p_0) + p(r|p_1) + p(r|p_2) + p(r|p_3)}{p(r|p_4) + p(r|p_5) + p(r|p_6) + p(r|p_7)}.$$

We obtain the result by plugging $p(r|p_0) = \frac{1}{\sqrt{2\pi\sigma^2}} \exp\left(-\frac{|r-p_0|^2}{2\sigma^2}\right)$ into the above. Similarly for LLR$(b_1|r)$, and LLR$(b_0|r)$.

Exercise 6-65: For BPSK, LLR $= \ln \frac{p(r|p_0)}{p(r|p_1)}$ for 2 point constellation. Show LLR $= \frac{2r}{\sigma^2}$.

Answer: Constellation points of BPSK, $p_0 = +1$, and $p_1 = -1$. Thus LLR $= \ln \frac{p(r|p_0)}{p(r|p_1)}$

$= \ln \frac{\exp\left(-\frac{|r-1|^2}{2\sigma^2}\right)}{\exp\left(-\frac{|r+1|^2}{2\sigma^2}\right)} = \frac{2r}{\sigma^2}$.

A demapper computation can be organized conveniently by a mapper table shown on far LHS of Figures 6-66 and 6-67 such that 'natural' binary sequence. In this way a mapper table is a list of constellation points and assigned bits are implicit.

Exercise 6-66: Confirm bit-to-symbol look up tables in Figures 6-66 and 6-67.

Exercise 6-67: Compute the average digital power of constellation of 8-DSQ in Figure 6-67. Hint: note that distance between constellation points is $2\sqrt{2}$. Answer: 10.

6.6.1.2 Performance of 8-DSQ and 8-PSK with LDPC Code

A simulation result is shown in Figure 6-68. The performance of BPSK is displayed as well for comparison. We can make a few observations;

a. No permutation (interleaver) used
b. Coded performance difference between 8-DSQ and 8-PSK is only 0.1dB
c. Required SNR for 8-DSQ (and 8-PSK) requires 6.7dB more than BPSK
d. Block error rate curve seems to shift horizontally, preserving the shape

We show, with other examples later, that there is no need of interleaver with AWGN channel. For time selective fading channel, the length of interleaver should be longer than the duration of fade so that the channel may appear to be AWGN.

Uncoded performance difference between 8-DSQ and 8-PSK is such that 8-DSQ is superior. With the same distance of 2 between constellation points the average power of 8-DSQ is 5.0 and that of 8-PSK is about 7.0, and thus the difference in dB is 1.46 dB (10 log (7/5)). SER performance of both shows the difference asymptotically at high SNR region. See Figure 6-69. Yet the coded performance, with LDPC, is very similar within 0.1dB. This indicates that a constellation may be chosen with other considerations; e.g., peak to average power ratio (PAPR) of 8-PSK is 1.0 while that of 8-DSQ is 1.8.

Required SNR of coded 8-PSK and 8-DSQ seems a horizontal shift from BPSK according to **c** and **d** above. This means that the use of BPSK during code development is well justified; decoupling code development and its application to high order constellation development. One quick prediction of the amount of horizontal shift is

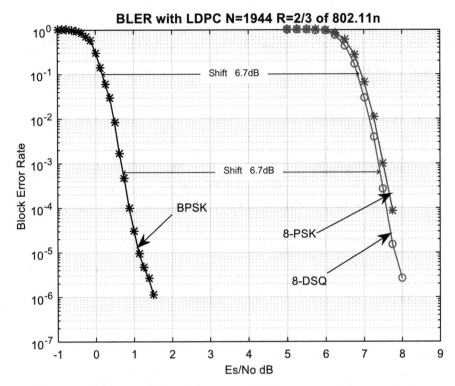

Figure 6-68: Block error rate performance of 8-DSQ, 8-PSK and BPSK with LDPC code (N=1944, R=2/3 of IEEE 802.11n). There is a horizontal shift 6.7dB compared to BPSK

using AWGN capacity formula; C=log$_2$ (1+SNR), and SNR = 2^C-1. With the current example 7.08 dB {=10 log [(2^(3*2/3) -1) / (2^(2/3)-1)]}, off by 0.38dB. This is quick but needs to be sharpened but predicts an approximate performance of higher order constellations from BPSK. For the capacity formula, see Appendix A.4 for additional details.

Exercise 6-68: Obtain an analytical expression of SER of 8-PSK and 8-DSQ in terms of Es/No. Hint: review Chapter 2 for SER.

Answer: x = sqrt(2*Es/No/5); SER =21/8*Q(x) - 2*Q(x)*Q(x); for 8DSQ
        x = sqrt(2*Es/No*sin(π/M)*sin(π /M)), M=8; SER = 2.0*Q(x); for 8PSK

6.6.1.3 More Coded Modulation Examples with 16QAM and 64QAM

We continue to use LDPC, N=1944 R=3/4 and R=1/2 of IEEE802.11n. In order to obtain 3 bits /Hz bandwidth efficiency, we use 16QAM with R=3/4, and 64QAM with R=1/2 code. The result of simulations is shown in Figure 6-70. We observe a few things;

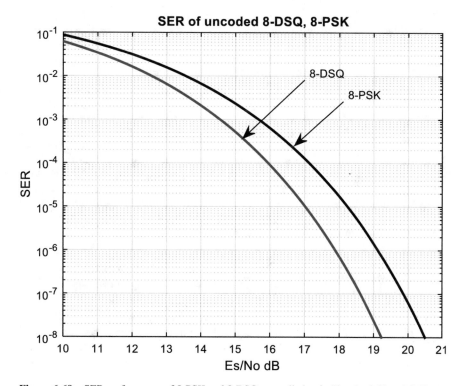

Figure 6-69: SER performance of 8-PSK and 8-DSQ constellation in Figures 6-66 and 6-67

- We used random permutation (interleaver); no or little difference.
- Horizontal shift of SNR; for 16QAM, $10 \log\{((2^{\wedge}(4*3/4)-1)/(2^{\wedge}(3/4)-1))\} = 10.1$ dB from capacity formula but actually 9.2dB, 0.9dB difference, and for 64QAM, $10\log\{((2^{\wedge}(6*1/2)-1)/(2^{\wedge}(1/2)-1))\} = 12.3$ dB predicted from capacity formula but actually 12.8 dB , 0.5dB difference, somewhat crude but well indicative in approximation.
- Mapper of 16QAM is the same as Figure 2-31 in Chapter 2 (Gray) and 64QAM mapper Gray as well.

From LDPC point of view, 16-QAM may be considered as 4 bit parallel channel and 64QAM as 6 bit parallel channel. In Figure 6-71, we display error performance of each bit (b3, b2, b1, b0) channel for 16QAM; they are unequal such that b2, b3 is much better than b1,b0. The combined performance is in-between as shown in Figure 6-71. This means that LDPC decoder averages out all bit channel performance.

Exercise 6-69: In 16-QAM, LDPC decoder uses 4 parallel channels of bit (b_0, b_1, b_2, b_3). As indicated in Figure 6-71, (b_2, b_3) bit channels are far better than (b_0, b_1). Confirm this is the case.

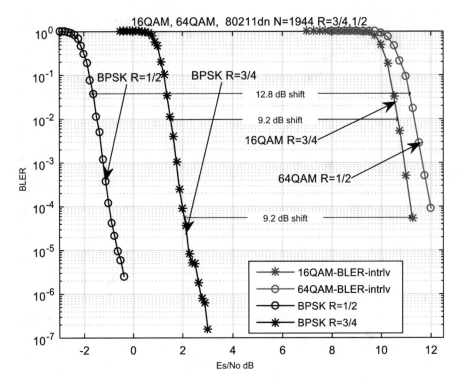

Figure 6-70:   3 bit /Hz obtained by 16QAM R=3/4 and 64QAM R=1/2

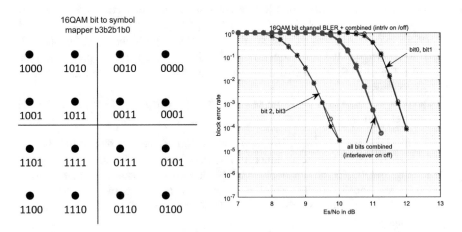

Figure 6-71: Mapper of 16QAM and each bit channel performance; b0, b1 much worse than b2, b3

Exercise 6-70: At 11.5dB of Es/No in Figure 6-71 (RHS), WER of (b_0, b_1) is about 1E-2, and that of (b_2, b_3) is very small, thus assume to be zero. When 4 parallel bit channels are used independently (i.e., no averaging effect of LDPC decoder), what is the expected combined WER? Answer: 0.5 * 1E-2, which is far worse than the all bits combined curve ($\approx$ 1E-5); LDPC decoder utilizes good bit channels of (b_2, b_3) well.

6.6.1.4 Gray Bit Assignment

We emphasize Gray bit assignment. This was discussed in Figure 2-33 of Chapter 2. This insight can be obtained looking at BER without coding. With Gray, BER is minimized. From LDPC point of view, BER is more important than SER.

Exercise 6-71: 1st quadrant of 4 bits is assigned by Gray and each quadrant is assigned by b4 b5 as shown RHS. Complete 64QAM bit assignment to be Gray.

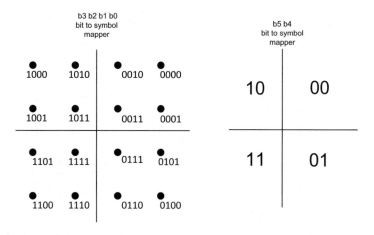

We repeat bit channel analysis for 64QAM, and it is shown in Figure 6-72.

6.6.1.5 On Permutation, Interleaver, and Scrambler

We use permutation and interleaver somewhat interchangeably. It may be desirable to distinguish them; in Turbo code context, permutation seems more appropriate on the other hand interleaver might be more so in case of handling time selective fading. In case of interleaver it can be done by modulation symbols rather than bits. We summarize it as in Figure 6-73.

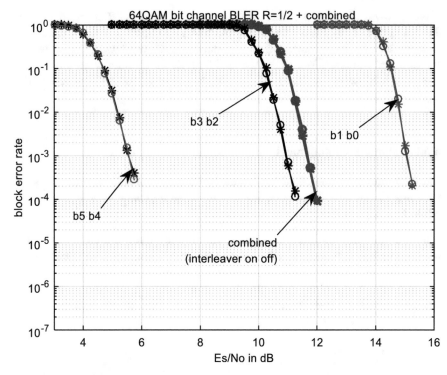

Figure 6-72: Performance of 64QAM bit channel, with Gray mapping; notice a large gap between b1 b0 with b4 b5 , around 9dB

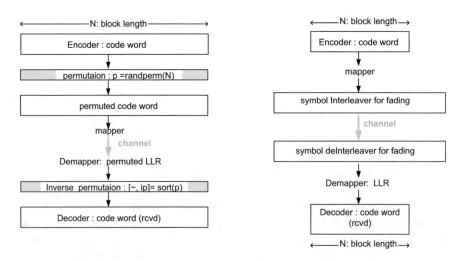

Figure 6-73: Different use of terminologies: permutation (LHS) and interleaver (RHS)

Figure 6-74: Scrambler
and descrambler

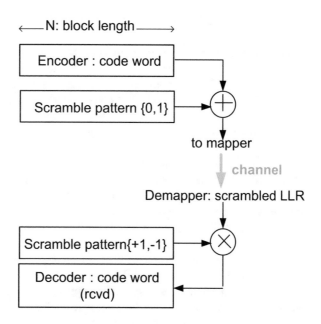

In order to be effective for time selective fading, the length of interleaver should be longer than average fade duration. In that case the block length of a code can be smaller than the fade duration.

Another device related with permutation and interleaver is a ***scrambler***; randomizing bits or modulation symbols. The output of encoder may be scrambled, i.e., adding (in mod 2) random bit pattern to the encoder output. The scrambled pattern can be descrambled at the receiver by inverting the sign of received samples (assuming BPSK) when scramble pattern is '1'. Even if an encoder generates zero codeword (it is a codeword for any linear code), actual transmission symbols are looking random and using all symbols available on constellation. This is summarized in Figure 6-74.

A scrambler may be useful for doing simulations without encoder, i.e., sending zero code word and for other things.

6.6.2 MLCM, TCM and BICM

So far we code all the bits, and a coupling between coding and modulation, similar to BICM, is through mapper and demapper, in particular soft bits, i.e., LLR of each bit. We noted that there is no need of interleaver (permutation) under AWGN channel. We now explore MLCM with higher order modulations. TCM is not covered here since MLCM can be used instead with block codes as well as convolutional codes. But, there is a vast literature on TCM if interested.

6.6.2.1 A Brief History of MLCM, TCM and BICM

In this subsection we review the application of coding to communication signals and systems. When HD was prevalent, mainly due to complexity, with algebraic code like BCH and RS, even convolution code, it was natural to separate coding and modulation; demodulator gives HD output to decoder. In this case entire bits are coded. When Ungerbeck published a paper on TCM scheme in 1982 [4], a convolutional code was tightly coupled with modulation, i.e., redundancy is accomplished by expanding constellation, and only selected bits of modulation symbol, after proper bit assignment with set partitioning, are coded. As this idea is well publicized, it was clear that there were similar, but not exactly the same, ideas. For example see MLCM of Imai [5] published in 1977. Another idea similar to MLCM, independently invented in-house to solve implementation speed requirements, has been implemented as a product [6]. TCM was extensively studied during 90s, and adapted as standards of voice band modems.

However, if it was applied to time selective fading channel, it was not very effective. BICM was a reaction to TCM in this context; after long enough interleaving in order to make time selective fading appear random, entire symbols are coded and Gray bit assignment rather than set partitioning assignment and soft decision. This coding was effective in handling time selective fading. We tend to use BICM as a synonym to coding all the bits but keep in mind that interleaving is for time selective fading and it can also be done by symbol interleaving as suggested in Figure 6-73 (RHS).

6.6.2.2 MLCM Idea

So far all the bits are coded, and a coupling between coding and modulation, similar to BICM, is through mapper and demapper, in particular soft bits, i.e., LLR of each bit. In MLCM, only selected bits in a symbol are coded. From Figure 6-72 (also from Figure 6-71), it is clear that each bit channel is dramatically different; b_4b_5 bit (MSB) channel is far better than b_1b_0 (LSB). Each bit channel may have a different code; b1b0 should have more coding, i.e., lower rate code and b_4b_5 may be coded lightly, i.e., higher rate code, and b_2b_3 may have intermediate code rate. In MLCM, b_4b_5 is not coded (code rate 1.0) and $b_0b_1b_2b_3$ are coded by LDPC. We describe it with a concrete example as shown in Figure 6-75.

6.6.2.3 MLCM Mapper (64QAM Example)

In BICM case, we emphasized Gray bit assignment. In MLCM we do not maintain Gray assignment when 16QAM is expanded to 64QAM but use a set partitioning so that the Euclidean distance, between constellation points (say ■) after expansion, must be maximum. This can be accomplished by translation of 16QAM into each

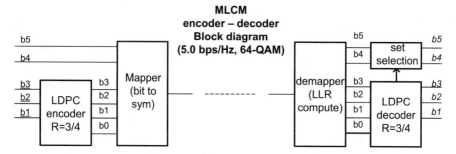

Figure 6-75: A block diagram of MLCM with 64QAM and 5 bit per sec /Hz bandwidth efficiency. Mapper will be different as well as demapper and decoding compared to BICM

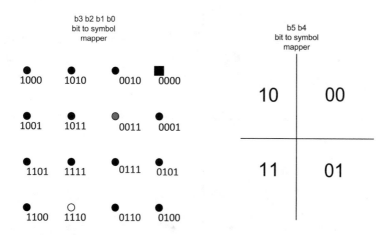

Figure 6-76: 16QAM will be expanded to 64QAM appropriate as MLCM mapper; 16QAM is Gray assigned and quadrant is assigned by Gray

quadrant. The set of b3 b2 b1 b0 is Gray assigned after expansion but overall it is not Gray due to the quadrant boundary. In other words coded bits (b3 b2 b1 b0) should be Gray after expansion, i.e., in 64QAM constellation. The result is shown in Figure 6-77.

The distance of constellation of uncoded bits is 4 times of adjacent symbols; 12 dB difference in SNR. When the constellation size of coded bits is QPSK rather than 16QAM it would be 6dB. The constellation size of coded bits can be 8 as well. The constellation size of coded block may be 4, 8, 16 even 32. Here we chose 16, which is typical when we use a powerful code like LDPC. It can be smaller when the underlying code is less powerful.

Exercise 6-72: When the constellation size of coded bits is 8, obtain SNR difference between coded bits and uncoded bits. Answer: 9 dB $= 10 \log (8)$, similarly for 12 dB with the size 16, and 6dB with the size 4.

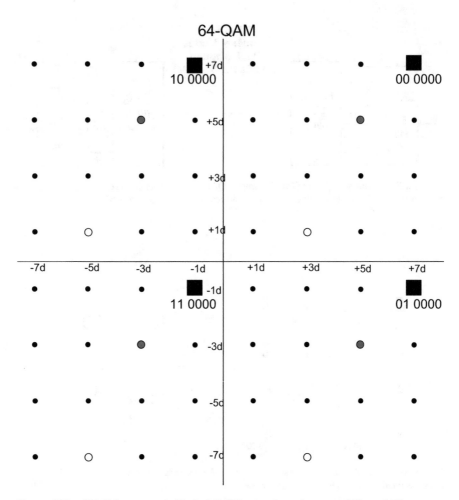

Figure 6-77: 64QAM mapper suitable for MLCM using the assignment of Figure 6-76

6.6.2.4   MLCM Implementation with N=1944 LDPC

With N=1944, R=3/4 code, 486 symbols of 64QAM can be transmitted in a code block since 1944/4= 486 as shown in Figure 6-78. These 64QAM symbols are through AWGN and become received samples.

In decoding side, LLRs of b3 b2 b1 b0 are computed from received samples with 64 point constellation, which are the input to LDPC decoder. The output of LDPC decoder will be used as 4 point constellation to compute LLRs of b4 b5, which will be used to make decision on b4 b5 bits.

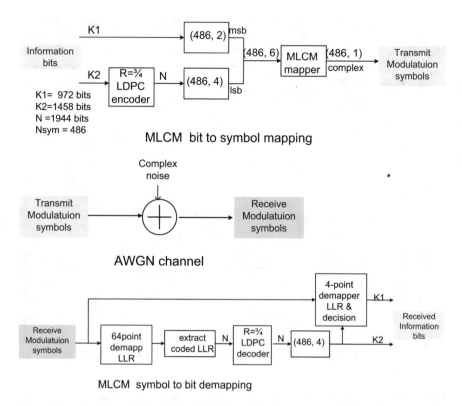

MLCM bit to symbol mapping

AWGN channel

MLCM symbol to bit demapping

Figure 6-78: With N=1944, R=3/4 code, 486 symbols of 64QAM can be transmitted in a code block since 1944/4= 486. Thus overall code rate is 5/6, i.e., 5 bits per symbol

6.6.2.5 Performance Comparison of MLCM with BICM

Note that here BICM means all the bits are coded with Gray assignment and no permutation (interleaver). Both have the same 5 bit per symbol bandwidth efficiency.

6.6.2.6 An Extension to 5.5 Bits/Symbol, 5.25 and 5.75 Bits/Symbol

By changing LDPC code rate, we can obtain 5.25, 5.5 and 5.75 bit /symbol system as shown in the table below; note that 5.5 bit/symbol system requires two consecutive symbols, and 5.25 and 5.75 requires 4 consecutive symbols.

LDPC Rate	3/4	13/16	7/8	15/16
Bit /symbol	5.0	5.25	5.5	5.75

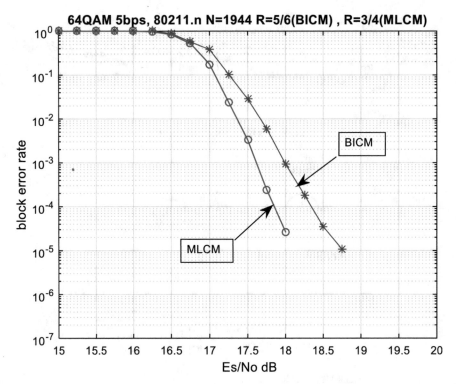

Figure 6-79: Performance of MLCM (64QAM with R=3/4 mapper as shown in Figure 6-77) compared with BICM (64QAM with R=5/6); there is net gain of 0.5dB or more

A block diagram of 5.5 bit /symbol with two consecutive symbols is shown in Figure 6-80. By changing code rate of LDPC a variety of bandwidth efficiency is doable but a number of consecutive symbols may increase.

Exercise 6-73: Using three consecutive symbols 4.67, 5.0, 5.33, and 5.67 bits / symbol system can be done. What are LDPC code rate necessary? Answer: 8/12, 9/12, 10/12 and 11/12.

6.6.2.7   An Example of MLCM to 128 DSQ

The idea of MLCM, in summary, is that constellation points are partitioned into two sets – coded set and uncoded set – and their relationship is a translation in two dimensional spaces, as an example of 64QAM in Figure 6-77 clearly shows. In each set, a bit assignment should be Gray so that the mix-up of adjacent constellations results in small bit errors. It may be called as a tessellation on plane.

We now extend this idea of MLCM to the constellation size of 128. With this size (2^7) of constellation — we may use 128 DSQ so that tessellation is possible. Two

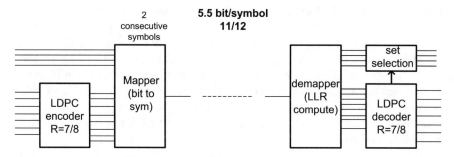

Figure 6-80: 5.5 bits / symbol system with two consecutive symbols

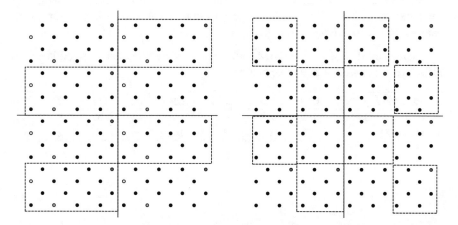

Figure 6-81: 128DSQ constellation is partitioned into two sets of coded block and uncoded block; coded block size of 16 on LHS and of 8 on RHS

examples of partition is shown in Figure 6-81. On LHS coded block size is 16, i.e. 4 bits of LSB, and on RHS coded block size is 8, i.e., 3 bits of LSB.

With 128 DSQ the bit assignment of coded sets cannot be completely Gray. That of uncoded sets can be completely Gray. It can be Gray as much as possible.

Exercise 6-74: The statement on the bit assignment being Gray above is true. Convince yourself it is true. Answer: try out all possible arrangement. You realize that some constellation points have too many neighboring constellation points. Gray assignment is possible for uncoded blocks.

Note that by rotating 128DSQ constellation by 45° counter-clock-wise it is a square constellation; constellation points on a diagonal will be on y-axis and overall in diamond shape.

Exercise 6-75: Try out 45° rotation of 128DSQ. Confirm that it is a square constellation.

6.6.2.8   An MLCM Mapper with 128 Points and its Extensions

Since the constellation block of 16 may be reasonable for powerful FECs like LDPC we fix its size and expand constellation size to 128. Then we need 8 of such blocks.

One possible MLCM mapper for 128 constellation points is shown in Figure 6-81; a coded block of 16QAM with Gray will be translated as shown in LHS of the figure to get an MLCM mapper. It is not unique; treating a shaded block as a constellation point (there are 16 of them). Another possibility is a coded block of 8 as 8-PSK and 8-DSQ.

Exercise 6-76: Try out other possible MLCM mapper of 128 constellation points with 16 QAM coded block as in Figure 6-82. Answer: All possibilities are huge but if we limit 2 per quadrant and allow only a certain symmetrical pattern, the combinations of 2 out of 4, thus 6. We need to choose one with desirable properties; the least power in average and in peak.

The idea of MLCM with 16QAM coded block can be extended to 512, 2048, and 8192. We summarize it as a table below;

Constellation size: Cn	8192	4096	2048	1024	512	256	128	64
No of 16QAM blocks	512	256	128	64	32	16	8	4
Per quadrant	128	64	32	16	8	4	2	1

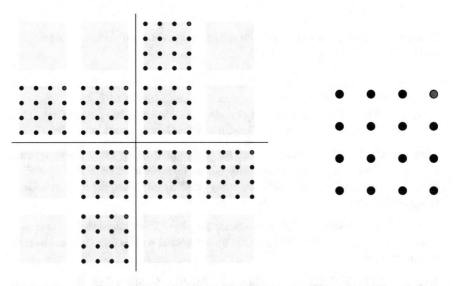

Figure 6-82: MLCM mapper for 128 constellation points; a coded block of 16QAM with Gray will be translated as shown in LHS to get an MLCM mapper. It is not unique; treating a shaded block as a constellation point, another possibility is 8PSK like and 8DSQ like choice

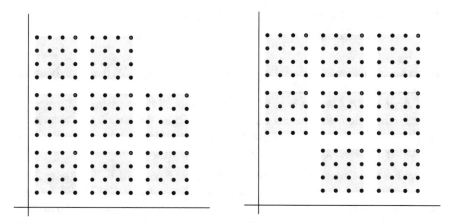

Figure 6-83: MLCM mapper for Cn=512 showing only 1st quadrant. Two possibilities are shown out of 9

For example in case of Cn = 512, 32 of 16 QAM blocks are treated as if a 32 point constellation and in each quadrant 8 are assigned. Thus the problem of MLCM mapper is now a translation of 16QAM blocks to 32 selected block-points.

Exercise 6-77: In one quadrant assignment of Cn = 512 case, how many possibilities are there? There are 8 of 16QAM blocks in a quadrant. Answer: if it is arranged compactly, then 3 x 3 = 9 square can be used. Since there are 8 of such blocks to be assigned to 9, there are 9, i.e., one empty spot can be any one of 9. For example two possible assignments of 1st quadrant are shown below.

Exercise 6-78: In Figure 6-83 find digital power of Cn = 512 constellation for both LHS and RHS. Answer: Hint: write a mapper and find average power; $\frac{1}{512} \sum_{p=1}^{512} |s_p|^2$.

In fact the idea can be extended to non-power of 2 constellation points but multiple of 16. It is preferable to be a multiple of 64 to have the same number per quadrant. For example Cn = 1563 = 16* 96 = 64* 24. It is not hard to get MLCM mapper for this case. 24 blocks in one quadrant are assigned to 5 x 5 squares emptying one block, perhaps one in corner; there are many more possibilities.

Exercise 6-79: Plan how to do mapping for Cn=1563. You need to use two consecutive symbols so that one symbol carries 10.5 bits /symbol.

Answer: It can carry more than 10.5 bits per symbol ($\log_2 (1536) = 10.585$), and thus 21 bits can be assigned to two consecutive symbols. Naming $I_1 I_2 I_3 \ldots I_{21}$ from LSB to MSB, $I_1 I_2 I_3 \ldots I_8$ are assigned to two 4 bits (16QAM), $I_9 I_{10} I_{11} \ldots I_{18}$ are two 5bits (32 constellation blocks of 16QAM), and $I_{19} I_{20} I_{21}$ are assigned to two ternary (A, B, C) symbols. There are 3 (A, B, C) of 32 constellation blocks of 16QAM, i.e., 32*3 = 96. Two consecutive symbols have nine (AA, BA, CA, AB, BB, CB, AC, BC, CC) combinations but need only 8, and thus CC is not used.

6.6.3   Channel Capacity of AWGN and of QAM Constellations

The concept of channel capacity is very useful since it provides the upper bound of performance, a reference point.

In Appendix A.4 we cover some basics of information theory including entropy, mutual information, and channel capacity. In particular we explain how to compute the channel capacity of AWGN and of DMC with constellation of equally likely probability distribution. Some results are shown in Figure 6-84.

Exercise 6-80: Our numerical examples of Figures 6-68, 6-70, 6-79 achieves 2 bits per sec /Hz, 3 and 5 respectively. Find Es/No dB required at 1E-4 BLER for each case. Then compare them to corresponding SNR required of channel capacity of 8QAM, 16QAM and 64QAM reading from Figure 6-84. How much deviation from the capacity? Answer: See table below.

Graph	constellation	BW efficiency	SNR at 1E-4 BLER from each figure	SNR required reading from Figure 6-84	Difference
Figure 6-68	8QAM	2 bits/ sec/Hz	7.5dB	≈5.1dB	2.4dB
Figure 6-70	16QAM	3 bits/ sec/Hz	11.0dB	≈9.1dB	1.9dB
Figure 6-79	64QAM (MLCM)	5 bits/ sec/Hz	17.7dB	≈16.1dB	1.6dB

Exercise 6-81: Compare SNR at 1E-4 BLER from Figures 6-68, 6-70, 6-79 to achieve 2, 3 and 5 bits per Hz with those of AWGN capacity. Answer: $10*\log_{10}(2^C-1)$ for C=2, 3, 5 bits per Hz will result in [4.77, 8.45, 14.9] dB.

Exercise 6-82*: Write a program to compute a channel capacity for constellations listed in Figure 6-84. Hint: study Appendix A.4 in the end of this chapter. The star * means a project exercise.

6.6.4   PAPR Reduction with Coding

When PAPR of a signal is large, PA should be able to handle the peak power (envelope) accordingly, i.e., linear, and thus it is less efficient in terms of power. If it is non-linear it will distort signals resulting in performance degradation and sideband spectral growth. This issue is important in practice and is covered in Chapter 8. Here we briefly demonstrate, by simulation results, that coding relieves the linearity requirement substantially. There are various techniques in handling PAPR reduction as surveyed in [8] but we focus on clipping. The idea of clipping is shown in Figure 6-85 and its CCDF (RHS) after clipping; PAPR after clipping is reduced 3dB at 1E-6 probability (Figure 6-86).

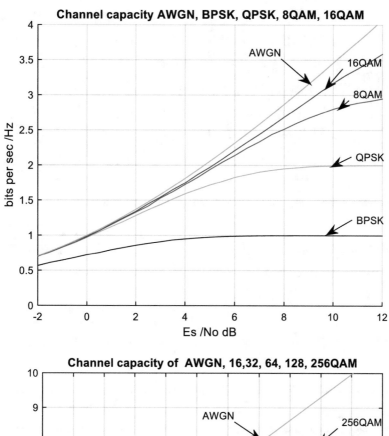

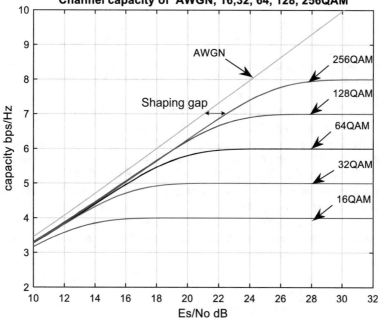

Figure 6-84: Channel capacity of AWGN, and with constellations of BPSK, QPSK, 8QAM, 16QAM to 256 QAM

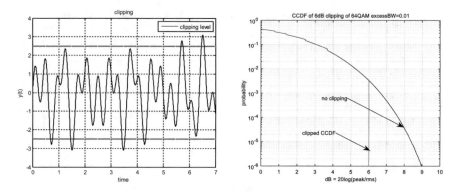

Figure 6-85: Clipping idea and CCDF after clipping and before clipping

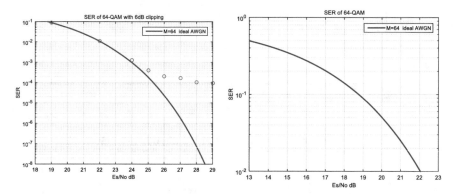

Figure 6-86: SER performance of 64QAM after clipping (circle) and before (solid line) on LHS; The error floor is formed around SER =1E-4 but not much impact below SER =1E-2

We use LDPC coding by IEEE 80211n N=1944, R=5/6 and its performance is not much impact down to BLER 1E-6 as shown in Figure 6-87.

We tried to apply more severe clipping of 4 dB and 5 dB and the result is shown in Figure 6-88. 4dB clipping results in substantial BLER performance degradation and with 5 dB clipping 0.2 dB degradation is observed (Figure 6-88).

In this particular example, 0.5% clipping can be handled by LDPC code with 0.2 dB BLER performance degradation; thus there is net power gain of 2-3 dB, i.e., PA can operate that much higher power. This seems very promising. But, more work is needed; for example we need to check it against real PA and check spectral regrowth.

6.6.5   Fading Channels

In Chapter 4 we covered fading channels in detail as multi-path fading; we showed, in summary, that it can be modeled as a linear time-varying frequency response

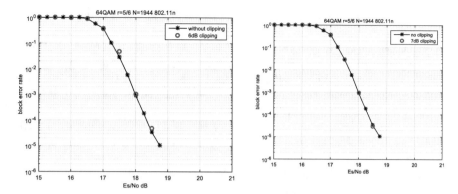

Figure 6-87: BLER of 64QAM with LDPC N=1944 R=5/6 of IEEE 802.11n with 6dB clipping and 7 dB clipping; very little impact

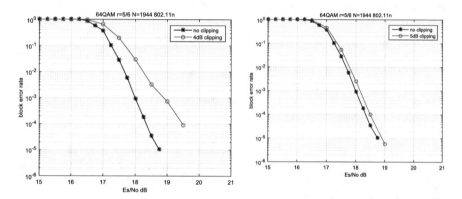

Figure 6-88: BLER of 64QAM with LDPC N=1944 R=5/6 of IEEE 802.11n with 4dB clipping and 5 dB clipping; with 4dB clipping degradation is substantial and with 5dB clipping about 0.2dB degradation

filter where the time-varying part is random in radio propagation environment. Specifically it is typical to model it as a tapped delay line with tap coefficients being random, e.g., PDF statistics of Rayleigh or Rice with a Doppler filter. This model can cover most of practical situations but sometimes may be too general so we categorize specific cases like time selective (narrow band flat frequency response), frequency selective, and both time and frequency selective, called doubly selective.

In this section we demonstrate that coding, especially with modern powerful FEC with SD, is very powerful in handling fading channels. Traditionally before the advent of such powerful coding, space diversity and frequency diversity were often used, may be still useful, and further such diversity may be combined with FEC. Note that one may interpret diversity as a form of repetition code.

6.6.5.1   Degradation of SER Performance Due to Rayleigh Fading

A brief comment on the SER curve follows in Figure 6-89. 2-QAM (same as BPSK) and FSK of both coherent (FSKC) and non-coherent (FSKNC) along with binary DPSK are shown. For DPSK a Rician channel with K=10 is also shown. It is clear that Rayleigh fading in general degrade a system tremendously such that it is not usable without additional remedies – some form of diversity or /and FEC.

It is useful in practice to understand how we obtain the curves in Figure 6-89, which in turn are obtained analytical expressions summarized in a table in Appendix A.5. In fact, even the degradation is tremendous but still the upper bound of performance due to the underlying assumptions – perfect interleaver and perfect channel estimation. We explain it next.

6.6.5.2   Fading as a Random Fluctuation of SNR

We first consider Rayleigh fading; time selective narrow band flat fading.

Rayleigh fading can be simulated by complex Gaussian process as shown in Figure 6-90. PDF of its power $(C_I^2 + C_Q^2)$ is exponential and that of its phase is uniform. We can normalize its power (γ) to be unity without loss of generality. The receiver processing requires the knowledge of channel (C_I, C_Q), which may be done by channel estimation using known pilots (i.e., sending some known signal occasionally). The other part is to handle Doppler frequency; the higher Doppler frequency the higher the faster fading. Typically, it is small relative to symbol rate. In order to handle this problem of fading rate, an interleaver is used when FEC is used.

Even if the channel estimation of (C_I, C_Q) is perfect and with a long enough interleaver, SNR at receiver is fluctuating according to PDF of $C_I^2 + C_Q^2$. This will result in performance degradation. This is shown in Figure 6-89 for binary modulations, and its analytical expressions of SER are tabulated in Appendix A.5.

In passing we make a comment on space diversity, say two diversity receiver combining $(m=2)$, the statistics of Rayleigh becomes that of Nakagami, which is shown (4-23) of Chapter 4. With SD combining SER performance improves and converges to that of AWGN as diversity order (m in Nakagami PDF) gets to infinity.

Exercise 6-83*: Study Figure 6-90 carefully. Write a program to generate the sequence of (C_I, C_Q) assuming that Doppler filter is approximated by a low pass filter and that $\gamma = 1.0$, i.e., the average power is unity. The star * means a project exercise.

We will make the above description more quantitative below by deriving the expression in Appendix A.5.

We denote SER of AWGN in terms of $\frac{Es}{No}$ as $p_e\left[\frac{Es}{No}|\text{AWGN}\right]$ and SER with Rayleigh fading as $p_e\left[\frac{Es}{No}|\text{Rayleigh}\right]$, and then $p_e\left[\frac{Es}{No}|\text{Rayleigh}\right] = E\left\{p_e\left[\frac{Es}{No}|\text{AWGN}\right]\right\}$ average over Rayleigh fading;

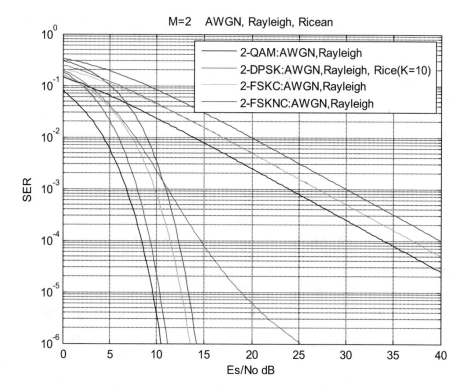

Figure 6-89: SER Performance with Rayleigh, Rice fading of BPSK, binary DPSK, and FSK coherent and non-coherent. These plots are based on analytical expressions of SER summarized in Appendix A.5

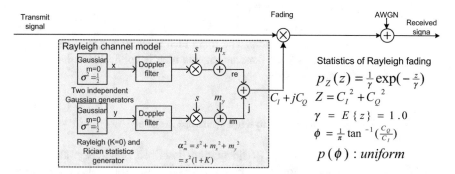

Figure 6-90: Rayleigh fading channel model (K=0) for simulations; $C_I + jC_Q$ random numbers are generated according to the description in dotted box, and PDF

$$p_e\left[\frac{Es}{No}\middle|\text{Rayleigh}\right] = \int_0^\infty p_e\left[\frac{Es}{No}z\middle|\text{AWGN}\right]P_z dz \qquad (6\text{-}57)$$

Using SER of M-PAM, $p_e\left[\frac{Es}{No}\middle|\text{PAM}\right] = \frac{2(M-1)}{M}Q\left[\sqrt{\frac{6}{M^2-1}\frac{Es}{No}}\right]$, $p_e\left[\frac{Es}{No}\middle|\text{Rayleigh}\right] =$
$\int_0^\infty \frac{2(M-1)}{M}Q\left[\sqrt{\frac{6}{M^2-1}\frac{Es}{No}z}\right]\frac{1}{\gamma}\exp\left(-\frac{z}{\gamma}\right)dz$. With the help of (5.1) to (5.6) on pp. 124 –
125 of [9], we obtain a closed form of the above integral as, $p_e\left[\frac{Es}{No}\middle|\text{Rayleigh}\right] =$
$\frac{M-1}{M}\left[1 - \sqrt{\frac{3Es/No\cdot\gamma}{M^2-1+3Es/No\cdot\gamma}}\right]$. When M=2 and γ=1.0 we obtain

$$p_e\left[\frac{Es}{No}\middle|\text{Rayleigh}, M=2\right] = \frac{1}{2}\left[1 - \sqrt{\frac{Es/No\cdot\gamma}{1+Es/No\cdot\gamma}}\right] = \frac{1}{2}\left[1 - \sqrt{\frac{Es/No}{1+Es/No}}\right]$$

The above is shown a table in Appendix A.5. Keep in mind the underlying assumptions; perfect phase tracking and infinite interleaver, which in practice must be approximated as part of receiver processing. Thus it is upper bound (best possible) performance, and may be a useful reference point.

Exercise 6-84: Given $p_e\left[\frac{Es}{No}\middle|\text{AWGN}\right] = Q\left[\sqrt{\frac{Es}{No}}\right]$ for coherent FSK, find the
corresponding $p_e\left[\frac{Es}{No}\middle|\text{Rayleigh}\right]$ Answer: See Appendix A.5 $\frac{1}{2}\left[1 - \sqrt{\frac{Es/No\cdot\gamma}{2+Es/No\cdot\gamma}}\right]$

Hint: $\int_0^\infty Q\left[\sqrt{\frac{6}{M^2-1}\frac{Es}{No}z}\right]\frac{1}{\gamma}\exp\left(-\frac{z}{\gamma}\right)dz = \frac{1}{2}\left[1 - \sqrt{\frac{3Es/No\cdot\gamma}{M^2-1+3Es/No\cdot\gamma}}\right]$ and apply M^2-
$1=6$.

Exercise 6-85: Given $p_e\left[\frac{Es}{No}\middle|\text{AWGN}\right] = \frac{1}{2}\exp\left(-\frac{Es}{No}\right)$ for 2-DPSK, find the
corresponding $p_e\left[\frac{Es}{No}\middle|\text{Rayleigh}\right]$. Answer: See Appendix A.5 $\frac{1}{2}\frac{1}{1+Es/No\cdot\gamma}$.

Hint: evaluate $\int_0^\infty \frac{1}{2}\exp\left(-\frac{Es}{No}\right)\frac{1}{\gamma}\exp\left(-\frac{z}{\gamma}\right)dz$.

In summary we derived analytical expressions of SER used in Figure 6-89, which is shown in Appendix A.5 as a table. The main idea is summarized in (6-57), i.e., treating Rayleigh fading as SNR fluctuation; this idea is applicable universally to other fading models such as Rice.

6.6.5.3   Channel Capacity of Rayleigh Fading Channel

Using the idea of fading as a SNR fluctuation described previously, along with AWGN channel capacity formula, $C = \log_2\left(1 + \frac{Es}{No}\right)$, we can evaluate a channel capacity of Rayleigh fading channel, C_{Rayleigh}, using (6-57).

$$C_{\text{Rayleigh}} = \int_0^\infty \log 2\left(1 + \frac{Es}{No}z\right)P_z dz \quad \int_0^\infty \log_2\left(1 + \frac{Es}{No}z\right)\frac{1}{\gamma}\exp\left(-\frac{z}{\gamma}\right)dz.$$

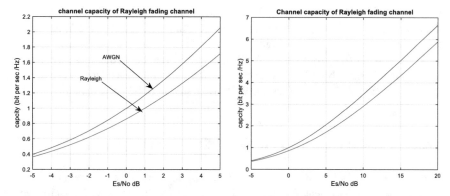

Figure 6-91: Channel capacity of Rayleigh fading channel using (6-57)

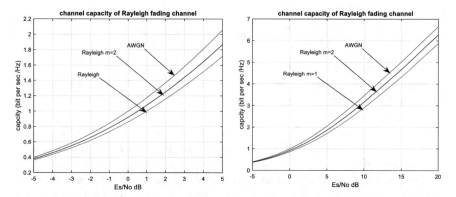

Figure 6-92: Channel capacity of Rayleigh fading with SD $m=2$

It seems that there is no closed form expression of the above integral but it can be evaluated numerically. The result of it is shown in Figure 6-91.

Exercise 6-86: From Figure 6-91, at capacity 1 bit per sec /Hz, what is SNR degradation of Rayleigh fading channel compared to AWGN? What about 5 bits per sec /Hz? Answer: $\approx$ 1dB, and $\approx$ 2.4dB. This degradation is far less than that of uncoded BPSK, which is $\approx$ 25dB at SER 1E-4 from Figure 6-89. This indicates that with coding SNR degradation due to time selective fading can be far less severe than without coding.

As mentioned before two space diversity combining ($m=2$) will change the underlying statistics from Rayleigh ($m=1$) to more benign so that we may expect that the degradation is less severe. This is shown in Figure 6-92 where we computed the channel capacity of $m=2$ Rayleigh (or Nakagami).

Exercise 6-87: From Figure 6-92, at capacity 1 bit per sec /Hz, what is SNR degradation of $m=2$ Rayleigh fading channel compared to AWGN? What about 5 bits per sec /Hz? Answer: $\approx 0.5\text{dB}$, and $\approx 1.2\text{dB}$.

We comment that from Figure 6-92, a common practical counter - measure of mitigation of fading, by SD followed by FEC, is well justified. As the order of SD, m, gets larger its statistics gets closer AWGN as shown in (4-23) of Chapter 4.

6.6.5.4   Channel Capacity of Frequency Selective Fading Channel

In time selective fading we assume the frequency response of a channel is flat, i.e., there is no frequency selective fading. Here we compute a channel capacity with frequency selective fading; this is shown in (6-58), where $\frac{(Es)_k}{No}$ is SNR in a subchannel.

$$C_{FS} = \frac{1}{L}\sum_{k=1}^{L} \log_2\left(1 + \frac{(Es)_k}{No}\right) \qquad (6\text{-}58)$$

This is shown in Figure 6-93.

Figure 6-93 is a discrete version and can be extended to continuous frequency response by changing a summation into an integral which we do not explore further here.

The computation of (6-58) can be done numerically for a given channel frequency response. When the frequency selective fading is time-varying, it can be averaged over time-varying statistics.

The problem of actually achieving a channel capacity of frequency selective fading is more complicated and its detailed exploration is beyond the scope of this section. Thus we describe some important ideas qualitatively here, summarized as Table 6-25;

For all of ISI and fading, slow and fast, a receiver needs to do proper channel estimations. We emphasize that when OFDM is used for doubly selective fading,

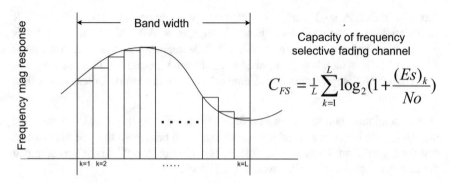

Figure 6-93: Channel capacity of frequency selective fading; sub channels are formed with equal frequency spacing for a given bandwidth and numbered from 1 to L

Table 6-25: Summary of ideas of fading counter-measure

Issue		Counter-measure	remarks
ISI		OFDM	Covered in Chapter5 in detail, see Figure 5-46
		Equalizer	Covered briefly in Chapter 8
Slow fading	Very slow (Cable)	Bit loading at transmit side for DSL (cable channel)	Mod-coding classes for each OFDM subchannel
	Slow (Wireless LAN)	Averaging with mod-coding for SNR, adaptively changing, SD or MIMO	OFDM is effective for slow frequency selective fading
Fast fading (vehicle)		Easier with wideband (Rice statistics) when there is LOS	Equalizer for ISI may be usable
		With OFDM each subchannel is close Rayleigh even when there is LOS	Additional measure is necessary to handle Rayleigh fading in subchannel as well as frequency selective averaging. See Figure 5-47.

both frequency and time, in each sub channel, there should be countermeasure to handle. See Figure 5-47 in Chapter 5 in order to understand the concept clearly.

6.6.5.5 Evaluate (6-57) When $p_e\left(\frac{Es}{No}|\text{AWGN}\right)$ is Available as Numerical Data

Exercise 6-88: In (6-57), $p_e\left[\frac{Es}{No}|\text{Rayleigh}\right] = \int_0^\infty p_e\left(\frac{Es}{No}z|\text{AWGN}\right) P_z(z)dz$, where $p_e\left[\frac{Es}{No}|\text{AWGN}\right]$ is given by numerical data, and $P_Z(z)$ may be given by an equation. It is useful to evaluate the integration under these circumstances. Write a program for it. Hint: $p_e\left[\frac{Es}{No}|\text{discrete points}\right]$ is interpolated, e.g., using $y_q = \text{spline}(x, y, x_q)$ where (x, y) data set and (x_q, y_q) are interpolated values. The integration is replaced by summation.

Exercise 6-89*: In Figure 6-84, the channel capacity of BPSK, QPSK, ..., and 256QAM with AWGN is computed. Using the idea of (6-57), evaluate the channel capacity of BPSK with Rayleigh fading by writing a program. Hint: BPSK channel capacity with AWGN is given numerically. It is interpolated, e.g., by spline as above, and the integration is done by summation.

6.7 References with Comments and Appendix

There is a vast amount of literature in coding [11]. Our reference is very limited and only for a very specific issue. This chapter emphasizes modern iteratively decodable codes [12], LDPC and Turbo, and its application to higher order modulations.

[1] L. Bahl, J. Cocke, F. Jelinek, and J. Raviv, "Optimal decoding of linear codes for minimizing symbol error rate," IEEE Trans. Inform. Theory,Vol. 20, No. 2, pp. 284-287, Mar 1974.

[2] C. Berrou, A. Glavieurx and P. Thitimajshima,"Near Shannon Limit Error Correcting coding and Decdoing: Tutbo Codes(1)", ICC '93 Geneva, Switzerland, May 1993, pp 1064-1070

[3] D. J. C.MacKay and R.M. Neal "Near Shannon Limit performance of low density parity check codes," Elec.Lett.,p.1645, August 1996

[4] G. Ungerbeck, "Channel coding with multilevel / phase signals," IEEETrans. Information Theorypp.58 -67, 1982

[5] H. Imai and S. Hirakawa, "A new multilevel coding method using error correcting codes," IEEE Trans. Information Theory, pp.371-377, May 1977

[6] S.M. Yang, Wei-Kang Cheng and Jun Shen "LSB-coded Modulation and its Application to Microwave Digital radios," ICC 96 IEEE

[7] IEEE Std 802.11n -2009 Annex R HT LDPC matrix definitions

[8] S.H. Han and J. H. Lee, "An Overview of Peak to Average power Ratio Reduction techniques for multi-carrier transmission" IEEE Wireless Communications, April 2005, pp.56 -65

[9] M.K. Simon, and M-S. Alouni, "Digital Communications over Fading Channels" Chapter 8, 2nd Ed., John Wiley 2005

[10] Larson, Knud J. "Short Convolutional Codes with Maximal free Distance for Rates ½, 1/3, and ¼" IEEE Trans. Information Theory , vol IT-19 pp.371-372 May 1973

[11] Lin, Shu and Costello, Daniel "Error Control Coding: Fundamentals and Applications" 2nd ed Pearson-Prentice Hall 2004

[12] Johnson, S. "Iterative Error Correction" Cambridge University Press 2010

Appendix 6

A.1   CM Decoding Example of Figure 6-36

A list of codewords with *H* in Figure 6-36; 8 codewords and their BPSK symbols

$$
\begin{vmatrix}
0 & 0 & 0 & 0 & 0 & 0 \\
1 & 0 & 0 & 1 & 0 & 1 \\
0 & 1 & 0 & 1 & 1 & 1 \\
1 & 1 & 0 & 0 & 1 & 0 \\
0 & 0 & 1 & 1 & 1 & 0 \\
1 & 0 & 1 & 0 & 1 & 1 \\
0 & 1 & 1 & 0 & 0 & 1 \\
1 & 1 & 1 & 1 & 0 & 0
\end{vmatrix}
\rightarrow BPSK \rightarrow
\begin{vmatrix}
1 & 1 & 1 & 1 & 1 & 1 \\
-1 & 1 & 1 & -1 & 1 & -1 \\
1 & -1 & 1 & -1 & -1 & -1 \\
-1 & -1 & 1 & 1 & -1 & 1 \\
1 & 1 & -1 & -1 & -1 & 1 \\
-1 & 1 & -1 & 1 & -1 & -1 \\
1 & -1 & -1 & 1 & 1 & -1 \\
-1 & -1 & -1 & -1 & 1 & 1
\end{vmatrix}
$$

r = [1.3802 2.2968 -0.5972 1.6096 1.2254 0.0753]
rbit = [0 0 1 0 0 0]

CM decoding (soft)

Computing correlation between codewords in BPSK symbols and **r**,

$$\text{symbol list} * \mathbf{r}^{\mathrm{T}} = \begin{bmatrix} 1 & 1 & 1 & 1 & 1 & 1 \\ -1 & 1 & 1 & -1 & 1 & -1 \\ 1 & -1 & 1 & -1 & -1 & -1 \\ -1 & -1 & 1 & 1 & -1 & 1 \\ 1 & 1 & -1 & -1 & -1 & 1 \\ -1 & 1 & -1 & 1 & -1 & -1 \\ 1 & -1 & -1 & 1 & 1 & -1 \\ -1 & -1 & -1 & -1 & 1 & 1 \end{bmatrix} \begin{bmatrix} 1.3802 \\ 2.2968 \\ -0.5972 \\ 1.6096 \\ 1.2254 \\ 0.0753 \end{bmatrix} = \begin{bmatrix} 5.9901 \\ -0.1401 \\ -4.4241 \\ -3.8147 \\ 1.5145 \\ 1.8227 \\ 2.4403 \\ -3.3887 \end{bmatrix}$$

Choose maximum correlation $5.9901 \rightarrow [1\ 1\ 1\ 1\ 1\ 1] \rightarrow [0\ 0\ 0\ 0\ 0\ 0]$

CM decoding (hard): minimum distance decoding

$$\text{Codewords list} \oplus \mathbf{rbit}^{\mathrm{T}} = \begin{bmatrix} 0 & 0 & 0 & 0 & 0 & 0 \\ 1 & 0 & 0 & 1 & 0 & 1 \\ 0 & 1 & 0 & 1 & 1 & 1 \\ 1 & 1 & 0 & 0 & 1 & 0 \\ 0 & 0 & 1 & 1 & 1 & 0 \\ 1 & 0 & 1 & 0 & 1 & 1 \\ 0 & 1 & 1 & 0 & 0 & 1 \\ 1 & 1 & 1 & 1 & 0 & 0 \end{bmatrix} \cdot \oplus \cdot \begin{bmatrix} 0 \\ 0 \\ 1 \\ 0 \\ 0 \\ 0 \end{bmatrix} \cdot = \cdot \begin{bmatrix} 1 \\ 4 \\ 5 \\ 4 \\ 2 \\ 3 \\ 2 \\ 3 \end{bmatrix}$$

Choose minimum $=1, \rightarrow [0\ 0\ 0\ 0\ 0\ 0]$ and it may be called minimum distance decoding. $\oplus$ means exclusive operation between two elements of multiplication.

This type of CM decoding quickly becomes impractical since a code with K=20 information bits has 1048576 codewords; and with K=200, no. of codewords are about 1.6E+60. However, in theory it is the best one can do in decoding. All error coding-decoding schemes are to overcome this practical problem.

A.2   The Computation of p_0 (p_1), and LLR for BPSK

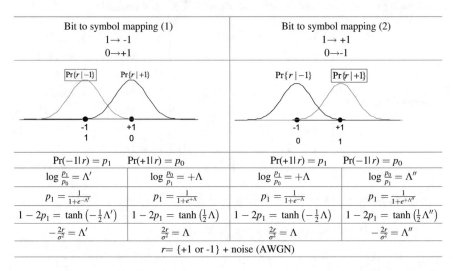

Bit to symbol mapping (1) $1 \to -1$ $0 \to +1$		Bit to symbol mapping (2) $1 \to +1$ $0 \to -1$	
$\Pr(-1\lvert r) = p_1$	$\Pr(+1\lvert r) = p_0$	$\Pr(+1\lvert r) = p_1$	$\Pr(-1\lvert r) = p_0$
$\log \frac{p_1}{p_0} = \Lambda'$	$\log \frac{p_0}{p_1} = +\Lambda$	$\log \frac{p_1}{p_0} = +\Lambda$	$\log \frac{p_0}{p_1} = \Lambda''$
$p_1 = \frac{1}{1+e^{-\Lambda'}}$	$p_1 = \frac{1}{1+e^{+\Lambda}}$	$p_1 = \frac{1}{1+e^{-\Lambda}}$	$p_1 = \frac{1}{1+e^{+\Lambda''}}$
$1 - 2p_1 = \tanh\left(-\frac{1}{2}\Lambda'\right)$	$1 - 2p_1 = \tanh\left(\frac{1}{2}\Lambda\right)$	$1 - 2p_1 = \tanh\left(-\frac{1}{2}\Lambda\right)$	$1 - 2p_1 = \tanh\left(\frac{1}{2}\Lambda''\right)$
$-\frac{2r}{\sigma^2} = \Lambda'$	$\frac{2r}{\sigma^2} = \Lambda$	$\frac{2r}{\sigma^2} = \Lambda$	$-\frac{2r}{\sigma^2} = \Lambda''$
$r = \{+1 \text{ or } -1\} + \text{noise (AWGN)}$			

The above table shows different possible choices of bit to symbol mapping and LLR Λ definition. For consistency and simplicity, we choose one shaded throughout. In particular, LLR$=\Lambda= \log \frac{p_0}{p_1}$ where $p_0=$ Pr($+1\lvert r$) and $p_1=$Pr($-1\lvert r$). Note that these probabilities can be obtained from Pr($r\lvert +1$) and Pr($r\lvert -1$), and assuming Pr($+1$) = Pr (-1)= ½. See Exercise 6-90 below.

Exercise 6-90: Find the relationships assuming Pr($+1$) = Pr(-1)= ½ and using conditional probabilities Pr($\pm1\lvert r$)= Pr($r\lvert \pm1$) Pr (±1)/Pr(r). Answer: Pr($\pm1\lvert r$)=Pr ($r\lvert \pm1$)/(Pr($r\lvert +1$)+Pr($r\lvert -1$)), and thus Λ=log (Pr($r\lvert +1$)/Pr($r\lvert -1$)).

Exercise 6-91: Show $p_1 = \frac{1}{1+e^{+\Lambda}}$ and $p_0 = \frac{1}{1+e^{-\Lambda}}$ from $\Lambda= \log \frac{p_0}{p_1}$ and $p_0=1-p_1$.

Exercise 6-92: Show $\frac{2r}{\sigma^2}=\Lambda$ assuming AWGN and Pr($+1$) = Pr(-1)= ½. Answer: recall that Pr($r\lvert +1$)$=\frac{1}{\sqrt{2\pi\sigma^2}} \exp\left(-\frac{(r-1)^2}{2\sigma^2}\right)$ and Pr($r\lvert -1$)$=\frac{1}{\sqrt{2\pi\sigma^2}} \exp\left(-\frac{(r+1)^2}{2\sigma^2}\right)$ and $\Lambda= \log$ (Pr($r\lvert +1$)/Pr($r\lvert -1$)).

Exercise 6-93: Show $1 - 2p_1 = \tanh\left(\frac{1}{2}\Lambda\right)$ using $\tanh\left(\frac{1}{2}\Lambda\right) = (e^{+\Lambda} - 1)/(e^{-\Lambda} + 1)$.

From the above, a probability may be assigned from the received sample value, r, as $p_0 = \frac{1}{1+e^{-\Lambda}} = \frac{1}{1+\exp\left(-2r/\sigma^2\right)}$ and $p_1 = 1 - p_o = \frac{1}{1+e^{+\Lambda}} = \frac{1}{1+\exp\left(+2r/\sigma^2\right)}$ assuming AWGN.

A.3 Different Expressions of Check Node LLR of LDPC

Different expressions of check node LLR

Other expressions of (6-53) is derived and given by (6-59), (6-60), and (6-61) below.
Starting from (6-53), we express it as

$$\Lambda(a_1) = 2\tanh^{-1}\log^{-1}\left[\log \prod_{k=2, k\neq 1}^{K} \tanh\left(\frac{1}{2}\Lambda(b_k)\right)\right]$$

$$= -\log\left\{\tanh\left(\frac{1}{2}\left[\sum_{k=2, k\neq 1}^{K} -\log\left\{\tanh\left(\frac{1}{2}\Lambda(b_k)\right)\right\}\right]\right)\right\}$$

which is the same as (6-59).
We used $y = -\log\tanh\left(\frac{x}{2}\right) = \log\frac{e^x+1}{e^x-1}$. Furthermore $e^y = \frac{e^x+1}{e^x-1}$, and $e^x = \frac{e^y+1}{e^y-1}$, i.e., its inverse is the same as original function.

$$\Lambda(a_1) = -\log\left\{\tanh\left(\frac{1}{2}\left[\sum_{k=2, k\neq 1}^{K} -\log\left\{\tanh\left(\frac{1}{2}\Lambda(b_k)\right)\right\}\right]\right)\right\} \qquad (6\text{-}59)$$

Additional manipulations are done for computational convenience to arrive at (6-60).

$$\Lambda(a_1) = \prod_{k=2, k\neq 1}^{K} sign(\Lambda(b_k))\left[-\log\left\{\tanh\left(\frac{1}{2}\left[\sum_{k=2, k\neq 1}^{K} -\log\left\{\tanh\left(\frac{1}{2}|\Lambda(b_k)|\right)\right\}\right]\right)\right\}\right]$$

$$\Lambda(a_1) = \prod_{k=2, k\neq 1}^{K} sign(\Lambda(b_k))\, \log\tanh 2\left[\sum_{k=2, k\neq 1}^{K} \log\tanh 2(\Lambda(b_k))\right] \qquad (6\text{-}60)$$

Define $\log\tanh 2(x) = -\log\left\{\tanh\left(\frac{1}{2}|x|\right)\right\}$

```
function y = logtanh2(x)
t = tanh(abs(x)/2);
y = -log(t);
```

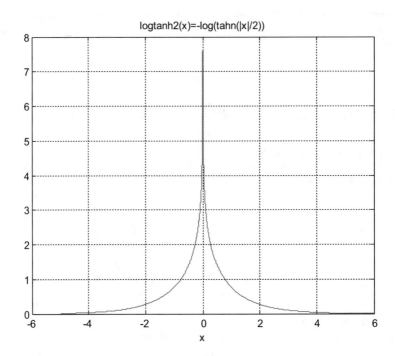

logtanh2(x)=-log(tahn(|x|/2))

The advantage of (6-60) is addition rather than multiplication being used. An approximation of (6-60) is given by

$$\Lambda(a_1) \approx \prod_{k=2, k\neq 1}^{K} sign(\Lambda(b_k)) \min_{k=2, k\neq 1}^{K} [\Lambda(b_k)] \tag{6-61}$$

For simplicity, (6-60) may be used in practice at the expense of 'small' performance degradation. (6-60) may be used with a look up table.

LLR of two input parity check and its iterative use
Another variations of (6-53) can be obtained by going back to Figure 6-39 where the computation of probability of two input parity check; two inputs are q_{14} and q_{15} and the outgoing message r_{11}, i.e., probability of two input parity check.

We repeat it here for convenience; we need to compute the probability of r_{11} being '0' and '1' with the condition of knowing the probability of q_{14} and q_{15} and of meeting a parity check.

$r_{11} = \; '0'$: if $\{q_{14} = \; '0'$ and $q_{15} = \; '0'\}$ or $\{\, q_{14} = \; '1'$ and $q_{15} = \; '1'\}$

$r_{11} = \; '1'$: if $\{q_{14} = \; '0'$ and $q_{15} = \; '1'\}$ or $\{\, q_{14} = \; '1'$ and $q_{15} = \; '0'\}$

With the notation of $p(q_{14} = '0')=1-p_1$, $p(q_{14} = '1')=p_1$, and $p(q_{15} = '0')=1-p_2$, p $(q_{15} = '1')=p_2$ the probabilities of $r_{11}='0'$ and $r_{11}='1'$ are given by

$$\mathrm{pr}\left(r_{11} = \; '0' \,|\, q_{14}, q_{15}\right) = p_1 p_2 + (1 - p_1)(\, 1 - p_2)$$

$$\mathrm{pr}\left(r_{11} = \; '1' \,|\, q_{14}, q_{15}\right) = (1 - p_1)\, p_2 + p_1 (\, 1 - p_2)$$

In terms of LLR, $\Lambda(r_{11})=$ LLR (r_{11}), $\Lambda_1=$ LLR(q_{14}) and $\Lambda_2=$ LLR(q_{15})

$$\Lambda(r_{11}) = \; \log \frac{p_1 p_2 + (1 - p_1)(1 - p_2)}{(1 - p_1)p_2 + p_1(1 - p_2)} \tag{6-62}$$

$$\Lambda(r_{11}) = \; \log \frac{1 + e^{\Lambda 1} e^{\Lambda 2}}{e^{\Lambda 1} + e^{\Lambda 2}} \tag{6-63}$$

(6-63) is obtained from (6-62) by using the relationship as shown below,

$$p_1 = \frac{1}{1 + e^{+\Lambda 1}} \; \text{and} \; 1 - p_1 = \frac{e^{+\Lambda 1}}{1 + e^{+\Lambda 1}}, \; \text{and} \; p_2 = \frac{1}{1 + e^{+\Lambda 2}} \; 1 - p_2 = \frac{e^{+\Lambda 2}}{1 + e^{+\Lambda 2}}.$$

These relationships can be found in *Appendix A.2*.

In general notation of two inputs, *a* and *b*, (6-63) can be denoted as

$$\Lambda(a \oplus b) = \; \log \frac{1 + e^{\Lambda(a)} e^{\Lambda(b)}}{e^{\Lambda(a)} + e^{\Lambda(b)}} \tag{6-64}$$

(6-64) can be further manipulated for ease of computation as follows.

$$\Lambda(a \oplus b) = \mathrm{sign}(\Lambda(a)) \; \mathrm{sign}(\Lambda(b)) \; \min \; [|\Lambda(a)|, |\Lambda(b)|] + C(a, b) \tag{6-65}$$

where $C(a, b) = \; \log (1 + e^{-\,|\,\Lambda(a) + \Lambda(b)|}) - \log (1 + e^{-\,|\,\Lambda(a) - \Lambda(b)|})$.

For the computation of $C(a,b)$, we may use a table lookup or a piecewise linear approximation of $g(x) = \log (1+e^{-\,|\,x|})$ which is shown below.

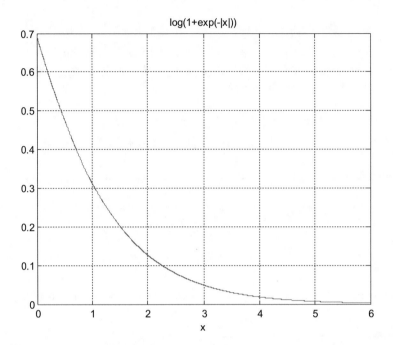

Another approximation of $C(a,b)$ is given by,

$$C(a,b) = g(x_1) - g(x_2) = \begin{cases} +c & \text{if } |x_1| < 2 \text{ and } |x_2| > 2 |x_1| \\ -c & \text{if } |x_2| < 2 \text{ and } |x_1| > 2 |x_2| \\ 0 & \text{otherwise} \end{cases}$$

c may be chosen to optimize performance e.g., c=0.55 or c=0.8 .

For more than two inputs, (6-64) can be used iteratively. This can be shown symbolically as in (6-66),

$$\Lambda(a \oplus b \oplus c) = \log \frac{1 + e^{\Lambda(a \oplus b)} e^{\Lambda(c)}}{e^{\Lambda(a \oplus b)} + e^{\Lambda(c)}} \tag{6-66}$$

If $C(a,b)$ is ignored in (6-65), then it can be simplified with 3 inputs as

$$\Lambda(a \oplus b \oplus c) = \text{sign}\,(\Lambda(a \oplus b))\,\text{sign}\,(\Lambda(c))\,\min\,[|\Lambda(a \oplus b)|, |\Lambda(c)|]$$
$$= \text{sign}\,(\Lambda(a))\,\text{sign}\,(\Lambda(b))\,\text{sign}\,(\Lambda(c))\,\min\,[|\Lambda(a)|, |\Lambda(b)|, |\Lambda(c)|].$$

(6-65) can be used iteratively in computer programming easily even though it is a bit cumbersome to denote in equations.

A.4 Computation of Channel Capacity

AWGN channel capacity formula

We used AWGN channel capacity formula $C = \log_2\left[1 + \frac{Es}{No}\right]$ in Sections 6.6.3 and 6.6.5. It is insightful to understand its meaning and underlying assumptions in its derivation. This is a precise mathematical answer to a question; how fast one can transmit data for a given SNR, due to AWGN, with 1 Hz bandwidth. The capacity is directly proportional to the bandwidth but logarithmically proportional to the power of signal.

The underlying assumptions, in $C = \log_2\left[1 + \frac{Es}{No}\right]$, are;

- Bandwidth of signal is the same as signaling rate; both I and Q channels are utilized. In other words, the matched filters are ideal LPFs.
- Channel impairment: AWGN only
- Constellation: continuous Gaussian distribution, not uniform (equally likely) distribution. In terms of FEC, infinitely long random codes
- Unit of C: bits per sec /Hz.

Thus it is the upper bound of performance under AWGN but is very useful as a reference point. This remarkable formula was obtained by C. Shannon in 1948.

Entropy: information measure

We define a measure of information, called entropy or Shannon entropy, H, in order to distinguish Boltzmann entropy in thermodynamics.

Exercise 6-94: we have a constellation of 4 points with equally likely probability. How many bits per symbol (constellation point) can transmit? Answer: log2 (4) =2.0.

Now each symbol (constellation point) is not uniformly distributed but by a probability distribution; say constellation points are number 1 to M; 1 with 0.1, 2with 0.25, 3 with 0.25 and 4 with 0.4.

We define entropy as H= 0.1 $\log_2(1/0.1)$+ 0.25 $\log_2(1/0.25)$ + 0.25 $\log_2(1/0.25)$+ 0.4 $\log_2(1/0.4)$ = 1.861 < 2.0; in general constellation X, with probability distribution Px,

$$H(X) = E\{\log_2(1/Px)\} = \sum_{i=1}^{M} p_i \log_2\left[\frac{1}{p_i}\right].$$

Mutual information I (X;Y)

Joint probability P(X;Y), and marginal probabilities P(X), P(Y) and conditional probabilities P(X|Y), P(Y|X) are related and shown in Venn diagram as in Figure 6-94 (LHSS);

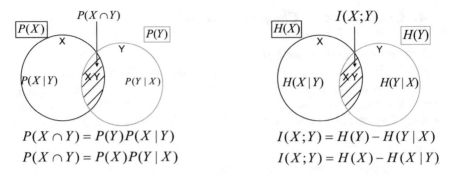

Figure 6-94: Joint probability and mutual information on Venn diagram.

Mutual information, I(X; Y), is graphically defined as shown in Figure 6-94 (RHS). Two extreme cases of mutual information are; I(X; Y) = 0 when X, Y are independent, i.e., $P(X \cap Y) = P(X)P(Y)$, and furthermore $I(X; Y) = H(X)$ when $P(Y|X)$=1.0, i.e., determistic case. In other words, mutual information is a measure of how random variable X and Y are correlated or dependent. It is symmetric; I(X; Y) = I(Y; X). Formally we define mutual information as,

$$I(X; Y) = \sum_{x \in X, y \in Y} P(x \cap y) \log_2 \frac{p(x \cap y)}{p(x)P(y)} = E\left\{ \log_2 \frac{P(x \cap y)}{p(x)p(y)} \right\}$$

A channel capacity is defined as the maximum mutual information optimized with input probability distribution P(x);

$$C(\text{bits}/\text{symbol}) = \max_{P(x)} I(X; Y)$$

Channel capacity of a constellation with uniform probability distribution; $p(x)$ =1/M.
In this case we can use $I(X; Y) = H(X) - H(X|Y) = \log_2 M - H(X|Y)$; there is no need of maximization.

$$I(X; Y) = E\left\{ \log_2 \frac{p(x \cap y)}{p(x)p(y)} \right\} = E\left\{ \log_2 \frac{1}{p(x)} \right\} + E\left\{ \log_2 \frac{p(y|x)p(x)}{p(y)} \right\}$$

where $P(y) = \sum_{x'} p(y|x')p(x') = \frac{1}{M}$ and $p(x) = \frac{1}{M}$. Thus

$$I(X;Y) = \log_2 M - E\left\{\log_2 \frac{1}{p(y|x)}\right\} = \log_2 M - \sum_{x,y}\left\{p(y|x)\log_2 \frac{1}{p(y|x)}\right\}.$$

Channel transition matrix

	y1	y2	y3	yK
a1	pr(y1la1)	pr(y2la1)	pr(y3la1)	pr(yKla1)
a2	pr(y1la2)	pr(y2la2)	pr(y3la2)	pr(yKla2)
a3	pr(y1la3)	pr(y2la3)	pr(y3la3)	pr(yKla3)
a4	pr(y1la4)	pr(y2la4)	pr(y3la4)	pr(yKla4)

e.g., pr(y1la1) $= \exp(-|y1-a1|^2/2\ \sigma^2)/sc$; sc such that $\sum_{i=1}^{4} pr\left(y_j|a_i\right) = 1.0$.

Computation of -H(X|Y) for equally likely constellation
pryx.*log2(pryx) in MATLAB computation

	y1	y2	yk
a1	pr(y1la1)*log$_2$\{ pr(y1la1)\}	pr(y2la1)*log$_2$\{ pr(y2la1)\}	pr(ykla1)*log$_2$\{ pr(ykla1)\}
a2	pr(y1la2)*log$_2$\{ pr(y1la2)\}	pr(y2la2)*log$_2$\{ pr(y2la2)\}	pr(ykla2)*log$_2$\{ pr(ykla2)\}
a3	pr(y1la3)*log$_2$\{ pr(y1la3)\}	pr(y2la3)*log$_2$\{ pr(y2la3)\}	pr(ykla3)*log$_2$\{ pr(ykla3)\}
a4	pr(y1la4)*log$_2$\{ pr(y1la4)\}	pr(y2la4)*log$_2$\{ pr(y2la4)\}	pr(ykla4)*log$_2$\{ pr(ykla4)\}

e.g. pr(y1la1) $\log_2$ \{ pr(y1la1) \} such that $0\ \log_2\{0\}= 0$ and $1\ \log_2\{1\} =0$.
-H(X|Y) = mean(sum(pryx.*log2(pryx)))

$$I(X;Y) = \log_2 M + \text{mean}(\text{sum}(\text{pryx}. * \log 2(\text{pryx})))$$

Thus we compute mutual information of constellations with equally likely distribution using the above from channel transition matrix (DMC).

A.5   SER Performance of Binary PSK, DPSK, FSK

For binary (M=2) case a summary is collected in a table below. The plot of Figure 6-89 is based on these expressions of SER.

Mod / Channel	symbol error rate for M=2			
	BPSK Coherent	2-DPSK Differentially coherent	FSK Orthogonal	FSK Non-coherent
AWGN	$Q\left(\sqrt{2\dfrac{E_s}{N_o}}\right)$	$\dfrac{1}{2}\exp\left(-\dfrac{Es}{No}\right)$	$Q\left(\sqrt{\dfrac{E_s}{N_o}}\right)$	$\dfrac{1}{2}\exp(-\dfrac{1}{2}\dfrac{E_s}{N_o})$
Rayleigh (amplitude)	$\dfrac{1}{2}\left(1-\sqrt{\dfrac{\gamma_s}{1+\gamma_s}}\right)$	$\dfrac{1}{2}\dfrac{1}{(1+\gamma_s)}$	$\dfrac{1}{2}\left(1-\sqrt{\dfrac{\gamma_s}{2+\gamma_s}}\right)$	$\dfrac{1}{(2+\gamma_s)}$
Rice (amplitude)	-	$\dfrac{1}{2}\dfrac{1+K}{1+K+\gamma_s}\exp\left[-\dfrac{K\gamma_s}{1+K+\gamma_s}\right]$	-	-

Note: $\overline{\gamma_s} = \overline{\alpha^2}\frac{E_s}{N_o}$ and without loss of generality let $\overline{\alpha^2}=1.0$. $K = \frac{m_1{}^2+m_2{}^2}{2\sigma^2}$ which means the power ratio of direct path vs. random path, and K=0 reduces to Rayleigh.

6.8 Problems

P6-1: A 4 bit repetition code uses two codewords {0000, 1111} to represent data {'0', '1'}respectively, and for transmission BPSK is used with bit to symbol mapping of $0\rightarrow+1$, $1\rightarrow-1$.

(1) What is the code weight (w_m) i.e., bit difference between two codewords {0000, 1111}?

(2) For a sequence of random data {'0', '1', '1', '0', '1', '0'}, find BPSK symbols to be transmitted, 'X', and express it 4 by 6 matrix.

(3) Obtain received samples, $y_i = X' + n_i$ when the noise samples at the receive side are given by

$$n_i = \begin{bmatrix} -0.1765 & -1.4491 & -0.1303 & -1.3617 & 0.5528 & 0.6601 \\ 0.7914 & 0.3335 & 0.1837 & 0.4550 & 1.0391 & -0.0679 \\ -1.3320 & 0.3914 & -0.4762 & -0.8487 & -1.1176 & -0.1952 \\ -2.3299 & 0.4517 & 0.8620 & -0.3349 & 1.2607 & -0.2176 \end{bmatrix}$$

(4) Make a decision for each data (all 6 of them) using soft decision CM decoding, i.e., adding 4 received samples and decide sign of it.

(5) Quantize received samples $[y_i] = sign\,(y_i)$ and then apply CM decoding in order to recover transmit data. Compare with the result of (4).

P6-2: CM (soft) decoding of 4 bit code is to add four received samples where two codewords {0000, 1111} for data {'0', '1'} and bit to symbol mapping of $0 \rightarrow +1$, $1 \rightarrow -1$ are used;

$$Y_s = \sum_{i=1}^{4} y_i = 4'X' + n_1 + n_2 + n_3 + n_4.$$

(1) What is the signal power and noise power of Y_s (added received samples)?
(2) Show that SNR of Y_s is increased by a factor of 4, which is the weight of the code; $(E_s/N_o)w_m$ where E_s/N_o is SNR of BPSK and $w_m=4$.
(3) Show that the word error probability is given by $P_w = Q\left(\sqrt{\frac{2E_s}{N_o} w_m}\right)$ and that it is equivalent to shifting BPSK symbol error rate performance by 6dB to the left.

P6-3: In a 4 bit repetition code, two codewords are chosen to be {0110, 1001} to represent {'0', '1'} respectively. A bit to symbol mapping of $0 \rightarrow +1$, $1 \rightarrow -1$ is used. Show that CM (soft) decoding rule needs to be modified $Y_s = y_1 - y_2 - y_3 + y_4$ and that the decision rule is that decide '0' when $Y_s > 0$ and '1' otherwise.

P6-4: The code rate is defined as $R = k/n$ where k is information bit and n is a number of bits in a codeword.

(1) What is R for 4 bit repetition code?
(2) Show that in a repetition code there is no <u>net</u> coding gain since the transmit power is increased by the same repetition factor.
(3) When a code rate is $R=2/3$ show that the power increase in transmit side is 1.76dB (called rate loss) assuming BPSK transmission. What about $R=5/6$, $R=3/4$, $R=1/2$, and $R=1/3$?

P6-5: Hadamard code, $(n, k) = (2^k, k)$, construction rule (i.e., codeword enumeration rule) is given by

$$H_{2m} = \begin{bmatrix} H_m & H_m \\ H_m & \overline{H}_m \end{bmatrix} \text{ with } m = 2 \text{ as } H_2 = \begin{bmatrix} 0 & 0 \\ 0 & 1 \end{bmatrix} \text{ and } \overline{H}_2 = \begin{bmatrix} 1 & 1 \\ 1 & 0 \end{bmatrix}.$$

(1) Construct Hadamard H_4, H_8, H_{16} and H_{32} by enumerating all codewords.
(2) Show that the minimum distance $d_{min} = n/2$ for all cases.
(3) Show generator matrices (G4) for H_4 by inspection that $G4 = \begin{bmatrix} 0 & 1 & 0 & 1 \\ 0 & 0 & 1 & 1 \end{bmatrix}$.
(4) Find generator matrices for H_8, and H_{16} (Use a computer program if necessary).

Hint: $G8 = \begin{bmatrix} 0 & 1 & 0 & 1 & 0 & 1 & 0 & 1 \\ 0 & 0 & 1 & 1 & 0 & 0 & 1 & 1 \\ 0 & 0 & 0 & 0 & 1 & 1 & 1 & 1 \end{bmatrix}$.

(5) Find generator matrices for H_{32}, H_{64}, and so on by induction.

P6-6: For linear block codes with codeword being equal probability, the average word error probability is given by $P_w = P_{e \mid m = 0}$ as shown in (6-2) where $P_{e \mid m = 0}$ is the probability of $\{m \neq 0 \mid m = 0 \text{ transmitted}\}$. Argue that in this case zero codeword instead of random data can be used for performance simulations. Choose a concrete code, say Hamming $(n, k) = (7, 4)$, and demonstrate it by using soft CM decoding.

P6-7: Show that $p_r(u_1 + u_2 + u_3) = 1 - p_r(\overline{u_1 + u_2 + u_3}) = 1 - p_r(\overline{u_1} \cdot \overline{u_2} \cdot \overline{u_3})$.

P6-8: Show that with the independence assumption, $1 - p_r(\overline{u_1} \cdot \overline{u_2} \cdot \overline{u_3}) = 1 - p_r(\overline{u_1}) \cdot p_r(\overline{u_1}) \cdot p_r(\overline{u_1}) = 1 - (1 - p_r(u_1))(1 - p_r(u_2))(1 - p_r(u_3))$. This is used in the text instead of so called 'union bound' $p_r(\text{union bound}) = \sum_{i=1}^{2^k-1} p_r(u_i)$.

P6-9: Obtain a codeword C_m with information u_m with a generator matrix

$$Gu_m = [1\ 0\ 1\ 1]\ G = \begin{bmatrix} 1 & 0 & 0 & 0 & 1 & 1 & 1 \\ 0 & 1 & 0 & 0 & 0 & 1 & 1 \\ 0 & 0 & 1 & 0 & 1 & 0 & 1 \\ 0 & 0 & 0 & 1 & 1 & 1 & 0 \end{bmatrix}. \text{ Then enumerate all the codewords.}$$

P6-10: A generator matrix may <u>not</u> be systematic, i.e., not in the form of $\mathbf{G} = [I_k | P]$. An example is shown below.

(1) Enumerate all the codewords for an example below, and make a code table from information bits to codewords.
(2) Compare this with Hadamard H_{16} in P6-5.
(3) How does one identify information bits from a recovered codeword?

$$G = \begin{bmatrix} 0 & 1 & 0 & 1 & 0 & 1 & 0 & 1 & 0 & 1 & 0 & 1 & 0 & 1 & 0 & 1 \\ 0 & 0 & 1 & 1 & 0 & 0 & 1 & 1 & 0 & 0 & 1 & 1 & 0 & 0 & 1 & 1 \\ 0 & 0 & 0 & 0 & 1 & 1 & 1 & 1 & 0 & 0 & 0 & 0 & 1 & 1 & 1 & 1 \\ 0 & 0 & 0 & 0 & 0 & 0 & 0 & 0 & 1 & 1 & 1 & 1 & 1 & 1 & 1 & 1 \end{bmatrix}$$

P6-11: Designate codeword bits as $[c_1\ c_2\ c_3\ c_4\ c_5\ c_6\ c_7\ c_8] = [u_1\ u_2\ u_3\ u_4\ p_1\ p_2\ p_3\ p_4]$, information bits as $[u_1\ u_2\ u_3\ u_4]$ and parity bits as $[p_1\ p_2\ p_3\ p_4]$ with an (8, 4) code whose parity check equations are given by

$$p_1 = u_1 + u_2 + u_3$$
$$p_2 = u_1 + u_2 + u_4$$
$$p_3 = u_1 + u_3 + u_4$$
$$p_4 = u_2 + u_3 + u_4$$

(1) Find the generator matrix and the parity check matrix.
(2) Show the code distance is 4.
(3) Is this true $P = P^T$ matrix form of parity equations?
(4) Is this self-dual code?

P6-12: Hamming code may be defined by its parity check matrix H, $(n - k) \times n$. In order to identify a single error, the columns must be distinct and include all possible combinations of $(n - k)$bit column other than zero. This is possible when $n = 2^m - 1$ and $(n - k) = m$ with $m \geq 3$. For a given $n = 2^m - 1$, there are many choices of H by placing different $(n - k)$bit column permutations. One example choice of H with $m = 3$ is used in Section 6.2.1, and corresponding G.

$$H = \begin{bmatrix} 1 & 0 & 1 & 1 & 1 & 0 & 0 \\ 1 & 1 & 0 & 1 & 0 & 1 & 0 \\ 1 & 1 & 1 & 0 & 0 & 0 & 1 \end{bmatrix} \quad G = \begin{bmatrix} 1 & 0 & 0 & 0 & 1 & 1 & 1 \\ 0 & 1 & 0 & 0 & 0 & 1 & 1 \\ 0 & 0 & 1 & 0 & 1 & 0 & 1 \\ 0 & 0 & 0 & 1 & 1 & 1 & 0 \end{bmatrix}$$

(1) How many such choices are possible?
(2) For a systematic code we choose $H = [P^T | I_{n-k}]$. How many choices of H?
(3) Show that all different choices lead to equivalent codes; $d_{min}=3$.

P6-13: Table 6-5 Hamming code is generated by using $c(X)=u(X)g(X)$ where a generator polynomial $g(X)=X^3+X+1$ and $u(X)=u_0+u_1X+u_2X^2+u_3X^3$, and transmitting c_0, first, then c_1, and c_6, last. Its non-systematic generator matrix is given by

$$G_{non\_sys} = \begin{bmatrix} g_0 & g_1 & g_2 & g_3 & 0 & 0 & 0 \\ 0 & g_0 & g_1 & g_2 & g_3 & 0 & 0 \\ 0 & 0 & g_0 & g_1 & g_2 & g_3 & 0 \\ 0 & 0 & 0 & g_0 & g_1 & g_2 & g_3 \end{bmatrix} = \begin{bmatrix} 1 & 1 & 0 & 1 & 0 & 0 & 0 \\ 0 & 1 & 1 & 0 & 1 & 0 & 0 \\ 0 & 0 & 1 & 1 & 0 & 1 & 0 \\ 0 & 0 & 0 & 1 & 1 & 0 & 1 \end{bmatrix}.$$

Show that the systematic G_{sys} and the parity check matrix H are given by.

$$G_{sys} = \begin{bmatrix} 1 & 0 & 0 & 0 & 1 & 1 & 0 \\ 0 & 1 & 0 & 0 & 0 & 1 & 1 \\ 0 & 0 & 1 & 0 & 1 & 1 & 1 \\ 0 & 0 & 0 & 1 & 1 & 0 & 1 \end{bmatrix} \quad H = \begin{bmatrix} 1 & 0 & 1 & 1 & 1 & 0 & 0 \\ 1 & 1 & 1 & 0 & 0 & 1 & 0 \\ 0 & 1 & 1 & 1 & 0 & 0 & 1 \end{bmatrix}$$

P6-14: Consider a common method of cyclic code generation with an example of Hamming (7, 4) code. The transmission sequence is u_0, u_1, u_2, u_3, c_4, c_5, c_6 by appending parity bits. The parity bits are generated by the remainder of X^{n-k} v $(X) / g(X) = c_4X^2+c_5X+c_6$ where X^{n-k} $v(X) = X^3(u_0 X^3+u_1 X^2+u_2X+u_3) = u_0 X^6+u_1 X^5+u_2X^4+u_3X^3$. And the codeword polynomial is given by $C(X) = c_0 X^6+c_1 X^5+c_2X^4+c_3X^3 + c_4X^2+c_5X+c_6$. The non-systematic generator matrix $\underline{G}$ for this code is

$$\underline{G} = \begin{bmatrix} g_3 & g_2 & g_1 & g_0 & 0 & 0 & 0 \\ 0 & g_3 & g_2 & g_1 & g_0 & 0 & 0 \\ 0 & 0 & g_3 & g_2 & g_1 & g_0 & 0 \\ 0 & 0 & 0 & g_3 & g_2 & g_1 & g_0 \end{bmatrix} = \begin{bmatrix} 1 & 0 & 1 & 1 & 0 & 0 & 0 \\ 0 & 1 & 0 & 1 & 1 & 0 & 0 \\ 0 & 0 & 1 & 0 & 1 & 1 & 0 \\ 0 & 0 & 0 & 1 & 0 & 1 & 1 \end{bmatrix}.$$

Find the parity check matrix.

P6-15: 7 bit repetition code may be generated by using a generator polynomial g (X) = (X^3+X+1) (X^3+X^2+1). Express g(X) in a polynomial form. Find the generator matrix and the parity check matrix.

P6-16: A single parity check code may be generated by g(X) = X+1 for $n=7$. Find the generator matrix and the parity check matrix.

P6-17: A maximal length code may be generated by g(X) = (X+1) (X^3+X+1) for $n=7$. Express g(X) in a polynomial form. Find the generator matrix and the parity check matrix.

P6-18: CRC-12 generator polynomial is given by g(X) = $X^{12}+X^{11}+X^3+X^2+X+1$. Typically CRC bits (i.e., parity bits) are appended at the end of data.

(1) Draw an encoder implementation diagram using shift registers.
(2) Its block length is 2^{12} = 4096 bits. What is the maximum number of data bits in a block?
(3) When the data are 3000 bits, how this can be handled? Is it OK?

P6-19: BCH code $n=15$, $k=11$, $t=1$ is listed with the generator polynomial g (X) octal notation 23 (use the same polynomial order as in Table 6-6).

(1) Express g(X) in a polynomial form.
(2) Draw error pattern detection circuit when a decoding scheme in Figure 6-13 is used. Hint: obtain H in a systematic form, and use the first column.

P6-20: A primitive polynomial p(X) = X^5+X^2+1 for $m=5$ is used to generate BCH codes $(n, k, t) = (31, 26, 1), (31, 21, 2), (31, 16, 3), (31, 11, 5), (31, 6, 7)$ which are listed in Table 6-6. Their generator polynomials in octal representation are given by [45], [3551], [107657], [5423325], [313365047] respectively.

(1) Draw an encoder block diagram for triple error correction code.
(2) Write a computer program (say Matlab) to implement it.
(3) (project*) Expand the program to implement all encoders.
(4) Write a program to find g(X) polynomial from octal representations.

P6-21: RS code is defined GF (2^3) with primitive polynomial p(X) = X^3+X+1. In particular, there is a code $(N, K) = (7, 3)$.

(1) List all the non-zero elements of GF (2^3) in power $(1, a, a^2, a^3, \ldots a^6)$ and corresponding 3-bit tuple.
(2) What is the number of bits in a block and how many parity bits?

(3) Find a generator polynomial with $g1(X) = (X+1)(X+a)(X+a^2)(X+a^3)$.
(4) Find a generator polynomial with $g2(X) = (X+a)(X+a^2)(X+a^3)\ (X+a^4)$.
(5) Draw a block diagram with $g1(X)$ and another one with $g2(X)$, and compare them.

P6-22: RS code is defined GF (2^{10}) with $p(X) = X^{10}+X^3+X+1$ (primitive polynomial).

(1) There is a code with $t=17$ symbol error correction. Find, block length (N), parity symbol length $(N-K)$, and the minimum distance (d).
(2) The block size is $N=1023$. The range of information size $K=1, 2,\ldots1023$. When $K=1$, what is the correctable symbols t in in a block? What about $K=1021$? What about $K=1023$?
(3) A shortened code (360, 326) is used in a standard. What is the length of shortening in symbols, and what is the minimum distance of the shortened code?
(4) Another shortened code (16, 14) is used on the same GF (2^{10}) with $p(X) = X^{10}+X^3+X+1$. Find a generator polynomial containing the factor of $(X+a)$.

P6-23: (project*) Write a program to encode binary BCH code with (m, t) using the list of primitive polynomial (function p=primpol.m, m up to 16) in sub-section 6.2.5.1.

P6-24: (project*) Write a program to decode binary BCH code with (m, t) using the list of primitive polynomial in P6-23. Use the decoding scheme described in Section 6.2.6.

P6-25: A recursive systematic convolution code (RSC) is defined by G$=16\ /\ 13 = 1110\ /\ 1011$. Its state transition table is shown in Figure 6-1.

(1) Draw an encoder block diagram.
(2) Draw a state transition flow diagram.
(3) Draw a trellis diagram with a block length 12 starting from zero state to terminating to zero state. How many trellis paths are there? Find the minimum distance for this case of a block code (Figure 6-95).

P6-26: A non-recursive convolution code is defined by G $= [16, 13] = [A1, A2]$.

(1) Generate a state transition table.
(2) Draw an encoder block diagram.
(3) Draw a state transition flow diagram.
(4) Draw a trellis diagram with a block length 12 starting from zero state to terminating to zero state.

P6-27: Rate 1/3 convolution code of $m=4$ is listed in Table 6-11 G $= [25, 33, 37]$ in octal form. Assuming non-recursive code, draw an encoder diagram.

P6-28: Rate 1/2 convolution code of $m=5$ is listed in Table 6-11 G $= [53, 75]$ in octal form. Assume non-recursive code.

		D¹	D²	D³	xin	A2	D¹	D²	D³		D¹	D²	D³	xin	A2	D¹	D²	D³
		0	0	0	0	0	0	0	0		0	0	1	0	0	1	0	0
		0	0	0	1	1	1	0	0		0	0	1	1	1	0	0	0
		1	0	0	0	1	0	1	0		1	0	1	0	1	1	1	0
		1	0	0	1	0	1	1	0		1	0	1	1	0	0	1	0
G = 16 / 13 = 1 1 1 0 / 1 0 1 1		0	1	0	0	1	1	0	1		0	1	1	0	1	0	0	1
		0	1	0	1	0	0	0	1		0	1	1	1	0	1	0	1
		1	1	0	0	0	1	1	1		1	1	1	0	0	0	1	1
		1	1	0	1	1	0	1	1		1	1	1	1	1	1	1	1

Fig. P6-1: G=16/13 RSC state transition table

(1) Draw an encoder diagram.
(2) A block length 25 starting from zero state to terminating to zero state code is generated. What is the code rate of this block code? How many termination bits are necessary?

P6-29: A table of input weight and output distance, A (w, d), is shown in Table 6-12. This is an enumeration of non-zero codewords with the input weight is 1, 2, 3, and so on. When a data block length is say 25, with $w=1$, there are 25 patterns. Enumerate all the codewords, count '1's and put it in the table. This will be repeated with $w=2$, $w=3$ and so on. Write a program to generate a table for G = [17, 13] non-recursive code.

P6-30: Consider a non-systematic convolutional code and its generator polynomial is given by [7 , 5] in octal form (see Figure 6-21). Find its impulse response by xin $= \delta[n]$ and A1[n]A2[n] where n=0, 1, 2, 3, ... And obtain the generator matrix.

P6-31: Show that a generator matrix for [7 , 5] non-systematic convolutional code when the block length is N=7; G=[11101100000000; 00111011000000; 00001110110000; 00000011101100; 00000000111011; 11000000001110; 10110000000011]. Compute the output for the input [1 0 0 0 1 1 0].

P6-32: For a systematic convolutional code [7/5], i.e., $G(D) = \frac{1+D+D^2}{1+D^2}$, find the impulse response , xout[n]A1[n] for xin $= \delta[n]$. Then obtain the corresponding generator matrix.

P6-33: (project*) m=6 non-systematic convolutional code is listed as [133, 171] in octal form (see Table 6-10). A block code is created by terminating to zero state with information bit being 22. Thus the block length is 22+6 with termination bits. What is the actual code rate?

(1) Write a computer program to generate a code state transition table.
(2) Write a computer program of encoder for random input bit stream.

P6-34: (project*) This problem is to write a program to simulate code performance using the P6-33 encoder. We assume BPSK modulation of bit to symbol mapping of 0→+1, 1→-1, and AWGN channel.

(1) Consider SNR in dB from -3.0dB to +2dB with 0.5dB step, and obtain the variance of Gaussian noise generator.
(2) Write a function routine to do CM branch metric computation when received samples are given by [a, b] for a branch of 2 bit.
(3) Write a Viterbi decoding routine, i.e., calculating state metric and branch metric.
(4) Write a trace back routine to select an optimum path, and recover transmit data.

P6-35: Assume BPSK with 0→+1, 1→-1 bit mapping. Soft CM branch metric of Rate ½ convolutional code can be represented as, where a, b are received samples.

Branch bits	00	01	10	11
Soft CM metric	$a+b$	$a-b$	$-a+b$	$-a-b$

(1) Find soft CM branch metric computation table when the code rate is 1/3.
(2) Obtain branch metric tables for both code rates (i.e., ½, 1/3) when the metric is Euclidean distance.

P6-36: A conditional probability can be expressed as $Pr(A|B)=Pr(AB)/Pr(B) = Pr(AB)/$scale factor. How does one determine the scale factor?

P6-37: DMC channel model is listed in Table 6-15. Obtain cross over probabilities of P_A, P_B, P_C and P_D when SNR (Es/No) is 2dB.

P6-38: G=7l5 non-recursive convolutional code is used as a block code of length 9 terminating to zero state, e.g., see Figure 6-26. Use CM branch metric.

(1) Draw a trellis diagram corresponding to this block code of length 9.
(2) How many total paths are there in the trellis diagram?
(3) After Viterbi decoding of state and branch metric computation in the forward direction, how many surviving paths are there in maximum?
(4) After backward traceback, there is a single surviving path left. How does one obtain transmit bits?

P6-39: (project*) Consider G=7l5 non-recursive convolutional code used as a block code of length 9 terminating to zero state which is the same as in P6-38. We consider a time reversed processing, i.e., the 9th samples are processed first, for backward Viterbi decoding of state and branch metric (CM) computation. Then trace back is done in forward direction. In order to decode this block code, remember that the entire block of received samples of 18 is necessary.

P6-40: (advanced *) Combine the results of P6-38 and P6-39, i.e., forward state metric computation (P6-38) and backward state metric computation (P6-39). Can you devise a scheme to generate soft outputs, LLR of transmit bits rather than {0, 1} decision comparable to BCJR in log domain? Hint: See J. Hagenauer and

P. Hoher,"A Viterbi Algorithm with Soft_Decision Outputs and its Applications", Proc. 1989 IEEE GLOBECOM'89, pp.47.1.1-47.1.7, Dallas, Texas, 1989.

P6-41: Given G1 and G2 generator matrices there is one way of forming a new generator matrix G, called *time sharing*, as shown below;

$$G_1 = [1\ 1\ 1\ 1\ 1] \quad G_2 = \begin{bmatrix} 1 & 0 & 0 & 0 & 1 & 0 & 1 \\ 0 & 1 & 0 & 0 & 1 & 1 & 1 \\ 0 & 0 & 1 & 0 & 1 & 1 & 0 \\ 0 & 0 & 0 & 1 & 0 & 1 & 1 \end{bmatrix} \quad \rightarrow \quad G = \begin{bmatrix} G_1 & 0 \\ 0 & G_2 \end{bmatrix}$$

Obtain a generator G explicitly and find its block size (n), information size (k).

P6-42: Given G1 and G2 generator matrices there is another way of forming a new generator matrix G, called *direct sums*, as shown below

$$G_1 = [1\ 1\ 1\ 1\ 1\ 1\ 1] \quad G_2 = \begin{bmatrix} 1 & 0 & 0 & 0 & 1 & 0 & 1 \\ 0 & 1 & 0 & 0 & 1 & 1 & 1 \\ 0 & 0 & 1 & 0 & 1 & 1 & 0 \\ 0 & 0 & 0 & 1 & 0 & 1 & 1 \end{bmatrix} \quad \rightarrow \quad G = \begin{bmatrix} G_1 \\ G_2 \end{bmatrix}$$

Obtain a generator G explicitly and find its block size (n), information size (k).

P6-43: One way of obtaining a low rate convolutional code such as rate=1/4, 1/6, so on is starting from rate=1/2 followed by repetition of 2, 3 and so on. For example, m=6 rate=1/2, generators (133, 171) in octal representation with the minimum free distance d_f =10, is listed in Table 6-10. With this code combining idea of repetition, rate=1/4 code generators are given by (133, 133, 171, 171). Similarly rate=1/6 code generators are (133, 133, 133, 171, 171, 171). Draw encoder diagrams for these low rate codes; the serial connection of the (133, 171) encoder and repetition code.

P6-44: A two dimensional product code was proposed early on as in Fig. P6-2 (LHS). Two codes of C1 (n_1,k_1,d_1) and C2 (n_2,k_2,d_2) are combined where $(n_1,k_1,$

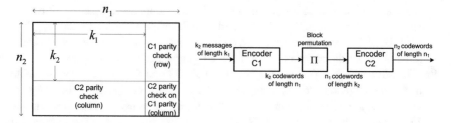

Fig. P6-2: Construction of product code C1 $\otimes$ C2 and its implementation with the cascaded connection of C1 and C2 through block permutation

d_1)=(block length, information length, minimum distance of C1), and similarly for C2. Another interpretation of the product code (LHS) is on the RHS of Fig. P6-2; C1 and C2 are connected serially through a block permutation.

(1) Show that the combined code C is given by $(n,k,d)=(n_1n_2,k_1k_2,d_1d_2)$. What is (n,k,d) of C when C1=C2=(7, 4, 3)?
(2) Compare C of (1) with a repetition code with the same information bits and minimum distance.
(3) Describe the block permutation on RHS for C1=C2= (7, 4, 3).

P6-45: Referring to Fig. P6-2, when C2= $(n_2,n_2,$ 1), a trivial code, the code C is a multiple (n_2 times) transmission of C1. Show this code C has the capability of burst error correction and find the length of correctable burst. Hint: Transmission sequence of (n_1, n_2) matrix is column by column.

P6-46: Referring to Fig. P6-2, C1 is a RS code (N, K, D) = (7, 5, 3) with GF (2^m) and C2= (7, 4, 3) code.

(1) What is (n, k, d) of the combined code C in bits? Hint: $m=3$.
(2) What is the size of the block permutation matrix?
(3) When C1=RS (255, 223, 33) with GF (2^m) of $m=8$, C2=convolutional code with G = (133, 171) of 64 states are connected directly, i.e., in cascade, what is the size of π matrix?

P6-47: Referring to Fig. P6-2, C = C1 $\otimes$ C2 product code can be decoded iteratively. Received samples are arranged as a matrix of (n_2, n_1).

(1) When C1 = C2 = (7, 4, 3), k_1k_2 =16 information bits are arranged in matrix (k_2 , k_1)=(4, 4), and then parity checks are added in rows and columns. Describe an iterative decoder with algebraic decoding of this product code in details; row by row decoding and followed by column by column decoding is repeated.

P6-48: One form of parity check matrix, H, is called IRA (irregular repeat accumulator) form for iterative decoding. An example of encoder is shown in Figure 6-44.

(1) Confirm that, column weight =[5 5 4 3 3] and row weight =[2 2 2 2 2 2 2 2 2] and perm = [11 18 1 15 2 6 3 12 7 19 8 16 4 13 5 9 10 17 14 20], H is given by

$$
H =
\begin{array}{c}
\xleftarrow{\quad \text{H1} \quad} \times \xrightarrow{\qquad\qquad \text{zigzag} \qquad\qquad} \\
\begin{array}{ccccc|cccccccccc}
0 & 0 & 1 & 0 & 1 & 1 & 0 & 0 & 0 & 0 & 0 & 0 & 0 & 0 & 0 \\
1 & 0 & 0 & 1 & 0 & 1 & 1 & 0 & 0 & 0 & 0 & 0 & 0 & 0 & 0 \\
1 & 1 & 0 & 0 & 0 & 0 & 1 & 1 & 0 & 0 & 0 & 0 & 0 & 0 & 0 \\
1 & 0 & 1 & 0 & 0 & 0 & 0 & 1 & 1 & 0 & 0 & 0 & 0 & 0 & 0 \\
0 & 1 & 0 & 0 & 1 & 0 & 0 & 0 & 1 & 1 & 0 & 0 & 0 & 0 & 0 \\
0 & 1 & 0 & 1 & 0 & 0 & 0 & 0 & 0 & 1 & 1 & 0 & 0 & 0 & 0 \\
1 & 0 & 1 & 0 & 0 & 0 & 0 & 0 & 0 & 0 & 1 & 1 & 0 & 0 & 0 \\
1 & 1 & 0 & 0 & 0 & 0 & 0 & 0 & 0 & 0 & 0 & 1 & 1 & 0 & 0 \\
0 & 1 & 0 & 1 & 0 & 0 & 0 & 0 & 0 & 0 & 0 & 0 & 1 & 1 & 0 \\
0 & 0 & 1 & 0 & 1 & 0 & 0 & 0 & 0 & 0 & 0 & 0 & 0 & 1 & 1 \\
\end{array}
\end{array}
$$

(2) Show that the generator matrix G is given by

$$
G =
\begin{array}{c}
\xleftarrow{\quad u \quad} \times \xrightarrow{\qquad\qquad \text{parity bits} \qquad\qquad} \\
\begin{array}{ccccc|cccccccccc}
1 & 0 & 0 & 0 & 0 & 0 & 1 & 0 & 1 & 1 & 1 & 0 & 1 & 1 & 1 \\
0 & 1 & 0 & 0 & 0 & 0 & 0 & 1 & 1 & 0 & 1 & 1 & 0 & 1 & 1 \\
0 & 0 & 1 & 0 & 0 & 1 & 1 & 1 & 0 & 0 & 0 & 1 & 1 & 1 & 0 \\
0 & 0 & 0 & 1 & 0 & 0 & 1 & 1 & 1 & 1 & 0 & 0 & 0 & 1 & 1 \\
0 & 0 & 0 & 0 & 1 & 1 & 1 & 1 & 1 & 0 & 0 & 0 & 0 & 0 & 1 \\
\end{array}
\end{array}
$$

In hardware implementation the use of the multiplication of information bit with generator matrix (i.e., uG) not used often but in MATLAB it is fast and easy. Hint: parity bits are computed $H1*u^T$ followed by accumulation. The generator matrix of accumulation is upper triangle.

P6-49: (project*) In P6-48 we change permutation to two cases below;
   permA = [12 20 2 14 10 15 8 13 18 1 9 5 17 11 7 16 6 3 19 4]
   permB = [1 6 11 16 2 7 12 17 3 8 13 18 4 9 14 19 5 10 15 20]

   (1) Find H1 corresponding to permA, and list all the 4-cycles. Hint: there are 3 of 4-cycles and one can find them by inspection.
   (2) Find H1 corresponding to permB, and list all the 4-cycles. Hint: there are 12 of 4-cycles, and write a program to list 4-ccyels for a given H.

P6-50: A circulant matrix is defined as a circular (left) shifted (square) identity matrix in column. Note that we chose a left shit of column even though it is possible to right shift or to row shift. When a circulant block size is 81, i.e., identity matrix (81, 81), there are 57 shifts. In MATLAB, it is circshift (eye (81), [0,-57]). Plot it. Hint: See Figure 6-47.

P6-51: (project*) In Figure 6-48, a prototype H (12, 24) is shown. It is lifted by block circulant size 81 to make an LDPC code (n, k) = (1944, 972) by defining the parity check matrix with assigning a block circulant for each prototype H as Figure 6-46. Write a program to generate a full H matrix size (972, 1944).

P6-52: A repetition code and single parity check code are dual codes; generator matrix G of repetition code is parity check matrix H of single parity check code.

Fig. P6-3: Product code (LHS) and one check bit punctured (RHS)

$$\begin{bmatrix} u_0 & u_1 & u_2 & p_9 \\ u_3 & u_4 & u_5 & p_{10} \\ u_6 & u_7 & u_8 & p_{11} \\ p_{12} & p_{13} & p_{14} & p_{15} \end{bmatrix} \qquad \begin{bmatrix} u_0 & u_1 & u_2 & p_9 \\ u_3 & u_4 & u_5 & p_{10} \\ u_6 & u_7 & u_8 & p_{11} \\ p_{12} & p_{13} & p_{14} & \end{bmatrix}$$

Fig. P6-4: Product code with transmission sequence (LHS), C1 parity checks (RHS)

$$\begin{bmatrix} u_0 & u_1 & u_2 & u_3 & p_{16} & p_{17} & p_{18} \\ u_4 & u_5 & u_6 & u_7 & p_{19} & p_{20} & p_{21} \\ u_8 & u_9 & u_{10} & u_{11} & p_{22} & p_{23} & p_{24} \\ u_{12} & u_{13} & u_{14} & u_{15} & p_{25} & p_{26} & p_{27} \\ p_{28} & p_{31} & p_{34} & p_{37} & p_{40} & p_{43} & p_{46} \\ p_{29} & p_{32} & p_{35} & p_{38} & p_{41} & p_{44} & p_{47} \\ p_{30} & p_{33} & p_{36} & p_{39} & p_{42} & p_{45} & p_{48} \end{bmatrix}$$

$$P^T = \begin{bmatrix} 1110 \\ 0111 \\ 1101 \end{bmatrix}$$

	Repetition code $(n, k, d) =$ (4,1,4)	Single parity check code (n, k, d) =(4,3,2)
Generator matrix G	G=[1 1 1 1]	$G= \begin{bmatrix} 1001 \\ 0101 \\ 0011 \end{bmatrix}$
Parity Check matrix H	$H= \begin{bmatrix} 1100 \\ 1010 \\ 1001 \end{bmatrix}$	H=[1 1 1 1]

(1) Consider repetition code $(n, k, d) = (5, 1, 5)$ and show its G and H.
(2) Consider a parity check code $(n, k, d) = (5, 4, 2)$ and show its G and H.

P6-53: A repetition code $(n, k, d) = (4, 1, 4)$ is *lifted* by an identity matrix size of 3, which increases it to $(n, k, d) = (12, 3, 4)$. Show that G of the lifted code is below;

$$G = \begin{bmatrix} 100 & 100 & 100 \\ 010 & 010 & 010 \\ 001 & 001 & 001 \end{bmatrix}$$. Find the corresponding H.

P6-54: A single parity check code $(n, k, d) = (4, 3, 2)$ is *lifted* by an identity matrix size of 3, which increases it to $(n, k, d) = (12, 9, 2)$. Find the generator matrix G and parity check matrix H for the lifted code. Hint: see P6-53.

P6-55: A product code $(n, k, d) = (16, 9, 4)$ of Fig. P6-3 (LHS) is formed by using a single parity check code $(n, k, d) = (4, 3, 2)$, i.e., C1=C2 = (4, 3, 2).

(1) Obtain H of product code $(n, k, d) = (16, 9, 4)$ code (LHS).

(2) Obtain the corresponding G. Hint: p_{15} check bit contains all information bits.

(3) Show that there is no 4-cycle in H.

P6-56: A code $(n, k, d) = (15, 9, 3)$ of Fig. P6-3 (RHS) is formed by first product code using a single parity check code $(n, k, d) = (4, 3, 2)$, i.e., C1=C2 = (4, 3, 2). Then puncture p_{15} check bit in order to obtain the code.

(1) Obtain H of product code $(n, k, d) = (15, 9, 3)$ code (LHS).

(2) Obtain the corresponding G.

(3) Confirm the minimum distance is 3 rather than 4 by inspecting G.

P6-57: (advanced*) We form a C = C1 $\otimes$ C2 product code with C1 = C2 = (7, 4, 3), $k_1 k_2 = 16$ information bits are arranged in matrix (k_2, k_1)=(4, 4), and then parity checks are added in rows and columns as shown in Fig. P6-4; transmission sequence is numbered in subscript, $u_0, u_1, \ldots .p_{48}$.

Find parity check matrix H (first) and then generator matrix G for this code.

P6-58: Consider Hamming code $(n, k, d) = (7, 4, 3)$ with $g(X) = 1+X+X^3$. Define $X^3 g(X^{-1}) = X3(1+X^{-1}+X^{-3}) = 1+X^2+X^3$.

(1) Show that X^7+1 is divisible by $X^3 g(X^{-1})$ by a long division.

(2) Find $h(X) = 1+X+X^2+X^4$ such that $X^7+1 = g(X)*h(X)$ by a long division.

(3) Show that a generator matrix G corresponding to $X^3 g(X^{-1})$ and a parity check matrix H corresponding to $h(X)$;

$$G = \begin{bmatrix} g_3 & g_2 & g_1 & g_0 & 0 & 0 & 0 \\ 0 & g_3 & g_2 & g_1 & g_0 & 0 & 0 \\ 0 & 0 & g_3 & g_2 & g_1 & g_0 & 0 \\ 0 & 0 & 0 & g_3 & g_2 & g_1 & g_0 \end{bmatrix} \qquad H = \begin{bmatrix} h_0 & h_1 & h_2 & h_3 & h_4 & 0 & 0 \\ 0 & h_0 & h_1 & h_2 & h_3 & h_4 & 0 \\ 0 & 0 & h_0 & h_1 & h_2 & h_3 & h_4 \end{bmatrix}$$

(4) Confirm that $GH^T=0$.

P6-59: Show that X^n+1 is divisible by $X^{n-k} g(X^{-1})$ for a given $g(X)$ in a cyclic code. c $(X)=(u_0+u_1 X+u_2 X^2+\ldots .u_{k-1}X^{k-1})$ $X^{n-k} g(X^{-1})$ generate the same codewords as $(u_0, u_1, u_2, \ldots .u_{k-1})$ followed by appending the remainder as parity bits as in (6-29).

P6-60: In a cyclic code for a given generator polynomial $g(X)$ and corresponding parity check polynomial $h(X)$ show that G and H are shown below such that $GH^T=0$;

$$G = \begin{bmatrix} g_{n-k} & \cdots & g_2 & g_1 & g_0 & 0 & 0 & \cdots & 0 \\ 0 & g_{n-k} & \cdots & g_2 & g_1 & g_0 & 0 & \cdots & 0 \\ & & \cdots & & & & & & \\ 0 & 0 & g_{n-k} & \cdots & & g_2 & g_1 & g_0 & 0 \\ 0 & 0 & 0 & g_{n-k} & \cdots & & g_2 & g_1 & g_0 \end{bmatrix}$$

$$H = \begin{bmatrix} h_0 & h_1 & h_2 & \cdots & h_{k-1} & h_k & \cdots & \cdots & 0 & 0 \\ 0 & h_0 & h_1 & h_2 & \cdots & h_{k-1} & h_k & \cdots & \cdots & 0 \\ & \cdots & \cdots & \cdots & & & & & & \\ 0 & 0 & \cdots & \cdots & h_0 & h_1 & h_2 & \cdots & h_{k-1} & h_k \end{bmatrix}$$

P6-61: In BPSK bit to symbol mapping '0'→+1, '1'→-1 is used. LLR is defined as $\Lambda = \log \frac{p_0}{p_1} = \log \frac{pr(+1|r)}{pr(-1|r)}$. Show $p_1 = \frac{1}{1+e^{+\Lambda}}$ and $p_0 = \frac{1}{1+e^{-\Lambda}}$. In BPSK, when a received sample r is given, what is LLR?

P6-62: In QPSK, $b_1 b_0 = 00 \to 1+j$, $01 \to 1-j$, $10 \to -1+j$, and $11 \to -1-j$ bit to symbol mapping is used. Generate an implicit mapping table (i.e., list only symbols). Derive explicit expressions of LLR of b_1 and b_0.

P6-63: In MLCM, a number of constellation points are Cn=3126. We may send 11.5 bits per symbol in the average, i.e., 23 bits per 2 symbols. The 23 bits are named as I_1, I_2, I_3, ...I_{23} from LSB to MSB. I_1 to I_8 are assigned to two 4 bits of 16QAM, and $I_9, I_{10}, \ldots I_{20}$ of 12 bits are assigned to 64 of 16-QAM blocks. How do you assign I_{21}, I_{22}, I_{23} of 3 bits into two symbols? Hint: There are 192 blocks of 16-QAM in the constellation.

P6-64: As information measure, Shannon entropy or simply entropy is defined as $H(X) = E\{ \log_2 (\frac{1}{P_X}) \} = \sum_{i=1}^{M} p_i \log_2 (\frac{1}{p_i})$. When we have a constellation of 64 points and the probability of each point is uniform, what is $H(X)$?

Chapter 7
Synchronization of Frame, Symbol Timing and Carrier

Contents

7.1	Packet Synchronization Examples	512
	7.1.1 PLCP Preamble Format of IEEE 802.11a	513
	7.1.2 RCV Processing of STS and LTS	516
	7.1.3 802.11$_a$ Synchronization Performance	525
	7.1.4 DS Spread Spectrum Synchronization Example	529
7.2	Symbol Timing Synchronization	532
	7.2.1 Symbol Timing Error Detector for PAM/QAM	533
	7.2.2 Known Digital Timing Error Detectors	539
	7.2.3 Numerical Confirmation of S-Curve of Timing Error Detectors	545
	7.2.4 Timing Detectors with Differentiation or with Hilbert Transform	550
	7.2.5 Intuitive Understanding of Timing Detectors	550
	7.2.6 Carrier Frequency Offset Estimation	552
	7.2.7 Embedding Digital TED Into Timing Recovery Loop	554
	7.2.8 Resampling and Resampling Control	556
	7.2.9 Simulations of Doppler Clock Frequency Shift	561
7.3	Carrier Phase Synchronization	565
	7.3.1 Carrier Recovery Loop and Its Components	566
	7.3.2 Phase Locked Loop Review	566
	7.3.3 Understanding Costas Loop for QPSK	567
	7.3.4 Carrier Phase Detectors	569
	7.3.5 All Digital Implementations of Carrier Recovery Loop	571
7.4	Quadrature Phase Imbalance Correction	572
	7.4.1 IQ Imbalance Model	573
	7.4.2 $\hat{\theta}$, φ_i, and φ_d Measurements	576
	7.4.3 2-Step Approach for the Estimation of $\hat{\theta}$, φ_i, and φ_d	577
	7.4.4 Additional Practical Issues	578
	7.4.5 Summary of IQ Phase Imbalance Digital Correction	579
7.5	References with Comments	580
Appendix 7		581
	A.1 Raised Cosine Pulse and its Pre-filtered RC Pulse	581
	A.2 Poisson Sum Formula for a Correlated Signal	582
	A.3 Review of Phase Locked Loops	583
	A.4 FIR Interpolation Filter Design and Coefficient Computation	588
7.6	Problems	594

© Springer Nature Switzerland AG 2020
S.-M. Yang, *Modern Digital Radio Communication Signals and Systems*,
https://doi.org/10.1007/978-3-030-57706-3_7

Abstract The literature on synchronization is vast and yet there seem no clear-cut fundamentals identified. Our approach here is to treat the topic from modern implementation point of view, namely the most of demodulation functions are implemented digitally, after sampling, rather than by analog signal processing. Another point is a fast synchronization, in fact, aiming as fast as theoretically possible with minimum variance. An extensive development for TED is based on 2 or 4 samples per symbol, which works with a large carrier frequency offset.

Key Innovative Terms physical frame synchronization with matched filters · TED list · TED for smaller excess BW · TED in sampling rate · carrier phase detector in symbol rate · digital quadrature phase imbalance correction with 2-step approach · simulation of Doppler clock frequency for wide band signals · weak signal initial code phase synchronization of GPS

General Terms all digital implementations · Costas loop, digital resampling, DS spread spectrum · Farrow structure, frame synchronizations · IEEE 802.11a preamble · STS · LTS · matched filter · Packet synchronization · Pre-filtered pulse · PLL · Polynomial interpolation · s-curve

List of Abbreviations

ADC	analog digital converter
AGC	automatic gain control
AWGN	additive white Gaussian noise
BETR	band edge timing recovery
BER	bit error rate
BPF	bandpass filter
CCW	counter clock wise
CDMA	code division multiple access
CP	cyclic prefix
CW	clock wise
DAC	digital analog converter
DC	direct current
DFT	discrete Fourier transform
DS	direct sequence
FIR	finite impulse response
FSK	frequency shift keying
GI	guard interval, see CP
GPS	global positioning system
IDFT	inverse discrete Fourier transform
IEEE	Institute of Electrical Electronics Engineers
IQ	in-phase quadrature phase
LEO	low Earth orbit
LNA	low noise amplifier
LO	local oscillator
LP, LPF	low pass, low pass filter

LTS	long-term training sequence
MPDU	MAC protocol data unit
OFDM	orthogonal frequency division multiplex
PAM	pulse amplitude modulation
PDI	post detection integrator
PER	packet error rate
PLL	phase locked loop
PLCP	physical layer convergence procedure
PN	pseudo random noise
ppm	part per million
PRN	pseudo random noise
PSDU	PLCP service data unit
QAM	quadrature amplitude modulation
QPSK	quadrature phase shift keying
RC	raised cosine
RCV	receive
SNR	signal to noise ratio
STS	short term training sequence
TED	timing error detector
VCO	voltage controlled oscillator

In this chapter we cover one of the most important topics in receiver signal processing; synchronization. The synchronization includes symbol clock frequency and sampling phase recovery, carrier frequency and phase estimation for coherent demodulation. Furthermore it includes physical layer frame detection related with OFDM symbol, with spread spectrum signal spreading sequence boundary, and with block error correcting code word boundary. Related to synchronization, we cover equalization and channel estimation briefly. The equalization tries to 'invert' a channel so that the frequency response of the overall channel including an equalizer should be flat. For fading channels, pilot symbols (or training symbols) are used for channel estimation.

We repeat the system block diagram in Figure 7-1 in order to see where this chapter topic belongs to an overall system. These functions, synchronization and equalizations, are not explicitly shown in this overall diagram but part of demodulation processing. Depending on the implementations carrier frequency local oscillator (LO) and sampling clock may also be involved, as indicated by shading in the figure.

The literature on synchronization is vast and yet there seem no clear-cut fundamentals identified. This is understandable because it should be adjusted to a signal format whenever it changes, sometimes taking advantage of peculiar features of it. Our approach here is to treat the topic from modern implementation point of view, namely the most of demodulation functions are implemented digitally (or numerically), after sampling, rather than by analog signal processing. Another point is a fast synchronization, in fact, aiming as fast as theoretically possible with minimum variance.

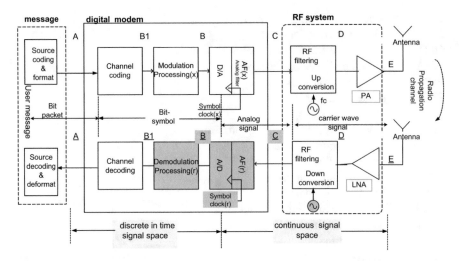

Figure 7-1: Digital radio system block diagram with emphasis of receiver signal processing – synchronization (symbol clock, carrier, and physical layer packet), equalization and channel estimation. These functions are not explicitly shown in this overall diagram but part of demodulation processing

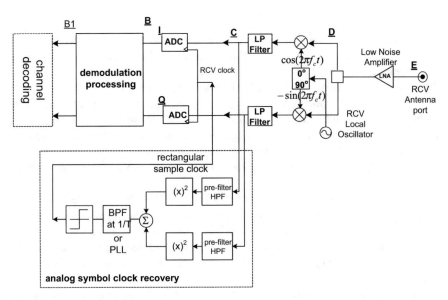

Figure 7-2: Modulation symbol clock recovery with analog signal processing using squaring and pre-filter (high-pass filter type) followed by a band-pass filter (or PLL) tuned to the symbol rate. We consider the variety of digital implementations based on the idea similar to this

For example, a symbol clock recovery may be done by purely analog processing as shown in Figure 7-2. This scheme is fast, independent of carrier frequency offset and of modulation symbol level, and relies on only the cyclostationary nature of a signal after squaring it. This will be elaborated in detail later in this chapter, where we consider variety of digital implementations based on the idea similar to this analog method.

We emphasize again that the scheme in Figure 7-2 utilizes only the cyclostationary nature after squaring, which comes from the fact that symbols are transmitted periodically with a symbol rate. (This has some similarity to signal power detection where no detailed features of a signal are necessary.) Thus it is applicable to any QAM (PAM) signal format, and even to dense constellation approaching analog one. This form of clock recovery is treated in detail here not only because of its applicability to any QAM (PAM) signal format but also because of the robustness under carrier frequency offset and channel fading. Furthermore it is amenable to digital implementations as well with variety of enhancements. It is practically well justified as well since it is fast.

Contrast to the symbol clock recovery in Figure 7-2, carrier phase synchronization inherently requires the details of signal constellations since the measurement of carrier phase deviation from an ideal constellation point should be done in digital domain. But actual phase correction may be done either in analog signal or in digital domain. We show an example in Figure 7-3, that the phase correction is done by changing the frequency of RCV local voltage controlled oscillator (VCO). Note that this scheme is similar to a PLL of sinusoidal signals, except for phase detection being elaborated with digital comparison against ideal constellation points in addition to mixing (i.e., quadrature demodulating) locally generated carrier with the incoming carrier modulated signal.

In modern implementations, demodulation processing, including both synchronization and equalization, is typically done digitally. Thus Figures 7-2 and 7-3 may be included inside demodulation processing as shown in Figure 7-4. This fully

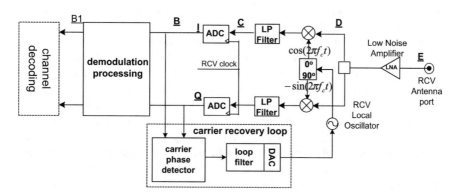

Figure 7-3: Carrier recovery loop; carrier phase detection must be inherently digital but loop filter and phase correction may be digital or analog. Here the phase correction is done by changing the frequency of LO in the form of voltage controlled oscillator (VCO)

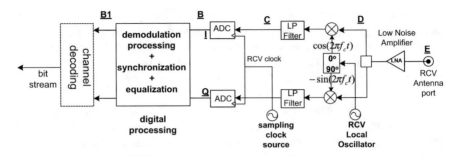

Figure 7-4: Fully digital implementation of demodulation processing; sampling clock and carrier frequency are free running, though both frequencies are close to those of transmitter, and all detection and necessary corrections are done digitally. We emphasize this full digital implementation scheme in this chapter

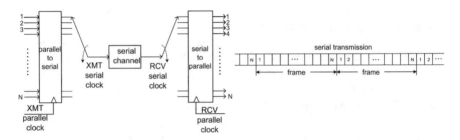

Figure 7-5: Frame (packet, block) synchronization problem arises due to serial transmission and the frame boundary must be recovered at the receive side, i.e., both serial clock and parallel clock must be recovered with correct boundary (phase)

digital implementation of synchronization (symbol clock recovery, carrier frequency and phase), equalization and demodulation will be our focus in this chapter. Sampling clock and carrier frequency are free running, though both frequencies are close to those of transmitter, and all detection and necessary corrections are done digitally. In this type of implementations, it is typical to have a sampling clock frequency higher than symbol rate, 2 or 4 times of it, being often used in practice.

Another class of synchronization problem may be represented symbolically as parallel to serial, serial transmission and back to parallel which is shown in Figure 7-5. It may be called frame synchronization; a frame can be a collection of modulation symbols, OFDM symbols, spread spectrum spreading sequences, and code words of a forward error correcting block code. It may be called packet synchronization.

A similar problem exists in higher layers (data link layer, network layer, application layer), where a block of bits forms a frame and the frame typically is marked by a special pattern known to receiver. And a receiver searches for this pattern by trial and error until found. Here we focus on the synchronization of physical layer frame, i.e., for a frame of OFDM symbols and spread spectrum sequence; examples from IEEE $802.11_{a,b,g,n,\ ac}$ signal format will be given in this chapter.

The equalization tries to 'invert' a channel so that the frequency response of the overall channel including an equalizer should be flat. There is a vast amount of literature on equalization, including adaptive equalization for bandlimited channels. In wideband wireless channel communications, an issue of sophisticated adaptive equalization is solved by utilizing an OFDM signal format, where each subcarrier channel is narrow enough so that it is close to flat in frequency domain. Thus we devote little pages on equalization in this book. However, even if it is flat there is a need of 1-tap equalizer. Moreover if it is not exactly flat enough, then additional measure of equalization beyond the 1-tap, using a small number of taps, might be necessary. In wireless environments, a multipath fading may create the need of channel estimation. For this problem, typically pilot symbols, known to receiver, are inserted at transmit side.

Example 7-1: Show the coefficients of 1-tap equalizer when a sampled channel impulse (end to end including all the filters) is represented by $h(n) = 0.7244 + j\,0.1941$ when $n = 0$, $h(n) = 0.0$ otherwise.

Answer: $c_{eq} = 1/h(n) = 1.2880 - j\,0.3451$. Note that obviously the channel with 1-tap equalizer $= 1/h(n) \times c_{eq} = 1.0 + j\,0.0$, and thus there is no cross-coupling between i-ch and q-ch. A real arithmetic implementation is shown in Figure 7-6.

Example 7-2: For a given $h(n) = h_{re} + jh_{im}$ when $n = 0$, $h(n) = 0.0$ otherwise, a corresponding 1-tap equalizer can be represented by two components; gain and phase rotation.

Answer: gain $= 1/\sqrt{h_{re}^2 + h_{im}^2}$ and phase rotation $= -\tan^{-1}(h_{im}/h_{re})$. Thus carrier phase rotation can be implemented as part of equalization. In practice, a channel impulse response is not directly available and thus it should be estimated.

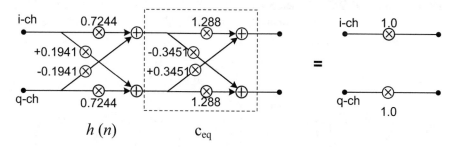

Figure 7-6: A numerical example of 1-tap equalizer, being the minimum required equalizer, which eliminates cross-coupling between i-ch and q-ch, and sets the gain to be 1.0. This can be decomposed as carrier phase rotation and gain. A precise phase rotation, eliminating cross coupling, is very important

7.1 Packet Synchronization Examples

In this section we give an example of physical layer packet synchronization using the signal format of IEEE 802.11$_a$. We use IEEE std 802.11^{TM} – 2012 which contains the standards of 802.11$_{a, g, n}$, and 802.11$_{ac}$ standard was published in 2014. IEEE 802.11$_{a, g, n, ac}$ is all based on OFDM with cyclic prefix (CP) and their base line is 802.11$_a$. Thus our example will be this basic packet format of it, called PLCP (physical layer convergence procedure) preamble. In passing we mention that 802.11$_{ac, n}$ contains the same packet format as legacy in addition new multi-antenna signaling preamble, and that 802.11$_g$ is essentially identical with 802.11$_a$ except for carrier frequency being in 2 GHz band. Another example of packet synchronization is direct sequence spread spectrum system in section 7.1.4, in particular, one from GPS.

A schematic representation of OFDM is shown in Figure 7-7.

Compare Figure 7-7 with Figure 7-5, and note the similarity of parallel to serial at the transmit side, serial transmission, and serial to parallel at the receive side. Our problem is to find the parallel symbol boundary at the receiver as well as the detection of the start of a packet since it is a burst transmission scheme, rather than continuous transmission.

OFDM signal parameters are described in the IEEE standard in detail and we summarize them briefly here. IDFT size $N = 64$, CP $v = 16$, and thus 80 samples per OFDM symbol are transmitted serially with the sampling rate $(N + v)/T = 20$ MHz.

OFDM symbol time T = 4.0 μsec of which CP is 0.8 μsec and data carrying portion is 3.2 μsec. Sub-carrier frequency spacing is 312.5 kHz (=1/3.2 μsec). 10 MHz, 5 MHz sampling rate schemes, corresponding to T = 8.0 μsec, and T =16.0 μsec with the same N and v, are specified in the IEEE standard but we focus only on the 20 MHz scheme.

Exercise 7-1: OFDM parameters are related. Find OFDM symbol time given N, v, and sampling rate above. Answer: $T = 80/20$ MHz $= 4$ μsec from $(N+v)/T = 20$ MHz.

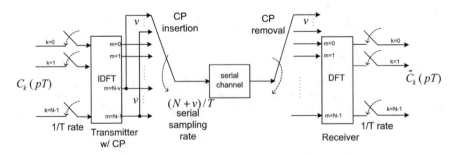

Figure 7-7: A schematic representation of OFDM transmission system with cyclic prefix (CP). CP is called guard interval (GI) in IEEE std 802.11^{TM} – 2012

Compare Figure 7-7 with Figure 7-5, and we emphasize again the similarity of parallel to serial at the transmit side, serial transmission, and serial to parallel at the receive side. Our problem of packet synchronization is to find the parallel symbol boundary at the receiver. In addition to the symbol boundary, the detection of the start of a packet is necessary in 802.11_a signaling scheme since it is a burst transmission scheme, rather than continuous transmission.

7.1.1 PLCP Preamble Format of IEEE 802.11a

We explain the PLCP preamble with the aid of Figure 7-8. Our focus is the preamble of 4 OFDM symbols (four of 4 μsec duration symbols – gray shade) in a packet.

IEEE 802.11_a physical layer packet is a sequence of OFDM symbols. Its preamble is added in addition to MAC data (MPDU) to be transmitted. Additional bits of 'service', 'tail', and 'pad' to MPDU are necessary to fit into concatenation of OFDM symbols. PLCP preamble consists of 4 OFDM symbols, fixed and known to receiver, – the first two are called short-term training sequence (STS) and the second two are called long-term training sequence (LTS). The preamble can be used to recover OFDM boundary and packet detection (start of a packet). The demarcation of 4 μsec in the figure is one OFDM symbol in time and each symbol contains 80 samples with 20 MHz sampling.

In passing we mention SIGNAL even though we do not use it for synchronization purpose since it carries the packet length information and rate (modulation class and

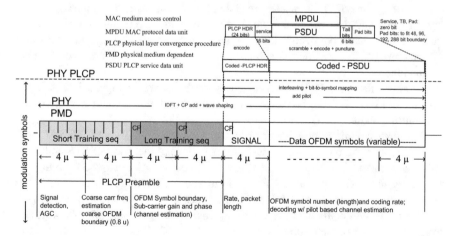

Figure 7-8: IEEE 802.11_a physical layer packet is a serial concatenation of OFDM symbols. Its preamble is added in addition to MAC data (MPDU) to be transmitted. PLCP preamble consists of 4 OFDM symbols, fixed and known to receiver, – the first two are called short-term training sequence (STS) and the second two are called long-term training sequence (LTS). Coded 48 bit symbol, called SIGNAL, contains rate (modulation and code rate) and data length in byte. The demarcation of 4 μsec is one OFDM symbol in time

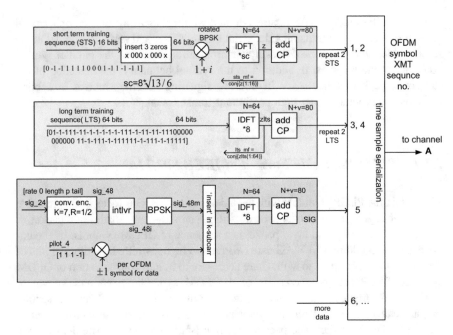

Figure 7-9: A physical layer packet generation by adding PLCP preamble of 2 OFDM symbols of STS, 2 OFDM symbols of LTS, and 1 symbol of SIGNAL. The first five OFDM symbols of a packet are sequentially transmitted as numbered. Additional data carrying OFDM symbols, numbered from 6, are followed

code rate). 24 bits of PLCP header are coded into 48 bits named as SIGNAL which contains rate and data length in byte. Without correctly decoding SIGNAL subsequent data cannot be decoded. Each data carrying OFDM symbol allocates 48 sub-carriers for data and 4 sub-carriers for pilots (will be used for channel estimation).

Another view of a PLCP packet generation is shown in Figure 7-9. A block diagram of Figure 7-9 may be viewed as a summary of digital implementations of the transmitting signal generation.

Below we explain Figure 7-9 one block at a time; STS, LTS, and SIGNAL.

We mention the convention of sub-carrier frequency index k. For STS and LTS, sub-carrier index, k, will be 0 to 63 where $k = 0$ being DC and $k = 31$ the highest frequency, $k = 63$ (equivalently $k = -1$) to $k = 32$ (equivalently $k = -32$) are negative frequencies. For SIGNAL, the k index of 4 pilot sub-carriers is {43, 57, 7, 21} or with equivalent negative frequency notation {−21, −7, +7, +2}, and the k index of 48 bit data sub-carrier is {48, 49, … 63, 1, 2, .. 26} or equivalently {−26, −25, …., −1, +1, +2, .., +26} except for pilots and DC ($k = 0$). Both indexing schemes are, of course, equivalent and it is a matter of convenience.

7.1.1.1   Short-Term Training Sequence (STS)

STS is in the top of Figure 7-9. A 16 digit pattern {0 −1 −1 1 1 1 1 0 0 0 1 −1 1 −1 −1 1}, BPSK modulated and no signal with zero, is expanded into 64 digits by inserting three 0s between digits, and rotated by 45 degree (multiplying 1 + i). Its corresponding sub-carrier index is $k = 0$ to 63. A scale factor, $\sqrt{13/6}$, is such that a symbol energy is 52 (the energy of every OFDM symbol in a packet is consistently 52). Additional scale factor, 8, is necessary to make the same energy per symbol after IDFT. (Note that IDFT definition has a scale factor 1/N while DFT scale factor is 1.0.) Due to the repeated insertion of 3 zeroes before IDFT, the time samples after IDFT will be periodic with the period of 16 samples.

One STS after IDFT and adding CP is shown in Figure 7-10. It is periodic with the period of 16 samples. Thus 16-sample pattern is repeated 5 times in one OFDM symbol. With one more repetition of STS, there are 10 such 16 sample patterns, as shown in Figure 7-8 (first two symbols of duration 8 μsec). Note that this 16 sample pattern in time may be obtained by taking IDFT with $N = 16$.

The 16 sample pattern in time is denoted as sts_mf (n) with $n = 1,..,16$. Its conjugate, sts_mf*(n) , with time reversal, will be used as a matched filter at the receiver to detect the 16 sample pattern.

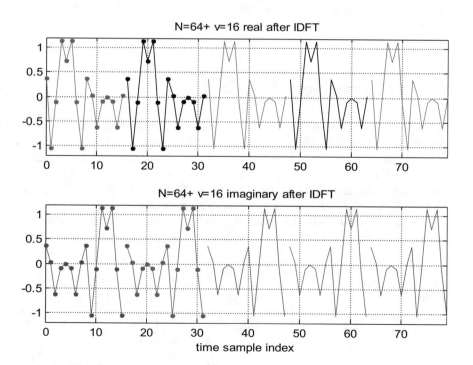

Figure 7-10: STS time samples after IDFT and periodic with the period of 16 samples

Example 7-3: $\{0 -1 -1 1 1 1 1 0 0 0 1 -1 1 -1 -1 1\}$ are expanded into 32 digit by inserting one '0' between digits. After IDFT with N = 32, the output will be periodic with 16 samples. Thus the output of N = 32 IDFT can be found by N = 16 IDFT of the original pattern.

Exercise 7-2: Show Example 7-3 using the definition of IDFT, and generalize it to two zero insertion, three zero insertion and so on.

7.1.1.2   Long-Term Training Sequence (LTS)

In LTS, 52 sub-carriers are filled with BPSK symbols, and sub-carrier locations with '0' out of 64 possible ones, DC and high frequencies, are not used. One OFDM symbol of LTS is shown in Figure 7-9 (middle portion) and given by,

$\{0 1 -1 -1 1 1 -1 1 -1 1 -1 -1 -1 -1 -1 1 1 -1 -1 1 1 -1 1 1 -1 1 1 1 1 1 1 0 0 0 0 0$
$0 0 0 0 0 0 1 1 - 1 -1 1 1 -1 1 -1 1 1 1 1 1 1 1 -1 -1 1 1 1 -1 1 1 - 1 1 1 1 1\}$

with sub-carrier index $k =$ to 63. Note that the energy of this symbol is 52, and thus there is no need of scaling. After IDFT, the same energy level is maintained with the scaling factor 8 as shown in Figure 7-9. And its 64 time samples is denoted as lts_mf (n), and its conjugate lts_mf $*(n)$, after time reversal, will be used as a matched filter to recover LTS boundary, i.e., OFDM symbol boundary.

7.1.2   RCV Processing of STS and LTS

By sending known, special OFDM symbol of STS and LTS, a receiver should detect the start of a packet, both 16-sample boundary and OFDM symbol boundary. Once this boundaries are found, i.e., synchronization is done, we can obtain carrier frequency & phase estimation, and sub-carrier channel phase and gain (channel estimation). These functions are listed in Figure 7-8 (bottom). Not listed in the figure, SNR in dB may also be estimated as part of STS processing.

This means that the synchronization should be searched under the condition of thermal noise and carrier frequency/phase offset as well as channel fading (maybe

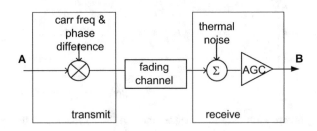

Figure 7-11: Impairments during synchronization recovery

both frequency selective and time selective fading) and sampling clock phase and frequency uncertainty. See Figure 7-11 for this channel model.

Yet the synchronization should be robust enough so that it should contribute negligibly to the packet error rate. Note that it is also true to SIGNAL symbol. When any of 5 preamble symbols, including SIGNAL symbol, cannot be recovered correctly, the rest of data symbols is not decodable and thus results in packet error.

In addition a part of STS (say the first symbol of a packet) may be used to set an optimum power level at the input of ADC (called AGC). Another choice is that AGC may be done, in RF side, without the use of STS. In order to focus on synchronization issue and to make the presentation simple, here we may set aside AGC problem. (See Chapter 8 for AGC detail.)

7.1.2.1   Matched Filter Algorithm

In order to detect a pattern of finite duration in time domain (e.g. STS or LTS), a matched filter, in the form of finite impulse response (FIR) digital filter, is used. A matched filter in this case is FIR with the filter coefficients of conjugated and time reversed version of the same pattern. For example of STS, the corresponding 16 sample time domain pattern, denoted as sts_mf(n) with $n = 1{:}16$, and its matched filter is sts_mf*(m) with $m = 16{:}{-}1{:}1$ and * means conjugation, and FIR implementation is shown in Figure 7-12.

We comment on the concept of matched filter. It is used in pulse shaping filter design, i.e., transmit filter and receive filter should be a pair of matched filters for optimum SNR. The matched filter concept is applicable to both digital and analog implementations, i.e., it may be implemented either in digital filter or in analog filter. (See Chapter 3 for details of matched filter.) In passing we also mention that the naming of matched filter is related with impedance matching of power amplifier; terminating impedance should be complex conjugation of power amplifier output impedance.

Example 7-4: We give a numerical example of matched filter output for the input of one STS symbol plus preceding zeroes (total 96 samples) in Figure 7-13. Look at them carefully. Top portions are real part of the input to the matched filter and bottoms are the output of it. Left hand side is with ideal channel and right hand side

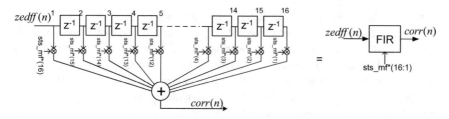

Figure 7-12: STS matched filter in FIR form, and its coefficients are sts_mf*(16:1)

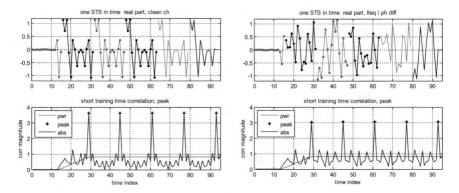

Figure 7-13: Matched filter outputs for one STS symbol under clean channel condition on LHS, and with frequency (1.25 subcarriers) and phase offset (−41 degree) on RHS. Top portion is the real part of input to the matched filter

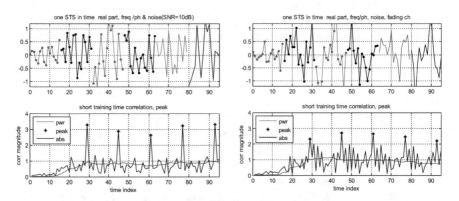

Figure 7-14: Thermal noise of 10 dB is added (LHS), and frequency selective fading channel is imposed (RHS) in addition to noise. There is noticeable degradation in correlation peaks

is with frequency and phase offset. Note that periodic nature of original STS is destroyed due to the frequency and phase offset, and peaks are reduced substantially even though they are clearly recognizable. The power level is computed with the averaging period of 16 samples and it reaches at 0.8125 (52/64) since symbol energy is set at 52 with 64 samples.

Example 7-5: Power level computation may be done by using the matched filter structure in Figure 7-12; samples stored, $sr(1:16)$ at a given time index n. Then power is given by *sum* $(sr(1:16).*$ conjugate$(sr(1:16)))/16$ where .* means term by term multiplication, and *sum* means add all 16 terms. When a longer averaging is needed, more storage is needed. With the same notation, matched filter computation is expressed as corr$(n) = $ *sum* $(sr(1:16).*$ conjugate$(sts\_mf(16:-1:1)))$ for a given input at n, and this is repeated with next input index.

AWGN is added (thus SNR 10 dB), in addition to the same carrier frequency and phase offset above, and a fading channel is imposed in Figure 7-14.

With a frequency fading channel the correlation peaks are reduced substantially and the degradation depends on the severity of fading.

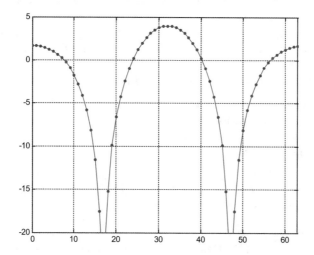

In this numerical example we use a digital filter with [1 −0.05 +0.8 −0.2] and the magnitude of its frequency response is shown on the left hand side.

As we can see in this example, the correlation peaks are reduced with noise, carrier frequency and phase offset and fading channels. However, in case of multiple peaks (5 in the example) additional improvement can be obtained by non-coherently combining multiple peaks, called post detection integration (PDI). The output of matched filter is partitioned into 16-sample segments and then the magnitude is added, i.e., non-coherently combined. Then find a peak after combining. Since there are two STS symbols in the preamble, up to 10 PDI is possible.

LTS will be processed similarly with a matched filter whose length is 64, denoted as lts_mf*(64:−1:1), again time reversal and conjugation of lts_mf(1:64).

Example 7-6: With the same notation as in the previous example, the matched filter output can be expressed as *sum* (*sr*(1:64).* conjugate (*lts_mf*(64:-1:1))) where .* means term by term multiplication, and *sum* means add all 64 terms.

A matched filter may be considered as a correlation implemented by filtering or convolution. When there is no degradation due to frequency and phase offset, fading and noise, it is an autocorrelation.

Example 7-7: Find an autocorrelation of real part of sts_mf (*n*); i.e., sum over *m* of {real (sts_mf (*m*)) real (sts_mf*(*m-n*))} and repeat for the imaginary part. Then compare them with a (complex) autocorrelation of sts_mf (*n*) and show that auto-correlation of sts_mf (*n*) = autocorrelation of real part of sts_mf (*n*) + autocorrelation of imaginary part of sts_mf (*n*).

7.1.2.2   STS Processing Detail

Here in this subsection we explain the processing of two STS symbols. Figure 7-15 summarizes the receiver signal processing of the preamble of 5 OFDM symbols; top is 2 STS symbols, middle is 2 LTS symbols and bottom SIGNAL symbol. A major task in preamble processing is to find OFDM symbol boundary in a robust manner with frequency offset, noise and fading channel. This is a synchronization problem. The processing of STS symbols have four distinct, but related, parts;

1) STS 16-sample boundary recovery
2) Frequency offset estimation
3) Carrier phase (common to all sub-carriers) may be estimated
4) SNR may be estimated.

The most critical part is 1) STS 16-sample boundary and the rest rely on the recovered boundary. OFDM symbol boundary is eventually found by LTS processing where this STS 16-sample boundary is used.

16-sample boundary recovery
Using Figure 7-15 (top block), the 16-sample boundary recovery of STS is explained. Here we assume that AGC is done so that the input level to ADC is optimum. If not, the first STS symbol may be dedicated to do AGC and the second STS symbol may be used for 16-sample boundary and thus PDI range is 5 rather than 10. Here we assume PDI range is 10.

A start of packet with power measurement: Power level is measured by averaging incoming samples stored in the matched filter, and is compared against a set threshold; the steady state power level is 52/64 (0.8125) and thus the threshold may be set say, 0.5 (this is a design parameter). Once the power level is above the threshold, the detection of a packet is declared. This works well when noise is not severely high. However, as noise is high (e.g., SNR 0 dB or worse), additional measure is necessary to make sure that the trigger is not due to noise. The idea is that power level must be above the threshold after initial trigger. Once confirmed, it is marked as a start of packet (denoted as pwr_index in the figure). This will be a reference point where all subsequent processing is done.

16 sample partition: The output of FIR, with matched filter coefficients of conjugate $(sts\_mf(16:-1:1))$, is stored (denoted in the figure as corr(1:Lt)) and parti-*tioned into 10 PDI sets of 16 samples beginning from the start of packet.*

PDI and peak locations: 10 PDI sets of 16 samples are, after magnitude squaring, added to make one set of 16 samples. This is called post detection integration or non-coherent combining, and then the location of the maximum after PDI is obtained ('peaks' in the figure). At the same time in each set of 16 samples peak locations are found, which may not be periodic due to noise and other impairments.

Combine PDI and peak locations: The peak location of PDI is periodically extended from the start of packet. Thus 10 STS boundary locations may be obtained.

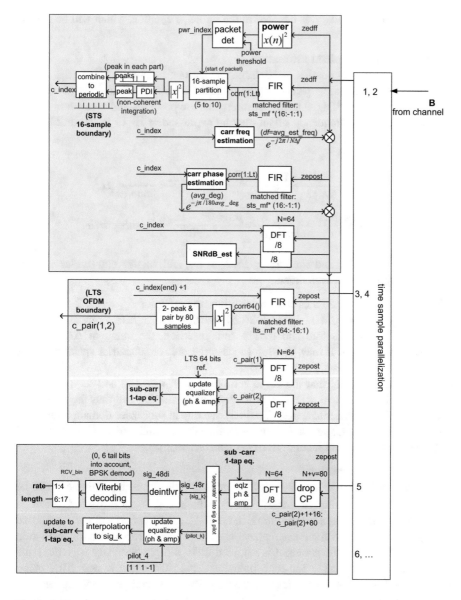

Figure 7-15: A block diagram of receiver processing of STS, LTS and SIGNAL. DFT/8 means DFT operation with scaling factor 1. In transmit side IDFT uses scaling factor 8, i.e., IDFT*8 since its standard definition has 1/N factor (here N = 64)

This should work well. However, additional improvement seems possible by utilizing the next peak of PDI common with peaks obtained before PDI if the peak of PDI has no common peak with those peaks before non-coherent combining. The result,

STS 16 sample boundary, is denoted in the figure as c_index, which will be used in carrier frequency offset estimation.

Carrier frequency offset estimation
Once 16-sample boundaries are determined, carrier frequency offset (Δf) can be estimated by comparing phase angles of matched filter output (corr(.)) at those boundaries. For 10 PDIs, 9 such measurements are possible, and take average over 9 estimations. The frequency offset may be expressed in Hz or in the unit of sub-carrier spacing;

$$\Delta f = \frac{ph2 - ph1}{2\pi(t2 - t1)} = \frac{angle2^o - angle1^o}{360^o} \frac{fs}{samples} [Hz]$$

$$\Delta f_k = \frac{ph2 - ph1}{2\pi} \frac{fs}{samples} \frac{N}{fs} = \frac{angle2^o - angle1^o}{90^o} [sub\_carrier] \qquad (7\text{-}1)$$

where $fs = 20$ *MHz* and *samples* $= 16$, $N = 64$, and, *angle2* and *angle1* are two consecutive phase measurements.

Exercise 7-3: With STS of 3 zero insertion, the range of $\Delta f_k = \pm 2$ [subcarrier spacing] or in terms of Hz, $\Delta f = \pm 2 \cdot 20$ *MHz*/64 $= 625$ *kHz*. Convince yourself that this statement is true and thus 4 zero insertion, the range of $\Delta f_k = \pm 3$ [subcarrier spacing]. It may be useful to use the unit of subcarrier spacing for the carrier frequency offset, rather than Hz.

Answer: remember that the angle is modulo $360°$ or $\pm180°$ and see (7-1) if the angle difference beyond $\pm180°$, it is ambiguous or aliasing and thus the maximum range is ±2 [subcarrier spacing]. You may see this in subcarrier domain after DFT. When the offset is ±2 subcarriers or more, then there is ambiguity. For example, +3 sub-carrier offset is not distinct from -1 subcarrier offset.

Carrier phase estimation
After the estimated frequency offset computed, the receiving samples will be compensated by multiplying $e^{-j2\pi \cdot \Delta f_k / N \cdot n}$ where Δf_k is an estimated offset in subcarrier spacing and n is the sample index [0,1,..,..]. This frequency compensated samples will be used for carrier phase estimation; as shown in Figure 7-15 (top block), the matched filter output is sampled at 16 sample boundary (c_index) and compute the phase angle for each boundary, and then take average. This phase estimation is denoted as avg_deg in the figure. The phase compensation can be done by multiplying $e^{-j2\pi \cdot avg\_deg / 360}$ to the entire samples. Note that this phase estimation is common to all subcarriers, and each subcarrier phase may not be the same and thus phase compensation for each subcarrier must be done precisely. Thus this phase estimation and compensation may be omitted. The carrier frequency offset is estimated before DFT as part of 16 sample boundary recovery since the frequency offset is common to all subcarriers. However, carrier phase should be estimated for each subcarrier with known reference symbols after DFT.

SNR estimation

After the 16-sample boundary is recovered, frequency offset is estimated and corrected as in Figure 7-15 (top portion), which are necessary steps before SNR estimation may be done.

SNR estimation relies on known constellation points in each subcarrier. Thus it should be done after DFT For STS its constellation points are known to receiver; $1 + j$, $-1-j$, or 0, with 0 being empty subcarrier and with the scaling factor $d = \sqrt{\frac{13}{6}}$ so that the energy of one OFDM symbol is 52.

The received points in constellation will deviate from the transmitted ones due to noise as well as due to fading and carrier frequency (residual less than one subcarrier frequency) and phase offset. See Figure 7-16. The error between transmit constellation point and receive one is measured $e_i = (A_i - a_i)$ where A_i is transmitted one and a_i being received for all subcarriers $i = 0$ to 63, and taking its squared magnitude $|A_i - a_i|^2$ for SNR estimation. Note that e_i is a complex number.

Then the estimated SNR, in ratio, is given by one OFDM symbol energy/sum of error squared magnitude in the symbol, i.e., $\text{SNR} = 52/\sum_{i=0}^{63}|A_i - a_i|^2$.

Note that to estimate SNR precisely, it is necessary to remove the impact to the error from fading, carrier frequency and phase offset; the error between the ideal and actual must be due only to the noise. This is important in practice if SNR should be measured accurately, in which case SNR estimation may be done later stage.

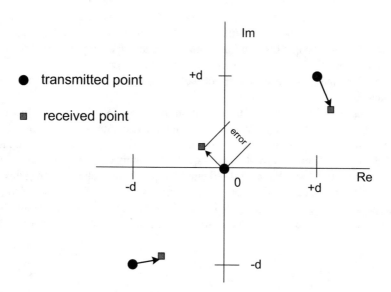

Figure 7-16: Constellation of STS and error measurements

7.1.2.3   LTS Processing Detail

Two additional OFDM symbols in PLCP preamble after STS, called LTS, will be processed as shown in Figure 7-15 (2nd block) to find OFDM symbol boundary with the matched filter, denoted as lts_mf(1:64) in the figure, and it is obtained by IDFT of LTS 64 bit sequence shown in Figure 7-9, then by conjugation and time reversal.

Additional processing of LTS is done after DFT in order to obtain one tap equalizer for each sub carrier (carrier phase and magnitude equalization). This is shown in the figure (a block of sub-carr 1-tap eq.) and it is part of equalization and channel estimation but not pursued further here.

OFDM symbol boundary from LTS processing
The computation of the matched filter in LTS processing can be simplified due to the fact that 16-sample boundary is known after STS processing; the problem is to resolve which 16-sample boundary corresponds to OFDM symbol boundary and thus the matched filter output is computed only at the 16-sample boundary. This means that the matched filter is computed 10 times for the duration of two LTS symbols rather than 80*2, reduction by the factor of 16. The uncertainty may occur if the detection of the start of packet is deviated more than 16-sample boundary (and thus the number of 16-sample boundary is less than the maximum 10 during two STS symbols). This is schematically shown in Figure 7-17.

The matched filter computation may be done at 16-sample boundary as well as at the mid-point between two 16-sample boundaries. This means that the matched filter is computed 20 times during two LTS symbols. This may be increased to 40 times, to 80 times and finally 160 times where there is no computational saving due to the knowledge of 16-sample boundaries. From the numerical examples (not shown here but experimented all possible cases), it seems that there is little gain by more computation, and thus we stick with the case of computation only at the 16-sample boundaries.

The matched filter outputs, 10 of them during two LTS symbols, are processed further in order to obtain OFDM symbol boundary. The output of the matched filter is in general a complex number and is converted to the squared magnitude (or magnitude). Then find the two peaks, OFDM boundaries, that should be separated by one OFDM symbol period (80 input samples to the matched filter); this idea can be implemented as follows. Find the maximum peak and then to see if there is corresponding peak among 9 outputs (except for the maximum out of 10 outputs). If the corresponding pair exists, that is the end of search. If not repeat for the next

Figure 7-17:   16-sample boundary (blue) and OFDM boundary (red)

maximum and then to see if there is corresponding peak among 8 outputs (exclude the maximum and the next maximum out of 10 outputs). Theoretically it can be repeated further. However, in practice, it is stopped at this point of two trials and the performance seems not degraded much. In a sense this process takes advantage of non-coherent combining (equivalent to post detection integration) and the knowledge that OFDM symbol boundaries are periodic.

Additional processing of LTS is to obtain the initial 1-tap equalizer coefficient after DFT. This issue is picked up in equalization but not pursued in synchronization context.

7.1.3   802.11$_a$ Synchronization Performance

The packet error rate (PER) performance is specified in IEEE std 802.11^{TM}-2012 as 10% for a packet length 4096 bytes (4096*8 bits). Roughly the corresponding bit error rate (BER) is 0.1/4096/8 = 3e-6. For this PER the minimum receiver input level sensitivity is specified depending on modulation and coding classes from BPSK, QPSK, 16-QAM to 64-QAM and coding rate from ½ , ¾, to 5/6. The minimum input level sensitivity specified in the standard is conservative and thus most can do nearly by 10 dB better. For example, BPSK ½ rate requires −82 dBm input level sensitivity but a typical commercial implementation can do around −90 dBm. The preamble of STS and LTS should be detectable even below this input level so that the impact of the synchronization error to PER is small.

7.1.3.1   Overall Performance Target

The importance of PLCP preamble detection including OFDM boundary recovery cannot be overemphasized; it must be most robust. A target for miss detection (plus false detection) should be much smaller than PER performance by order of magnitude, i.e., for 10% target PER, the probability of wrong detection is 1% or 1e-2 or less. In this way the contribution to PER by synchronization error is negligibly small. Note that once the synchronization is lost, i.e., OFDM boundary is lost; no amount of FEC, or diversity can help to improve the performance.

The control signal after LTS, called SIGNAL, should have a similar impact to PER since without correctly decoding it (i.e., without knowing modulation and coding class, and data length, which are contained in it), the rest of data is not decodable. SIGNAL uses BPSK ½ convolution code with K = 7, and its BER performance is well-known in the literature. In order to set the target performance of packet/OFDM symbol boundary detection, a numerical simulation program (Matlab) has been written for SIGNAL error performance and the result is shown in Figure 7-18. Note SNR scale, x-axis, where the rate loss correction, code rate ½ (3 dB) and the usage of 52 subcarriers out of 64, 52/64 (0.9 dB), is NOT included for convenience. Where a typical textbook includes 3 dB loss to emphasize the net gain

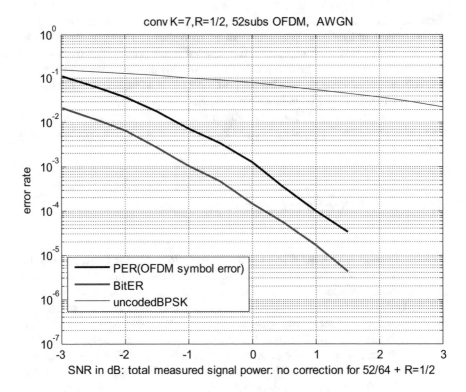

Figure 7-18: SIGNAL with convolution code R = ½, K = 7 using 52 subcarriers out of 64. Note that ½ (3 dB) and 52/64 (0.9 dB) rate power correction is NOT included for convenience

of coding, i.e., 3 dB more transmit power necessary for parity bits. And 0.9 dB is peculiar to this particular case since out of 64 subcarriers only 52 are used.

Around 1.5 dB SNR, the performance of SIGNAL BER is 3e-6 from the figure. With 20 MHz bandwidth the thermal noise is -103 dBm and with noise figure 5 dB it becomes -98 dBm. Thus for SIGNAL the minimum input level sensitivity is -96.5 dBm, which is much lower than -90 dBm of the minimum input level sensitivity for typical implementation. The choice of the modulation and coding of SIGNAL seems good. With this Wi-Fi example, we are not designing the preamble signal itself since it is already standardized. So we analyze it. SIGNAL error rate performance seems reasonably good. In passing we mention that in Figure 7-15, the receiver processing of SIGNAL, an OFDM symbol, is shown schematically. And the simulation result in Figure 7-18 is done for flat thermal noise channel.

From this discussion in this section, the target for packet and OFDM symbol boundary detection performance, through STS and LTS processing, should be 1% error (miss detection and false detection) with SNR 1.5 dB or less. In other words, for given 1% of the packet detection error, the required SNR should be less than 1.5 dB, much less for better margin.

7.1.3.2 STS and LTS Performance for Flat Channel

We established the target performance of detection error probability for STS and LTS. Now we examine the performance using simulations.

The detection algorithm of STS and LTS is essentially based on matched filter and post detection integration and combining, which was described in the sections 7.1.2.1, 7.1.2.2, and 7.1.2.3. It is possible to have many simulation cases with frequency offset, phase error and thermal noise as well as channel models (flat, frequency selective and time selective). We divide the cases with flat channel and frequency selective channel. We treat flat channel case in this subsection and frequency selective channel next. By assuming slowing fading flat channel performance may represent that of time selective channel.

We choose frequency offset (1.25 subcarrier frequency spacing) and phase error (−41 degree), which is the same condition as in Figure 7-13 where the output of matched filter is displayed. The thermal noise is varied from SNR = −5 dB to +3 dB. The simulation result is shown Figure 7-19. The result exceeds far better than 1.5 dB SNR for the target 1% detection error rate; −4 dB SNR for LTS, and less than −5 dB for STS to have 1% detection error.

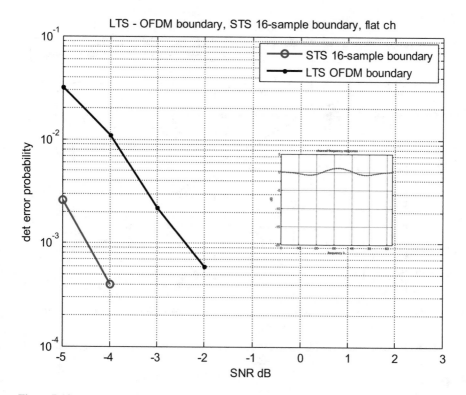

Figure 7-19: Simulation result for detection error probability of STS and LTS under thermal noise (SNR −5 dB to +3 dB) with frequency offset (1.25 subcarriers) and phase error (−41 degree) same as in Figure 7-13

Exercise 7-4* (project): Write a computer program (e.g., Matlab) to obtain simulation result in Figure 7-19 based on the algorithm described in Figure 7-15. This is a small project but worth trying out in order to have in-depth understanding.

7.1.3.3 STS and LTS Performance for Frequency Selective Channel

We repeat the simulation with the same condition of Figure 7-19, but with frequency selective channel, somewhat severe which is shown as insert in Figure 7-20.

The frequency selective channel degrades the performance a great deal; 4 to 5 dB. For a given 1% detection error, the required SNR is about 0.2 dB, from the figure, for both STS and LTS, which still meets our requirement of 1.5 dB with the margin of 1.3 dB.

We conclude the algorithm based on matched filter and PDI combining in IEEE 802.11a seems adequate with some margin. From this analysis we can learn different packet detection symbol design (e.g., more symbols and different STS and LTS patterns) for different requirements or for further refinement of performance.

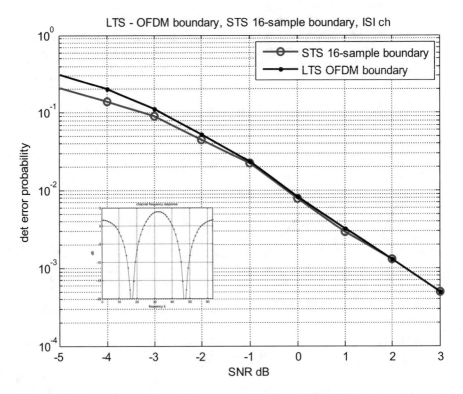

Figure 7-20: Simulation result for detection error probability of STS and LTS under the same as in Figure 7-13 plus frequency selective channel shown as insert

7.1.4   DS Spread Spectrum Synchronization Example

A brief summary of direct sequence (DS) spread spectrum system is introduced in Figure 7-21; for data $d(m)$, the transmit signal, $c(n)$, is generated by 'multiplying' spreading code $s(n)$, and modulo 2 addition (exclusive OR) is multiplication for binary $\{0,1\}$ signal. In the receive the same spreading code is 'multiplied', with the correct phase, in order to recover the data $d(m)$.

DS spread spectrum signaling scheme is often used in cellular, known as coded division multiple access (CDMA) where multiple users share the same frequency and time, and each user is assigned with different spreading code phase.

With the example in Figure 7-21, the physical layer synchronization issue is the same as shown in Figure 7-5, where the synchronization problem is summarized as recovering parallel clock with correct phase from serial transmission. This can be seen below figure. For simplicity we use the period of $s(n)$ is 15 since it is a maximal length sequence with 4 bit memory (shift register).

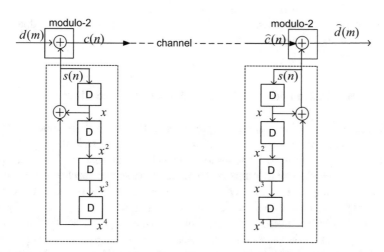

Figure 7-21: A baseband digital representation of DS spread spectrum sequence signaling; for data $d(m)$, the transmit signal, $c(n)$, is generated by 'multiplying' spreading code $s(n)$, and modulo 2 addition (exclusive OR) is multiplication for binary $\{0,1\}$ signal. In the receive the same spreading code is 'multiplied' in order to recover the data $d(m)$

Exercise 7-5: Write a program (e.g., Matlab) to generate c(n) with generating polynomial $x^4 + x + 1$ with $d(m) = 0$ and with all one state. Answer: {0 1 0 1 1 0 0 1 0 0 0 1 1 1 1}

In GPS (global positioning system) DS spread signal, called GPS C/A signal, is broadcast from GPS satellites toward GPS receivers on the earth. One of the issues is to find the boundary of spreading code; in particular under the condition of weak signals.

In this subsection we use this GPS C/A as an additional example of physical layer packet synchronization, which is interesting and appropriate since the synchronization is one of central issues in spread spectrum signaling. It should be recovered first before carrier frequency offset and phase correction under low SNR. In GPS environment, the frequency offset is due to the movement of devices on earth, called Doppler frequency.

Now C/A signal is summarized. The spreading code is 10 stage (memory) maximum length sequence, also called pseudo random number (PRN), and 1023 bit long. It is transmitted with 1.023 MHz, and one period is 1 msec. Its carrier frequency, called L1 is $1.023*1540 = 1575.42$ MHz, i.e., the transmit carrier frequency is an integer multiple (1540) of the chip rate 1.023 MHz. It uses a form of BPSK; it is on Q-channel, while I-channel is not empty but used by another, called P(Y) code. In typical BPSK signaling, only one quadrature channel (I-channel) is used and the other is empty. This signal carries data of *Navigation Message*; 1 bit data takes 20 msec i.e., 20 periods of PRN.

Exercise 7-6: What is the data rate of the navigation message? Answer: 50 bps (bit per sec). A frame of the message contains a total of 1500 bits. How long does it take to send a frame? Answer: 30 seconds.

7.1.4.1 Acquisition and Tracking with Analog Matched Filter

Traditionally the synchronization of spread spectrum (SS) code is divided into two parts; initial code phase *acquisition* and subsequent *tracking*. In terms of parallel and serial clocks, *acquisition* corresponds to parallel clock and *tracking* does to sampling phase of serial clock. The literature on the issue of DS spread spectrum signal is vast with different implementation schemes. Here our approach is based on modern implementations, i.e., digital implementations after sampling. Consistently with the previous section and for the clarity of understanding, we stick with matched filter as algorithm basis, rather than serial search type which may be simpler than the former. Consequently the partition of acquisition and tracking is not clear-cut since both happen nearly at the same time even though acquisition is still slightly ahead.

We first show the algorithmic concept of PRN code acquisition assuming matched filter (time-reversed, conjugated PRN code) can be implemented in analog circuits. This is shown in Figure 7-22.

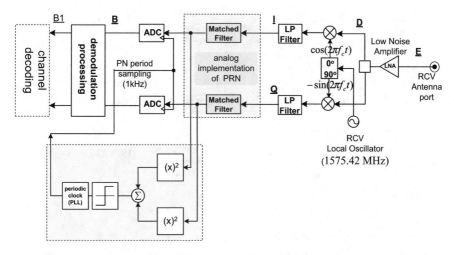

Figure 7-22: Matched filter PRN code phase acquisition; matched filter (time-reversed PRN code of one period), and squaring/summing, must be implemented in analog signal which might be problematic. This diagram shows clearly the concept of algorithm

Compare Figure 7-22 with the modulation symbol clock recovery in Figure 7-2. If LP filter + PRN matched filter is seen as a single matched LP filter in Figure 7-22, then the two figures use essentially the same algorithm (except for pre-filter HPF).

7.1.4.2   Digital Matched Filter Implementation

Digital implementation of PRN matched filter is straightforward, using FIR structure, and addition signal processing of squaring can be done precisely as well. With digital approach there is a variety of possible implementations to choose from, which may be adapted to particular situations even though we consider and show all digital implementations.

Exercise 7-7* (project): Write a simulation program (Matlab) for GPS C/A signal assuming 10 stage maximum length polynomial is $x^{10} + x^3 + 1$ with the initial phase is all ones (11...11) and with carrier frequency offset close to zero. Repeat with carrier frequency offset 1 kHz.

25 PRN periods carry 1 bit and thus the output of matched filter will be added for 25 periods coherently to recover data. The PRN code boundary will be used for coherent addition.

7.1.4.3   Weak Signal Initial Code Phase Synchronization

It is of great practical interest to recover initial PRN code boundary under very weak signal condition, which may happen when GPS devices are inside building, or high

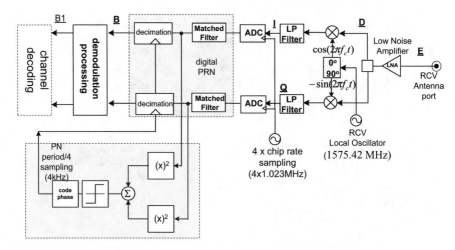

Figure 7-23: Digital matched filter implementation of PRN code acquisition; $4\times$ chip rate sampling results in $4\times$ samples per PRN period, which should be decimated to one sample per PRN period. $4\times$ chip rate sampling phase adjustment may be made from the code phase (4 kHz). This will be explained in section 7.2 symbol clock synchronization

rise building canyon inside city. In the literature of GPS, carrier to noise density C/No-Hz is often used and typically is 45 dB C/No-Hz, which corresponds to, with 1.023 MHz signaling rate, C/N is 15 dB.

The target weak signal to recover is 30 dB lower than typical 45 dB C/No-Hz, 15 dB C/No-Hz, which is equivalent to C/N is -15 dB, well below 0 dB. We do not pursue it here further but it is an interesting technical challenge.

7.2   Symbol Timing Synchronization

This section and next will cover modulation symbol timing synchronization and carrier phase recovery respectively. All digital schemes will be emphasized, and Figure 7-4: fully digital implementation of demodulation processing will be re-drawn with some additional detail in Figure 7-24. This is our base line of synchronization scheme that the timing recovery and carrier phase synchronization are deliberately 'de-coupled'; the timing is processed in sampling rate, higher than symbol rate, ahead of carrier frequency and phase recovery while carrier phase is processed in symbol rate. The choice of this scheme is born out of design practice; it is fast, even under substantial carrier frequency offset, and universally applicable to most modulations, easy to understand, and good performance close to theoretical optimum (maximum likelihood estimation). This scheme may also be a departure point for understanding various other schemes.

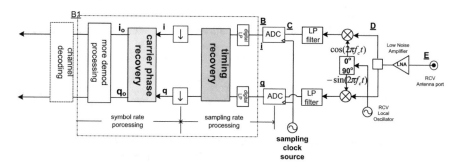

Figure 7-24: A fully digital implementation of symbol timing and carrier phase recovery; the timing recovery and phase synchronization are deliberately 'de-coupled' so that the timing is processed in sampling rate, higher than symbol rate, and carrier phase is processed in symbol rate

7.2.1 Symbol Timing Error Detector for PAM/QAM

A received signal after a matched filter will be represented as,

$$z(t - \tau) = \sum_{k=-\infty}^{+\infty} a_k h(t - \tau - kT) + n(t) \tag{7-2}$$

where $h(t)$ is an end-to-end impulse response with $h(t) = h_T(t) \otimes h_R(t)$ where $\otimes$ denotes convolution, and $h_T(t)$ is an impulse response of transmit side and $h_R(t)$ one of receive side. T is a symbol period, and τ is the delay through transmission and needs to be estimated at the receiver. In the following, however, τ is not explicitly used, rather a variable, t-τ, is treated like t for notational convenience.

$\{a_k\}$ is a sequence of digital symbols to be transmitted. For example, $\{a_k\} = \{+1, -1\}$ for BPSK and $\{a_k\} = \{+1, -1, +j, -j\}$ for QPSK ($45°$ rotated version). However, for the purpose of this section argument of timing error detector derivation, discrete level assumptions are not necessary and indeed it can be allowed to be even analog, e.g., $\{a_k\} = \{$normal distribution with unity variance$\}$ or $\{a_k\} = \{$uniform distribution $-\sqrt{3}$ to $+\sqrt{3}$ with unity variance$\}$. This is one of major points of this timing error detector development. $\{a_k\}$ is assumed to be uncorrelated and without loss of generality we may set the average power being unity, i.e., $E\{|a|^2\} = 1.0$ where $E\{\}$ denotes an average operation. To make a presentation simple, $\{a_k\}$ is real, i.e., PAM. However, it can be extended to the case of $\{a_k\}$ being complex, i.e., QAM. One method is by considering two, I and Q, real symbols, and adding them together, as shown in Figure 7-2, after squaring.

Correlation of $z(t)$ with $z(t \mp \Delta T)$

Consider a multiplication of a received signal $z(t)$ with its shifted version $z(t \mp \Delta T)$ and taking an expectation, which may be called auto-correlation, as

$$E\{z(t)z(t \mp \Delta T)\} = \sum_{k=-\infty}^{+\infty} \sum_{l=-\infty}^{+\infty} E\{a_k a_l\} h(t - kT) h(t \mp \Delta T - lT) + E\{n^2(t)\}$$

where the terms of noise and data correlation are eliminated due to the lack of it. Furthermore using the assumption that data are uncorrelated, it is reduced to (7-3) below.

$$E\{z(t)z(t \mp \Delta T)\} = E\{|a_k|^2\} \sum_{k=-\infty}^{\infty} h(t - kT)h(t \mp \Delta T - kT) + \sigma^2 \qquad (7\text{-}3)$$

where $E\{|a_k|^2\}$ signal power and σ^2 noise power, and note that being $\Delta = 0$ then it becomes a squaring. The symbol timing synchronizer in Figure 7-2 with analog signal processing is based on squaring. The shift is denoted as $\mp \Delta T$ with $0 \leq \Delta < 1$ and $\Delta = 0, \frac{1}{4}, \frac{1}{2}$ is of particular interest for digital implementations, which will be clear later why we choose this peculiar delay for auto-correlation.

Symbol timing synchronization scheme, i.e. timing error detector, is essentially based on the periodic nature of autocorrelation of $z(t)$ in (7-2) above, which is called cyclostationary in the literature. The nature of $z(t)$ in (7-2), PAM/QAM, is such that its correlation of $z(t)$ with $z(t \mp \Delta T)$ is cyclostationary under the assumption of $\{a_k\}$ being uncorrelated. Note that the discreteness of $\{a_k\}$ is not assumed as we mentioned earlier.

Exercise 7-8: Prove that (7-3) is periodic with T. Hint: $t \rightarrow t+T$ in (7-3) and then absorb into summation index k. Answer: replace t with $t + T$, and (7-3) does not change.

7.2.1.1   Closed Form Expression of Timing Detector for Band-Limited Signals

$S(\mp\Delta T)$ or S_Δ is a timing error detection curve without noise and it is called S-curve. It is periodic with a period of T.

$$S(\mp\Delta T) = \sum_{k=-\infty}^{\infty} h(t - kT)h(t \mp \Delta T - kT) \qquad (7\text{-}4)$$

A variety of timing error detection algorithms can be found by examining $S(\mp\Delta T)$, and thus we focus on finding explicit closed form expressions for (7-4). A detailed examination is based on the identity, so called Poisson sum formula

$$\sum_{n=-\infty}^{n=\infty} h(t - nT)h(t - \Delta T - nT) = \sum_{m=-\infty}^{m=\infty} \left[\frac{1}{T} \int_{v=-\infty}^{v=\infty} H\left(\frac{m}{T} - v\right) H(v) e^{-j2\pi\Delta Tv} dv \right] e^{\frac{j2\pi}{T}mt} \qquad (7\text{-}5)$$

where $H(f)$ is a Fourier transform of $h(t)$, and for convenience this identity is derived in Appendix A.2 at the end of this chapter.

$$S(\mp\Delta T) = \sum_{m=-\infty}^{\infty} \left[\frac{1}{T} \int_{v=-\infty}^{v=\infty} H\left(\frac{m}{T} - v\right) H(v) e^{\mp j2\pi\Delta Tv} dv \right] e^{j\frac{2\pi}{T}mt} \qquad (7\text{-}6)$$

It is a Fourier series expansion which is possible because it is periodic with T, and the integral expression inside the parenthesis [ ] is the coefficients of Fourier series. The coefficients are a function of Δ for a given m (harmonic number). This Fourier series expansion is very useful when $H(f)$, frequency response of a shaping pulse, is band-limited.

With the assumptions that $H(f)$ is band limited and $h(t)$ is real and even symmetry, which means that $H(-f) = H(f)$ and only $m = 0, \pm 1$ terms need to be considered. This assumption is well justified in practice since in typical band-width efficient transmission system, the bandwidth of $H(f)$ is less than two times of symbol rate.

Exercise 7-9: $h(t)$ is real and even symmetry, and then $H(-f) = H(f)$. Show this from the definition of Fourier transform.

Assuming the band limited $H(f)$ and $h(t)$ real, we obtain S-curve in a closed form by evaluating (7-6) for $m = 0, 1, -1$ case as,

$$S(\mp\Delta T) = \frac{2}{T} \int_0^{\frac{1}{T}} |H(v)|^2 \cos(2\pi\Delta Tv) dv$$

$$+ \left[\frac{2}{T} \int_0^{\frac{1}{T}} H\left(\frac{1}{T} - v\right) H(v) \cos(2\pi\Delta Tv) dv \right] \cos\left(\frac{2\pi t}{T}\right)$$

$$+ \left[\frac{2}{T} \int_0^{\frac{1}{T}} H\left(\frac{1}{T} - v\right) H(v) \sin(2\pi\Delta Tv) dv \right] \sin\left(\frac{2\pi t}{T}\right).$$

$$S(\mp\Delta T) = C_0(\Delta) + A_1(\Delta) \cos\left(\frac{2\pi t}{T}\right) + B_1(\Delta) \sin\left(\frac{2\pi t}{T}\right) \qquad (7\text{-}7)$$

Exercise 7-10: Fill the details of getting (7-7) from (7-6) with only three of $m = -1, 0, 1$.

Another form of expression of (7-7) is also useful. By changing integration variable $y = \left(vT - \frac{1}{2}\right)$, and assuming that if $h(t)$ is even symmetrical in time domain, $H\left(\frac{1}{T} - v\right) H(v)$ is even symmetrical around $\frac{1}{2T}$ (or even in the new variable y around zero), it can be reduced to

$$S(\mp\Delta T) = C_0(\Delta) + \left[\frac{2}{T^2} \int_{-0.5}^{0.5} H\left(\frac{y+0.5}{T}\right) H\left(\frac{-y+0.5}{T}\right) \cos(2\pi\Delta y) dy \right] \cos\left(\frac{2\pi t}{T} \mp \pi\Delta\right)$$

$$S(\mp\Delta T) = C_0(\Delta) + C_1(\Delta) \cos\left(\frac{2\pi t}{T} \mp \pi\Delta\right) \qquad (7\text{-}8)$$

Note that the integrand of C_1 is even function of y and Δ.

Exercise 7-11: Obtain (7-8) from (7-7) by changing integration variable $y = vT - 0.5$.

Evaluation of (7-8) with specific Δ

It is useful to express (7-8) with specific values of Δ.

$\Delta = 0, \pm 1, \pm 2, \ldots \ldots$	$S(\mp\Delta T) = C_0(\Delta) + C_1(\Delta)(-1)^{\lvert\Delta\rvert}\cos\left(\frac{2\pi t}{T}\right)$
$\Delta = \pm\frac{1}{2}, \pm\frac{3}{2}, \pm\frac{5}{2}, \ldots \ldots$	$S(\mp\Delta T) = C_0(\Delta) + C_1(\Delta)\,\mathrm{sgn}\,(\Delta)(-1)^{\lvert\Delta\rvert-\frac{1}{2}}\sin\left(\frac{2\pi t}{T}\right)$
$\lvert\Delta\rvert\le\frac{1}{2}$	$S(\mp\Delta T) = C_0(\Delta) + C_1(\Delta)\cos\left(\frac{2\pi t}{T}\mp\varepsilon\pi\right) \quad 0\le\varepsilon\le\frac{1}{2}$
$\frac{1}{2}\le\lvert\Delta\rvert\le 1$	$S(\mp\Delta T) = C_0(\Delta) + C_1(\Delta)\,\mathrm{sgn}\,(\Delta)\sin\left(\frac{2\pi t}{T}\mp\varepsilon\pi\right) \quad 0\le\varepsilon\le\frac{1}{2}$

Exercise 7-12: Confirm the expression in the above table of (7-8) with specific values of Δ.

Now comparing (7-7) with the specific evaluation of (7-8) above, we can relate A_1 and B_1 with $C_1(\Delta)$. For example, with $\Delta = 0$, then $A_1 = +C_1(\Delta)$ and $B_1 = 0$, and with $\Delta = \pm\frac{1}{2}$, then $A_1 = 0$ and $B_1 = \pm C_1(\Delta)$, and so on.

From (7-7) when t is replaced by $t\mp\frac{T}{2}$ then A_1 and B_1 become negative. (Convince yourself that this is the case.) Again from (7-7) when t is replaced by $t\mp\frac{T}{4}$ then A_1 and B_1 are exchanged i.e., sin( ) term becomes cos( ) and vice versa. (Convince yourself this is the case as an exercise.) In other words, one with $t \to t \mp\frac{T}{2}$ and the other $t \to t \mp\frac{T}{4}$, which means a half symbol and a quarter symbol delay to a received signal. The half symbol delay negates and the quarter symbol delay exchanges cos() with sin() and vice versa. We arrange this relationship in Table 7-1.

Table 7-1: S-curve with different Δ and half and quarter delay

$$\sum_{n=-\infty}^{n=\infty} h(t - nT)h(t + \Delta T - nT) = C_0 + A_1\cos\left(\frac{2\pi t}{T}\right) + B_1\sin\left(\frac{2\pi t}{T}\right)$$

$\Delta = +$ means delay, $-$ means early

	$t \to t$			$t \to t\mp\frac{T}{2}$ $+, -$ not matter, negation			$t \to t\mp\frac{T}{4}$ $A_1 \leftrightarrow B_1$		
Δ	C_0	A_1	B_1	C_0	A_1	B_1	C_0	A_1	B_1
0	$+$	$+$	0	$+$	$-$	0	$+$	0	$\pm$
± 1	$-$	$-$	0	$-$	$+$	0	$-$	0	$\mp$
± 2	$+$	$+$	0	$+$	$-$	0	$+$	0	$\pm$
± 3	$-$	$-$	0	$-$	$+$	0	$-$	0	$\mp$
$+\frac{1}{2}$	0	0	$-$	0	0	$+$	0	$\pm$	0
$-\frac{1}{2}$			$+$			$-$		$\mp$	
$+\frac{3}{2}$	0	0	$+$	0	0	$-$	0	$\mp$	0
$-\frac{3}{2}$			$-$			$+$		$\pm$	

Table 7-1 will be used to form different timing error detection algorithms in the next section.

Concrete examples of $C_1(\Delta)$ values

In order to compute the coefficient $C_1(\Delta)$ we need to specify $H(f)$, a pulse shape. We computed $C_1(\Delta)$ for two specific cases of $h(t)$: one is a raised cosine pulse with an excess bandwidth β and its pre-filtered raised cosine pulse. For $C_1(\Delta)$ computation, we use a definite integral from a mathematical table,

$$\int_0^{\frac{\pi}{2}} \cos{}^{m-1} x \cos axdx = \frac{\pi}{2^m mB\left(\frac{m+a+1}{2}, \frac{m-a+1}{2}\right)} \quad \text{where } B(u, v) = \frac{\Gamma(u)\Gamma(v)}{\Gamma(u+v)}$$

.

Exercise 7-13:   Confirm the result of Table 7-2. For a given $H(f)$, use(7-8) and reduce it so that the above integration in a table is used.

The expressions in Table 7-2 are numerically evaluated and displayed in Figure 7-25 with total 10 cases of $\Delta = \{0, 0.5, 1, 1.5, 2, 2.5, 3, 3.5, 4, 4.5\}$. The top figure is for raised cosine and the bottom is for pre-filtered pulse. As can be seen in the figure, it is noted that there is essentially no sign change except for when the value of $C_1(\Delta)$ is near zero with $\beta > 0.6$ for $\Delta > 3.5$. This property will be used to increase the timing error detector gain; especially it is important when β is small say less than 0.4. A variety of digital symbol timing error detector can be found by applying the result of this section, which will be discussed in the next section.

Importance of pre-filtering

In Figure 7-25 the timing detector gain $C_1(\Delta)$ is displayed for raised cosine and its pre-filtered pulse. A rough approximation of pre-filtered pulse can be done a high-pass filter or differentiator. However, we define it precisely in Appendix A.1. It is clear from the figure the detector gain is bigger with pre-filtering. This will be confirmed further with different timing detectors summarized in Table 7-3 later. The benefit of pre-filtering was heavily emphasized for high order modulations with small excess bandwidth for highly bandwidth efficient systems in the literature. See [5].

Table 7-2: $C_1(\Delta)$ for raise cosine and pre-filtered raised cosine	raised cosine	$2C_1^{RC}(\Delta) = \frac{\beta}{2\Gamma(2+\Delta\beta)\Gamma(2-\Delta\beta)}$
	pre-filtered RC	$2C_1^{pRC}(\Delta) = \frac{6\beta}{\Gamma(3+\Delta\beta)\Gamma(3-\Delta\beta)}$

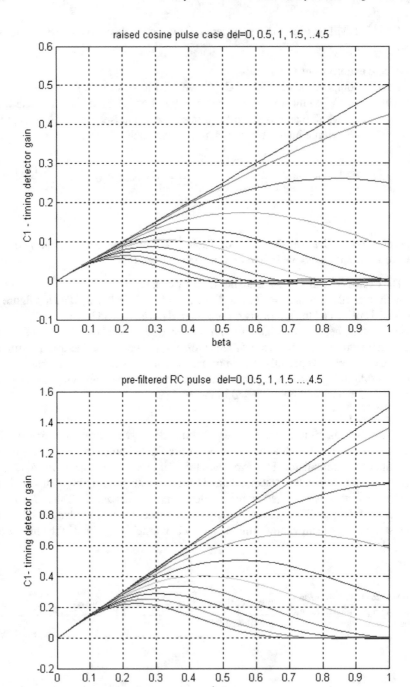

Figure 7-25: Numerical calculation of $2C_1$ as a function of ($\Delta = $ del, $\beta = $ excess bandwidth)

7.2.2 Known Digital Timing Error Detectors

A list of known timing detectors is summarized in Table 7-3. The derivation of S-curve is done by using Poisson sum formula and concrete examples are raised cosine pulse and pre-filtered raised cosine pulse.

Example 7-8: Find S-curve expression for the timing detector algorithm, the second row in the table $u(\tau) = z^2(\tau) - z^2(\tau - T/2)$. In Table 7-1, look at $\Delta = 0$ row and combine $t \to t$ column and $t \to t \mp \frac{T}{2}$. The result is $2A_1$, i.e., $2C_1(0)$. Thus S-curve is given by $2C_1(0) \cos\left(\frac{2\pi\tau}{T}\right)$. The computation of $2C_1(0)$ can be done using Table 7-2 for raised cosine and pre-filtered raised cosine.

Example 7-9: Find S-curve expression for the timing detector algorithm, the third row in the table $u(\tau) = z^2\left(\tau - \frac{T}{4}\right) - z^2\left(\tau - \frac{3T}{4}\right)$. Compared with the previous detector there is a quarter of symbol period delay. In Table 7-1, look at $\Delta = 0$ row and $t \to t \mp \frac{T}{4}$. The result is $2B_1$, i.e., $2C_1(0)$. Thus S-curve is given by $2C_1(0) \sin\left(\frac{2\pi\tau}{T}\right)$.

Exercise 7-14: Derive S-curves in Table 7-3. Hint: Use (7-8) and Table 7-1 for additional combination, and Table 7-2 for $2C_1(\Delta)$ computation, following the example above.

Examine Table 7-3 and we make following comments. It is convenient that there is no DC (the first row in the table) and that the error should be zero at the correct sampling point, i.e., sin() rather than cos(). For the first two rows S-curve is a cosine function. The third row, wave difference method or early late method, seems to be the most convenient since it is a sine function but the quarter symbol delay is necessary. In digital sampled implementation this means sampling 4 times of symbol rate to implement the quarter delay easily.

7.2.2.1 Extensions of Known Timing Detectors

The existing timing detectors listed in Table 7-3 can be extended for further enhancement for higher detector gain and/or hang-up free locking behavior.

Two timing detectors listed in Table 7-3 (2nd and 3rd), $u(\tau) = z^2(\tau) - z^2(\tau - T/2)$ and $u(\tau) = z^2\left(\tau - \frac{T}{4}\right) - z^2\left(\tau - \frac{3T}{4}\right)$ are combined into one. This is implementable with 4 times symbol rate sampling. This timing error detector (shaded) is embedded in digital synchronizer scheme as shown in Figure 7-26. Additional components in the figure, re-sampling, decimation from sampling rate to symbol rate, and digital loop filter, are shown to see where the timing error detector belongs. These will be treated later, and for now, our focus is on timing error detector itself.

Table 7-3: A summary of known timing detector algorithm and their S-curves

Algorithm Description $z(t)$ is a received signal $z(t)$ as $z(t) = \sum_{k=-\infty}^{\infty} a_k h(t-kT) + n(t)$	S-Curve without noise $h(t)$-band limited in frequency domain, and even symmetry in time domain			Comments
	General	Raised Cosine	Pre-filtered Raised cosine [5][13]	
$u(\tau) = z^2(\tau)$	$\sum_{k=-\infty}^{\infty} h^2(\tau - kT)$ $= 2D_0 + 2D_1 \cos\left(\frac{2\pi\tau}{T}\right)$ $D_m = \frac{1}{T}\int_0^{1/T} H\left(\frac{m}{T} - v\right) H(v)\,dv$ $m = 0\ or\ 1$	$1 - \frac{\beta}{4} + \frac{\beta}{4}\cos\left(\frac{2\pi\tau}{T}\right)$	$\frac{3\beta}{4} + \frac{3\beta}{4}\cos\left(\frac{2\pi\tau}{T}\right)$	Squaring [4] [12]
$u(\tau) = z^2(\tau)$ $- z^2(\tau - T/2)$	$\sum_{k=-\infty}^{\infty} h^2(\tau - kT)$ $- \sum_{k=-\infty}^{\infty} h^2\left(\tau - \frac{T}{2} - kT\right)$ $= 2C_1(0)\cos\left(\frac{2\pi\tau}{T}\right)$	$\frac{\beta}{2}\cos\left(\frac{2\pi\tau}{T}\right)$	$\frac{3\beta}{2}\cos\left(\frac{2\pi\tau}{T}\right)$	Wave Difference [3] or [2]
$u(\tau) = z^2\left(\tau - \frac{T}{4}\right)$ $- z^2\left(\tau - \frac{3T}{4}\right)$	$\sum_{k=-\infty}^{\infty} h^2\left(\tau - \frac{T}{4} - kT\right)$ $- \sum_{k=-\infty}^{\infty} h^2\left(\tau - \frac{3T}{4} - kT\right)$ $= 2C_1(0)\sin\left(\frac{2\pi\tau}{T}\right)$	$\frac{\beta}{2}\sin\left(\frac{2\pi\tau}{T}\right)$	$\frac{3\beta}{2}\sin\left(\frac{2\pi\tau}{T}\right)$	Wave Difference [3] or Early-late [8] [10]

$u(\tau) = z(\tau - T/2)$ $\{z(\tau) - z(\tau - T)\}$	$\sum\limits_{k=-\infty}^{k=\infty} h\left(\tau - \frac{T}{2} - kT\right) h(\tau - kT)$ $- \sum\limits_{k=-\infty}^{k=\infty} h\left(\tau - \frac{T}{2} - kT\right) h(t - T - kT)$ $= 2C_1\left(\frac{1}{2}\right)\sin\left(\frac{2\pi\tau}{T}\right)$	$\dfrac{\beta}{2\Gamma\left(2+\frac{\beta}{2}\right)\Gamma\left(2-\frac{\beta}{2}\right)}$ $\sin\left(\frac{2\pi\tau}{T}\right)$ $\approx 0.43\beta\sin\left(\frac{2\pi\tau}{T}\right)$	$\dfrac{6\beta}{\Gamma\left(3+\frac{\beta}{2}\right)\Gamma\left(3-\frac{\beta}{2}\right)}$ $\sin\left(\frac{2\pi\tau}{T}\right)$ $\approx 1.35\beta\sin\left(\frac{2\pi\tau}{T}\right)$	Gardner [1] and its use in [6] [7]
$u(\tau) = z(\tau)$ $[z(\tau+\varepsilon T) - z(\tau - \varepsilon T)]$	$\sum\limits_{k=-\infty}^{k=\infty} h(\tau - kT)[h(\tau+\varepsilon T - kT)$ $-h(\tau - \varepsilon T - kT)]$ $= -2C_1(\varepsilon)\sin(\pi\varepsilon)\sin\left(\frac{2\pi\tau}{T}\right)$	$-\dfrac{\beta\sin(\pi\varepsilon)}{2\Gamma(2+\varepsilon\beta)\Gamma(2-\varepsilon\beta)}$ $\sin\left(\frac{2\pi\tau}{T}\right)$ $\approx -0.34\beta$ $\sin\left(\frac{2\pi\tau}{T}\right)$	$-\dfrac{6\beta\sin(\pi\varepsilon)}{\Gamma(3+\varepsilon\beta)\Gamma(3-\varepsilon\beta)}$ $\sin\left(\frac{2\pi\tau}{T}\right)$ $\approx -1.0\beta$ $\sin\left(\frac{2\pi\tau}{T}\right)$	$\varepsilon = \frac{1}{4}$
$u(\tau) = z(\tau)$ $\left\{\frac{d}{d\tau}z(\tau)\right\}$	$S_{Diff} = -2C_{Diff}\sin\frac{2\pi\tau}{T}$ $C_{Diff} = \frac{2\pi}{T}$ $\int_0^+ H\left(\frac{1}{T} - v\right)H(v)dv$			[8] and ML
$u(\tau) = z(\tau)$ $\{\hat{z}(\tau)\}$	$S_{Hilbert} = +2D_1\sin\left(\frac{2\pi\tau}{T}\right)$ $D_m = \frac{1}{T}\int_0^{1/T} H\left(\frac{m}{T} - v\right)H(v)dv$ $\hat{z}(\tau) = H\{z(\tau)\}$ Hilbert	$-\dfrac{\beta}{4}\sin\left(\frac{2\pi\tau}{T}\right)$	$-\dfrac{3\beta}{4}\sin\left(\frac{2\pi\tau}{T}\right)$	[4] and [13] BETR

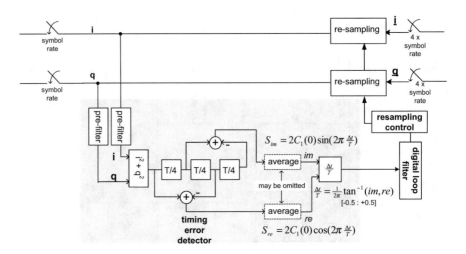

Figure 7-26: All digital symbol timing recovery scheme with timing detector (wave difference method). Re-sampling, decimation from sampling rate to symbol rate, and digital loop filters are shown as well

7.2.2.2 Timing Detectors Related with One in Figure 7-26

It is interesting and insightful that the timing detector of Figure 7-26 can be obtained from seemingly different approaches in the literature; [2], [7], [10], [11].

The first method of derivation is based on Reference [2], and a timing detector algorithm is described as,

$$\varepsilon_m = -\frac{1}{2\pi}\arg\left(\sum_{n=mLN}^{(m+1)LN-1} |z_n|^2 e^{-j\frac{2\pi \cdot n}{N}}\right) \qquad (7\text{-}9)$$

where $\varepsilon_m \in \{-0.5, 0.5\}$ is a timing error, and z_n is obtained by sampling a received signal (7-2). N samples per symbol and L symbols of averaging in the mth time segment. This algorithm is derived in [2] and more than 4 samples per symbol may be used. As argued in [2], the simplest implementation is possible by using 4 samples per symbol ($N = 4$), which results in the same structure as in Figure 7-26. The sign of detector gain is opposite. This particular implementation is shown in [11].

Exercise 7-15: Observe that the main computation of (7-9) is a discrete Fourier transform of $|z_n|^2$ evaluated at $f = 1/T$ symbol rate. Then in order to get timing error the phase in radian (arg) is computed.

Answer: $e^{-j\frac{2\pi \cdot n}{N}}$ is obtained from $e^{-j2\pi f \cdot \Delta t \cdot n}$ with sample period $\Delta t = \frac{T}{N}$ and frequency $f = 1/T$. n is the time index of $|z_n|^2$.

The second related timing detector to Figure 7-26 is based on so-called early-late gate approximation of ML (maximum likelihood) timing detector u

$(\tau) = z^2(\tau + \varepsilon T) - z^2(\tau - \varepsilon T)$, in Figure 1 of Reference [7] where the absolute value is used rather than squaring, but here it is modified for easy comparison. See also Reference [10] page 319.

Exercise 7-16: Find the S-curve of the early-late timing error detector u $(\tau) = z^2(\tau + \varepsilon T) - z^2(\tau - \varepsilon T)$.

Answer: One can compute S-curve for this case by considering $\Delta = 0$ and $t \rightarrow t \pm \varepsilon T$. With $\Delta = 0$ (squaring), $S(\mp\Delta T) = C_0(0) + C_1(0) \cos\left(\frac{2\pi t}{T}\right)$ using the table in section 7.2.1.1. By applying $t \rightarrow t + \varepsilon T$, $S(t + \varepsilon T) = C_0(0) + C_1(0) \cos\left(\frac{2\pi t}{T} + 2\pi\varepsilon\right)$ and applying $t \rightarrow t - \varepsilon T$, $S(t - \varepsilon T) = C_0(0) + C_1(0) \cos\left(\frac{2\pi t}{T} - 2\pi\varepsilon\right)$, and then by subtracting $S(t + \varepsilon T) - S(t - \varepsilon T)$ we obtain

$$S = -2C_1(0) \sin(2\pi\varepsilon) \sin\left(\frac{2\pi\tau}{T}\right) \qquad (7\text{-}10)$$

Note that the detector gain is maximum when $\varepsilon = \frac{1}{4}$ and is identical with one (3rd row) in Table 7-3. It is sensible to choose $\varepsilon = \frac{1}{4}$ with 4 samples per symbol and the other hand in analog implementation it does not need to be precisely $\varepsilon = \frac{1}{4}$.

In light of the fact that the early – late algorithm is approximation to ML detector, so is the timing detector in Figure 7-26.

7.2.2.3   Gardner's Timing Detector and its Extension

Gardner detector in Reference [1] is a timing detection algorithm, and its S-curve is shown in Table 7-3 (4th row). This can be derived by using Table 7-1. Take $\Delta = -\frac{1}{2}$ and $\Delta = +\frac{1}{2}$, and take difference, then both terms time substituted $t \rightarrow t \mp \frac{T}{2}$ by a half of symbol period. The result gives the same detector structure derived in [1], and shown in Table 7-3.

Exercise 7-17: Show that S-curve of the timing error detector (Gardner detector) u $(\tau) = z(\tau - T/2)\{z(\tau) - z(\tau - T)\}$ is given by $S = 2C_1\left(\frac{1}{2}\right) \sin\left(\frac{2\pi\tau}{T}\right)$. Hint: follow the above discussion as a guide. Note that $2C_1(\frac{1}{2}) < 2C_1(0)$. Gardner detector gain is slightly smaller than that of wave difference.

By looking at the sign pattern of Table 7-1, variety of new timing detection algorithms can be formed by properly combining terms.

Gardner's algorithm can be extended by adding more terms as,

$$u_{new1}(\tau) = z\left(\tau - \frac{3T}{2}\right)[z(\tau) - z(\tau - T) + z(\tau - 2T) - z(\tau - 3T)]$$

And its S-curve is given by

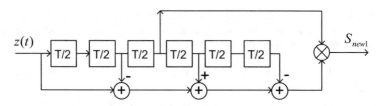

Figure 7-27: New1 timing detector: extension of Gardner's detector for higher gain

$$S_{new1} = -\left[2C_1\left(\tfrac{1}{2}\right) + 2C_1\left(\tfrac{3}{2}\right)\right]\sin\left(\tfrac{2\pi\tau}{T}\right)$$

It is easy to extend Figure 7-27 in the above to have more terms in summation part. Note that the detector gain may be increased by adding more terms. This algorithm resembles a digital calculation of a derivative, but is not exactly the same since a digital derivative is given by $h_d(n) = \tfrac{1}{n}(-1)^n, n = \pm1, \pm2, ..$ where $h_d(n) = 0$ with $n = 0$.

While two samples per symbol may be enough to implement Figure 7-27, other combinations are possible if 4 samples per symbol are available. Wave difference and Gardner algorithms are combined as shown below.

$$u_{new2}(\tau) = z\left(\tau - \tfrac{T}{2}\right)[z(\tau) - z(\tau - T)] + z\left(\tau - \tfrac{T}{4}\right)\left[z\left(\tau - \tfrac{T}{4}\right) - z\left(\tau - \tfrac{5T}{4}\right)\right]]$$

Its S-curve is given by $S_{new2} = \left[2C_1(0) + 2C_1\left(\tfrac{1}{2}\right)\right]\sin\left(\tfrac{2\pi\tau}{T}\right)$.

The above two examples are found with the motivation to increase a timing detector gain, at least asymptotically without noise and sampled digital implementation in mind with two samples per symbol or four samples per symbol. It is also shown that any detector based on two samples per symbol can be extended to four sample per symbol one by 'delaying' a quarter of symbol period $(t \to t \mp \tfrac{T}{4})$ to the detector, for example shown in Figure 7-26. By using both cosine and sine terms the detector curve is a straight line and an "instantaneous timing error phase" can be measured with the four samples. Being the detector characteristic a straight line up to $\tfrac{T}{2}$, Reference [2] showed that this timing detector is free of hang up and fast.

Exercise 7-18: Show S_{new1} for New1 timing detector in Figure 7-27.

Answer: Make a table for each summation term to organize the computation.

Δ	−3/2	−1/2	+1/2	+3/2
$t \to t$	0	−T	−2T	−3T
sign	−	+	−	+
$C_1(\Delta)$	$C_1\left(\tfrac{3}{2}\right)$	$C_1\left(\tfrac{1}{2}\right)$	$C_1\left(\tfrac{1}{2}\right)$	$C_1\left(\tfrac{3}{2}\right)$

$-T, -2T, -3T$ are the same as 0 due to the S_{new1} is periodic. Look up Table 7-1 for sign.

Figure 7-28: New2 timing
detector: combining wave
difference and Gardener

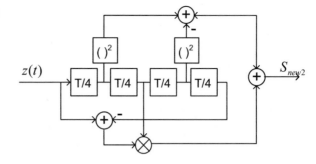

Exercise 7-19: Show S_{new2} for New2 timing detector in Figure 7-28.
Answer: Again make a table and look up Table 7-1 for sign.

Δ	$-1/2$	$+1/2$	$+0$	0
$t \to t$	0	$-T$	$-1/4$	$-3/4$
sign	$+$	$-$	$+$	$-$
$C_1(\Delta)$	$C_1(\frac{1}{2})$	$C_1(\frac{1}{2})$	$C_1(0)$	$C_1(0)$

7.2.3 *Numerical Confirmation of S-Curve of Timing Error Detectors*

We obtain some closed form expression of $S(\mp\Delta T)$ in (7-4) for $h(t)$ with the end to
end pulse shape, bandlimited in frequency, even symmetry and real in time. This
gives us some insight into the improvement of timing error detector as well as its
understanding. However, the numerical evaluation of (7-4) is another way to confirm
the description so far, and its computation is straightforward; we need to compute

$$S(\mp\Delta, t \to t \mp \varepsilon T) = \sum_{k=-\infty}^{\infty} h(t - kT)h(t \mp \Delta T - kT) \text{ where } \varepsilon = 0, \tfrac{1}{4}, \tfrac{1}{2} \text{ for a given}$$

$h(t)$. Since it is periodic the range of t is 0 to T, i.e., one period. The time span of $h(t)$
is 2 for time limited pulses (e.g., triangle pulse), 16 or so may be sufficient for
practical pulses (e.g., raised cosine with excess bandwidth beta); thus k ranges -8 to
$+8$. Two loops in t and in k are necessary while Δ, $\mp\varepsilon$ are scalar parameters. This
numerical approach allows any practical pulse shape as well.

Example 7-10: S-curve computation with Matlab program for raised cosine pulse
and triangle time limited pulse. Samples per symbol period is *Nsam*, and pulse span
is limited to $-Lsym$ to $+Lsym$; Δ is denoted by *del*, and $ph = \varepsilon$ means $t \to t \mp \varepsilon T$.

```
Nsam = 64; Lsym = 8;              % samples per symbol, symbols of span
rc =0;                           % 1= RC,, 0=Prefiltered, 2=triangle
beta = 0.31;                      % excess bandwidth
n= (0:1:Nsam)';                  % S_crive time index
nL = length(n);
k =(-Lsym:1:+Lsym)';             % pulse span vector
del= 0;                          % del 0: squaring, delta
ph = 0;                          % t → t
```

```
% raised cosine pulse
for jn = 1: nL
  ndNk= ndN(jn)-k + ph;   % k(thus ndNk, ht, htd) is a vector
  ht = sin(pi*(ndNk+eps))./(pi*(ndNk+eps)).*cos(beta*pi*(ndNk))./(1-
4*(beta*(ndNk)).*(beta*(ndNk)));
                          %eps for numerical stability
  htd = sin(pi*(ndNk+del+eps))./(pi*(ndNk+del+eps)) .*cos(beta*pi*(ndNk+del))
./(1-4*(beta*(ndNk+del)).*(beta*(ndNk+del))));
  s_rc(jn) = ht'*htd;        %multiply and summing
end
```

```
% triangle pulse  spanning two symbols
for jn = 1: nL
  ht = zeros(length(k),1);
  for jk = 1: length(k)
   ndNk = ndN(jn)-k(jk) + ph;
   if ndNk >-1 && ndNk <=0,   ht(jk) = ndNk+1;   end
   if ndNk >0 && ndNk < 1,      ht(jk) = -ndNk+1;   end
  end
  htd = zeros(length(k),1);
  for jk = 1: length(k)
   ndNk = ndN(jn)+del -k(jk) + ph;
   if ndNk >-1 && ndNk <=0,   htd(jk) = ndNk+1;   end
  if ndNk >0 && ndNk < 1,      htd(jk) = -ndNk+1;   end
  end
  s_rc(jn) = ht'*htd;
end
```

Exercise 7-20:   Write a Matlab function, using the above, with parameters of Δ, $\mp\varepsilon$, pulse span($-Lsym$ to $+Lsym$), symbol periods, samples per symbol, and pulse shape.
Answer: function s_rc = smy_synT2a(del, ph, k, nL, ndN, beta, rc).

Example 7-11: Numerical computation of S-curve of timing detector in Figure 7-26; this is listed in Table 7-3 (2nd and 3rd row) as $u(\tau) = z^2(\tau) - z^2(\tau - T/2)$ and $u(\tau) = z^2\left(\tau - \frac{T}{4}\right) - z^2\left(\tau - \frac{3T}{4}\right)$.

Once the function in the above is written the rest of computation is organized as 4 function calls.

```
del=0;          % del 0: squaring
ph =0;          src0  = smy_synT2a(del, ph, k, nL, ndN, beta, rc);
ph =-1/2;       src12 = smy_synT2a(del, ph, k, nL, ndN, beta, rc);
supper = src0 - src12;
ph =-1/4;       src14 = smy_synT2a(del, ph, k, nL, ndN, beta, rc);
ph =-3/4;       src34 = smy_synT2a(del, ph, k, nL, ndN, beta, rc);
sbottom = src14- src34;
```

The result of this computation is shown in Figure 7-29. We show two symbol periods to see the periodicity. This is precisely the same as in Table 7-3 for raised cosine pulse. The detector gain is proportional to β excess bandwidth.

Now the pulse is changed to a triangle, and we repeat the same Matlab function with different pulse shape parameter. The result is displayed in Figure 7-30.

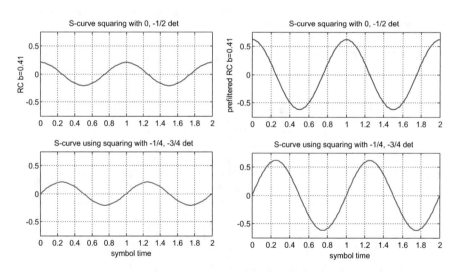

Figure 7-29: S-curve with wave difference timing detector; $u(\tau) = z^2(\tau) - z^2(\tau - T/2)$ (top part) and $u(\tau) = z^2\left(\tau - \frac{T}{4}\right) - z^2\left(\tau - \frac{3T}{4}\right)$ (bottom part). Left side is for raised cosine and right side is for pre-filtered RC. Compare with Table 7-3 (2nd and 3rd row)

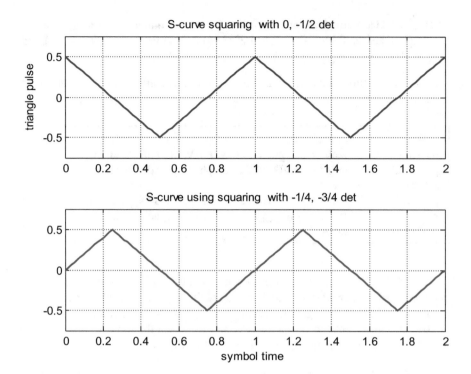

Figure 7-30: S-curve of wave difference algorithm with triangle pulse. The detector gain, 0.5, is similar to $\beta = 1.0$ of RC pulse

Example 7-12: Gardner and New1 timing detector (Figure 7-27) comparison

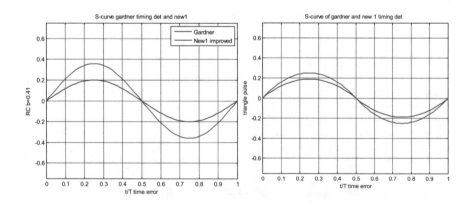

New1 timing detector in Figure 7-27, compared to Gardener's (4th row listed in Table 7-3), is greater in the detector gain. The improvement seems to be relatively less as the excess bandwidth gets close to 1.0. This is not surprising if one looks at C_1

(Δ, β) in Figure 7-25; only when $\Delta = 0$ the gain increases monotonically and as$\Delta > 1.0$, it gets smaller.

With a triangle pulse Gardner's detector is not impressive in detector gain (upper right figure of Figure 7-31). In fact, it is much worse than that of raised cosine with $\beta = 0.991$. The wave difference (squaring) method in Figure 7-30 works well with the triangle pulse.

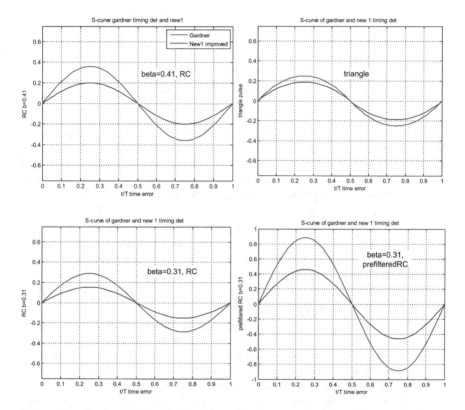

Figure 7-31: Gardner and New1 timing detector S-curve comparison; four different pulse shapes. It is obvious that New1 detector gain is always bigger. Detector gain of triangle pulse is much smaller than that of wave difference detector

7.2.4   Timing Detectors with Differentiation or with Hilbert Transform

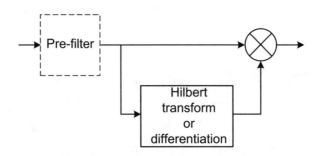

Historically early on, in voice band modem development for telephone lines, so called band edge timing recovery (BETR) was discovered and used extensively [4], which was initially described in [13]. In BETR a received signal is band-pass filtered at $\frac{1}{2T}$ and in parallel filtered at $-\frac{1}{2T}$, then the former and the latter after conjugation are multiplied in order to generate timing detection signal. The band pass filter is an approximation to pre-filtering. This can be done on real signal of IF frequency or equivalently in base band with complex envelope signal which may be considered as IF frequency is zero. The upper side band is represented as $\{z(t) + j\hat{z}(t)\}$ while the lower side band is $\{z(t) - j\hat{z}(t)\}$ where $\hat{z}(t)$ is a Hilbert transform of $z(t)$. Thus BETR timing signal is formed by $im\{[z(t) + j\hat{z}(t)][z(t) - j\hat{z}(t)]^*\} = 2\{z(t)\hat{z}(t)\}$ which is shown in the figure above. Note that $re\{[z(t) + j\hat{z}(t)][z(t) - j\hat{z}(t)]\} = z^2(t) + \hat{z}^2(t)$ can be used, but it will be the same as squaring.

In Table 7-3 in the last row, S-curve timing detector with Hilbert transform is listed and their detector curves are shown. $S_{Hilbert}$ has the same gain as the squaring except for sin( ) function rather than cos( ). There is no DC component. It is interesting to relate BETR as multiplication of signal and its Hilbert transform, and thus its S-curve can be computed readily.

Instead of Hilbert transform **differentiation** may be used for timing detector. This timing detector is listed in Table 7-3 as well. Its detector gain is expressed in the integral form and no explicit formula for the integration yet found.

7.2.5   Intuitive Understanding of Timing Detectors

In passing we mention Reference [9] where a timing recovery based on one sample per symbol is described. See problems of P7-14 and P7-15 for additional detail.

We give two concrete numerical examples so that intuitive understanding of timing detector is possible.

Example 7-13: BPSK with the end to end impulse response being triangle – transmit pulse is rectangular, and its matched filter is rectangular too. The timing detector uses differentiation, i.e., a received signal is multiplied by its derivative. And the resulting waveform is periodic in symbol time with occasional gaps. Note that the periodic wave happens when there is a transition from 1 to -1 or vice versa and there is a gap when there is no transition, i.e., consecutively 1s or -1s. A specific illustration is shown in Figure 7-32.

Example 7-14: BPSK with the end to end impulse response being a RC with excess bandwidth $\beta = 0.981$. Timing detector uses squaring, i.e., a received signal is multiplied by itself. This is listed in Table 7-3 (1st row). Again see that the periodic wave (cosine like in a symbol) happens when there is a transition from 1 to -1 or vice versa and there is a gap when there is no transition, i.e., consecutively 1s or -1s. This is shown in Figure 7-33.

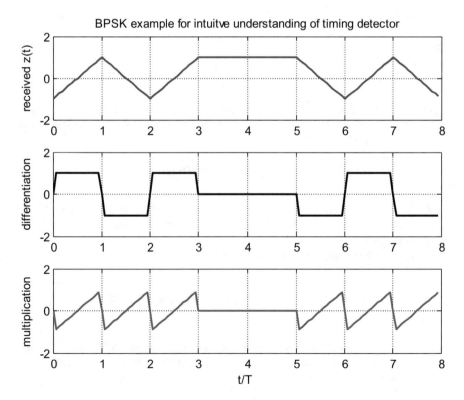

Figure 7-32: Triangle pulse with differentiation timing detector (received wave form multiplied by its derivative)

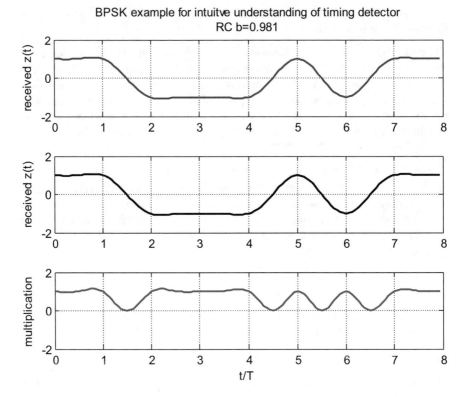

Figure 7-33: RC = 0.981 pulse with squaring timing detector; note that transition occurs 4 times and 'cosine' wave occurs 4 times, and no transition no 'cosine' wave . As shown in Table 7-3, there is DC, which shifts the waveform upward

7.2.6 Carrier Frequency Offset Estimation

A carrier frequency offset may be estimated from receive samples. This idea is shown in Figure 7-34. A shaded block processes receive samples to obtain a carrier frequency offset.

   A signal to be processed is represented by

$$z(t) = \left\{ m(t)e^{j\phi(t)} \sum_{k=-\infty}^{\infty} a_k h(t - kT) + n(t) \right\} e^{j2\pi f_e t}$$

where $m(t)e^{j\phi(t)}$ represents a fading which is assumed to be slow, thus little change during ΔT. We consider $E\{z(t)z^*(t \mp \Delta T)\}$, explicitly with the conjugation as, ignoring noise,

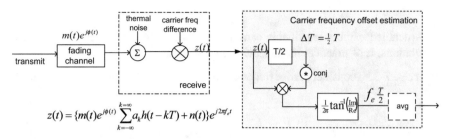

Figure 7-34: Carrier frequency offset estimation from z (t); carrier frequency offset is introduced after thermal noise, as part of down conversion LO

$$E\{z(t)z^*(t \mp \Delta T)\} = e^{-j2\pi f_e(\mp \Delta T)} \left\{ |m(t)|^2 \sum_k \sum_i E(a_i a_k^*) h(t - iT) h(t \mp \Delta T - kT) \right\}.$$

Note that the quantity inside parenthesis of $\{\}$ is real, assuming that data is uncorrelated

$$|m(t)|^2 \left\{ \sum_k \sum_i E(a_i a_k^*) h(t - iT) h(t \mp \Delta T - kT) \right\} =$$

$|m(t)|^2 E\left(|a_k|^2\right) S(\mp \Delta T)$. Thus

$$E\{z(t)z^*(t \mp \Delta T)\} = e^{-j2\pi f_e(\mp \Delta T)} \; |m(t)|^2 E\left(|a_k|^2\right) S(\mp \Delta T)$$

Exercise 7-21: $E\{z(t)z^*(t - \frac{1}{2}\Delta T)\} = e^{-j2\pi f_e\left(\frac{1}{2}T\right)} \; |m(t)|^2 E\left(|a_k|^2\right) S(-\frac{1}{2}\Delta T)$ and thus from its phase we can obtain frequency offset as $f_e\left(\frac{1}{2}T\right) = \frac{1}{2\pi} \tan^{-1}\left(\frac{im}{re}\right)$. This is shown in Figure 7-34.

Example 7-15: The carrier frequency estimate of Figure 7-34 with the sampling rate of $4\times$ symbol rate is shown below. Complex multiplication is explicit with real multiplication.

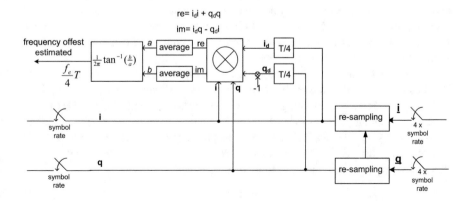

A detailed exploration of this carrier frequency estimation, such as numerical simulations, is of practical interest but not pursued here.

Exercise 7-22: consider carrier frequency offset due to fading and thus z(t) is given by

$$z(t) = \left\{ m(t)e^{\,j2\pi f_e t} \sum_{k=-\infty}^{\infty} a_k h(t - kT) + n(t) \right\}$$

Compute $E\{z(t)z^*(t \mp \Delta T)\} = e^{-j2\pi f_e(\mp \Delta T)} |m(t)|^2 E\Big(|a_k|^2\Big) S(\mp \Delta T)$ ignoring the noise. This is the same as before, and thus frequency offset due to fading can also be estimated with the scheme of Figure 7-34.

7.2.7 Embedding Digital TED Into Timing Recovery Loop

Now we explore those components in detail. Before getting into a full digital implementation of Figure 7-26, we first consider only TED is digital but a loop filter and voltage controlled clock are analog as shown in Figure 7-35.

Exercise 7-23: The components of loop in Figure 7-35 are summarized below.

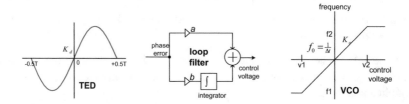

When K_d, a, b, and Ko are positive, the loop is in negative feedback mode. For example, if the output of TED is positive, the frequency of VCO is increased, and sampling occurs earlier and thus TED phase error will be reduced. What happens when the output of TED is negative? What about around zero? If it is around zero, it is called in locks or in tracking mode.

We explain Figure 7-35; the sampling rate is 4 times of symbol rate to be concrete but it can be 2 times or higher. In order to show that this timing recovery loop is in fact to synchronize the sampling clock with the transmit symbol clock $(1/T)$ with optimum sampling phase of symbols, a signal chain from transmit side is reminded in faded line. Overall it is a PLL except for its phase detector being replaced by TED and 4× sampling (ADC). For an overview of PLL see A.3 for refreshing its basics; this is strongly recommended if a topic of PLL is new.

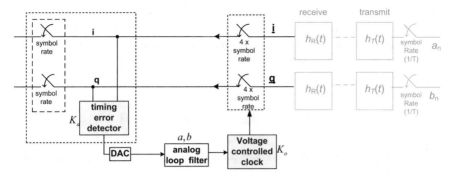

Figure 7-35: Embedding digital TED into an analog loop with analog loop filter and voltage controlled clock oscillator. Overall it is PLL except that TED replaces phase detector in PLL

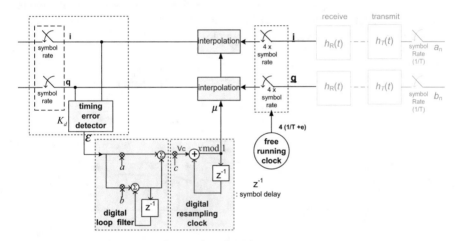

Figure 7-36: Embedding digital TED into digital recovery loop

Exercise 7-24: Compare Figure 7-35 (TED is digital, and the rest of timing recovery is analog) with Figure 7-2 (a complete analog timing recovery scheme), and discuss. Hint: In Figure 7-2, the role of PLL is a very sharp BPF, and it is a feedforward system. On the other hand, in Figure 7-35, it is a negative feedback system, i.e., once the loop is converged (or in tracking mode), the output of TED is around zero, and sample rate to symbol rate reduction is to select one sample out of 4.

Now a full digital implementation of Figure 7-35 will be outlined.

In Figure 7-36 we show a possible digital implementation of Figure 7-35. Note that digital resampling clock is an integrator with modulo 1 since a sampling phase must be within $-\frac{1}{2}$ to $\frac{1}{2}$ (or within 0 to 1.0). A digital loop filter, similar to analog counterpart, has two paths – direct and integration with a and b parameters respectively. Additional parameter, c, may be used as digital resampling clock adjustment speed equivalent to K_o gain of VCO. The free running sampling clock must be approximately 4 times of symbol rate denoted as $4(1/T +e)$ in the figure. We explain resampling block next section.

7.2.8 Resampling and Resampling Control

Conceptually resampling can be understood as shown in Figure 7-37; interpolation and resample (decimation).

However, in timing recovery context, the phase of resampling is adjusted each symbol; it is not a fixed interpolation and decimation as in sample rate change. This is illustrated in Figure 7-38. An impulse response $h(t)$ which is evaluated at $t = -2 + \mu$, $t = -1 + \mu$, $t = +\mu$, $t = 1 + \mu$, $t = 2 + \mu$, $t = 3 + \mu$, as in the figure and then the output is $x(n + \mu)$. In addition to FIR filter calculation, $h(t)$ should be evaluated for a given delay, μ at every sample.

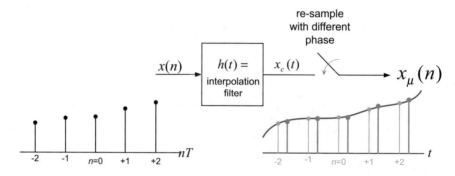

Figure 7-37: Conceptually resampling is an interpolation and followed by decimation

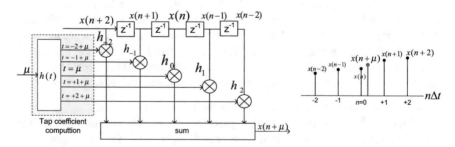

Figure 7-38: Resampling block of 5-tap FIR filter with an impulse response $h(t)$ which is evaluated at $t = -2 + \mu$, $t = -1 + \mu$, $t = +\mu$, $t = 1 + \mu$, $t = 2 + \mu$, as shown; then the output is x $(n + \mu)$ shown on RHS

Exercise 7-25: In Figure 7-38, if one needs to compute $x(n - \mu)$, how does one get the tap coefficients? Answer: evaluate $h(t)$, at each t with $\mu \rightarrow -\mu$, i.e., $t = -2-\mu$, $t = -1 - \mu$, $t = -\mu$, $t = 1 - \mu$, $t = 2 - \mu$, $t = 3 - \mu$.

In summary given a resample phase $(\pm\mu)$ the output of resampler is given by x $(n \pm \mu)$ if a resampling interpolation filter, $h(t)$, is evaluated at $h(k\Delta t \pm \mu)$ where $k = \ldots -3, -2, -1, 0, +1, +2, \ldots$, and Δt is a sample period, not explicit, but $\Delta t = 1$ for notational simplicity.

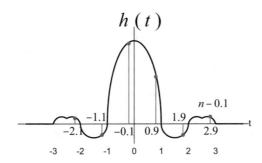

Impulse Response of Resampling Interpolation Filter

We describe the required characteristics of an interpolating impulse response $h(t)$ in this context of resampling.

1. $h(n\Delta t) = \delta(n)$, $n = 0, \pm 1, \pm 2, \ldots \pm L$ a Nyquist pulse
2. $h(t) = h(-t)$ even symmetry around zero
3. $h(t)$ is finite in time for practicality

An example of $h(t)$ with 6 symbol period span is shown at LHS.

For a given pair of Fourier transform $h(t) \leftrightarrow H(f)$. What are the requirements of $H(f)$? It should be a low pass filter; the magnitude should be flat (due to 1. above) and phase should zero (due to 2. above), and its cutoff frequency is $0.5/\Delta t$ (due to 3. this can only be approximate).

In appendix A.4, we discuss how to design interpolating FIR filters based on the constraint of 1,2, and 3 above.

Exercise 7-26: Two sample (linear) interpolator using FIR filter shown below. Tap coefficients are computed, for a given μ, as $h_{-1} = h(-1 + \mu)$, $h_0 = h(0 + \mu)$, and $h_1 = h(1 + \mu)$. This FIR filter (triangle $h(t)$) is equivalent to two point interpolation, i.e., $x(n + \mu) = (1 - \mu)x(n) + (\mu)x(n + 1)$. Show $x(n - \mu) = (1 - \mu)x(n) + (\mu)x(n - 1)$.

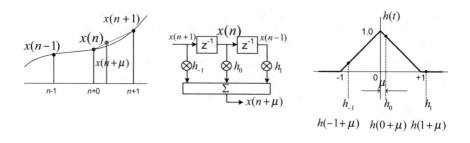

$$h(-1+\mu) \quad h(0+\mu) \quad h(1+\mu)$$

Exercise 7-27: The evaluation of $h(k + \mu)$, where $k = ..,-3, -2, -1, 0, +1, +2, ...$ and $\mu = [-0.5, +0.5]$, can be done by a lookup table. Obtain a lookup table size if the span of interpolation is 6 and the resolution of μ is $\frac{1}{128}$. Answer: 6×128.

Digital resampling clock- overflow and underflow

Exercise 7-28: Find the range of μ in Figure 7-38. Answer: $\mu = -1/2$ to $+1/2$. Thus we use the notation of x modulo 1 in Figure 7-36. Note that $x (n + 0.6)$ can be $x (n + 1 - 0.4)$.

In Figure 7-36, digital resampling clock generates resample phases μ but, no indication of possibility of overflow ($u > 0.5$) or of underflow ($u < -0.5$) in the figure. The details of the operation of u mod 1 are shown in Figure 7-39.

Exercise 7-29: Discuss the meaning of overflow and underflow. Hint: remember that μ being positive needs to increase the frequency of resampling clock, and thus $u > 0.5$ means the increasing resampling frequency is too large. Answer: see table.

	u < −0.5	u > +0.5
Situation	Resampling clock frequency < incoming clock frequency Loop tried to increase resampling clock frequency	Resampling clock frequency > incoming clock frequency Loop tried to decrease resampling clock frequency
Additional measure	More samples of the resampler output	Skip samples, i.e., less samples
Remarks	In analog loop VCO frequency is changed to match with the incoming clock frequency. This is done in digital loop by a buffer - skipping and inserting samples.	

Example 7-16: Resampler and its control when $u > +0.5$ occurs. In the figure below with resample phase $u > +0.5$, a resampler generates $x(2 + u) = x(3 + (u - 1))$. Thus x (2) is not used (skipped).

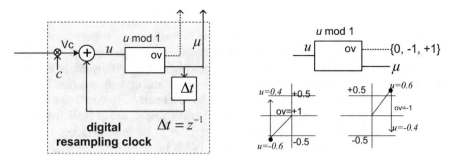

Figure 7-39: Overflow and underflow of digital resampling clock, with examples

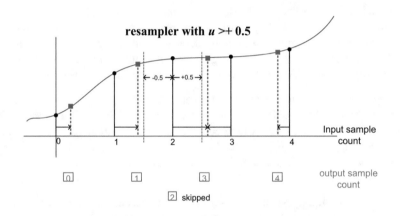

Example 7-17: Resampler and its control when $u < -0.5$ occurs. In the figure below with resample phase $u < -0.5$, a resampler generates $x(2-|u|) = x(1+|(u+1)|)$. Thus $x(1)$ is used twice generating additional sample.

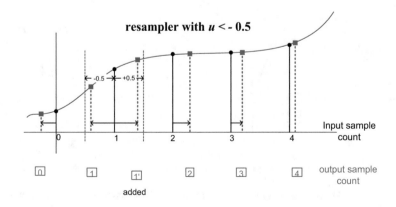

In most practical situations this overflow and under flow may occur when there is a clock frequency offset or with very high Doppler clock frequency shift. Thus, it is important to understand the boundary conditions where the overflow/underflow may occur. Thus an exploration of how to utilize overflow and underflow would be very interesting and important in practice to handle those situations. It should be a form of numerical simulations for a given model of Doppler channel, and of clock frequency offset. We leave this as *advanced project exercises* to interested readers with the above discussion as a starting point, at the end of Subsection 7.2.8 and 7.2.9.

Sample to symbol decimation

We need to explain the difference between Figure 7-35, analog loop with voltage controlled clock oscillator (VCO), and Figure 7-36, digital loop with digital resampling clock (mod 1 integrator), which may be called DCO. In analog loop, we mentioned that in order for PLL to be negative feedback system, K_d, a, b, and Ko are all positive. On the other hand in digital loop, K_d, must be negative for the loop to be a negative feedback system. This TED difference is shown in Figure 7-40.

Exercise 7-30: Confirm the above statement. Answer: In Figure 7-35, when the output of analog loop filter is positive, the frequency of VCO increases. Thus sampling occurs earlier $(-\mu)$, which compensates later sampling of positive TED. In Figure 7-36, the output of digital loop filter is positive, DCO output increases. Thus it is not negative feedback and the sign must be changed to negative feedback. One way is to negative slope of TED.

Once symbol timing synchronization is achieved we need one sample per symbol time, e.g., one out of 4 samples per symbol. This is generally true. However, in practice, one sample out of 4 or 2 is designated as a symbol sampling point ('0') and thus the digital loop is updated every symbol.

We so far outlined symbol timing recovery, and detailed all its necessary components -TED and the loop components - with a special emphasis on all digital implementations as shown in Figure 7-36. This is a necessary step to develop a

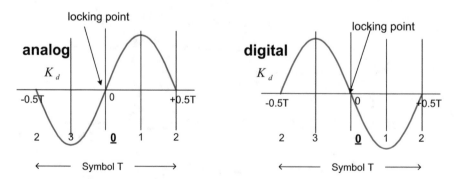

Figure 7-40: TED for analog loop (LHS) of Figure 7-35, and TED for digital loop (RHS) of Figure 7-36

symbol timing synchronization subsystem but may not be quite sufficient in practice – we need to do additional experiments and numerical simulations, from transmit to receive, i.e., end to end. We leave this as a project to the interested as below.

Exercise 7-31* (project): Write a Matlab program to do simulations of Figure 7-36.

7.2.9 Simulations of Doppler Clock Frequency Shift

Our focus will be on obtaining channel simulation models, in particular with a large Doppler clock frequency shift. This topic is not often treated in the literature because Doppler frequency shift of symbol clock is typically small enough so that a timing recovery may handle it well. However, there are situations that it may not be the case, e.g., low Earth orbit (LEO) satellite channel with wide bandwidth, say, 500 MHz.

7.2.9.1 Doppler Frequency Shift

Doppler frequency shift is due to the relative movement between transmitter and receiver; the velocity (v) is positive when two are approaching while it is negative when two are moving away. Its Doppler frequency shift is given by $f_D = f_o \frac{v}{c}$ where c is speed of light. Thus the frequency is given by $f = f_o\left(1 + \frac{v}{c}\right)$. Commonly such Doppler shift can be heard by an emergency vehicle with siren, approaching and moving away; when it is approaching the pitch of siren is getting higher while it getting lower when it is moving away.

Example 7-18: When a phone in vehicle is moving away from base station with speed of 200 km per hour, what is Doppler frequency shift in ppm (part per million) ? Answer: 0.185 ppm; $\frac{v}{c}$ = 200/3600/300000 = 1.85E-7.

Example 7-19: In LEO satellite when the altitude of satellite is 1150 km and the elevation angle is 50°, then Doppler velocity relative to Earth is shown to be +4.73 km/sec. What is Doppler frequency shift in ppm (part per million)? What about when Doppler velocity −4.73 km/sec due to orbiting? Answer: 15.8 ppm; $\frac{v}{c}$ = 4.73/ 300000 = 1.58E-5, and −15.8 ppm when it is −50°.

Doppler shift happens to both carrier frequency and symbol clock, i.e., affecting an entire signal with the same percentage shift (e.g., ppm). If a signal is expressed by $x(t) \leftrightarrow X(f)$ a Fourier transform pair, its Doppler shifted version of it is given by time scale change $t \rightarrow \frac{t}{(1+\varepsilon)}$ or by frequency scale change $f \rightarrow f(1 + \varepsilon)$ where $\varepsilon(t) = \frac{v(t)}{c}$, i.e., the velocity may be a function of time.

7.2.9.2 Simulation Model of Doppler Clock Frequency Shift

We try to obtain a simulation model of Doppler clock frequency shift in this section. In order to get insight we try a few examples.

Example 7-20: Suppose that a Doppler shift is represented by a small frequency jump, i.e., $\varepsilon=$ constant $= \varepsilon_p$. A sampled transmit signal x $(n\Delta t) = \sum_k a(k)h(n\Delta t - kM\Delta t)$ where Δt is sample period and there are M sample per symbol. Now $\Delta t \rightarrow \frac{\Delta t}{(1+\varepsilon_p)}$ due to Doppler shift. For both before and after Doppler shift, signal samples are calculated by $x(n) = \sum_k a(k)h(n - kM)$, and the difference is in time scale as shown in the figure above when $\varepsilon_p \geq 0$. This is the same as a symbol period is decreased by $\frac{1}{(1+\varepsilon_p)} \leq 1$ as shown in the figure.

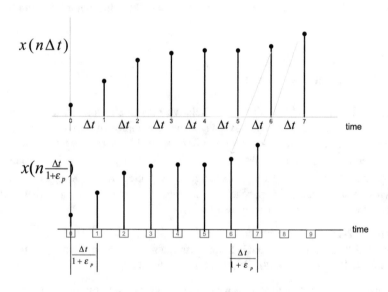

Exercise 7-32: When a Doppler shift is such that $\varepsilon_p \leq 0$ what happens to the sample period? It is obviously increased. Draw $x\left(n\frac{\Delta t}{(1+\varepsilon_p)}\right)$ for this case.

Now we consider cases with Doppler shift is a function of time, $\varepsilon(t)$.

Example 7-21:

$$\varepsilon(t) = \varepsilon_p \cos\left(\frac{2\pi t}{T_L}\right) = \varepsilon_p \cos\left(\frac{2\pi n\Delta t}{L\Delta t}\right).$$

Example 7-22: $\varepsilon(t) = \varepsilon_p \, sgn\left(\sin\left(\frac{2\pi t}{T_L}\right)\right)$

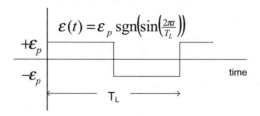

We need to compute a time phase $n\frac{\Delta t}{(1+\varepsilon)}$ for each n for a given $\varepsilon(n\Delta t)$; for example, $\varepsilon(n\Delta t) = \varepsilon_p \, sgn\left(\sin\left(\frac{2\pi n\Delta t}{L\Delta t}\right)\right)$. Notice that $n\frac{\Delta t}{(1+\varepsilon)}$ is no longer integer multiple of Δt but with fraction of it when $\varepsilon(n\Delta t) \neq 0$ as in this example. This can be implemented by DAC as shown in Figure 7-41 where the output of DAC is impulse train with unequal time spacing and zeroes between impulses.

A digital implementation of Figure 7-41 is shown in Figure 7-42 and will be explained here. Assuming $M\Delta t = T$, i.e., there are M samples per symbol time (T), x $(n\Delta t) = \sum_k a(k)h(n\Delta t - kM\Delta t)$ are generated by interpolating data symbols (a) with transmit filter ($h\,(t)$). These samples are interpolated by a factor L by insertion of zeros between samples as shown in Figure 7-42, and are assigned to time positions due to Doppler frequency shift. The time positions are computed by tracking f-state and n-cntr as shown in Figure 7-42 (bottom). Depending on the overflow/underflow, ov $=\{-1, 0, +1\}$ of 'mod 1' , n-cntr is {no increment, +1 increment, +2 increment} to be adjusted Doppler frequency change. For 'mod 1' operation, it is shown in the figure.

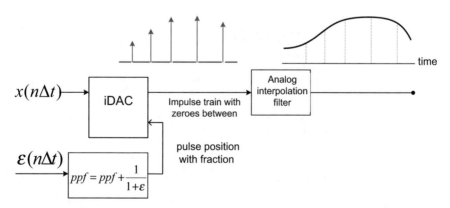

Figure 7-41: Generation of samples with Doppler clock frequency shift using a DAC and analog interpolation filter; an impulse is generated at pulse position and zeroes between impulses

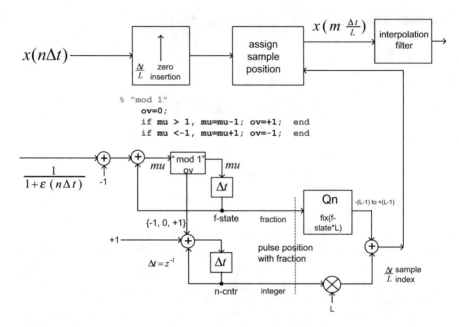

Figure 7-42: Digital implementation of Doppler clock shift where sample rate is increased by L interpolation factor; a fraction and integer part is tracked separately and combined into a time position (index)

Note that the interpolation factor L will determine the time position resolution; the higher the L, the higher the resolution.

Example 7-23: A numerical example of Figure 7-42 is shown in Figure 7-43. Doppler frequency shift is given by $\varepsilon(n\Delta t) = \varepsilon_p \, sgn\left(cos\left(\frac{2\pi n\Delta t}{L\Delta t} \right)\right)$ with $L = 18$, $\varepsilon_p = 0.2501$ which is rather large but to see the effect easily.

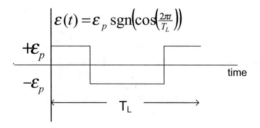

Exercise 7-33* (project): Write a Matlab program to generate Figure 7-43, and then extend it to handle practical Doppler frequency shift.

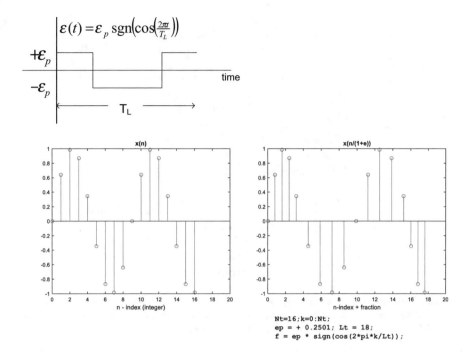

Figure 7-43: A numerical example of Figure 7-42 implementation

7.3   Carrier Phase Synchronization

Carrier phase synchronization inherently requires the details of signal constellations since the measurement of carrier phase deviation from an ideal constellation point should be done in digital domain. However, actual phase correction may be done either in analog signal or digitally. In modern implementations, demodulation processing, including both synchronization and equalization, is typically done digitally. This fully digital implementation of symbol timing, carrier phase synchronization is shown in Figure 7-24 where sampling clock and carrier frequency are free running, and all detection and necessary corrections are done digitally. In particular, symbol timing recovery is decoupled from carrier frequency and phase, and performed in sampling rate, which is typically higher than symbol rate by 2 or 4 times. On the other hand all-digital carrier loop is done in symbol rate. This is our base line of synchronization of symbol timing and carrier phase, and it is fast, applicable universally to any digital modulation schemes.

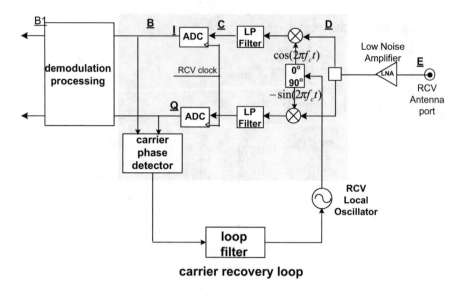

Figure 7-44: Analog carrier recovery loop, which was used often in the past, is a starting point of our development of all digital implementations. Carrier phase detector requires the details of signal constellation. Shading indicates that it is part of phase detector

7.3.1   Carrier Recovery Loop and Its Components

Our starting point of carrier recovery loop will be re-captured here in Figure 7-44 again for convenience even though we showed it in Figure 7-3 before. This analog scheme was used often in practice. Note that this scheme is similar to a PLL of sinusoidal signals, except for phase detection being elaborated with digital comparison against ideal constellation points in addition to mixing (i.e., quadrature demodulating) locally generated carrier with the incoming carrier modulated signal. Furthermore, in this analog scheme, the phase correction is done by changing the frequency of RCV local voltage controlled oscillator (VCO). Historically it was developed, for QPSK, purely in analog form, known as Costas loop. This will be explained further below

7.3.2   Phase Locked Loop Review

See A.3 Review of phase locked loops at the end of this chapter. It is important to refresh it at this point if not done earlier.

First we compare this carrier recovery loop with a phase locked loop (PLL) of sinusoidal input signal. This is shown in A.3. There is one to one correspondence of three components – phase detector, loop filter and VCO – between PLL and carrier recovery loop.

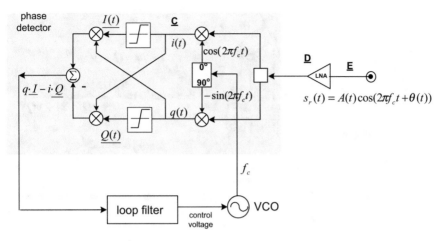

Figure 7-45: Costas loop for QPSK carrier phase recovery; a concrete example of Figure 7-44. i $(t) \rightarrow \underline{I(t)}$ and $q(t) \rightarrow \underline{Q(t)}$ are a decision constellation point

7.3.3   Understanding Costas Loop for QPSK

Carrier phase synchronization scheme, called Costas loop, was invented early on, which is shown in Figure 7-45. It may be considered as a concrete example of Figure 7-44. We explain the phase detector of Costas loop in Figure 7-45. The equation $q\,\underline{I} - i\,\underline{Q}$ is the phase measurement against a constellation point $(\underline{I(t)},\underline{Q(t)})$. This will be explained using Figure 7-46. Note that a constellation point (I, Q) is by decision from a received sample (x, y), called decision-directed, and that (i, q) is related to the sample (x, y) by scaling factor c, which may be done by gain control so that $i^2 + q^2 = I^2 + Q^2$. As shown in Figure 7-46, (i, q) is on the same circle as (I, Q).

Exercise 7-34:   Show that, from $i + jq = (I + jQ)e^{j\theta}$, $\sin(\theta)=(qI - iQ)/(I^2 + Q^2)$, and $\tan(\theta)=(qI - iQ)/(iI + qQ) = (yI - xQ)/(xI + yQ)$.

Answer: Using Euler identity, $(i + jq)/(I + jQ) = \sin(\theta) + j\cos(\theta) = (iI + qQ)/(I^2 + Q^2) + j(qI - iQ)/(I^2 + Q^2)$.

Exercise   7-35: From   Figure   7-46   derive   the   relationship $\sin(\theta) = (qI - iQ)/(I^2 + Q^2) = (yI - xQ)/\sqrt{(x^2 + y^2)(I^2 + Q^2)}$.

Answer: Using $q = cy$, $i = cx$, $i^2 + q^2 = I^2 + Q^2$, $\sin(\theta) = (qI - iQ)/(I^2 + Q^2)$ can be expressed in terms of x, y, I, Q.

In Figure 7-45 of Costas loop, the phase measurement is done by $s = qI - iQ$ without denominator $\sqrt{(I^2 + Q^2)}$ which is a constant and may be absorbed as part of phase detector gain.

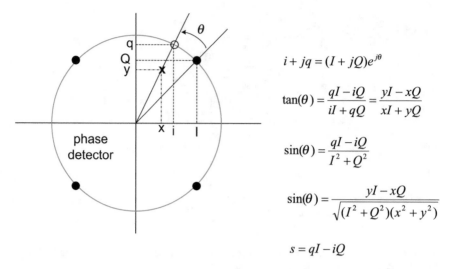

$$i + jq = (I + jQ)e^{j\theta}$$

$$\tan(\theta) = \frac{qI - iQ}{iI + qQ} = \frac{yI - xQ}{xI + yQ}$$

$$\sin(\theta) = \frac{qI - iQ}{I^2 + Q^2}$$

$$\sin(\theta) = \frac{yI - xQ}{\sqrt{(I^2 + Q^2)(x^2 + y^2)}}$$

$$s = qI - iQ$$

Figure 7-46: Phase error measurement of QPSK; (i, q) and (I, Q) are on the circle. $I = [x]$ and $Q = [y]$,e.g., a received sample in the first quadrant is decided to be $(I, Q) = (+1, +1)$

From Figure 7-46, $s = qI - iQ$ is a sine function with different phase error $\sqrt{(I^2 + Q^2)}$ being a constant. This, called s-curve, was computed and shown in Figure 7-47.

Thus it is clear that the Costas loop of Figure 7-45 is an example of Figure 7-44 for QPSK. The shaded region computes the phase error using $s = qI - iQ$.

Note that it is periodic in 90°, i.e., every 90° rotation the constellation looks identical. This is due to the fact that decision-directed determination of $I = [x]$ and $Q = [y]$, i.e., receiver does not know a transmitted symbol. This periodicity can be extended to 180° with BPSK.

Exercise 7-36: Convince yourself that the above statement is correct. Hint: consider the decision boundary of BPSK.

Exercise 7-37: In some situations, known symbols are transmitted for carrier recovery, called pilot. In this pilot scheme, it is possible that the range of phase error measurement can be extended to 360°.

The phase ambiguity of 90° and 180° can be handled digitally by using differential encoding or phase invariance encoding at transmit side and corresponding decoding at receiver, which was discussed in Chapter 2 (section 2.7.2). In this chapter we are not concerned with the resolution of the phase ambiguity.

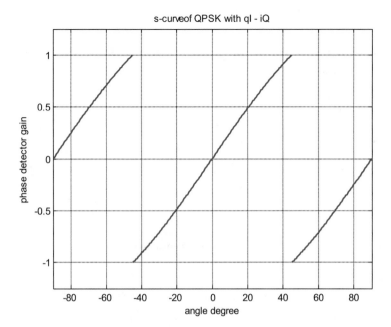

s-curve of QPSK with qI - iQ

Figure 7-47: s-curve of QPSK using $s = qI - iQ$ (See Figure 7-46)

7.3.4 Carrier Phase Detectors

We now consider carrier phase error detectors with QAM/PSK constellation in general. As mentioned before the detection of carrier phase requires the details of constellation of a signal. We start with 16-QAM example whose constellation is shown in Figure 7-48 with different decision boundaries.

We first consider how to compute s-curve of carrier phase error detector, without noise, for a given constellation (e.g., 16QAM here) and its decision boundaries (decision directed case) as dotted lines in Figure 7-48.

The idea of getting s-curve is simple; for a given carrier angle θ , compute a received sample by $i + jq = (I + jQ)e^{j\theta}$ for I, Q of a constellation point, and then quantize $I_h = [i]$ and $Q_h = [q]$ using decision boundaries, and compute angle estimate $s = (qI_h - iQ_h)/(iI_h + qQ_h)$. This is repeated for all constellation points. Sum them and average (i.e., divide by a number of constellation points.)

The procedure for programming is as follows:

1. Set the range of angle is $-45^o < \theta < 45^o$ for decision directed case
2. Compute a receive sample by $i + jq = (I + jQ)e^{j\theta}$
3. Using decision boundaries (two cases in Figure 7-48), make a decision on i, q into $I_h = [i]$ and $Q_h = [q]$.
4. Angle measure: $s_k = (qI_h - iQ_h)/(iI_h + qQ_h)$ for constellation point k.
5. $\frac{1}{Nc}\sum_{k=1}^{Nc} s_k$ where $Nc =$ a number of constellation points

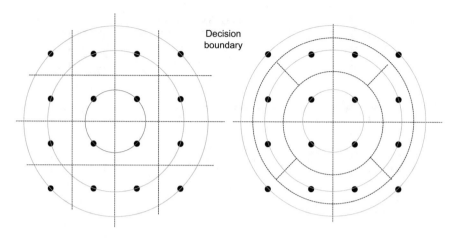

Figure 7-48: Phase detectors for 16-QAM with different decision boundary

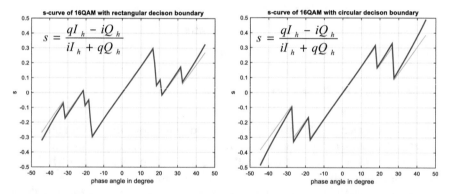

Figure 7-49: s-curves of 16QAM with two different decision boundaries (rectangular, circular LHS of Figure 7-48 and RHS of it respectively)

6. Set different angle θ and repeat 1–5 for $-45° < \theta < 45°$ (decision directed case)

For 16QAM with two different decision boundaries s-curves are shown in Figure 7-49 using the computational procedure above.

Note that the circular decision boundaries (RHS of Figure 7-48) result in a better s-curve; it has higher detector gain when the angle is $|\theta| > 20°$.

Modified decision boundary

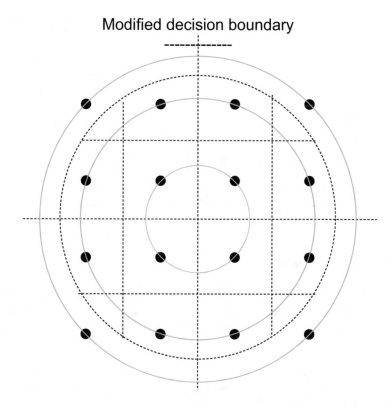

Exercise 7-38: Can you give reasons why circular decision boundary is better than rectangular one? Hint: consider a decision on i, q into $I_h = [i]$ and $Q_h = [q]$ when $|\theta| > 10°$, and there will be a decision error for some constellation points. Consider modifying decision boundary as shown LHS for 4 'outer corner' constellation points.

Exercise 7-39: Simplify the computation of $s_k = (qI_h - iQ_h)/(iI_h + qQ_h)$ for QPSK. Hint: consider $s_k = (qI_h - iQ_h)$ as in Costas loop. Generalize it to QAM constellations.

7.3.5 All Digital Implementations of Carrier Recovery Loop

A complete digital implementation of carrier recovery loop is shown in Figure 7-50; digital phase detector, digital loop filter, digital VCO, and phase de-rotation.

Note that Figure 7-50 has a one-to-one correspondence with Figure 7-44 of analog carrier recovery loop, and with Figure 7-45 of QPSK carrier recovery loop.

We explain each component in this carrier recovery loop. First the phase detector was covered in the previous section of 7.3.4 in detail. A digital loop filter is a digital

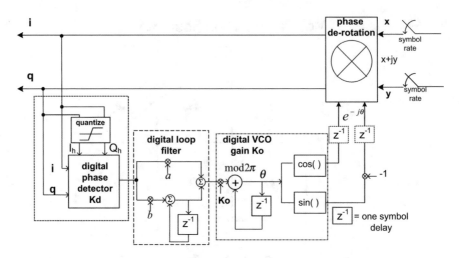

Figure 7-50: All digital implementation of carrier recovery loop

version of analog loop filter in 2nd order as shown in Figure 7-62 of A.3 Review of phased locked loops. A digital VCO is a digital version of analog VCO. In digital VCO a complex exponential in the form of $e^{-j\theta}$ is produced as $\cos(\theta) - j \sin(\theta)$. The adder in the integrator may be of mod 2π. The negation of angle θ is necessary in order to de-rotate the phase of incoming sample, i.e., we need the angle difference. In other words, the loop is in negative feedback. The phase de-rotation operation is to get de-rotated samples (i, q) by $i + jq = (x + jy) e^{-j\theta}$. Real arithmetic will be used for obtaining (i, q) in practice, as shown in the figure.

In passing we mention that the output of digital VCO is delayed by one symbol, which is shown in Figure 7-50 with dotted line. This is not essential and thus may be omitted.

7.4   Quadrature Phase Imbalance Correction

Closely related with carrier phase recovery, yet a distinct problem of quadrature phase error (or IQ phase imbalance) will be discussed here. A quadrature modulator (demodulator) needs two carrier frequencies with 90° separation, i.e., $\cos(2\pi f_c t)$ and $\sin(2\pi f_c t)$. IQ phase imbalance problem, i.e., quadrature phase error, will occur when two are not exactly 90° apart. The implementation issues of quadrature up-conversion and down-conversion are discussed in Chapter 8, where quadrature phase error problem is solved by calibration. Here we will discuss the digital correction of quadrature phase error adaptively at a receiver.

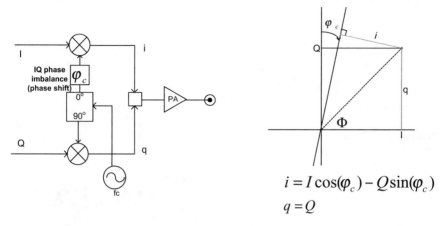

$$i = I\cos(\varphi_c) - Q\sin(\varphi_c)$$
$$q = Q$$

Figure 7-51: IQ phase imbalance representation in quadrature modulator; a phase shifting block is added (LHS) on $0°$ branch of $90°$ splitter

7.4.1   IQ Imbalance Model

IQ phase imbalance may be modeled conveniently by adding a phase shift block as shown in Figure 7-51, and its mathematical model is shown on RHS of the figure. The imaginary axis on a complex plane of constellation is rotated by φ_c (clock-wise). A point on the plane (I, Q) will be changed to (i, q), and the computation of (i, q) is done by $i = I \cos (\varphi_c) - Q \sin (\varphi_c)$ and $q = Q$. This can be seen as $i = \sqrt{I^2 + Q^2} \sin (90° - \Phi - \varphi_c) = \sqrt{I^2 + Q^2} \cos (\Phi + \varphi_c)$.

A phase shifting block may be added on $90°$ side as shown in Figure 7-52.

Exercise 7-40: Confirm the (i, q) as shown in RHS of Figure 7-52. Hint: express q as $q = \sqrt{I^2 + Q^2} \sin (\Phi - \varphi_s)$.

With both φ_c and φ_s, we have $i = I \cos (\varphi_c) - Q \sin (\varphi_c)$ and $q = -I \sin (\varphi_s) + Q \cos (\varphi_s)$ .This is summarized in Figure 7-53.

In passing note that IQ imbalance is not the same as carrier rotation, which may be represented by both real and imaginary axis of the same amount of angle and with the same direction.

Exercise 7-41: From Figure 7-53, express θ_r and φ terms of φ_c and φ_s. And when $\varphi_c = 6°$ and $\varphi_s = 10°$, find θ_r and φ.

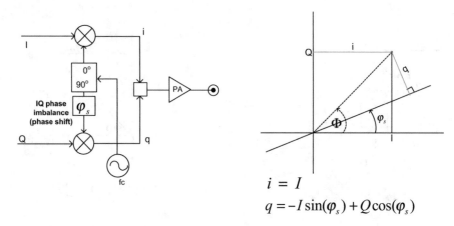

$$i = I$$
$$q = -I\sin(\varphi_s) + Q\cos(\varphi_s)$$

Figure 7-52: IQ phase imbalance representation in quadrature modulator; a phase shifting block is added (LHS) on 90° branch of 90° splitter

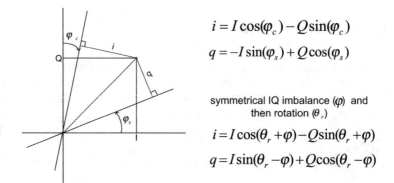

$$i = I\cos(\varphi_c) - Q\sin(\varphi_c)$$
$$q = -I\sin(\varphi_s) + Q\cos(\varphi_s)$$

symmetrical IQ imbalance (φ) and then rotation (θ_r)

$$i = I\cos(\theta_r + \varphi) - Q\sin(\theta_r + \varphi)$$
$$q = I\sin(\theta_r - \varphi) + Q\cos(\theta_r - \varphi)$$

Figure 7-53: (i, q) with both φ_c and φ_s and its decomposition into rotation (θ_r) and symmetrical IQ phase imbalance (φ)

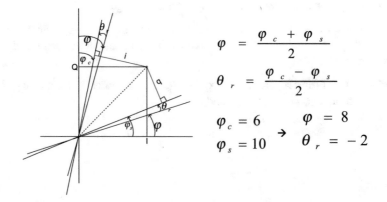

$$\varphi = \frac{\varphi_c + \varphi_s}{2}$$

$$\theta_r = \frac{\varphi_c - \varphi_s}{2}$$

$$\begin{array}{c} \varphi_c = 6 \\ \varphi_s = 10 \end{array} \rightarrow \begin{array}{c} \varphi = 8 \\ \theta_r = -2 \end{array}$$

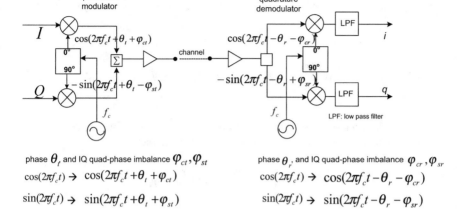

phase θ_t and IQ quad-phase imbalance $\varphi_{ct}, \varphi_{st}$

$\cos(2\pi f_c t) \rightarrow \cos(2\pi f_c t + \theta_t + \varphi_{ct})$

$\sin(2\pi f_c t) \rightarrow \sin(2\pi f_c t + \theta_t + \varphi_{st})$

phase θ_r and IQ quad-phase imbalance $\varphi_{cr}, \varphi_{sr}$

$\cos(2\pi f_c t) \rightarrow \cos(2\pi f_c t - \theta_r - \varphi_{cr})$

$\sin(2\pi f_c t) \rightarrow \sin(2\pi f_c t - \theta_r - \varphi_{sr})$

Figure 7-54: Carrier phase (θ_t, θ_r) and IQ imbalance φ_{ct}, φ_{st}, φ_{cr}, φ_{cr} at transmit and receive

Answer: $\varphi = 8°$, $\theta_r = -2°$

Remember that $\varphi_c = 6°$ cw and $\varphi_s = 10°$ ccw. Thus $\theta_r = -2°$ means ccw. Study the figure in detail.

We need to consider the case of both carrier rotation and IQ phase imbalance simultaneously at both transmitter and receiver. This is shown in Figure 7-54.

Now we need to express (i, q) in terms of (I, Q) and of $(\theta_t, \theta_r, \varphi_{ct}, \varphi_{st}, \varphi_{cr}, \varphi_{cr})$. It is straightforward manipulations of trigonometric equations, and at receive, LPFs eliminate high frequency components.

$$i = I \cos\left(\theta_t + \theta_r + \varphi_{ct} + \varphi_{cr}\right) - Q \sin\left(\theta_t + \theta_r - \varphi_{st} + \varphi_{cr}\right)$$

$$q = I \sin\left(\theta_t + \theta_r + \varphi_{ct} - \varphi_{sr}\right) + Q \cos\left(\theta_t + \theta_r - \varphi_{st} - \varphi_{sr}\right)$$

We introduce φ_i, φ_d and $\widehat{\theta}$ as below

$$\varphi_i = \tfrac{1}{2}\left(\varphi_{ct} + \varphi_{st} + \varphi_{cr} + \varphi_{sr}\right)$$
$$\varphi_d = \tfrac{1}{2}\left(\varphi_{ct} + \varphi_{st} - \varphi_{cr} - \varphi_{sr}\right) \qquad \rightarrow$$
$$\widehat{\theta} = \theta_t + \theta_r + \tfrac{1}{2}\left(\varphi_{ct} - \varphi_{st}\right) + \tfrac{1}{2}\left(\varphi_{cr} - \varphi_{sr}\right)$$

$$\widehat{\theta} + \varphi_i = \theta_t + \theta_r + \varphi_{ct} + \varphi_{cr}$$
$$\widehat{\theta} - \varphi_d = \theta_t + \theta_r - \varphi_{st} + \varphi_{cr}$$
$$\widehat{\theta} + \varphi_d = \theta_t + \theta_r + \varphi_{ct} - \varphi_{sr}$$
$$\widehat{\theta} - \varphi_i = \theta_t + \theta_r - \varphi_{st} - \varphi_{sr}$$

In terms of new variables of φ_i, φ_d and $\widehat{\theta}$, we express (i, q) as

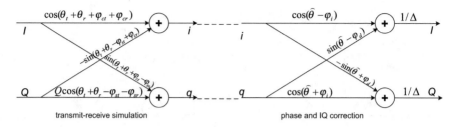

transmit-receive simulation   phase and IQ correction

Figure 7-55: Carrier phase and IQ phase imbalance can be recovered by if φ_i, φ_d and $\widehat{\theta}$, and thus transmit symbol (I, Q) can be recovered if if φ_i, φ_d and $\widehat{\theta}$ are known

$$i = I\cos\left(\widehat{\theta} + \varphi_i\right) - q\sin\left(\widehat{\theta} - \varphi_d\right) \tag{7-11}$$

$$q = I\sin\left(\widehat{\theta} + \varphi_d\right) + Q\cos\left(\widehat{\theta} - \varphi_i\right) \tag{7-12}$$

We obtain (I, Q), from the above two equations, in terms of (i, q) and of φ_i, φ_d and $\widehat{\theta}$.

$$I\Delta = +i\cos\left(\widehat{\theta} - \varphi_i\right) + q\sin\left(\widehat{\theta} - \varphi_d\right)$$

$$Q\Delta = -i\sin\left(\widehat{\theta} + \varphi_d\right) + q\cos\left(\widehat{\theta} + \varphi_i\right)$$

$$\Delta = \cos\left(\widehat{\theta} - \varphi_i\right)\cos\left(\widehat{\theta} + \varphi_i\right) + \sin\left(\widehat{\theta} - \varphi_d\right)\sin\left(\widehat{\theta} + \varphi_d\right)$$

At the receiver we can summarize the result so far in Figure 7-55; transmit symbol (I, Q) can be recovered from (i, q), if φ_i, φ_d and $\widehat{\theta}$ are known (or estimated correctly).

Exercise 7-42: Confirm the correction of the phase and IQ phase imbalance in Figure 7-55, i.e., obtain (I, Q), in terms of (i, q) and of φ_i, φ_d and $\widehat{\theta}$.

7.4.2 $\widehat{\theta}$, φ_i , and φ_d Measurements

From (7-11) and (7-12), we can find $\widehat{\theta}$, φ_i , and φ_d as

$$\tan\left(\widehat{\theta}\right) = \frac{qI - iQ}{iI + qQ} \quad \text{when} \quad \varphi_i = 0 \text{ and } \varphi_d = 0 \tag{7-13}$$

$$\cos\left(\varphi_i\right) = \frac{iI - qQ}{I^2 - Q^2} \quad \text{when} \quad \widehat{\theta} = 0 \tag{7-14}$$

$$\sin(\varphi_d) = \frac{qI - iQ}{I^2 - Q^2} \quad \text{when } \widehat{\theta} = 0 \tag{7-15}$$

Exercise 7-43: Obtain (7-13)–(7-15), the measurements of $\widehat{\theta}$, φ_i, and φ_d, in terms of (i, q) received samples, and (I, Q) transmitted symbols. Hint: in order to obtain $\widehat{\theta}$ apply the condition of $\varphi_i = 0$ and $\varphi_d = 0$ first to (7-11) and (7-12), and similarly for φ_i, and φ_d.

Keep in mind that a total phase rotation of $\widehat{\theta}$ is given by $\widehat{\theta} = \theta_t + \frac{1}{2} \times (\varphi_{ct} - \varphi_{st}) + \theta_r + \frac{1}{2}(\varphi_{cr} - \varphi_{sr})$ where $\frac{1}{2}(\varphi_{ct} - \varphi_{st})$ is due to IQ imbalance of transmit side, and $\frac{1}{2}(\varphi_{cr} - \varphi_{sr})$ is due to that of receive side.

In order to use (7-13)–(7-15), we did numerical experiments with 16-QAM constellation shown in Figure 7-56. With $\theta_t + \theta_r = -5°$, $\varphi_{ct} = -10°$, $\varphi_{st} = -10°$, $\varphi_{cr} = -2°$, $\varphi_{sr} = 8°$, we obtain $\widehat{\theta} = -10°$, $\varphi_i = -7°$, $\varphi_d = -13°$. We try to numerically compute $\widehat{\theta}$, φ_i, and φ_d using (7-13)–(7-15). The result is summarized in Figure 7-56 where the black numbers on diagonal are the measurement of $\widehat{\theta}$, red on off-diagonal for φ_d, and blue on off-diagonal for φ_i.

For φ_i, and φ_d, only off-diagonal points are used, which is of necessity due to the fact that $I^2 - Q^2 = 0$ with diagonal points.

For the measurement of $\widehat{\theta}$, only diagonal constellation points are used. Though all constellation points may be used but the estimation using diagonal points are more accurate when $\varphi_i \neq 0$ and $\varphi_d \neq 0$.

Keep in mind that numerically computation of $\widehat{\theta}$ is done when $\varphi_i \neq 0$ and $\varphi_d \neq 0$, i.e., ignoring the condition of (7-13). It turns out that averaging two diagonal points we can measure it accurately as shown in Figure 7-56; $(-9.0213) + (-11.0801) = -20.1014$. The average is -10.0507.

For the measurement of φ_d with $\widehat{\theta} \neq 0$, similarly averaging four off-diagonal points, we can measure it accurately as shown in Figure 7-56; $(-24.6553 - 1.4846 - 26.6733 + 0.3343)/4 = -13.1197$.

For the measurement of φ_i with $\widehat{\theta} \neq 0$, it turns out that we got a very poor estimation of it even with averaging four off-diagonal points; $(28.7148 + 22.4730i + 34.5010 + 28.9966i)/4 = 15.8040 + 12.8674i$ while its true value is $-7°$. Note that there are imaginary numbers in the estimation of φ_i. This means $|iI - qQ| > |I^2 - Q^2|$ in (7-14) which is not possible with $\cos(\varphi_i)$. The sign is wrong but it does not matter, which will be clear later.

7.4.3 2-Step Approach for the Estimation of $\widehat{\theta}$, φ_i, and φ_d

In order to solve the problem of measurement of φ_i we use the 2-step approach shown in Figure 7-57; first measure $\widehat{\theta}$, and then φ_i, and φ_d.

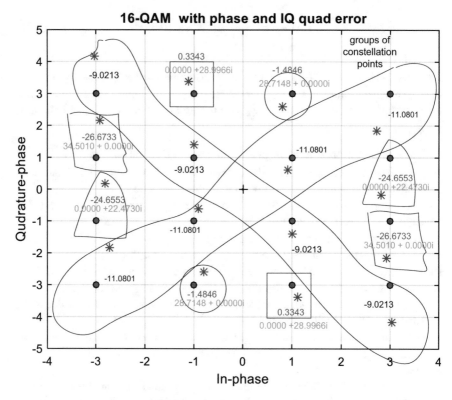

Figure 7-56: Measurements of $\widehat{\theta}$, φ_i, and φ_d.; the black numbers on diagonal are the measurement of $\widehat{\theta}$, red on off-diagonal for φ_d, and blue on off-diagonal for φ_i

We tried to apply the 2-step approach to the same IQ phase imbalance as in Figure 7-56, and the result is very good.

After the first step of $\widehat{\theta}$ estimation, its corrected constellation is shown in Figure 7-58 (RHS). With this corrected constellation, the condition of phase error being zero is fulfilled when we estimate φ_i, and φ_d, and thus accurate estimations are possible. The constellation after the correction of φ_i, and φ_d is shown in Figure 7-59.

7.4.4 Additional Practical Issues

For a decision-directed iteration for the estimation of $\widehat{\theta}$, φ_i, and φ_d we need to estimate (I, Q) at the receiver. In our presentation we assumed (I, Q) are known at the receive side which is equivalent to a pilot scheme; known symbols are sent.

In our presentation we ignore the noise in order to develop the algorithm of 2-step approach. In practice this must be simulated, which is beyond our current scope.

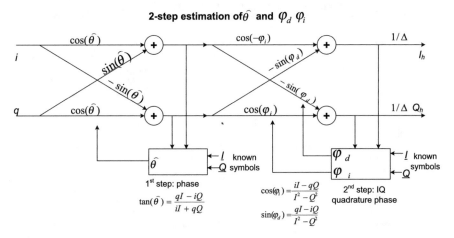

Figure 7-57: A 2-step approach of estimation of $\widehat{\theta}$, φ_i, and φ_d; first measure $\widehat{\theta}$ and second measure φ_i, and φ_d

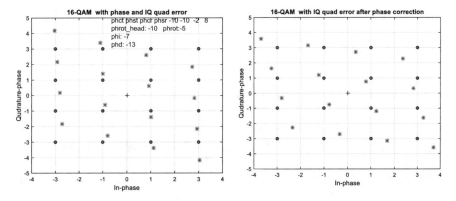

Figure 7-58: The first step of $\widehat{\theta}$ estimation with the same parameter as in Figure 7-56, and its result is on RHS

7.4.5 Summary of IQ Phase Imbalance Digital Correction

In this section we described how to represent, measure and correct quadrature phase imbalance in digital domain. This can be done by decision-directed or pilot method even though we did not pursue it further. In particular, the 2-step method is practically useful not only because of accuracy but it can be combined with a separate carrier recovery followed by IQ phase imbalance correction.

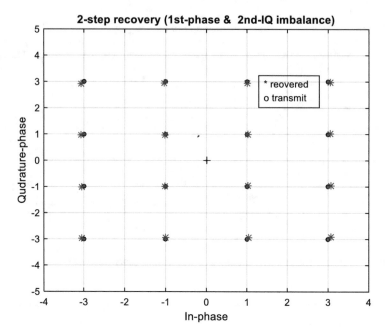

Figure 7-59: 2-step corrected constellation

7.5   References with Comments

For digital symbol timing synchronization

[1] F.M. Gardner, "A BPSK/QPSK Timing-error Detector for Sampled Receivers," IEEE Trans. Commun., vol. COM-34, pp.423-429, May 1986

[2] M. Oerder and H. Meyer, "Digital Filter and Squaring Timing Recovery," IEEE Trans. Commun., vol. COM-36, pp.605-612, May 1988

[3] O. Agazzi, C.-P.J. Tzeng, D.G. Messerschmitt, and D.A. Hodges, "Timing recovery in Digital Subscriber loops," IEEE Trans. Commun., vol. COM-33, pp.558-569, June 1985

[4] N.K. Jablon, " Joint Blind Equalization, Carrier Recovery, and Timing Recovery for High-Order QAM signal Constellations," IEEE Trans. Signal Processing , vol.40, No.6, pp.1383 - 1398, June 1992

[5] B. Lankl and G. Sebald, "Jitter-reduced digital timing recovery for multilevel PAM and QAM systems," Conference Record, IEEE International Conference on page(s): 804 - 810 vol.2, May 1993 Geneva, Switzerland 1993

[6] A. D'Andrea and M. Luise, "Optimization of Symbol Timing Recovery for QAM Data Demodulators," IEEE Trans. Commun., vol. COM-44, pp.399-406, March 1996

[7] F. Harris and M. Rice, "Multirate Digital Filters for Symbol Timing Sychronization in Software Defined Radios," IEEE Journal on Selected Areas in Commun., vol. 19 No.12, pp. 2346-2357 December 2001

[8] A.A. D'Amico, A.N. D'Andrea and R. Reggiannimi, " Efficient Non-Data-Adided Carrier and Clock Recovery for Satellite DVB at Very Low Signal-to-Noise Ratios," IEEE Journal on Selected Areas in Commun., vol. 19 No.12, pp. 2320-2330 December 2001

[9] K.H. Mueller and M. Muller, "Timing Recovery in all digital synchronous data receivers," IEEE Trans. Comm. , vol. COM-24, No.5, pp.516 –531, May 1976

Textbook chapters on synchronization

[10] J.G. Proakis and et. al, "Digital Communications" , pp. 319 (early-late), 5th edition, McGraw Hill 2008

[11] Heinrich Meyer et. al., "Digital Communication Receivers: synchronization, channel estimation and signal processing" pp.291, John Wiley & Sons Inc. 1998.

Early papers on synchronization

[12] L.E. Franks and J.P. Bubrouski, "Statistical properties of timing jitter in a PAM timing recovery system," IEEE Trans. Comm., vol. COM-22, No.7, pp. 913-920, July 1974.

[13] D.N. Godard, "Passband Timing Recovery in All digital Modem Receiver," IEEE Trans. Comm. , vol COM-26, No.5, pp.517 –523, May 1978

Appendix 7

A.1 Raised Cosine Pulse and its Pre-filtered RC Pulse

A raised cosine pulse is often used in digital data transmission for base band pulse shaping. We also consider its pre-filtered pulse, which is motivated by eliminating self-noise or pattern dependent noise. This becomes serious when the modulation order is high, using a band limited pulse, less than 1/T with say $\beta \ll 1$, in order to be of high bandwidth efficiency. The impulse response confined in time ,e.g. rectangular, may eliminate self-noise at the expense of using more bandwidth.

The pre-filtering is defined as, $h_{pf}(t) = k\, h_{RC}(t) \otimes h_{RC}(t)\, \cos \frac{2\pi t}{T}$ where k is a scaling factor and $\otimes$ denotes convolution in time domain. In other words, in frequency domain, $H_{pf}(f) = \frac{k}{2} H_{RC}(f)\left[H_{RC}\left(f - \frac{1}{T}\right) + H_{RC}\left(f + \frac{1}{T}\right)\right]$.

Raised cosine pulse and its pre-filtered one

	Raised Cosine Pulse	Pre-filtered Raised cosine pulse	
Time domain	$h_{RC}(t) = \dfrac{\sin\left(\frac{\pi t}{T}\right)}{\frac{\pi t}{T}} \dfrac{\cos\left(\frac{\beta \pi t}{T}\right)}{1-4\left(\frac{\beta t}{T}\right)^2}$	$h_{pf}(t) = \dfrac{\sin\left(\frac{\beta \pi t}{T}\right)}{\frac{\pi t}{T}} \dfrac{\cos\left(\frac{\pi t}{T}\right)}{1-\left(\frac{t}{T}\right)^2}$	
Frequency domain	$H_{RC}(f) = T$ $= \dfrac{T}{2}\left\{ 1 + \cos \dfrac{\pi T}{\beta}\left(f - \dfrac{1-\beta}{2T}\right)\right\}$ $= \dfrac{T}{2}\left\{ 1 + \cos \dfrac{\pi T}{\beta}\left(f + \dfrac{1-\beta}{2T}\right)\right\}$	$H_{pf}(f) = 0$ $= T\cos^2 \dfrac{\pi T}{\beta}\left(f - \dfrac{1}{2T}\right)$ $= T\cos^2 \dfrac{\pi T}{\beta}\left(f + \dfrac{1}{2T}\right)$	$\|f\| < \dfrac{1-\beta}{2T}$ $\dfrac{1-\beta}{2T} < f < \dfrac{1+\beta}{2T}$ $-\dfrac{1+\beta}{2T} < f < -\dfrac{1-\beta}{2T}$

Exercise 7-44: Raised cosine pulse and its pre-filtered RC pulse with $\beta = 0.41$ are shown below.

Exercise 7-45: Sketch the frequency response of RC and pre-filtered RC. Hint: use the argument to define pre-filtering.

Note that the pre-filtered pulse is not used for transmission but used for only timing recovery. It can be approximately done by a high pass filter and for digital implementation see Reference [5] as an example.

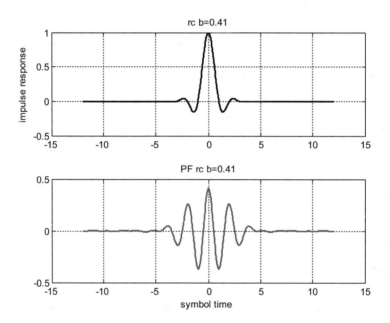

A.2 Poisson Sum Formula for a Correlated Signal

In this appendix, we like to show the identity, used extensively in the text

$$\sum_{n=-\infty}^{n=\infty} h(t - nT)h(t \mp \Delta T - nT) = \sum_{m=-\infty}^{m=\infty} \left[\frac{1}{T} \int_{v=-\infty}^{v=\infty} H\left(\frac{m}{T} - v\right) H(v) e^{\mp j2\pi\Delta Tv} dv \right] e^{j\frac{2\pi}{T}mt}$$

This identity is quickly seen by applying three identities used in Fourier Transform. Firstly multiplication in time domain can be expressed by convolution in frequency domain and vice versa, and secondly periodic extension of a signal in time domain, i.e., periodic signal, can be expressed as a sampling in frequency domain. Thirdly a delayed time domain signal transform can be expressed by multiplying a delay factor ($e^{\mp j2\pi f\Delta T}$ where ΔT is delay.) to the original transform.

Define Fourier transform pair as

$$u(t) = \int_{-\infty}^{\infty} U(f)e^{j2\pi ft}df \quad \Leftrightarrow \quad U(f) = \int_{-\infty}^{\infty} u(t)e^{-j2\pi ft}dt.$$

Multiplication   Convolution

$$u(t) = v(t)w(t) \quad \Leftrightarrow \quad U(f) = \int_{-\infty}^{\infty} V(f-v)W(v)dv$$

Periodic extension

$$\sum_{k=-\infty}^{k=\infty} u(t-kT) = \frac{1}{T} \sum_{m=-\infty}^{m=\infty} U\left(\frac{m}{T}\right)e^{j2\pi\frac{m}{T}t}$$

Delayed signal transform

$$u(t \mp \Delta T) = \int_{f=-\infty}^{f=\infty} \left[U(f)e^{\mp j2\pi f\Delta T}\right]e^{j2\pi ft}df$$

By choosing $v(t) = h(t)$, and $w(t) = h(t \mp \Delta T)$, and applying the above three identities, the result is obtained.

Exercise 7-46: Fill the additional details for the above derivation.

A.3 Review of Phase Locked Loops

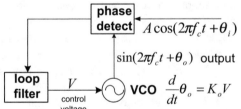

The phase locked loops (PLL) are extensively used in synchronization subsystems - carrier and timing - , and in frequency synthesizers, e.g., frequency variable oscillators, of RF systems. We cover its basics here, to be concrete and simple in describing, starting with one of frequency and phase synchronization of sine wave as shown on the left. The key components of PLL - VCO, phase detector, and loop filter -will be explained, and then the linear model of PLL will be derived below. A thorough understanding of PLL is important and useful as well.

VCO
The model of voltage controlled oscillator is shown in Figure 7-60. The characteristics of control voltage to output frequency is shown RHS of the figure; control voltage ranges from v1 to v2 and its corresponding output frequency. Its slope is given as a loop parameter K_o. In fact, it is used in Chapter 2 as part of FSK signal generation. See Figure 2-56 if you are curious.

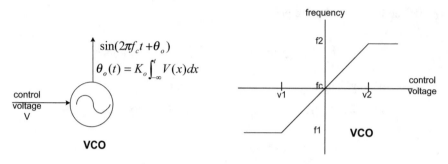

Figure 7-60: VCO and its input and output relationship

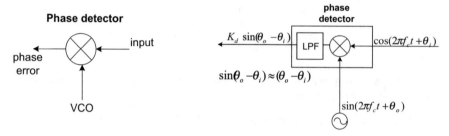

Figure 7-61: Phase detector symbol (LHS) and its implementation in case of sinusoids

Phase detector

In Figure 7-61, its symbol is on LHS and its actual implementation is on RHS when the input is sinusoids. We use the relationship of $\cos(2\pi f_c t + \theta_i) \ \sin(2\pi f_c t + \theta_o) = \frac{1}{2}\sin(\theta_o - \theta_i) + \frac{1}{2}\sin(4\pi f_c t + \theta_o + \theta_i)$, and the higher harmonic is low pass filtered. The gain of phase detector is denoted by K_d. Note that $\sin(\theta_o - \theta_i) \approx (\theta_o - \theta_i)$ when *the phase difference is small*.

Loop filter

For the second order loop we often use a filter with direct path and integration path as shown in Figure 7-62. When there is no integration path ($b = 0$) it becomes a first order loop. A frequency offset, if available, can be inserted as shown in dotted line.

Linear model of PLL is useful to see its behavior when it is in lock or tracking mode, i.e., $\sin(\theta_o - \theta_i) \approx (\theta_o - \theta_i)$ being small or the approximation works well. Its model is in Figure 7-63 and its transfer function $H(s) = \frac{\theta_o(s)}{\theta_i(s)}$ is shown as well. Note that VCO is modeled as an integrator, $K_o \frac{1}{s}$.

Exercise 7-47: Confirm the transfer function $H(s)$ using the flow graph representation in Figure 7-63. Hint: $\theta_o(s) = K_o K_d \, a\{\theta_i(s) - \theta_o(s)\}\frac{1}{s} + K_o K_d \, b\{\theta_i(s) - \theta_o(s)\}\frac{1}{s^2}$. And $H(s) = \frac{\theta_o(s)}{\theta_i(s)} = \frac{sK_o K_d a + K_o K_d b}{s^2 + sK_o K_d a + K_o K_d b}$.

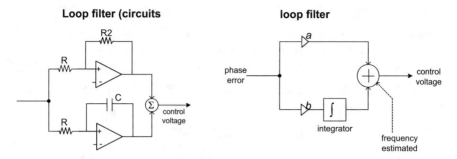

Figure 7-62: Loop filter in circuits (LHS) and its flow diagram (RHS); direct path with gain a and integration path with gain b. Dotted line indicates the addition of frequency offset possible.

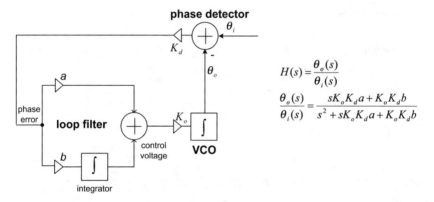

$$H(s) = \frac{\theta_o(s)}{\theta_i(s)}$$

$$\frac{\theta_o(s)}{\theta_i(s)} = \frac{sK_oK_da + K_oK_db}{s^2 + sK_oK_da + K_oK_db}$$

Figure 7-63: Linearized model of PLL for the behavior in lock mode (tracking mode) and its transfer function is shown as well

Exercise 7-48: Show by letting, $K_oK_d\,a = A$, $K_oK_d\,b = B$, $H(s) = \frac{\theta_o(s)}{\theta_i(s)} = \frac{sA+B}{s^2+sA+B}$. When $b = 0$, $H(s) = \frac{A}{s+A}$

In the literature $H(s)$ is often expressed in terms of ζ and ω_n.

Exercise 7-49: The transfer function is often expressed as $H(s) = \frac{s2\zeta\omega_n+\omega_n^2}{s^2+s2\zeta\omega_n+\omega_n^2}$. Show $\omega_n = \sqrt{B}$, $\zeta = \frac{A}{2\sqrt{B}}$, called natural (radian) frequency and damping factor respectively. Answer: $B = \omega_n^2$, $A = 2\zeta\omega_n$ then $\omega_n = \sqrt{B}$, $\zeta = \frac{A}{2\sqrt{B}}$.

Exercise 7-50: When the damping factor $\zeta = \frac{1}{\sqrt{2}}$, express B in terms of A. Answer: $B = \frac{1}{2}A^2$. Try other cases of $\zeta = \frac{1}{2}$ and $\zeta = 1$; $B = A^2$ and $B = \frac{1}{4}A^2$.

Now we try to understand the behavior of Figure 7-63, linearized model of PLL through examples and exercises. We state two theorems from Laplace transform.

The <u>final value theorem</u> of Laplace transform states that $\lim_{t \to \infty} y(t) = \lim_{s \to 0} sY(s)$.
Hint: Search Internet, like *Wikipedia*, to get the proof and the conditions under which it is valid. At least understand it so that one can use it.

The <u>initial value theorem</u> of Laplace transform states that $\lim_{t \to 0} y(t) = \lim_{s \to \infty} sY(s)$.

We define the phase error as $\theta_e(s) = \theta_i(s) - \theta_o(s) = \frac{s^2}{s^2 + As + B}\theta_i(s)$ Laplace transform domain. Using the final value theorem we obtain following;

Input	Error; $\theta_e(s) = \theta_i(s) - \theta_o(s)$
phase jump θ_i $\Delta\theta$ ⎯⎯⎯⎯ 　　　　time	$\lim_{t\to\infty}\theta_e(t) = \lim_{s\to 0} s\left(\frac{s^2}{s^2+As+B}\theta_i(s)\right) = 0$; i.e., tracks out the input phase jump where $\theta_i(s) = \frac{\Delta\theta}{s}$ (step).
frequency jump i.e., phase ramp θ_i ／ $\Delta\omega t$ ／ 　　　time	$\lim_{t\to\infty}\theta_e(t) = \lim_{s\to 0} s\left(\frac{s^2}{s^2+As+B}\theta_i(s)\right) = 0$; i.e., tracks out the input frequency jump where $\theta_i(s) = \frac{\Delta\omega}{s^2}$.

Exercise 7-51: For the first order loop (i.e., $b = 0$), $\lim_{t\to\infty}\theta_e(t) = \lim_{s\to 0} s\left(\frac{A}{s+A}\theta_i(s)\right) = 0$ when there is a phase jump, i.e., $\theta_i(s) = \frac{\Delta\theta}{s}$. On the other hand if there is a frequency jump (i.e., phase ramp), i.e., $\theta_i(s) = \frac{\Delta\omega}{s^2}$ then $\lim_{t\to\infty}\theta_e(t) = \frac{1}{A}$. This is the reason the second order loop is often used in practice.

In the above we have shown that a 'small' input jump of both phase and frequency can be tracked out when the loop is in lock (a linear model assumes it). Next two exercises extend it further to see the time behavior of a linearized model of PLL. This may be skipped in the first reading.

Exercise 7-52: In Figure 7-63, there is a 'small' input phase jump, K_i, at time 0; the jump is 'small' enough that the linear model is valid. Find the initial output phase and final phase; $\theta_o(t = 0^+)$ and $\theta_o(t = +\infty)$. And find the initial VCO frequency and final VCO frequency; $\frac{d}{dt}\theta_o(t)\big|_{t=0^+}$ and $\frac{d}{dt}\theta_o(t)\big|_{t=+\infty}$.

Answer: In terms of s, the output phase is represented by

$$\theta_o(s) = \frac{sK_oK_da + K_oK_db}{s^2 + sK_oK_da + K_oK_db}\theta_i(s) = \frac{sK_oK_da + K_oK_db}{s^2 + sK_oK_da + K_oK_db}K_i\frac{1}{s}$$

VCO frequency $\frac{d}{dt}\theta_o(t) \leftarrow$ Laplace Transform $\rightarrow \frac{sK_oK_da + K_oK_db}{s^2 + sK_oK_da + K_oK_db}K_i$

We use that the Laplace transform of differentiation is $\frac{dx(t)}{dt} \leftarrow \rightarrow s\,X(s)$. Using the initial and final value theorems in Laplace transform, we obtain;

$$\theta_o(t = 0^+) = \lim_{s \to \infty} s\theta_o(s) = 0 \text{ and } \theta_o(t \to \infty) = \lim_{s \to 0} s\theta_0(s) = K_i$$

$$\frac{d}{dt}\theta_o(t = 0^+) = \lim_{s \to \infty} s^2\theta_o(s) = K_oK_daK_i, \text{ and}$$

$$\frac{d}{dt}\theta_o(t \to \infty) = \lim_{s \to 0} s^2\theta_o(s) = 0$$

As part of intuitive understanding PLL, in particular a linear model of it, we try to see the above result without resorting to the final and initial value theorem of Laplace transform. Under lock condition of PLL where $\theta_o(t) \approx 0$, if $\theta_i(t) = K_iu(t)$ is applied where $u(t)$ is a unit step what will happen in a linear model of PLL shown in Figure 7-63?

$\theta_o(t = 0^+)$ cannot change instantaneously due to the presence of integrator, and thus must be the previous value, i.e., zero. The input to VCO will change instantaneously, through the direct path, with the gain of K_daK_i and thus VCO frequency will be $K_o$times gain (K_daK_i). When PLL is settled, the phase error must be zero $\theta_e(t) = \theta_i(t) - \theta_o(t)$, and thus $\theta_o(t \to \infty) = \theta_i(t \to \infty) = K_i$.

Exercise 7-53: In the previous exercise we looked for the initial and final output phase and VCO frequency. In this exercise we try to obtain both phase change in time and VCO frequency change in time.
Answer:

$$\frac{sK_oK_da + K_oK_db}{s^2 + sK_oK_da + K_oK_db}K_i\frac{1}{s} = \frac{sA + B}{s^2 + sA + B}K_i\frac{1}{s} \leftarrow \text{Laplace transform}$$

$$\rightarrow \text{output phase in time}$$

$$\frac{sK_oK_da + K_oK_db}{s^2 + sK_oK_da + K_oK_db}K_i = \frac{sA + B}{s^2 + sA + B}K_i \leftarrow \text{Laplace transform}$$

$$\rightarrow \text{VCO frequency in time}$$

For actual inverse Laplace transform of the above, we need to classify them into three categories – over damped $(\zeta > 1.0)$, critically damped $(\zeta = 1.0)$, and under damped $(\zeta < 1.0)$ cases with $A = 2\zeta\omega_nB = \omega_n^2$. We leave it as a small project below.

Exercise 7-54*(project): Find the impulse response for three categories above depending on damping factor, and then the unit step responses. Hint: look for a

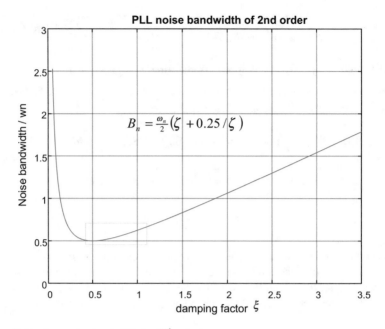

PLL noise bandwidth of 2nd order

$$B_n = \tfrac{\omega_n}{2}\left(\zeta + 0.25/\zeta\right)$$

damping factor ζ

Figure 7-64:   loop noise bandwidth for 2nd order loop

good textbook on signals and systems such as Oppenheim (Oppenheim, Alan V., et. al. ,"Systems and Signals", 2nd ed 1996, Prentice Hall) or on Laplace transform.

Noise bandwidth of the second order loop

Noise bandwidth of PLL is defined by $B_n = \frac{1}{\max\limits_{f} |H(f)|^2} \int_0^\infty |H(f)|^2 df$.

When it is applied to $H(s)$ of the 2nd order loop where $s = 2\pi f$ we obtain $B_n = \frac{\omega_n}{2}\left(\frac{1}{4\zeta}+\zeta\right) = \frac{1}{4}\left(A+\frac{B}{A}\right)..$

Exercise 7-55* (project): Show the above statement, i.e., noise bandwidth is expressed as $B_n = \frac{\omega_n}{2}\left(\frac{1}{4\zeta}+\zeta\right) = \frac{1}{4}\left(A+\frac{B}{A}\right)$.

Hint: Note that $\max\limits_{f} |H(f)|^2 = 1.0$. $2B_n = \int_{-\infty}^{+\infty}|H(f)|^2 df = \int_{-\infty}^{+\infty}|h(t)|^2 dt$.

Using the above result we show the minimum, $B_n = \frac{\omega_n}{2}$, is achieved when $\zeta = 0.5$. This is plotted in Figure 7-64.

A.4 FIR Interpolation Filter Design and Coefficient Computation

In order to implement digital timing recovery, a digital interpolator filter is necessary. An example of interpolation (resampling) filter is shown in Figure 7-38. In this appendix we discuss how to design such interpolation filter impulse response $h(t)$. In particular our focus is on a finite duration $h(t)$ with specific properties; $h(n\Delta t) = \delta(n)$,

$n = 0, \pm 1, \pm 2, \ldots \pm L$ a Nyquist pulse with a finite span ($Q = 2L$) and $h(t) = h(-t)$ even symmetry around zero.

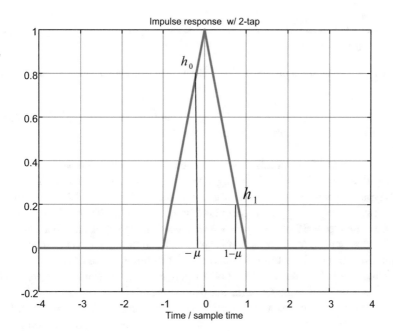

An example of such impulse response with $L = 1$, is shown on the left, symmetrical around zero and triangular. The coefficients of its sampled impulse response, with delay $(-\mu)$, h_o and h_1, are shown as well.

A topic of interpolation in general is well covered in the literature of numerical analysis. Reference [A4-1] may be useful if interpolation is new to a reader since it is written in digital filtering point of view. However, our need of it is specific to symbol timing recovery signal processing, where typically the span of impulse response may be limited to $L < 5$.

An impulse response of a low pass filter with vestigially symmetrical transition from passband to stopband around 0.5*sampling frequency ($1/\Delta t$) is shown below where β is a transition bandwidth. It is a Nyquist pulse with even symmetry around zero (i.e., $h_{VLP}(t) = h_{VLP}(-t)$).

$$h_{VLP}(t) = \frac{\sin\left(\pi \frac{t}{\Delta t}\right)}{\pi \frac{t}{\Delta t}} w\left(\frac{t}{\Delta t}\beta\right).$$

A transition bandwidth β must be matched with underlying pulse shape used in generating a signal. The choice of $w\left(\frac{t}{\Delta t}\beta\right)$ is wide open but commonly a raised cosine case $w\left(\frac{t}{\Delta t}\beta\right) = \cos\left(\pi \frac{t}{\Delta t}\beta\right) / \left(1 - \left(2\beta \frac{t}{\Delta t}\right)^2\right)$. Another possible choice is

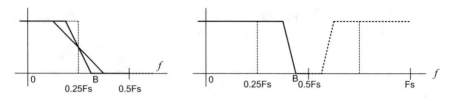

Figure 7-65: Signal spectrum sketch; LHS typically used shape in data transmission

given by $w\left(\frac{t}{\Delta t}\beta\right) = \sin\left(\frac{1}{2}\beta\pi\frac{t}{\Delta t}\right)/\left(\frac{1}{2}\beta\pi\frac{t}{\Delta t}\right) \otimes \sin\left(\frac{1}{2}\beta\pi\frac{t}{\Delta t}\right)/\left(\frac{1}{2}\beta\pi\frac{t}{\Delta t}\right)$ which is triangular in frequency domain. Note that when $\beta = 0.0$, it becomes an ideal lowpass
filter.

The time span of the impulse response in the above is not finite but it can be made
finite with a span of $Q = 2L$ by using a rectangular window of length $2L$.

A typical spectrum shape in data transmission is shown in LHS of Figure 7-65
where two samples per symbol period are used. In general the spectrum of a signal
may be similar to RHS of the figure where L needs to be larger than that LHS. Keep
in mind that our use of interpolation is specifically in timing recovery context where
a signal spectrum is bandlimited close to 0.25 of sampling rate (Fs).

As the cutoff frequency of interpolator gets smaller, the span of L gets smaller.
Thus β can be used to match with the bandwidth of signal; the smaller the cutoff
frequency the larger the excess bandwidth (β). For example when $B = 0.3Fs$, β may
be chosen to be 0.4. But in case of $B = 0.4Fs$, β may be 0.2. This discussion gives an
insight how the span of interpolation L is impacted by the bandwidth of a signal B in
Figure 7-65 and it can also be used to design an interpolator.

A useful numerical computation based essentially the idea of matching signal
spectrum is done in Reference [A4-2]. Actual computation may be done using interp.
m of MATLAB as **[h4,C]=interp(h0,r,L,alpha)** where C is filter coefficients, L is span, alpha is cutoff frequency relative to 0.5sampling frequency, and r is
interpolation rate.

An example design is shown in Figure 7-66. The match between the numerical
design and Nyquist pulse based design is good. $L = 3$ ($Q = 6$ point interpolation)
and cutoff frequency of a signal is assumed to be $0.3Fs$. A Nyquist pulse is given by
$\beta = 0.4$ with $w\left(\frac{t}{\Delta t}\beta\right) = \sin\left(\frac{1}{2}\beta\pi\frac{t}{\Delta t}\right)/\left(\frac{1}{2}\beta\pi\frac{t}{\Delta t}\right) \otimes \sin\left(\frac{1}{2}\beta\pi\frac{t}{\Delta t}\right)/\left(\frac{1}{2}\beta\pi\frac{t}{\Delta t}\right)$.

For interpolator filter design for timing recovery, two methods discussed above
may be good enough. Traditionally polynomial based approximation of fractional
delay is also used in timing recovery, in particular so called Farrow structure, in
order to simplify hardware implementation. Polynomials used there are Lagrange
type. For our purpose we chose only odd degree polynomials to make sure to be
linear phase (zero phases). This topic is well covered in the literature as well as in
[A4-1].

We list the base polynomial coefficients in Table 7-4 $Q = 2, 4, 6, 8$. Note than
polynomial based interpolator has only one parameter of Q, span of impulse
response.

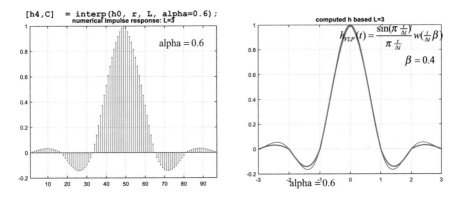

Figure 7-66: $L = 3$ $Q = 6$ interpolation $r = 16$. Numerical design (LHS) and Nyquist pulse based design (RHS). Match is good with predictable excess bandwidth 0.4

The computation of the coefficients is done by a matrix inversion method. For example, with $Q = 4$, let $l_0(t) = c_3 t^3 + c_2 t^2 + c_1 t + c_0$ and imposing four conditions; $l_0(0) = 1, l_0(-1) = l_0(1) = l_0(2) = 0$. Similarly $l_k(t)$ where $k = -1, 1, 2$ can be found. This is the definition of Lagrange base polynomials. A numerical example of Lagrange base polynomials is shown in Figure 7-67, where four base polynomials are shown on LHS. The corresponding impulse response, shown on RHS, is obtained from the base polynomials by evaluating the range of $0 \le t \le 1$ and shifting.

Denote the samples values as $y_k = y(t)|_{t = k}, k = -1, 0, 1, 2$ with $Q = 4$. An interpolated value at μ, $y(\mu)$, where $0 \le \mu \le 1$, is given by

$$y(\mu) = y_{-1} l_{-1}(\mu) + y_0 l_0(\mu) + y_1 l_1(\mu) + y_2 l_2(\mu).$$

In order to express $y(\mu)$ in terms of convolution with $h(t)$

$$y(\mu) = y_{-1} h_1(\mu) + y_0 h_0(\mu) + y_1 h_{-1}(\mu) + y_2 h_{-2}(\mu).$$

Thus we obtain $h(t)$ as shown in RHS of Figure 7-67.

Farrow structure implementation is not pursued here but can be done from the listed coefficients of Lagrange base polynomials.

[A4-1] R. W. Schafer and L. R. Rabiner, "A digital signal processing approach to interpolation," *Proc.* IEEE, vol. '61, pp. 692-702, June 1973.

[A4-2] Oetken, G., Thomas W. Parks, and H. W. Schüssler. "New results in the design of digital interpolators." IEEE® Transactions on Acoustics, Speech, and Signal Processing. Vol. ASSP-23, 1975, pp. 301–309.

Table 7-4:   Approximate envelope shape of h(t) and Lagrange base polynomial

L	Q FIR span	Approximate envelope shape of impulse response $h(t)$: non-zero span of Q sample period	Lagrange base polynomial $l_k(t)$ with Q such ones, $0 \le t \le 1$
1	2	L=1, Q=2	$l_0(t) = -t + 1$ $l_1(t) = +t$
2	4	L=2, Q=4	$l_{-1}(t) = -\frac{1}{6}t^3 + \frac{1}{2}t^2 - \frac{1}{3}t$ $l_0(t) = \frac{1}{2}t^3 - t^2 - \frac{1}{2}t + 1$ $l_1(t) = -\frac{1}{2}t^3 + \frac{1}{2}t^2 + t$ $l_2(t) = \frac{1}{6}t^3 - \frac{1}{6}t$
3	6	L=3, Q=6	$l_{-2}(t) = \frac{-1}{120}t^5 + \frac{1}{24}t^4 - \frac{1}{24}t^3 - \frac{1}{24}t^2 + \frac{1}{20}t$ $l_{-1}(t) = \frac{1}{24}t^5 - \frac{1}{6}t^4 - \frac{1}{24}t^3 + \frac{2}{3}t^2 - \frac{1}{2}t$ $l_0(t) = -\frac{1}{12}t^5 + \frac{1}{4}t^4 + \frac{5}{12}t^3 - \frac{5}{4}t^2 - \frac{1}{3}t + 1$ $\quad l_1(t) = \frac{1}{12}t^5 - \frac{1}{6}t^4 - \frac{7}{12}t^3 + \frac{2}{3}t^2 + t$ $l_2(t) = -\frac{1}{24}t^5 + \frac{1}{24}t^4 + \frac{7}{24}t^3 - \frac{1}{24}t^2 - \frac{1}{4}t$ $l_3(t) = \frac{1}{120}t^5 - \frac{1}{24}t^3 + \frac{1}{30}t$

4	
8	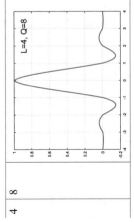 L=4, Q=8

$L_{-3}(t) = \left[\frac{-1}{5040}, \frac{1}{720}, \frac{-1}{720}, \frac{-1}{144}, \frac{-1}{90}, \frac{1}{180}, \frac{1}{105}, 0\right]$ $L_{-2}(t) = \left[\frac{1}{720}, \frac{-1}{120}, \frac{-1}{360}, \frac{1}{12}, \frac{-71}{720}, \frac{-3}{40}, \frac{1}{10}, 0\right]$

$L_{-1}(t) = \left[\frac{-1}{240}, \frac{1}{48}, \frac{3}{80}, \frac{-13}{48}, \frac{1}{15}, \frac{3}{4}, \frac{-3}{5}, 0\right]$

$l_0(t) = \left[\frac{1}{144}, \frac{-1}{36}, \frac{-7}{72}, \frac{7}{18}, \frac{49}{144}, \frac{-49}{36}, \frac{-1}{4}, 1\right]$

$l_1(t) = \left[\frac{-1}{144}, \frac{1}{48}, \frac{-13}{48}, \frac{-11}{18}, \frac{3}{4}, 1, 0\right]$

$l_2(t) = \left[\frac{1}{240}, \frac{-1}{120}, \frac{-3}{40}, \frac{89}{240}, \frac{-3}{40}, \frac{-3}{10}, 0\right]$

$l_3(t) = \left[\frac{-1}{720}, \frac{1}{720}, \frac{17}{720}, \frac{-1}{144}, \frac{-4}{45}, \frac{1}{180}, \frac{1}{15}, 0\right]$

$l_4(t) = \left[\frac{1}{5040}, 0, \frac{-1}{360}, 0, \frac{7}{720}, 0, \frac{-1}{140}, 0\right]$

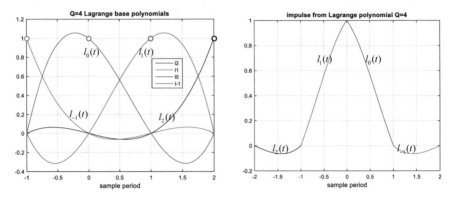

Figure 7-67: Lagrange base polynomial and its impulse response

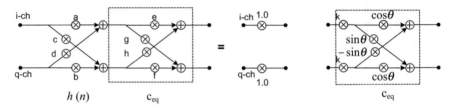

Fig. P7-1: Complex baseband representation of $h(n)$ and C_{eq} and combined result

7.6 Problems

P7-1: When a sampled channel response is given by $h(n) = (0.7244 - j*0.1941)$, i.e., a 1-tap. We are interested in removing cross-coupling between i-ch and q-ch as shown in Fig. P7-1.

(1) Find a complex baseband representation of $h(n)$; a, b, c, and d.
(2) Find a complex baseband representation of C_{eq}; e, f, g, and h.
(3) Find a complex baseband representation of C_{eq} with gain (k) and angle (θ). Carrier recovery is to find angle (θ) assuming gain (k) is found. Thus a carrier recovery is considered as a single tap equalizer.

P7-2: Referring to Figure 7-7, IDFT size $N = 64$, CP $v = 16$, and thus 80 samples per OFDM symbol are transmitted serially with the sampling rate $(N+v)/T = 10$ MHz.

(1) What is OFDM symbol period (T) in sec?
(2) What is CP period (v) in sec?
(3) What is subcarrier frequency spacing (10 MHz/N) in Hz?
(4) With sub-carrier numbering $k = 0, 1,..., 63$, '0' corresponds to DC. What is the highest positive frequency k number?

P7-3: Referring to Figure 7-8, how many OFDM symbols in PLCP preamble? How many OFDM symbols in SIGNAL?

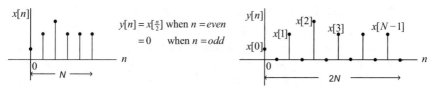

Fig. P7-2:   Form y[n] from x[n] by inserting one zero between samples

P7-4: In Fig. P7-2, $y[n]$ is generated from $x[n]$ inserting one zero between samples. This insertion of zeros can be extended to adding two zeros between samples and three zeros between samples, and so on.

(1) For a given DFT of $x[n]$, $X[k]$, show that $Y[k] = X[k]$ where $k = 0, 1, 2, \ldots$, $2N-1$, i.e., $Y[k]$ is the same as $X[k]$ repeated twice. The length of $y[n]$ becomes $2N$.

(2) When two zeros are inserted between samples, show that $Y[k]$ is the same as $X[k]$ repeated three times. The length of $y[n]$ becomes $3N$.

(3) When $L-1$ zeros are inserted between samples, show that $Y[k]$ is the same as $X[k]$ repeated L times. The length of $y[n]$ becomes LN.

P7-5: $Y[k]$ is generated from $X[k]$ inserting one zero between frequency samples. This insertion of zeros can be extended to adding two zeros between frequency samples and three zeros between frequency samples, and so on. Express $y[n]$ in terms of $x[n]$. Hint: See P7-4 and this is the dual case of time and frequency index change. STS in Figure 7-10 uses the case with three zeros between frequency samples.

P7-6: A short term training sequence, with rotation, is given by

$$\text{sts}[n] = \{0, -1, -1, 1, 1, 1, 1, 0, 0, 0, 1, -1, 1, -1, -1, 1\}(1+j).$$

80-sample time domain STS signal can be found by taking 16 sample-IDFT ($N = 16$) and repeat 5 times with a proper scaling using the result of P7-5. Show the scaling factor is $2\sqrt{\frac{13}{6}}$ rather than $8\sqrt{\frac{13}{6}}$ when IDFT size is 64 then adding CP.

P7-7: A matched filter, $g[n]$, to a sequence $h[n]$ is defined as , * being complex conjugation,

$$g[n] = h * [-n].$$

(1) Find a matched filter when $h[n] = \{1, 1, -1, 0, 0, 1, -1, -1\}'$.

(2) $h[n]$ is rotated by $(1 + j)$, i.e., $h[n](1 + j)$, and find the corresponding matched filter.

P7-8: Channel impairments of carrier frequency offset, carrier phase offset, and fading channel are shown in Figure 7-11. In particular, frequency offset and phase offset is modeled as a multiplication factor to a signal in discrete-time.

(1) When frequency offset is 1.3 relative to subcarrier spacing of OFDM, find the exponent (Δf) of multiplication factor as in $e^{\,j2\pi\frac{\Delta f}{N}n}$.

(2) A digital filter is used to model fading channel (frequency selective fading) ; its impulse response is $[1, -0.05, +0.8, -0.2]$. Find its frequency response such that the power gain through the filter is unity.

P7-9: Carrier frequency offset can be measured by comparing phase differences at the peak output of STS matched filter since $\Delta\varphi = 2\pi\Delta f(t_2 - t_1)$[rad], i.e., phase difference is frequency offset times time difference. Time difference $(t_2 - t_1)$ can be measured by a number of samples/f_s. This is shown in (7-1). How many such measurements are available at the output of STS matched filter with 10 PDI?

P7-10: A carrier frequency offset measured in Hz can be expressed by a number of subcarrier spacing. Confirm that Δf_k in (7-1) when $N = 64$, samples $= 16$. This measurement can be used to compensate the frequency offset by multiplying a factor to receiving samples. Show the factor is $e^{-j2\pi\cdot\Delta f_k/N\cdot n}$.

P7-11: A simple example of DS spread spectrum is given in Figure 7-21.

(1) When 4 bit state is $[1\ 1\ 1\ 1]$, show $s[n] = [0\ 1\ 0\ 1\ 1\ 0\ 0\ 1\ 0\ 0\ 0\ 1\ 1\ 1\ 1]$.

(2) When $d[m] =0$, then $[0\ 1\ 0\ 1\ 1\ 0\ 0\ 1\ 0\ 0\ 0\ 1\ 1\ 1\ 1]$ is transmitted. When $d[m]$ $=1$, what is the transmitted pattern?

(3) This DS system can be interpreted as a repetition code. What are the code words?

P7-12: A received signal is represented by $z(t) = \sum_{k=-\infty}^{+\infty} a_k h(t - kT) + n(t)$ where $\{a_k\}$ is a sequence of digital symbols to be transmitted and $h(t)$ is an end to end impulse response.

(1) When a_k is uncorrelated, i.e., $E\{a_k a_l\} = 0$, when $k \neq l$, $E\{a_k a_l\} = \sigma_s^2$ when $k = l$, ignoring noise, show $E\left\{|z(t)|^2\right\} = \sigma_s^2 \sum_{k=-\infty}^{+\infty} |h(t - kT)|^2$

(2) Show that $\sum_{k=-\infty}^{+\infty} |h(t - kT)|^2$ is periodic with the period of T.

(3) From   (2)   its   Fourier   series   can   be expressed $\sum_{k=-\infty}^{+\infty} |h(t - kT)|^2 = \sum_{m=-\infty}^{+\infty} C_m e^{\,j2\pi\frac{m}{T}t}$.

(4) Express C_m in terms of $H(f)$. Hint: use (7-6) and Table 7-3.

P7-13: With BPSK transmit pulse (filter) is rectangular as shown in the figure below. Receive pulse (filter) is the same as the transmit one as a matched filter.

(1) Show the end to end pulse is triangular and draw it.

(2) Transmitted 12 symbols are given by
$[1\ 1\ 1\ -1\ 1\ -1\ -1\ 1\ -1\ -1\ -1\ 1\ 1]$. Draw received waveforms $z(t)$ for the 12 symbols.

(3) Draw $\frac{d}{dt}z(t)$ and the output of timing detector of $z(t)\frac{dz(t)}{dt}$.

(4) Draw the output of timing detector of $z^2(t)$.

P7-14: In P7-13, BPSK $z(t)$ is defined. Denote $z_k = z(kT)$, $z_{k-\frac{1}{2}} = z\left(kT - \frac{1}{2}T\right)$ with $k = 0, 1, 2, ..11$. Gardner PD is defined as $z_{k-\frac{1}{2}}(z_k - z_{k-1})$, i.e., samples of mid-point between two data points are multiplied. This PD is listed in Table 7-3. Apply this PD to the received wave form with transmitted 12 symbols are given by $[1\ 1\ 1\ -1\ 1\ -1\ -1\ 1\ -1\ -1\ -1\ 1\ 1\ ]$.

(1) Obtain PD numerical values when sampling phase offset is zero.
(2) Obtain PD numerical values when sampling phase offset is +0.2 (late).
(3) Obtain PD numerical values when sampling phase offset is −0.2(early).

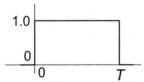

P7-15: One sample per symbol timing PD is described in Reference [9], called Mueller Muller (MM) PD. A sample at kT is denoted as y_k, and given by

$$y_k = \sum_{m=-\infty}^{+\infty} A_m h(kT + \tau - mT)$$

where A_m transmit symbols, $h(t)$ is an end to end impulse response, and τ is a sampling phase at kT. Then a phase measurement at kT, PD_k, is given by

$$PD_k = y_k A_{k-1} - y_{k-1} A_k.$$

(1) Show that $E(PD_k)$, an average PD_k, called 'S-curve', is given by

$$E(PD_k) = E\left(|A_k|^2\right)\{h(\tau + T) - h(\tau - T)\}..$$

(2) Obtain $E(PD_k)$ if $h(t)$ is a triangular pulse described in P7-13.
(3) Obtain numerical values of PD_k Transmitted 12 symbols are given by $[1\ 1\ 1\ -1\ 1\ -1\ -1\ 1\ -1\ -1\ -1\ 1\ 1\ ]$ in P7-13 if τ is 0, 0.1, 0.25, 0.375, 0.5 assuming $E(|A_k|^2)=1.0$.

P7-16: A slight variation of MMPD in P7-15, with defining $e_k = y_k - A_k$, is given by

$$PD_k = y_k A_{k-1} - y_{k-1} A_k = e_k A_{k-1} - e_{k-1} A_k.$$

(1) Show it is correct.

(2) Sometimes in order to eliminate multiplications, a quantized version $qPD_k = \text{sgn}(e_k)\,\text{sgn}(A_{k-1}) - \text{sgn}(e_{k-1})\,\text{sgn}(A_k)$ is used where sgn $(x) = 1, 0, -1$ when $x > 0$, $x = 0$, and $x < 0$ respectively. What are the possible values of qPD_k?

(3) Obtain $E(qPD_k)$ using the result of P7-15.

P7-17: In Figure 7-34, carrier frequency offset estimation method is described using $E\{z(t)z^*(t \mp \Delta T)\} = e^{-j2\pi f_e(\mp\Delta T)}\{|m(t)|^2 E(|a_k|^2)S(\mp\Delta T)\}$. One key assumption is that $\{|m(t)|^2 E(|a_k|^2)S(\mp\Delta T)\}$ is real, not complex. It is a practical issue. Discuss the conditions to make { } is real.

P7-18: In analog PLL, VCO is often used and its input (control voltage) and output, (frequency of sinusoidal wave or rectangular wave) characteristic is shown below.

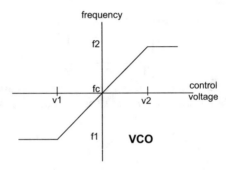

When $v_1 = -1.0$ volt, $v_2 = +1.0$ volt, and $f_1 = 9.9$ MHz, $f_2 = 10.1$ MHz in the figure.

(1) What is the output frequency when $v_c = -0.5$ volt?
(2) What is the output frequency when $v_c = +0.5$ volt?
(3) If a symbol rate is 4 MHz and control voltage is zero, assuming a square wave of peak voltage is 0.5 volt, sketch output waveform for 3 symbol period.

P7-19: In a linear model of PLL, the phase error of phase detector is proportional to the input phase and modeled as $\theta_e = k_d(\theta_i - \theta_o)$. Sketch its input – output characteristic for the timing phase range $-0.5T \le \theta_i \le +0.5T$.

P7-20: A practical phase detector is often a mixer or a multiplier as shown in Fig. P7-3. Show its linear model (RHS) assuming $\sin(\theta_i - \theta_o) \approx (\theta_i - \theta_o)$, i.e., in lock.

P7-21: An analog loop filter in PLL is shown in Fig. P7-4 (LHS).

A digital loop filter is obtained from analog counterpart by replacing an analog integrator by a digital one. Show that they are equivalent.

P7-22: A digital linear PLL model for timing recovery may be obtained from its analog counterpart. It is shown in Fig. P7-5.

(1) Confirm that a linear model for digital PLL is obtained by $s \to (1 - z^{-1})$

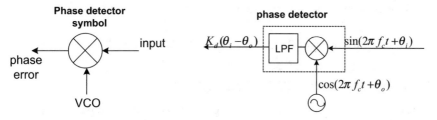

Fig. P7-3: PD symbol and sinusoidal phase detector

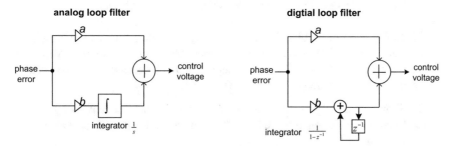

Fig. P7-4: Analog loop filter and digital loop filter

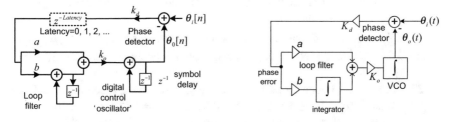

Fig. P7-5: A linear model for digital loop (LHS) from its analog counter part

(2) Show that $H(s) = \frac{\theta_o(s)}{\theta_i(s)} = \frac{sK_oK_da + K_oK_db}{s^2 + sK_oK_da + K_oK_db}$ becomes, with latency $=$
0, $H(z^{-1}) = \frac{\theta_o(z^{-1})}{\theta_i(z^{-1})} = \frac{\left(\widehat{k_p} + \widehat{k_i}\right) - \widehat{k_p}z^{-1}}{\left(\widehat{k_p} + \widehat{k_i} + 1\right) - \left(\widehat{k_p} + 2\right)z^{-1} + z^{-2}}$ where $\widehat{k_p} = k_ok_da$, $\widehat{k_i} = k_ok_db$.

(3) Find $H(z^{-1})$ when $b=0$.

(4) Find $H(z^{-1})$ when Latency $=1, 2, 3\ldots D$. Note that this latency may come
from digital design implementation of pipelining and parallel processing.

P7-23: Using the result of P7-22, when $\widehat{k_p} = 0.25$, $\widehat{k_i} = 5/64$ (0.0781), $D = 0$ show
that $H(z^{-1}) = \frac{0.2471 - 0.1882z^{-1}}{1.000 - 1.694z^{-1} + 0.7529z^{-2}}$.

P7-24: For $H(z^{-1})$ given in P7-23, obtain and plot

(1) An impulse response by evaluating the output when the input is $\delta[n]$.

(2) A step response by evaluating the output when the input is $u[n]$.

(3) A frequency response by letting $z = e^{j2\pi f}$ and evaluating $H(e^{-j2\pi f})$

P7-25: Using the result of P7-22, when $\hat{k}_p=0.25$, $\hat{k}_i=5/64$ (0.0781), $D=3$ show that

$$H(z^{-1}) = \frac{0.3281z^{-3}-0.2500z^{-4}}{1-2z^{-1}+z^{-2}+0.3281z^{-3}-0.2500z^{-4}}$$

P7-26: For $H(z^{-1})$ given in P7-25, obtain and plot

(1) An impulse response by evaluating the output when the input is $\delta[n]$.
(2) A step response by evaluating the output when the input is $u[n]$.
(3) A frequency response by letting $z = e^{j2\pi f}$ and evaluating $H(e^{-j2\pi f})$
(4) Note that this system is nearly unstable. How do you fix it?

P7-27: In Fig. P7-5 of problem P7-22, DCO (digital control 'oscillator') is a digital integrator. The input to the integrator increases the output 'phase' increases. This behavior is opposite to VCO with the input-output characteristic being a positive slope (see the figure in problem P7-18).

(1) Show that the input- (frequency) output of DCO has a negative slope.
(2) In order to be a negative feedback PLL, show that PD should have a negative slope as well.

P7-28: Using a MATLAB function [h4, C] = interp (h0,8,3,0.6) with h0 = [0 0 0 0 0 1 0 0 0 0] obtain an interpolating filter C, and plot its frequency response. Compare with a filter using Nyquist pulse method using $h_{VLP}(t) = \sin\left(\pi\frac{t}{\Delta t}\right)/\pi\frac{t}{\Delta t} \cdot w\left(\frac{t}{\Delta t}\beta\right)$ with $\beta = 0.4$ where $w\left(\frac{t}{\Delta t}\beta\right) = \sin\left(\frac{1}{2}\beta\pi\frac{t}{\Delta t}\right)/\left(\frac{1}{2}\beta\pi\frac{t}{\Delta t}\right) \otimes \sin\left(\frac{1}{2}\beta\pi\frac{t}{\Delta t}\right)/\left(\frac{1}{2}\beta\pi\frac{t}{\Delta t}\right)$. Note that the pulse span is 5 sample period (L = 3). Hint: See Figure 7-66.

P7-29: Find Q = 3 Lagrange polynomials, $l_0(t)$, $l_1(t)$, $l_2(t)$. By definition of base polynomial $l_0(0) = 1$, $l_0(1) = 0$, $l_0(2) = 0$ where $l_0(t) = c_2t^2 + c_1t + c_0$. It is clear $c_0 = 1$ from $l_0(0) = 1$. From $l_0(1) = 0$, $c_2 + c_1 = -1$, and from $l_0(2) = 0$, $4c_2 + 2c_1 = -1$. Solving two simultaneous equations we have $l_0(t)=\frac{1}{2}t^2 - \frac{3}{2}t + 1$. Apply $l_1(1) = 0$, $l_1(1) = 1$, $l_1(2) = 0$, and obtain $l_1(t)$. Similarly obtain $l_2(t)$.

(1) Plot $l_0(t)$, $l_1(t)$, $l_2(t)$ with $0 \le t \le 2$.
(2) Obtain FIR filter $h(t)$ with $-2 \le t \le 1$. Is this linear phase?

P7-30: Received samples of 5 are observed in QPSK system (4 constellation points; $1 + j$, $-1 + j$, $-1 - j$, $1 - j$) as [1.185 − 0.773i, −1.197 + 0.771i, 0.764 + 1.193i, 0.763 + 1.199i, −0.775 − 1.186i]. With the decision directed recovery of transmit symbol, find angle measurements for each sample using the relationship in Figure 7-46. Then take an average of 5 measurements.

P7-31: Compare two carrier phase detectors for 16-QAM shown in Figure 7-48. Their decision boundaries are different. Consider 4 constellation points in the quadrant.

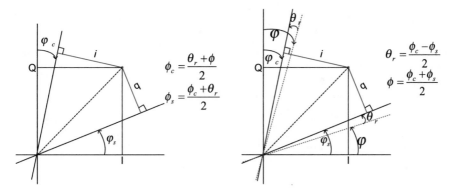

Fig. P7-6: IQ phase imbalance representations

(1) Show that two constellation points on the diagonal can be used for measuring angle 45°. This is true for both LHS and RHS decision boundaries.
(2) For two constellation points on the off-diagonal, the range of angle measurement is different for LHS and RHS decision boundaries. Estimate the range of angle measurements for each constellation point.

P7-32: IQ phase imbalance, two asymmetrical components, represented in Fig. P7-6 (LHS) can be represented by one symmetrical IQ phase imbalance and phase rotation (RHS).

(1) When $\phi_c = 10°$, $\phi_s = 6°$, find θ_r and ϕ.
(2) Is this relationship valid when $\phi_c = -10°$?

Chapter 8
Practical Implementation Issues

Contents

8.1 Transceiver Architecture 605
     8.1.1 Direct Conversion Transceiver 606
     8.1.2 Heterodyne Conversion Transceiver 608
     8.1.3 Implementation Issues of Quadrature Up-Conversion 610
     8.1.4 Implementation Issues of Quadrature Down-Conversion 612
     8.1.5 SSB Signals and Image Cancellation Schemes 614
     8.1.6 Transceiver of Low Digital IF with Image-Cancelling 619
     8.1.7 Calibration of Quadrature Modulator I Demodulator 649
     8.1.8 Summary of Transceiver Architectures 650
8.2 Practical Issues of RF Transmit Signal Generation 651
     8.2.1 DAC 652
     8.2.2 Transmit Filters and Complex Baseband Equivalence 658
     8.2.3 TX Signal Level Distribution and TX Power Control 660
     8.2.4 PA and Non-linearity 662
     8.2.5 Generation of Symbol Clock and Carrier Frequency 670
     8.2.6 Summary of RF Transmit Signal Generation 673
8.3 Practical Issues of RF Receive Signal Processing 673
     8.3.1 ADC 675
     8.3.2 RX Filters and Complex Baseband Representation 678
     8.3.3 RCV Dynamic Range and AGC 678
     8.3.4 LNA, NF, and Receiver Sensitivity Threshold 682
     8.3.5 Re-generation of Symbol Clock and Carrier Frequency 685
     8.3.6 Summary of RF Receive Signal Processing 685
8.4 Chapter Summary and References with Comments 686
     8.4.1 Chapter Summary 686
     8.4.2 References with Comments 686
8.5 Problems 687

Abstract This chapter covers transceiver architectures – direct conversion, heterodyne conversion, and low digital IF image cancelling transceiver -, transmit RF
signal generation and receiver RF signal processing. We examine the required
number of bits, and the optimum power level for DAC and ADC. A particular
emphasis is on low digital IF image cancelling transceiver; it is described in great
detail so that one skilled in the art of transceiver can implement. The quality

© Springer Nature Switzerland AG 2020   603
S.-M. Yang, *Modern Digital Radio Communication Signals and Systems*,
https://doi.org/10.1007/978-3-030-57706-3_8

measurement of quadrature modulator (demodulator) related with DC offset and quadrature imbalance is explained in detail so that they can be calibrated as well.

Key Innovative Terms low digital IF image cancelling transceiver · digital correction of quadrature phase imbalance (in chapter 7 as well) · calibration of DC offset and quadrature phase and amplitude imbalance

General Terms complex baseband equivalence · control loop · DC offset · dynamic range · image cancellation · optimum level to ADC · optimum level from DAC · PAPR · phase noise · quadrature amplitude imbalance · quadrature phase imbalance · quantization noise · receiver threshold · sensitivity · SSB · transceiver

List of Abbreviations

ADC	analog digital converter
AGC	automatic gain control
AM	amplitude modulation
CMOS	complementary metal oxide silicon
DAC	digital analog converter
DC	direct current
DFT	discrete Fourier transform
DNL	differential non-linearity
DSP	digital signal processing
FIR	finite impulse response
FM	frequency modulation
FSK	frequency shift keying
FT	Fourier transform
IDFT	inverse discrete Fourier transform
IEEE	institute of electrical and electronics engineers
IF	intermediate frequency
IIR	infinite impulse response
IMD	intermodulation distortion
I, Q	in-phase, quadrature phase
ISI	inter symbol interference
LHS	left hand side
LNA	low noise amplifier
LO	local oscillator
LP(F)	low pass filter
LSB	lower side band
LSB	least significant bit
LTE	long term evolution
MSB	most significant bit
NF	noise figure
OFDM	orthogonal frequency division multiplex

PA	power amplifier
PAM	pulse amplitude response
PAPR	peak to average power ratio
PDF	probability density function
PSK	phase shift keying
ppm	part per million
QAM	quadrature amplitude modulation
RC	resistor, capacitor
RCV	receive
RHS:	right hand side
RF	radio frequency
RMS	root mean square
RSC	recursive systematic (convolution) code
RX	receive
SIR	signal to interference ratio
SINR	signal to interference and noise ratio
SNqR	signal to quantization noise ratio
SSB	single side band
SAW	surface acoustic wave (filter)
TV	television
TX	transmit
USB	upper side band
VGA	variable gain amplifier
XMT	transmit

Here we focus our attention to practical implementation issues related with continuous signal space shown in Figure 8-1, which include RF system and part of digital modem, in particular, including DAC and ADC. We emphasize as 'implementation issues' because their implementations are done by analog circuits, rather than digitally (or numerically). Remember that Figure 8-1 is a block diagram showing an overall picture, and each block in detail requires experts on it to implement it properly. Our goal is to understand the issues in detail, from system engineer point of view who put together a communication system. This means that our representations are mostly not circuits but mathematical operations.

8.1 Transceiver Architecture

In Figure 8-1, signal processing blocks of B $\rightarrow$C $\rightarrow$D $\rightarrow$E in the transmit side and $\underline{E}$ $\rightarrow$$\underline{D}$ $\rightarrow$$\underline{C}$ $\rightarrow$$\underline{B}$ in the receive side are collectively called a *transceiver*. In some cases DAC and ADC may not be part of a transceiver. DAC and ADC are implemented by mostly analog circuits as well even though they are the interface between discrete-in-

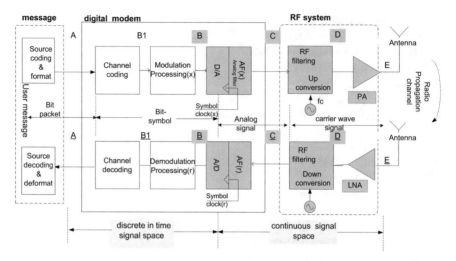

Figure 8-1: Digital radio communication system block diagram with emphasis on DAC, up conversion and PA in the transmit direction and LNA, down conversion and ADC in the receive direction as highlighted in shade

time (digital) signal space and continuous (analog) signal space. For our purpose we include them as part of a transceiver, but will be treated in another section.

LHS of (B - $\underline{B}$) in Figure 8-1 is abstracted as a source of modulated symbols, getting into DAC in the transmit side, and getting out from ADC in the receive side, and denoted as bit-to-symbol mapping and symbol-to-bit demapping. In the transmit side the output of a transceiver is at the XMT antenna port and the input at RCV is antenna port in the receive side. As an example, see Figure 8-2.

8.1.1   Direct Conversion Transceiver

Baseband (I and Q) signals, after DAC and LPF (low pass filter), are up-converted directly to a carrier frequency in the transmit side, and RF signals are down converted, after LP filter and ADC sampling, to I and Q baseband signals directly. 'Direct' means that there is no intermediate frequency (IF) translation. This is shown in Figure 8-2.

In the transmit side, the signal at D in Figure 8-2, after frequency translation and combining, is represented by

$$s_x(t) = i(t) \cos\left(2\pi f_c t\right) - q(t) \sin\left(2\pi f_c t\right) \tag{8-1}$$

And the signal is amplified by PA for transmission but not explicitly part of (8-1).

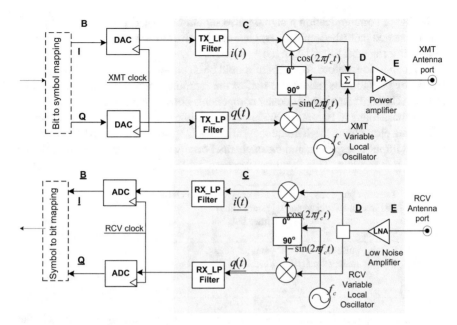

Figure 8-2: Transceiver with direct conversion. For multiple channels, the frequency of local carrier (f_c) is variable

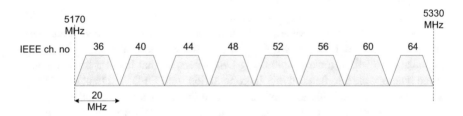

Figure 8-3: IEEE 802.11a example of frequency channel plan with 20 MHz channel bandwidth. IEEE uses channel number system (36, 40...)

In the receive side, the same signal (if we ignore noise and any other impairment) will be demodulated down to baseband to generate I and Q signals. Twice carrier frequency should be eliminated by RX_LP filter.

In many wireless communications, RF is divided into channels and each user occupies a designated channel. For example, IEEE 802.11a standard uses 5 GHz band with 20 MHz channels shown in Figure 8-3. In this multi-channel environment, a transmit signal should be band-limited enough so that it occupies only its own channel but no spill over to adjacent channels. In the receive side a receiver should have a good channel selectivity so that the presence of neighboring signals should not interfere and degrade the reception.

In digital communication a signal pulse should be both ISI free (Nyquist pulse) and a matched pair (the impulse response of a receiver filter should be time-reversed to that of a transmit filter, i.e., $h_R(t) = h_X(-t)$).

In receive side, before ADC, there should be an anti-aliasing filter to remove high frequency components, above the half of the sampling frequency. In transmit side, after DAC, there are high frequency components above sampling frequency.

Low pass filters in Figure 8-2 are used for multiple purposes – anti-aliasing before ADC and eliminating high frequency after DAC, pulse shaping for ISI free and matched impulse response, and clean channel occupancy at transmitter and channel selectivity at receiver. This is summarized as a table below.

TX_LP	Transmit pulse shaping (ISI free and matched pair)	Filtering (smoothing) after DAC	Out-band rejection for clean channel occupancy *RF filter (center frequency variable) may be added before or after PA
RX_LP	Receive pulse shaping (ISI free and matched pair)	anti-aliasing before ADC	Out-band rejection for channel selectivity *RF filter (center frequency variable) may be added before or after LNA

The direct conversion transceiver is conceptually simple but not used often in the past due to implementation issues –DC offset, IQ imbalance, flicker noise and even order nonlinearity requirement. These issues are more difficult with multi-channel environment due to the change of channel.

A major reason for not using direct conversion seems to be related with filters – no baseband filters for adequate channel selectivity, but in IF, such filters were available, e.g., SAW filters. The use of direct conversion was supposed to be a major technical innovation, for example using DSP filters, so that it has a fancy name of homodyne opposed to heterodyne. In conjunction with OFDM signal, 802.11a/g/n was implemented with direct conversion due to the motivation of highly integrated RF transceiver chip made of CMOS without external SAW filters. We will be back to these issues after discussing heterodyne transceiver.

8.1.2   Heterodyne Conversion Transceiver

The use of IF stage was a major innovation in broadcasting channels like AM, FM, and TV. By adding a sharp IF filter (e.g., SAW filter), the channel selectivity was improved considerably. Most of radios used IF frequency, say 70 MHz when the implementation were mostly done by analog discrete components. Overall still most radios have more than a single (direct) conversion.

IQ imbalance problems are there but slightly easier since IF frequency is fixed. The flicker noise problem near DC is essentially gone. DC offset can be removed easily by DC blocking since it is not part of the signal.

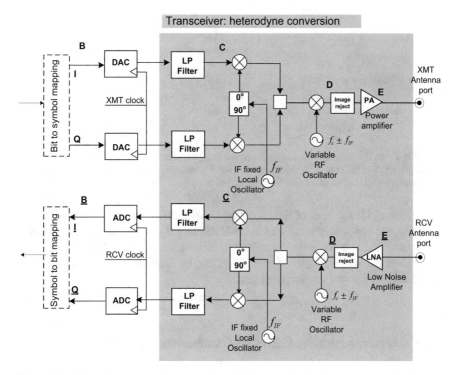

Figure 8-4: Heterodyne conversion transceiver. IF frequency is fixed and common for all channels, and channel selection is done by variable RF oscillator. DC offset and IQ imbalance problems are there but easier since IF frequency is fixed. A new problem is the need of image reject filter

Figure 8-5: Spectrum at D in Figure 8-4 with image frequency (dotted). The part A shows with LO frequency being $f_c + f_{IF}$ and the part B shows with $f_c - f_{IF}$. The frequency difference between two images of spectrum is always $2f_{IF}$

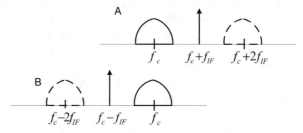

A new problem for heterodyne transceiver is the need of image reject filter. The transmit side is shown in Figure 8-5. The image may appear above RF LO frequency if it is $f_c + f_{IF}$ (A in Figure 8-5) or below if it is $f_c - f_{IF}$ (B in Figure 8-5). An image reject filter removes the image. Only one side band of IF signal will be transmitted, upper sideband with LO frequency $f_c - f_{IF}$ or lower sideband with LO frequency $f_c + f_{IF}$. Note that the image reject filter can be a lowpass or a highpass filter depending on which side band is rejected.

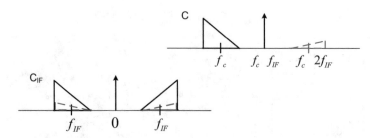

Figure 8-6: In the receive side the signal at the image location (red) can be interference once down-converted to IF. Thus before down conversion the image must be filtered or image cancelling conversion (which will be discussed later) should be used

In receive side the need of filtering image is essentially the same as transmit side. However, the image is not its own but could be other signals as shown in Figure 8-6, and hence a signal at the image location can become interference. Thus before down conversion, the image frequency should be clean by filtering or by image cancelling conversion which will be discussed later.

8.1.3   Implementation Issues of Quadrature Up-Conversion

There are practical issues in implementing a transceiver as in Figure 8-2 and in Figure 8-4. In particular, we address quadrature up- and down- conversion circuits in this section. They are DC-offset in I-rail and in Q-rail, IQ phase imbalance, and IQ amplitude imbalance, which are shown in Figure 8-7 for the transmit side.

In order to implement the equation (8-1), the amplitude of I-rail and Q-rail should be perfectly balanced (the same) and any difference in gain (imbalance) results in degradation. I-rail (or Q-rail) means the entire component from DAC, Filter, VGA, and mixer to the combiner as well as LO port. Any gain mismatch in the entire chain will be amplitude imbalance.

For I-side, LO should be cosine and for Q-side should be sine, i.e., 90° phase difference, called quadrature. Any deviation from 90° difference (phase imbalance) degrades the signal.

The equation (8-1) is modified with these degradations as,

$$
\begin{aligned}
s_x(t) = {} & (1 + \Delta A)(i(t) + DC_i) \cos\left(2\pi f_c t\right) \\
& - \left(q(t) + DC_q\right) \sin\left(2\pi f_c t + \Delta\varphi\right)
\end{aligned}
\tag{8-2}
$$

where ΔA is IQ amplitude imbalance, $\Delta\varphi$ is quadrature phase imbalance, and DC_i and DC_q are DC offset in I-rail and Q-rail respectively.

All combined effects in (8-2) are a bit hard to see, and thus we may consider only DC offset, ignoring IQ amplitude and quadrature phase imbalance. DC offset will result in carrier leakage which wastes signal power.

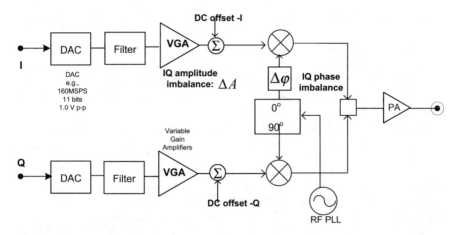

Figure 8-7: Implementation issues of quadrature up-conversion; DC offset (I and Q) = LO leakage, IQ amplitude imbalance, and IQ Phase imbalance

Example 8-1: Consider only DC offset and see that it will manifest as a carrier leakage, also called LO leakage. $(i(t) + DC_i) \cos (2\pi f_c t) - (q(t) + DC_q) \sin (2\pi f_c t)$ can be expressed as in (8-3),

$$[i(t) \cos (2\pi f_c t) - q(t) \sin (2\pi f_c t)] + (a) \cos (2\pi f_c t + ph) \qquad (8\text{-}3)$$

where for carrier leakage term $a = \sqrt{DC_i^2 + DC_q^2}$ and $ph = \tan^{-1}\left(\frac{DC_q}{DC_i}\right)$.

Example 8-2: Carrier leakage is often expressed in dB relative to signal power. For example, -10 dB LO leakage means that the LO leakage power is 1/10 of signal power. What is the power loss of a signal [in dB] due to LO leakage? The answer is 0.42 dB. Answer in general is: signal power loss [dB] $= 10*\log \left(\frac{1}{1+10^{-LOdB/10}}\right)$.

LO leakage 0 dB: signal power loss $= 3$ dB, and LO leakage -20 dB: signal power loss $= 0.043$ dB.

Example 8-3: Assuming that the bandwidth of a signal is 10 MHz and spectrum resolution bandwidth 100 kHz, show a transmit spectrum with LO leakage -15 dB.

Answer: signal spectrum is down by 20 dB (10log(10e6/100e3)) from the reference point of RMS signal power. LO leakage power is down 15 dB compared to RMS signal power. Thus LO tone is 'sticking out' by 5 dB above continuous signal spectrum. When LO leakage is -20 dB, then there is no 'sticking out', i.e., buried in the signal spectrum.

In order to see the impact of IQ imbalance, we drop DC offset from (8-2);
$s_x(t) = (1 + \Delta A)i(t) \cos (2\pi f_c t) - q(t) \sin (2\pi f_c t + \Delta\varphi)$. By rearranging it, we show as

$$s_x(t) = [(1 + \Delta A)i(t) - q(t)\sin(\Delta\varphi)]\cos(2\pi f_c t) - [q(t)\cos(\Delta\varphi)]\sin(2\pi f_c t).$$

Note that I–side has Q signal component, i.e., $q(t)\sin(\Delta\varphi)$, and this is noise to I signal; i.e., Q signal is spilled over to I, called *crosstalk between I and Q*. And signal to interference ratio (SIR) is $20\log(\sin(\Delta\varphi))$ dB. For example, $\Delta\varphi = 1°$, SIR $= 35$ dB. In quadrature modulation signaling scheme, I-Q channel cross-talking interference is most damaging to system performance.

For amplitude imbalance $(1 + \Delta A)$, the signal power is reduced by $20\log(1 + \Delta A)$ assuming it is less than 1.0. For example, $(1 + \Delta A) = 0.95$, SNR is reduced by 0.44 dB.

Example 8-4: A system SNR is given by 40 dB. Consider the phase imbalance $1°$ and amplitude imbalance $(1 + \Delta A) = 0.95$. What is a degraded SNR due to phase imbalance? What is a degraded SNR due to amplitude imbalance?
Answer: 1/SINR due to phase imbalance $= 10*\log(10^{-40/10} + 10^{-35/10}) = -33.8$ dB, and SNR due to amplitude imbalance $= 40 - 0.44 = 39.56$ dB. Note that phase imbalance is more damaging. What about the case of SNR 30 dB? Answer: 1/SINR $= -28.8$ dB and SNR $= +29.56$ dB.

8.1.4   Implementation Issues of Quadrature Down-Conversion

We consider the receive side. Figure 8-8 sketches the corresponding issues of DC offset, IQ amplitude imbalance, and IQ quadrature phase imbalance.

DC offset and its impact to ADC effective bits
The effect from transmit side is cumulative to the receive side. LO leakage at the transmit side will become after demodulation DC offset. It may cancel or worsen depending on the sign of DC. The impact to system performance is to shift (offset) the receive signal constellation but DC offset may be removed digitally completely, thus no degradation of performance.

DC offset may reduce ADC dynamic range so that its effective bits may be reduced or with excessive DC offset there may be clipping (overload distortion). This ADC impact can be eliminated only by reducing DC offset itself.

Example 8-5: A numerical example is given on the impact to ADC of DC offset reducing effect number of bits of ADC, i.e., reducing quantization noise. ADC has a reference voltage $+0.5$ to -0.5 [V] with 50 ohm termination, and 10 bits (1 sign bit + 9 bits of magnitude) to cover the range. A signal (say OFDM) has a peak to average power ratio $+12$ dB.

1) What is optimum RMS input voltage to ADC? Or power in dBm?
2) What is the quantization noise degradation if DC offset is 10% of 0.5 V (peak), i.e., 0.05 V? What about 0.1 V case?
3) Optimum means to minimize quantization noise without overloading (clipping). What is the optimum signal to quantization noise?

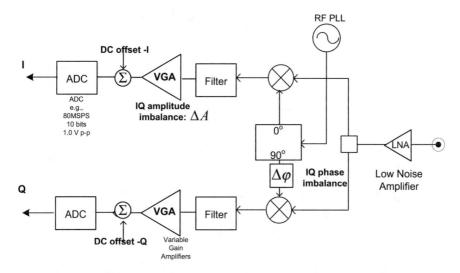

Figure 8-8: Implementation issues of quadrature down conversion: DC offset (I and Q), IQ amplitude imbalance, and IQ Phase imbalance

Answer: 1) The peak power ADC can handle is 7.0 dBm (= 10*log (0.5²/ 50*1000)). Note that 0 dBm means 1 miliwatt of power. The optimum (maximum) input power with no overloading (clipping) is −5 dBm (= 7 dBm – 12 dB).

In terms of voltage, 12 dB power back-off means 1/15.8 (= $10^{-12/10}$) in ratio. RMS voltage is given by 0.1258 $\left(= \frac{0.5}{\sqrt{15.8}}\right)$.

2) The input power to ADC is maintained to be −5 dBm even with DC offset, and thus the net signal power is reduced by the power due to DC offset; 0.05^2 +net_sig² = 0.1258^2, and thus net_sig² = 0.1154^2 = 0.0133. The power of net_sig with 50 ohm termination is given by −5.75 dBm (= 10*log (0.1154²/50*1000)). Thus 0.75 dB is quantization noise increase. For 0.1 V DC offset, 0.1^2 +net_sig² = 0.1258^2 and net_sig² = 0.076^2 = 0.0058, in dBm, −9.33 dBm (= 10*log (0.076²/50*1000)). Thus quantization noise increase is 4.33 dB. Note that the quantization noise increases very rapidly as DC offset is close to RMS. See Figure 8-9 where x-axis is normalized by 0.1258 V (signal RMS).

3) From 1), signal power is given by −5 dBm. We need to find the quantization noise power. The quantization range is +1/2 q to −1/2 q where q = 0.5/ 2^{10-1} = 0.98 mV in this example. The quantization noise power depends on the probability density function of the quantization noise. A uniform distribution from −1/2q to +1/2q results in variance (power) of $\frac{q^2}{12}$, and it is −58.0dBm $\left(= 10\log \left(\frac{0.5}{2^9}\right)^2 \frac{1}{12} 1/50 * 1000\right)$. Thus signal to quantization noise ratio (SN$_q$R) is 53 dB. When the error is concentrated in two points (−1/2q, +1/ 2q) with discrete probability, the variance is $\frac{q^2}{4}$ and it is

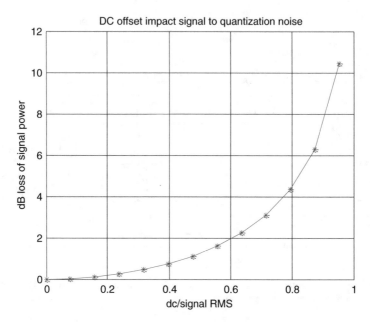

Figure 8-9: Signal to quantization noise reduction due to DC offset

-53.2dBm $\left(= 10 \log \left(\frac{0.5}{2^9}\right)^2 \frac{1}{4} 1/50 * 1000 \right)$, and thus signal to quantization noise is 48.2 dB . Note that in this example we assumed peak to average power is +12 dB.

If this example is new and thus a bit hard, you may revisit it after reading Section 8.3.1 on ADC.

IQ amplitude and phase imbalance

Like DC offset, IQ amplitude imbalance and phase imbalance of transmit side and those of receive side are cumulative; may cancel or make worse again. This is shown in Figure 8-10 for amplitude imbalance, and in Figure 8-11 for phase imbalance.

8.1.5 SSB Signals and Image Cancellation Schemes

In digital communication context, SSB (single side band) is not used; instead two basebands (I and Q) are modulated by quadrature modulator as in Figure 8-2, generating a transmit signal, $s_x(t)$ given by (8-1). In analog voice communication context, SSB was a major technical innovation in 1930s and used extensively in telephony until digital transmission took over in 1970s. In terms of bandwidth efficiency quadrature modulation is as good as SSB since the same RF channel is used twice by two quadrature signals, i.e., I and Q channels are orthogonal due to their 90° phase difference. I/Q modulation is convenient since the spectrum of digital

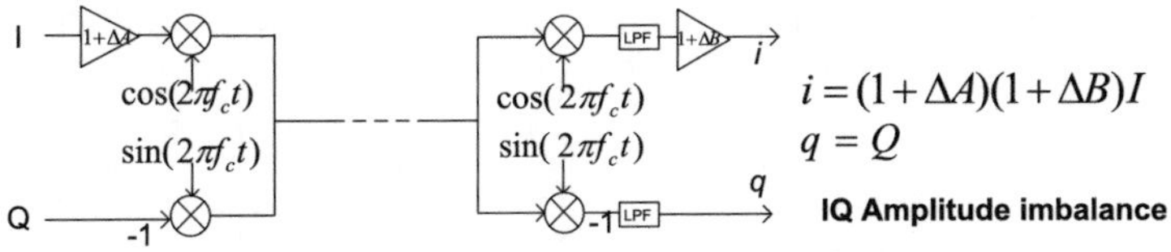

$$i = (1+\Delta A)(1+\Delta B)I$$
$$q = Q$$

IQ Amplitude imbalance

Figure 8-10: IQ amplitude imbalance is cumulative

$$i = I - Q \sin(-\varphi_s')$$
$$q = I\sin(-\varphi_s') + Q\cos(\varphi_s^i - \varphi_s')$$

IQ phase imbalance

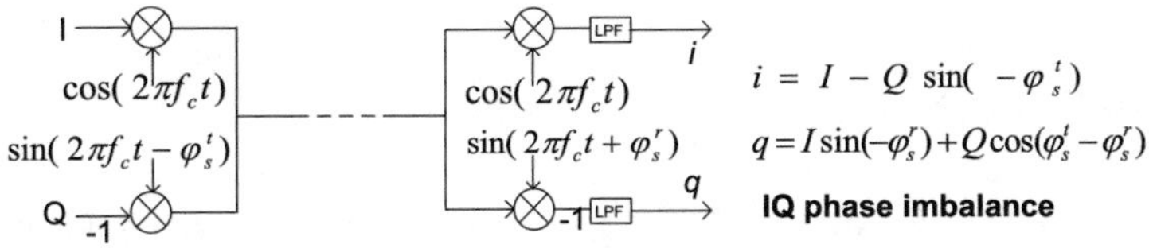

Figure 8-11: IQ phase imbalance is cumulative

communication signals at DC is often not zero but tends to be the largest whereas for the spectrum of voice signals near DC is null. Here our starting point is quadrature modulation and it will be related with image cancellation schemes since the same mechanism applies to both SSB and image cancelling. *This will be used when we discuss image cancelling transceiver next section.* Thus it is important to understand it thoroughly.

SSB signals with a single tone

The Fourier Transform (FT) of (8-1), is given by $S_x(f)$,

$$2S_x(f) = I(f - f_c) + jQ(f - f_c) + I(f + f_c) - jQ(f + f_c) \qquad (8\text{-}4)$$

where FT pair $i(t) \leftrightarrow I(f)$ and $q(t) \leftrightarrow Q(f)$.

   This can be seen easily from $\cos(2\pi f_c t) = \frac{1}{2}\left(e^{j2\pi f_c t} + e^{-j2\pi f_c t}\right)$ and $\sin(2\pi f_c t) = \frac{1}{2j}\left(e^{j2\pi f_c t} - e^{-j2\pi f_c t}\right)$ with FT pairs of $i(t)e^{j2\pi f_c t} \leftrightarrow I(f - f_c)$, $i(t)e^{-j2\pi f_c t} \leftrightarrow I(f + f_c), jq(t)e^{j2\pi f_c t} \leftrightarrow jQ(f - f_c), jq(t)e^{-j2\pi f_c t} \leftrightarrow jQ(f + f_c)$.

   The first two terms of (8-4) is positive frequency and we can see that $I(f)$ and Q (f) are added in frequency domain by $90°$ (or j). Similarly for the negative frequency terms, i.e., 3rd and 4th terms in (8-4), are quadrature with $-90°$ (or $-j$).

Example 8-6: We consider two single tones of the same frequency; i $(t) = A_m \cos(2\pi f_m t)$, $q(t) = A_m \sin(2\pi f_m t)$. Consider complex (not real) signals in time, called analytical signals, $i(t) + jq(t)$, $i(t) - jq(t)$ and find their spectra.

   Answer: See the figure below.

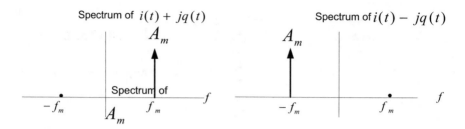

We obtain the above as follows.

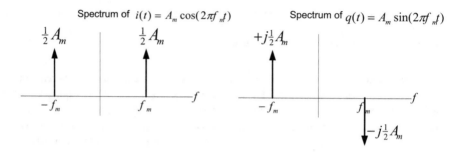

Exercise 8-1: Consider another combinations $ji(t) + q(t)$ and $-ji(t) + q(t)$ from Example 8-6 and sketch their spectra.

Answer:

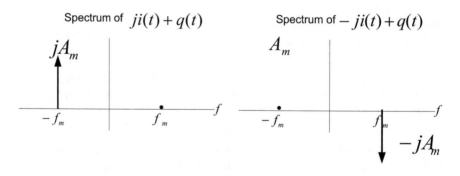

−90° phase shifter (Hilbert transform)

In order to generate a SSB signal in the above example, we need to choose $i(t) = A_m \cos(2\pi f_m t)$, $q(t) = A_m \sin(2\pi f_m t)$. In other words, $i(t)$ and $q(t)$ are related by $-90°$ phase shift since $A_m \cos(2\pi f_m t - 90°) = A_m \sin(2\pi f_m t)$.

We generalize this $-90°$ phase shifter, and its transfer function is given as in Figure 8-12; $\angle H(f) = -j\,\mathrm{sgn}\,(f)$, and apply not only to sinusoids to any signal.

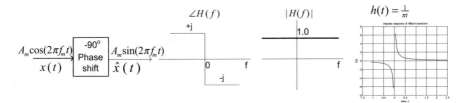

Figure 8-12: $-90°$ phase shift circuits and its frequency response or Hilbert transformer; $\angle H(f) = -j \,\text{sgn}(f)$ and $|H(f)| = 1.0$ and its impulse response is $h(t) = \frac{1}{\pi \cdot t}$

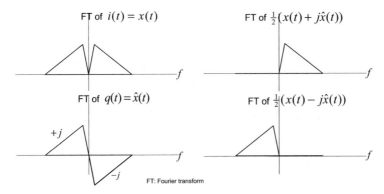

Figure 8-13: For a given $i(t) = x(t)$, $q(t) = \hat{x}(t)$, i.e., $-90°$ shifted version of $i(t) = x(t)$. Then we form analytical signals $\frac{1}{2}(x(t) + j\hat{x}(t))$ for upper side band, and $\frac{1}{2}(x(t) - j\hat{x}(t))$ for lower sideband. The scale change to ½ here is necessary for easy visual comparison. FT stands for Fourier Transform

Table 8-1: Examples of Hilbert transform pairs

$x(t)$	$\hat{x}(t)$	$x(t)$	$\hat{x}(t)$
$\cos(2\pi ft)$	$\sin(2\pi ft)$	$e^{\,j2\pi f_c t}$	$-je^{\,j2\pi f_c t}$
$\sin(2\pi ft)$	$-\cos(2\pi ft)$	$e^{-j2\pi f_c t}$	$je^{-j2\pi f_c t}$
$m(t)\cos(2\pi ft)$	$m(t)\sin(2\pi ft)$	$\delta(t)$	$\frac{1}{\pi \cdot t}$
$m(t)\sin(2\pi ft)$	$-m(t)\cos(2\pi ft)$	$\frac{\sin(t)}{t}$	$\frac{1-\cos(t)}{t}$

This may be intuitively justified that any (practical) signal can be decomposed into many single frequency components.

We use Hilbert transformer to generate a SSB signal as shown in Figure 8-13.

By choosing $q(t) = \hat{i}(t)$, i.e., Hilbert transform of $i(t)$, a quadrature modulator will generate SSB signals. In terms of equation, it is given by (8-5), for upper side band,

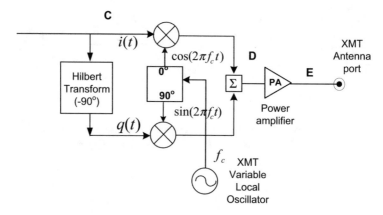

Figure 8-14: SSB signal generation using Hilbert transform with quadrature modulator and it can be used for image cancelling mixers

$$s_x(t) = \mathrm{Re}\left\{\left(i(t) + j\hat{i}(t)\right)e^{j2\pi f_c t}\right\} = i(t)\cos\left(2\pi f_c t\right) - \hat{i}(t)\sin\left(2\pi f_c t\right) \quad (8\text{-}5)$$

This SSB generation mechanism is the same as image-cancelling modulator (mixer).

Example 8-7: Consider the inverse Hilbert transform. From Figure 8-13 one can see $H(\hat{x}(t)) = -x(t)$; the inverse is the same Hilbert transform but negation. This statement can be verified by looking at Table 8-1 as well.

With the inverse Hilbert transform we can do the receive side of Figure 8-14 as shown in Figure 8-15 below. In the figure Hilbert transform block is denoted as $+90°$ phase shifter, rather than $-90°$.

Exercise 8-2: Another way of generating and recovering SSB signals is to use a filter, probably sharp one, as shown Figure 8-16. But our focus is to use Hilbert transform since we discuss SSB related with digital image cancelling application next.

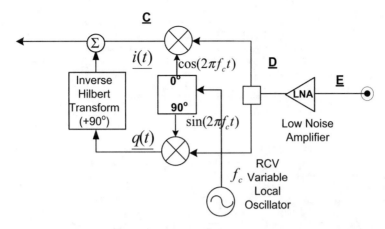

Figure 8-15: SSB signal demodulation, which is the inverse of SSB generation in Figure 8-14. The inverse Hilbert transform is the same as Hilbert transform with negation. Note that the recovery of $i(t)$ may not need both quadrature demodulator and inverse Hilbert transform. We keep the structure for later use as well as for the symmetry

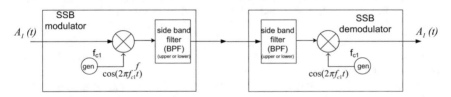

Figure 8-16: SSB generation and reception using sharp filters

8.1.6   Transceiver of Low Digital IF with Image-Cancelling

We discussed SSB signal generation and detection using $-90°$ phase shifter (or Hilbert transformer), and the same mechanism is applicable to image cancelling. In this section we propose a transceiver with digital image cancelling capability and elaborate it in great detail.

As transceiver architectures, we discussed direct and heterodyne (IF) conversion types, and their practical implementation issues; DC offset and carrier leakage, IQ phase and amplitude imbalance and flicker noise for direct conversion, and image cancelling requirements for heterodyne conversion architecture in addition to all problems of direct conversion except for flicker noise. When IF frequency is fixed, it is easier to deal with DC offset and IQ imbalance than when it is variable. However, it may still need to meet stringent requirements. In order to meet such cases some form of calibrations may be used. In this section we suggest transceiver architecture in order to alleviate all these problems. It is a low IF heterodyne scheme with digital implementation of image cancelling. A low IF scheme in the literature is often

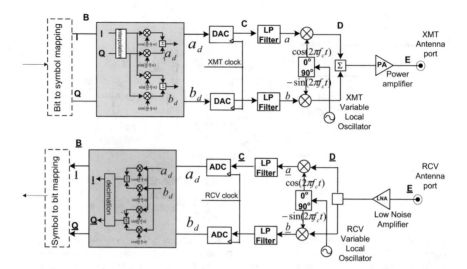

Figure 8-17: Digital image-cancelling transceiver architecture; I and Q are digitally quadrature modulated with the symbol rate as a digital IF carrier frequency, and followed by quadrature modulation by RF carrier frequency, and both TX side and RX side are shown

condemned to be useless due to the problem of image rejection filtering requirement being so severe, especially in broadband applications.

Due to digital implementation of this scheme, IQ phase and amplitude balance of digital IF is nearly perfect within digital computation accuracy, and no DC offset. RF quadrature modulator and demodulator may have amplitude and phase imbalance but the image may be suppressed further by filtering without impacting the main signal. It also may have DC offset but out of band, i.e., there is no overlap with the main signal thus easy to use DC blocking. A major price for this scheme is the complexity of ADC and DAC – doubling the sampling speed. This architecture is schematically shown in Figure 8-17, and will be elaborated in great detail.

In Figure 8-17, note that RHS of (a_d, b_d) is the same as direct conversion shown in Figure 8-2. If $b_d = H(a_d)$, where $H(\ )$ is Hilbert transform, i.e., b_d is $-90°$ phase shifted version of a_d, then it is the same as Figure 8-14 and Figure 8-15 of SSB signal generation and recovery. Thus one sideband of the spectrum of a_d will be cancelled as shown in Figure 8-13. We explain how to generate a_d and b_d, their analog version a and b after DAC and low pass filter. In order to generate digital IF signal (a_d) and its Hilbert transform (b_d), I and Q must be interpolated 4 times, i.e., sampling frequency is 4 times of symbol rate. In terms of equation it is given by, $\frac{\Delta t}{4}n$ the time index of interpolated sequence with $\frac{1}{\Delta t}$ symbol rate, and $\frac{4}{\Delta t}$ sampling rate, as

$$a_d + jb_d = \left[i\left(\frac{\Delta t}{4}n\right) + jq\left(\frac{\Delta t}{4}n\right) \right] e^{j\frac{2\pi\Delta t}{\Delta t\,4}n} \qquad (8\text{-}6)$$

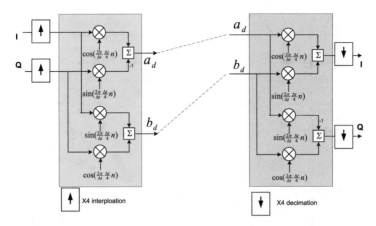

Figure 8-18: Digital IF generation from (I, Q) to (a_d, b_d) generation and vise versa. The digital IF carrier frequency is the same as symbol rate ($\frac{1}{\Delta t}$) and the sampling is 4 times of symbol rate

The equation (8-6) is for upper sideband case. (For lower sideband, see Figure 8-13.) The interpolated sequence is denoted as $i\left(\frac{\Delta t}{4}n\right)$ and $q\left(\frac{\Delta t}{4}n\right)$, a slight abuse of notation. The interpolation digital filter will be discussed later. The equation (8-6) can be 'implemented' by a block diagram, with real arithmetic computation, as in Figure 8-18.

This is a brief summary of the transceiver architecture with low IF digital image-cancelling. We now explain each part of the scheme in detail. First in Section 8.1.6.1 we explain the image cancelling for the transmit side, and next in Section 8.1.6.2 we do it for the receive side. These two sections are a bit lengthy but the point is to see the mechanism of image cancelling in both transmit and receive directions.

8.1.6.1   TX Signal Generation of Digital Image-Cancelling Transceiver

I, Q are sequences sampled at symbol rate ($1/\Delta t$), and interpolated by a factor of 4 times. Digital quadrature modulators produce a_d and b_d with the IF carrier frequency same as symbol rate ($1/\Delta t$) where b_d is Hilbert transform of a_d. By forming an analytical signal of $a_d + jb_d$ a lower sideband is cancelled. A particular choice of IF carrier frequency, $1/\Delta t = 1/T_s$, and interpolation factor of 4, will be generalized and explained later. However, we mention that it is a practically useful choice simplifying the implementations. DACs and LP filters are essentially transparent to the signal here, and will be elaborated later.

The action of quadrature modulator with the carrier frequency of f_c is shown as being frequency shifted by $e^{j2\pi f_c t}$ with no negative frequency component, and taking real part in time domain. This is captured schematically in Figure 8-19.

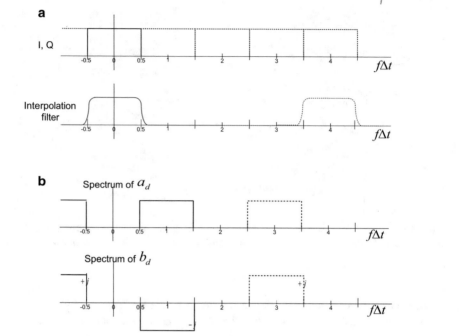

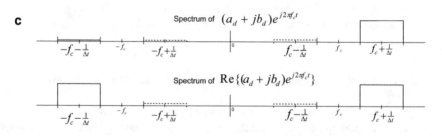

Figure 8-19: Sketch of frequency spectrum in signal processing of digital image cancelling TX signal generation from I, Q. DACs and LP filters are transparent to signals. a) I, Q samples and interpolation filter. b) Digital IF quadrature modulation and upper sideband selection. c) RF quadrature modulator action

Exercise 8-3: It is noted that the signal processing of Figure 8-19 can be summarized as in (8-7),

$$\text{Re}\left\{(i+jq)e^{\,j2\pi f_{IF}t}e^{\,j2\pi f_c t}\right\} = \text{Re}\left\{(i+jq)e^{\,j2\pi(\,f_{IF}+f_c)t}\right\} \qquad (8\text{-}7)$$

Show that the above statement is true, and the implementation may be simplified. However, the structure developed here will be useful and the reasons will be clarified later.

8.1.6.2 RX Signal Processing of Digital Image-Cancelling Transceiver

The receive side signal processing is essentially the inverse of the transmit side; quadrature RF carrier demodulating, after low pass filtering and ADC, followed by digital IF demodulator to generate I, Q samples. See Figure 8-17. However, it is different from TX side due to the presence of interfering image signals, not its own image but foreign signals as interference at the image frequency. In this section we show in detail that this digital image-cancelling scheme is able to eliminate the interfering image signals and to recover I,Q samples correctly even in the presence of the interfering image signals.

In order to see how RX signal processing works, it is convenient to decompose a RX signal into in-phase and quadrature phase components even though it is a single time domain signal, as shown in Figure 8-20. In-phase is denoted as $r(t)$ and quadrature as $\bar{r}(t)$. We treat each one at a time, and the result will be combined in order to obtain I, Q samples.

First we consider only the in-phase, and it is demodulated down to digital IF frequency. The sketch of the signal spectrum is shown in Figure 8-21. Note that twice the carrier frequency component is filtered out. At this point, the interfering image is mixed up with the target signal (upper sideband).

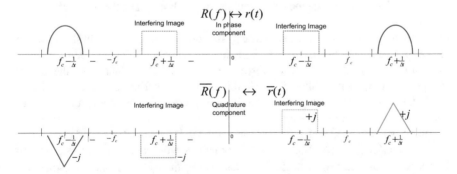

Figure 8-20: A RX signal (frequency spectrum) is represented by two frequency components – in-phase and quadrature phase components. It is real in time domain and thus the even symmetry for in-phase and odd symmetry for quadrature phase component. Red means negative frequency and blue positive frequency. $+j$, $-j$ denotes the phase of 90°, −90°

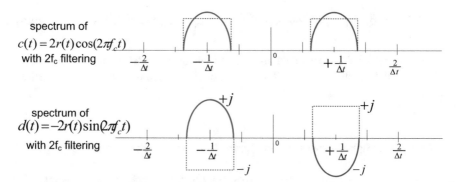

Figure 8-21: $c(t)$ is (in-phase) part of $a(t)$, $d(t)$ is (in-phase) part of $b(t)$. Note that twice the carrier frequency component is filtered out. At this point, the interfering image is mixed up with the target signal

$c(t)$ is a part of $a(t)$ and $d(t)$ a part of $b(t)$, corresponding to in-phase part of RF signal, $r(t)$. Later the quadrature part of RF, $\bar{r}(t)$ will be processed and the other part of $a(t)$ and $b(t)$, denoted as $\bar{c}(t)$ and $\bar{d}(t)$ respectively, will be obtained.

$c(t)$ and $d(t)$ are processed by digital IF demodulator, shown in Figure 8-18 in order to obtain a part of I and Q. Later $\bar{c}(t)$ and $\bar{d}(t)$ are processed similarly by digital IF demodulator in order to obtain the other part of I and Q.

Let us here proceed with $c(t)$ and $d(t)$ first.

In order to obtain I signal (part of), as shown in Figure 8-18, we need to compute $\frac{2}{4}c(t)\cos\left(2\pi\frac{t}{\Delta t}\right) + \frac{2}{4}d(t)\sin\left(2\pi\frac{t}{\Delta t}\right)$ and their frequency domain spectrum is shown schematically in Figure 8-22 in order to see the cancellation. In the last part of the figure shows the result.

In order to obtain Q signal (part of), we need to compute $-\frac{2}{4}c(t)\sin\left(2\pi\frac{t}{\Delta t}\right) + \frac{2}{4}d(t)\cos\left(2\pi\frac{t}{\Delta t}\right)$ and their frequency domain spectrum is shown schematically in Figure 8-23 in order to see the cancellation, and the last part of the figure shows the result.

Secondly we consider the quadrature phase component of RF signal, shown in the lower part of Figure 8-20, and it is demodulated down to digital IF frequency. The sketch of the signal spectrum is shown in Figure 8-24. Note that twice the carrier frequency component is filtered out. At this point, the interfering image is mixed up with the target signal (upper sideband). The process is essentially the same as in-phase component processing.

In order to obtain Q signal (part of), as shown in Figure 8-18, we need to compute $-\frac{2}{4}\bar{c}(t)\sin\left(2\pi\frac{t}{\Delta t}\right) + \frac{2}{4}\bar{d}(t)\cos\left(2\pi\frac{t}{\Delta t}\right)$. In order to see the cancellation, a step by step process is shown in Figure 8-25. In the last part of the figure shows the result.

In order to obtain I signal (part of), we need to compute $\frac{2}{4}\bar{c}(t)\cos\left(2\pi\frac{t}{\Delta t}\right) + \frac{2}{4}\bar{d}(t)\sin\left(2\pi\frac{t}{\Delta t}\right)$. In order to see the cancellation, again a step by step process is shown in Figure 8-26. In the last part of the figure shows the result.

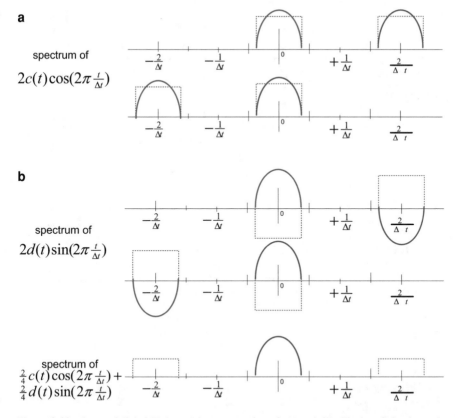

a

spectrum of

$2c(t)\cos(2\pi\frac{t}{\Delta t})$

b

spectrum of

$2d(t)\sin(2\pi\frac{t}{\Delta t})$

spectrum of

$\frac{2}{4}c(t)\cos(2\pi\frac{t}{\Delta t}) + \frac{2}{4}d(t)\sin(2\pi\frac{t}{\Delta t})$

Figure 8-22: I part of digital IF demodulator processing of $c(t)$ and $d(t)$. The interfering image is cancelled around DC and the main upper sideband (I-signal) is cancelled around twice of digital IF carrier frequency. The image amplitude is reduced by ½ compared to I-signal

Let us summarize the above. I and Q signal spectrum is shown in Figure 8-27. The interfering image signal should be removed by filtering. The decimation filter will accomplish it.

Example 8-8: consider $c(t)$ and $d(t)$ in Figure 8-21, and $\bar{c}(t)$ and $\bar{d}(t)$ in Figure 8-24. Form analytical signals $c(t) + jd(t)$, and $\bar{c}(t) + j\bar{d}(t)$, and sketch their spectra.

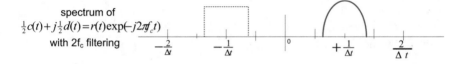

spectrum of

$\frac{1}{2}c(t) + j\frac{1}{2}d(t) = r(t)\exp(-j2\pi f_c t)$

with $2f_c$ filtering

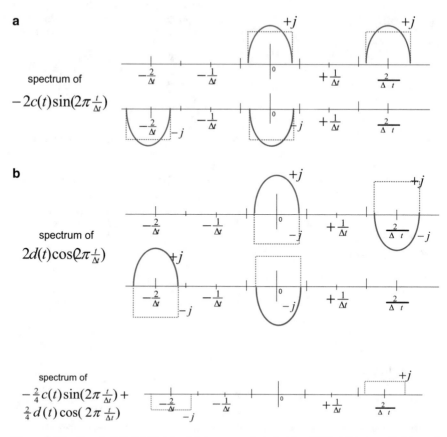

Figure 8-23: Q part of digital IF demodulator processing of $c(t)$ and $d(t)$. All is cancelled around DC and the interfering image is around twice of digital IF carrier frequency. The image amplitude is reduced by ½

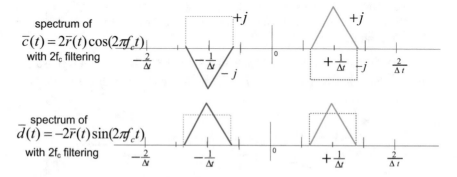

Figure 8-24: $\bar{c}(t)$ is (quadrature) part of $a(t)$, and $\bar{d}(t)$ (quadrature) part of $b(t)$. Note that the twice of carrier frequency is filtered out. At this point, the interfering image (rectangle shape) is mixed up with the target (triangle shape)

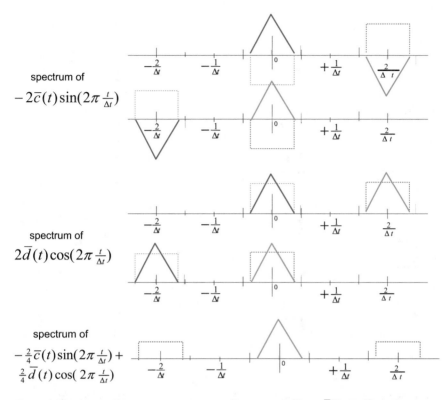

spectrum of

$-2\overline{c}(t)\sin(2\pi\frac{t}{\Delta t})$

spectrum of

$2\overline{d}(t)\cos(2\pi\frac{t}{\Delta t})$

spectrum of

$-\frac{2}{4}\overline{c}(t)\sin(2\pi\frac{t}{\Delta t})+$
$\frac{2}{4}\overline{d}(t)\cos(2\pi\frac{t}{\Delta t})$

Figure 8-25: Q part of digital IF demodulation processing of $\overline{c}(t)$ and $\overline{d}(t)$. The interfering image is cancelled around DC and the main upper sideband (Q-signal) is cancelled around twice of digital IF carrier frequency. The image amplitude is reduced by ½ compared to Q-signal

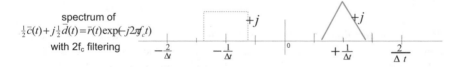

spectrum of
$\frac{1}{2}\overline{c}(t)+j\frac{1}{2}\overline{d}(t)=\overline{r}(t)\exp(-j2\pi f_c t)$
with $2f_c$ filtering

Exercise 8-4: Obtain the result of Figure 8-27 by considering $(c(t)+jd(t))e^{\frac{-j2\pi}{\Delta t}t}$, and $(\overline{c}(t)+j\overline{d}(t))e^{\frac{-j2\pi}{\Delta t}t}$.

8.1.6.3   Filtering Considerations for Digital Image Cancelling Transceiver

In the previous section we focused on the image cancelling aspect of the transceiver architecture shown in Figure 8-17. In this section our attention will be on the filtering, which will be performed by interpolation filters, DACs and LP filters in

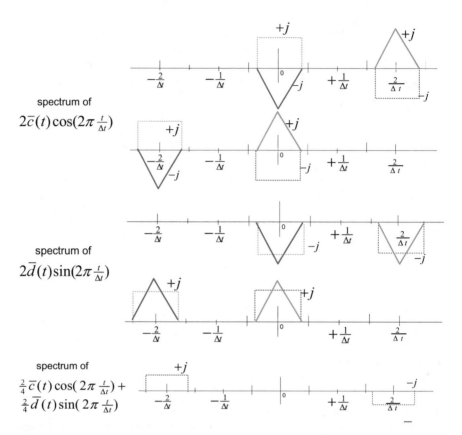

spectrum of
$$2\overline{c}(t)\cos(2\pi\tfrac{t}{\Delta t})$$

spectrum of
$$2\overline{d}(t)\sin(2\pi\tfrac{t}{\Delta t})$$

spectrum of
$$\tfrac{2}{4}\overline{c}(t)\cos(2\pi\tfrac{t}{\Delta t})+$$
$$\tfrac{2}{4}\overline{d}(t)\sin(2\pi\tfrac{t}{\Delta t})$$

Figure 8-26: I part of digital IF demodulator processing of $\overline{c}(t)$ and $\overline{d}(t)$. All is cancelled around DC and the interfering image is around twice of digital IF carrier frequency. The image amplitude is reduced by ½

the transmit direction and by decimation filters, and LP filters in the receive direction. Note that DAC can be considered as a filter whose impulse response is a rectangular pulse. The pulse width is the inverse of conversion rate.

As shown in Figure 8-3, radio channels are band limited and the transmit signal must be fit into a designated channel cleanly without spilling over to adjacent channels. It may be called *clean channel occupancy*. Regulatory specifications in the form of spectrum mask must be met. In the receive direction a selected channel must be processed so that the performance is good even with adjacent channel presence and interference. This is called *channel selectivity*. We discuss here how this filtering requirement can be met.

For the transmit side the interpolation filters (I and Q) eliminate frequency components from about 0.5 to 2.0 of symbol rate (1/2 of sampling rate) as shown in Figure 8-28 (the first one). After digital IF modulation a_d, b_d are generated and will be converted to analog and low pass filtered shown in second and third figure in

a

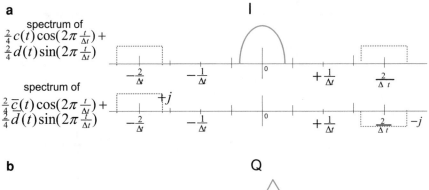

b

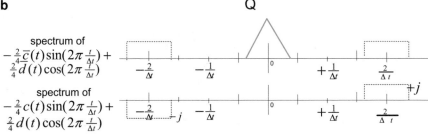

Figure 8-27: I and Q signal after digital IF demodulation. The interfering image signal should be removed by filtering. The decimation filter will accomplish it

Figure 8-28. The combined frequency response of DAC and LPF should be flat in the passband.

In practical design of LPF, it should compensate for DAC frequency response so that the combined frequency response with DAC is flat. The main objective of LPF is to remove higher digital harmonics.

For clean channel occupancy and band limiting the interpolation filters play a major role here. A pulse shaping filter can be done by twice symbol rate in direct conversion transceiver but, here the x4 interpolation filter can be a shaping filter as well. In practice RF filter may be added to further clean up but the clean channel occupancy filter is mainly done at the baseband.

Now consider the receive side filtering.

If there is no adjacent signal interference, the signals before ADC should have a spectrum centered on digital IF frequency (i.e., symbol rate). As shown before, the spectra of $c(t)$ and $d(t)$ in Figure 8-21, and of $\bar{c}(t)$ and $\bar{d}(t)$ in Figure 8-24 are centered on digital IF frequency. This is shown in Figure 8-29 (first part) and LPF needs to eliminate the frequency above ½ sampling rate (i.e., 2 times symbol rate). If there is adjacent channel interference, this LPF should be sharpened so there is no aliasing due to the interference. Unlike LPF in the transmit side, this LPF is important for channel selectivity. Together with decimation filter it mainly determines the channel selectivity. It is anti-aliasing and at the same time responsible for all out of band rejection.

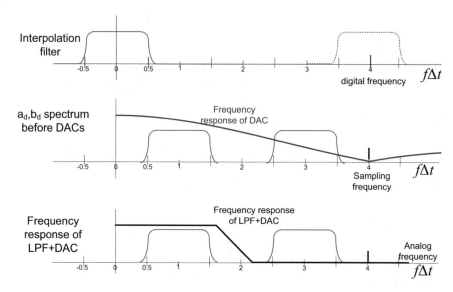

Figure 8-28: Filtering in the transmit side; interpolation filter, DAC and LPF +DAC frequency response

The decimation filter can be the other half of pulse shaping filter together with interpolation filter at the transmit side.

In passing we also mention that OFDM signals do not need a shaping filter and thus there is additional freedom in choosing interpolation and decimation filters. On the other hand there is no reason the same type of shaping filter cannot used for OFDM case. Out of band rejection requirement is the same and only the transition from pass-band to stop-band may be different. This will be elaborated further in Section 8.1.6.10.

8.1.6.4   Digital IF Modulation and Demodulation

In this section we explain digital IF modulation and demodulation shown in Figure 8-18 in some detail. As discussed briefly, denoting symbol time as Δt and sampling time as $\frac{\Delta t}{4}$ (i.e., four samples per symbol period), and interpolated samples as $i\left(\frac{\Delta t}{4} n\right)$ and $q\left(\frac{\Delta t}{4} n\right)$, digital IF modulation computation is given by (8-6), and demodulation is the inverse. For a quick reference we repeat (8-6) here; $a_d + jb_d = \left[i\left(\frac{\Delta t}{4} n\right) + jq\left(\frac{\Delta t}{4} n\right)\right] e^{j\frac{2\pi\Delta t}{\Delta t 4}n}$ and its inverse is $i\left(\frac{\Delta t}{4} n\right) + jq\left(\frac{\Delta t}{4} n\right) = [a_d + jb_d] e^{-j\frac{2\pi\Delta t}{\Delta t 4}n}$.

By choosing digital IF carrier frequency the same as symbol rate and 4 times interpolation the computation is greatly simplified since

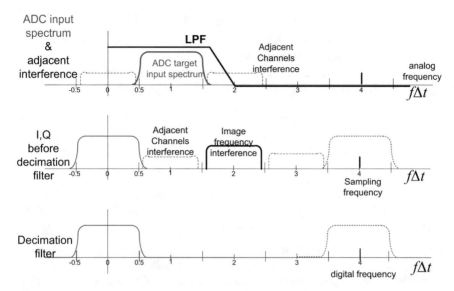

Figure 8-29: Receive side filtering LPF before ADC is for anti-aliasing and should remove all frequency above 2 times of symbol rate (i.e., 1/2 of sampling rate). After digital IF demodulation the all adjacent channel interference, including image frequency interference, is removed by decimation filter

n	0	1	2	3	4	5	6
$\cos\left(\frac{2\pi}{\Delta t}\frac{\Delta t}{4}n\right)$	1	0	-1	0	1	0	-1
$\sin\left(\frac{2\pi}{\Delta t}\frac{\Delta t}{4}n\right)$	0	1	0	-1	0	1	0

$\cos\left(\frac{2\pi}{\Delta t}\frac{\Delta t}{4}n\right)$ and $\sin\left(\frac{2\pi}{\Delta t}\frac{\Delta t}{4}n\right)$ are 0, 1, or -1. Furthermore numerically it is perfect implementation thus no errors unlike analog circuit implementation.

Digital IF carrier frequency and interpolation factor

There is flexibility in the choice of digital IF carrier frequency and interpolation factor. Let us work out the following example.

Example 8-9: If digital IF carrier frequency ($f_{IF} = F_k$) is chosen to be 2 times symbol rate ($\frac{F_k}{\Delta t} = \frac{2}{\Delta t}$), what is the interpolation factor (I_k) so that $\cos\left(\frac{2\pi \cdot F_k}{\Delta t}\frac{\Delta t}{I_k}n\right)$ and $\sin\left(\frac{2\pi \cdot F_k}{\Delta t}\frac{\Delta t}{I_k}n\right)$ are 0, 1, -1? Answer: 8. If f_{IF} is 3 times symbol rate, the interpolation factor is 12.

Another interesting set of choice of (F_k, I_k), so that $\cos\left(\frac{2\pi \cdot F_k}{I_k} n\right)$ and $\sin\left(\frac{2\pi \cdot F_k}{I_k} n\right)$ are $0, 1, -1$, is that (digital IF carrier, interpolation factor); $(0.5, 2)$, $(1.5, 6)$, $(2.5, 10)$ etc.. Note that the ratio I_k / F_k should be 4.

Summarizing the above the choice of digital IF carrier an interpolation factor so that $\cos\left(\frac{2\pi \cdot F_k}{I_k} n\right)$ and $\sin\left(\frac{2\pi \cdot F_k}{I_k} n\right)$ are $0, 1, -1$ is put into a table.

Digital IF carrier ($\frac{F_k}{\Delta t}$)	0.5	1	1.5	2	2.5	3	$I_k / F_k = 4$
Interpolation factor ($\frac{\Delta t}{I_k}$)	2	4	6	8	10	12	

If we allow additional computations, there are more choices. One interesting choice might be $F_k = 0.5$ and $I_k = 4$ with $I_k / F_k = 8$. This choice requires multiplications since $\cos\left(\frac{2\pi \cdot F_k}{I_k} n\right) = \cos\left(\frac{2\pi}{8} n\right) = 1, \frac{1}{\sqrt{2}}, 0, \frac{1}{\sqrt{2}}, -1, \frac{1}{\sqrt{2}}, 0, \frac{1}{\sqrt{2}}$ with $n = 0, 1, 2, 3, 4, \ldots, 7.$, and similarly for sine part. This choice is interesting because it relaxes the analog LPF i.e., the transition from passband to stopband has much more room. On the other hand it takes a full advantage of digital image cancelling, i.e., filtering the image this close would be very difficult. We redraw Figure 8-28 to see the filtering requirement for transmit side of this case and similarly we do

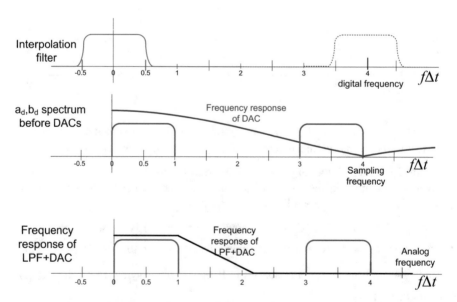

Figure 8-30: Filtering in the transmit side with digital IF frequency 0.5 times symbol rate. Compare to Figure 8-28. LPF transition from passband to stopband is much gentler. Note that RF LO frequency will be offset by 0.5 symbol rate (0.5 channel bandwidth)

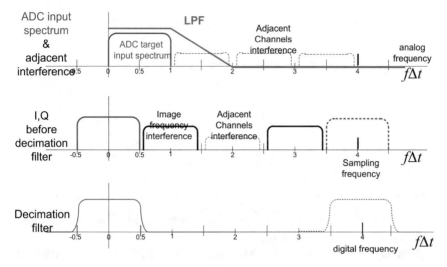

Figure 8-31: Compared to Figure 8-29, the transition, from passband to stopband, of anti-aliasing LPF can be much gentler. Decimation filter will remove both image frequency interference and adjacent channel interference

Figure 8-29 for receive side. They are shown in Figure 8-30 and Figure 8-31 respectively.

8.1.6.5 Interpolation and Decimation Digital Filters

We consider the implementations of interpolation and decimation digital filters. In the context of interpolation and decimation, finite impulse response (FIR) filters are often used to take an advantage of special structures of interpolation and decimation, which will be clear as we show how to organize the computation. The duration of impulse response of FIR is finite, and thus for any FIR the computation can be expressed as a convolution, n, k being the sequence index,

$$y(n) = \sum_{k=0}^{N-1} h(k)x(k-n) \tag{8-8}$$

where the input sequence is $x(n)$, output sequence $y(n)$ and $h(k)$ is an impulse response of a filter. $n = 0, 1, 2, \ldots \infty$. Its frequency response is given by

$$H(f) = \sum_{k=0}^{N-1} h(k)e^{-j2\pi f \cdot k} \tag{8-9}$$

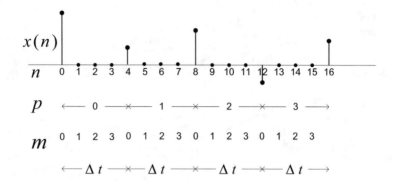

Figure 8-32: Index relationship and zeros are inserted at the input of interpolation filter

Note that we dropped the time interval (Δt or $\frac{\Delta t}{4}$) between samples for convenience, and we will insert it as necessary.

First we consider interpolation. With the interpolation factor 4, one input sample will generate 4 output samples. Formally this can be done by inserting zeros on the input (3 zeros in case of factor 4 and thus 4 input samples with zeros), then use (8-8) to generate 4 output samples. Then the index should be changed $n = 4p+m$ with $m = 0, 1, 2, 3$ or by a pair of index (p, m). The time interval between samples is changed from Δt to $\frac{\Delta t}{4}$. This is shown in Figure 8-32.

$$x(n) \rightarrow \boxed{h_0, h_1, h_2, \dots} \rightarrow y(n)$$

An efficient computation can be organized by taking advantage of multiplication by zero. We will show this. First FIR filter is represented using shift registers (one sample delay per shift register) in Figure 8-33 and an abbreviated representation is shown below where $h_0 = h(0)$, $h_1 = h(1)$,...

It is not difficult to see that a poly-phase structure, instead of a straightforward FIR filter in Figure 8-33, can be used. When a new input comes, say $x(0)$, then we need to compute 4 output samples $y(0)$, $y(1)$, $y(2)$, $y(3)$. Note that the computation of $y(0)$ involves only a set of filter coefficients, namely $h_0, h_4, h_8, \dots$, and $y(1)$ involves $h_1, h_5, h_9, \dots$, and $y(2)$ involves $h_2, h_6, h_{10}, \dots$, and $y(3)$ involves $h_3, h_7, h_{11}, \dots$. This can be put into a block diagram shown in Figure 8-34.

The poly-phase structure developed for interpolation can be used for decimation. It uses the same convolution computation of (8-8) and the structure is essentially the inverse; for 4 input samples, 1 output sample will be computed. This is shown in Figure 8-35. The filter coefficients of a decimation filter is in general not the same as that of an interpolation filter.

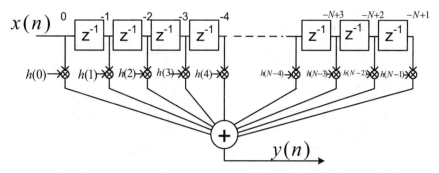

Figure 8-33: FIR filter with shift register implementation. Input $x(n)$ is shifted into the registers by one sample and all the stored samples are shifted, and then multiply and add to generate an output. This computes the equation (8-8). A snap shot $n = 0$ is shown. The abbreviated notation is shown below

Figure 8-34: Poly-phase structure of interpolation (x 4) digital filter

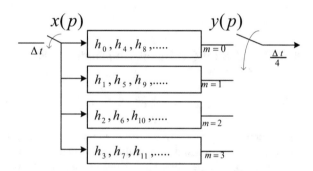

Figure 8-35: Poly-phase structure of decimation filter with a factor of 4. The filter coefficients of a decimation filter is in general not the same as those of a matching interpolation filter

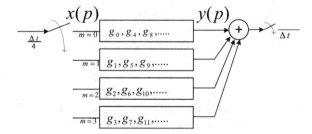

8.1.6.6 Implementation Example of Digital IF Image Cancelling Blocks

With the poly-phase structure and digital IF combined, we give a specific implementation example of the blocks in Figure 8-18. In the figure, digital IF frequency is the same as symbol rate and the interpolation factor is 4.

In passing we note that Figure 8-18 works for different choice of IF carrier frequency $\left(\frac{F_k}{\Delta t}\right)$ and interpolation factor (I_k) by changing to $\cos\left(\frac{2\pi \cdot F_k}{I_k} n\right)$ and $\sin\left(\frac{2\pi \cdot F_k}{I_k} n\right)$ from $\cos\left(\frac{2\pi}{\Delta t}\frac{\Delta t}{4} n\right)$ and $\sin\left(\frac{2\pi}{\Delta t}\frac{\Delta t}{4} n\right)$. The figure shows the case of upper side band. For lower side band appropriate sign change is necessary.

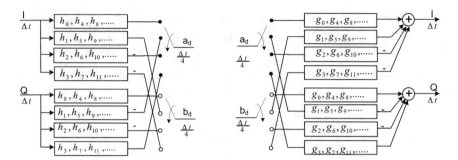

Figure 8-36: Poly-phase implementation of interpolation and decimation of Figure 8-18. Due to the ratio of digital IF frequency and interpolation factor to be 4, there is no multiplication. Upper sideband will be selected with the sign

Exercise 8-5: Work out this sign change in detail. Hint: start with (8-6).

8.1.6.7 Effects of Quadrature Amplitude and Phase Imbalance to Image Cancellation

We discussed practical issues of quadrature modulators and demodulators of transceiver architectures in Section 8.1.3 and 8.1.4. They may be summarized as DC offset related problems and IQ amplitude imbalance and phase imbalance, as shown in Figure 8-7 for transmit side and Figure 8-8 for receive side.

In this section we will examine the impact of IQ amplitude and phase errors to image cancelling performance. We will use SSB signal generation and recovery schemes depicted in Figure 8-14 and Figure 8-15 respectively. Compare them with Figure 8-17. In the transmit direction RHS of a_d, b_d in Figure 8-17 (upper part) is considered to be quadrature modulator and we view it as Figure 8-14. Remember that b_d is Hilbert transform of a_d, i.e., $-90°$ phase shifted version. In the receive direction RHS of a_d, b_d in Figure 8-17 (lower part) is considered to be quadrature demodulator and we view it as Figure 8-15. I and Q can be recovered by going through inverse Hilbert transform, i.e., $+90°$ phase shift. Digital IF modulator and demodulator may be considered perfect $90°$ phase shifters due to its digital implementations, in particular multiplication free implementation if digital IF carrier frequency (F_k) and interpolation factor (I_k) is chosen to be $I_k/F_k = 4$. See Section 8.1.6.4 for details. In order to see the impact to image cancelling performance, we use a single frequency tone as a test signal, i.e., $A_m \cos(2\pi f_m t)$ and its Hilbert transform $A_m \sin(2\pi f_m t)$. This gets into a quadrature modulator with in-phase carrier $A_c \cos(2\pi f_c t)$, and quadrature carrier $A_c(1 + \varepsilon) \sin(2\pi f_c t)$ with the imbalance $1 + \varepsilon$. (The imbalance in dB is given by $20 \log(1 + \varepsilon)$. 1.0 dB imbalance is equivalent to $\varepsilon = 0.122$.) Then in-phase component after modulation is given by $A_c A_m \cos(2\pi f_m t) \cos(2\pi f_c t)$ and quadrature component by $-A_c(1 + \varepsilon) A_m \sin(2\pi f_m t) \sin(2\pi f_c t)$. The sum of two will be transmitted. The sum contains

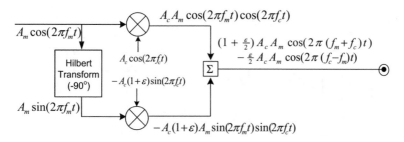

Figure 8-37: Image cancellation degradation due to amplitude imbalance

the lower sideband due to the imbalance and may disappear as $\varepsilon \rightarrow 0$, and it is given by

$$\left(1 + \frac{\varepsilon}{2}\right)A_cA_m \cos\left(2\pi(f_m + f_c)t\right) - \frac{\varepsilon}{2}A_cA_m \cos\left(2\pi(f_c - f_m)t\right) \tag{8-10}$$

The image rejection in dB is defined by the power ratio of upper sideband (desired) and lower sideband (image), and from (8-10) it is given by

$$\left(\frac{\varepsilon}{2 + \varepsilon}\right)^2 \tag{8-11}$$

$$\text{IR [dB]} = 10\log\left(\frac{\varepsilon}{2 + \varepsilon}\right)^2 \approx 10\log\left(\frac{\varepsilon^2}{4}\right) \tag{8-12}$$

Example 8-10: 0.1 dB amplitude imbalance results in $\varepsilon = 0.012$, and thus IR [dB] $= -45$ dB. 1.0 dB amplitude imbalance will result in -25 dB image rejection. Note that the amplitude imbalance can occur due to the gain difference between in-phase path and quadrature phase path and the effect is the same. This can be seen from Figure 8-37 and also from (8-10).

With the same setup as Figure 8-37 we consider quadrature phase imbalance. This is represented by $\Delta\varphi$ and quadrature carrier is now given by $-A_c \sin(2\pi f_c t + \Delta\varphi)$.

In order to see the degradation due to quadrature phase imbalance we need to express $A_cA_m \cos(2\pi f_m t)\cos(2\pi f_c t) - A_cA_m \sin(2\pi f_m t)\sin(2\pi f_c t + \Delta\varphi)$ in terms of USB and LSB components. They are, after straightforward but tedious manipulations, shown in Figure 8-38, and given by (8-13).

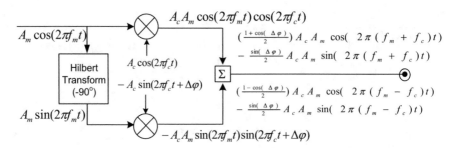

Figure 8-38: Image cancellation degradation due to quadrature phase imbalance. In addition to image component there is interference term; $-\frac{\sin(\Delta\varphi)}{2}A_cA_m\sin(2\pi(f_m+f_c)t)$ in upper side band.

$$\left(\frac{1+\cos(\Delta\varphi)}{2}\right)A_cA_m\cos(2\pi(f_m+f_c)t)-\frac{\sin(\Delta\varphi)}{2}A_cA_m\sin(2\pi(f_m+f_c)t)+$$

$$\left(\frac{1-\cos(\Delta\varphi)}{2}\right)A_cA_m\cos(2\pi(f_m-f_c)t)-\frac{\sin(\Delta\varphi)}{2}A_cA_m\sin(2\pi(f_m-f_c)t)$$

$$(8\text{-}13)$$

LSB power is given by $\left(\frac{1-\cos(\Delta\varphi)}{2}\right)^2+\left(\frac{\sin(\Delta\varphi)}{2}\right)^2$ and USB interference power is given by $\left(\frac{\sin(\Delta\varphi)}{2}\right)^2$, and desired USB power by $\left(\frac{1+\cos(\Delta\varphi)}{2}\right)^2$ where we set $A_cA_m=1$.

$$\text{Image cancelling}:\quad\frac{(1-\cos(\Delta\varphi))^2+\sin^2(\Delta\varphi)}{(1+\cos(\Delta\varphi))^2}\approx\frac{\sin^2(\Delta\varphi)}{4}\qquad(8\text{-}14)$$

$$\text{In-band interference}:\quad\frac{\sin^2(\Delta\varphi)}{(1+\cos(\Delta\varphi))^2}\approx\frac{\sin^2(\Delta\varphi)}{4}\qquad(8\text{-}15)$$

$$\text{IR [dB]}\approx10\log\left(\frac{\sin^2(\Delta\varphi)}{4}\right)\qquad(8\text{-}16)$$

Example 8-11: 1 degree phase imbalance will degrade image cancelling -41 dB and at the same time the interference to signal in USB is -41 dB. The phase imbalance impact is severe in the sense that not only image generation but also interference.

Example 8-12: When both amplitude imbalance and phase imbalance are present, an approximate image rejection would be adding both (8-11) and (8-14). What is image cancelling in dB when both 0.1 dB amplitude imbalance and 1 degree phase imbalance are present? $10\log(10^{-45/10}+10^{-41/10})=-39.5$ dB using the IR

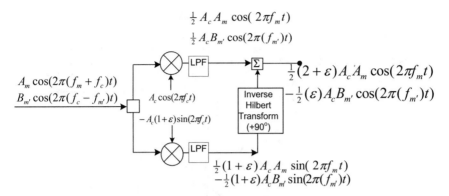

Figure 8-39: Image cancellation degradation due to receive side amplitude imbalance. LPFs eliminate twice carrier frequency components. Un-cancelled image signal, $-\frac{1}{2}(\varepsilon)A_c B_{m'}$ $\cos(2\pi(f_{m'})t)$, is present as interference after demodulation. For inverse Hilbert transform use Table 8-1 from $\hat{x}(t)$ to $x(t)$ direction

[dB] calculation done before. Make sure that this method is equivalent to adding (8-11) and (8-14) and then converting to dB.

Now we consider the impact of quadrature demodulator amplitude and phase imbalance to image cancelling. As expected it is the similar to transmit side analysis except that a signal at the image location should be considered. Thus we use $A_m \cos(2\pi(f_m + f_c)t) + B_{m'} \cos(2\pi(f_c - f_{m'})t)$ as a receive test signal; the first term is desired USB tone and the second term is a signal at the image location. $f_{m'}$ means that the image signal is not exactly the same frequency since it is not its own image, but other signal. It could be represented by $B_{m'} \sin(2\pi(f_c - f_{m'})t)$.

After quadrature demodulation and combining both in-phase and quadrature components the output becomes, as shown in Figure 8-39,

$$\frac{1}{2}(2 + \varepsilon)A_c A_m \cos(2\pi f_m t) - \frac{1}{2}(\varepsilon)A_c B_{m'} \cos(2\pi(f_{m'})t) \qquad (8\text{-}17)$$

Compare (8-17) with (8-10) it is identical except for $B_{m'}$ in the second term. As $B_{m'}$ becomes small, the interference gets small, which can be done by filtering. As $\varepsilon \to 0$, i.e., no gain imbalance, the image can be cancelled completely even $B_{m'}$ is not zero.

IR in dB is given by

$$10 \log \left[\left(\frac{B_{m'}}{A_m}\right)^2 \left(\frac{\varepsilon}{2 + \varepsilon}\right)^2 \right] \qquad (8\text{-}18)$$

With the same receive test signal we consider quadrature phase imbalance, and its impact to image cancellation. This is summarized in Figure 8-40. In this case after quadrature demodulation and combining both in-phase and quadrature components the output contains, after somewhat tedious calculations, 4 terms, and is given by

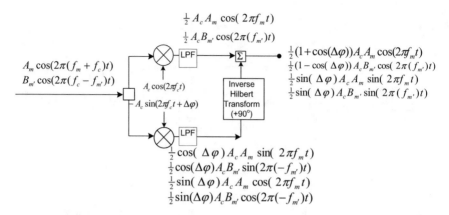

Figure 8-40: Image cancellation degradation due to quadrature phase imbalance

$$\frac{1}{2}(1 + \cos{(\Delta\varphi)})A_cA_m \cos{(2\pi f_mt)} + \frac{1}{2}(1 - \cos{(\Delta\varphi)})A_cB_{m'} \cos{(2\pi(\ f_{m'})t)}$$

$$+\frac{1}{2}\sin{(\Delta\varphi)}A_cA_m \sin{(2\pi f_mt)} + \frac{1}{2}\sin{(\Delta\varphi)}A_cB_{m'} \sin{(2\pi(\ f_{m'})t)}$$

$$(8\text{-}19)$$

The first term is desired signal and the rest are interference due to the phase imbalance ($\Delta\varphi$). This is shown in Figure 8-40. Compare (8-19) with (8-13) and they are identical except for $B_{m'}$.

Then image rejection is given by

$$\text{Image cancelling}: \frac{(1 - \cos{(\Delta\varphi)})^2 + \sin^2(\Delta\varphi)}{(1 + \cos{(\Delta\varphi)})^2}\left(\frac{B_{m'}}{A_m}\right)^2$$

$$\approx \frac{\sin^2(\Delta\varphi)}{4}\left(\frac{B_{m'}}{A_m}\right)^2 \qquad (8\text{-}20)$$

$$\text{In-band interference}: \frac{\sin^2(\Delta\varphi)}{(1 + \cos{(\Delta\varphi)})^2} \approx \frac{\sin^2(\Delta\varphi)}{4} \qquad (8\text{-}21)$$

IQ phase imbalance problem

Note that the phase imbalance creates in-band interference in addition to generating image signal even though there is no image signal. This can be seen in Figure 8-40; there are four terms in the output of demodulator. The in-band interference is $\frac{1}{2}\sin{(\Delta\varphi)}A_cA_m \sin{(2\pi f_mt)}$, which cannot be suppressed by filtering while the image signals can be suppressed further by filtering, i.e. $B_{m'} \to 0$. This can also be seen from (8-20) and (8-21) for receive side as well as from (8-15) and (8-16) for transmit side. The digital IF block can suppress the image signal, but cannot do anything to reduce the in-band interference due to the phase imbalance ($\Delta\varphi$). Thus it is necessary to reduce the phase imbalance of an analog quadrature modulator and demodulator to acceptable level. However, $\Delta\varphi$ can be measured as part of digital signal processing, and then counter-measured (thus cancelled the phase imbalance further). *A fast and effective, believed to be new, scheme will be explained in conjunction with carrier phase recovery process, separately in detail in Chapter 7.*

The analysis method is directly applicable to the case with digital IF block. In order to see it note that the digital IF block is essentially to translate the frequency by digital IF frequency (f_{IF}). Assuming digital IF frequency translation is perfect, the test signal is the same single tone except for the frequency change from f_m to $f_m + f_{IF}$ at the input to analog quadrature modulator and at the output of analog demodulator.

Exercise 8-6: In Figure 8-18, apply a single tone and obtain a_d and b_d.

8.1.6.8   DC Offset – Filtering and Cancellation

DC offset in the baseband becomes a carrier leakage in RF signal, and vice versa. Carrier (local oscillator) leakage in analog quadrature modulator is not removed by inserting digital IF image cancelling stage. In fact, we use the same quadrature modulator as in direct conversion transceiver. Major difference is that DC offset is out of band with digital IF whereas it is in-band with direct conversion. In other words, RF spectrum in Figure 8-17 shows that carrier leakage at f_c is not overlapped with the main signal. Thus carrier leakage filtering or DC cancellation at the baseband can be done easily since there is no signal there for digital IF case but the same operation should be very carefully done, not to damage the main signal, for direct conversion transceiver. Baseband gain block does not need to be DC coupled, and a simple DC blocking or DC cancellation scheme would be used. DC coupled amplifiers are prone to DC offset unless DC is cancelled carefully.

In RF as shown in Figure 8-17, carrier frequency can be filtered by a notch filter tuned to the frequency. In baseband it can be filtered by a high pass filter, a simple RC filter blocking DC as shown Figure 8-44. DC offset can be filtered by a digital

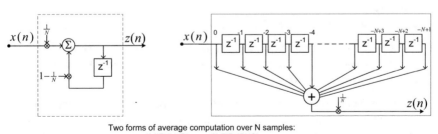

Two forms of average computation over N samples:
1) IIR (LHS)   2) FIR (RHS)

Figure 8-41: Two forms of average computation for N samples

Figure 8-42: DC
cancellation with average
(DC offset)

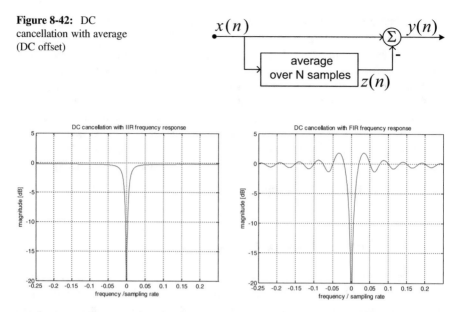

Figure 8-43: Frequency response of DC cancellation with IIR and FIR DC average, $N = 20$

filter. In one design, all of these three methods may be deployed or one method may
be good enough. These are choices that a system designer may use appropriately.

We discuss a digital cancellation first. When there are N, signal samples in digital
form we can estimate DC by computing average of those samples. This can be
represented by a FIR filter. A similar moving average can be computed by a simple
IIR filter. It is shown in Figure 8-41.

When DC, i.e., an average is estimated, then it is subtracted from the main signal.
This is shown in Figure 8-42, and a transfer function of the cancellation using IIR
DC estimation is given by $1 - \frac{1-a}{1-az^{-1}}$ where $a = 1 - \frac{1}{N}$, and the transfer function
using FIR filter is given by $1 - \frac{1}{N}\frac{1-z^{-N}}{1-z^{-1}}$ where we use the identity
$1 - z^{-N} = (1 - z^{-1})(1 + z^{-1} + z^{-2}.. + z^{-N+1})$. Frequency responses, for example,
are shown in Figure 8-43.

As the sample size (N) gets larger, the corner frequency gets closer to DC.

Exercise 8-7: A sampling frequency is 20 MHz and the required corner frequency
is 100 kHz. What is N, a rough sample size? $N = 200$. Change the corner frequency
to 1 MHz, and then $N = 20$.

Now consider analog and RF filtering of DC / carrier frequency. A simple analog
filter can be RC shown in Figure 8-44, and its frequency response is shown next. A
simple LC circuit can be used as a RF notch filter as in the same figure. Depending
on the Q of a coil its frequency response can be obtained.

Note that a simple DC blocking circuit in Figure 8-44 may be interpreted as a DC
cancellation using a RC low pass filter as average estimation block since the
canceller transfer function is given by

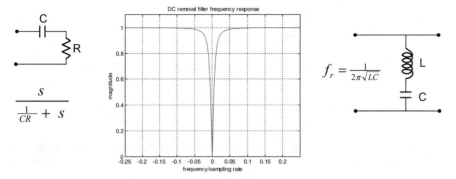

$$\frac{S}{\frac{1}{CR} + s}$$

$$f_r = \frac{1}{2\pi\sqrt{LC}}$$

Figure 8-44: RC high-pass filter, transfer function, and its frequency response. RF notch filter can be implemented by an LC

$$1 - \frac{\frac{1}{CR}}{\frac{1}{CR} + s} = \frac{s}{\frac{1}{CR} + s} \qquad (8\text{-}22)$$

Example 8-13: Work out the detail of (8-22), and draw a block diagram where ▷ a unity gain amplifier is for isolation.

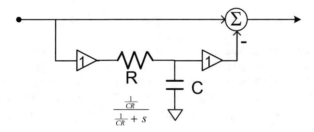

$$\frac{\frac{1}{CR}}{\frac{1}{CR} + s}$$

From (8-22), a digital filter can be derived by approximating differentiation by difference $s = 1 - z^{-1}$. Substituting it into (8-22), it becomes

$$a\frac{1 - z^{-1}}{(1 - az^{-1})} \text{ with } a = \frac{CR}{1 + CR} \qquad (8\text{-}23)$$

Exercise 8-8: Derive transfer a function of digital filters in Figure 8-45 and confirm that they are the same.

Exercise 8-9: We wish to change the gain of a filter to be unity at $z^{-1} = -1$ in Figure 8-45. Find the gain factor necessary. Answer $= \frac{1+a}{2a}$.

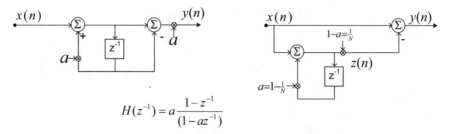

$$H(z^{-1}) = a\frac{1-z^{-1}}{(1-az^{-1})}$$

Figure 8-45: Digital filter from analog DC blocking (RC filter) in Figure 8-44, and it is the same as DC cancellation from Figure 8-42 with IIR average estimation. a is a filter parameter to determine a cutoff frequency

Exercise 8-10: In Figure 8-45, show that $a = 0.95$ corresponds to $N = 20$. Thus the cutoff frequency is related with the length of samples for averaging.

Additional comments on DC cancellation (filtering) are in order. The standards of IEEE 802.11a, 11g, 11n, and 11ac use OFDM signaling format. In order to make DC offset filtering easy, one sub-channel at DC (subcarrier index $k = 0$) is not used for data transmission. Emptying subcarrier index $k = 0$ is not necessary for digital IF transceiver since the sub-channel is at digital carrier frequency after digital IF modulation. The sub-channel of index $k = 0$ is well usable with digital IF. Direct conversion transceiver is used for OFDM signaling without using DC sub-channel. However, calibrations to cancel DC during power up seem to be used to handle DC filtering without damaging nearby sub-channels. With digital IF, DC filtering can be done easily without impacting the signal of interest.

Most digitally modulated signals with a single carrier as opposed to multiple sub-carriers like OFDM have a spectrum with high energy content near DC, often the largest. DC filtering introduces a form of ISI (inter-symbol interference) and thus degrades performance. The cutoff frequency of DC filtering should be small enough so that ISI should be negligible. It can be more demanding as modulation levels go higher, e.g., 256-QAM.

8.1.6.9 Different Transceiver Architectures with Digital IF

A basic architecture, with digital IF image cancelling, is shown in Figure 8-17. In this section we discuss different transceiver architectures with digital IF conversion. One variation is shown in Figure 8-46.

Compared to Figure 8-17 of digital image cancelling, this architecture is much simpler but requires sharp image rejecting filters. In order to make them in one frequency, rather than frequency variable, one may add an analog IF stage. (See Figure 8-47 as an example.) Implementing a sharp IF/RF filter inside chip may be challenging. Assuming it is available the architecture is attractive; in addition to simplicity, IQ amplitude and phase imbalance problems are absent. However, RF

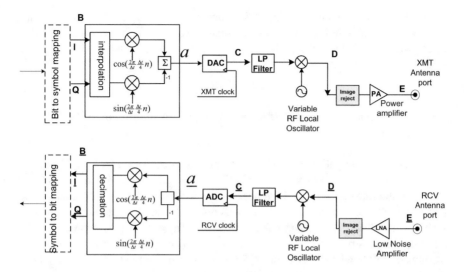

Figure 8-46: Digital IF conversion architecture without digital image cancelling but using image rejecting filters

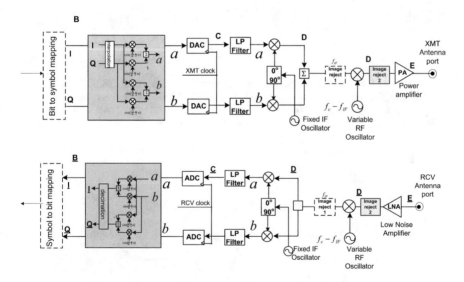

Figure 8-47: Fixed IF stage insertion to a transceiver architecture. Image filters are shown with dotted lines, which means that they are not essential but will improve the rejection further. Image reject 2 are necessary but can be simple depending on IF frequency, which should be reasonably high

CMOS filter technology may still need to be developed in the future. *In this architecture we may rely heavily on a good filter technology in RFIC.* Currently it seems not available. When there is no need of RF translation, e.g., backplane trace interconnection, this low digital IF scheme in Figure 8-46 is well usable.

An IF stage can be added to Figure 8-17 and resulted in Figure 8-47 as it is added to direct conversion in Figure 8-2 and resulted in Figure 8-4. We showed the image filters with dotted line, which means that they are not essential but will improve further on image rejection.

8.1.6.10 Alternative Form of Interpolation and Decimation of OFDM Signals

We consider an alternative form of interpolation and decimation for OFDM signals.

In Figure 8-17 the interpolation to 4 times of symbol rate in general, $1/\Delta t$, is shown in the transmit side, and the decimation, inverse of interpolation, should be performed at the receive side. In sub-section 8.1.6.5, detailed implementation schemes are discussed, in particular taking advantage of FIR structure to make into poly-phase structure as shown in Figure 8-34 (interpolator) and in Figure 8-35 (decimator). When the interpolator of Figure 8-34 is applied to OFDM signal, it can

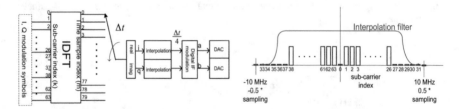

Figure 8-48: OFDM signal generation with interpolation after IDFT. IEEE802.11algl with 20 MHz sampling rate as a concrete example, i.e., $\Delta t = 1/20e6$. The frequency response of an interpolation filter is shown (RHS)

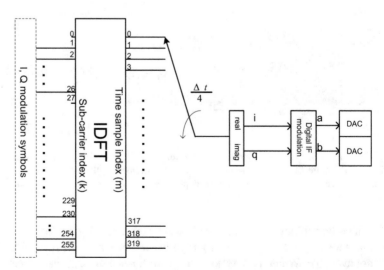

Figure 8-49: IDFT is used for interpolation, example of 4 times from IDFT size 64 to 256. This figure is equivalent to Figure 8-48 implementation

be shown as in Figure 8-48. The frequency response of an interpolation filter may look like one in Figure 8-48 RHS. Note that actual use of subcarriers is 52 out of 64, and that cyclic prefix of 16 is added to 64.

Alternatively the interpolation may be combined with IDFT by increasing the size of IDFT from 64 to 256, a factor of 4 (the same as interpolation factor). This is shown in Figure 8-49.

Exercise 8-11: Similarly the decimation can be combined into DFT at the receiver by increasing the size of DFT by a factor of 4 (the same as decimation factor). This is left as an exercise as it is not hard to see it.

8.1.6.11 Multi-channel Processing with Digital IF

Here we show that a digital IF transceiver, with image cancelling, can be used for multi-channel receiver processing. In sub-section 8.1.6.2, we showed in great detail that indeed the digital IF stage proposed in Figure 8-17 can be capable of cancelling unwanted sideband (or image). In step by step we started from Figure 8-20, RF spectrum of a receiver, and arrived at Figure 8-27. In the process we saw the lower sideband cancellation. However, cancelling upper sideband instead of lower sideband requires a small change in digital IF demodulation. See Figure 8-13.

This can be seen easily by considering complex exponential rather than sinusoidal processing. See Example 8-8, which is captured as Figure 8-50 for convenience, and subsequent Exercise.

We can obtain the upper sideband as shown in Figure 8-27 by considering $(c(t) + jd(t))e^{-j\frac{2\pi}{\Delta t}t}$, and $(\bar{c}(t) + j\bar{d}(t))e^{-j\frac{2\pi}{\Delta t}t}$. We can obtain the lower sidebands by $(c(t) + jd(t))e^{+j\frac{2\pi}{\Delta t}t}$, and $(\bar{c}(t) + j\bar{d}(t))e^{+j\frac{2\pi}{\Delta t}t}$. We may visualize the processing moving the spectrum to shift right on the frequency axis to get LSB. This can be translated into a sign change in sinusoidal processing. For mid-band channel, digital IF carrier frequency is zero.

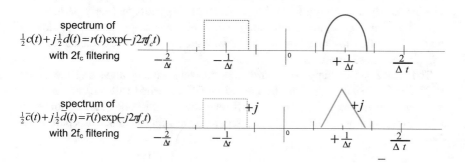

Figure 8-50: Analytical representation of $c(t) + jd(t)$ and $\bar{c}(t) + j\bar{d}(t)$

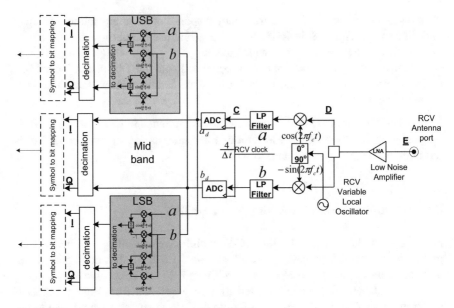

Figure 8-51: Multi-channel processing with digital IF

Thus we can process three channels as shown in Figure 8-51. As the sampling rate gets higher more channels can be processed. For example, 5 channels with 8 times of symbol rate sampling, and 7 channels with 12 times of symbol rate sampling.

8.1.6.12   Summary of Digital Image Cancelling Transceiver

We proposed a low digital IF transceiver architecture with image cancelling. We showed its cancellation capability, somewhat counter-intuitive since two sidebands are mixed before digital IF conversion to baseband. We still rely on the precision of analog quadrature modulator I demodulator balancing amplitude and phase between in-phase circuit and quadrature phase circuit even if we assume digital IF is perfect. (We can safely assume digital IF implementation is perfect or nearly perfect.) IQ phase imbalance generates a cross-talk noise between in-phase and quadrature phase, in addition to degrading image suppression. IQ amplitude imbalance only degrades image suppression.

The image can be suppressed further by filtering without impacting the signal itself. For LO leakage, this can be filtered or DC cancelled (filtered) at baseband again without impacting the signal itself, and thus easily done. This contrasts to direct conversion where DC filtering should be done with care since it is part of the signal.

8.1.7 Calibration of Quadrature Modulator | Demodulator

In Section 8.1.3 and 8.1.4, we discussed the implementation issues of quadrature modulator and demodulator addressing gain and phase imbalance and DC offset (LO leakage) as shown in Figure 8-7 and Figure 8-8. Here we discuss the same issues but with specific test signals, rather than modulated traffic signals, in order to measure the quality of quadrature modulator and demodulator. This quality measurement can be extended to calibrations.

We first consider the receive side, i.e., quadrature demodulator; the test signal is a single RF tone above carrier frequency; $f_c + f_m$, and after demodulation we observe baseband signals, which will be cosine in in-phase and sine in quadrature phase. This is shown Figure 8-39 and 8-40. When there is LO leakage, there is DC offset in the baseband. When there is amplitude and phase imbalance in quadrature demodulator, there will be amplitude imbalance and crosstalk between in-phase (cosine) and quadrature (sine) path. If the in-phase is applied to horizontal input and the quadrature phase is applied to vertical input to an oscilloscope (if analog) to digital display, these can be captured as shown in Figure 8-52. The gain difference shows up as elongation of a circle, the phase imbalance blurs circle due to interference, and the center of the circle is shifted away from an origin due to DC offset. Calibration process will make the circle centered at the origin, without elongation and without blurring.

We now consider the transmit side, i.e., quadrature modulator; here test signals are a single tone with 90° phase difference, i.e., cosine and sine. When the quadrature modulator is perfect, it will generate a single RF tone at $f_c + f_m$ (i.e., we choose upper sideband).

However, when there is the gain and phase imbalance in quadrature modulator, there will be unsuppressed image and the cross-talk between in-phase (cosine) path

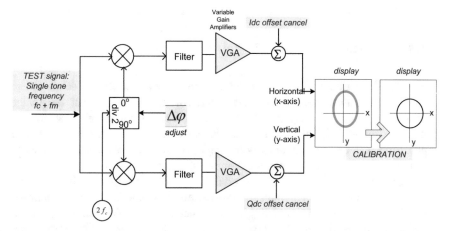

Figure 8-52: Quality measurement of quadrature demodulator and possible correction with calibrations (the arrow indicates). Display of a circle is visual purpose and may be useful for analog signals but for digital signals, numeric computations are easy to do than display

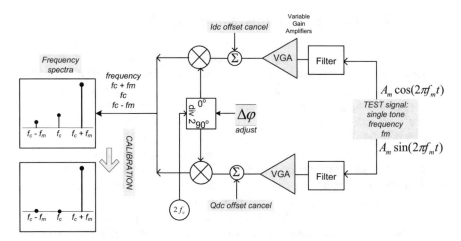

Figure 8-53: Quality measurement and calibration of quadrature modulator

and quadrature phase (sine) path as shown in Figure 8-37 and 8-38. If the frequency spectra are displayed, the unsuppressed sideband is at $f_c - f_m$. DC offset shows up as LO leakage at f_c. This is shown in Figure 8-53. The calibration process will suppress the spectra at f_c and $f_c - f_m$ as shown in the figure.

Exercise 8-12: Alternative to spectra display in measuring the quality of quadrature modulator is to use a calibrated demodulator. Connect the output of transmit RF to the input of a perfect (or well calibrated) demodulator, and then observe in-phase and quadrature phase baseband as in receive side. Then observe a circle display shown in Figure 8-52. Understand this scheme by working out the details with the aid of Figure 8-39 and 8-40.

8.1.8   Summary of Transceiver Architectures

We explored low digital IF image cancelling transceiver architecture in detail. Image cancelling mechanism was clearly demonstrated. It was shown that the image cancelling performance may rely on the amplitude and phase imbalance of RF quadrature mixers, and the relationship between their requirements and image suppression was clearly explained with single tone test signals. And their quality measurement scheme, extendible to calibrations, was also shown.

   Along the exploration, as basic material, direct conversion, heterodyne conversion and other variations of architectures are reviewed as well. SSB signal generation method is utilized for understanding image cancellations since both are the same mechanism. IQ phase imbalance will generate crosstalk (interference) between in-phase and quadrature phase signals in addition to degrading image suppression. A new digital removal method of the IQ phase imbalance at the receiver is elaborated in Section 7.4 of Chapter 7.

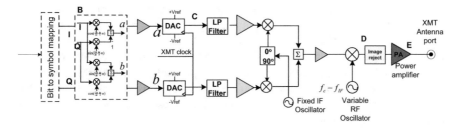

Figure 8-54: Transmit signal chain with explicit gain blocks; digital gain (blue), baseband analog gain (green), IF gain (beige) and RF gain (red). Transmit power is referenced at the antenna port. This figure emphasizes the signal level changes along the path. Transmit power is specified in terms of dBm (e.g., 0 dBm = 1 miliwatt), referenced at the antenna port

8.2   Practical Issues of RF Transmit Signal Generation

Both in this section and the next we will cover practical issues in a transceiver; signal level distribution along the signal chain, how many bits of digital representation in DAC and ADC, and dynamic range of devices (mixers, filters, and amplifiers) used in the chain. This section covers the transmit side and the next section does the receive side.

When we discussed transceiver architectures (e.g., low digital IF image canceling architecture with IF stage shown in Figure 8-47), gain blocks were implicit. In Figure 8-54, we added explicit gain blocks; digital gain (blue), baseband analog gain (green), IF gain (beige) and RF gain (red) in order to show the signal level change along the chain. The power (P), expressed in dBm, can be related with the voltage (V) by specifying terminating resister (e.g., $R = 50$ ohm); $P = V^2/R$ [watt]. In RF / IF signal, it is convenient to use the power, but in baseband, the voltage may be convenient. By specifying the terminating resister R, a signal level can be specified by a power, even in baseband; a voltage is converted to a power assuming a (hypothetical) termination with R.

Analog baseband signals after DAC should be amplified through the chain, and the power at the antenna port should be a desired power level. Transmit power is typically referenced at the antenna port.

Example 8-14: When a desired power at the antenna port is +20 dBm, and the signal level at one of DACs (both I and Q respectively) is −5 dBm, what is the overall gain in dB necessary?

Answer: +22 dB since I and Q combined power is −2 dBm. Note that the gain of 22 dB will be distributed as analog baseband, IF gain and RF gain including PA.

How do you distribute the gain or maintain signal levels along the path to maximize the signal quality? And how many bits in DAC are necessary to satisfy the signal quality requirements? We will answer these questions from system design point of view.

Signal quality may be measured by signal to noise (and interference) ratio. Here the noise is not thermal noise, but 'circuit noise', quantization noise, overload distortion, and inter-modulation noise due to non-linearity. The term 'circuit noise' may be related with physical design (circuit partition, isolation, layout, clock distribution etc....), and detailed characterization of it would be in the realm of physical circuit design and beyond the scope of system design level. Ideally 'circuit noise' should be below quantization noise but probably above thermal noise. We may assume 'circuit noise' is smaller than or comparable to quantization noise. Thus here we compute quantization noise explicitly but the estimation of 'circuit noise' in detail is beyond the scope.

8.2.1 DAC

A DAC converts a digital number into an analog voltage. 3 bit DAC example is shown in Table 8-2. It is 2's complement digital representation and its reference voltage is given as $\pm$Vref. Sign-magnitude representation is shown as well in the table.

Note that the output voltage is given by fractional value times Vref (reference voltage). The fractional value may be obtained by integer value $/2^{b-1}$ where b is the number of bits of DAC. In 2's complement representation, MSB (most significant bit) is, interpreted as negative value of 2^{b-1}, i.e., -2^{b-1} with integer digital value and this is equivalent to the value of -1 with fractional digital value if '1', otherwise ('0') zero.

Exercise 8-13: In sign-magnitude, MSB is the sign; '0' means positive, '1' means negative. In 2's complement, MSB carries the value of -2^{b-1} (or -1 with fractional value) if '1' and zero if '0'. See these statements are true from Table 8-2.

Hint: Examine the table carefully.

Table 8-2: 3 bit 2's complement DAC example with voltage range $= \pm$Vref

INPUT	2's complement digital value		OUTPUT	sign-magnitude digital value	
digital representation (3 bit)	integer	fraction	analog voltage with reference voltage *Vref	integer	fraction
000	+0	+0.00	0.00*Vref	+0	+0.00
001	+1	+0.25	0.25*Vref	+1	+0.25
010	+2	+0.50	0.50*Vref	+2	+0.50
011	+3	+0.75	0.75*Vref	+3	+0.75
100	−4	−1.00	−1.00*Vref	−0	−0.00
101	−3	−0.75	−0.75*Vref	−1	−0.25
110	−2	−0.50	−0.50*Vref	−2	−0.50
111	−1	−0.25	−0.25*Vref	−3	−0.75

Figure 8-55: 3 bit 2's complement DAC and its input-output relationship. Note that only discrete points (black dots) are defined. The output voltage may deviate from ideal points (black dots) within the range (vertical red line) of −0.125Vref to +0.125Vref from an ideal point. This output voltage uncertainty (in general −1/2^{b}Vref to +1/2^{b}Vref, i.e., ±LSB) will create quantization noise

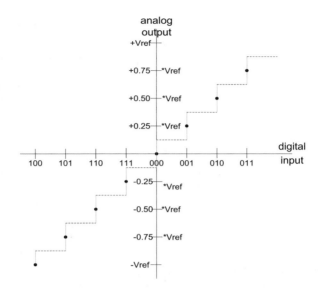

The input – output relationship may be seen graphically as in Figure 8-55. Note that only discrete points (black dots) are defined for an ideal DAC. However, a practical DAC may generally have a voltage uncertainty shown in vertical red lines.

Exercise 8-14: Suppose that the voltage uncertainty is in general −1/2^{b}Vref to +1/2^{b}Vref with b bits. Show that, with $b = 10$, the uncertainty range is about ±0.1024 % of Vref, i.e., ±LSB (least significant bit).

Exercise 8-15: Consider 8 bit 2's complement DAC with Vref = 0.5 V. What is the analog voltage corresponding to 0100001, and 1111000?
   Answer: (+1/2 +1/64)0.5 = 0.2578 V, and (−1+1/2+1/4+1/8+1/16) 0.5 = −0.0313 V

Error voltage of DAC and its PDF
The quantization noise is due to the output voltage uncertainty. A statistical treatment of the voltage uncertainty requires a probability density distribution function (PDF). In practice getting a PDF is not trivial but a plausible assumption, perhaps confirmed by experience, may be used.

A uniform distribution is often used in practice; all values, within the range of uncertainty, are equally likely with the average corresponding to an ideal point. This means the average is zero. Another PDF, somewhat pessimistic, may be most likely values being around the two edges with the equal probability. Thus 2 point discrete PDF with the probability ½. These are shown Table 8-3.

Exercise 8-16: Another 2 point discrete PDF is shown in the third column. Confirm the power of error voltage in Table 8-3. Hint: start with the definition of variance.

Table 8-3: Different PDFs of DAC error voltage

PDF of error voltage	Uniform	2 point discrete	±½ LSB error
Graph $q = \frac{1}{2^b} Vref$			
power of error voltage	$\sigma^2 = q^2/3$	$\sigma^2 = q^2$	$\sigma^2 = q^2/4$
RMS of error voltage	$rms = q/\sqrt{3} \approx 0.577q$	$rms = q$	$rms = q/2$
Vref / error voltage RMS	$2^b\sqrt{3}$	2^b	$2^b\,2$

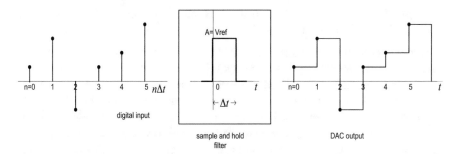

Figure 8-56:   DAC is a filter and its impulse response is a rectangular pulse (sample and hold) with pulse width Δt and with amplitude A = Vref when the input binary values are fractional. The normalized magnitude of frequency response is called sin(x)/(x) where $x = \pi \Delta t f$

Conversion rate and frequency response
In addition to resolution (a number of bits), a conversion rate of a DAC is an important parameter to be specified. A conversion rate is the inverse of conversion time (Δt). The analog output is maintained during the duration of conversion time, called sample and hold. This is equivalent to a filter with rectangular pulse shape. Its pulse width is Δt and amplitude is A = Vref if input binary samples have fractional value representation.

Exercise 8-17:   Find a frequency response of a filter of rectangular pulse shown in Figure 8-56. Answer: $H(f) = V_{ref} \Delta t \, e^{-j\pi\Delta t f} \frac{\sin(x)}{(x)}$ where $x = \pi\Delta t f$. Try out starting from the definition of Fourier Transform. Note that the frequency response has a null at the conversion rate and its integer multiples.

Peak to average power ratio (PAPR) of a signal
One of signal characterizations of a signal in time domain is peak to average power ratio. A single tone sine wave has 2 (i.e., 3 dB) since its power is $\frac{1}{2} A^2$ where A is the peak amplitude.

Example 8-15:   Consider PAPR of N tones of equal amplitude sine wave. Its power is $N \, 1/2A^2$ and the largest possible peak is $N*A$. Thus PAPR is $10*\log(2N)$ in dB.

Example 8-16:   A signal with 2 point discrete distribution (e.g., error voltage of DAC), with the range [−q to +q], has PAPR of $10 \log 1 = 0$ dB since its power is q^2. A random pulse train of rectangular shape with two possible values of −1 V to +1 V (100% duty cycle) is another example of 0 dB PAPR. A trivial example of 0 dB PAPR is DC

Example 8-17:   A signal with uniform distribution (e.g., error voltage of DAC) has PAPR of $10 \log 3 = 4.77$ dB since its power is $q^2/3$ with the range [−q to +q].

In general PAPR of a signal requires the probability distribution function of its amplitude in order to know how often the peak happens.

Example 8-18: A signal has a normal (Gaussian) distribution with mean m = 0. What is PAPR of the signal? It could be infinity with approaching zero probability.

Clipping at 3σ (overload distortion), PAPR is $10\log(3^2) = 9.5$ dB, and clipping happens with the probability 0.0027 (= Q(3)*2), i.e., 0.27%.

Clipping at $3.5\ \sigma$ (overload distortion less than 3σ), PAPR is $10\log(3.5^2) = 10.9$ dB, and clipping happens with the probability 4.6e-004 (= Q(3.5)*2).

Clipping at $4\ \sigma$ (overload distortion less than 3.5σ), PAPR is $10\log(4^2) = 12.0$ dB, and clipping happens with the probability 6.3e-005 (= Q(4)*2).

Note: $Q(x) = \frac{1}{\sqrt{2\pi}}\int_x^\infty e^{-\frac{y^2}{2}}dy$ i.e., $Q(x)$ is the tail end probability of zero mean and unity variance Gaussian.

Computation of the amplitude distribution of a signal is to categorize each sample of a signal into the categories of range, and to count the occurrence frequency in each range category, called histogram. It is the same as PDF function calculation from samples. It is straightforward but tedious. This amplitude PDF is useful to decide how much back off is necessary for power amplifier as well, which will be discussed later in this chapter.

Signal to quantization noise ratio at the output of DAC

It is convenient to represent digital samples by the same number of bits as with the resolution bits of DAC (perhaps obvious) and the digital values are in fraction; b-bit fixed point representation. Thus, with this fixed point representation, the maximum possible peak power is 1.0 (unity, i.e., 0 dB).

The peak power that a DAC can handle is the square of reference voltage; V^2_{ref}, which corresponds to the digital peak power 1.0. A digital power (variance of a signal) is converted to analog power by multiplying together; $P_{analog} = P_{digital} * V^2_{ref}$. In terms of voltage, it is $V_{analog} = V_{digital} * V_{ref}$ where $V_{digital} = \sqrt{P_{digital}}$ and $V_{analog} = \sqrt{P_{analog}}$, i.e., the analog voltage is proportional to digital voltage by V_{ref}.

The scaling of b-bit fixed point representation of digital samples should be chosen to maximize signal to quantization noise ratio (SN_qR). The peak power should be 0 dB or digital power should be backed off by PAPR dB;

$$P_{digital}[dB] = 0 - PAPR\ [dB] \qquad (8\text{-}24)$$

$$P_{analog} = V^2_{ref}/PAPR \qquad (8\text{-}25)$$

Example 8-19: PAPR of a signal is known to be 12 dB, and $V_{ref} = 0.5$ V, and termination resistance 50 ohm. What is the optimum power after DAC in dBm?

Answer: $P_v = 0.5^2 * 10^{-12/10} = 0.25 * 0.0631 = 0.0158$, $V_{analog} = \sqrt{P_v} = 0.1256$ [V], P_{analog} [dBm] $= 10\log(0.0158/50*1000) = -5.01$ dBm.

We can calculate SN_qR for <u>uniform distribution</u>.

Figure 8-57: Integral non-linearity (INL) within $\pm\frac{1}{2}$ LSB (inside green shade)

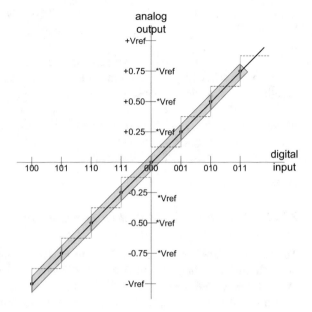

$$\mathrm{SN_qR} = \frac{P_{analog}}{P_{error}} = \frac{2^{2b}}{PAPR}3 \tag{8-26}$$

$$\mathrm{SN_qR\ [dB]} = b6.02 + [4.77] - \mathrm{PAPR\ [dB]} \tag{8-27}$$

With 2 point discrete distribution (LSB error), it is reduced by the factor 3 (or 4.77 dB) compared to uniform distribution. With 2 point discrete distribution (1/2 LSB error), it is increased by the factor 4/3 (1.23 dB) over uniform distribution. The quality of DAC, even with the same resolution bit (b) influences $\mathrm{SN_qR}$ differently. In (8-27), the first term is due to resolution bits, the second term due to the quality of DAC and the third signal dynamic range expressed by PAPR.

Exercise 8-18: With 2 point distribution of quantization error (LSB error), and PAPR $=$ 12 [dB], $\mathrm{SN_qR}$ [dB] $=$ 42 dB is required. How many bits of DAC is necessary? Answer: From (8-27), with 4.77 dB reduction, $42 = b\,6.02 - 12$. Solving it we obtain $b = 9$ bits.

Exercise 8-19: In the above exercise how many bits are allocated for overload distortion? Answer: 12 dB PAPR corresponds 2 bits, and thus 7 bits out of total 9 bits are used for quantization noise.

In practice, PAPR can be a design parameter; if less overload distortion is desired, it should allow it to be larger, and then more bits in DAC are necessary.

Impairments of DAC, and testing
A systematic deviation from the ideal points (black dots) degrades DAC performance. In Figure 8-57, the integral non-linearity (INL) within $\pm\frac{1}{2}$ LSB (inside green

shade) is shown. Differential non-linearity (DNL) is DAC voltage difference between adjacent codes. If it is LSB for every code, then DNL is zero. In Figure 8-57, it is 1 LSB.

DC offset is another possible impairment of DAC.

The pulse shape may deviate from a clean rectangular pulse, such as overshooting, thus the frequency response may deviate from sin(x)/(x) type.

Testing of DAC can be done by a digital single tone with different frequency or by random patterns. For time domain an oscilloscope trace can be observed with horizontal triggering by conversion clock. For frequency domain – e.g., harmonic distortion, a spectrum analyzer can be used.

8.2.2 Transmit Filters and Complex Baseband Equivalence

The output of DAC, staircase signal due to DAC's sample and hold, are smoothed out by a low pass filter. This LPF should compensate for the frequency response of DAC so that overall frequency response should be flat in the passband or should meet transmit pulse shaping with proper transition to stopband. The stopband should have enough attenuation to remove higher digital harmonics and thus to meet clean channel occupancy requirement. This is shown in Figure 8-28. Additionally IF /RF filter may be used in order to meet the clean channel occupancy.

IF / RF filter and their complex baseband equivalence

One can find a complex baseband representation of IF/RF filters, which can be cascaded with baseband filters to see the impact of all filters in the chain. We explain how to find a complex baseband from IF/RF filter frequency response. Without loss of generality we use the center frequency of IF/RF filters as f_{IF} .

Step 1. Move a filter frequency response centered at f_{IF}, $H_{IF}(f)$, to zero frequency (DC); the frequency response of negative frequency should be set zero first, and then move. This is easy visually as shown below. The baseband frequency response is represented by $H_b(f)$.

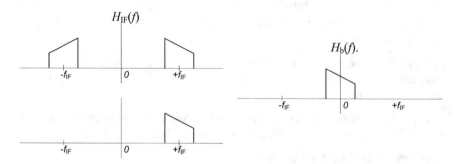

Step 2. The baseband frequency response, $H_b(f)$, is converted to time domain, i.e., the impulse response. Note that this impulse response $h_b(t)$, is not real but

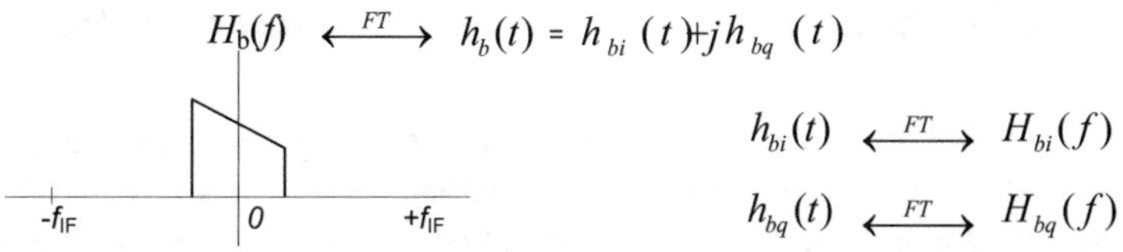

$$H_b(f) \xrightarrow{\ FT\ } h_b(t) = h_{bi}(t) + j h_{bq}(t)$$

$$h_{bi}(t) \xrightarrow{\ FT\ } H_{bi}(f)$$

$$h_{bq}(t) \xrightarrow{\ FT\ } H_{bq}(f)$$

Figure 8-58:   Complex baseband frequency response of IF/RF filters, and corresponding baseband frequency response

Figure 8-59: Overall filter response from baseband to IF/RF filters. Notice there is cross coupling between in-phase and quadrature-phase

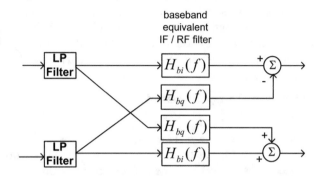

complex, and it is represented by $h_{bi}(t) + j\ h_{bq}(t)$, where $h_{bi}(t)$ and $h_{bq}(t)$ are real. Their corresponding frequency responses are $H_{bi}(f)$ and $H_{bq}(f)$ respectively.

Step 3. Obtain overall frequency response from baseband to IF/RF. This is shown in Figure 8-59.

Example 8-20: Consider an example of a RF filter whose magnitude is even symmetry and phase is odd symmetry around the RF carrier frequency. This is a typical filter frequency response. Determine $H_{bq}(f)$.

Answer: In Figure 8-58, the frequency response is symmetric in magnitude and asymmetric in phase and thus the impulse response is real-valued. Thus $h_{bq}(t)$, must be zero and $H_{bq}(f)$ be so as well. In this case there is no cross coupling. No cross coupling is desired for good performance.

Example 8-21: Consider the impulse response corresponding to $H_{IF}(f) \xleftrightarrow{FT} h_{IF}(t)$. Show that it can be expressed as $h_{IF}(t) = \mathrm{Re}\ \{[\ h_{bi}(t) + j\ h_{bq}(t)]\ e^{\ j2\pi\ f_{IF}t}\ \}$.

Answer: $[\ h_{bi}(t) + j\ h_{bq}(t)]$ is the baseband complex impulse response. $[\ h_{bi}(t) + j\ h_{bq}(t)]\ e^{\ j2\pi\ f_{IF}t}$ is a frequency translation of it to the 'right' only, with no negative frequency part. By taking real part of it, we arrive at the starting point of Step 1 in the above. Note that any IF/RF filter can be represented by the impulse response $\mathrm{Re}\ \{[\ h_{bi}(t) + j\ h_{bq}(t)]\ e^{\ j2\pi\ f_{IF}t}\ \}$, whose Fourier transform is the filter frequency response.

8.2.3 TX Signal Level Distribution and TX Power Control

The signal level of TX chain is high enough so we can ignore thermal noise in TX signal context. Thus the signal quality at the output of DAC plus LPF, expressed by SN_qR, may be degraded further due to 'circuit noise'. The detailed characterization of 'circuit noise' is in the realm of circuit design and thus beyond the scope here. We use it as a kind of background noise. It degrades the signal quality as the analog signal goes through the chain toward antenna, starting from DAC. This means that the signal level should be as high as possible with baseband gain. On the other hand the high level signal may overload some devices along the chain, which creates non-linear distortion, called inter-modulation (IMD) noise. Thus the signal level should be maintained as high as possible without causing IMD noise. A good design should be such that most IMD is due to PA. This will be elaborated further later in this chapter.

Open loop TX power control

Example 8-22: Power at the antenna port is specified to be +24 dBm, and the signal level at one of DACs (both I and Q respectively) is −5 dBm, the overall gain in dB necessary is +26 dB since I and Q combined power is −2 dBm.

This gain of 26 dB (Example 8-22) will be distributed among analog baseband, IF gain and RF gain including PA. A task is how to distribute this. Open loop TX power control means that insertion loss of each device and gain of each amplifier are assigned to meet the required TX power and to maintain the signal quality.

Figure 8-60 shows how signal level should be distributed. However, answers are not unique depending on device characteristics (insertion loss, overload signal level,

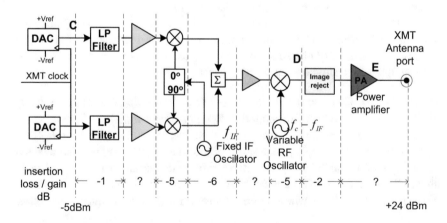

Figure 8-60: An example of signal level fluctuation and gain distribution with open loop control. −5 dBm at DAC (each) should increase to +24 dBm with given insertion loss. Distribute the gain. Total gain necessary 42 dB = 26 dB + 18 dB (compensating insertion losses)

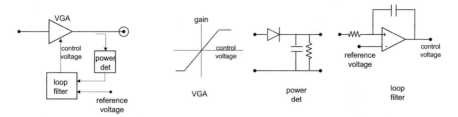

Figure 8-61: Closed loop power control: overall loop and loop components are variable gain amplifier (VGA), power detector, loop filter and reference voltage (to set power level)

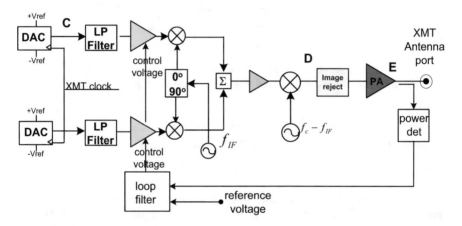

Figure 8-62: Closed loop power control when VGA is baseband amplifier rather than PA

gain range of amplifiers). In practice, a computer program (or spread sheet) may be used.

Closed loop TX power control

In open loop power control, TX power is set at a certain level and may vary slightly due to the change of gain and insertion loss of each device, and due to temperature and supply voltage. If TX power level should be controlled tightly within a small specified range regardless of temperature and device variations, it can be controlled by a closed loop. One possible implementation is shown in Figure 8-61 where there are four components necessary; variable gain amplifier which is characterized by control voltage vs. gain, power detector (power vs. output voltage), loop filter (time constant; how fast the control settles), and reference voltage (setting power level). Ultimately the stability of power level relies on the stability of reference voltage. VGA can be PA itself in Figure 8-61 or baseband amplifier as shown in Figure 8-62.

In some cases TX power level should change because of system requirements. For example, TX power should be controlled to reduce interference when the same

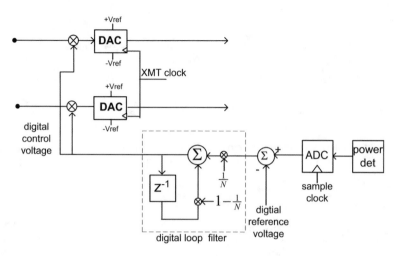

Figure 8-63: Digital closed loop power control; 'digital control voltage' is a scaling of digital number. Digital reference voltage is related with ADC reference voltage

frequency is re-used in cellular environment. In this case 'reference voltage' should be changed to meet the required TX power level.

Example 8-23: From Figure 8-62, VGA (gain variation) is to be done digitally and control voltage, reference voltage and loop filter will be implemented digitally as well. The power detector output voltage should be sampled and converted to a digital number by using ADC. Draw a block diagram by modifying Figure 8-62. How is loop time constant (i.e., how fast the loop settles to a desired state) determined? Hint: take ADC sampling rate into account.

   Answer: It is straightforward to modify Figure 8-62 for overall block diagram and it is left as an exercise. A detailed block diagram for digital control part is shown in Figure 8-63; for digital reference voltage, digital loop filter, digital control voltage, and digital samples being scaled, there are corresponding analog counter-parts. In particular, digital multiplication by 'digital control voltage' corresponds to VGA.

8.2.4   PA and Non-linearity

A design of PA itself requires perhaps experts dedicated the whole of one's career. Our treatment is mostly its external behavior and characteristics in a system. Every device (baseband amplifiers, IF/RF amplifiers, mixers and filters) in the transmit chain has a dynamic range of its own, which is the range of the largest signal from the smallest signal that a device can handle. A device may experience non-linearity when a signal is large. It is PA which delivers a transmit power to the antenna, which means that it deals with a large signal. By properly distributing signal levels, other

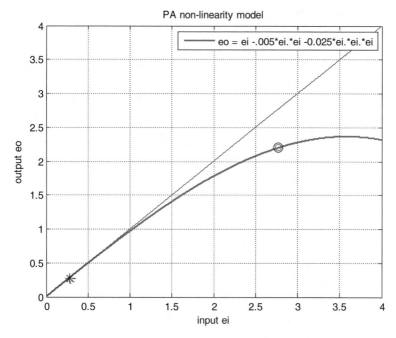

Figure 8-64: PA non-linearity model with a three term polynomial (eo = ei -.005*ei² −0.025*ei³); a small circle (o) indicates 1 dB compression point, which means the power gain is reduced by 1 dB compared to linear gain. A star (*) indicates operating point with 10 dB backoff

devices should experience little non-linearity. This in general is a good system design practice, which implies that 'circuit noise' must be small.

Example 8-24: With 50 ohm load (terminating resistance), what is the output voltage swing (peak to peak) if the output power is +30 dBm (1 watt) and if an input signal is a pure tone? What about +20 dBm (100 mWatt)? +10 dBm (10 mWatt)?

Answer: For 30 dBm, $V_{peak}^2/50/2 = 1.0$ watt, $V_{peak} = \sqrt{100} = 10$ V. Thus the peak to peak swing is 20 V. For 20 dBm, $V_{peak}^2/50/2 = 10^{+20/10}/1000 = 0.1$, $V_{peak} = \sqrt{10} = 3.16$ V, the peak to peak is 6.32 V. For 10 dBm, $V_{peak} = 1.0$ V, thus the peak to peak is 2 V.

PA non-linearity model

The non-linearity can be modeled by many different ways. We give an example with a polynomial model in Figure 8-64.

1 dB compression point – PA power handling capacity

In Figure 8-64, a small circle indicates 1 dB compression point, which means the power gain is reduced by 1 dB compared to linear case (blue line). Note that e_o in the figure is normalized; an actual output power of PA can be obtained by e_o multiplying its power gain. Similarly e_i is normalized by actual power level, say in watt. 1 dB

Table 8-4: PA non-linearity example with three tone test signals

Transfer function Input: 3 tones	$e_o = a_1 e_i + a_2 e_i^2 + a_3 e_i^3$ $e_i = A \cos(\alpha t) + B \cos(\beta t) + C \cos(\gamma t)$	
	Out of band undesired	In-band desired + intermodulation terms
dc	$\frac{1}{2} a_2 (A^2 + B^2 + C^2)$	
1st order		$a_1 A \cos(\alpha t) + a_1 B \cos(\beta t) +$ $a_1 C \cos(\gamma t)$
		$\frac{3}{4} a_3 A (A^2 + 2B^2 + 2C^2) \cos(\alpha t) +$ $\frac{3}{4} a_3 B (B^2 + 2C^2 + 2A^2) \cos(\beta t) +$ $\frac{3}{4} a_3 C (C^2 + 2A^2 + 2B^2) \cos(\gamma t)$
2nd order	$\frac{1}{2} a_2 A^2 \cos(2\alpha t) +$ $\frac{1}{2} a_2 B^2 \cos(2\beta t) +$ $\frac{1}{2} a_2 C^2 \cos(2\gamma t) +$	
	$a_2 AB \cos((\alpha + \beta)t) +$ $a_2 BC \cos((\beta + \gamma)t) +$ $a_2 CA \cos((\gamma + \alpha)t) +$ $a_2 AB \cos((\alpha - \beta)t) +$ $a_2 BC \cos((\beta - \gamma)t) +$ $a_2 CA \cos((\gamma - \alpha)t)$	
3rd order	$\frac{1}{4} a_3 A^3 \cos(3\alpha t) +$ $\frac{1}{4} a_3 B^3 \cos(3\beta t) +$ $\frac{1}{4} a_3 C^3 \cos(3\gamma t) +$	
	$\frac{3}{4} a_3 A^2 B \cos((2\alpha + \beta)t) +$ $\frac{3}{4} a_3 A^2 C \cos((2\alpha + \gamma)t) +$ $\frac{3}{4} a_3 B^2 A \cos((2\beta + \alpha)t) +$ $\frac{3}{4} a_3 B^2 C \cos((2\beta + \gamma)t) +$ $\frac{3}{4} a_3 C^2 A \cos((2\gamma + \alpha)t) +$ $\frac{3}{4} a_3 C^2 B \cos((2\gamma + \beta)t) +$	$\frac{3}{4} a_3 A^2 B \cos((2\alpha - \beta)t) +$ $\frac{3}{4} a_3 A^2 C \cos((2\alpha - \gamma)t) +$ $\frac{3}{4} a_3 B^2 A \cos((2\beta - \alpha)t) +$ $\frac{3}{4} a_3 B^2 C \cos((2\beta - \gamma)t) +$ $\frac{3}{4} a_3 C^2 A \cos((2\gamma - \alpha)t) +$ $\frac{3}{4} a_3 C^2 B \cos((2\gamma - \beta)t)$
	$\frac{3}{2} a_3 ABC \cos((\alpha + \beta + \gamma)t)$	$\frac{3}{2} a_3 ABC \cos((\alpha + \beta - \gamma)t) +$ $\frac{3}{2} a_3 ABC \cos((\alpha - \beta + \gamma)t) +$ $\frac{3}{2} a_3 ABC \cos((\alpha - \beta - \gamma)t) +$

compression point with actual PA output power (e.g., +35 dBm) is often used to indicate the PA power capacity. 1 dB compression power may be considered to be a peak power that a PA can handle. Other indicator is PA saturation power where no further power gain is possible.

Non-linearity creates roughly two types of non-linear distortion –intermodulation (IMD) noise and harmonics. IMD is an in-band (or near) noise due to the non-linearity (mildly deviating slightly from linear relationship). Harmonics are out of band and occur at the integer multiple frequencies of the signal carrier frequency, which is possible only through non-linearity. Since the harmonics are out of band it can be filtered out. On the other hand IMD noise is mixed up with the desired signal and thus once mixed it is not easy to separate it out. The two types of non-linear distortion can be seen easily when we consider pure sine wave tones as input test signal. In Table 8-4, three tone test signals are used as a test signal and the output after passing a non-linear transfer are displayed with the two columns of harmonic terms and intermodulation terms.

Example 8-25: We give a numerical example. See Table 8-4 and Figure 8-64. Note that the intermodulation terms are in-band and harmonic terms are out of band

Transfer function Input: 3 tones	$e_o = e_i - 0.005e_i^2 - 0.025e_i^3 \; ei = \cos(\alpha t) + \cos(\beta t) + \cos(\gamma t)$ Three tone frequency : [1001 MHz, 1002 MHz, 1003 MHz]	
	Out of band undesired harmonics	In-band desired + intermodulation terms
dc	−0.075 [0 MHz]	
1st order		0.9062 [1001, 1002, 1003 MHz]
2nd order	−0.0025 [2002, 2004, 2006 MHz]	
	−0.005 [2003, 2005, 2004 MHz] −0.005 [1, 1, 2 MHz]	
3rd order	−0.063 [3003, 3006, 3009 MHz]	
	−0.0188 [3004, 3005, 3005,3007, 3007, 3008 MHz]	−0.0188 [1000, 999, 1001, 1003, 1005, 1004 MHz]
	−0.0375 [3006 MHz]	−0.0375 [1000, 1002, 1004 MHz]

Exercise 8-20: Try another numerical example. The three tone are the same frequency, but increase the amplitude A = B = C = 1.5. Hint: Use Table 8-4 to compute the amplitudes since the tones due to nonlinearity are the same.

What is the impact to the continuous spectrum signal, i.e., digitally modulated, data carrying, signals? We examined the impact of non-linearity to sinusoidal tones in frequency domain. In addition to the original tones, new tones are created due to non-linearity which we categorize them into two – harmonics away from the original tones and intermodulation distortion of in band and near in band. In the same way, new spectrum will be created for the continuous spectrum. Intermodulation distortions are mixed up with desired signal in band and adjacent bands. This is sometimes called spectral growth.

PAPR of a signal and its impact to PA backoff
For a given PA, with 1 dB compression power which may be considered to be a peak power that it can handle, what is an average operating power that it can handle? We define PA backoff in dB as, with the peak power being 1 dB compression power,

$$\text{Average operating power [dBm]} = \text{peak power [dBm]} \\ - \text{PA backoff [dB]} \qquad (8\text{-}28)$$

When we choose PA backoff [dB] the same as PAPR of an input signal, it is very close to linear operation. If we allow small amount of nonlinearity, PA backoff [dB] may be smaller than PAPR so that operating power can be higher. This is a

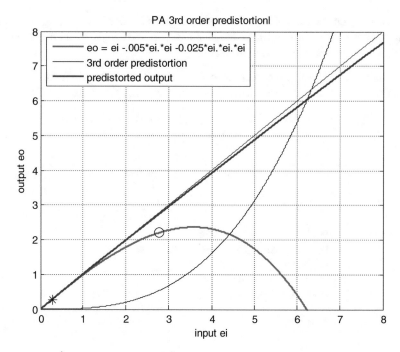

Figure 8-65: 3rd order non-linearity is compensated by pre-distortion (black). In practice typically the improvement is not as dramatic as shown in the figure due to PA model being more complex and due to adjustment of predistorter being not exact

system design parameter. Note that this problem is similar to DAC expressed by (8-24). However, in DAC case, it is backed off exactly by PAPR so that there is very little overload distortion. In PA case, PA backoff [dB] may be smaller than PAPR of a signal and thus overall system IMD is dominated by PA.

Exercise 8-21: The 1-dB compression of a PA is +35 dBm and PAPR of a signal is 12 dB. A system designer decided to use PA backoff 10 dB. What is the operating PA output? How much non-linearity is allowed?

Answer: operating PA output power is +25 dBm ($= 35 - 10$). In Figure 8-64, it is indicated by a star (*). Backoff is 2 dB less than PAPR.

PA non-linearity improvement

In order to increase an average operating power, PA itself should have a large peak power, e.g., 1-dB compression power being high. Inside PA many different techniques are used and beyond the scope here. Sometimes for a given PA, a form of pre-distortion circuits in IF or in baseband may be added to improve operating output power. Even though it is outside of PA it can be treated as effective peak power handling capacity of PA.

In Figure 8-65, see 3rd order pre-distortion (black), output with pre-distortion (thick blue), without pre-distortion (red). Since main distortion is due to mostly 3rd

Figure 8-66: Vestigial symmetric frequency response of ISI free end to end pulse

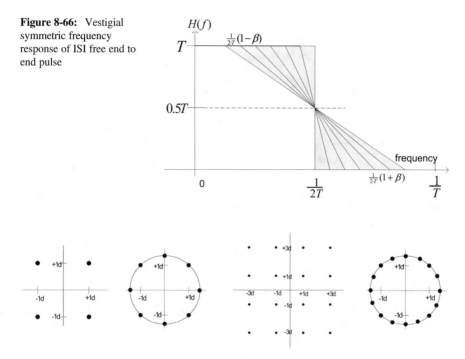

Figure 8-67: QPSK, 8-PSK, 16-QAM, 16-PSK; signal constellation examples

order nonlinearity, the result seems extremely good. In practice, 2 dB improvement is typical since PA nonlinearity is more complex than just the 3rd order and obtaining an exact model (i.e., obtaining correct polynomial coefficient) is limited.

PAPR of a signal and its reduction

Another way to increase an average operating power is to reduce PAPR of a signal itself so that the backoff is less. This can be seen from (8-28). In order to reduce PAPR we need to understand how PAPR of a signal can be related with signal construction – pulse shape (TX filter), signal constellation and modulation (e.g., QAM, PSK, FSK), and single carrier vs. multicarrier (OFDM). We explain briefly these parameters.

Figure 8-66 is a set of frequency responses for pulse shaping – end to end pulse. Intersymbol interference (ISI) free end to end pulse requires the vestigial symmetry around half of symbol rate (Nyquist frequency). A raised cosine pulse is another set of such family.

The square root of this frequency response will be used in transmit side. A design parameter is β, called excess bandwidth; as β gets close zero, PAPR increases, but, less bandwidth is required.

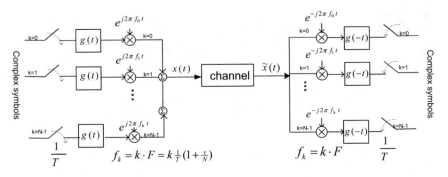

Figure 8-68: Multicarrier signal generation and reception in analog representation. This can be implemented using IDFT and DFT when subcarrier frequency is related with 1/T symbol rate. $g(t)$ is typically a rectangular pulse and then is called OFDM

PAPR dependency on constellation and modulation will be shown below. In Figure 8-67, different constellations are shown; QPSK, 8-PSK, 16-QAM and 16-PSK.

Exercise 8-22: Show that PSK constellations have PAPR $= 1.0$. Answer: Peak power $= 2d^2$ for all constellation points. Average power is $2d^2$ $1/4*4 = 2d^2$ for QPSK, $2d^2$ $1/8*8 = 2d^2$ for 8-PSK and $2d^2$ $1/16*16 = 2d^2$ for 16-PSK.

Exercise 8-23: Find PAPR for 16-QAM. Answer: Peak power $= 2*3^2$ $d^2 = 18$ d^2 for 4 corner constellation points. Average power $= 1/16$ $(4*2d^2 + 4*18d^2+8* 10d^2) = 10$ d^2. We assume all constellation points are equally likely. PAPR $= 9/5$ (or 2.56 dB).

Exercise 8-24: Find PAPR for 64-QAM. Answer: Peak power $= 2* 7^2$ d^2. Average power $= 2(M^2-1)/3$ $d^2 = 2*21$ d^2 (work out the formula). PAPR $= 7/3$ (or 3.68 dB).

16-PSK is less often used in practice. 16-QAM performs better than 16-PSK for a given average signal power to ratio (SNR). If PAPR reduction is more important in a system design, 16-PSK will be chosen against 16-QAM. One may consider FSK modulation against QAM since its PAPR is the same as sine wave. Sometimes 32-QAM or 128-QAM may be chosen in the consideration of PAPR since it does not have outer corner constellation point and hence less PAPR compared to square constellations (e.g., 64-QAM, 256-QAM).

We consider PAPR of multicarrier signal, i.e., OFDM signals.

A multicarrier signal such as OFDM is used by most wideband systems (e.g., IEEE 802.11a, n, g, ac and 4G LTE). This trend will continue. The essence of OFDM can be understood readily from its analog equivalent representation shown in Figure 8-68. A single carrier system uses only one branch. In practice it is implemented using IDFT and DFT when subcarrier frequency spacing is related with 1/T symbol rate; $f_k = k\frac{1}{T}\left(1 + \frac{v}{N}\right)$ where k is the subcarrier index, and N is the total number of subcarriers, and v is guard period (or cyclic prefix) $g(t)$ is typically a rectangular pulse.

PDF distribution of amplitude of an OFDM signal can be obtained from a summation of identical PDF distribution of each subcarrier. Due to the central limit theorem, it can be well approximated by a normal (Gaussian) distribution regardless of a subcarrier PDF distribution. Thus PDF of an OFDM signal is approximately a normal distribution. This assertion can be confirmed experimentally by measuring (or computing) the distribution function of an OFDM signal.

Example 8-26: Using a normal distribution, how much backoff in dB is necessary if 1% clipping (overload), 0.1%, and 0.01% are allowed.

Answer: $2*Q(2.577) = 0.01$, 20log $(2.577) = 8.2$ dB; $2*Q(3.277) = 0.001$, 20log $(3.277) = 10.3$ dB; $2*Q(3.89) = 0.0001$, 20log $(3.89) = 11.8$ dB. Hint: We computed, by trial and error, the amplitude so that 2 times tail probability to be 1%. Similarly we can compute it for 0.1 % and for 0.01%.

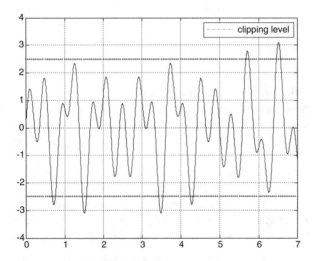

For OFDM signals, there are many suggestions for PAPR reduction. It is basically a form of digital clipping as shown on the left figure; above a certain threshold a signal is 'clipped'. A simple clipping is hard, i.e., maintaining the same level (rectangle). This non-linearity creates spectral regrowth around the band edge and in-band. The idea is to soften this hard clipping so that the regrowth is less. Windowing is rectangular form of clipping into triangle or similar shapes by 'windowing'. Peak cancelling is to generate another legitimate signal to cancel the peak, and this canceling signal is created by using reserved subcarriers; some subcarriers are reserved for it by using data carrying capacity. This idea is extended further to do even 'smart clipping', which is possible by using reserved subcarriers judiciously. The implementation details of these algorithms are beyond the scope here but it is covered in Chapter 6 in conjunction with powerful forward error correcting codes.

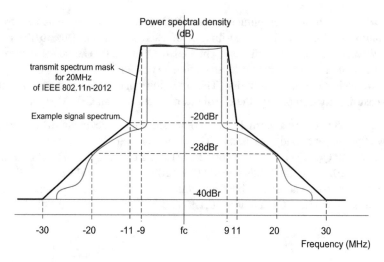

Figure 8-69: Spectrum mask of 802.11n for 20 MHz

Spectrum mask - regulatory requirements

A transmit signal must fit into a designated channel cleanly without spilling over to adjacent channels, called *clean channel occupancy*. Regulatory specification in the form of spectrum mask must be met. One example is spectrum mask of IEEE 802.11n-2012 is shown in Figure 8-69. Transmit filtering is applied for the clean channel occupancy. However, due to non-linearity, in particular IMD, there will be spectral re-growth around in-band and adjacent channels. In this mask of Figure 8-69, the reference point is the maximum of a spectrum around carrier frequency. In many spectrum mask specifications, a total power is specified as a reference point at carrier frequency, and spectrum analysis resolution bandwidth is specified.

Example 8-27: If the spectrum mask of Figure 8-69 uses a total power in 20 MHz specified as a reference point with the resolution bandwidth 100 kHz, how should be the mask modified while maintaining the same clean occupancy specification (i.e., power spill over to adjacent channels) as the figure?

Answer: -20 dBr (in the figure) should be changed to -20 dBr $- 10 \log$ (20 MHz/ 100 kHz) $= -43$ dBr from the reference, i.e., it is 23 dB higher vertically. This happens if the carrier frequency, a single tone, is of full power without data modulation.

8.2.5   Generation of Symbol Clock and Carrier Frequency

There are two frequency sources in generating a transmit signal; XMT clock of DAC and RF / IF carrier frequency. These frequency sources are, of course, periodic in time and a frequency is the inverse of the period. XMT clock is related with a symbol

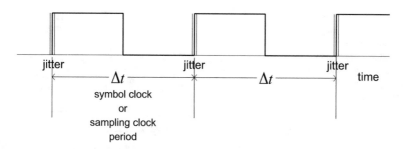

Figure 8-70: Jitter of symbol clock or XMT clock (sampling) of DAC; only the rising edge is used

clock by an integer, and it is rectangular in time. A carrier frequency is sinusoidal in time.

If the frequency source in practice is not perfectly periodic, the frequency may have a tolerance, which may be specified by part per million (ppm), i.e., 10^{-6}. This 'small' fluctuation in frequency must be tracked in the receive side.

For symbol clock, related with DAC sampling clock, this frequency fluctuation is called jitter as shown in Figure 8-70. Note that only the rising edge is used.

Exercise 8-25: A sampling clock is 20 MHz and its frequency tolerance is specified ± 20 ppm. What is the frequency tolerance in Hz? Answer: $\pm 400[\text{Hz}]$ ($=20e^6 * 20e^{-6}$).

For carrier frequency, in addition to frequency tolerance being specified, ideally it should be perfectly sinusoidal. In practice, it is not quite so. This can be seen clearly in the frequency domain, and thus it is often called spectral purity problem. In frequency domain a perfect sinusoid is an impulse. A slightly 'impure' sinusoid has a spectrum blurred in frequency domain. This blurred sideband is called phase noise. The internal circuit mechanisms for the imperfect sinusoid may be complex and beyond the scope here. However, fortunately, this impurity can be modeled and understood as a frequency modulation (FM) by noise. Or it can be modeled by AM for amplitude fluctuation but here we focus on the phase nose due to FM.

Exercise 8-26: A carrier frequency modulated by a sinusoid is considered. Find out the spectrum of $Vo = A\cos\left(2\pi f_o t + 2\pi \int\limits_{-\infty}^{t} x_{ph}(\tau)d\tau\right)$ when $x_{ph}(t)$ is a cosine wave.

Answer: V_o is represented in the form of a complex envelope multiplied by $e^{j2\pi f_o t}$, as $V_o = A\operatorname{Re}\left\{(a(t)+jb(t))e^{j2\pi f_o t}\right\}$ where $a(t)+jb(t) = c(t) = e^{j2\pi \int_{-\infty}^{t} x_{ph}(\tau)d\tau}$ and thus the complex envelope of V_o is obtained. In order to find the spectrum of V_o we need to find the spectrum of its complex envelope and to frequency translate by f_o. Given $x_{ph}(t) = A_m\cos(2\pi f_m t)$, $c(t) = e^{j\frac{A_m}{f_m}\sin(2\pi f_m t)}$. In order to find the spectrum we need to find the Fourier Transform of $c(t)$. In fact, Fourier series since it is periodic.

Figure 8-71: Carrier
frequency generation using
PLL and phase noise is
modeled as a frequency
modulation by noise-like
signal $x_{ph}(t)$. VCO is of pure
sinusoid

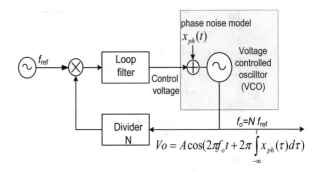

Figure 8-72: Spectral
density of phase noise
measurement

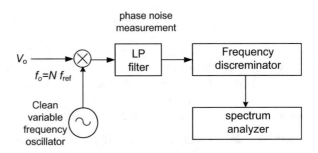

A carrier is generated by using PLL from a reference frequency. This reference is typically provided in a system as a very stable timing source by a crystal oscillator or by a timing source synchronized with GPS (global positioning system). VCO is a pure cosine wave and its control voltage is modulated by $x_{ph}(t)$ modeling phase

noise,   and   thus   the   output   of   VCO   is   given   by $Vo =$

$A\cos\left(2\pi f_o t + 2\pi \int\limits_{-\infty}^{t} x_{ph}(\tau)d\tau\right)$ as shown in Figure 8-71.

The output frequency f_o is given by $f_o = N f_{ref}$ where N is the factor of a divider.

Actual measurement of spectral density around a carrier is essentially FM demodulation as shown in Figure 8-72. The output of an actual carrier frequency generator will go through FM demodulation – mixing down, low-pass filter, frequency discriminator and display of the power measured.

The impact of the phase noise (impurity) into a system is explained in the receiver signal processing when a carrier recovery is discussed; this is left as a project exercise.

8.2.6   Summary of RF Transmit Signal Generation

In order to maximize signal to quantization noise (SN_qR), the scaling of b-bit fixed point representation of digital samples before and after DAC, the peak power in digital number should be 1.0 or 0 dB (i.e., samples are in fraction), and thus average power should be backed off by PAPR [dB] as shown in (8-24). The corresponding optimum analog power at DAC should be amplified in order to deliver a desired power level at the antenna port. Along the TX chain signal level must be distributed so the signal quality is maintained. A good design is such that IMD distortion is dominated by PA. In other words, IMD contribution from all other device should be negligibly small.

PAPR characterization of a signal was discussed. The frequency response of baseband filters and RF/IF can be obtained by complex baseband representation, and a detailed method was described.

8.3   Practical Issues of RF Receive Signal Processing

A receive signal chain with explicit gain blocks is shown in Figure 8-73; digital gain (blue), baseband gain (green), IF gain (beige) and RF gain (red) are inserted. Receive power is specified in terms of dBm at the antenna port. Its dynamic range of the largest signal to the smallest signal power difference is tremendously large; typically around 60 dB, say -30 dBm to -90 dBm. Yet the signal power at the input of ADC should be constant (optimum level). A receiver chain should provide this variable gain while maintaining the signal quality along the chain; this function is called AGC (automatic gain control). This aspect is distinctly different from transmit side. How many bits of ADC are necessary to satisfy the signal quality requirements? How to distribute the gain to accommodate the whole dynamic range of received signal?

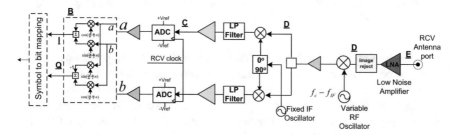

Figure 8-73: Receive signal chain with explicit gain blocks; digital gain (blue), baseband gain (green), IF gain (beige) and RF gain (red). This figure emphasizes the signal level change along the path. Receive power is specified in terms of dBm at the antenna port. There is one-to-one correspondence with Figure 8-54 of transmit side

Table 8-5: 3 bit 2's complement ADC example

INPUT		OUTPUT
Analog input voltage / Vref	Mid-point value / V_{ref}	3 bit in 2's complement
−0.125 to +0.125	+0.00	000
+0.125 to +0.375	+0.25	001
+0.375 to +0.625	+0.50	010
+0.625 to +∞	+0.75	011
−0.875 to +∞	−1.00	100
−0.625 to −0.875	−0.75	101
−0.375 to −0.625	−0.50	110
−0.125 to −0.375	−0.25	111

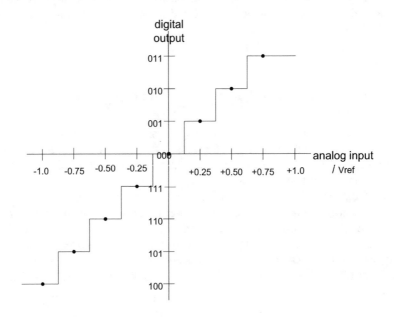

Figure 8-74: 3 bit 2's complement ADC graphical representation; each digit has the range of ±LSB. A dot is the midpoint in a range

In the receive side signal processing, the thermal noise is important as a signal level can be very low, close to thermal noise level. In addition to thermal noise, 'circuit noise', quantization noise, overload distortion and IMD noise due to non-linearity should be considered. However, overload distortion and IMD due to non-linearity is not as severe as those of transmit side since the signal level is typically low. Note that thermal noise is unavoidable while others may be designed out by using bigger device or by good physical design.

8.3.1   ADC

An ADC converts an analog voltage as input to a digital number as output. 3 bit ADC example is shown in Table 8-5. The output is 2's complement digital representation and its reference voltage is given by $\pm V_{ref}$, which is the peak voltage that ADC can handle. The error voltage in each digital representation is $\pm 1/2^{b} V_{ref}$, i.e., $\pm$LSB; a digital representation corresponds to a midpoint analog voltage while the analog input may within the range of $-1/2^{b} V_{ref}$ to $+1/2^{b} V_{ref}$ as the mid-point in the middle. This creates the uncertainty in the input voltage, which is called a quantization error.

The input – output relationship of ADC can be represented graphically as shown in Figure 8-74.

Exercise 8-27: In order to have less than ± 0.1 % of V_{ref} voltage range that can be distinguished (i.e., error voltage) in digital representation, how many bits of ADC is necessary? Answer: Find b such that $1/2^{b} < 0.001$, and $b = 10$. Actual quantization error with 10 bits is ± 0.098 %.

Exercise 8-28: Consider 8 bit 2's complement ADC and $V_{ref} = 0.5$ V. Find a digital representation of input voltages of +0.23 V, -0.4 V and -0.6 V.

   Answer: $+0.23 \rightarrow 0.23 /0.5 = 0.46 \rightarrow \underline{00111010}$ (back to analog $\rightarrow 0.4531$ $\rightarrow 0.2266$, error $= 0.0034$).

   $-0.4 \rightarrow -0.4/0.5 = -0.8 \rightarrow 1- 0.8 = 0.2 \rightarrow 0011001 \rightarrow \underline{10011001}$ (back to analog $-0.8047 \rightarrow -0.4023$, error $= 0.0023$).

   $-0.6 \rightarrow \underline{10000000}$ since it is beyond V_{ref} with some saturation logic (back to analog $-1.0 \rightarrow -0.5$, error $= - 0.1$ V).

Quantization error of ADC and its PDF

In order to compute quantization noise, we need a PDF of quantization error voltage. We note that, with b bit ADC, the range of quantization error is $\pm 1/2^{b} V_{ref}$. It seems plausible to assume that the errors are distributed uniformly within the range, if ADC is designed and behaves well. Using the notation $q = \pm 1/2^{b} V_{ref}$, its variance (power of error voltage) is given by $q^{2}/3$ and thus rms value is $q/\sqrt{3} \approx q0.577$. (See the 1st column of Table 8-3.) Note that the uniform distribution is due to the dynamic

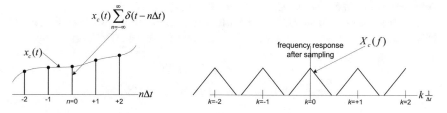

Figure 8-75: Sampling (ADC) creates a periodic frequency response. If not bandlimited there will be aliasing (i.e., folding back high frequency parts)

variability of input signal, i.e., at the sampling instant it can be any value within the error range.

Conversion rate of ADC and frequency response

In addition to resolution bits, a conversion rate (sampling rate) of an ADC is an important parameter. A conversion rate is the inverse of conversion time (Δt). A conversion rate must be at least twice faster than the bandwidth of a signal (sampling theorem). In practice an anti-aliasing (analog) filter will be placed before sampling. ADC (sampling) can be represented by the multiplication of an incoming signal $x_c(t)$ by a periodic impulse train $s(t) = \sum_{n=-\infty}^{+\infty} \delta(t - n\Delta t)$ whose Fourier Transform is $S(f) = \frac{1}{\Delta t}\sum_{k=-\infty}^{+\infty} \delta\left(f - \frac{k}{\Delta t}\right)$. Thus the frequency response after sampling is $X_c(f)$ * $S(f)$ where * denotes convolution. This frequency response is periodic and is shown in Figure 8-75. Note that if $X_c(f)$ is not bandlimited there will be aliasing.

Optimum input power level to ADC

In order to find an optimum power level, we need to know PAPR of a received signal. This was discussed in conjunction with DAC. For convenience we repeat partially here. For example, PAPR of a single tone is 3 dB.

Exercise 8-29: A signal has a normal (Gaussian) distribution with mean m = 0. What is PAPR of the signal? It could be infinity with approaching zero probability. Thus we need to limit a peak, say 4σ (four times of rms).

Clipping at 3σ(overload distortion), PAPR is $10\log(3^2) = 9.5$ dB, and clipping happens with the probability 0.0027 (= Q(3)*2), i.e., 0.27%.

Clipping at 3.5σ(overload distortion less than 3σ), PAPR is $10\log(3.5^2) = 10.9$ dB, and clipping happens with the probability 4.6e-004 (= Q(3.5)*2).

Clipping at 4σ(overload distortion less than 3.5σ), PAPR is $10\log(4^2) = 12.0$ dB, and clipping happens with the probability 6.3e-005 (= Q(4)*2).

Note: $Q(x) = \frac{1}{\sqrt{2\pi}}\int_x^\infty e^{-\frac{y^2}{2}}dy$ i.e., Q(x) is the tail end probability of zero mean and unity variance Gaussian.

In practical system design situation, PAPR of a signal, perhaps with interference as well, may be computed or measured experimentally. In order to know the probability of a peak, it is necessary to find out PDF or cumulative distribution of a signal of interest.

Exercise 8-30: With $V_{ref} = 0.5$ and $R = 50$ ohm what is the peak power in dBm? Answer: $10\log(0.5^2/50 *1000) = +6.99$ dBm.

The peak power that an ADC can handle, with a termination resistance R, is given by $P_{\text{ADC peak}} = V_{ref}^2 / R$. Thus the optimum input power of ADC is given by

$$P_{\text{ADC opt}} = P_{\text{ADC peak}}/\text{PAPR}$$
$$P_{\text{ADC opt}}[\text{dBm}] = P_{\text{ADC peak}}[\text{dBm}] - \text{PAPR [dB]}$$

Signal to quantization noise ratio
With uniform distribution of quantization noise, we can calculate signal to quantization noise ratio (SN_qR) of ADC, and it is given by

$$SN_qR = \frac{P_{ADC \cdot opt}}{P_{quantization}} = 3 \frac{2^{2b}}{PAPR} \tag{8-29}$$

$$SN_qR \text{ [dB]} = b\,6.02 + [4.77] - PAPR \text{ [dB]} \tag{8-30}$$

Exercise 8-31: With uniform distribution of quantization error, and PAPR = 12 [dB], SN_qR [dB] = 40 dB is required. How many bits of ADC are necessary? Answer: From (8-30), $40 = b\,6.02 + [4.77] - 12$. Solving it we obtain $b = 7.85$ thus one can use $b = 8$ bit ADC.

Exercise 8-32: In the above example how many bits are allocated for overload distortion? Answer: 12 dB PAPR corresponds 2 bits, and thus 6 bits out of total 8 bits are used for quantization noise.

Effect of DC offset to ADC
DC offset, typically due to baseband analog amplifiers before ADC, may reduce the dynamic range of ADC so that its effective bits may be reduced or with excessive DC offset there may be a clipping. Note that if there a DC offset it increases a peak of a signal one sided depending on the sign of DC, and thus it reduces the effective bits of ADC. This can be eliminated only by reducing DC offset itself.

Example 8-28: A numerical example is given on the impact to ADC of DC offset reducing effect number of bits of ADC, i.e., reducing quantization noise. ADC has a reference voltage +0.5 to −0.5 [V] with 50 ohm termination, and 10 bits (1 sign bit + 9 bits of magnitude) to cover the range. A signal has a peak to average power ratio +12 dB.

What is the quantization noise degradation if DC offset is 10% of 0.5 V (peak), i.e., 0.05 V? What about DC offset of 0.1 V case?

1) The peak power ADC can handle is 7.0 dBm (= 10*log (0.5²/50*1000)). The optimum (maximum) input power with no overloading (clipping) is −5 dBm (= 7 dBm − 12 dB).
2) The input power to ADC is maintained to be −5 dBm even with DC offset, and thus the net signal power is reduced by the power due to DC offset; 0.05^2 +net_sig² = 0.1258², and thus net_sig² = 0.1154² = 0.0133. The power of net_sig with 50 ohm termination is given by −5.75 dBm (= 10*log (0.1154²/ 50*1000)). Thus 0.75 dB is quantization noise increase. For 0.1 V Dc offset, 0.1^2 +net_sig² = 0.1258² and net_sig² = 0.076² = 0.0058, in dBm, −9.33 dBm (= 10*log (0.076²/50*1000)). Thus quantization noise is 4.33 dB.

A precision DC supply or an analog single tone with different frequency may be used for testing ADC. Any systematic deviation from a transfer graph in Figure 8-74

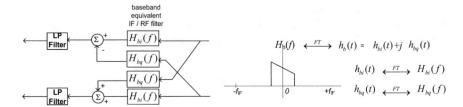

Figure 8-76: Overall filter response from RF/IF to baseband. The cross-coupling between in-phase and quadrature phase degrades a system performance

degrades ADC performance; e.g., DC offset, missing codes, skewed distribution (not uniform distribution of error voltage) etc...

8.3.2   RX Filters and Complex Baseband Representation

A signal received at RX antenna port will go through RF/IF filters, baseband analog filters and digital filters in order to meet channel selectivity requirements as well as to meet an end-to-end pulse shaping. Most filtering requirements will be met by analog baseband filters and digital filters but RF/IF filters are crucial to provide enough stop band attenuation. In Section 8.2.2 we discussed how we can find an overall filter response using complex baseband equivalence of RF/IF filters. Here we repeat it briefly for convenience.

In order to find complex baseband equivalence, move a RF/IF filter frequency response to baseband from left to right (i.e., only positive frequency), then take an inverse Fourier transform which in general results in complex impulse response. Separate the complex impulse response into real and imaginary. For each real and imaginary part, take Fourier transform. An overall frequency response is a cascade of RF/IF equivalence and LP filters. This process is shown in Figure 8-76.

Exercise 8-33: What are the conditions that there is no cross coupling in complex baseband?

Answer: In order to eliminate cross-coupling, the complex baseband impulse response must be real. If an impulse response is real in time, its frequency response should be, around DC, even symmetric in magnitude and odd symmetric in phase, i.e., $H_b(f) = H_b^*(-f)$ where * means complex conjugation. (This is also called Hermitian symmetry.)

8.3.3   RCV Dynamic Range and AGC

The signal level at receive antenna port changes due to radio propagation media and movement and typically its dynamic range can be 50 to 60 dB. One important task of

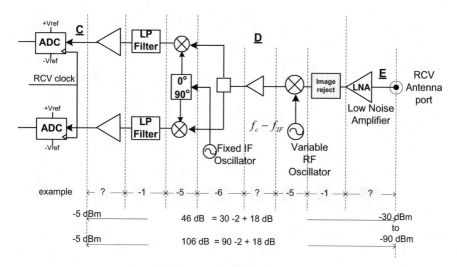

Figure 8-77: Estimation of required gain for a given receive input range

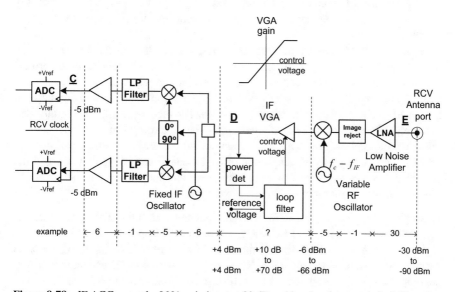

Figure 8-78: IF AGC example; LNA gain is set at 30 dB and baseband amp gain is 6 dB each

a receive chain is that this varying signal level must be constant before sampling at ADC input by AGC.

An example is shown in Figure 8-77 with the insertion loss of each device in the chain; overall gain should be 46 dB when the received signal is at overload point (−30 dBm), and 106 dB when it is the smallest signal (−90 dBm), called a receiver threshold. The overload point is the largest signal that an LNA can handle without

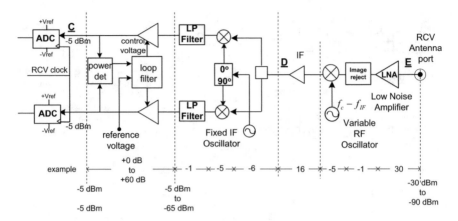

Figure 8-79: Baseband AGC with analog control; input to baseband amp ranges from −5 dBm to −65 dBm and, providing 0 to 60 dB gain, the output level become −5 dBm

poor error rate performance due to overload distortion. The threshold is the minimum signal level that can meet the required bit (or symbol, packet) error rate, say 10^{-3}. This dynamic range must be equalized to a constant level before sampling, say −5 dBm.

Sometimes LNA may have a few gain settings so that the receiver dynamic range can be wider; typically, in case of a very strong signal, LNA gain is set to zero dB and otherwise it is set to a fixed gain and thus extending overload point. By having a high enough fixed gain of LNA, a system noise figure (NF) can be made close to LNA's NF. Thus the threshold can be improved. The receive threshold is limited due to thermal noise at the front end of a receiver. NF is an indicator of additional thermal noise due to amplification. This will be elaborated further later in this section.

AGC can be done by IF amplifiers or by baseband amplifiers or both, and the control voltage, along with power measurement and loop filter, can be generated digitally or by analog method. We give examples.

Example 8-29: See Figure 8-78 IF AGC example carefully. To be concrete the insertion loss is assigned and LNA and baseband amp gain have a fixed value.

Due to LNA gain of 30 dB and mixer and image filter loss, input to IF amp is varying from −6 dBm to −66 dBm. By providing +10 to +70 dB gain automatically the output level of IF amp becomes +4 dBm. With baseband gain 6 dB and LP filter and other insertion loss, the signal level before sampling becomes −5 dBm.

IF AGC requires four components; IF VGA, power detector, loop filter and reference voltage. IF variable gain amplifier exhibits control voltage vs. gain sketched in the figure. Power detector may be implemented by rectifier followed by a low pass filter. Loop filter is an integrator. Reference voltage sets the output power level. Figure 8-78 shows graphical diagrams. We emphasize that these four components are essential and that they interact. The required dynamic behavior, e.g., how fast it settles after receive level change, depends on system requirement. Accordingly IF AGC must be designed.

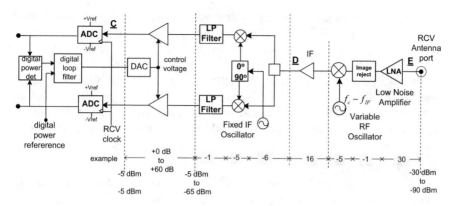

Figure 8-80: Baseband AGC with digital control. VGAs are the same as in Figure 8-79 but controlled digitally. Digital control voltage will be converted to analog, perhaps as part of VGA

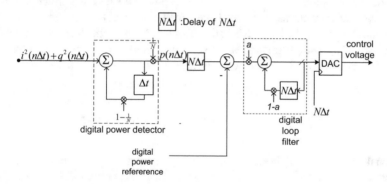

Figure 8-81: Implementation of digital control generation; four components of power detection, loop filter, reference power and VGA (not shown)are clearly visible

Example 8-30: Baseband AGC with analog control. See Figure 8-79 carefully.

Example 8-31: Baseband AGC with digital control. See Figure 8-80 carefully.

Example 8-32: An implementation of digital control voltage generation is shown in Figure 8-81. Even though actual mechanism of computing power detection, power reference, loop filter and control voltage may be different from analog counter parts, clearly four components are identified. The averaging period of power detection is specified by a number of samples (N) and sampling period (Δt), i.e., $N\Delta t$. This detected average power, in digital number representation, is compared with reference power level in digital number representation; the comparison is to subtract reference power from detected average power. This difference is integrated by a loop filter and its time constant is denoted by a. This digital control voltage is converted to analog voltage by DAC. In practice, VGA may be designed so that a given digital control in bits may correspond to a VGA gain.

In digital control the outputting control voltage (equivalently sending control digits) does not need to be periodically regular. One control register digits can be held for a long time or can be updated regularly. This essentially makes a system time-varying, i.e., loop filter time constant changing. It is flexible.

Since the power computation is done after ADC, i.e., samples from ADC, for the duration of AGC settling time, the input power to ADC may be not optimum. In particular, ADC should be able handle overload properly, e.g., using a saturation logic or using a mechanism to detect overload condition. One should be careful when one takes advantage of flexible loop time constant change.

Simulation of AGC

Exercise 8-34*(project): A subsystem of AGC can be isolated and simulated in computer to predict its dynamic behavior and performance. This simulation may be important when a digital control is used since one should be careful with ADC overload. Hint: rather than a concrete numerical example here we emphasize the importance of simulation of a subsystem reflecting real situation correctly. In order to do realistic simulations, one must have a realistic model of four components plus other components in the loop. 'Realistic' does not mean all the details of circuits but rather capturing and abstracting important parameters – one key system engineer skill. Test signal must be 'realistic' as well.

8.3.4 LNA, NF, and Receiver Sensitivity Threshold

Thermal noise is present due to thermal random fluctuation of electrons once a signal is converted to electric current and to voltage due to a current through a resistor. When this noise is much smaller than a signal, it can be ignored. It is not the case at the front end of a receiver chain since a signal can be so small that this thermal noise has a major impact to the performance of a system; a fundamental fact of life that cannot be avoided. In addition to thermal noise, a system may add its own noise, with different origin including thermal, during signal amplification. This is conveniently specified by a noise figure (NF) of an amplifier and of a system. LNA should have a small NF, thus is called low noise amplifier. A minimum signal level, that a receiver can have a desired system error rate performance, is called receiver sensitivity threshold typically measured in dBm.

Exercise 8-35: The thermal noise is characterized by its power spectrum with $S_n(f) = kT$ [watt/Hz] where k is Boltzmann constant 1.38×10^{-23} [joule / Kelvin degree] and T is temperature in Kelvin ($°K = °C + 273$). It is flat, also called white, for all radio frequencies. Thus the thermal noise per Hz at room temperature (20 °C) is, $293 \times 1.38 \times 10^{-23}$ [watt /Hz], which is -174 dBm / Hz. When the bandwidth of a system is 1 MHz, then the noise power is -114 dBm / MHz .

Noise figure (NF) is defined how much additional noise is added after amplification and is measured as the ratio of the input signal to noise (SNR $_{in}$) over the output signal to noise (SNR $_{out}$), i.e., NF = SNR $_{in}$ / SNR $_{out}$. The input noise (N_i) is

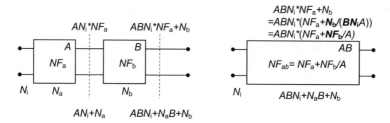

Figure 8-82: NF of two cascaded amp in terms of each NF and gain. A is gain, NF_a is NF of the first amp, N_a is the noise added by amp, N_i is the input noise

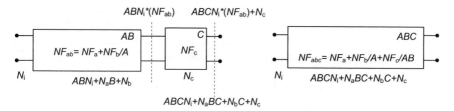

Figure 8-83: NF of three cascaded amplifiers is given by $NF_{abc} = NF_a + NF_b/A + NF_c/AB$ with the repeated application of the same argument

amplified by power gain A at the output, and addition noise (N_a) is added due to the amplifier. Thus the total output noise is $N_i A + N_a$. The power gain the ratio of the output power over input power $A = S_{out}/S_{in}$. NF is commonly used in dB and NF [dB] is given by,

$$\text{NF [dB]} = 10 \, \log\, (\text{SNR}_{in}/\text{SNR}_{out}) = 10 \, \log\left(1 + \frac{N_a}{AN_i}\right) \qquad (8\text{-}31)$$

Exercise 8-36: Show that (8-31) is true. What does it mean NF being 3 dB?

Answer: $\text{NF} = (S_{in}/N_i)/(S_{out}/(N_i A + N_a)) = (N_i A + N_a)/(N_i A) = 1 + N_a/(N_i A)$. 3 dB means that $\text{NF} = 2$, thus $N_a/(N_i A) = 1$. The added noise is the same as the amplified input noise.

Example 8-33: NF and gain are usually specified for a given amplifier. We need to find NF of two cascaded amplifiers, and each amp is specified with NF and gain. Note that the NF of first amp is critically important. This is shown in Figure 8-82, and further extended to Figure 8-83.

A is gain, NF_a is NF of the first amp, N_a is the noise added by amp, N_i is the input noise. Similarly for the second amp B is gain NF_b is NF of the second amp, N_b is the noise added by amp. The input noise to the second amp, i.e., the output noise of the first amp is expressed as $AN_i * NF_a$. Similarly the output noise of the second amp is

Figure 8-84: A numerical example of receiver threshold, NF, SNR and receiver dynamic range

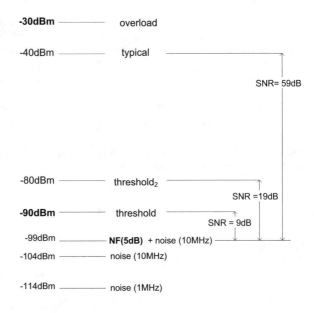

given by $ABN_i*NF_a+N_b$ and is the same as $ABN_i*(NF_a+NF_b/A)$. Thus the overall NF is given by,

$$NF_{ab} = NF_a + NF_b/A \qquad (8\text{-}32)$$

Exercise 8-37: $A = 20$ dB, $NF_a = 2$ dB and $B = 20$ dB and $NF_b = 10$ dB are given. What is the overall NF_{ab} ? Answer: use (8-32) after changing dB to ratio, then back to dB. $10 \log(10^{0.2} +10^{1.0}/10^2) = 2.27$ dB. Note that this is close to NF of the first amp. This shows that the importance of low NF of LNA.

Receiver sensitivity threshold

A receiver sensitivity threshold can be a complex function of many parameters – bandwidth, modulation and coding and implementation issues as well. This may be summarized as theoretically SNR required plus implementation loss to meet performance requirements.

Example 8-34: We show the relations of the sensitivity threshold, SNR and NF. A numerical example of a receiver dynamic range is given by -30 dBm to -90 dBm in Figure 8-84. In 10 MHz bandwidth the thermal noise power is -104 dBm.

If NF is 5 dB, the effective noise power is -99 dBm. Thus SNR available is 9 dB. If the required SNR is 19 dB, then the minimum threshold may be adjusted to -80 dBm. If NF can be improved by say 2 dB, then the threshold can be improved by 2 dB. If the required SNR can be improved by say 3 dB, then the threshold can be changed by the same amount. This relationship can be visualized graphically as in Figure 8-84. When symbol error rate is performance criterion, then the noise power measuring bandwidth is the same as symbol rate. (Explain why.) In this example

10 MHz should be the same as symbol rate. Note that NF is less critical when the receiver input signal is typical (say −40 dBm), i.e., much larger than the threshold level. For example, 10 dB NF in this case makes SNR 5 dB worse, and SNR = 54 dB, which is still very high and little impact to system.

8.3.5 Re-generation of Symbol Clock and Carrier Frequency

In receive signal processing we also need a sampling clock, RCV clock for ADC and RF/IF carrier frequency. The quality requirements of both RCV clock and carrier frequency are identical to those of transmit side as we discussed in Section 8.2.5. For example, the spectral impurity of RCV carrier creates blurred sideband called phase noise and degrades the system performance. In terms of frequency stability there is no need of regulatory requirements specified but it is required for system performance.

We use the word 're-generation' of symbol clock and carrier frequency rather than 'generation'. The reason is that in the receiver, not only generation but also symbol clock and carrier must be synchronized with those of transmit. For coherent demodulation the local carrier phase must be synchronized with the incoming signal as well. For sampling clock, the sampling phase must be optimized so that sampling should occur at optimum clock phase. This synchronization issue is the topic of Chapter 7 and integral part of digital modem design. In practice this synchronization processing occupies the substantial part of the complexity of receiver demodulation processing.

8.3.6 Summary of RF Receive Signal Processing

In order to maximize signal to quantization noise (SN_qR), the scaling of b-bit fixed point representation of digital samples before and after ADC, the average power should be backed off by PAPR [dB] as shown in (8-29). The corresponding optimum analog power at ADC should be delivered after AGC of the dynamic range of received signal typically 60 dB or so. When the receive signal level is close to the receiver threshold, the thermal noise may impact a system substantially. NF of LNA is important to minimize.

AGC may be done in IF or in baseband or both. AGC control signal may be generated digitally or in analog method.

8.4   Chapter Summary and References with Comments

8.4.1   Chapter Summary

We covered low digital IF transceiver architecture in great detail, from justifying and proving the validity of its method to actual implementation using DSP in the first part of chapter. The method of low IF transceiver was condemned as not a viable scheme in the past due the severe requirement of sharp filtering of image. However, in this chapter a saving grace is digital image cancelling with simplicity thus the image filtering requirement is essentially eliminated. Additional benefit of it is that DC offset is out of band. In order to keep multiplication free image canceller, a frequency of digital IF can be 0.5 times of symbol rate with interpolation factor of 2 whereas typically digital IF is the same as symbol rate and interpolation factor is 4.

After discussing the transceiver architecture, we discussed the issues of transmit side and those of receive side in two subsequent sections.

8.4.2   References with Comments

For this chapter a very limited reference is provided for additional extension of the topics discussed here. We did not use these references in our work but provide them so that an interested reader may find additional references from these.

Brief comments on each: [1] is most closely relevant to our chapter as the title of it indicates – transceivers for wireless communications. However, our discussion is centered on a low IF digital image cancelling architecture. [2] and [3] are focused on circuit level design even though they discuss some issues of system level as well. [3] is more so on the details of circuit design of key RF components – amplifiers, oscillators and synthesizers, and mixers.

[1] Qizheng Gu, "RF system design of transceivers for wireless communications", Springer 2005

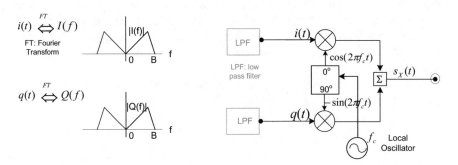

Fig. P8-1: Quadrature modulator (RHS) to generate a carrier modulated transmit signal $s_X(t)$

[2] Behzad Razavi," RF Microelectronics", 2nd ed, Prentice Hall 2012
[3] Thomas H. Lee "The Design of CMOS Radio-Frequency Integrated Circuits"
     2nd ed, Cambridge University Press 2004

8.5   Problems

P8-1: A quadrature modulator shown in Fig. P8-1(RHS) is used in order to generate
RF signal $s_X(t) = i(t) \cos (2\pi f_c t) - q(t) \sin (2\pi f_c t)$ where $i(t)$ and $q(t)$ are analog
baseband signals shown in Fig. P8-1 LHS.

(1) When $q(t) = 0$ in Fig. P8-1, it is called double side band (DSB). Sketch the
     magnitude of its spectrum of $s_X(t)$.
(2) When $i(t) = 0$ in Fig. P8-1, it is called double side band (DSB). Sketch the
     magnitude of its spectrum of $s_X(t)$.
(3) When $q(t) = i(t)$ in Fig. P8-1, sketch the magnitude of its spectrum of $s_X(t)$. In
     this case the same signal is transmitted twice.
(4) When, $i(t)$ and $q(t)$ are two independent baseband signals in Fig. P8-1, sketch
     the magnitude of its spectrum of $s_X(t)$. In this case two independent baseband
     signals are transmitted at the same time. What is the overall transmission rate?
     Compare with the above (1) – (3).

P8-2: What, in P8-1, is an RF channel bandwidth required in order for a signal to
occupy cleanly within the RF bandwidth? Consider all the cases of (1) to (4).

P8-3: A quadrature modulator in Fig. P8-1 (RHS) is universal in the sense that all
different analog modulations can be, in principle, implemented with it by choosing $i(t)$ and $q(t)$ properly. For example, an AM (amplitude modulation), used in
AM broadcasting, can be done by choosing $q(t) = 0$ and $i(t) = DC + \widehat{i}(t)$ where
$\widehat{i}(t)$ is a baseband signal and DC (constant) is chosen to the peak of $\widehat{i}(t)$.
Assuming DC $= 1.0$, $\widehat{i}(t) = \sin (2\pi 1000 t)$, and $f_c = 50$KHz, draw $s_X(t)$.

P8-4: Another example of quadrature modulation is to generate SSB. By choosing q
$(t) =$ Hilbert transform $(i(t))$, SSB signal can be generated. This idea is also called
carrierless amplitude and phase modulation and was used.

Fig. P8-2: FSK angle
generation from PAM with a
pulse g(t)

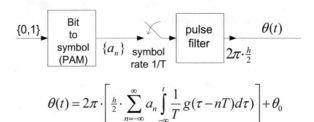

$$\theta(t) = 2\pi \cdot \left[\frac{h}{2} \cdot \sum_{n=-\infty}^{\infty} a_n \int_{-\infty}^{t} \frac{1}{T} g(\tau - nT) d\tau \right] + \theta_0$$

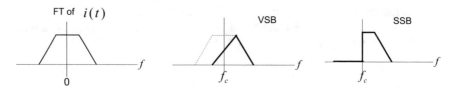

Fig. P8-3: SSB and VSB comparison

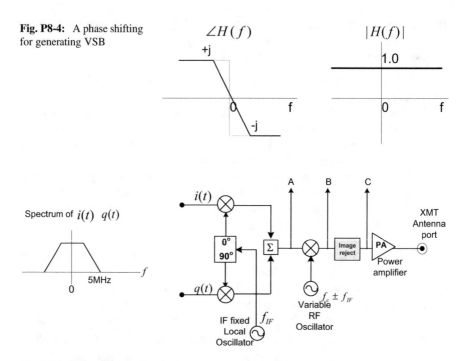

Fig. P8-4: A phase shifting for generating VSB

Fig. P8-5: Heterodyne conversion transmit side

(1) Choose $i(t) = A_m \cos(2\pi f_m t)$, $q(t) = A_m \sin(2\pi f_m t)$. Obtain $s_X(t)$ with this example, and sketch its spectrum.

(2) Choose $i(t) = m(t) \cos(2\pi f_{IF} t)$, $q(t) = m(t) \sin(2\pi f_{IF} t)$ where $m(t) = A_m \cos(2\pi f_m t)$. Obtain $s_X(t)$ and sketch its spectrum.

(3) Choose $i(t) = m(t) \cos(2\pi f_{IF} t)$, $q(t) = m(t) \sin(2\pi f_{IF} t)$ where $m(t)$ is a bandlimited baseband signal whose spectrum is shown in Fig. P8-1 (LHS). Sketch its spectrum.

P8-5: An FSK signal covered in Chapter 2 can be generated by using a quadrature modulator. This is shown in Section 2.8.3.

(1) Show that any FSK signal can be generated by choosing $i(t) = \cos(\theta(t))$ and

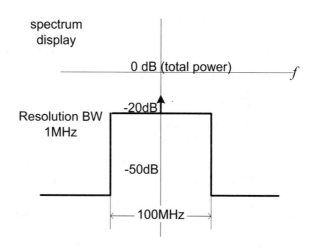

Fig. P8-6: Unrealistic looking power spectral density measured

$q(t) = \sin(\theta(t))$ where $\theta(t)$ is generated as in Fig. P8-2.

(2) Can we claim 'any' FSK? Consider different levels of $\{a_n\}$, pulse shape of g (t) and choice of modulation index h.

(3) Are $i(t)$ and $q(t)$ independent? What can we say about the dimension of FSK, one dimension or two?

P8-6: A vestigial sideband (VSB) with 8 levels is used in digital TV in US. The vestige of other sideband is transmitted, i.e., DC part of spectrum is transmitted. A comparison with SSB is shown in Fig. P8-3.

One method of generating a VSB signal is to use a highpass filter (for upper sideband) with a stopband to passband shape – vestigial symmetry after carrier modulation by $\cos(2\pi f_c t)$. Devise a method using a quadrature modulator to generate VSB signals. Hint: similar to SSB, a phase shifting circuit may be used as shown Fig. P8-4.

P8-7: Heterodyne conversion means there is an intermediate frequency (IF) conversion, and then a signal is moved to a carrier frequency. This is shown in Fig. P8-5. IF frequency is 140 MHz while baseband signal bandwidth is bandlimited to 5 MHz and carrier frequency is 1 GHz.

(1) Sketch signal spectrum at A

(2) Sketch signal spectrum at B

(3) Sketch signal spectrum at C when RF oscillator frequency is $f_c + f_{IF}$ where the image reject filter is a low pass.

(4) Sketch signal spectrum at C when RF oscillator frequency is $f_c - f_{IF}$ where the image reject filter is a high pass.

(5) If RF channel spacing is 10 MHz, a bandpass filter is needed in (3) and (4) above. Sketch the frequency magnitude of it.

P8-8: The impact of DC offset; $i(t)$ contains DC offset DC_i, i.e., $i(t) + DC_i$ and $q(t)$ contains DC offset DC_q, i.e., $q(t) + DC_q$ in Fig. P8-5. Express the signal at A as given by $s_A(t) = (i(t) + DC_i) \cos(2\pi f_{IF}t) - (q(t) + DC_q) \sin(2\pi f_{IF}t)$. Confirm a carrier leakage term; $a \cos(2\pi f_{IF}t + \phi)$ where $a = \sqrt{DC_i^2 + DC_q^2}$ and $\phi = \tan^{-1}\left(\frac{DC_q}{DC_i}\right)$.

P8-9: Carrier leakage (also called LO leakage) is often expressed in dB relative to signal power. For example 0 dB LO leakage means signal power loss 3 dB. Show that signal power loss $[dB] = 10 * \log\left(\frac{1}{1+10^{-LOdB/10}}\right)$. What is LO leakage if power loss is less than 0.1 dB?

P8-10: A spectrum of transmit signal $s_x(t)$ is shown in Fig. P8-6 (unrealistically clean look!), and LO leakage is shown at carrier frequency 'sticking out' above 8 dB from -20 dB. Spectrum analyzer resolution bandwidth is 1 MHz and signal bandwidth is 100 MHz, and thus spectrum is -20 dB below reference power (total power). What is LO leakage in dB? What is the power loss due to this LO leakage?

P8-11: Quadrature phase imbalance, i.e., cosine and sine are not $90°$ apart, can be captured as $\Delta\phi$ in $s_x(t) = i(t) \cos(2\pi f_c t) - q(t) \sin(2\pi f_c t + \Delta\phi)$. It can be expressed as

$s_x(t) = [i(t) - q(t) \sin(\Delta\phi)] \cos(2\pi f_c t) - [q(t) \cos(\Delta\phi)] \sin(2\pi f_c t)$. Note that I side has Q signal component $q(t) \sin(\Delta\phi)$, which is interference to I signal. Show that signal to interference ratio in dB is given by $20 \log(|\sin(\Delta\phi)|)$. To have 40 dB SIR, obtain how accurate quadrature phase balance, in degree, should be.

P8-12: A system SNR is given by 40 dB.

(1) Quadrature phase imbalance is $1°$. What is overall SINR?
(2) SIR due to quadrature phase imbalance is 40 dB. What is overall SINR?

P8-13: An ADC uses reference voltage ± 0.5V. That means that it can handle the peak voltage of 0.5 V. RMS input level to ADC, without clipping (peak distortion) is known to be 0.1257 V due to a large peak to average power ratio.

(1) Assuming 50Ω termination, what is the peak power in dBm that ADC can handle?
(2) What is the peak to average power ratio of a signal used here in dB?
(3) If there is 0.05 V DC offset (10% of ADC reference voltage), what is RMS voltage without peak distortion?
(4) What about 0.1 V DC offset case?

P8-14: Establish a Fourier transform pair from its definition as;

$$e^{j2\pi \cdot f_m \cdot t} \overset{FT}{\Leftrightarrow} \delta(f - f_m).$$

(1) Draw a frequency response.

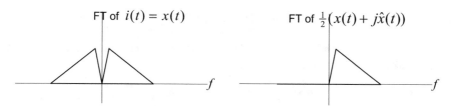

Fig. P8-7: Analytical signal or SSB

(2) Draw a frequency response of $e^{-j2\pi \cdot f_m \cdot t}$
(3) Draw a frequency response of $\cos(2\pi f_m t)$
(4) Draw a frequency response of $\sin(2\pi f_m t)$

P8-15: Hilbert transform is defined in frequency domain as a phase shifting transfer function; $H(f) = -j \operatorname{sgn}(f)$. Using this establish some of Hilbert transform pairs in Table 8-1; for example,

(1) $\cos(2\pi ft) \overset{HT}{\Leftrightarrow} \sin(2\pi ft)$
(2) $\sin(2\pi ft) \overset{HT}{\Leftrightarrow} -\cos(2\pi ft)$
(3) Show that $H(\hat{x}(t)) = -x(t)$

P8-16: Consider a Fourier transform pair,

$$\frac{1}{\pi t} \overset{FT}{\Leftrightarrow} -j \operatorname{sgn}(f).$$

(1) Show its validity.
(2) Show Hilbert transform can be expressed in terms of convolution;

$$\hat{x}(t) = x(t) * \frac{1}{\pi t}$$

P8-17: Find Fourier transform of $(x(t) + j\hat{x}(t)) \cdot e^{j2\pi \cdot f_c \cdot t}$ for a given $x(t)$ shown in Fig. P8-7. Then find FT of $\operatorname{Re}\left\{(x(t) + j\hat{x}(t)) \cdot e^{j2\pi \cdot f_c \cdot t}\right\}$ and sketch it.

P8-18: With digital IF carrier frequency being symbol rate $(\frac{1}{\Delta t})$, upper sideband can be generated by, where m is integer index,

$$a_d + jb_d = [i(m) + jq(m)]e^{j\frac{2\pi}{\Delta t}m}.$$

(1) Sketch the spectrum of $(a_d + jb_d)$.
(2) Show that lower sideband can be obtained by

$$c_d + jd_d = [i(m) + jq(m)]e^{-j\frac{2\pi}{\Delta t}m}$$

(3) Obtain c_d and d_d, and draw a block diagram for lower side band case.
(4) Sketch the spectrum of $c_d + jd_d$.

P8-19: In Figure 8-18 (RHS), $i(m)$ and $q(m)$ are recovered from a_d and b_d by demodulation processing. Derive the demodulation processing block from,

$$a_d + jb_d = [i(m) + jq(m)]e^{j\frac{2\pi}{\Delta t}m}.$$

P8-20: The figures from Figure 8-20 to Figure 8-27 show receiver processing of digital IF image cancelling mechanism. It is straightforward but tedious. It may be simplified by using analytic signals as suggested in Example 8-8 and Exercise 8-4. Propose the same argument using this idea.

P8-21: (advanced*) A RF channel bandwidth is given by 80 MHz. Consider 'clean channel occupancy' and 'channel selectivity'. What is the maximum symbol rate of baseband signal to fit into this channel? If the excess bandwidth is 20%, what is the maximum symbol rate?

P8-22: In an image cancelling transceiver, digital IF frequency ($\frac{F_k}{\Delta t}$) relative to symbol rate is chosen to be $F_k = 0.5$, and interpolation factor ($\frac{\Delta t}{I_k}$) relative to symbol rate can be chosen to be $I_k = 2$ in order to minimize digital IF modulation since $\cos\left(\frac{2\pi \cdot F_k}{\Delta t} \frac{\Delta t}{I_k} n\right) = \cos\left(\frac{\pi}{2} n\right) = \{1, 0, -1, 0, 1, 0, -1, 0, ...\}$ and $\sin\left(\frac{2\pi \cdot F_k}{\Delta t} \frac{\Delta t}{I_k} n\right) = \sin\left(\frac{\pi}{2} n\right) = \{0,1, 0,-1,0,1, 0,1,....\}$ where $n = 0, 1, 2, 3,.....$

(1) One practical issue is that a signal should have a zero excess bandwidth. Explain it.
(2) If $F_k = 0.75$ is chosen, what is I_k so that $\frac{F_k}{I_k} = 4$? What is the minimum sampling rate in this case?
(3) Draw a spectrum diagrams similar to Figure 8-30 at transmit side for being $F_k = 0.75$, and $I_k = 4$.

P8-23: (advanced*) In an image cancelling transceiver, x4 interpolation filter is shown in Figure 8-34. This interpolation filter can be implemented by a cascade of two x2 interpolation filters.

(1) Draw a block diagram for this cascade implementation.
(2) The 2nd stage interpolation may be simpler than the 1st stage one. Explain why. In fact the 2nd stage interpolation may be done by a half power filter whose cutoff frequency is 0.25 times sampling rate.
(3) A digital filter of ideal low pass with the cutoff frequency, f_o, has an impulse response of, where $n = \{...., -2,-1, 0, 1, 2, ...\}$,

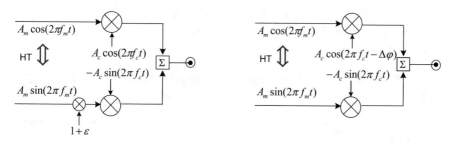

Fig. P8-8: Quadrature amplitude imbalance (LHS) and quadrature phase imbalance (RHS)

$$h(n) = \frac{\sin(2\pi f_o n)}{\pi n}.$$

Obtain $h(n)$ when $f_o = 0.25$ for $n = \{-4 \text{ to } +4\}$.

P8-24: (advanced*) In an image cancelling transceiver, x4 decimation filter is shown in Fig. P8-8. This decimation filter can be implemented by a cascade of two x2 decimation filters.

(1) Draw a block diagram for this cascade implementation.
(2) The 1st stage decimation may be simpler than the 2nd stage one. Explain why. In fact the 1st stage decimation may be done by a half power filter whose cutoff frequency is 0.25 times sampling rate.
(3) Compare with P8-23, and describe the difference.

P8-25: A quadrature modulator amplitude imbalance is modeled as shown in Fig. P8-8 (LHS) where the gain of quadrature branch is imbalance by $1 + \varepsilon$.

(1) Show that image rejection [dB] is given by $IR[dB] = 10\log\left(\frac{\varepsilon}{2+\varepsilon}\right)^2$ as in (8-12).
(2) If the required $IR = -30$ dB, what is the amplitude imbalance required in dB, i.e., $20\log(1+\varepsilon)$?

P8-26: A quadrature phase imbalance is modeled as shown in Fig. P8-8 (RHS) where the phase imbalance is $\Delta\varphi$.

(1) IR [dB] is modeled as the ratio of unwanted LSB power / desired USB power, and it may be shown as in (8-16).
(2) Quadrature phase imbalance generates the interference between I and Q channels and it may be represented by the USB interference power /desired USB power. Show that it can be represented by (8-15).
(3) For a required IR [dB] $= -30$ dB, what is the allowed quadrature phase imbalance $\Delta\varphi$ in degree?

P8-27: Two forms of average computation - IIR and FIR - is shown in Figure 8-41.

(1) Find a transfer function of IIR DC estimation.

(2) Find a transfer function of FIR DC estimation.

(3) Explain how one chooses the length of average N.

P8-28: We consider 8-bit 2's complement representation. Fill the blanks in a table below.

Binary	integer	fraction	Binary	integer	fraction
0000 0000	0	0.0	1000 1001		
1000 0000	-128	-1.0	0001 1100		
0000 1001	9		1001 1100		

P8-29: Consider 8-bit 2's complement DAC with $V_{ref} = \pm 0.5$ V. What are the analog voltage corresponding to binary data shown in P8-28 Table.

P8-30: We consider 8-bit sign magnitude representation. Fill the blanks in a table below.

Binary	integer	fraction	Binary	integer	fraction
0000 0000	0	0.0	1000 1001	-9	
1000 0000	0	-0.0	0001 1100		
0000 1001	9		1001 1100		

P8-31: Consider 8-bit sign magnitude DAC with $V_{ref} = \pm 0.5$V. What are the analog voltage corresponding to binary data shown in P8-30 Table.

P8-32: The conversion rate of a DAC is 1 GHz and its impulse response is represented by sample-hold. What is the duration of sample-hold pulse (rectangular pulse)? Assuming $V_{ref} = \pm 1.0$V draw its impulse response and its frequency response.

P8-33: Consider a 8-bit DAC and compute signal to quantization ratio (SN_qR) for a signal with PAPR (peak to average power ratio) is 1 dB if the error voltage distribution is uniform. What happens to SN_qR if PAPR is 3 dB?

P8-34: PA should handle power. Consider a single tone test signal. What is the current swing, with 50 ohm termination, if the output power is +30 dBm (1 watt). If the maximum current swing is limited by 1 mA, what is the output power with 50 ohm termination?

P8-35: PA power handling capacity is described by 1 dB compression point or by saturation power. Explain the terms.

P8-36: In Table 8-4, three tone test signals (e_i) are used for a non-linearity of up to 3rd order terms. In practice, two-tone signals are often used. Generate 3rd order in-band harmonics using the table. Obtain numerical results with $e_o = e_i - 0.005e_i^2 - 0.025e_i^3$ and $e_i = \cos(2\pi 1001e6t) + \cos(2\pi 1002e6t)$, and sketch spectrum.

P8-37: Clipping is used in order to reduce the required backoff. If PAPR of a signal with normal amplitude distribution is 10 dB, what is the power gain if 1%

clipping is used? $Q(x) = \frac{1}{\sqrt{2\pi}} \int_x^\infty e^{-\frac{y^2}{2}} dy$ i.e., $Q(x)$ is the tail end probability of zero mean and unity variance Gaussian. For example, $Q(0) = 0.5$, $Q(2.577) = 0.01/2$.

P8-38: Referring to Figure 8-71, a simple sinusoidal phase noise (jitter) is considered. There is a problem of finding Fourier transform (series) of

$$c(t) = e^{j\frac{Am}{fm}\sin(2\pi f_m t)} = e^{j\beta\sin\left(2\pi\frac{t}{T}\right)} = \sum_{k=-\infty}^{\infty} C_k e^{j2\pi\frac{k}{T}t}.$$

Note that $c(t)$ is periodic with the period of T since $c(t) = c(t+T)$. Thus we need to expand $c(t)$ with a Fourier series in order to find spectrum of $c(t)$.

(1) Show that C_k can be expressed by Bessel function of first kind, $C_k = J_k(\beta)$ where it is defined as

$$J_k(\beta) = \frac{1}{2\pi} \int_{-\pi}^{+\pi} e^{j(\beta\sin\theta - k\theta)} d\theta.$$

(2) It is known that $J_{-k}(\beta) = (-1)^k J_k(\beta)$ and $J_0^2(\beta) + J_1^2(\beta) + \ldots = 1.0$.
(3) MATLAB has a function to compute $J_k(\beta) = \text{besselj}(k, \beta)$. Plot $J_0(x)$.

P8-39: Consider 8 bit sign magnitude ADC and $V_{ref} = 0.5$ V. Find a digital representation of input voltages of +0.23 V, −0.4 V and −0.6 V. Compare with 2's complement representation.

P8-40: A PDF of ADC quantization error voltage is uniform and, using the notation of $q = \pm 1/2^b V_{ref}$, show that its variance (power of error voltage) is $q^2/3$. When ADC $V_{ref} = \pm 0.5$V with 10bits, sketch its PDF, and obtain rms (root mean square) voltage of error.

P8-41: A periodic impulse train $s(t)$ whose Fourier Transform $S(f)$ is a pair of FT. $s(t) = \sum_{n=-\infty}^{+\infty} \delta(t - n\Delta t) \overset{FT}{\Leftrightarrow} S(f) = \frac{1}{\Delta t}\sum_{k=-\infty}^{+\infty} \delta\left(f - \frac{k}{\Delta t}\right)$.
Obtain the pair by computing Fourier series coefficients since $s(t)$ is periodic.

P8-42: RF/IF filter frequency response can be represented by a complex baseband, which is summarized in Figure 8-76. I and Q channels are cross-coupled and interfere each other due to the cross-branches. In order to remove the cross-coupling, an impulse response of a complex baseband must be real. Then in frequency domain there is conjugate symmetry, i.e., even symmetry for magnitude and odd symmetry for phase.

(1) When there is a constant phase (even symmetry) in a complex baseband, there is a cross-coupling between I and Q. Is it true? If so how much cross-coupling?
(2) When there is a slope in magnitude (odd symmetry), there is a cross-coupling between I and Q. Is it true? Explain it.

Chapter 9
Review of Signals and Systems, and of Probability and Random Process

Contents

9.1 Continuous-Time Signals and Systems 699
   9.1.1 Impulse Response and Convolution Integral – Time Domain 699
   9.1.2 Frequency Response and Fourier Transform 702
   9.1.3 Signal Power and Noise Power 705
9.2 Review of Discrete-Time Signals and Systems 713
   9.2.1 Discrete-Time Convolution Sum and Discrete-Time Unit Impulse 713
   9.2.2 Discrete Fourier Transform Properties and Pairs 715
9.3 Conversion Between Discrete-Time Signals and Continuous-Time Signals 717
   9.3.1 Discrete-Time Signal from Continuous-Time Signal by Sampling 717
   9.3.2 Continuous-Time Signals from Discrete-Time Signal by De-sampling
      (Interpolation) 719
9.4 Probability, Random Variable and Process 721
   9.4.1 Basics of Probability 721
   9.4.2 Conditional Probability 722
   9.4.3 Probability of Independent Events 724
   9.4.4 Random Variable and CDF and PDF 724
   9.4.5 Expected Value (Average) 725
   9.4.6 Some Useful Probability Distributions 726
   9.4.7 Q(x) and Related Functions and Different Representations 726
   9.4.8 Stochastic Process 730
   9.4.9 Stationary Process, Correlation and Power Density Spectrum 731
   9.4.10 Processes Through Linear Systems 732
   9.4.11 Periodically Stationary Process 733
9.5 Chapter Summary and References with Comments 733
   9.5.1 Chapter Summary 733
   9.5.2 References with Comments 734
9.6 Problems 734

Abstract This chapter is a collection of review material from signals and systems – continuous and discrete, and conversion between them – and from probability, random variable, and stochastic process.

General Terms average · BPSK · CDF · continuous-time · continuous to discrete · CW · discrete-time · discrete to continuous · expected value · Fourier transforms frequency response · impulse response · interpolation · noise power · PAM · PDF ·

© Springer Nature Switzerland AG 2020   697
S.-M. Yang, *Modern Digital Radio Communication Signals and Systems*,
https://doi.org/10.1007/978-3-030-57706-3_9

probability · random variable · sampling · signal power · signals and systems · spectral density · stationary · stochastic process · unit impulse · unit step

Keywords ■■■

List of Abbreviations

ADC	analog digital conversion
CDF	cumulative density function
DAC, D/A	digital analog conversion
LHS	left hand side
LNA	low noise amplifier
LPF	low pass filter
MAP	maximum a posterior probability
ML	maximum likelihood
PDF	probability density function
PSK	phase shift keying
RHS	right hand side
SER	symbol error rate
SNR	signal to noise ratio

We review the important results from signals and systems. We cover continuous-time signals and discrete-time signals, and deterministic signals and random processes (signals). We provide Fourier transform properties and useful example pairs in a table form. There are two separate tables; one for continuous-time signals and the other for discrete –time signals. For more detailed expositions, there are a number of good textbooks, e.g., [1] and [2].

We also review the topics of probability, random variables and random (stochastic) processes. We list useful probability density functions that are useful in the context of digital communications. For more detailed expositions, there are a number of good textbooks as well, e.g., [3] and [4].

We emphasize again that this collection of relevant material, from signals and systems, probability, random variable and process useful to digital communication systems, is mainly for summarized reference. When detailed exposition is necessary, it is important to consult a relevant textbook.

9.1 Continuous-Time Signals and Systems

9.1.1 Impulse Response and Convolution Integral – Time Domain

We now explain one of the most important results from the study of signals and systems, in digital communications systems point of view, which is summarized in Figure 9-1. In linear time-invariant systems, a system, as a black box, can be completely specified by its impulse response, $h(t)$ or its Fourier Transform $H(f)$. The meaning of linear time-invariant is summarized as; linear means that $x(t) = 0 \rightarrow y(t) = 0$, and $a \cdot x(t) \rightarrow a \cdot y(t)$, and time-invariant means that $x(t - D) \rightarrow y(t - D)$ or $\delta(t - D) \rightarrow h(t - D)$.

The definition of unit impulse $\delta(t)$ is $\delta(t) = 1.0$ when $t = 0$, $\delta(t) = 0.0$ when t $\neq$ 0. It may be defined as a derivative of unit step function $u(t)$ as $\delta(t) = \frac{d}{dt}u(t)$. This is shown in Figure 9-2.

Exercise 9-1: Be familiar with following useful relationships. Hint: use [1] if you need additional help.

$$\int_{-\infty}^{+\infty} \delta(\tau)d\tau = 1.0.$$

$$\int_{0}^{+\infty} \delta(t - \tau)d\tau = u(t).$$

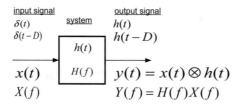

Figure 9-1: Linear time-invariant systems with input and output; when the input is a unit impulse its output is its impulse response. For any input, the output is expressed as a convolution integral of the input with the impulse response

Figure 9-2: Unit impulse defined; it can be related with unit step function

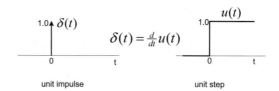

$$\int_{-\infty}^{+\infty} x(\tau)\delta(t-\tau)d\tau = x(t) \text{ i.e.}, x(t) = x(t) \otimes \delta(t)$$

Note that $x(t) = x(t) \otimes \delta(t)$ for any continuous-time signal, where $\otimes$ means convolution.

From the definition of the impulse response and the fact that any input signal can be considered as a train of impulses (with different amplitude), the output of a system with $h(t)$ can be shown as a convolution integral.

$$y(t) = \int_{-\infty}^{+\infty} h(\lambda)x(t-\lambda)d\lambda \tag{9-1}$$

It is also denoted as

$$y(t) = x(t) \otimes h(t) \tag{9-2}$$

By changing integration variable it can be shown that

$$y(t) = \int_{-\infty}^{+\infty} x(\lambda)h(t-\lambda)d\lambda \tag{9-3}$$

It is important to be thoroughly familiar with the convolution integral.

Exercise 9-2: Time reversal and shift. Find x(−t) and x(−t + 0.5) for a given x(t).

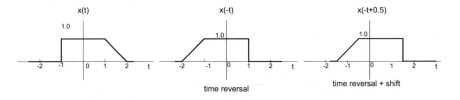

Exercise 9-3: Time scale and shift. Given x(t) find x(2t) and x(2t − 1).

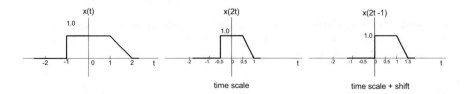

Exercise 9-4: Time reversal, scale and shift. Given x(t) find x(−2t) and x(−2t − 1).

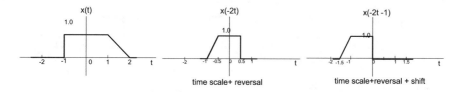

Example 9-1: Find y(t) with h(t) a rectangular pulse width T and amplitude A, and x (t) is the same as h(t).

Example 9-2: A system impulse response is rectangular as in Example 9-1, and the input is two unit impulses separated by T/2 (half of the width of rectangular pulse). Find the output.

Exercise 9-5: A system impulse response is rectangular as in Example 9-1, and the input is two unit impulses separated by T (the width of rectangular pulse) and the 2nd is negative, i.e., $\delta(t) - \delta(t - T)$. Find the output and sketch it.
Answer:

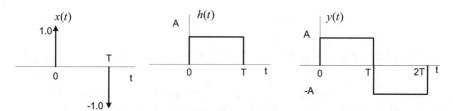

Example 9-3: A system impulse response is triangular and the input shown below (3 unit impulses) in Figure 9-5. Find the output and sketch.
   The process of computing the output, a convolution integral, is shown Figure 9-5 as well. Work out the details for yourself.

Exercise 9-6: Write a computer program (e.g., Matlab) for computation of Example 9-3.

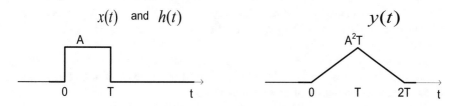

Figure 9-3: Solution to Example 9-1; a convolution of rectangular pulses become a triangular pulse

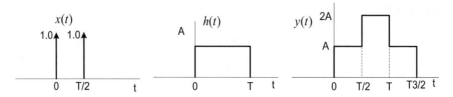

Figure 9-4: Example 9-2 solution

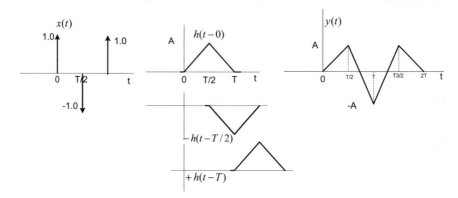

Figure 9-5: Example 9-3 and it shows the process of convolution integral to obtain the output

9.1.2   Frequency Response and Fourier Transform

Another useful characterization of a signal and of a linear time-invariant system, is the frequency response through Fourier transform pair. It is defined by

$$X(f) = \int_{-\infty}^{+\infty} x(t)e^{-j2\pi ft} dt \qquad (9\text{-}4)$$

$$x(t) = \int_{-\infty}^{+\infty} X(f)e^{+j2\pi ft} df \qquad (9\text{-}5)$$

Here we use frequency f in [Hz], or cycles per second, rather than radian frequency which is related with Hz frequency as $\omega = 2\pi f$ [rad/sec]. With this choice of frequency f, (9-4) and (9-5) are symmetrical, and in practice, [Hz] frequency is often used.

In Table 9-1, we list the important properties of Fourier transform and some useful transform pairs. Each pair can be confirmed from the definition of (9-4) and (9-5). We leave the details of proof as exercises.

Now we provide examples to show how to use Table 9-1.

Table 9-1: Fourier Transform Properties and Example pairs

Useful Fourier transform properties			
Time domain	Frequency domain	comments	
$x(t) = \int_{-\infty}^{+\infty} X(f)e^{+j2\pi ft}df$	$X(f) = \int_{-\infty}^{+\infty} x(t)e^{-j2\pi ft}dt$	Definition	Row no.
$x(at)$	$\frac{1}{\|a\|}X\left(\frac{f}{a}\right)$	$a \neq 0$ special case: $a = -1$,	1
$x^*(t)$	$X^*(-f)$	conjugation	2
	For real $x(t)$, $X(f) = X^*(-f)$. Thus $\|X(-f)\| = \|X(f)\|$, and $\angle X^*(f) = -\angle X(f)$		3
$x^*(-t)$	$X^*(f)$	Combine above two (1,2)	4
$x(t-D)$	$e^{-j2\pi fD}X(f)$	delay	5
$e^{\pm j2\pi f_o t}x(t)$	$X(f \mp f_o)$	modulation	6
$x(t) \circledast y(t)$	$X(f)Y(f)$	Time convolution	7
$x(t)y(t)$	$X(f) \circledast Y(f)$	Time multiplication	8
$\int_{-\infty}^{+\infty}\|x(t)\|^2 dt = \int_{-\infty}^{+\infty}\|X(f)\|^2 df$		Energy measure	9
$\int_{-\infty}^{+\infty} x(t)y^*(t)dt = \int_{-\infty}^{+\infty} X(f)Y^*(f)df$		Parseval's theorem	10
Useful Fourier transform pairs			
$x(t) = \delta(t)$	$X(f) = 1.0$	Flat spectrum	11
$x(t) = 1.0$	$X(f) = \delta(f)$	DC	12
Rectangle (t = −1/2, t = +1/2) amplitude = 1.0	$\frac{\sin(\pi f)}{\pi f}$	Rectangle pulse in time domain	13
$\frac{\sin(\pi t/T)}{(\pi t/T)}$	Brick filter (f = −1/2T, f = +1/2T) amplitude = T	Ideal low pass filter impulse response	14
$\cos(2\pi f_o t)$	$\frac{1}{2}\delta(f - f_o) + \frac{1}{2}\delta(f + f_o)$	cosine	15
$\sin(2\pi f_o t)$	$\frac{1}{2j}\delta(f - f_o) - \frac{1}{2j}\delta(f + f_o)$	sine	16
$e^{-\pi t^2}$	$e^{-\pi f^2}$	Gaussian	17
$e^{-a\|t\|}$	$\frac{2a}{a^2+(2\pi f)^2}$	exponential	18
$e^{-at}u(t)$	$\frac{1}{a+j2\pi f}$		
$sgn(t)$	$\frac{1}{j\pi f}$	$sgn(t) = 1, \ t > 0$ $= 0, \ t = 0$ $= -1, \ t < 0$	19
$\frac{1}{\pi t}$	$-j \ sgn(f)$	Hilbert transform	20
$\sum_{n=-\infty}^{\infty} \delta(t - nT)$	$\sum_{k=-\infty}^{\infty} \frac{1}{T}\delta\left(f - \frac{k}{T}\right)$	Impulse train, Fourier series coefficient: $a_k = \frac{1}{T}$	21
$x(t) \cdot \sum_{n=-\infty}^{\infty} \delta(t - nT)$	$\sum_{k=-\infty}^{\infty} \frac{1}{T}X\left(f - \frac{k}{T}\right)$	Impulse train sampling	
$p(t) = \sum_{n=-\infty}^{\infty} h(t - nT)$	$\sum_{k=-\infty}^{\infty} \frac{1}{T} H\left(\frac{k}{T}\right)\delta\left(f - \frac{k}{T}\right)$	$p(t)$ is periodic with T	22
$p(t) = \sum_{k=-\infty}^{\infty} a_k e^{j2\pi \frac{k}{T}t}$ $a_k = \frac{1}{T}\int_0^T p(t)e^{-j2\pi \frac{k}{T}t}$	$\sum_{k=-\infty}^{\infty} a_k\delta\left(f - \frac{k}{T}\right)$	Periodic pulse $p(t)$ with Fourier series expansion	23
$s(t) = \sum_{n=-\infty}^{\infty} x(t - nT) = $	$\sum_{k=-\infty}^{\infty} \frac{1}{T} X\left(\frac{k}{T}\right) e^{j2\pi \frac{k}{T}t}$	Poisson sum formula of periodic pulse- 22 and 23 combined	24

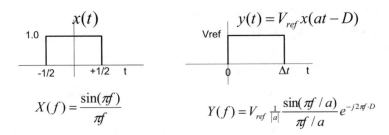

$$X(f) = \frac{\sin(\pi f)}{\pi f}$$

$$Y(f) = V_{ref} \frac{1}{|a|} \frac{\sin(\pi f / a)}{\pi f / a} e^{-j2\pi f \cdot D}$$

Figure 9-6: Example 9-4 solution

Example 9-4: Find a Fourier transform of a rectangular pulse with amplitude V_{ref}, and the pulse width from 0 to Δt using Table 9-1.

In row 13 of Table 9-1 a special form rectangular pulse $x(t)$ and its transform pair is listed. The rectangular pulse in Example 9-4, $y(t)$, can be given in terms of $x(t)$ as $y(t) = V_{ref} x(at - D)$ with $a = 1/\Delta t$, $D = \Delta t/2$. Convince yourself that it is true. You may want to review the previous exercises on time scale and time shift.

We can obtain Fourier transform of $y(t)$ by using Row 1 (time scale) and Row 5 (delay) of Table 9-1. The solution is shown in Figure 9-6.

Certainly this can be confirmed by direct use of (9-4) applied to $y(t)$.

Example 9-5: Find a Fourier transform of a triangular pulse in Example 9-1 using Table 9-1.

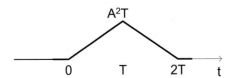

First find Fourier transform of a rectangular pulse in Example 9-1. Using time scale (row 1) and delay(row 5), from $X(f) = \frac{\sin(\pi f)}{\pi f}$ we obtain $AT \frac{\sin(\pi fT)}{\pi fT} e^{-j\pi fT}$; this can be also obtained from Example 9-4 by letting $\Delta t = T$ and $V_{ref} = A$. The triangle pulse above is a convolution of two identical rectangular pulses. Using row 7 in Table 9-1,

$$Y(f) = AT \frac{\sin(\pi fT)}{\pi fT} e^{-j\pi fT} \; AT \frac{\sin(\pi fT)}{\pi fT} e^{-j\pi fT} = \left[AT \frac{\sin(\pi fT)}{\pi fT} \right]^2 e^{-j2\pi fT}.$$

Example 9-6: Find Fourier transform of unit step pulse, $u(t)$, in Figure 9-2 (RHS).

Express $u(t) = \frac{1}{2} sgn(t) + \frac{1}{2}$ where $sgn(t) = 1.0$, if $t > 0$, $sgn(t) = 0.0$, if $t = 0$, $sgn(t) = -1.0$, if $t < 0$. See row 19 of Table 9-1. Using row 12 and row 19, $U(f) = \frac{1}{j2\pi f} + \frac{1}{2}\delta(f)$.

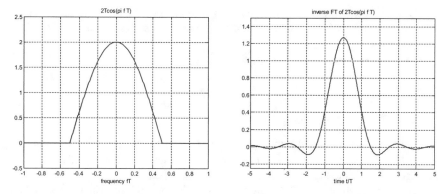

Figure 9-7: Example 9-7, a pair of $Y(f) = 2T \cos(\pi fT)$ and its inverse FT $y(t)$

Example 9-7: Given $Y(f) = 2T \cos(\pi fT)$ if $|f| < 1/2T$, $Y(f) = 0.0$ otherwise, find its inverse Fourier transform $y(t)$.

Use the identity of $2T \cos(\pi fT) = Te^{+j\pi fT} + Te^{-j\pi fT}$, and then compute $T \int_{-\frac{1}{2T}}^{+\frac{1}{2T}} e^{+j\pi fT + j2\pi ft} df$ and $T \int_{-\frac{1}{2T}}^{+\frac{1}{2T}} e^{-j\pi fT + j2\pi ft} df$.

The integrations can be changed in the form of $K \int_{-a}^{+a} e^{j\theta} d\theta$.

$$T \int_{-\frac{1}{2T}}^{+\frac{1}{2T}} e^{+j\pi fT + j2\pi ft} df = \frac{T}{j2\pi[t + 0.5T]} \int_{-\frac{1}{2T}}^{+\frac{1}{2T}} e^{+j2\pi f(t+0.5T)} d(\ f \cdot j2\pi[t + 0.5T])$$

$$= \frac{T}{j2\pi[t + 0.5T]} \int_{-j\pi(\frac{t}{T}+0.5)}^{+j\pi(\frac{t}{T}+0.5)} e^{+\theta} d(\theta) = \frac{\sin\left(\pi \frac{t + 0.5T}{T}\right)}{\pi \frac{t + 0.5T}{T}}$$

Similarly other terms may be integrated.

The result is given by

$$y(t) = \frac{\sin\left(\pi \frac{t+0.5T}{T}\right)}{\pi \frac{t+0.5T}{T}} + \frac{\sin\left(\pi \frac{t-0.5T}{T}\right)}{\pi \frac{t-0.5T}{T}} \tag{9-6}$$

Exercise 9-7: Write a program to plot (9-6). Hint: See Figure 9-7 (RHS).

9.1.3   Signal Power and Noise Power

Consider a continuous wave (CW) signal represented by

$$s_x(t) = A \cos(2\pi f_c t + \theta) \tag{9-7}$$

where the information is carried by amplitude (A) and phase (θ) or frequency (time derivative of phase $\frac{d\theta}{dt}$). In this section we compute the signal power of (9-7) in various ways.

In general the power of a CW signal, P_{s_x}, is defined as,

$$P_{s_x} = \lim_{\tau \to \infty} \frac{1}{2\tau} \int_{-\tau}^{+\tau} s_x^2(t)dt \tag{9-8}$$

Pause to think about (9-8); the square of $s_x(t)$, which is real, integrated (summed) over time interval ($-\tau$ to $+\tau$) and then divided by the time interval, i.e., taking an average. The physical unit of power is [jules/sec] = [watt].

We may approximate CW signal as deterministic, rather than random when $B \ll f_c$. If the bandwidth (B) of the information carrying signal is much smaller than the carrier frequency (f_c), i.e., $B \ll f_c$, which is typically the case, then it is close to a deterministic cosine (or sine) wave. When s_x is very close to a cosine wave (i. e. , $A(t) = A$, $\theta(t) = $ constant for a measuring period), its power is approximately $\frac{A^2}{2}$. For a pure sine wave, the power definition is simplified to measure for one period ($T_c = \frac{1}{f_c}$) since it is periodic. And its power is precisely $\frac{A^2}{2}$. If voltage amplifier gain is k, then the power after PA will be $k^2 \frac{A^2}{2}$.

Even when $A(t)$ is not constant, the power can be expressed exclusively by the power of complex envelope as,

$$P_{s_x} = \lim_{\tau \to \infty} \frac{1}{2\tau} \int_{-\tau}^{+\tau} \frac{1}{2} A^2(t)dt \tag{9-9}$$

using that $\lim_{\tau \to \infty} \frac{1}{2\tau} \int_{-\tau}^{+\tau} \frac{1}{2} A^2(t) \cos(4\pi f_c t + 2\theta)dt = 0$, which holds if $B \ll f_c$, i.e., the bandwidth of complex envelope signal is much smaller than the carrier frequency.

Exercise 9-8: Convince yourself that claiming that the equation (9-9) is correct by using $\cos^2(x) = (1 + \cos 2x)/2$.

9.1.3.1   Computation of CW Signal Power from Baseband Signal

Now the complex envelope is not a deterministic signal but a random process (or signal) because the information to be transmitted is random. A random process needs to be specified by probability distribution or probability density function. Then it is necessary to introduce the average of (9-9) above where the average is taken with the underlying probability density function (PDF) of the random process. The probability and random process (signal) will be reviewed in this chapter later.

Furthermore the power can be related to the symbol time if $A(t)$ is generated synchronously by converting discrete-time symbols to analog signals. We first show the result in equation form as in (9-10).

$$\cdot \lim_{\tau \to \infty} \frac{1}{2\tau} E\left[\int_{-\tau}^{+\tau} \frac{1}{2}A^2(t)dt\right] = \cdot \frac{1}{T}E\left[\int_{\varepsilon}^{\varepsilon+T} \frac{1}{2}A^2(t)dt\right] \qquad (9\text{-}10)$$

RHS of (9-10) is equivalent to the power of a periodic signal with its period T except that the expectation is necessary for a random signal. Remember that the randomness of signal $A(t)$ is due to that of data and that $A(t)$ is generated synchronously with the period of T, symbol time.

Intuitively (9-10) can be understood as follows. First RHS of (9-10) is seen in time domain; dividing the time axis into a number of symbol period (T) segments and then take an integration of $A^2(t)$, a particular instantiation in time (ε). This segmenting and integration may be repeated for many symbol periods. The whole process of instantiation in time will be averaged to obtain average power. Secondly the average can be obtained from an ensemble of possible data in one symbol period. Then averaging over the ensemble of data can be performed before the integration. Thus the result of (9-10) above is obtained as in (9-11); i.e., assuming the average power per symbol period is the same, P_{s_x} can be computed from the complex envelope as

$$P_{s_x} = \frac{1}{T}\int_0^T E\left[\frac{1}{2}A^2(t)\right]dt = \frac{1}{2}E\left[A^2(t)\right] \text{ per } T \qquad (9\text{-}11)$$

In summary, the power of synchronous transmission signal represented by CW signal in (9-7) can be obtained from baseband complex envelope $A(t)$ as in (9-11) when $B \ll f_c$. The randomness of $A(t)$, due to data, can be taken into account by averaging. As mentioned briefly if any signal gain k in RF, then the power, after PA, will be $\frac{k^2}{2}E(A^2)$.

9.1.3.2 Computation of Signal Power from Complex Envelope

$s_x(t) = A\cos(2\pi f_c t + \theta)$ of (9-7) can be written in the form

$$s_x(t) = Re\left\{[A_I + jA_Q]e^{j2\pi f_c t}\right\} \qquad (9\text{-}12)$$

where $A_I = A\cos(\theta)$ and $A_Q = A\sin(\theta)$, and $A^2 = A_I^2 + A_Q^2$.
Furthermore (9-12) can be written as

$$s_x(t) = A_I\cos(2\pi f_c t) - A_Q\sin(2\pi f_c t) \qquad (9\text{-}13)$$

using the relationship of $e^{j2\pi f_c t} = \cos(2\pi f_c t) + j\sin(2\pi f_c t)$, and the complex envelope, $C = A_I + jA_Q$. Since A_I and A_Q are real, $|C|^2 = A^2$. In other words, the power of CW signal (9-7), equivalently (9-12), can be computed from the power of its complex envelope, which is random because the data that are contained are random.

9.1.3.3   Power Gain Through a Filter and Spectral Density

A_I and A_Q defined in (9-12) are generated with random data, discrete in time, passing through a filter. In practice, data to be transmitted go through a digital to analog converter (DAC) and followed by an analog filter. Here we treat the cascading DAC and an analog filter as a filter. In many practical situations we are interested in the power and spectrum of A_I and A_Q. We present the result first here without detailed justification.

The power gain through a filter, which is represented by its impulse response $h(t)$, is simple when the input data symbols are uncorrelated. It is shown in Figure 9-8, and it is most useful for a majority of applications. But more general cases of discrete data with correlation will be treated as we need them. We remark that the correlation of data symbols generates a non-flat frequency spectrum, through Fourier transform of the correlation. Thus the overall spectrum will be changed accordingly, i.e., by multiplication of discrete spectrum and continuous spectrum.

The above result is justified through an example below.

Example 9-8: Binary phase shit keying (BPSK). {0,1} bit is mapped to {+d, −d} symbols. {0,1} are equally likely; the probability of {0} and {1} is ½. The transmit pulse shape is rectangular with the width of symbol (i.e., bit) period T and the amplitude 1.0 [volt]. What is the average symbol power per symbol period at the output of the shaping filter? Express it in terms of d and T.
Answer: discrete-time symbol power (variance) $\sigma^2 = \frac{1}{2}(+d)^2 + \frac{1}{2}(-d)^2 = d^2$. This

is the definition of variance. If it is not clear, Section 9.4 probability, random variable and process, in particular 9.4.5 Expected value (average), may help.

When these symbols of discrete random process (signal) go through a shaping pulse (e.g., rectangular pulse), what is the equivalent power gain?

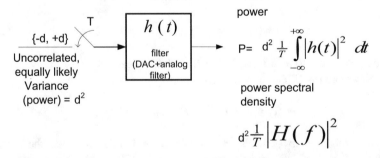

Figure 9-8: Power and spectral density from BPSK, and it is generalized to pulse amplitude modulation (PAM)

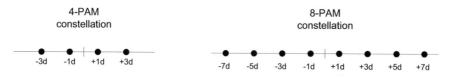

Figure 9-9: M-level constellations of PAM with M = 4, M = 8 examples

The 'power gain' through a shaping filter, in this case of a rectangular pulse, can be obtained by computing $\frac{1}{T}\int_{-\infty}^{+\infty}|h(t)|^2\,dt = 1$.

Sometimes it is convenient to use $\frac{1}{T}\int_{-\infty}^{+\infty}|H(f)|^2\,df$ which is Parseval's theorem where $H(f)$ is Fourier transform of $h(t)$. See row 9 of Table 9-1. Note that the power gain through the shaping filter can be normalized to be unity by scaling the pulse properly if it is not unity. With the rectangular pulse the power gain is unity. Then the power is determined by discrete-time symbols. This is a convenient view when we analyze the signal to noise ratio and the error performance.

The symbol energy (E_s), i.e., power delivered for the duration of symbol time T, can be expressed as $E_s = P\,T = d^2T$, the average power (P) times the symbol time. In this BPSK the symbol energy is also bit energy (E_b) as well.

The above example result can be generalized into pulse amplitude modulations (PAM). Figure 9-8 shows also this generalization too; digital power is σ^2 regardless of underlying modulation levels. We give a few examples how to compute digital power for a given constellation.

Example 9-9: Digital power of PAM. 4 level PAM's constellation and 8-level PAM are shown in Figure 9-9. Assume the symbols are equally likely, i.e., the probability is ¼ (in general 1/M). Find the digital power (variance).

$$\sigma^2 = \tfrac{1}{4}(+3d)^2 + \tfrac{1}{4}(+d)^2 + \tfrac{1}{4}(-d)^2 + \tfrac{1}{4}(-3d)^2 = \tfrac{1}{2}(1^2 + 3^2)d^2 = 5d^2 \ \text{ for } \ M = 4.$$

$$\sigma^2 = \tfrac{1}{4}(1^2 + 3^2 + 5^2 + 7^2)d^2 = 21\,d^2 \ \text{ for } \ M = 8$$

Exercise 9-9: For M = 10, show that $\sigma^2 = \tfrac{1}{5}(1^2 + 3^2 + 5^2 + 7^2 + 9^2)\,d^2 = 35\,d^2$

It is generalized to M-level PAM as, with M being even,

$\sigma^2 = \tfrac{2}{M}(1^2 + 3^2 + 5^2 + 7^2 + \cdots + (M-1)^2)\,d^2 = \tfrac{M^2-1}{3}\,d^2$. In order to see it, from a math table, one may look up the sum S of $1^2 + 3^2 + 5^2 + 7^2 + \cdots + (M-1)^2$. And it is given by $S = \tfrac{(M-1)M(M+1)}{6}$.

Example 9-10: QPSK constellation and 8-PSK are shown in Figure 9-10. Compute the digital power assuming all symbols are equally likely.

Answer: $\sigma^2 = \tfrac{1}{4}(d^2 + d^2)\,4 = 2d^2$ for QPSK.

$\sigma^2 = \tfrac{1}{8}(d^2)\,8 = d^2$ for 8-PSK.

An example below shows that the signal spectrum may change due to the presence of a digital filtering. It also changes the probability distribution as well.

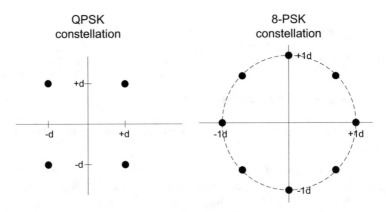

Figure 9-10: QPSK and 8-PSK constellations and its digital power: $2d^2$ and d^2 respectively

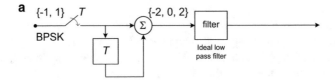

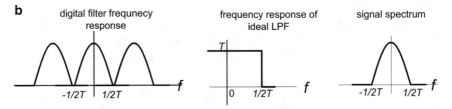

Figure 9-11: Digital filter $1 + z^{-1}$ is added after BPSK modulation and followed by an ideal -low pass filter, and its frequency response is the same as in Example 9-7 (Figure 9-7)

Example 9-11: A simple digital filter is added as in Figure 9-11 and the analog filter is an ideal low pass filter. And its signal spectrum is given by $2T \cos(\pi fT)$ as in Example 9-7 (see Figure 9-7). A digital filter is used for shaping signal spectrum.

Exercise 9-10: Show that the symbol $\{-1, +1\}$ becomes $\{-2, 0, +2\}$. Possible two symbols before addition are $\{+1, +1\}$, $\{+1, -1\}, \{-1, +1\}$, $\{-1, -1\}$ and after addition they are $\{+2, 0, 0, -2\}$.

  Digital power after addition is $\sigma^2/d^2 = \frac{1}{4}(-2)^2 + \frac{1}{2}(0)^2 + \frac{1}{4}(+2)^2 = 2$. Note that the probability distribution is changed.

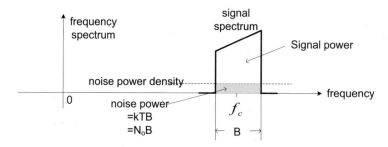

Figure 9-12: Thermal noise density, noise power with bandwidth B

9.1.3.4 Noise Power and Noise Power Spectral Density

A thermal noise is added at the receiver front end, i.e., at the input of low noise amplifier (LNA). It is characterized by its flat spectral density, and its underlying probability density function (PDF) is Gaussian. As the frequency is as high as optical frequency range then the assumption of flat spectrum no longer holds. In this book unless otherwise stated the thermal noise has the flat spectral density. The thermal noise power is proportional to the bandwidth; $P_n = kTB$ [Watt], where $k = 1.38 \times 10^{-23}$ [JK^{-1}], the Boltzmann constant and J = joule, T = temperature in Kelvin and B is the bandwidth in Hz. For example with B = 1.0 [Hz], T = 300 °K, noise power is -174 dBm, and with B = 1.0 [MHz] noise power is -114 dBm where 0 dBm denotes 1.0 miliwatt in dB scale.

Signal to noise ratio (SNR) is signal power / noise power for a given bandwidth. Noise density is symbolically denoted by N_o which is given by $N_o = kT$ for thermal noise with one-sided spectrum (only positive frequency). Then the noise power is given by $P_n = N_o B$. RF signal in time domain is real as shown in the equation (9-12), and one-sided spectrum is sufficient since the spectrum is symmetrical in frequency at zero. The noise is also real and thus one sided spectrum is sufficient. This is shown in Figure 9-12.

Often the double sided (both positive and negative frequency) spectral density is defined as $\frac{N_o}{2} = \frac{1}{2}kT$ and the double sided bandwidth is $2B$ for both positive and negative frequencies. Thus the noise power for real signal in time (RF or IF) is equivalent to one-sided spectrum. The double sided noise spectral density $\frac{N_o}{2}$ is useful when the time domain representation of noise signal is not real but complex (e.g., complex envelope, i-channel and q-channel). Then its frequency response may not be symmetric and hence it is necessary to consider the both positive and negative frequencies. However, the complex signal in time domain is represented by combining two real signals (i and q) into a complex signal. For example the complex envelope, C in (9-13), is represented by $A_I(t) + jA_Q(t)$ where both $A_I(t)$ and $A_Q(t)$ are real. Thus the one-sided noise spectral density $\frac{N_o}{2}$ may be applied for each real channel.

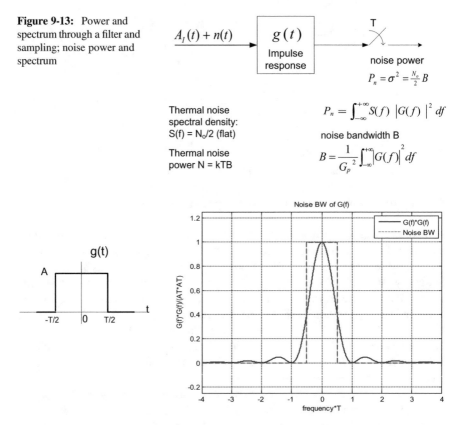

Figure 9-13: Power and spectrum through a filter and sampling; noise power and spectrum

$$A_I(t) + n(t)$$

$$g(t)$$
Impulse response

T

noise power

$$P_n = \sigma^2 = \frac{N_o}{2} B$$

Thermal noise spectral density:
S(f) = N_o/2 (flat)

$$P_n = \int_{-\infty}^{+\infty} S(f) \; |G(f)|^2 \, df$$

noise bandwidth B

Thermal noise power N = kTB

$$B = \frac{1}{G_p^2} \int_{-\infty}^{+\infty} |G(f)|^2 \, df$$

Noise BW of G(f)

g(t)

A

-T/2   0   T/2   t

G(f)*G(f)/(AT*AT)

frequency*T

Figure 9-14: Rectangular pulse and its frequency response (magnitude square) explained in Example 9-12. Noise BW is shown and B = 1/T

9.1.3.5   Noise Power After Passing Through a Receive Filter

In the receive direction a signal plus noise will go through a receive filter and will be sampled for further processing. We are interested in signal power and noise power after sampling. What type of receive filter should one use in order maximize signal to noise power ratio (SNR)? A short answer is a matched filter, which will be greatly elaborated with a dedicated chapter, Chapter 3, since it is very important in digital communication signal design. Here our focus will be on the thermal noise passing through a receive filter.

It is useful to define the noise bandwidth of a receiver filter, which is shown in Figure 9-14 and it is given by

$$B = \frac{1}{G_p{}^2} \int_{-\infty}^{+\infty} |G(f)|^2 df \tag{9-14}$$

where $G_p{}^2$ is the peak of $|G(f)|^2$. It is double sided noise bandwidth.

Example 9-12: For the rectangular pulse of the width T, show that the noise bandwidth is $B = \frac{1}{T}$.

A rectangular pulse $g(t)$ is shown in Figure 9-14 (LHS) and its Fourier transform is given by $G(f) = AT \frac{\sin(\pi f T)}{(\pi f T)}$. Thus when $f = 0$, $G_p{}^2 = (AT)^2$. In this case the integration is easy in time domain using Parseval's theorem (Table 9-1, row 9); $\int |g(t)|^2 dt = A^2 T$. We now obtain $B = 1/T$.

In fact, this is the minimum noise bandwidth with the signaling rate $1/T$. If the transmit pulse is rectangular, it is a matched filter as well and thus optimum. In the literature this pair of rectangular pulses is often used from the beginning of digital communication signal development, and in particular the receive side is implemented as 'integrate and dump'.

9.2   Review of Discrete-Time Signals and Systems

A discrete-in-time signal is a sequence of numbers or symbols. The duration between the numbers may not have time concept in general and discrete series of events. However, in our applications, a sequence is generated in a regular time interval when it is converted to continuous-time signal, or the sequence is obtained from a continuous-time signal by sampling it. And the time is called a sampling period or symbol time (T). In this book, we consider only the discrete-in-time signal with a constant interval between symbols unless otherwise stated. We use the abbreviation; $x(nT) = x(n)$ i.e., often times we omit the underlying sampling (symbol) period.

9.2.1   Discrete-Time Convolution Sum and Discrete-Time Unit Impulse

$\delta(n)$ is a unit impulse where $\delta(n) = 1.0$ when $n = 0$ and $\delta(n) = 0.0$ when $n \neq 0$.

A discrete time-invariant system, like its continuous-time counterpart, can be completely specified by its unit impulse response $h(n)$, i.e., the system output when the input is $\delta(n)$, or its Fourier transform $H(e^{j2\pi f})$. Rather than using simple f, the use of $e^{j2\pi f}$ represents the periodic nature of digital frequency and its period is 1.0 normalized by $1/T$ (sampling frequency). Note that the 'time-invariant' is the overuse of terms for simplicity; to be precise it should be 'shift-invariant' since discrete-time signals are defined only on discrete time index (n) being integers.

Figure 9-15: Discrete-time linear shift-invariant systems with input and output; when the input is a unit impulse, its output is the impulse response of a system $h(n)$. For any input $x(n)$, the output $y(n)$ is expressed as a convolution sum

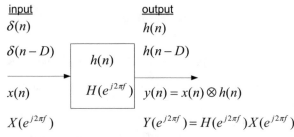

input

$\delta(n)$

$\delta(n-D)$

$x(n)$

$X(e^{j2\pi f})$

output

$h(n)$

$h(n-D)$

$H(e^{j2\pi f})$ $y(n) = x(n) \otimes h(n)$

$Y(e^{j2\pi f}) = H(e^{j2\pi f})X(e^{j2\pi f})$

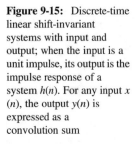

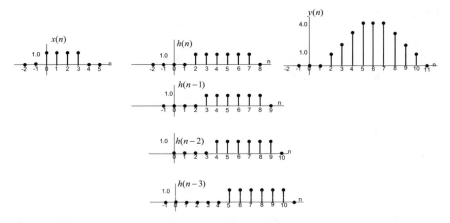

Figure 9-16: x(n) and h(n) are given as in the figure, and y(n) is shown

$$y(n) = \sum_{k=-\infty}^{\infty} h(k)x(n-k) \tag{9-15}$$

The convolution sum, (9-15), is denoted by using the convolution symbol $\otimes$ as,

$$y(n) = x(n) \otimes h(n) \tag{9-16}$$

By changing the summation index of (9-15), it can be summed as,

$$y(n) = \sum_{k=-\infty}^{\infty} h(n-k)x(k) \tag{9-17}$$

Example 9-13: Convolution of two finite sequences.

The process of computation of a convolution sum is in Figure 9-16. $y(n) = \sum_{m=0}^{3} h(n-m)x(m)$ is used in the figure. Alternatively $y(n) = \sum_{k=2}^{7} h(k)x(n-k)$ may be used. Note that the summation index is finite due to the fact that both input and system impulse response are finite in time.

Exercise 9-11*: $h(n)$ is finite in duration (N), and $x(n)$ is semi-infinite. Write a subroutine of a computer program to compute the output for any given input, $y(n)$ $n = 0,1,2,\ldots\ldots\ldots$ Hint: The use of (9-15), rather than (9-17) is more convenient to organize a program. The input $x(n)$ is stored in memory and compute (9-15). Then for next input $x(n + 1)$, all the memory contents are shifted $x(n - k)$ and the new x $(n + 1)$ is stored in the memory, and then compute (9-15) for the output again. This is repeated for each new input. The star * means a project exercise.

9.2.2   Discrete Fourier Transform Properties and Pairs

The Fourier transform of a discrete-time signal is periodic in frequency domain and its frequency is $1/T$ (sampling rate). Thus the Fourier transform of $x(nT)$ is denoted as $X(e^{j2\pi fT})$ in the literature to emphasize the fact that it is periodic.

The Fourier transform of $x(nT) = x(n)$ is given by,

$$X\left(e^{j2\pi fT}\right) = \sum_{n=-\infty}^{\infty} x(nT)\, e^{-j2\pi\, nfT} \tag{9-18}$$

And its inverse transform is given by

$$x(nT) = \int_{-\frac{1}{2T}}^{+\frac{1}{2T}} X\left(e^{j2\pi fT}\right) e^{j2\pi\, fT\,\, n} dfT \tag{9-19}$$

By making $T = 1$, it can be implicit. We are using a frequency, rather than radian frequency and they are related as $\omega = 2\pi f$.

Example 9-14: Find a frequency response of a system with unit sample response (impulse response) $h(nT) = h(n)$ as shown in Figure 9-17, similar to a rectangular pulse except it is discrete in time.

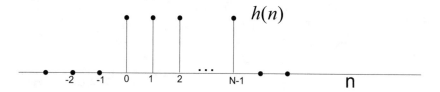

Figure 9-17: Example of unit sample response; discrete-time rectangular pulse

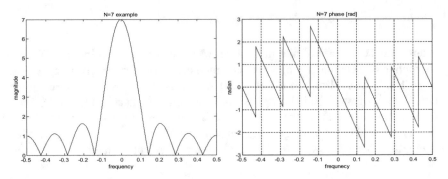

Figure 9-18: Amplitude (LHS) and phase response of Fourier transform of $N = 7$ rectangular pulse

Table 9-2: Discrete Fourier Transform Properties and Example pairs

Time domain	Frequency domain	
Useful discrete Fourier transform properties and pairs		
Time domain	Frequency domain	
$x(n) = \int_{-1/2}^{+1/2} X\left(e^{j2\pi f}\right) e^{+j2\pi fn} df$	$X\left(e^{j2\pi f}\right) = \sum_{n=-\infty}^{n=+\infty} x(n)e^{-j2\pi fn}$	Definition
$x(nT)$	$X(e^{j2\pi fT})$	
$x^*(n)$	$X^*(e^{-j2\pi f})$	conjugation
	For real $x(n)$, $X(e^{j2\pi f}) = X^*(e^{-j2\pi f})$. Thus $\lvert X(e^{-j2\pi f}) \rvert = \lvert X(e^{j2\pi f}) \rvert$, and $\angle X^*(e^{j2\pi f}) = -\angle X(e^{j2\pi f})$	
$x^*(-n)$	$X^*(e^{j2\pi f})$	
$x(n - n_D)$	$e^{-j2\pi f n_D} X\left(e^{j2\pi f}\right)$	delay
$e^{j2\pi f_o n} x(n)$	$X\left(e^{j2\pi(\, f - f_o)}\right)$	modulation
$x(n) \circledast y(n)$	$X(e^{j2\pi f})Y(e^{j2\pi f})$	Time convolution
$x(n)y(n)$	$X(e^{j2\pi f}) \circledast Y(e^{j2\pi f})$	Time multiplication
$\sum_{n=-\infty}^{n=+\infty} \lvert x(n) \rvert^2 = \int_{-0.5}^{+0.5} \lvert X(f) \rvert^2 df$		Energy measure
$\sum_{n=-\infty}^{n=+\infty} x(n)y^*(n) = \int_{-0.5}^{+0.5} X\left(e^{j2\pi f}\right) Y^*\left(e^{j2\pi f}\right) df$		Parseval's theorem
Useful Fourier transform pairs		
$x(n) = \delta(n)$	$X(e^{j2\pi f}) = 1.0$	
$x(n) = 1.0$	$X(f) = \sum_{k=-\infty}^{k=+\infty} \delta(\, f - k)$	$\delta(f)$ and repetition in frequency
$x(n) = 1.0, 0 \le n \le M$ $x(n) = 0.0$, otherwise	$\frac{\sin\left(\pi f(M+1)\right)}{\sin(\pi f)} e^{-j\pi fM}$	Discrete-time rectangle shape
$\frac{\sin(\pi 2 f_c n)}{\pi n}$	Brick filter ($f = -1/2 f_c$, $f = +1/2 f_c$)	Ideal low pass filter impulse response
$2\cos(2\pi f_o n)$	$\sum_{k=-\infty}^{k=+\infty} \delta(\, f - f_o + k) + \delta(\, f + f_o + k)$	Repetition $\delta(f - f_o) + \delta(f + f_o)$
$2j\sin(2\pi f_o n)$	$\sum_{k=-\infty}^{k=+\infty} \delta(\, f - f_o + k) - \delta(\, f + f_o + k)$	$\delta(f - f_o) - \delta(f + f_o)$
$\dfrac{\sin^2\left(\dfrac{\pi n}{2}\right)}{\left(\dfrac{\pi n}{2}\right)} \quad n \ne 0$ $\qquad\quad 0 \quad n = 0$	$-j\,\text{sgn}(f)$	Hilbert transform

Solution:

$$H\left(e^{j2\pi fT}\right) = \sum_{n=0}^{N-1} e^{-j2\pi nfT} = \frac{1 - e^{-j2\pi NfT}}{1 - e^{-j2\pi fT}} = \frac{\sin\left(\pi fTN\right)}{\sin\left(\pi fT\right)} e^{-j(N-1)\pi fT}$$

And the last part manipulation of $H(e^{j2\pi fT})$ is to see the magnitude and phase separately. Example magnitude and phase with N = 7 are displayed in Figure 9-18. We use the formula for the summation of a series;

$$\sum_{n=0}^{N-1}\left[e^{-j2\pi fT}\right]^n = \frac{1 - \left[e^{-j2\pi fT}\right]^N}{1 - \left[e^{-j2\pi fT}\right]} \tag{9-20}$$

9.3   Conversion Between Discrete-Time Signals and Continuous-Time Signals

9.3.1   Discrete-Time Signal from Continuous-Time Signal by Sampling

For a Fourier transform pair of a continuous-time signal $x_c(t) \leftrightarrow X_c(f)$, we will express the Fourier transform of the sampled $x_s(t) \leftrightarrow X_s(f)$ and discrete-time version as shown in Figure 9-19. It is given by

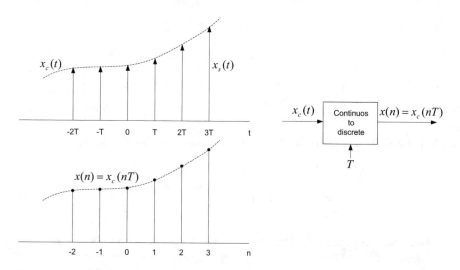

Figure 9-19: Sampling a continuous-time signal to generate a discrete –time signal

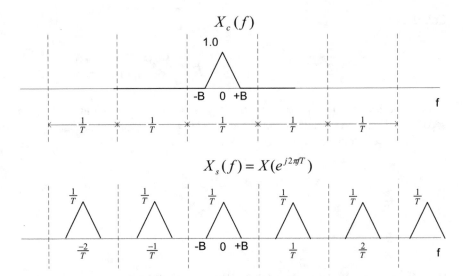

Figure 9-20: Frequency response of sampled continuous-time signals without aliasing

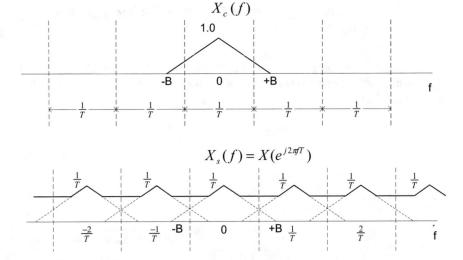

Figure 9-21: Frequency response of sampled continuous-time signals with aliasing

$$X(e^{j2\pi fT}) = X_s(f) = \frac{1}{T} \sum_{k=-\infty}^{k=+\infty} X_c\left(f - \frac{k}{T}\right) \qquad (9\text{-}21)$$

Note that it is periodic in frequency with the frequency period 1/T. The equation can be visualized as follows. The frequency response of the continuous-time signal

$X_c(f)$ is divided by a segment of $\frac{1}{T}$. Then add all of them up with a scale factor of $\frac{1}{T}$. Examples are shown in Figure 9-20 and in Figure 9-21.

In Figure 9-20, there is no aliasing since the continuous-time signal is band limited less than the sampling frequency and in Figure 9-21, the sampling frequency is not high enough compared to the bandwidth of the signal and thus there is aliasing distortion. Once this distortion happens, it is not possible to remove it. In order to avoid the aliasing distortion, it is important to limit the bandwidth by low pass filtering.

9.3.2   Continuous-Time Signals from Discrete-Time Signal by De-sampling (Interpolation)

Discrete-time signals may be converted to continuous-time signals. In practical situations digital-to-analog (D/A) converters are employed, which can be modeled

$$x_c(t) = \sum_{n=-\infty}^{n=+\infty} x(n) \cdot h(t - nT) \qquad (9\text{-}22)$$

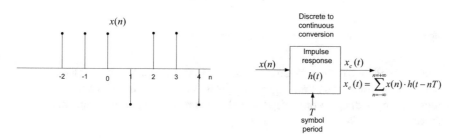

Figure 9-22:   Discrete-time signal to continuous-time signal conversion with an interpolating filter $h(t)$

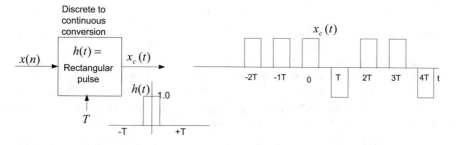

Figure 9-23:   Discrete-time signal to continuous-time signal conversion with a rectangular pulse interpolating filter

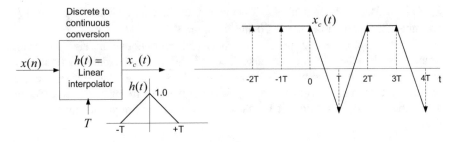

Figure 9-24: Linear interpolation filter (triangular pulse filter)

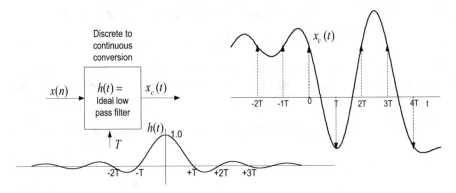

Figure 9-25: Ideal low pass filter is used for interpolation

as a linear filter (often a rectangular pulse). Here in order to relate discrete-time signals with continuous-time signal, we treat it as essentially a linear filtering or convolution of discrete-time signals with an impulse response of the filter. It is shown in Figure 9-22, and is denoted as,

$$x_c(t) = \sum_{n=-\infty}^{n=+\infty} x(n) \cdot h(t - nT) \tag{9-22}$$

We give some examples. we consider the case of $h(t) = \delta(t)$ where $\delta(t)$ is sometimes called Dirac delta function, i.e., $x_c(t) = \sum_{n=-\infty}^{n=+\infty} x(n) \cdot \delta(t - nT)$. In this case the unit sample in discrete time is replaced by impulse trains. Due to the impulse function, it appears to be discrete but it is considered as an analog signal.

More examples of different interpolating filters are given below; Figure 9-23, Figure 9-24, and Figure 9-25. The filter may be an ideal low pass filter, a brick wall in frequency domain (i.e., a rectangle) as shown in Figure 9-25. Its impulse response is given by $h(t) = \frac{\sin(\pi t/T)}{\pi t/T}$. Of all the example filters in the above, the ideal low pass filter provides the minimum bandwidth for the continuous-time signal.

9.4 Probability, Random Variable and Process

We give a summary of the results useful to our study from probability, random variable and stochastic process. The understanding of these topics is very important for modern digital radio communication signals and systems. Data message itself is a random variable and once it is converted to continuous-time signal it is called a random (stochastic) process. Its power and power spectrum should be defined by the use of average. The performance of digital communication systems is often expressed in terms of error probability of symbols, bits, and packets (a collection of bits). The radio communication channels of noise, interference and fading are understood by random processes with underlying probability density functions. The receiver processing aims to achieve the optimization criteria expressed as maximum likelihood or maximum a posterior probability.

In this section we list a collection of important items in probability, random variable and random process. For detailed exposition, there are many good textbooks, e.g., [3] and [4].

9.4.1 Basics of Probability

The probability of mutually exclusive events (i.e., $A \cap B = \varnothing$), is given by $P(A \cup B) = P(A) + P(B)$. We use 'OR', '+' for $\cup$ (union) and 'AND', '.' for $\cap$ (intersection). Note that this property is taken to be 'obvious' or an axiom in probability. If it is not mutually exclusive, the probability A OR B is given by $P(A \cup B) = P(A) + P(B) - P(A \cap B)$. This is shown diagrammatically in Figure 9-26.

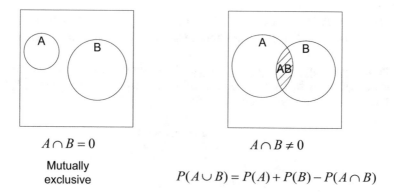

$A \cap B = 0$ $A \cap B \neq 0$

Mutually
exclusive $P(A \cup B) = P(A) + P(B) - P(A \cap B)$

Figure 9-26: Mutually exclusive event and its probability is P(A) + P(B)

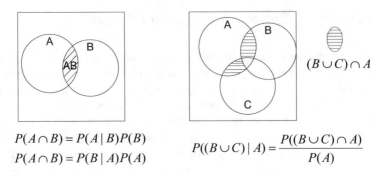

$$P(A \cap B) = P(A\,|\,B)P(B)$$
$$P(A \cap B) = P(B\,|\,A)P(A)$$

$$P((B \cup C)\,|\,A) = \frac{P((B \cup C) \cap A)}{P(A)}$$

Figure 9-27: Examples of conditional probability

9.4.2   *Conditional Probability*

The concept of conditional probability is important in our applications to modern digital communications and thus its thorough understanding is crucial.

Definition of conditional probability: It is the probability of intersection 'normalized' by the probability of the condition.

Two graphical examples are shown in Figure 9-27.

Example 9-15: A dice has 6 faces and each face is represented by a number from a set of $\{1, 2, 3, 4, 5, 6\}$. We assume the probability of each face is 1/6; $P(i) = \frac{1}{6}$ for $i = 1, 2, 3, 4, 5, 6$. We let $B = \{even\} = \{2, 4, 6\}$ and $A = \{2\}$. What is the probability of $P(A\,|\,B)$, the probability of A given B?

Answer:

$P(A \cap B) = P(A) = \frac{1}{6}$ and $P(B) = \frac{3}{6}$. And thus the solution is given by $P(A\,|\,B) = \frac{P(A \cap B)}{P(B)} = \frac{\frac{1}{6}}{\frac{3}{6}} = \frac{1}{3}$.

Exercise 9-12: Let now $C = \{2, 3\}$. What is the conditional probability of C for a given B?

Hint: Note that $P(C) = 1/6 + 1/6 = 1/3$, and $P(C \cap B) = 1/6$ and $P(B) = 1/2$. $P(C\,|\,B) = 1/3$. Since set C contains an odd number, 3, the condition B (even) does not help.

Conditional probability as Bayes' Theorem: It is useful to express the conditional probability in terms of total probability with mutually exclusive events $\{A_k\}$ for $k = 1, 2, \ldots,$ n and it is given by

$$P(A_i\,|\,B) = \frac{P(B|A_i)P(A_i)}{\sum_{k=1}^{n} P(B|A_k)P(A_k)}.$$

This is shown pictorially in Figure 9-28.

Example 9-16: There are two jars, and the first jar contains 8 white balls and 2 red balls and the second jar contains 1 white ball and 9 red balls. Randomly select a jar and take out a ball. It was a white ball. What is the probability that the ball (white) is taken from the first jar?

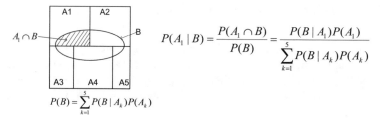

$$P(A_1 \mid B) = \frac{P(A_1 \cap B)}{P(B)} = \frac{P(B \mid A_1)P(A_1)}{\sum\limits_{k=1}^{5} P(B \mid A_k)P(A_k)}$$

$$P(B) = \sum_{k=1}^{5} P(B \mid A_k)P(A_k)$$

Figure 9-28: Bayes Theorem; graphical representation

Figure 9-29: Example 9-16 graphical representation

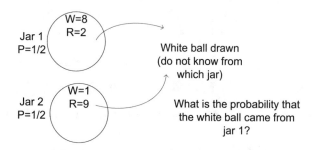

Jar 1
P=1/2
W=8
R=2

White ball drawn
(do not know from
which jar)

Jar 2
P=1/2
W=1
R=9

What is the probability that
the white ball came from
jar 1?

Answer: This is an example of computing a conditional probability. It is often said a posterior probability since it is the probability after the white ball is observed (known). We need to compute the probability of the first jar after observing a white ball;

$$P(A_1 = \mathit{'firstjar'})|B = \mathit{'white\ ball'}).$$

$$P(A_1 = \mathit{'first\ jar'}) = \frac{1}{2} \text{ and } P(A_2 = \mathit{'\ sec\ ond\ jar'})$$

$$= \frac{1}{2} \text{ due to the random selection.}$$

$$P(B = \mathit{'whte\ ball'}) = \sum_{k=1}^{2} P(B|A_k)P(A_k) = \frac{8}{10}\frac{1}{2} + \frac{1}{10}\frac{1}{2}$$

$$P(B = \mathit{'whte\ ball'} \cap A_1 = \mathit{'firstjar'}) = P(B|\ A_1)P(A_1) = \frac{8}{10}\frac{1}{2}$$

$$P(A_1 = \mathit{'firstjar'})|B = \mathit{'white\ ball'}) = \frac{\frac{8}{10}\frac{1}{2}}{\frac{8}{10}\frac{1}{2} + \frac{1}{10}\frac{1}{2}} = \frac{8}{9}.$$

Exercise 9-13: Find the probability of the white ball that is taken from the second jar. (Answer = 1/9). Find the conditional probability if the red ball was taken.

This example can be cast into digital communication problem. Jar 1 and Jar 2 represents '0' and '1' bits, and white and red ball represent error behavior of a channel. We will elaborate this amply in this book.

9.4.3   Probability of Independent Events

The statistical independence is an important concept and it arises when repeated trials of a single experiment are considered. Two events A and B are independent if and only if the probability $P(A \cap B) = P(A)P(B)$. This can be understood that in general $P(A \cap B) = P(A)P(B|A)$ but due to the independence $P(B|A) = P(B)$ and thus the result is obtained.

A good example is a repeated coin tossing (or tossing multiple identical coins at the same time). Consider three repetitions of coin tossing. What is the probability of all three being a head? Since each tossing is independent, it is ½ ½ ½ = 1/8.

For the _independent_ events, since $P(A \cap B) = P(A)P(B)$ the probability of union P $(A \cup B)$ can be expressed as

$$P(A \cup B) = P(A) + P(B) - P(A)P(B)$$

$$P(\overline{A} \cap \overline{B}) = P(\overline{A})P(\overline{B}) = (1 - P(A))(1 - P(B))$$

$$P(A \cup B) = 1 - P(\overline{A \cup B}) = 1 - (1 - P(A))(1 - P(B))$$

In the above we use, so called De Morgan's law

$$\overline{A \cup B} = \overline{A} \cap \overline{B}.$$

9.4.4   Random Variable and CDF and PDF

We define a function $\mathbf{x}(s)$ whose variable 's' will not be a number but an outcome of a probability experiment. For example an experiment may be the rolling of a dice with each face designated as $f_1, f_2, f_3, f_4, f_5, f_6$. We make the value $\mathbf{x}(f_k) = k$. This is a random variable $\mathbf{x}$.

For a given number x, we use the notation $\{ \mathbf{x} \le x \}$ to represent a set of all outcomes 's' and the set $\{ \mathbf{x} \le x \}$ is an event for any real number x. Its probability $P\{ \mathbf{x} \le x \}$ is a number depending on x; that is a function of x. The function is called distribution function (or cumulative distribution function CDF), $F_{\mathbf{x}}(x)$ for $-\infty < x < +\infty$.

$$F_{\mathbf{x}}(x) = P\{ \mathbf{x} \le x \}.$$

And its derivative of $F_{\mathbf{x}}(x) = F(x)$ is called probability density function (PDF), f (x),

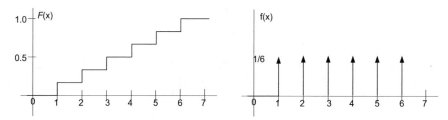

Figure 9-30: Example of CDF and PDF

$$f(x) = \frac{dF(x)}{dx}.$$

Example 9-17: fair dice experiment we define the random variable **x** by $x(f_k) = k$ and since $P(f_k) = 1/6$ it is a staircase function as in Figure 9-30. Its density function is an impulse as shown in the same figure.

Properties of CDF
(a) $F(-\infty) = 0$ $F(+\infty) = 1$
(b) It is a non-decreasing function of x

We can see the properties of CDF from an example and it is generally true.

9.4.5 *Expected Value (Average)*

The expected value of a (real) random variable **x** is the integral

$$E\{\mathbf{x}\} = \int_{-\infty}^{+\infty} xf(x)dx.$$

where $f(x)$ is the probability density function of **x**.
If **x** is of discrete type, taking values x_n with the probability p_n, then

$$E\{\mathbf{x}\} = \sum_n x_n P\{\mathbf{x} = x_n\} = \sum_n x_n p_n.$$

A random variable **x** in Example 9-17 is a fair dice tossing, and its expected value is

$$E\{\mathbf{x}\} = \frac{1 + 2 + 3 + 4 + 5 + 6}{6} = 3.5.$$

Expected value of $g(\mathbf{x})$

$$E\{g(\mathbf{x})\} = \int_{-\infty}^{+\infty} g(x)f(x)dx.$$

The mean (expected value), m, of a random variable places the center of gravity of $f(x)$. Another important parameter is its variance σ^2 defined by

$$\sigma^2 = E\left\{(\mathbf{x} - m)^2\right\} = \int_{-\infty}^{+\infty} (x - m)^2 f(x)dx$$

If $\mathbf{x}$ is of discrete type, taking values x_n, then

$$\sigma^2 = \sum_n (x_n - m)^2 \, P\{\mathbf{x} = x_n\}$$

A useful relationship is obtained from the definition

$$\sigma^2 = E\left\{(\mathbf{x})^2\right\} - E^2\{\mathbf{x}\}.$$

Exercise 9-14: Derive $\sigma^2 = E\{(\mathbf{x})^2\} - E^2\{\mathbf{x}\}$ from $E\{(\mathbf{x} - m)^2\}$. Hint: expand $E\{(\mathbf{x} - m)^2\}$ with $E\{\mathbf{x}\} = m$.

9.4.6   Some Useful Probability Distributions

One can obtain CDF from PDF by using $f(x) = \frac{dF(x)}{dx}$ or CDF is an integral of PDF.

9.4.7   Q(x) and Related Functions and Different Representations

We often use the Q(x) function for error rate expression, and it is a tail end of zero mean and unit variance Gaussian PDF and it is given by,

$$Q(x) = \frac{1}{\sqrt{2\pi}} \int_x^{+\infty} e^{-\frac{z^2}{2}} dz.$$

$erfc(x) = \frac{2}{\sqrt{\pi}} \int_x^{+\infty} e^{-z^2} dz$ is used sometimes and it is related with Q(x) as,

Table 9-3: Probability density function (PDF) examples

	PDF $f(x)$	$m = E\{\mathbf{x}\}$	σ^2	Remarks
Uniform	$\frac{1}{b-a}$	$\frac{1}{2}(a+b)$	$\frac{1}{12}(a-b)^2$	Function ranges from a to b inclusive, and zero otherwise
$\mathbf{y} = \sin(\mathbf{x} + \emptyset)$	$f(y) = \frac{1}{\pi\sqrt{1-y^2}}$	0		$\mathbf{y} = \sin(\mathbf{x} + \emptyset)$, the range of $\mathbf{x} = \{a = -\pi$ and $b = +\pi\}$ and the range of $\mathbf{y} = (-1,1)$
Gaussian	$\frac{1}{\sigma\sqrt{2\pi}} e^{\frac{(x-m)^2}{2\sigma^2}}$	m	σ^2	$Q(y) = \frac{1}{\sqrt{2\pi}}\int_y^{+\infty} e^{-\frac{(x)^2}{2}}\,dx$ Gaussian or normal and its derivatives.

(continued)

Table 9-3: (continued)

	PDF $f(x)$	$m = E\{x\}$	σ^2	Remarks
Rayleigh	$\frac{1}{2\sigma^2} e^{-\frac{x}{2\sigma^2}}$ $m = 2\sigma^2 = 1.0$	$2\sigma^2$	Power = mean	$X = Y_1^2 + Y_2^2$ and Y_1 and Y_2 are zero mean independent Gaussian (central chi distribution)
Rice	$\frac{1}{2\sigma^2} e^{-\frac{(s^2+x)}{2\sigma^2}} I_0\left(\frac{s}{\sigma^2}\sqrt{x}\right)$ $I_0(y)$: 0th order Bessel $s^2 = m_1^2 + m_2^2$ $m = 2\sigma^2 = 0.25$ $s^2 = 0.75$ Dotted line: Rayleigh	$2\sigma^2 + s^2$	Power = mean	The same as Rayleigh except for non-zero mean Gaussian (non central chi distribution). It can be obtained by adding two square of Gaussian with mean and variance of m and σ^2.
Lognormal	$\frac{1}{\sigma\sqrt{2\pi}} e^{-\frac{(x-m)^2}{2\sigma^2}}$	m [dB]	σ^2	σ (not square) and m in dB scale
discrete (2 points)	$p\delta(x - A) + (1 - p)\delta(x + A)$	0	A^2	$p = 1/2$: uniform extendible with many points

Binomial	$\sum_{k=0}^{n} \frac{n!}{k!(n-k)!} p^k (1-p)^{n-k} \delta(x-k)$	np	$np(1-p)$	Sum of n independent and identical random variables of coin flipping with $P\{head\} = p$
Poisson	$\sum_{k=0}^{+\infty} e^{-a} \frac{a^k}{k!} \delta(x-k)$	a	a	

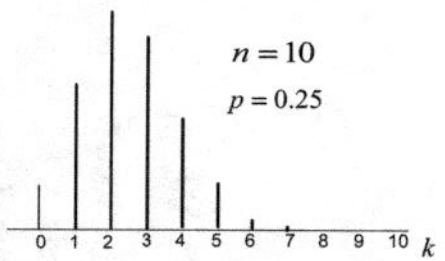

$n = 10$
$p = 0.25$

0 1 2 3 4 5 6 7 8 9 10 k

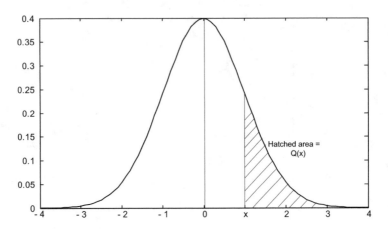

Figure 9-31: Definition of Q(x) graphical representation

$$Q(x) = \frac{1}{2} erfc\left(\frac{x}{\sqrt{2}}\right).$$

And erfc(x) can be expressed by Q(x) as, $erfc(x) = 2Q(\sqrt{2}x)$.

The tail end probability of Gaussian distribution, as shown in Figure 9-31 (hatched), with mean m and variance σ is given by $Q\left(\frac{x-m}{\sigma}\right)$.

Craig's representation of Q(x) is given by,

$$Q(x) = \frac{1}{\pi} \int_0^{\pi/2} \exp\left(-\frac{x^2}{2\sin^2\theta}\right) d\theta \quad \text{with } x \geq 0.$$

This expression may be useful for numerical evaluations of Q(x) and certain error rate expressions related with fading. The representation of Q(x) in the above may be derived from PSK SER derivation; see pp. 284 of [5].

9.4.8   Stochastic Process

We consider a carrier wave signal given by $s_x(t) = A \cos(2\pi f_c t + \theta)$. When A, f_c, and θ are fixed parameters, then it is a deterministic signal. However, in order to carry information, A and θ should modulate according to the information and then the signal becomes a stochastic process (signal). A and θ are random variables according to data. A set of data represented by A and θ is known, and then it is a time function. There are many such set, and thus it is a family of time functions. When time is fixed at t, it is a function of two, A and θ, random variables.

The mean $m(t)$ of a random process $\mathbf{x}(t)$ is the expected value of the random variable $\mathbf{x}(t)$ at t (fixed parameter):

$$m(t) = E\{\mathbf{x}(t)\} = \int_{-\infty}^{+\infty} xf(x;t)dx.$$

The autocorrelation $R(t_1, t_2)$ of a random process $\mathbf{x}(t)$ the joint moment of the random variables $\mathbf{x}(t_1)$ and $\mathbf{x}(t_2)$:

$$R(t_1, t_2) = E\{\mathbf{x}(t_1)\mathbf{x}(t_2)\} = \int_{-\infty}^{+\infty} x_1 x_2 f(x_1, x_2; t_1, t_2)dx_1 dx_2$$

Example 9-18: The process $\mathbf{x}(t)$ is given by

$$\mathbf{x}(t) = a \cos(2\pi f_c t) + b \sin(2\pi f_c t)$$

where a and b are two independent random variables with each zero mean and each variance σ^2 and f_c is constant.

<u>Solution:</u> It is clear that $E\{\mathbf{x}(t)\} = 0$. And since $E\{a\,b\} = 0$ (due to independence),

$$
\begin{aligned}
R(t_1, t_2) &= E[a \cos(2\pi f_c t_1) + b \sin(2\pi f_c t_1)][a \cos(2\pi f_c t_2) + b \sin(2\pi f_c t_2)] \\
&= E\{a^2\} \cos(2\pi f_c t_1) \cos(2\pi f_c t_2) + E\{b^2\} \sin(2\pi f_c t_1) \sin(2\pi f_c t_2) \\
&= \sigma^2 \cos(2\pi f_c(t_1 - t_2))
\end{aligned}
$$

9.4.9 Stationary Process, Correlation and Power Density Spectrum

A stochastic process is stationary (<u>in the strict sense</u>) if the statistics are not affected by a shift in the time origin. This means that the two processes $\mathbf{x}(t)$ and $\mathbf{x}(t + \varepsilon)$ have the same statistics for any ε. In particular we must have PDF $f(x; t) = f(x; t + \varepsilon)$ and thus $f(x; t) = f(x)$. A consequence of the stationary definition is

(a) $E\{\mathbf{x}(t)\} = m = $ constant
(b) $E\{\mathbf{x}(t_1)\mathbf{x}(t_2)\} = E\{\mathbf{x}(t)\mathbf{x}(t + (t_1 - t_2))\}$
(c) $R(\tau) = E\{\mathbf{x}(t)\mathbf{x}(t + \tau)\}$

The autocorrelation depends only on time difference $(t_1 - t_2) = \tau$.

When the above (a) and (b) hold we say it is stationary <u>in the wide sense</u>. If $\mathbf{x}(t)$ is normal and stationary in the wide sense, then it is stationary also in the strict sense since the normal processes are uniquely defined in terms of its mean and autocorrelation.

Some properties of the autocorrelation of stationary processes are:

- $R(\tau) = R(-\tau)$
- $R(0) = E\{|\mathbf{x}(t)|^2\} \geq 0$
- $|R(\tau)| \leq R(0)$

The power spectrum density (or simply spectral density) of stationary process $\mathbf{x}(t)$ is defined as the Fourier transform of its autocorrelation function :

$$S(f) = \int_{-\infty}^{+\infty} R(\tau)e^{-j2\pi f\tau}d\tau$$

Its inverse is given by

$$R(\tau) = \int_{-\infty}^{+\infty} S(f)e^{+j2\pi f\tau}df$$

By setting $\tau = 0$, we obtain $R(0) = \int_{-\infty}^{+\infty}S(f)df = E\{|\mathbf{x}(t)|^2\}$. This means that the total area of $S(f)$ is the average power.

9.4.10   *Processes Through Linear Systems*

We now consider a linear system represented by its impulse response $h(t)$ as shown in Figure 9-32. Input/output is stochastic processes (signals) rather than deterministic. In Figure 9-1, we used deterministic signals. Essentially the same result is applied to the stochastic signals. The output process $\mathbf{y}(t)$ is a convolution of input process $\mathbf{x}(t)$ and the impulse response of the system $h(t)$.

However, the output autocorrelation R_Y is the convolution of input correlation R_X with the impulse responses of $h(t) \otimes h^*(-t)$. Since $h^*(-t)$ has the Fourier transform of $H^*(f)$ the output power spectrum is given $|H(f)|^2 S_{XX}(f)$.

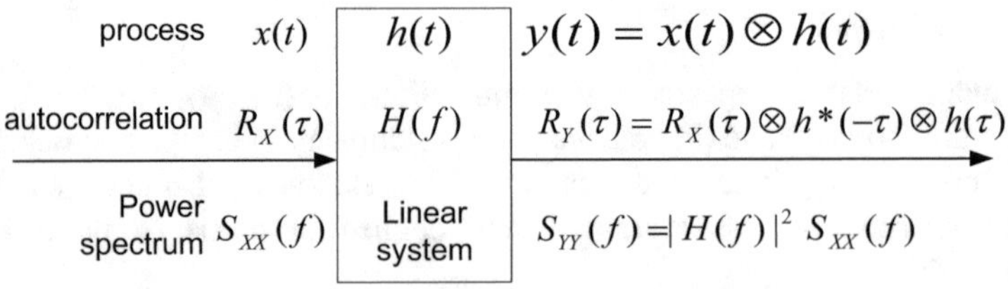

process	$x(t)$	$h(t)$	$y(t) = x(t) \otimes h(t)$
autocorrelation	$R_X(\tau)$	$H(f)$	$R_Y(\tau) = R_X(\tau) \otimes h^*(-\tau) \otimes h(\tau)$
Power spectrum	$S_{XX}(f)$	Linear system	$S_{YY}(f) = |H(f)|^2 \, S_{XX}(f)$

Figure 9-32: Stochastic process through linear systems

9.4.11   Periodically Stationary Process

$\mathbf{x}(t)$ is periodically stationary (or cyclostationary) with period T in the wide sense if

(a) $E\{\mathbf{x}(t)\} = m(t) = E\{\mathbf{x}(t + kT)\}$
(b) $E\{\mathbf{x}(t_1)\mathbf{x}(t_2)\} = E\{\mathbf{x}(t_1 + kT)\mathbf{x}(t_2 + mT)\}$

Example 9-19: The process $\mathbf{x}(t)$ is given by

$$\mathbf{x}(t) = \sum_{n=-\infty}^{n=+\infty} a_n h(t - nT)$$

where $E\{a_n\} = 0$ and $E\{a_i a_j\} = 0$ if $i \neq j$ and $E\{a_i^2\} = \sigma^2$ if $i = j$. Find the autocorrelation of $\mathbf{x}(t)$ and show it is periodically stationary.
Solution: It is clear that $E\{\mathbf{x}(t)\} = 0$ and the autocorrelation, $R(t_1, t_2)$, is given by

$$
\begin{aligned}
R(t_1, t_2) &= E\left\{ \sum_{n=-\infty}^{n=+\infty} a_n h(t_1 - nT) \sum_{m=-\infty}^{m=+\infty} a_m h(t_2 - mT) \right\} \\
&= \left\{ \sum_{n=-\infty}^{n=+\infty} \sum_{m=-\infty}^{m=+\infty} E\{a_m a_n\} h(t_1 - nT) h(t_2 - mT) \right\} \\
&= E\{a_i^2\} \left\{ \sum_{k=-\infty}^{k=+\infty} h(t_1 - kT) h(t_2 - kT) \right\}
\end{aligned}
$$

Adding multiple T to t_1 and to t_2 does not change $R(t_1, t_2)$ and thus it is periodic with the period of T. In some cases using the variables of $t_1 = t$ and $t_2 = t + \tau$ may be more convenient without losing generality.

9.5   Chapter Summary and References with Comments

9.5.1   Chapter Summary

We reviewed the important results from signals and systems – both continuous and discrete in time. We expanded to include deterministic signals and random processes (signals). Communications signals may be considered random signals because the information to carry is random (or probabilistic). Its underlying statistics are described by random variables and their expected values (mean and variance). The concept of probability is the basis for error performance, and in particular that of the conditional probability is important for optimum decoding schemes such as maximum a posterior (MAP) and maximum likelihood (ML) detection.

9.5.2 References with Comments

Our reference is very limited to some well- known textbooks for signals and systems of both continuous and discrete-time [1] and [2], probability and random processes [3] and [4]. Some old books have a new edition available. There are many fine textbooks available on the market.

[1] Oppenheim, Alan V., Willsky, Alan S. with Nawab S. Hamid," Signals and Systems", 2nd ed 1996, Prentice Hall, Upper Saddle River, New Jersey 07458

[2] Oppenheim, Alan V. and Schafer, Ronald W., "Discrete-time Signal Processing " 3rd ed, Prentice Hall, Pearson Education Limited 2014

[3] Papoulis, Athansios and Pillai, S. Unnikrishna " Probability, Random Variables, and Stochastic Process" 3rd ed 1984 McGraw-Hill, and 4th ed 2002

[4] Ash, Robert B., "Basic Probability" John Wiley &Sons 1970, and later edition 2008 Dover Books on Mathematics

[5] Proakis, J. and Salehi,M., "Digital Communications" 5th ed, McGraw Hill 2008, pp.284

9.6 Problems

P9-1: From the definition of unit impulse $\delta(t) = 1.0$ when $t = 0$, $\delta(t) = 0$ when $t \neq 0$, and using $\int_{-\infty}^{+\infty} \delta(\tau)d\tau = 1.0$, show that $\delta(2t) = \frac{1}{2}\delta(t)$, and $\delta(-t) = \delta(t)$.

P9-2: A linear system is defined as input $x(t) = 0$ produces its output $y(t) = 0$, and $a\,x(t)$ generates $a\,y(t)$ where a is constant. Determine if the system is linear;

(1) $y(t) = x(t) + 1$

(2) $y(t) = x(\sin(t))$

P9-3: $x(t)$ is shown in Fig. P9-1. Sketch three cases; $x(-t)$, $x\left(\frac{1}{2}t - 1\right)$, $x\left(-\frac{1}{2}t - 1\right)$.

P9-4: For $x(t) = \sin(2\pi t)$, draw $x(t)$, $x(-t)$, $x\left(\frac{1}{2}t - 1\right)$, $x(2t - 1)$ for $-2 < t < 2$.

P9-5: Show that two exponential functions, $e^{-t}u(t)$ and $2e^{-2t}u(t)$, are convolved and the result is $2(e^{-t} - e^{-2t})u(t)$. Sketch the result.

P9-6: The impulse response of a system is $3e^{-3t}u(t)$. What is the output for the input of $u(t)$?

Fig. P9-1: The graph of x(t)

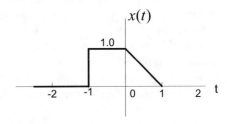

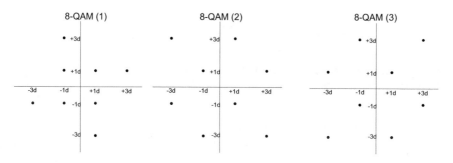

Fig. P9-2: Different 8-QAM constellations are obtained from 16-QAM

P9-7: Show, from the definition of Fourier transform pair $x(t) \Leftrightarrow X(f)$, that

(1) $x(-t) \Leftrightarrow X(-f)$
(2) $\delta(t) \Leftrightarrow 1$
(3) $1 \Leftrightarrow \delta(f)$
(4) $x^*(t) \Leftrightarrow X^*(-f)$

P9-8: Show Fourier transform pair; $\sum\limits_{n=-\infty}^{\infty} \delta(t - nT) \Leftrightarrow \sum\limits_{k=-\infty}^{\infty} \frac{1}{T}\delta\left(f - \frac{k}{T}\right)$.

P9-9: $\sum\limits_{n=-\infty}^{\infty} \delta(t - nT) = \sum\limits_{k=-\infty}^{\infty} \frac{1}{T}e^{+j2\pi\frac{k}{T}t}$, i.e., periodic in T and can be expanded in complex exponential.

P9-10: Plot $y(t) = \dfrac{\sin\left(\pi\frac{t+0.5T}{T}\right)}{\pi\frac{t+0.5T}{T}} + \dfrac{\sin\left(\pi\frac{t-0.5T}{T}\right)}{\pi\frac{t-0.5T}{T}}$ using a computer program.

P9-11: A discrete impulse response, $h[n]$, can be scaled in three ways; find the scale factors

(1) Power gain of a system is unity
(2) Peak of the impulse response is unity
(3) The frequency response at DC is 0 dB (i.e, unity)

P9-12: The impulse response is $h[n] = u[n] - u[n - 5]$. For the input $u[n]$, obtain the output, and sketch it.

P9-13: The system impulse response is $\delta[n - 3]$. Find the output when the input is given by $\sin\left(2\pi\frac{1}{3}n\right)u[n]$ and sketch it.

P9-14: Three different 8-QAM constellations are shown in Fig. P9-2.
Compute the digital power of three 8-QAM(m) m = 1, 2, 3. Compare them which one is the least power when the distance between points is the same.

P9-15: 0 dBm denotes 10^{-3} watts in dB scale.

(1) Express -174 dBm in watts.
(2) Express $+30$ dBm in watts.

P9-16: Show a discrete Fourier transform pair;

$$\sum_{k=-\infty}^{\infty} \delta[n - kN] \Leftrightarrow \frac{1}{N} \sum_{k=-\infty}^{\infty} \delta\left[f - \frac{k}{N}\right].$$

Compare with the continuous case in P9-8.

P9-17: For periodic square wave

$$x[n] = 1, \quad n \mid \leq M$$
$$= 0, \quad M < \mid n \mid \leq N/2 \quad \text{and } x[n+N] = x[n]$$

Find the Fourier series coefficient a_k so that its discrete Fourier transform is given

by $\sum_{k=-\infty}^{\infty} a_k \delta\left[f - \frac{k}{N}\right]$. Hint: $a_k = \frac{\sin[(2\pi k/N)(M+0.5)]}{N \sin(\pi k/N)}$ $k \neq 0, \pm N, \pm 2N \dots$ and

$a_k = \frac{(2M+1)}{N}$ $k = 0, \pm N, \pm 2N \dots$.

Sketch when $N = 10$, $M = 4$.

P9-18: Find the discrete Fourier transform of $a^n u[n]$ when $|a| < 1$.

P9-19: Continuous signal conversion to discrete-time signal is mathematically

modeled by multiplying an impulse train, $\sum_{n=-\infty}^{\infty} \delta(t - n\Delta t)$, to the continuous

input, $x_c(t)$, as $x(n\Delta t) = x_c(t) \cdot \sum_{n=-\infty}^{\infty} \delta(t - n\Delta t)$ where Δt is a sampling period.

And then omitting Δt (or set $\Delta t = 1$) we denote it as $x(n\Delta t) = x[n]$.

(1) Show that Fourier transform pair is given by $x(n\Delta t) \Leftrightarrow \sum_{k=-\infty}^{\infty} \frac{1}{\Delta t} X_c\left(f - \frac{k}{\Delta t}\right)$.

(2) Rather than using an impulse train, one can use a periodic pulse as

$\sum_{n=-\infty}^{\infty} h(t - n\Delta t)$. Show that the sampled frequency response is given by

$\sum_{k=-\infty}^{\infty} \frac{1}{\Delta t} H\left(\frac{k}{\Delta t}\right) X_c\left(f - \frac{k}{\Delta t}\right)$

(3) A physical implementation of the continuous signal to discrete one is called A/D. Before A/D, there is typically a low pass filter. Why?

P9-20: Discrete-time signal conversion to continuous one is modeled as an interpo-

lation; $x_c(t) = \sum_{n=-\infty}^{\infty} x(n\Delta t) g(t - n\Delta t)$ where Δt is a sample period. In practical

D/A implementation, the interpolation filter ($g(t)$) may be done by two steps – first a rectangular pulse (rec (t)) and then followed by an analog filter (A(t)). Express the overall interpolation filter $g(t)$ in terms of these two step impulse responses.

P9-21: A Venn diagram is shown in Fig. P9-3, where $A \cap B = AB$ is hatched.

(1) Draw a diagram of $A \cup B = A + B$ with hatch.
(2) Draw a diagram of $\overline{A}$ by hatching
(3) Draw a diagram of $\overline{A} \cap B = \overline{A}B$ by hatching.
(4) $B = (AB) + (\overline{A}B)$

Fig. P9-3: Venn Diagram

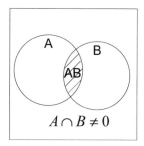

$$A \cap B \neq 0$$

(5) $A \cup B = A + B = A + \overline{A}B$

(6) Show that $P(A \cup B) = P(A) + P(B) - P(A \cap B)$.

P9-22: From Fig. P9-2, hatch $\overline{A + B}$ and $\overline{A} \cdot \overline{B}$. Show De Morgan law $\overline{A + B} = \overline{A} \cdot \overline{B}$ and similarly its dual $\overline{AB} = \overline{A} + \overline{B}$.

P9-23: An experiment of tossing of a coin twice consists of four events; $hh, ht, th,$ and tt. Assuming the probabilities of the events are given by $P\{hh\} = p^2, P\{ht\} = pq, P\{th\} = pq,$ and $P\{tt\} = q^2$ where $p + q = 1$, find the probability of heads at first tossing $P\{ht, hh\}$ and find the probability of heads at second tossing $P\{th, hh\}$. The intersection of two events is $\{hh\}$ and $P\{hh\} = p^2$. Can you conclude that two events are independent? Hint: yes

P9-24: Is it possible two events are independent and mutually exclusive? Hint: draw such a Venn diagram.

P9-25: For the independent events, the probability of union $P(A \cup B)$ can be expressed as $P(A \cup B) = 1 - (1 - P(A))(1 - P(B))$.

P9-26: A box contains three white balls and four red balls, $w_1, w_2, w_3, r_1, r_2, r_3, r_4$. We remove from it at random two balls in succession. What is the probability that the first ball is red and the second is white? Hint: use conditional probability.

P9-27: A coin, with head probability p and tail one $q = 1 - p$, is tossed n times. Show that the probability of heading being shown k times is given by

$$P\{k \text{ heads in } n \text{ tossings}\} = p_n(k) = C_k^n p^k q^{n-k} \text{ where } C_k^n = \frac{n!}{k!(n-k)!}.$$

P9-28: A binomial probability density function is related with Gaussian probability density function, known as DeMoivre-Lapace Theorem,

$$p_n(k) = \frac{n!}{k!(n-k)!} p^k q^{n-k} \approx \frac{1}{\sqrt{2\pi npq}} e^{-(k-np)^2/2npq}.$$

(1) Plot both binomial and Gaussian when $n = 10, p = 1/2, q = 1/2$

(2) Plot both binomial and Gaussian when $n = 10, p = 2/5, q = 3/5$

Index

0-9, and Symbols

1+D system, 84, 85, 87–91, 127, 128
10GBase-T, 433, 435
1-tap equalizer, 263, 274, 282, 298, 511, 525
2-ray model, 192–194
2's complement, 652, 653, 674, 675
2-step approach, 577–579
32-QAM, 56, 57, 60–62, 125, 126, 668
5G, 5
8-QAM, 55, 56, 61, 62, 67, 125, 126
-90° phase shifter, 618

A

Acquisition, 530–531
Additive white Gaussian noise (AWGN), 8, 13, 16, 18, 37, 44, 45, 47–52, 54, 59, 64, 66, 107, 118, 119, 122, 123, 133, 135, 142, 143, 172, 173, 242, 304, 306–308, 323, 324, 335, 336, 339, 341, 344, 385, 390–392, 397–399, 411, 412, 420, 442, 443, 450, 451, 455, 456, 461, 464, 470, 471, 474, 476–479, 482, 483, 487, 496, 518
Address commutating, 267, 270–272, 275, 278, 279
Advanced mobile phone service (AMPS), 250
Algebraic codes, 14, 323, 462
Algebraic decoding, 366–369, 499
Algebraic decoding process, 366, 367
All digital implementations, 531, 560, 566, 571–572
Altitude, 236–238, 561
Ambiguity zone detection (AZD), 88–90, 119

AM broadcasting, 9, 27, 250
Amplitude phase shift keying (APSK), 62–67, 435
Amplitude shift keying (ASK), 11
Analytical performance, 331
Antenna
　beam angle, 190–191
　gain, 22, 23, 186–191, 199–201, 204, 205, 238, 241
　patterns, 189, 191, 205, 240
Aperture, 189–191
Aperture area, 190, 191, 199
A posterior probability (APP), 391–393, 396–397, 399, 400, 406, 412, 417–419, 438, 441, 442, 444, 451
APP of branches, 399, 441, 442
Applications, 5, 25–27, 119, 142, 216, 254, 265, 309, 323, 385, 438, 450, 451, 453–455, 462, 479, 480, 498, 510, 618, 620, 683
Approximate cycle extrinsic message node degree (ACE), 430
Arc angle, 189
Autocorrelation, 109, 223, 519, 534, 731–733
Automatic gain control (AGC), 517, 520, 673, 678–682, 685
Average, 42–44, 55, 56, 60, 72, 74, 75, 93, 107–110, 121, 124, 125, 170, 194, 199, 202, 209, 214, 219, 222, 305, 325, 455, 457, 461, 468, 469, 474, 480, 492, 503, 522, 533, 569, 577, 612, 614, 641, 642, 644, 653, 655, 665–668, 673, 677, 681, 685, 706–709, 721, 725–726, 732

© Springer Nature Switzerland AG 2020
S.-M. Yang, *Modern Digital Radio Communication Signals and Systems*,
https://doi.org/10.1007/978-3-030-57706-3

B

Band diagonal, 421, 433, 435
Band edge timing recovery (BETR), 541, 550
Band edge vestigial symmetry, 149–151
Bandlimited channels, 511
Baseband signals, 8–12, 111, 299, 606, 649,
    651, 706–707
Bayes' Theorem, 722
Beam angle, 190
Beam forming, 23, 226, 242
Belief propagation (BP), 412
Bessel function, 158, 204
Binary phase shift keying (BPSK), 12, 13, 17,
    18, 21, 37–44, 47–49, 52–55, 63, 64, 67,
    107, 110, 118, 120–124, 135, 170, 172,
    224, 324, 325, 327, 335, 336, 339–341,
    343, 366, 380, 384, 385, 389, 390, 392,
    397, 411, 418, 420, 442, 443, 450–453,
    455, 456, 461, 471, 474, 475, 477, 479–
    482, 490, 491, 496, 497, 503, 515, 516,
    525, 530, 533, 551, 568
Binary symmetric channel (BSC), 8, 323, 392,
    395, 397–400, 412, 450
Bit error rate (BER), 48, 66–72, 82, 118, 126,
    199, 325, 327–331, 340–343, 365, 366,
    459, 525, 526
Bit interleaved coded modulation (BICM), 242,
    461–463, 465, 466
Bit loading, 28, 256, 304, 305, 479
Bit node
    updating, 412–414, 417, 419
Bit-to-symbol
    mapping, 11, 26, 606
    tables, 455
Block codes, 20, 328, 329, 331, 332, 334, 336,
    338–348, 353, 369, 370, 378–381, 384,
    403–406, 461, 492, 495–497, 510
Block transmission, 252–264, 268, 273, 280,
    303
Bluetooth, 4, 5, 102
Bose, Chaudhuri, Hocquenghem (BCH), 14,
    323, 331, 358, 359, 361–369, 422, 433,
    462, 494, 495
Branch metrics, 13, 385–391, 395, 397–399,
    406, 438, 441–444, 448, 497
Brickwall filter, 139

C

Calibration of quadrature modulator, 649–650
Carlson's rule, 94
Carrier frequency
    offset, 509, 516, 518, 519, 522, 523, 530–
    532, 552, 554
    offset estimation, 522, 553

Carrier phase
    detection, 39, 509, 565, 569
    detectors, 566
    error detectors, 569
Carrier phase synchronization, 509, 532,
    565–572
Carrier recovery loop, 509, 566, 571–572
Cellular frequency reuse, 229–230
Cellular networks, 5, 20, 26–28
Cellular systems, 228, 233–235, 250
Central limit theorem, 204, 669
Channel
    binary symmetric, 8, 395, 412
    capacity, 14, 25, 27, 28, 228, 250, 304, 308,
    323, 329, 330, 433, 437, 470, 471,
    476–479, 487, 488
    coding, 7, 8, 13, 14, 20, 26, 28, 36,
    47, 66, 199, 256, 304, 305, 307, 309,
    323–488
    discrete memoryless, 8, 397
    dispersion, 22, 254, 262–265, 297, 298
    emulators, 210
    erasure, 8, 392, 450, 451
    estimations, 19–20, 282, 309, 474, 478, 508,
    511, 514, 516, 524
    model with scattering description, 214–215
    selectivity, 607, 608, 628, 629, 678
    sounding, 220–225
    transition matrix, 489
Check node
    updating, 415, 417, 419
Chirp signal, 141, 221, 224–225
Chi-square, 207
Circulant matrix, 434, 500
Clean channel occupancy, 608, 628,
    629, 658, 670
Clipping, 470, 472, 473, 612, 613, 656, 669,
    676, 677
Cluster sizes, 230–232
Co-channel distance, 232
Co-channel interference, 230, 233–234
Code division multiple access (CDMA)
    code division multiplex, 20
Coded modulation (CM), 13, 46, 47, 52, 123,
    124, 305–307, 324, 325, 327, 330, 331,
    336, 339, 341–344, 347–349, 384–388,
    390, 391, 395, 410, 456–459, 480, 481,
    490–492, 497
Coding error performance, 322
Coherent demodulation, 19, 21, 49, 106, 242,
    507, 685
Common platform
    with commutating filters, 21, 314
Community antenna TV (CATV), 304
Commutating

filters, 21, 22, 252, 265, 274, 275, 278, 280, 287, 288, 302, 303, 309
polyphase filters, 265
Complex baseband envelope, 34
Complex baseband equivalence, 658–659, 678
Complex conjugation, 137, 140, 517, 678
Complex envelopes, 8–12, 14–17, 24, 26, 35–42, 72, 97, 98, 103, 104, 106, 111, 118, 129, 130, 133, 141, 204, 206, 216, 224, 250, 253, 269, 299, 301, 550, 671, 706–708, 711
Complimentary cumulative density function (CCDF), 74, 75, 127, 470, 472
Conditional probabilities, 46, 122, 123, 392, 393, 396, 397, 399, 412, 413, 415, 443, 482, 487, 497, 722–723, 733
Constellations, 10–12, 17, 18, 37–43, 53, 55–62, 65–69, 74, 87, 89–91, 98, 106, 118, 120–128, 280, 293–297, 435, 451, 453–457, 461–464, 466–471, 487–489, 503, 509, 523, 565–569, 571, 573, 577, 578, 580, 612, 667, 668, 709
Constituent codes, 371, 437, 448, 449
Continuous phase frequency shift keying (CPFSK), 93–108, 116, 129
Continuous-time, 8, 698, 700, 713, 717–720
Continuous to discrete, 697
Continuous wave (CW)
signals, 5, 8, 9
Control loop, 660, 661
Conversion rate of ADC, 676
Convolution
codes, 13, 307, 370–372, 375, 392, 403, 439, 447–450, 495, 525, 526
integral, 269, 289, 699–702
Convolutional codes, 28, 323, 370–407, 438, 461, 462, 480, 496–499
Convolutional code tables, 375–378
Coordinate system, 189, 231, 232
Cordless phones
digital, 4
Correlation
by convolution, 136
decoding, 13, 14, 336, 341, 385, 410
used as branch metric, 385–386, 391, 497
Correlation metric (CM)
correlation metric decoding, 13, 14, 324, 327, 341–344, 410
decoders, 324, 327
Cosine modulated multi-tone (CMT), 303
Costas loop, 566–568, 571
Crest factor, 223

Cumulative distribution function (CDF), 74, 127, 214–217
Cycles, 93, 101, 120, 128, 250, 425–430, 432, 433, 655
Cyclic codes, 338, 349–361, 366, 433, 493, 502
Cyclic prefix, 21, 139, 142, 251, 255–256, 276, 512, 647, 668
Cyclic redundancy check (CRC)
polynomial, 357, 358, 494
Cyclostationary, 109, 509, 534, 733

D

DC cancellation, 641, 642, 644
DC offset, 608–614, 619, 620, 636, 641–644, 649, 650, 658, 677, 678, 686
Decimation filters, 625, 628–631, 633–635
Deep space communication systems, 4
Delay power profile, 213, 214, 216
Delay spread, 24, 200, 221, 222, 225, 241, 254, 262–264, 282, 297–298, 305
Demapper, 91, 418, 451, 453–455, 461–463
Demodulation, 14–23, 37, 49, 96, 106, 118–120, 249, 254, 260, 262, 289, 507–510, 532, 565, 612, 619, 627, 629–633, 639, 647, 649, 672, 685
Differential coding, 67, 68, 75, 76, 80–84, 110, 111, 118
Differential phase shift keying (DPSK), 52, 62–66, 474, 475, 489
Diffraction, 4, 186, 191–202
Digital analog converter (DAC)
conversion rate, 655, 694
error voltage, 653–655
integral non-linearity, 657, 658
Digital European cordless telephone (DECT), 102
Digital IF
carrier frequencies, 620, 621, 625–628, 630, 631, 636, 644, 647
modulations, 299, 620, 628, 630, 644
Digital image-cancelling transceiver, 620–627
Digital loop filters, 539, 542, 555, 560, 571, 662
Digital modulations, 10, 11, 14, 18, 26, 28, 33–130, 223, 302, 565
Digital multi-tone (DMT), 22, 28, 252–265, 268, 271–274, 277, 280, 282, 303, 309
Digital resampling, 555, 556, 558–560
Digital signal processing (DSP), 3, 4, 20, 139, 171, 299, 354, 591, 608, 641, 686
Digital subscriber loop (DSL), 4, 5, 22, 24, 25, 28, 252, 255, 256, 303, 304, 479

Dirac delta function, 720
Direct conversion transceiver, 606–609, 629,
      641, 644
Direct RF pulse, 221–222, 224
Direct sequence (DS)
   spread spectrum, 20, 77, 529–531
Directivity, 189–190, 226
Discrete Fourier transform (DFT), 21, 22, 112,
      115, 153, 252–258, 260, 262–265, 268–
      274, 278–280, 282, 284, 289, 290, 296,
      300–302, 309, 515, 521–525, 542, 647,
      668
Discrete memoryless channel (DMC), 8, 323,
      395, 397, 398, 450, 470, 489, 497
Discrete-time
   signals and systems, 698, 713–717
Discrete to continuous, 719, 720
Distributed antennas, 235
DOCSIS 3.1, 62, 435
Doppler
   clock frequency shift, 560–564
Doppler filters, 206, 208, 210, 213, 214, 218,
      219, 473, 474
Doppler frequency
   maximum, 24, 221
Doppler spectrum, 205–206, 209, 210, 214–
      216, 219–221, 225, 227, 228, 241, 307
Double sideband (DSB), 9, 10
Double square (DSQ), 61–62
   128-DSQ, 126, 466, 467
   8-DSQ, 451, 454–457, 468
Double squared QAM, 451
Doubly selective fading
   channels, 216, 305–308
Down-conversion, 15, 16, 553, 572, 606, 610,
      612–614
DSQ constellations, 62, 126
DS spread spectrum
   synchronization, 529–532
Dual codes, 334, 404, 500
Duo-binary, 24, 84–86, 90–91, 449
DVB-S2, 65, 430, 433, 435
Dynamic range, 612, 651, 657, 662, 673, 677,
      678, 680, 684, 685

E
Early–late algorithm, 540, 542, 543
Edge tables, 418–420
Effective isotropic radiated power (EIRP), 190,
      191
Elevation angle, 237–240, 561
Empirical path loss models, 195, 226

End-to-end pulse, 17, 26, 88, 138, 141–143,
      145–146, 148, 153–156, 164, 173, 272,
      545, 667, 678
Equalization
   adaptive, 511
Erasures, 8, 432, 433, 450, 451
Error correction, 14, 26, 65, 323, 325, 327, 331,
      334, 336, 357, 359, 366, 432, 480, 494,
      495, 499
Error location polynomial, 367–369
Error patterns, 336, 359–361, 494
Excess bandwidth, 74, 85, 144, 146, 149–151,
      155, 157, 264, 276, 277, 285, 286, 294,
      295, 309, 537, 538, 545, 547, 548, 551,
      590, 591, 667
Excision, 24
Expected value, 708, 725–726, 730
Extending, 73, 366, 369–370, 680
Extrinsic information, 437, 448
Eye patterns, 156, 167–168, 280–282, 293–297

F
Fading, 4, 13, 14, 22–24, 37, 174, 195–198,
      200–204, 209, 210, 212, 214–220, 223,
      227, 236–238, 240–242, 279, 305–308,
      323, 348, 390, 391, 448, 455, 459, 461,
      462, 472–480, 507, 509, 517–520, 523,
      552, 554, 721, 730
Fading channel
   counter measures, 479
   counter-measure with coding, 322
Fade margin, 197–201, 242
Farrow structure, 590, 591
Fast fading, 479
Fast Fourier transform (FFT), 114–116, 130,
      206, 251, 279, 283
Filter bank multicarrier (FBMC), 252, 303
Filter design
   analog and digital hybrid, 132
Filter method
   of CP, 265, 268, 274
   of windowing, 157–159, 265, 272
Filtered multi-tone (FMT), 252, 303
Filtered OFDMs, 264, 274–284, 303,
      308, 309
FIR filter design
   Parks-McClellan algorithm, 158
First order loop, 584, 586
Flat channel, 118, 527–528
Flow diagrams, 374, 376, 495, 585
Forward–backward algorithm, 393–396
Forward error correcting code (FECs), 407, 468

Forward error correction (FEC), 13, 14, 21, 65, 66, 69, 242, 304, 306, 323, 407, 473, 474, 478, 487, 525

Fourier transforms, 108–110, 114, 117, 128, 137, 147, 148, 153, 155, 205, 286, 534, 535, 557, 561, 582, 583, 615, 617, 655, 659, 671, 676, 678

Frame boundary, 20, 510

Frame synchronizations, 510

Free space loss, 23, 24, 186–187, 195, 199, 241

Free space propagation, 186, 187

Frequency division multiplex (FDM), 5, 9, 22, 26, 28, 112, 250, 252, 265, 266, 274, 276, 280, 283–284, 303, 308

Frequency domain FIR filter, 132

Frequency hopping (FH), 20–22, 301–302, 309

Frequency modulation (FM), 9, 10, 26, 27, 93–108, 112, 116, 129, 250, 608, 671, 672

Frequency responses, 4, 24, 43, 84, 85, 110, 112, 118, 128, 135, 137–139, 148, 150, 151, 155, 162, 163, 213, 254, 256, 258, 259, 298, 304, 472, 473, 478, 507, 511, 519, 535, 582, 617, 629, 630, 633, 642, 643, 646, 647, 655, 658, 659, 667, 673, 675, 676, 678, 702–705, 710–712, 715, 718

Frequency reuse, 228–231

Frequency selective
channels, 23, 241, 304–305, 309, 478, 516, 528

fading, 209–213, 241, 304, 305, 478, 479, 516, 527

fading channels, 209, 305, 478, 518

Frequency shift keying (FSK), 11, 21, 52, 66, 93–108, 115, 116, 118, 119, 128–130, 474–476, 489, 583, 667, 668

Frequency tolerance, 671

Fresnel zones, 194–195

Friis equation, 186, 191

Fundamental Theorem of transformation, 205

G

Galois field (GF), 361–366, 408, 424, 427, 428, 431, 432, 494, 495, 499

Gardner
TED, 506, 554

Gaussian elimination, 332, 352, 354, 408, 421

Gaussian MSK (GMSK), 102, 103, 105, 107

Gaussian pulse, 102–103, 107

Generator matrix
non-systematic, 332, 493, 496

Generator polynomials, 350–352, 354, 355, 357–359, 361, 363, 365, 366, 422, 493–496, 502

Girth, 425, 432

Global positioning system (GPS), 6, 225, 512, 530–532, 672

Global system for mobile communications (GSM), 5, 102, 103, 227, 228

GMR, 435

Gray bit assignment, 67, 68, 70, 71, 81, 459, 462

Gray coding, 11, 63, 67, 83, 84

Ground reflection, 192–194

GSM channels, 227

Guard interval, 512

H

Hadamard, 326, 327, 330, 491, 492

Hamming codes, 14, 334–339, 351, 360, 369, 370, 493, 502

Hard decision (HD)
decoding, 13, 325, 335–336

Heterodyne conversion transceiver, 608–610

Hexagonal shape, 229

Hilbert transform
pairs, 617

I

Ideal LPFs, 139, 140, 145, 146, 487

IDFT-DFT pair, 22, 255, 260

IEEE 802.11a
preamble, 513

IF sampling, 284, 299–302

Image cancellations, 614–618, 636–641, 650

Image-cancelling, 610, 615, 618–648, 650, 686

Image cancelling transceiver, 615, 627–630, 648, 650

Impulse responses, 4, 11, 12, 22, 76–78, 85, 86, 128, 135–140, 146, 150–154, 156, 164–166, 168, 169, 173, 210, 211, 214, 216, 218, 220–222, 224, 251, 258, 262–264, 266, 274, 280–282, 293–299, 304, 370, 496, 511, 517, 533, 551, 556, 557, 581, 587, 589–592, 594, 608, 617, 628, 633, 655, 658, 659, 678, 699–701, 703, 708, 713–716, 720, 732

In-band interference, 640

Independence, 329–330, 344, 346, 348, 349, 380, 382, 383, 414, 441, 448, 492, 724, 731

Independent events, 724
In-door, 23, 241
Input weight and output distance, 380–384, 496
Integral non-linearity (INL), 657
Integrate and dump, 138, 144, 713
Inter-channel interference (ICI), 264, 293, 298
Interference
    limited system, 228
Interleaver, 28, 223, 306, 437, 447–449, 453,
        455, 457, 459–461, 465, 474, 476
Inter-modulation distortion (IMD), 111, 660,
        664, 666, 670, 673, 674
Interpolation
    polynomial, 600
Intersymbol interference (ISI)
    free, 667
    free condition, 133
Inverse discrete Fourier transform (IDFT), 22,
        112, 116, 250–255, 260, 265, 267, 272,
        274, 275, 277, 280, 284, 287, 288, 301,
        309, 512, 515, 516, 521, 524, 646, 647,
        668
IQ amplitude imbalance, 610–615, 636, 648
IQ imbalance model, 573–576
IQ Phase imbalance
    corrections, 572, 576, 579
I-rail, 610
Iridium, 283
Iterative computations, 394–396, 399, 412,
        418, 439
Iterative decoding, 242, 391, 437, 499

J
Jakes spectrum, 205–206, 227, 228, 241
Jitter, 671

K
Kaiser window, 157, 158

L
Large scale fading, 23, 202, 240
Layered
    approach, 26–27
    models, 26
L-carrier, 27
Lift
    circulant block, 437
Line of sight (LOS), 5, 6, 23, 24, 186, 194–195,
        203, 204, 208–209, 479

Line of sight microwave radios, 4, 56
Linear binary block codes, 13, 331–370
Linear channels, 169, 172–174, 298
Linear codes, 328, 340, 341, 350, 370, 371,
        373, 377, 381, 391, 431, 461, 480
Link, 5, 20, 26, 27, 194, 195, 199, 200, 233,
        510
Link budget, 199–202, 244
Local area network (LAN), 4, 5, 139, 186, 252,
        254, 435, 479
Log likelihood ratios (LLR), 49, 411, 450
    BPSK, 49
    check node of LDPC, 483–486
    computations, 453–455
LO leakage, 611, 612, 648–650
Long term evolution (LTE), 5, 28, 252, 254,
        255, 257, 264, 668
Long-term training sequence (LTS), 513, 514,
        516–528
Loop filters, 509, 554, 555, 560, 566, 572, 583–
        585, 661, 662, 680–682
Low density parity check code (LDPC), 14, 28,
        29, 65, 69, 242, 307, 323, 340, 381,
        407–437, 450–453, 455–457, 459,
        462–468, 472, 473, 479,
        483–486, 500
    encoder, 420–425
    low density parity code, 14, 407
Low digital IF, 619–648, 650, 651, 686
Low Earth orbit (LEO)
    fading channel model, 236
    satellite, 236–240, 561
Low noise amplifier (LNA), 3, 16, 20, 24, 29,
        36, 37, 71, 134, 185, 199, 606, 608, 679,
        680, 682–685, 711
Low pass filter (LPF), 39, 85, 87, 88, 96, 114,
        128, 135, 139, 158, 159, 206, 211, 241,
        258, 277, 299, 300, 474, 557, 589, 606,
        620, 629–633, 642, 658, 660, 680

M
MAC protocol data unit (MPDU), 513
Magnitude response, 137, 139
Majority rule, 13
Matched filter
    digital, 28, 88, 134, 153, 155, 156, 517,
        530–532, 713
    frame, 13, 20, 26, 506, 507, 510, 530
Maximal length code, 337, 339, 351, 494
Maximum a posterior probability (MAP)
        45–52, 123

Maximum likelihood (ML), 45–52, 89, 119, 123, 174, 433, 532, 541–543, 733
Maximum likelihood sequence detector (MLSD), 174, 370, 371, 385, 391, 407
Maximum likelihood sequence detection (MLSE), 88
Mean, 3, 4, 6, 16, 17, 19, 22, 24, 45, 46, 64, 72, 74–76, 102, 107, 108, 111, 122, 135, 137, 140, 150, 153–156, 165, 170, 174, 190, 195–197, 207–209, 215–217, 222, 227, 229, 233, 250, 273, 283, 305, 324, 325, 328, 329, 331, 340, 350, 359, 371, 380, 383, 385, 387, 390, 397, 407, 408, 410, 420, 421, 430, 432–437, 439, 441, 446, 447, 449, 455, 457, 465, 470, 474, 481, 489, 490, 656, 676, 726, 728, 730, 731, 733
Message passing decoding, 407, 412, 420, 433
Message-passing (MP), 411, 412, 433
Microwave, 5, 6, 23, 119, 186, 480
Minimum distance, 46–48, 51, 123, 325, 327, 331, 334–336, 338, 340, 341, 358, 361, 365, 370, 385, 431–433, 481, 491, 495, 499, 502
Minimum free distance, 377, 378, 382, 498
Minimum shift keying (MSK), 11
   frequency, 93–108
Minimum weights, 334, 377
M-level constellations, 709
Mobile apps, 26, 27
Mobility, 4, 23, 305
Modulation index, 9, 97, 98, 100, 102, 103, 116, 129
Modulation symbols, 11, 12, 26, 35, 36, 38, 39, 133, 253, 307, 451, 459, 461, 462, 508–510, 531, 532
Monte Carlo method, 214, 218–220
M-PAM, 44, 45, 47, 48, 53–55, 72, 98, 125, 135, 168, 476
M-QAM, 55, 57, 59, 60, 125
M-sequence, 223
Multi input multi output (MIMO), 22, 23, 28, 226, 257, 479
Multilevel coded modulation (MLCM) mapper, 461–469
Multi-level constellations, 453–455
Multipath fading, 23, 199, 200, 202–220, 227, 228, 511
Mutual information, 470, 487–489
Mutually exclusive, 329–330, 380, 721, 722

N
Naming of OFDM signals in use, 29, 112–114, 249–309
Nakagami
   fading, 207
NASA CCSDS, 435
Net link budget, 199–202
Networks, 4–6, 24, 26, 27, 224, 510
Noise bandwidth, 19, 28, 44, 45, 54, 71, 72, 88, 107, 122, 126, 142–145, 154, 211, 264, 588, 712, 713
Noise figure (NF), 71, 126, 199, 201, 526, 680, 682–685
Noise power, 18, 19, 44–45, 48, 71, 72, 88, 89, 118, 122, 126, 137, 142–144, 163, 173, 174, 199, 233, 491, 534, 613, 682, 684, 705–713
Noise power spectral density, 711
Noncentral chi-square, 208
Non-systematic code, 351, 375, 383, 386, 449
Non-white Gaussian noise, 135
Non-white noise, 172–174
Numerically controlled oscillator (NCO), 95, 302
Nyquist
   conditions, 17, 28, 146, 153, 211
   criteria, 17, 26, 28, 133, 146–149, 154, 164, 173
   pulses, 28, 131–181, 557, 589–591

O
Offset QAM, 37, 72–75, 101, 118, 303
One tap equalizer, 28, 262, 296, 524
On-off keying (OOK), 11, 25, 37–43, 48–52, 67, 89, 118–122, 221
Open System Interconnection (OSI), 26, 27
Optimization of square root Nyquist filter, 155, 160, 161
Optimum level from DAC, 604
Optimum level to ADC, 676
Orthogonal, 5, 11, 37–43, 48–52, 67, 118, 121, 242, 614
Orthogonal frequency division multiple access (OFDMA), 20
   with staggering, 73, 284–297
   symbol, 22, 116, 130, 225, 250–252, 254, 255, 257, 260–261, 264–268, 270, 271, 273, 282, 283, 286, 305, 306, 308, 313–316, 318, 510, 512–516, 520, 523–526, 594

Orthogonal frequency division multiple access (OFDMA) (*cont.*)
  symbol boundary, 225, 260–261, 264, 268, 271, 273, 282, 314, 513, 516, 520, 524, 525
Orthogonality, 101, 250, 251, 254, 263, 265, 269, 270, 274, 276, 284–286, 293, 294, 302–304
Outage requirement, 198, 199
Outdoor mobile, 241
Overload distortion, 612, 651, 656, 657, 666, 674, 676, 677, 680

P
Packet synchronization, 510, 512–532
Parallel clock, 510, 529, 530
Parity
  bit, 14, 324–328, 331, 339, 341, 344, 346, 348, 350, 358, 404, 410, 412, 421, 429–431, 493, 494, 500, 502
  check matrix, 332–334, 336, 337, 356, 357, 366, 369, 370, 403, 407–410, 412, 420, 425, 428, 431, 433, 437, 453, 492–494, 499–502
Parseval's theorem, 144, 703, 709, 713, 716
Partial response signalling (PRS), 84, 90–93, 118, 128
Partial response system (PRS), 84
Passband edge, 158, 160, 161
Passband ripple, 158, 161
Path loss
  exponent, 23, 187, 194, 195, 233, 241
  radio channel, 186–188
Peak distortion, 156, 165–166, 168, 169, 171
Peak frequency deviation, 93–95, 97, 103
Peak to average power ratio (PAPR)
  reduction, 323, 470–472, 668, 669
  reduction with coding, 470–472
Peak to RMS ratio, 223
Periodically stationary process, 733
Permutation
  matrices, 423, 424, 434, 435, 499
  random, 434, 448, 457, 461
Phase ambiguity, 75, 76, 80, 82, 568
Phased array antenna, 238–240
Phase detector
  gain, 567, 584
Phase invariance coding, 80–84
Phase jump, 107, 586
Phase locked loop (PLL), 95, 508, 509, 554, 555, 560, 566, 583–588, 672
Phase noise, 185, 671, 672, 685
Phase response, 137, 716

Phase shift keying (PSK), 53–66, 107, 118, 489, 569, 667, 668
Phase trajectory, 98, 100, 101
Physical, 5, 25–27, 46, 71, 113, 205, 305, 508, 512–514, 529, 530, 652, 674, 706
Physical layer convergence procedure (PLCP) preamble, 512–516, 524, 594
Physical layer frame, 507, 510
Pilot symbols, 282, 507, 511
Poisson sum formula, 534, 539, 582–583, 703
Polar coordinate, 188
Polarization, 22, 191
Poly-phase implementation, 636
Poly-phase structure, 634, 635, 646
Post detection filter, 107
Post detection integration (PDI), 519–521, 525, 527, 528
Power amplifier (PA)
  1 dB compression, 663–666
  non-linearity, 65, 663–671
  power amplifier, 72
Power density spectrum, 731–732
Power detector, 661, 662, 680
Power distribution, 114, 205, 207, 209
Power gain, 19, 42, 122, 208, 472, 663, 664, 683, 708–710
Power line channel, 25
Power line communication, 25
Power spectral density (PSD), 43–44, 103, 105, 107–119, 260
Pre-filtered
  pulses, 537, 539, 581–582
  RC, 547, 582
  RC pulse, 581–582
Pre-filtering, 537, 550, 581, 582
Primitive polynomials, 76, 78, 79, 127, 362, 494, 495
PRN code acquisition, 530, 532
Pseudo-noise (PN) sequence, 20
Pseudo random (PN) noise, 21, 76
Pulse coded modulation (PCM), 24, 27, 84
Pulse mismatch
  mismatch loss, 142, 163, 166, 264, 272
Pulse shaping filters, 11–12, 14, 16, 17, 20, 39, 101, 126, 154, 265, 274, 302–304, 517, 629, 630
Puncturing, 369–370, 439, 442, 443

Q
Q-rail, 610
Quad phase shift keying (QPSK), 10, 11, 17, 19, 55, 57, 63, 72, 257, 283, 463, 471, 479, 503, 525, 533, 566–569, 571, 667, 668

Quadrature amplitude imbalance, 610, 693
Quadrature amplitude modulation (QAM), 11, 53–66, 72–75, 97, 107, 115, 118, 126, 242, 250, 253, 281, 295, 303, 468–471, 509, 533–537, 569, 571, 667, 668
Quadrature modulations
  digital, 299–301, 621
Quadrature partial response system (QPRS), 90, 91
Quadrature phase imbalance, 579, 610, 612, 637–640
Quadrature phase imbalance correction, 572–579
Quality measurement of quadrature demodulator, 649
Quantization noise, 612–614, 651–653, 657, 673–675, 677, 685

R

Radiation, 189, 190
Radio communication signals, 3–29, 721
Radio propagation channels, 5, 8, 16, 23–24, 185
Raised cosine, 74, 146, 149, 211, 276, 285, 537, 539, 540, 545, 547, 549, 581–582, 589, 667
Raised cosine filters, 39, 276
Random variables, 197, 204, 205, 207, 208, 214–218, 488, 698, 708, 721–733
Range estimation, 199–202
Rayleigh fading, 20, 22, 23, 204–209, 305–309, 474–479
RCV dynamic range, 678–682
Receiver noise BW, 71
Receiver sensitivity threshold, 682–685
Receiver threshold, 21, 90, 199, 200, 679, 685
Rectangular pulses, 11, 12, 16, 17, 22, 24, 38, 39, 41, 42, 93, 98–100, 105, 112, 116, 118, 121, 122, 138–142, 144, 162–168, 171, 211, 222, 254, 256, 258, 265, 270, 272, 274, 277, 280, 309, 628, 655, 658, 668, 701, 704, 708, 709, 712, 713, 715, 716, 719, 720
Recursive systematic code (RSC), 370–375, 379, 383, 437–439, 442, 449, 450, 495, 496
Reed-Muller codes, 331, 334–339
Reference voltage, 612, 652, 656, 661, 662, 675, 677, 680
Reflections, 186, 191–202
Regular LDPC, 409, 435
Regular polygon, 229

Reliability, 198, 199
Remainder computation, 354–356
Remainders, 350, 351, 353–357, 364, 493, 502
Repeat accumulate (RA), 14, 227, 228, 421–425, 429, 430, 435
  AR4JA, 424, 435
Repeat-accumulator (RA) code, 421
Repeaters, 24, 235
Repetition, 249, 254, 260, 268, 270, 275, 323–328, 331, 340, 491, 498, 515
Repetition codes, 13, 14, 20–22, 80, 308, 323–326, 332, 338, 340, 350, 351, 404, 473, 490, 491, 494, 498–501
Repetitions, 14, 22, 221–224, 242, 716
Resampler, 557–559
Resampling
  control, 556–561
  digital, 555, 560
RF systems, 16, 20, 185, 583, 605
RFIC, 645
Rician fading, 208, 219, 220
Riemann zeta function, 234
RM codes, 338
Root Nyquist
  filter, 155, 160, 161
  pulse, 28, 133–181, 557, 589–591, 600, 608
Root raise cosine filter, 144
Rotational invariance, 76, 82–84, 118
Rotation invariance, 82, 83
Row distance, 429, 430
RS code
  Reed Solomon, 14
Run lengths, 77

S

Sampled channel model, 211, 214, 218–219
Sampling, 16, 17, 19, 37, 107, 116, 130, 133–138, 140, 145, 148, 156, 165–169, 172, 206, 211, 218, 221, 251, 252, 256, 260, 262, 266, 271, 273, 278, 280–283, 289, 290, 293–295, 299, 301, 304, 507, 510, 512, 513, 517, 530, 532, 533, 539, 542, 553–556, 560, 565, 582, 589, 590, 606, 608, 620, 621, 628–631, 642, 646, 648, 662, 671, 675, 676, 679–681, 685, 703, 712, 713, 715, 717–719
Sampling clock
  free running, 510, 556, 565
Satellite communication systems, 4, 65
Satellite motion, 184
SAW filters, 608

Scattering, 22, 186, 191–202, 206, 214, 216,
    219, 227
Scattering description, 214
Scrambler, 68, 76–80, 118, 127, 223, 337,
    459–461
S-curve, 534–536, 539, 540, 543–550, 568–570
The second order loop, 584, 586, 588
Sectored antennas, 234
Self-synchronizing, 78, 79, 127
Sensitivity, 39, 199, 525, 526, 684
Serial
    clocks, 510, 530
    to parallel, 510, 512, 513, 530
    transmission, 167, 260, 265–274, 280, 510,
        512, 513, 529
Serial, de-serial (SERDES), 5, 25
Serial transmission–OFDM, 265–274
Session layer, 27
Sgn (*t*), 704
Shadowing, 23, 195, 197, 227
Shannon, 28, 433, 480, 487, 503
Shaping pulse, 38, 40, 42, 102, 103, 133,
        153–161, 266, 279, 280, 535, 708
Shaping pulse design, 133, 154–162
Shift register circuits, 354–358
Shortening, 139, 142, 270, 358, 368–370, 495
Short-term training sequence (STS), 513–528
Signal power
    CW, 706–707
Signals
    continuous-time, 698–713, 717–721
    discrete-time, 8, 11, 118, 133, 134, 249,
        698, 713–720
    and systems, 1–32, 247–319, 462, 588,
        697–733
Signal to interference and noise ratio (SINR),
    612
Signal to quantization noise ratio (SNqR), 613,
    656, 657, 660, 673, 677, 685
Sign-magnitude representation, 652
Simulation of Doppler clock frequency
    561–565
Single sideband (SSB), 9–11, 26, 27, 250,
    614–620, 636, 650
Slow fading, 479
Small scale fading, 23, 24, 202, 203, 240, 241
SNR estimation, 523
SNR loss due to pulse mismatch, 141–143
SNR penalty of pulse mismatch with CP, 264
Soft decision (SD), 13, 14, 324–327, 330, 334,
    340, 342–349, 370, 386–391, 412, 462,
    473, 474, 477–479, 490
Software defined radio (SDR), 3, 134

Solid angle, 189–191
Space diversity, 22, 23, 207, 242, 305, 473,
    474, 477
Spatial channels, 22
Spatial reuse, 5, 250
Spectral density, 43, 111, 137, 143, 672,
    708–711, 732
Spectrum analyzer, 103, 112–115, 224, 258,
    658
Spectrum mask, 154, 162, 265, 628, 670
Spectrum of OFDM, 21–22, 256, 258,
    276–277, 285, 286, 510, 644
Spectrum shaping, 75–93
Spreading gain, 20, 21
Spread spectrum (SS), 20–22, 80, 221–224,
    507, 510, 512, 529–532
Squaring, 67, 160, 508, 509, 520, 531, 533,
    534, 540, 543, 549–552
Staggered QAM, 72–75, 126, 279, 284–297
State diagram, 371–376
State probability, 395, 399, 401, 406, 438,
    444, 445
State transitions, 371, 372, 375–378, 388, 389,
    397, 404, 439–441, 495, 496
Stationary, 109, 731, 733
Stationary process, 731–732
Statistical models, 202–220
Stochastic process, 721, 730–732
Stopband attenuation, 158
Stopband edge, 158, 160
Stopping sets, 425, 428, 429
Sub-carrier, 21, 22, 28, 116, 242, 250–252, 257,
    258, 262, 269, 283, 512, 514–516,
    522–524
Sub-channel
    permutation, 305, 306
    spacing, 250–252, 266, 286
Sum-product algorithm (SPA), 412
Survived state, 387
Symbol
    clock, 19, 20, 73, 101, 280, 294,
        507, 509, 532, 554, 555, 561, 565,
        670–672, 685
    clock recovery, 20, 508–510, 531
    energy, 19, 44, 48, 71, 122, 125, 126, 515,
        516, 518, 523, 709
    error rate, 37, 66, 87, 118, 339, 366
    timing synchronization, 532–564
Symbol error rate (SER), 19, 46–54, 57–61,
    63–71, 73–74, 87–89, 91, 92, 106–108,
    118, 124–126, 135, 169–172, 365, 384,
    391, 451, 455–457, 459, 472, 474–477,
    480, 489, 491, 684

Synchronization
    performance, 20, 21, 37, 66, 525–528, 532
Syndrome, 335–337, 356–360, 367, 368, 418

T
Tail-biting termination, 379, 381
Tamed FM, 106
T-carrier, 27
Terrestrial mobile systems, 186
Tessellation, 229, 466
The final value theorem, 586
The initial value theorem, 586
Thermal noise, 16, 18, 19, 24, 37, 44, 45, 66,
        71, 96, 118, 122, 126, 133, 143, 185,
        199–200, 228, 233, 516, 518, 526, 527,
        553, 651, 652, 660, 674, 680, 682, 684,
        685, 711, 712
Time division duplex (TDD)
    smaller excess BW, 506
Time reversal and conjugation, 136, 141, 519
Timing error detector (TED), 533–537, 539,
        543, 545, 554–556, 560
Timing recovery loop, 554–556
Tone test signals, 650, 664, 665
Traceback, 387, 497
Tracking, 240, 476, 530–531, 554, 555, 563,
        584, 585
Training symbols, 507
Transceiver, 29, 605–610, 619–648, 651, 686
Transceiver architectures, 605–651, 686
Transfer function, 80, 85, 110, 137, 371,
        373–377, 382, 584, 585, 616, 642, 643,
        664, 665
Transmit (TX)
    power control, 660–662
    signal level distribution, 660–662
Transport, 6, 27
Trapping set, 425
Trellis, 371, 373, 375, 377–380, 386, 387,
        389–392, 400, 403, 404, 406, 438, 439,
        441, 495, 497
Trellis coded modulation (TCM), 242, 461, 462
Trellis representation, 373, 377, 385, 390, 391,
        403–406
Turbo code
    decoding, 424, 438, 447
    iterative decoding, 28
Twisted pair, 5, 24, 28, 62, 256

U
Underwater acoustic channels, 186
Underwater communication, 25

Uniform scattering, 204–206, 219
Unit impulse, 76, 77, 127, 134, 146, 173, 699,
        713–715
Unit step, 587, 699, 704
Up-conversion, 9, 16, 21, 572, 606, 610–612
Use of repeaters and distributed antennas, 235

V
Variable gain amplifier (VGA), 610, 661, 662,
        680, 681
Variance, 17–19, 24, 44, 46, 74, 108, 110,
        120–122, 124, 168, 170, 195–200, 208–
        210, 215, 222, 442, 497, 507, 533, 613,
        653, 656, 675, 676, 708, 709, 726, 728,
        730, 731, 733
Vestigial sideband (VSB), 9, 10, 303
Vestigial symmetry, 144, 149, 150, 154, 160,
        667
Viterbi decoding
    algorithm, 385–388, 390
Voltage controlled oscillator (VCO), 94, 95,
        106, 128, 129, 224, 509, 554, 556, 558,
        560, 566, 571, 572, 583, 584, 586, 587,
        672

W
Wave difference, 539, 540, 542–545, 547–549
Wavelength division multiplexing (WDM), 25,
        28
Weak signal code phase synchronization of
        GPS, 531–532
WER *vs.* Es/No, 322
Wide-sense stationary, 108
Wiener-Khinchin theorem, 108, 117
Wi-Fi, 4, 5, 435, 526
Window, 112, 115, 157–159, 162, 280, 309,
        590
Window approach for digital filter design, 152
Windowing, 112, 113, 115, 151, 156, 157, 161,
        254, 260, 265, 267, 272, 273, 280, 282,
        303, 309, 669
Windowing method, 156–158
Wireless microwave access (WiMAX), 4, 28,
        186, 252, 254, 255, 257, 435

Z
ZigBee, 5

Printed in the United States
by Baker & Taylor Publisher Services